To the memory of
Vladimir Vladimirovich Beloussov

Space would fail, were I to attempt to write the history of this discussion, which is as old as our science itself.

—Eduard Suess, 1888

Contents

Acknowledgments

I owe thanks to many people for helping to create this book. First and foremost, I would have never hit upon the idea of writing it had Richard Ernst and Ken Buchan not invited its various ancestors and encouraged me all along the way. I am also grateful to them for agreeing to contribute the Foreword. Professor Abhijit Basu, GSA Books Science Editor, deserves credit for encouraging and superintending its metamorphosis from an overgrown paper to a small book. I am grateful to Professors William R. Dickinson, Gerald M. Friedman and W.A.S. Sarjeant for their very useful reviews. In addition to a fine technical review, Professor Sarjeant kindly took it onto himself to edit my entire text linguistically and stylistically. Without his editing, the Teutonic tendencies in my English would have been far more apparent in the text than they now are. I am deeply saddened that he died before I could place a published copy of this book into his hands.

Many of my Russian books I acquired and read with the help of my (very patient) friend and colleague Professor Boris A. Natal'in. I am particularly grateful to Mrs. Nadezhda Aleksandrovna Serper, the able head librarian of the Library of the All-Russian Institute of Economy of Mineral Resources in Moscow, for finding, among others, the proper reference to the extraordinarily rare 1903 paper by Pavlov in which the term syneclise was first defined. Dr. V.N. Sholpo kindly sent me the most recent literature on our common friend, the late Professor Beloussov. Beloussov's photograph accompanying the dedication I owe to the kindness of his widow, Mrs. Natalia Beloussova. Professor Fuat Sezgin, the great Arabist scholar and historian of science and technology of Frankfurt am Main, has held my hand as I ventured into the barely charted forest of the history of the Islamic earth sciences. Mrs. Maxine Dobrin gave me her late husband Professor Milton Dobrin's copy of Osmond Fisher's second edition of the *Physics of the Earth's Crust* as a memento of my late regretted friend and teacher. I owe to Dan McKenzie's generosity the first edition (Osmond Fisher's personal copy!) of the same book. Dan, among others, also taught me much about mantle plumes and gravity observations. I owe to John F. Dewey's dedicated and uniquely competent teaching whatever rigor my tectonic thinking may possess. Kevin Burke has been and remains my guide in all matters pertaining to African geology and the vertical motions of the lithosphere. Walter Pitman III was instrumental for my learning more about flood legends than I ever thought I would. Xavier Le Pichon enabled me to use the library facilities of Paris along with much encouragement. Dr. Alfred Schedl of the Geologischen Bundesanstalt, Vienna, alerted me to the existence of H. Wettstein's book and kindly allowed me to examine his copy of it. I owe much of my knowledge of the geology of the Cordillera of the western United States to conversations, discussions and literature exchange with B. Clark Burchfiel, the late lamented Peter J. Coney, Gregory A. Davis, William R. Dickinson, Gordon P. Eaton, Warren B. Hamilton, Brian Wernicke, and Mary Lou Zoback. Clement G. Chase, Gordon P. Eaton, George A. Thompson, and Brian Wernicke very kindly updated my knowledge of the Cainozoic uplift problem of the western United States while I was writing this book. Professors Halil İnalcık and Robert Dankoff helped me with the identification of the place names in Evliya Çelebi's *Seyahatname*.

I am deeply indebted to the kindness of the faculty and members of the Division of Geological and Planetary Sciences of the California Institute of Technology. By inviting me as a Moore Distinguished Scholar for a whole calendar year, they not only bestowed upon me a great honor, but also granted me the complete freedom necessary to devote myself entirely to my research. They also generously supported a good part of the research I undertook as their guest. My friends and colleagues Kerry Sieh and Brian Wernicke have advised me on numerous occasions on the current interpretations and supplied the relevant literature on parts of the American west, to the geological understanding of which they themselves have contributed so gloriously. Brian exhibited yet another act of great generosity by giving me his collection of the valuable volumes of the Geological Survey of New York. Another friend, Professor G.J. Wasserburg, informed me of T.J.J. See's ideas and generously presented me with his own copies of his publications. Mr. Jim O'Donnel, the great librarian of the division, patiently and cheerfully tracked down all the obscure references I placed before him and the competent Interlibrary Loan Office of Caltech put me under a great obligation by obtaining them all. Without the selfless assistance of Dr. Anke Friedrich (now of the Universität Potsdam, Germany) I would have been unable to locate and print many of the maps illustrating my account of the history of ideas in American tectonics. She also did much to dampen my incurable nostalgia for Europe, despite the great acts of wonderful hospitality I was shown while I resided in Pasadena. When she left for Germany, she handed some of my uncompleted business to Dr. Eric Cowgill, who attended to them with skill and devotion. Dr. Nathan A. Niemi (now of the University of California at Santa Barbara) not only discussed with me the cratonic stratigraphy of the Plateau Country, but also introduced me to the magical world of antiquarian book buying on the Internet (notwithstanding the threat of being murdered by my wife!). Mrs. Vala Hjörleifsdóttir of Caltech kindly assisted me in reading Urban Hiärne's wonderful book, a fine copy of which had been miraculously found for my collection by my friend Ed Rogers. Miss Joanne M. Giberson kindly helped me generate the topographic base for Figures 68 and 78, which were then drafted by Mrs. Jan Main. Mr. Jian Xin "Ken" Ou has been instrumental in reconciling me to various computers and programs at Caltech, which I would have been unable to use otherwise. Mr. Wayne Waller of the Digital Media Center of the Sherman Fairchild Library of Engineering and Applied Sciences in Caltech is responsible for the excellent quality of all the illustrations of Chapter XII and some others in other chapters, except Figure 119 and profile 2 in Figure 92, which were shot by Mr. John Sullivan of the Huntington Library in Pasadena. Mr. Waller also prepared the computer base for the dust jacket illustration. Messrs. Erden Soysal and Süleyman Kaçar in Istanbul are responsible for the rest. Mr. Brian Ebersole, the able science librarian of the Pomona College, provided a photographic reproduction of Figure 55 from an original in the Woodford Collection. Mrs. Loreta Young, my able secretary while in Caltech, served me with competence and cheerfulness and very substantially contributed to the various phases of the production of Chapter XII.

A special debt I owe to six great book dealers in the United States and to one great map dealer: Mr. Dan Cassidy of the Five Quail bookshop in Phoenix, Arizona, Michael Ginsburg of Michael Ginsburg Books in Sharon, Massachusetts, Mr. Robert D. Haines Jr. of the Argonaut bookshop in San Francisco, California, Mr. Richard Lafon of Sam Weller's bookshop in Salt Lake City, Utah, Mr. Paul Mahoney of the Antique Map Gallery in Denver, Colorado, my former schoolmate Mr. Ed Rogers of Poncha Springs, Colorado, and Mr. C.E. Van Norman of Wantagh Rare Book Company in Neversink, New York, have performed wonders in finding for my collection a surprising number of rare treasures of the western United States geological and geographical literature for most reasonable prices during the short time I was in Caltech. In fact, being a well-trained geologist and a fine scholar of the history of the subject himself, I owe to Ed not a few suggestions on the literature I needed to read.

Last, but not least, I am deeply grateful to my colleagues and students in the Istanbul Technical University and to my family. The former have tolerated my frequent absences from the office and from the classroom, and my wife, Oya, and son, Asım, my cluttering up the house with endless piles of books and sheets of maps while writing. Oya also helped me use my library in Istanbul while I was in Caltech. I do not dare ask any of them whether the result justifies their efforts and forbearance.

All uncredited illustrative material in this book is taken from books and documents in my private collection. My gratitude to my parents' unbounded generosity that enabled me to amass that collection cannot be adequately cast into words. They not only financed the collection but encouraged that it be created.

This research was also supported in part by the İstanbul Technical University and the Turkish Academy of Sciences.

Foreword

Broad domal uplifts of the lithosphere are associated with short-term geological events that are among the most dramatic in Earth's history, namely the eruption of millions of cubic kilometers of flood basalts and associated intrusions over 1–5 m.y. Important examples include the 65 Ma Deccan Traps of India, originally covering 1.8 million km^2, and the 120 Ma Ontong Java oceanic plateau of the Pacific basin, at 40 million km^3. It has been proposed that many of these magmatic events were associated with abrupt changes in climate and sea level, with break-up of supercontinents, and more speculatively, with global-scale extinctions and magnetic superchrons. Large plumes rising from deep within the mantle are now widely considered as the cause of such domal uplifts and their associated magmatism. Whereas mantle plumes from throughout the Mesozoic and Cenozoic have been relatively easy to identify and study because of their association with continental flood basalt provinces and oceanic plateaus, in earlier times the volcanic record of mantle plumes has been largely erased by erosion and other tectonic processes. Among the important features related to plumes that may persist in this older record are those associated with uplift above the plume. These features include triple junction rifting (or rift stars) and the sedimentary effects of domal topography. Identification of such features related to domal uplift is critical for assembling a complete catalogue of mantle plumes through time.

Mantle plumes were first recognized in the 1960s, and their association with broad-scale uplift shortly thereafter. However, the concept of uplift is woven through the entire history of geological and pre-geological thought. Indeed, for most of history, movement of the Earth's surface was viewed in a vertical sense (as in flooding or sea level changes). It was only with the plate tectonic revolution that large-scale horizontal movements gained prominence. In recent years, however, growing recognition of the importance of mantle plumes has reasserted the role of broad vertical movement.

In this volume, Celâl Şengör leads us on a fascinating journey that follows the development of ideas concerning large wavelength lithospheric deformation that forms broad uplifts and basins. The journey begins millennia ago with Middle Eastern and Asian mythology and ends with the plate tectonic revolution in the mid-20th century. In between, Şengör has assembled a wealth of detail from many parts of the world. The reader is treated to a multitude of legends, observations, and theories along with a host of characters who have explored this subject, from Plato and Aristotle, through Élie de Beaumont and Suess, to Cloos, Wilson and Burke. In order to tell that story, Celâl Şengör has consulted an immense number and variety of sources, many from his personal large collection of rare geological and historical texts.

We are especially pleased to have had a small role in the inception of this important historical review. The concept originally formed a part of papers on uplift and rifting that Şengör contributed to the volume on identifying old mantle plumes (*Mantle Plumes: Their Identification Through Time*, 2001, Geological Society of America Special Paper 352). As the review grew in size and scope, it became clear that it deserved its own volume.

Whether you read this volume as a geologist or as a historian, we know that you will have an enjoyable journey tracing the connections between ancient mythology and modern concepts of large wavelength deformation.

Richard Ernst and Ken Buchan
Geological Survey of Canada
Ottawa, Canada

Preface

This book reviews, in summary fashion, what I think to be the main stages of the evolution of ideas concerning structures commonly known as *epeirogenic*—one of the most neglected subjects in the historiography of tectonics. In the first chapter, I develop geological reasons for avoiding this term and introduce a new terminology.

This is not a scholarly book in the usual sense, much less a comprehensive one, because not all aspects of the subject matter have been reviewed and documented with the same thoroughness. Indeed, some, such as the tectonic interpretation of gravity observations that has critical importance for our understanding of long-wavelength structures of the lithosphere, has hardly been touched at all. I develop the history more thoroughly before the beginning of the twentieth century because the earlier history is generally less well known. But even the pre-twentieth century history that I discuss is much less complete than would be contained in a comprehensive treatise on the history of tectonics. That is why I subtitled the book as "materials for a history" rather than "a history."

I hardly enter into the biographical details of the personalities involved in the development of ideas on long-wavelength lithospheric structures. I assume that the reader is familiar with the *Geographers: Bio-bibliographic Studies*, published for the International Geographical Union, Commission on the History of Geographical Thought by Mansell (London), with the *Dictionary of Scientific Biography* (edited by C.C. Gillispie), with William A.S. Sarjeant's monumental *Geologists and the History of Geology* and its supplements, with Dietmar Henze's *Enzyklopädie der Entdecker und Erforscher der Erde*, with Numa Broc's *Dictionnaire Illustré des Explorateurs Français du XIXe Siècle*, and with Pauly's *Real-Encyclopädie der Classischen Altertumswissenschaft*, re-edited by Georg Wissowa.

As sources of biographical information, they are underused and not actively referenced in this study since they are readily available in major libraries. Rather, I refer to what I myself have used for this work, hoping that the interested reader will wander off to these sources too. Because the *Dictionary of Scientific Biography* includes summaries of widely used biographies, I indicate herein where I found its entries to be inadequate.

In all direct quotations, words written in *italics* and between square brackets [...] represent my own commentary. Transliteration has been a major problem for me, because this book involves sources from a number of languages using different alphabets and so many different sources. The solution I have found is an easy one for me, but not a satisfactory one for the reader: I have copied transliterations as I found them in my sources and transliterated my own readings as they sound to me. The only excuse I can offer for this procedure is that it enables my reader to recognize the names I transliterate in my sources and to be able to read them. Formal transliterations involve so many diacritical marks that for the uninitiated they are often unreadable. My procedure led to some inconsistencies and in one or two cases to different transliterations for the same word. But in such cases, the transliterated word is so obvious that I chose to remain faithful to the original orthography as used in my original source.

My emphasis in this book is on identifying the reasons why certain theories arose, why some were subsequently abandoned, and why others proved more resilient to falsification. Although the psychology of the individuals involved, and the sociology of the associated communities, in generating, falsifying and defending theories that answer the question "why happened" are of great importance, my interest is more on "how happened," which is more safely and readily answered by studying the logic employed in theory-building and theory-destruction.

The narrative in this book is provided in leaps and bounds, in the style of a research article rather than that of a book, because its purpose is to highlight only the significant episodes in the development of a very specialized topic in the history of tectonics. Because ideas of one or two personalities dominated each such episode, it is mainly on their work that I concentrate.[1] The research efforts of such dominant personalities almost invariably benefited from interaction with numerous others who were less prominent. This book cannot, and does not attempt to, tell the story of entire communities involved in the development of the observations and the theories that are treated in it, although I am aware that it is the collective work of those scientific communities within a social milieu that constitutes the proper subject of the history of science. (However, I disagree with Livingstone {1992, p. 11} that the significance of what he calls "minor figures" are more important than that of major figures in casting light on the history of a science {in Livingstone's case, geography} as a social phenomenon). As Bertrand Russell once said, history is admittedly more than the record of individual men, however great. Nietzsche would have thus classified my history with his "monumental" and "critical" histories, but certainly not with his "antiquarian" histories. Like the great science historian Charles Coulston Gillispie (1960, p. 521), however, I believe that history is made by men, not by causes or forces, and, like him, I therefore have tried to write with due attention to the intellectual personalities who have borne the battle.

The peculiarities of this book arise from its evolution. It originated in the modest intention of writing a small introduction to a paper on establishing geological criteria for identifying former mantle plumes. The paper had been invited by Richard Ernst and Ken Buchan, both of the Geological Survey of Canada, for a Geological Society of America symposium concerning the recognition of inactive plumes (Şengör, 2001a). The small introduction grew under my hand, however, as I progressively discovered my own ignorance of the subject, to become what I thought was appropriate as an independent paper for the same symposium. Upon receipt of my bulky typescript, my editors dutifully sent it out for refereeing notwithstanding its disproportionate size. After having received the advice of the referees that it was simply too large and too historical for inclusion in a technical volume, they suggested, following the advice of all three of the referees, that it might perhaps better be submitted as an independent book, though still related to our symposium. I sent my typescript plus the reviews to Professor Abhijit Basu, the Geological Society of America (GSA) book editor, who suggested that I might consider enlarging to make it more suitable for an independent book format. I exploited this welcome opportunity by adding to the text in a number of places and enriching the biographical information (and pictures) and other peripheral material in the endnotes, which I could not sensibly have considered putting into a paper. A further enlargement was made later when, as a Moore Distinguished Scholar at the California Institute of Technology, I decided to add the history of ideas on long-wavelength structures that had developed in the western United States.

Some readers may feel that a disproportionately long and detailed account is given of the development of ideas on long-wavelength lithospheric structures in the western United States. The reason for this focus is that there does not exist either a single satisfactory synthetic account or relevant summaries of parts of this subject to which I could have referred my readers. Also, the history of ideas on the falcogenic structures in the western United States illustrates better than most accounts in the history of geology the conjectural nature of science and the relations of ideas to observations.

The reader may well wonder how far the interpretation of mythology, a largely unwritten corpus of ideas and observations, or even the recovery of its very content, may be reliable for the purpose for which I employ it in the beginning of the historical narrative. There is little that I can say in the defense of my usage that would be satisfactory to any degree and that I might find entirely acceptable. Instead, in defense of using the mythology to understand patterns and ways of thinking, I here quote the words of Sir James Frazer (1919, p. ix–x, from the Preface to the third edition of his monumental *The Golden Bough*, part IV):

> The longer I occupy myself with questions of ancient mythology the more diffident I become of success in dealing with them, and I am apt to think that we who spend our years in searching for solutions of these insoluble problems are like Sisypus perpetually rolling his stone up hill only to see it revolve again into the valley, or like the daughters of Danaus doomed for ever to pour water into broken jars that can hold no water. If we are taxed with wasting life in seeking to know what can never be known, and what, if it could be discovered, would not be worth knowing, what can we plead in our defence? I fear very little. Such pursuits can hardly be defended on the ground of pure reason. We can only say that something, we know not what, drives us to attack the great enemy Ignorance wherever we see him, and that if we fail, as we probably shall, in our attack on his entrenchments, it may be useless but it is not inglorious to fall in leading a Forlorn Hope.

I am actually more optimistic than Sir James, however. I believe, with Alexander von Humboldt, that "Nobody would be in a position to deal with the history of a philosophical viewpoint, if he buries the times of the Age of Heroes entirely in oblivion. The myths of peoples, mixed with history and geography, do not entirely belong to the ideal world" (von Humboldt, 1835, p. 110). Arnold Toynbee echoes von Humboldt when he says that anyone who starts reading the *Iliad* as history will find that it is full of fiction but, equally, anyone who starts reading it as fiction will find that it is full of history (Toynbee, 1947, p. 44). This, I think, must be true for any mythology. See Brillante (1990) and Greene (1992) for recent optimistic analyses of the myth-history relationships and Haussig (1992, ch. 1 {p. 3-10}) for a remarkable employment of mythology to pursue a number of historical relationships.

Finally, a word about the dedication. My relationship with the late Professor Vladimir Vladimirovich Beloussov began under stormy circumstances. In 1979, following submittal of a paper to *Eos* about his ideas on global tectonics, Beloussov complained to the editor, Fred Spilhaus, in his cover letter (which, as I remember, also conveyed his Christmas greetings) that he was lately having difficulty getting his papers published in western journals. Spilhaus sent the paper to Kevin Burke, asking for comment. Kevin responded that the best way to pursue was perhaps to publish a paragraph by paragraph rebuttal together with Beloussov's text in adjacent columns of *Eos*. Spilhaus agreed, and Kevin suggested that I write a first draft, as many of Beloussov's criticisms against western scientists were built on a basis of the history of tectonics as it had evolved in continental Europe. In the end, Kevin thought that the rebuttal reflected more of me than of him and placed my name first in the authorship.[2] I had not met Beloussov before but recalled that my doctoral thesis advisor, Professor John F. Dewey, spoke with him on the telephone always with great deference.

I shall never forget my first encounter with Professor Beloussov during the International Geological Congress the following year in Paris. I walked up to him during a break and introduced myself. He did not quite catch my Turkish name and did not have his spectacles on and could not read my badge. With a courteous smile, he politely shook me by the hand and hurriedly put on his glasses, and for an instant, an expression—a mixture of surprise and discomfort—came upon him as if he had involuntarily swallowed something unpleasant. An instant later, the courteous gentleman was back—but not quite the smile—and we had a brief conversation. He then excused himself, saying he had a meeting to attend.

Later he became my guest in İstanbul. When he came to my house, my wife and I had before us a gentleman of the first water—carrying an air of Imperial Russia, engaging in pleasant and very learned conversation, now on the history of geology in Russia, now on art, now on the politics of the Soviet Union. Our correspondence lasted until his death, during the course of which he kindly presented me with a number of his books, including the hard-to-find early editions, all generously autographed. I learned much from Professor Vladimir Beloussov, including things I put in this book—hence, in part, the dedication, as an earnest expression of my admiration for the man and the scientist. But the most valuable thing he taught me was how totalitarian regimes, whatever their nature, sooner or later break the spirit of scientists who need the atmosphere of freedom and trust to flourish. (On this aspect of Beloussov's life, see especially chapter 7, entitled "Vladimir Vladimirovich Beloussov—a fate of a great scientist in an epoch of totalitarian regime" in the memoirs of his friend, the great Russian geologist, Victor Khain, 1997).

A.M.C. Şengör
İstanbul, 13 November 2000 and Pasadena, 3 March 2002

The large-wavelength deformations of the lithosphere: Materials for a history of the evolution of thought from the earliest times to plate tectonics

A.M.C. Şengör
İTÜ Maden Fakültesi, Jeoloji Bölümü, and Avrasya Yerbilimleri Enstitüsü, Ayazağa 34469, İstanbul, Turkey

ABSTRACT

Today, the geodynamic deformations of the lithosphere manifest themselves in two main categories: *structures of small wavelength* and *structures of large wavelength*—"wavelength of structure" being defined as the distance between two amplitude crests of cogenetic structures belonging to the characteristic size category within a field of deformation. I call structures of small wavelength *copeogenic* (because they *cut* the lithosphere) and the structures of long wavelength *falcogenic* (because they *bend* the lithosphere). This book traces the rise of the awareness of long-wavelength structures with the objective of understanding their essential features.

The subdued expression and enormous size of long-wavelength structures have been joint impediments to the recognition and the understanding of their nature, yet many have known of their existence from the earliest times—mainly on the basis of observations of sea-level change. Change of level has been inferred so early that the origin of this inference is lost among mythic speculations. Vertical motions of the rocky surface with respect to a reference fixed to the earth have been much harder to recognize because of the difficulty of finding an appropriate point of reference and the selection of gauges showing distance to that point of reference in the past. The earliest explanatory models were based on observations that, in some areas, land was actively gaining on the sea and that in others in the past, some of the present land areas had been covered by marine waters, as shown by fossils.

These early models involve now long-abandoned mechanisms invented from few and disconnected observations, but they helped to make a clear distinction between structures of small wavelength and structures of large wavelength. It was already implicitly understood that the former could be investigated on a scale ranging from single outcrops to individual mountains, whereas the study of the latter necessitated a regional approach. Small-wavelength structures were thought to form quickly, even catastrophically. Large-wavelength structures seemed to evolve slowly, but belief in such legends as Atlantis, the continent that allegedly had become submerged in one day and night, blurred the picture for a long time. Distinctions based on size, geometry, and timing of evolution remained disputed as long as means of observation of large-wavelength structures remained inadequate. Only with the development of biostratigraphy in the late eighteenth century and of geomorphological methods of slope investigation in the early twentieth century was the presence of large-wavelength structures eventually recognized beyond doubt. These methods have also helped us understand their evolution. In particular, the detailed topographic investigations carried out in the United States west of the Mississippi River since the beginning of the nineteenth century made the presence of large-wavelength structures indisputable. When those topographic data became combined with geological investigations from

the middle of the nineteenth century, it became obvious that older inferences concerning the relative rate at which such immense structures grow were correct.

However, that understanding was complete only after the period of development of geophysical methods to investigate what underlies long-wavelength structures. The latter part of that developmental period included the recognition of the mantle-plume generated uplifts. Between 1800 and 1960, geologists tried to accommodate the large-wavelength structures within the framework of all-encompassing global tectonic theories that were not nearly detailed enough for the purpose. Plate tectonics provided for the first time a comprehensive and detailed theory into the framework of which J. Tuzo Wilson placed his hypothesis of mantle plumes. It is now clear that mechanical loading, thermal changes in the mantle, and intracrustal flow events dominate the origin and evolution of long wavelength structures. Mantle plumes are the most significant non-plate-boundary generators of long-wavelength structures, and it is these structures and the fills of the associated lithogenetic environments that constitute their most faithful record.

CHAPTER I

INTRODUCTION

The purpose of this book is to trace, in barest outline, the rise of the awareness of structures commonly called "epeirogenic." Their subdued expression and enormous size have been joint impediments to the recognition of their presence and the understanding of their nature. Yet many have known, or sensed, their existence from the earliest times. Although these structures have long eluded explanation, they have nevertheless formed a significant part of almost every tectonic theory since Plato. The theory of mantle plumes in our times has offered, I believe for the first time, a testable and robust explanation for their most elusive subclass, domal uplifts, termed cymatogens[3] by King (1959, p. 117; 1962, p. 200; 1967a, p. 205). It now seems that such structures are the most prominent and the least ambiguous earmarks of mantle plume activity (Şengör, 2001a).

The following history aims at crystalizing the essential features of large-wavelength structures and at exposing the thicket of prejudice that has grown around them for millenia. This approach is a bit like reviewing the impressions of every blind Indian who has ever touched an elephant in order to find out the essential features of elephants. This may appear silly, when the obvious thing to do is to go and look at an elephant, but regrettably, we only have the backs of elephants available for inspection; we are nearly as blind as the other Indians, and we have grown up with their accounts of elephants. Even when we do our own "looking" (e.g., Şengör, 2001a), we must do so being aware of our own blinkers. One purpose of the following account is to point out where our blinkers may be and the extent to which they may restrict our field of vision.

My historical account is not homogeneous, with detail and quotations provided for the first 90% of this book (the relevance of which may not be immediately obvious), and rather short narrative text for the last 10%. The reason for this structure is that the terrain traversed in the earlier history is essentially virgin. Despite the work devoted to the subject since the nineteenth century, the evolution of geological thought and knowledge in antiquity and in the Middle Ages is very scantily known; its setting within the general history of thought has hardly been outlined[4], yet its impact on later developments has been immense. Because prejudices and observations are what we are after, the citing of passages that are perhaps longer than common in the numerous precursorist historical preludes in many scientific books was unavoidable. The limited space available naturally dictated the selection of highlights only. Connections between them had to be established by generalized passages that could not be documented in any detail. Some background information and some peripheral, but necessary, points are stored in endnotes. Some of these may seem superfluous to geologists, and others to historians. This was unavoidable because the book is addressed to them both. I apologize for any annoyance this may cause.

Two Kinds of Deformation of the Earth's Lithosphere

Today the geodynamic deformations of the lithosphere[5] manifest themselves in two main categories: *structures of small wavelength* and *structures of large wavelength* (Schmidt-Thomé, 1972, p. 1)[6]—"wavelength of structure" being defined as the distance between two amplitude crests of cogenetic structures within a field of deformation (Fig. 1 herein). Structures of small wavelength are those whose spatial repeat distances range

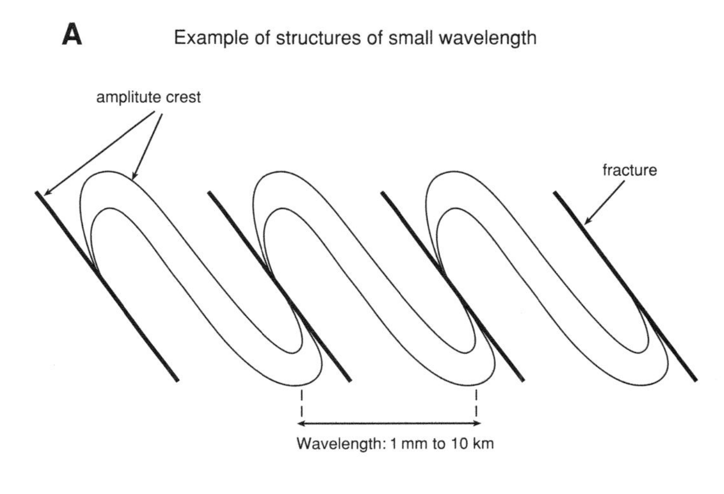

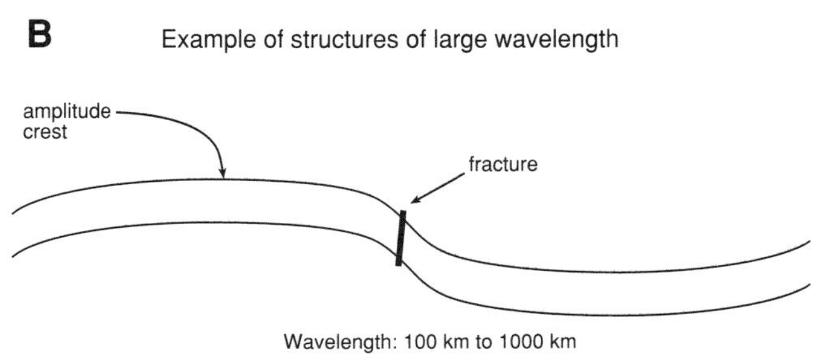

Figure 1. Concept of wavelength of structure. A: A folded and regularly imbricated bed illustrates a small-wavelength structure. This picture could apply to crenulation cleavage as well as to a series of imbricated folds of mountain-size (e.g., those in the Säntis Range in eastern Switzerland {Heim, 1905, Plate III}). Even if no folds were present, the imbricated structure would still define a group of cogenetic structures disposed with small repeat distances between them (an ideal example being in the Assynt district of the Scottish Highlands, where the section from the basal quartzite through Pipe-Rock, Fucoid Beds, Serpulite Grit, and to the limestone is repeated many times beneath the Glencoul thrust-plane, forming a *Schuppenstruktur* {Peach et al., 1907, p. 515}). Here the designation "wavelength" naturally presupposes no identity of distances between consecutive structures. It is employed only to call to mind an image of periodic recurrence with no assumption of regularity of period. B: Illustration of a large-wavelength structure that is *necessarily* of lithospheric dimensions.

from below a millimeter-scale to 10-km-scale[7]. In many, the amplitude exceeds the wavelength. They are commonly associated with *fracture* or other kinds of *structural discontinuities* on a scale similar to the size of the main structure; in some, such discontinuities constitute the only evidence of deformation. *Orogens* (Kober, 1921, p. 21), a bundling together of structures resulting from shortening; *taphrogens* (Şengör, 1995, p. 54, on the basis of Krenkel, 1922, p. 181 and footnote), a grouping of structures of extension into one large domain; and *keirogens* (Şengör and Natal'in, 1996, p. 490 and 639, note 8), belts along which wrench faults crowd, are the three dominant families of small-wavelength structures on earth. Orogens, taphrogens, and keirogens have map dimensions that range within similar limits, and the dominant type of strain along them corresponds with the character of the deformation along the three types of lithospheric plate boundaries. In fact, if an orogen, a taphrogen, or a keirogen itself is not the expression of a present or a past plate boundary or of a plate boundary zone, then it must at least be now (or have been in the past) part of a deformation field associated with one (see especially Şengör, 1999a; also Şengör, 1990). That is why it is wholly inappropriate to speak of "intraplate" tectonism in places where rates of displacement exceed a centimeter per year and where a considerable family of structures of small wavelength take up the deformation (e.g., Davis, 1980). Orogens, taphrogens, and keirogens have both pure end-members and transitional types (Woodcock, 1986)—such as transpressional orogens, or transtensional taphrogens, or transpressional or transtensional keiroges, or keirogenic orogens, or orogenic keirogens. Some of these transitional types can generate bewilderingly complex strain histories (e.g., Dewey, 2002).

Plate interiors, by contrast, are characterized by the dominance of large-wavelength structures having wavelengths from hundreds to thousands of kilometers (i.e., mostly megascopic structures; e.g., Bally et al., 1980; Hinze et al., 1980; Brown and Reilinger, 1986; Park, 1988, p. 188–209). Amplitudes of such structures are *always* only a small fraction of their wavelength. In these structures, fracturing is *always* subordinate to the bending of the lithosphere (cf. Şengör, 2001a), despite some persistent claims to the contrary (Cloos, 1939; Burke and Whiteman, 1973; Burke and Dewey, 1973; Ernst et al., 1995a, 1995b; Şengör and Burke, 1978; Şengör, 1995; Baragar et al., 1996; Ernst et al., 1996; Ernst and Buchan, 1997), at least on earth (cf. McKenzie, 1994). In fact, structures having large wavelengths commonly appear as large domes (*cymatogens* of King, 1959, p. 117; also see above; in part equivalent to *geotumors* of Haarmann, 1930, p. 13–14) or *downwarps*[8] such as those that characterize the present geomorphology of Scandinavia, Canada, Patagonia (e.g., Schütte, 1939; Mörner, 1979, 1980; Peltier, 1980; Grønlie, 1981; Fjeldskaar and Cathles, 1991), and Africa (Krenkel, 1922, p. 176 ff.; 1925, fig. 4 and plate I; 1957, p. 454–455, 457; King, 1962, especially p. 288 ff.; Holmes, 1965, fig. 763; Schmidt-Thomé, 1972, fig. 21–49; Burke, 1996; and the various contributions in Selley, 1997) or the basement of the United States in the mid-continent region (e.g., Hinze et al., 1980; Collinson et al., 1988; Fisher et al., 1988; Gerhard and Anderson, 1988; Sloss, 1988a; Leighton et al., 1990; Leighton, 1996)[9], and of the Russian (Nalivkin, 1976) and Angaran cratons (Bazanov et al., 1976). Other structures that fall under the large-wavelength category include large peripheral molasse basins appearing as both foredeeps and backdeeps[10] (or "retroarc basins" in Dickinson's 1974 terminology) and parallelling compressional arc orogens or collisional mountain belts (van Houten, 1969; Dickinson, 1974; Allen and Homewood, 1986; Şengör, 1990; Macqueen and Leckie, 1992; Dorobek and Ross, 1995; Van Wagoner and Bertram, 1995) or compressional segments of large strike-slip faults (e.g., Steel et al., 1985) and coupled lithospheric "outer rise" bulges on the continent side (e.g., Warsi and Molnar, 1977; Lyon-Caen et al., 1985) or their oceanic equivalents (e.g., Wilson, 1965). They are similar in origin to halos of depression and uplift couples surrounding point loads such as seamounts (Lambeck and Nakiboğlu, 1981; Watts et al., 1985; Nakada and Lambeck, 1986), and elastic and visco-elastic flexing of margins of sedimentary basins of diverse types (Beaumont, 1978; Beaumont et al., 1982; Dewey, 1982; Watts et al., 1982). Large-wavelength flexing is also the cause of the amplifications of structural relief of large- and small-wavelength structures by intraplate compressive stresses (Cloetingh, 1988; Cloetingh et al., 1985, 1989; Etheridge et al., 1991; cf. Stille's concept of "synorogenesis" {Stille, 1919, p. 205–206; 1924, p. 16}; also see Solomon, 1987). The swelling and shrinking of mid-ocean ridges—depending on spreading velocity (Hays and Pitman, 1973; Pitman, 1978; Turcotte and Burke, 1978) and thermal upheaval and subsidence of large "superswells" within the oceanic lithosphere (Davies and Pribac, 1993)—are further evidence of structural evolution with a large wavelength. Thermal subsidence of Atlantic-type continental margins (Pitman, 1978; Pitman and Golovchenko, 1991) also produces long-wavelength lithospheric structures. Both Bally and Snelson's (1980) "basins on rigid lithosphere" (also St. John et al., 1984) and Helwig's (1985) "flexural basins" are manifestations of the same. In addition, lateral variations in viscosity of the lithosphere (Karpytchev, 1997) contribute to generating large-wavelength geoid anomalies and also contribute to plate motions. An elastic/brittle upper crust adjusts itself to flow in the mid-crust, commonly away from orogenic welts (Royden, 1996; Royden et al., 1997; McQuarrie and Chase, 2000) or intra-taphrogenic highs (Block and Royden, 1990; Wernicke, 1990; Kaufman and Royden, 1994; McKenzie et al., 2000). The motions of the upper crust in such situations are also comprehended under large-wavelength movements.

As the short and incomplete list reviewed above shows, the recognition of the great variety of large-wavelength structures has grown rapidly in the last three decades (see especially Menard, 1973; cf. Stille, 1924 or King, 1955). This recognition results, in part, from the vastly improved technologies such as digital topography and gravity imaging provided by satellite observations (for background, see Anderson and Cazenave, 1986 and Lambeck, 1988) and seismic reflection profiling (as examples among many throughout the world, see Yilmaz, 1987, for methods, and Blundell et al., 1992; Meissner et al., 1992; and Gee and Zeyen, 1996, for some of the European results), but also in part from the recognition of the plate structure of the lithosphere. Such structures are not only easier to image now, but they are also easier to understand because we know so much more than before about the thermal and mechanical behavior of the lithosphere and the properties of the underlying mantle (Menard, 1973).

Despite that, in most recent textbooks on tectonics, large-wavelength structures receive only scant treatment (e.g., van der Pluijm and Marshak, 1997, p. 466–470[11]), while in the overwhelming majority of such textbooks, large-wavelength structures are not even recognized as a class (e.g., Bally et al., 1985; Kearey and Vine, 1990, 1996; Moores and Twiss, 1995; Mercier and Vergely, 1992), not even in scientific communities in which the concept of epeirogenic vertical motions used to be considered of central importance (e.g., Chain and Michajlov, 1989, especially p. 32–34; Frisch and Loeschke, 1990; Eisbacher, 1991; Miller, 1992; Khain and Lomize, 1995). In the *Encyclopedia of Structural Geology and Plate Tectonics* (Seyfert, 1987), not only is there no entry under "epeirogeny," but neither this nor any derivative term even appear in the index! This negligence has been instrumental in leading to misinterpretations of the geological record implicit in some recent popular models of sea-level change (e.g., Vail et al., 1977; Haq et al., 1988), as Pitman (1978) and Pitman and Golovchenko (1991) have documented.

It is, however, remarkable how much already had been learned about these structures before the 1960s, despite the fact that neither most of our present tools nor our models were available then. That knowledge was the basis from which we have so rapidly expanded our horizons since the advent of plate tectonics. It is useful, therefore, to know something about that former knowledge and its sources if we are to understand where we stand now. Our knowledge of very large-wavelength structures has also been burdened by a variety of biases resulting from faulty observations, unjustified inferences, and empty assertions, all much compounded by the great difficulties in observing them. We can only peel these biases off if we understand where the deficiencies in our knowledge come from and where, in our concepts and nomenclature, they now hide. Accordingly, this book, the last in a series of three publications on the recognition of mantle plume activity in the geological record (for the others, see Şengör, 2001a; Şengör and Natal'in, 2001), presents a brief synopsis of the evolution of thought on *large-wavelength structures that we now know always to crown active plume heads.*

But first, we must generate a shorthand so as not to have to repeat the cumbersome triplets *small-wavelength structures* and *large-wavelength structures* and also hopefully to create a firmer image of the two structure families in the minds of earth scientists.

Copeogenic and *Falcogenic* Structures: Definition

In the period 1872–1880, while studying the lacustrine terraces around the ancient Lake Bonneville (Pyne, 1980, p. 135; also see Hunt, 1980a, 1982), the great American geologist Grove Karl Gilbert (Fig. 2)[12] noticed bending of the terraces with very large wavelength (~300 km; see Gilbert, 1882, plates XLII and XLIII; 1890, plate L, reproduced as fig. 1.5 in Hunt, 1982; also see Ellenberger, 1989, fig. 3) resulting from what he thought were "broader displacements causing continents and oceans." Gilbert coined the somewhat inappropriate term *epeirogeny*[13] for this kind of structure generation and contrasted it with *orogeny*[14], a process that habitually creates narrower, more closely crowded structures (Gilbert, 1890, p. 340). Gilbert subordinated both epeirogeny and orogeny to *diastrophism*[15], a term apparently[16] invented by John Wesley Powell to designate the deformation of the earth's outer rocky rind in general. I should have followed Gilbert in using *epeirogeny* and *orogeny* to designate the large and small wavelength structures, respectively, or "to contrast the phenomena of the narrower geographic waves with those of the broader swells," as he put it himself (Gilbert, 1890, p. 340), but for two reasons: (1) The term *orogeny* has been burdened with the idea of morphogenic mountain-building since at least the middle of the nineteenth century, as the etymology of the term indeed implies, and with the implication of compressive deformation, especially in current usage (Sengör, 1990); and (2) The term *epeirogeny* has also been loaded with all sorts of contradictory tectonic interpretations[17]. It is thus necessary to create new short-hand notations for the two categories of structures. I believe, with Gillispie

Figure 2. The great American geologist, Grove Karl Gilbert (1843–1918), who coined the term *epeirogeny*.

(1960, p. 77), that "Humanists may complain of the jargon of the specialties, sometimes with justice. But no science can flourish until it has its own language in which words denote things or conditions and not qualities, all loaded with vague residues of human experience." As he pointed out, "To name is to know, not essences, nor totalities, but what we can know." (Gillispie, 1960, p. 172; also see p. 233 for Lavoisier on scientific nomenclature). For the large-wavelength structures, I thus propose the term *falcogenic* from the classical Greek φάλκης[18], the bent rib of a ship, for these structures generally *bend* the outer rocky rind of the earth. By contrast, the small wavelength structures *cut up* the lithosphere and therefore I propose to group them under the term *copeogenic* structures, from the Greek κοπή, to cut up, slaughter[19]. All plate boundaries are copeogenic structures or *copeogens*. All intraplate deformation is falcogenic and generates *falcogens*, but not uncommonly plate boundaries themselves do get caught up in falcogens (e.g., the axial depressions and culminations along spreading ridges, such as the Australian-Antarctic Depression {Veevers, 1984} or fluctuating basement depths in subduction trenches {e.g., see fig. 1 in Dewey, 1980}).

Figure 3 illustrates a breakdown of tectonic phenomena into the two major classes of *plate boundary* and *plate interior* events. The plate boundary processes, or copeogenic events, consist mainly of horizontal phenomena and readily divide into the events taking place on the three principal types of plate boundaries. By contrast, the plate interior processes, or falcogenic events, are all ultimately isostasy-related vertical phenomena, as Dutton (Fig. 4)[20] recognized more than a century ago (Dutton, 1880, p. 20–21). These plate interior processes fall into five groups, namely: (1) those processes related to the heating or cooling of the lithosphere or of the mantle below the lithosphere (*thermal isostasy*[21]: Crough, 1979, 1983; McGetchin et al., 1980; Heestand and Crough, 1981; Turcotte and Angevine, 1982; McKenzie, 1984; White and McKenzie, 1989; McKenzie, 1994); (2) those processes related to the loading of the lithosphere by sediments, such as at deltas (*sedimentary isostasy*: e.g., Fisk and McFarlan, 1955, especially fig. 11; Doust and Omatsola, 1989; Driscoll and Karner, 1994; also see Hall et al., 1982, fig. 8); (3) those processes related to the loading or unloading of the lithosphere through structural processes, such as thrusting or extensional detachment faulting (*structural isostasy*: e.g., Beaumont et al., 1982; Johnson and Beaumont, 1995; Wernicke, 1985; Wernicke and Axen, 1988); (4) those processes triggered by the formation of continental glaciers (*glacio-isostasy*, which is really a part of sedimentary isostasy, if one remembers that ice is first, when it is still snow, a sedimentary, then a metamorphic rock; see Chappell, 1974; Peltier, 1980; Sabadini et al., 1991); and finally (5) those processes brought about by the formation and disappearance or waxing and waning of water bodies (*hydro-isostasy*: Gilbert, 1890; Chappell, 1974). All of these isostatic mechanisms involve a flexing of the lithosphere (Forsyth, 1979) that does not alter the fabric of the rocks at outcrop. Influencing all of these also is the state of stress within the lithospere (Gay, 1980; Cloetingh, 1988; Cloetingh et al., 1985, 1989; Etheridge et al., 1991; Zoback and Zoback, 1989, 1997; Zoback et al., 1989).

Under extreme horizontal stress conditions, the lithosphere may buckle, even fracture, away from plate boundaries, creating structures between falcogenic and copeogenic in nature. As almost all such structures result from convergent plate boundary activities, or from the convergent component of transform-fault boundaries (they are all germanotype[22] in character: Ellenberger, 1989), I find no harm in using for them Stille's old designation *synorogenic* (meaning "at the same time as orogeny" {Stille, 1919, p. 205–206; 1924, p. 16}). For instance, the area of present deformation and seismicity between the Chagos Bank and the 90E Ridge in the Indian Ocean (Wiens et al., 1985; Wiens, 1985/86; Karner and Weissel, 1990) is a synoro-

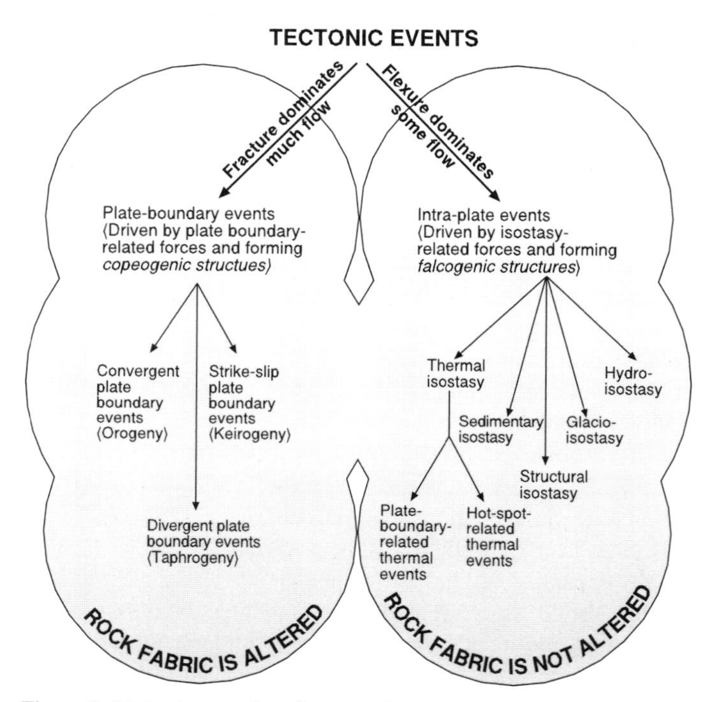

Figure 3. Main classes of earth-sourced tectonic events on earth. Impact of major bolides are capable of generating tectonic edifices as spectacular (or more) as those formed by the indigenous processes (endogenous and exogenous) of the rocky planets, but this classification does not show them. Fracture and flow are meant to imply those only within the lithosphere. Although the plate boundary events are divided into three end-member types, it is clear that only short stretches of the worldwide plate boundary network can be fitted into one or other of these end-member types (e.g., Woodcock, 1986, fig. 1; also, after him, Şengör, 1990, fig. 27). Of the intraplate events, varieties are shown only for thermal isostatic processes. Note that plate boundary events are indicated to alter the fabric of rocks, whereas plate interior processes are shown not to. These cases of alteration of fabric or non-alteration of fabric must be qualified with the word "dominantly." For instance, any fracture or fault atop a falcogenic dome does alter the local rock fabric there (let us say, 150,000 km^3 in volume for a fault zone that is 500 km long, 1 km wide, and 30 km deep), yet the doming itself, affecting volumes (within the lithosphere only!) on the order of 50–100 million km^3, does not alter, at least not visibly.

genic structure resulting from the activity of the convergent plate boundary in the Himalaya (Şengör, 1987). If that zone eventually evolves into a subduction zone, a synorogenic structure will have turned into an orogenic one.

As pointed out above, the copeogenic and falcogenic structures correspond *generally* with plate boundary and plate interior structures. Because copeogenic structures cut the lithosphere, they are by definition plate boundary structures. But falcogenic structures only gently bend the lithosphere. In places, they bend a whole plate boundary, as illustrated by the depression of the Antarctic/Indian plate boundary south of Australia creating the Australian-Antarctic Depression (Veevers, 1984), which is a falcogenic event taking place smack in the middle of an active plate boundary. Therefore, falcogenic events do not *just* create plate interior structures.

Figure 4. Sharp observer, imaginative theorist, accomplished writer, and immortal scientist in one mortal frame: the man who paved the way for Gilbert's distinction of *epeirogeny* from *orogeny* and developed the theory of isostasy: Clarence Edward Dutton (1841–1912), from de Margerie (1954, p. 633).

Using the labels "plate boundary" and "plate interior" to distinguish copeogenic and falcogenic structures often leads to another kind of trouble because many structures form thousands of kilometers away from the *usually depicted* plate boundaries (e.g., Lake Baykal from the Himalaya: see Dewey, 1977 or 1987; Mattauer, 1986; Şengör and Natal'in, 1996, fig. 21.9 and p. 609), yet they are within the *plate boundary zone* of the same plate boundary (Şengör, 1987; Şengör and Natal'in, 1996). Plate boundary zones are difficult to draw accurately because their boundaries against cratons are commonly marked, not necessarily by individual structures of significant strain and/or displacement, but by a zone of strain and/or displacement gradient that in places may be hundreds of kilometers wide (e.g., consider the whole of the British Isles as the *edge* of an Alpide plate boundary zone: Dewey and Windley, 1988).

In the following pages, I review the gross outlines of the somewhat checkered history of the recognition of falcogenic structures. One of my aims is to build a foundation for the discussions of the structure and evolution of mantle-plume-related falcogenic domes. These grand structures, together with their stratigraphic and structural record, may constitute the best evidence for identifying past plumes in the history of our planet (e.g., Şengör, 2001a). However, this book contains materials on a history of all kinds of falcogenic structures.

CHAPTER II

FLOODING AND DESICCATION—EARLIEST RECORDS AND INTERPRETATIONS

From Myth to Science[23]: Exogenic Geodynamics

Rise of land from the sea has been the theme of the oldest mythologies of mankind of which we have record. It was probably conceived by farmers living on deltaic plains[24], where the flatland was witnessed to emerge welcoming a new cycle of life after every major flood[25]. Until the sixth century B.C., the record shows that it was universally believed that the water level had dropped to give birth to dry land (Fig. 5). In the sixth century B.C., we have the first clear statements of differential land uplift. This dichotomy of opinion has initiated a controversy that has been settled only in the last decade of the twentieth century with a compromise favoring both differential land movement and the movements of the hydrosphere stemming from diverse mechanisms, both tectonic and atectonic, as the cause of the relative water/land movements (see especially Dewey and Pitman, 1998).[26]

Mythological Accounts Giving the Waters the Active Role in Inundating or Desiccating the Land

Middle Eastern Myths

The earliest myths clearly let the waters swell and subside over an immobile foundation, and for this, they had a ready example in the river floods of the Nile or the great Mesopotamian rivers, the Tigris and the Euphrates. Table 1 outlines the five oldest Middle Eastern representatives of the flood myths, namely the Sumerian (fourth to third millenium B.C.), two Akkadian (third to second millenium B.C.), the Judaic (first millenium B.C.), and the Greek (first millenium B.C.) in the form of a correlation chart. I think that the correlation chart shows beyond a reasonable doubt that in the Sumerian, in the two versions of the Akkadian, in the Judaic, and in the Greek flood stories, we are looking at different recensions of *the very same story*[27]. The theme, the sequence of recitation, the characters are the same (or, at least, of the same type), and there are amazing agreements of detail.

I highlighted in Table 1 some of the physical characteristics of the flood. First and foremost is the ubiquitous association with the wind; those passages containing a mention of wind are highlighted in Table 1 with **boldface type**. It is extraordinary that in all accounts the flood is accompanied by a wind *and it seems that the waters are at the command of the wind.* The violence thus comes from the wind, the thunder, and the thunderbolt. It seems as if it were an *exodynamic source* that unleashed the flood. In other words, *it is the waters that are believed to have moved and not the land.* There is, however, a detail that may be interpreted as a complication: in all the six recensions, including the Biblical, the god who unleashes the disaster is a

Figure 5. The Jebel Djudi, or "Deluge Mountains," as viewed from the south, on 12 June 1873, by George Smith (1840–1876), the man who first discovered the pre-Biblical flood legends involving an ark in the Middle East (from Smith, 1875, facing p. 108).

TABLE 1. A CORRELATION CHART OF FIVE DIFFERENT RECENSIONS OF THE ORIGINALLY MESOPOTAMIAN FLOOD LEGEND

SUMERIAN DELUGE MYTH[1]	AKKADIAN ATRA-HASÎS MYTH[2]	AKKADIAN GILGAMESH MYTH[3]	JUDAIC NOAH MYTH[4]	GRECIAN DEUCALION MYTH[5]	GRECIAN DEUCALION MYTH Alternative account
	Tablet II, Column viii				
		8Utnapishtim[6] said to him, to Gilgamesh:			
		9"I will reveal to thee, Gilgamesh, a hidden matter			
		10And a secret of the gods will I tell thee:			
		11Shurippak—a city which thou knowest,			
		12[(And) which on Euphrates' [banks] is situate —			
		13That city was ancient, (as were) gods within it,			
	32The Assembly ...[... 33Do not obey ...[... 34The gods commanded total destruction, 35Enlil[7] did an evil deed on the peoples.	14When their heart led to the great gods to produce the flood.	*Genesis 6:11:* And God said to Noah "I have determined to make an end of all flesh; for the earth is filled with violence through them; behold, I will destroy them with the earth.	[Jove had decided to punish mankind for their violence and declares this to the pantheon] When he had done, some proclaimed their approval of his words, and added fuel to his wrath, while others played their parts by giving silent consent. And yet they all grieved over the threatened loss of the human race, and asked what would be the state of the world bereft of mortals. Who would bring incense to their altars? Was he planning to give over the world to the wild beasts to despoil? As they thus questioned, their king bade them be of good cheer (for the rest should be his care), for he would give them another race of wondrous origin far different from the first.	
		15*[There] were* Anu[8], their father,			
		16Valient Enlil, their counselor,			
		17Ninurta, their assistant[9],			
		18Ennuge, their irrigator[10],			
		19Ninigiku-Ea was also present with them;			

				And now he was in act to hurl his thunderbolts 'gainst the whole world; but he stayed his hand in fear lest perchance the sacred heavens should take fire from so huge a conflagration, and burn from pole to pole. He remembered also that 'twas in the fates that a time would come when sea and land, the unkindled palace of the sky and the beleaguered structure of the universe should be destroyed by fire. He preferred a different punishment, to destroy the human race beneath the waves and to send down rain from every quarter of the sky.	
	Tablet III *Column i*				
	.. 11Atra-hasîs opened his mouth 12And addressed his lord, 13'Teach me the meaning [of the dream] 14[...] .. that I may seek its outcome.'				And Prometheus had a son Deucalion. He reigning in the regions about Phthia, married Pyrrha, the daughter of Epimetheus and Pandora, the first woman fashioned by the gods.
154*By the wall* I will say a word to thee, [take my word], 155[Give] ear to my instruction[11]:	15[Enki][12] opened his mouth 16And addressed his slave, 17'You say, "What am I to seek?" 18Observe the message that I will speak to you:	20Their words he repeats to the reed-hut:			
	20Wall, listen to me!	21'Reed-hut, reed-hut! Wall, wall!			And when Zeus would destroy the men of the Bronze Age, Deucalion by the advice of Prometheus
	21Reed wall, observe all my words!	22Reed-hut, hearken! Wall, reflect!			
		23Man of Shuruppak, son of Ubar-Tutu,			
	22Destroy your house, build a boat, 33Let the pitch be tough, and so give (the boat) strength.	24Tear down (this) house, build a ship!	Make yourself an ark of gopher wood; make rooms in the ark, and cover it inside out with pitch.		constructed a chest, and having stored it with provisions, he embarked in it with Pyrrha.
	23Spurn property and save life	25Give up possessions, seek thou life.			
		26Foreswear (worldly) goods and keep the soul alive!			

(continued)

TABLE 1. A CORRELATION CHART OF FIVE DIFFERENT RECENSIONS OF THE ORIGINALLY MESOPOTAMIAN FLOOD LEGEND *(continued)*

SUMERIAN DELUGE MYTH[1]	AKKADIAN ATRA-HASÎS MYTH[2]	AKKADIAN GILGAMESH MYTH[3]	JUDAIC NOAH MYTH[4]	GRECIAN DEUCALION MYTH[5]	GRECIAN DEUCALION MYTH Alternative account
	25The boat which you build 26....] be equal [(..)] (Two lines missing)	28The ship that thou shalt build, 29Her dimensions shall be to measure. 30Equal shall be her width and her length.	This is how you are to make it: the length of the ark three hundred cubits, its breadth fifty cubits, and its height, thirty cubits.		
	29Roof it over like the Apsû[13]. 30So that the sun shall not see inside it	31Like the Apsu thou shalt ceil her.	Make a roof for the ark, and finish it to a cubit above;		
	31Let it be roofed over above and below[14].		and set the door of the ark in its side;		
	32The tackle should be very strong,		make it with lower, second and third decks.		
156*By our* ... a flood[15] [*will sweep*] over the cult centres;			For behold, I will bring a flood of waters upon the earth,		
157To destroy the seed of mankind ..., 158Is the decision, the word of the assembly [of the gods]. 159*By* the word commanded by Anu (and) Enlil ..., Its kingship, its rule [*will be put to an end*] (Approximately 40 lines destroyed)			to destroy all flesh in which is the breath of life from under heaven;		
			everything that is on the earth shall die.		
			But I will establish my covenant with you;		
			and you shall come into the ark,		
			you, your sons, your wife, and your sons' wives with you.		

		27Aboard the ship take thou the seed of all living things.	And of every living thing of all flesh, you shall bring two of every sort into the ark, to keep them alive with you; they shall be male and female. Of the birds according to their kinds, and of the animals according to their kinds, of every creeping thing of the ground according to its kind, two of every sort shall come in to you, to keep them alive. Also take with you every sort of food for you and for them." Noah did this; he did all that God commanded him. *Genesis 7:* Then the Lord said to Noah, "Go into the ark, you and all your household, for I have seen that you are righteous before me in this generation. Take with you seven pairs of all clean animals, the male and his mate; and a pair of the animals that are not clean, the male and his mate; and seven pairs of the birds of the air also, male and female, to keep their kind alive upon the face of all the earth.		Lucian records that Deucalion had taken a pair of all animals into his vessel (see Cuvier, 1827, p. 147, footnote).
	36He opened the water-clock and filled it; 37He announced to him the coming of the flood[16] for the seventh night.		For in seven days I will send rain upon the earth forty days and forty nights;		
			and every living thing that I have made I will blot out from the face of the ground.		
	38Atra-hasîs received the command,	32I understood, and I said to Ea, my lord: 33'[Behold], my lord, what thou hast thus ordered, 34I will be honored to carry out. 35[But what] shall I answer the city, the people, and elders?'			

(continued)

TABLE 1. A CORRELATION CHART OF FIVE DIFFERENT RECENSIONS OF THE ORIGINALLY MESOPOTAMIAN FLOOD LEGEND *(continued)*

SUMERIAN DELUGE MYTH[1]	AKKADIAN ATRA-HASÎS MYTH[2]	AKKADIAN GILGAMESH MYTH[3]	JUDAIC NOAH MYTH[4]	GRECIAN DEUCALION MYTH[5]	GRECIAN DEUCALION MYTH Alternative account
		36Ea opened his mouth to speak, 37Saying to me his servant:			
	39He assembled the elders to his gate. 40Atra-hasîs opened his mouth 41And addressed the elders,	38'Thou shalt then thus speak unto them:			
	42My god [does not agree] with your god, 43Enki and [Enlil] are angry with one another. 44They have expelled me from [my house (?)], 45Since I reverence [Enki], 46[He told me] of this matter.	39"I have learned that Enlil is hostile to me,			
	47I can [not] live in [your ...], 48I cannot [set my feet on] the earth of Enlil.	40So that I cannot reside in your city, 41Nor set my f[oo]t in Enlil's territory,			
	49With the gods .. [... 50[This] is what he told me [...	42To the Deep I will therefore go down, To dwell with my lord Ea,			
	34I will rain down upon you here 35An abundance of birds, a profusion of fishes *Column ii*	43[But upon] you he will shower down abundance, 44[The *choicest*] birds, the *rarest* fishes. 45[*The land shall have its fill*] of harvest riches. 46[He who at dusk orders] the husk-greens, 47Will shower down upon you a rain of wheat." 48With the first glow of dawn, 49The land was gathered [about me]. (three lines, too fragmentary for translation)			
	13 [The child carried] the pitch, 10 The elders [... 11 The carpenter [carried his axe] 12 The reed-worker [carried his stone]. 14 The poor man [brought what was needed]	54The little ones [carr]ied bitumen, 55While the grown ones brought [all else] that was needful.			
	15. [...	56On the fifth day I laid her framework.			

	16. He/They .. [...	57One (whole) acre was her floor space, Ten dozen cubits the height of each of her walls,	
	17 . [...	58Ten dozen cubits each edge of the square deck.	
	18 Arta-hasîs (10 lines missing)	59I laid out the contours (and) joined her together.	
		60I provided her with six decks,	
		61Dividing her thus into seven parts.	
		62The floor plan I divided into nine parts.	
		63I hammered water-plugs into her.	
		64I saw to the punting-holes and laid in supplies.	
		65Six'sar' (measures) of bitumen I poured into the furnace,	
		66Three sar of asphalt [I also] poured inside	
	29 Bringing [...	67Three sar of oil the basket-bearers carried,	
		68Aside from the one sar of oil which the *calking* consumed.	
		69And the two sar of oil [which] the boatman stowed away.	
		70Bullocks I slaughtered for the [people], 71And I killed sheep every day. 72Must, red wine, oil, and white wine 73[I gave] the workmen [to drink], as though river water, 74That they might feast as on New Year's Day. 75I op[ened .	
		76[On the sev]enth [day] the ship was completed.	
		77[*The launching*] was very difficult, 78So that they had to shift the floor planks above and below, 79[*Until*] two thirds of [*the structure*] [*had g*]one [*into the water*].	

(continued)

TABLE 1. A CORRELATION CHART OF FIVE DIFFERENT RECENSIONS OF THE ORIGINALLY MESOPOTAMIAN FLOOD LEGEND *(continued)*

SUMERIAN DELUGE MYTH[1]	AKKADIAN ATRA-HASÎS MYTH[2]	AKKADIAN GILGAMESH MYTH[3]	JUDAIC NOAH MYTH[4]	GRECIAN DEUCALION MYTH[5]	GRECIAN DEUCALION MYTH Alternative account
	30Whatever he [had ... 31Whatever he had [...	80[Whatever I had] I laded upon her: 81Whatever I had of silver I laded upon her; 82Whatever I [had] of gold I laded upon her;	And Noah did all that the lord had commanded him. Noah was six hundred years old when the flood waters came upon the earth.		
		84All my family and kin I made go aboard the ship.	And Noah and his sons and his wife and his sons' wives with him went into the arc, to escape the waters of the flood.		
	32Clean (animals) . [..........] 33Fat (animals) [...........] 34He caught [and put on board] 35The winged [birds of] the heavens. 36The cattle (?) [...............] 37 The wild [creatures(?) ....] . 38..[...........] he put on board	83Whatever I had of all the living beings I [laded] upon her. 85The beasts of the field, the wild creatures of the field,	Of clean animals, and of animals that are not clean, and of birds, and of everything that creeps on the ground, two and two, male and female, went into the ark with Noah, as God had commanded Noah.		
	39 ...] the moon disappeared. 40 ...] he invited his people 41 ...] to a banquet 42 ...] . he sent his family on board, 43 They ate and they drank. (One line missing) 45But he was in and out: he could not sit, could not crouch, (One line missing) 47For his heart was broken and he was vomiting gall.				
		86Shamash had set me a stated time: 87When he who orders unease at night, Will shower down a rain of blight, 89Board thou the ship and batten up the entrance!' That stated time had arrived: 90'He who orders unease at night, showers down a rain of blight.' 91I watched the appearance of the weather.			

	48The appearance of weather changed,	92The weather was awesome to behold.	In the six hundredth year of Noah's life, in the second month, on the seventeenth day of the month, on that day all the fountains of the great deep burst forth, and the windows of the heavens were opened. And rain fell upon the earth forty days and forty nights. On the very same day Noah and his sons, Shem and Ham and Japheth, and Noah's wife and the three wives of his sons with them entered the ark, they and every beast according to its kind, and all the cattle according to their kinds, and every creeping thing that creeps on the earth according to its kind, and every bird according to its kind, every bird of every sort. They went into the ark with Noah, two and two of all flesh in which there was the breath of life. And that they entered, male and female of all flesh, went in as God commanded him; and the Lord shut him in.	
	50As soon as he heard Adad's voice 51Pitch was brought for him to close his door 52After he had bolted his door	93I boarded the ship and battened up the entrance. 94To batten down the (whole) ship, to Puzur-Amurri, the boatman, 95I handed over the structure together with its contents.		

(continued)

TABLE 1. A CORRELATION CHART OF FIVE DIFFERENT RECENSIONS OF THE ORIGINALLY MESOPOTAMIAN FLOOD LEGEND *(continued)*

SUMERIAN DELUGE MYTH[1]	AKKADIAN ATRA-HASÎS MYTH[2]	AKKADIAN GILGAMESH MYTH[3]	JUDAIC NOAH MYTH[4]	GRECIAN DEUCALION MYTH[5]	GRECIAN DEUCALION MYTH Alternative account
201All the windstorms, exceedingly powerful attacked as one.	53Adad was roaring in the clouds **54The winds became savage as he arose,** 55 He severed the hawser and set the boat adrift. (Three lines missing) *Column iii*	96With the first glow of dawn, 97A black cloud rose up from the horizon. 98Inside it Adad thunders,	The flood continued for forty days upon the earth; and the waters increased, and bore up the ark, and it rose high above the earth. The waters prevailed and increased greatly upon the earth; and the ark floated on the face of the waters.	**Straightway he shuts the North-wind up in the cave of Aeolus, and blasts soever that put the clouds to flight; but he lets the South-wind loose. Forth flies the South-wind with dripping wings, his awful face shrouded in pitchy darkness. His beard is heavy with rain; water flows in streams down his hoary locks; dark clouds rest upon his brow; while his wings and garments drip with dew. And, when he presses the low-hanging clouds with his broad hands, a crashing sound goes forth; and next the dense clouds pour forth their rain. Iris, the messenger of Juno, clad in robes of many hues, draws up water and feeds it to the clouds. The standing grain is overthrown; the crops which have been the object of farmers' prayers lie ruined; and the hard labor of the tedious year has come to naught.**	
	4 ...]. .. 5 ...] ... **the storm** 6 ...] .**were yoked** 7[Zû with] his talons [rent] the heavens. (One line missing)	**99While Shullat and Hanish go in front,** **100Moving as hera**lds **over hill and plain.** 101Erragal tears out the posts; 102Forth comes Ninurta and causes the dykes to follow. 103The Anunnaki lift up the torches,		The wrath of Jove is not content with the waters from his own sky; his sea-god brother aids him with auxiliary waves. He summons his rivers to council. When these have assembled at the palace of their king, he says: "Now is no time to employ a long harangue. Put forth all your strength, for there is need. Open wide your doors, away with all restraining dykes, and give full rein to your river steds." So he commands, and the rivers return, uncurb their fountains' mouths, and in unbridled course go racing to the sea.	

	9[He ...] the land	**104Setting the land ablaze with their glare.** **105Consternation over Adad reaches to the heavens,** **106Who turned to blackness all that had been light.**			
		127Six days and [six] nights **128Blows the flood wind, as the south storm sweeps the land.**			
202At the same time, the flood sweeps *over the cult-centres.*[17]	10And shattered its noise [like a pot] 11[...] the flood [set out]	**107[The wide] land was shattered like [a pot]!** **108For one day the south-storm [blew],** **109Gathering speed as it blew, [submerging the mountains],**	And the waters prevailed so mightily upon the earth that all the high mountains under the whole heaven were covered; the waters prevailed above the mountains, covering them fifteen cubits deep. And all flesh died that moved upon the earth, birds, cattle, beasts, all swarming creatures that swarm upon the earth, and every man; everything on the dry land in whose nostrils was the breath of life died. He blotted out every living thing that was upon the face of the ground, man and animals, and creeping things and birds of the air; they were blotted out from the earth.	Neptune himself smites the earth with his trident. She trembles, and at the stroke flings open wide a way for the waters. The rivers overleap all bounds and flood the open plains. And not alone orchards, crops and herds, men and dwellings, but shrines as well and their sacred contents do they sweep away. If any house stood firm and has been able to resist that huge misfortune undestroyed, still do the overtopping waves cover its roof, and its towers lie hid beneath the flood. And now the sea and land have no distinction. All is sea, and a sea without a shore.	All men were destroyed, except a few who fled to the high mountains in the neighborhood. It was then that the mountains in Thessaly parted and that all the world outside the Isthmus and Peloponnese was overwhelmed.

(continued)

TABLE 1. A CORRELATION CHART OF FIVE DIFFERENT RECENSIONS OF THE ORIGINALLY MESOPOTAMIAN FLOOD LEGEND *(continued)*

SUMERIAN DELUGE MYTH[1]	AKKADIAN ATRA-HASÎS MYTH[2]	AKKADIAN GILGAMESH MYTH[3]	JUDAIC NOAH MYTH[4]	GRECIAN DEUCALION MYTH[5]	GRECIAN DEUCALION MYTH Alternative account
	12Its might came upon the peoples [like a battle array] 13One person did[not] see another, 14They were [not] recognizable in the destruction.	110Overtaking the [people] like a battle. 111No one can see his fellow, 112Nor can the people be recognized from heaven.		Here one man seeks a hill-top in his flight; another sits in his curved skiff, plying the oars where lately he has plowed; one sails over his fields of grain or the roof of his buried farmhouse, and one takes fish caught in the elm-tree's top. And sometimes it chanced that an anchor was embedded in a grassy meadow, or the curving keels brushed over the vineyard tops. And where but now slender goats had browsed, the ugly sea calves rested. The Nereids are amazed to see beneath the waters groves and cities and the haunts of men. The dolphins invade the woods, brushing against the high branches, and shake the oak-trees as they knock against them in their course.	
				The wolf swims among the sheep, while tawny lions and tigers are borne along by the waves. Neither does the power of his lightening store avail the boar, nor his swift limbs the stag, since both are swept away by the flood; and the wandering bird, after long searching for a place to alight, falls with weary wings into the sea. The sea in unchecked liberty has now buried all the hills, and strange waves now bent upon mountain-peaks. Most living things are drowned outright. Those who have escaped the water slow starvation at last o'ercomes through lack of food.	

	15[The flood]bellowed like a bull, 16[Like] a whinnying wild ass the winds [howled]. (One line missing) 18The darkness[was dense], there was no sun.	
	30The Anunnaki, the great gods, 31Were sitting in thirst and hunger.	115The gods cowered like dogs Crouched against the outer wall.
	25[Enki] was beside himself, 26[Seeing that] his sons were thrown down before him (One line missing)	
	28Nintu, the great lady, 29Her lips were covered with feverishness 32The goddess saw it as she wept, 33The midwife of the gods, the wise Mami. 34(She spoke,) "Let the day become dark, 35Let it be gloom again.	116Ishtar cried out like a women in travail, 117The sweet-voiced mistress of the [gods] moans aloud: 118"The olden days are alas turned to clay, 119Because I bespoke evil in the Assembly of the gods.
	36In the assembly of the gods 37How did I, with them, command total destruction?" (One line missing)	120How could I bespeak evil in the Assembly of the gods, 121Ordering battle for destruction of my people,
	39Enlil has had enough of bringing about an evil command, 40Like that Tiruru, he uttered abominable evil. (One line missing)	
	(One line missing) 42As a result of my own choice 43And to my own hurt I have listened to their noise.	
	44My offspring—cut off from me—have become like flies.	122When it is I myself who gave birth to my people! 123Like the spawn of the fishes they fill the sea!"

(continued)

TABLE 1. A CORRELATION CHART OF FIVE DIFFERENT RECENSIONS OF THE ORIGINALLY MESOPOTAMIAN FLOOD LEGEND *(continued)*

SUMERIAN DELUGE MYTH[1]	AKKADIAN ATRA-HASÎS MYTH[2]	AKKADIAN GILGAMESH MYTH[3]	JUDAIC NOAH MYTH[4]	GRECIAN DEUCALION MYTH[5]	GRECIAN DEUCALION MYTH Alternative account
	(One line missing) 46And as for me, like the occupant of a house of lamentation My cry has died away 48Shall I go up to heaven 49As if I were to live in a treasure house? (One line missing) 51Where has Anu the president gone, 52Whose divine sons obeyed his command? 53He who did not consider but brought about a flood 54 And consigned the peoples to destruction (One line missing)				
	Column iv (Three lines missing) 4Nintu was wailing [... 5"What? Have they given birth to the [rolling (?)] sea? 6They have filled the river like dragon flies! (One line missing) 8Like a raft they have put in to the edge, 9Like a raft ... they have put in to the bank 10I have seen and wept over them 11I have ended my lamentation for them" 12She wept and eased her feelings 13Nintu wailed and spent her emotion (One line missing)		Only Noah was left, and those that were with him in the ark.	The land of Phocis separates the Boeotian from the Oetean fields, a fertile land, while still it was land. But at that time it was but a part of the sea, a broad expanse of sudden waters. There Mount Parnassus lifts its two peaks skyward, high and steep, piercing the clouds. When here Deucalion and his wife, borne in a little skiff, had come to land—for the sea had covered all things else—they first worshipped the Corycian nymphs and the mountain deities, and the goddess fate revealing Themis, who in those days kept the oracles. There was no better man than he, none more scrupulous of right, nor than she was any woman more reverent of the gods.	

	15The gods wept with her for the land 16She was surfeited with grief and thirsted for beer. (One line missing) 18Where she sat, they sat weeping (One line missing) 20Like sheep they filled the trough. 21Their lips were feverishly athirst 22They were suffering cramp from hunger (One line missing)	124The Anunnaki gods weep with her, 125The gods, all humbled, sit and weep, 126Their lips drawn tight, [...] one and all.	*Genesis 8:* But God remembered Noah and all the beasts and all the cattle that were with him in the ark.	When now Jove saw that the world was all one stagnant pool, and that only one man was left from those who were but now so many thousands, and that but one woman was left, both innocent and both worshippers of God,	
203After, for seven days (and) seven nights, **204The flood had *swept over* the land[18],** **205(And) the huge boat[19] had been tossed about by the windstorms on the great waters**	24For seven days and seven nights 25Came the deluge, the storm, [the flood] 26Where it . [... 27Was thrown down [... (25 or 26 lines missing to the end of column)	**127Six days and [six] nights** **128Blows the flood wind, as the south storm sweeps the land.**	**And God made a wind blow over the earth,** **and the waters subsided;** **the fountains of the deep and the windows of the heavens were closed,** **the rain from the heavens was restrained,** **and the waters receded from the earth continually.**	he rent the clouds asunder, and when these had been swept away by the North-wind he showed the land once more to the sky, and the heavens to the land. Then too the anger of the sea subsides, when the sea's great ruler lays by his three-pronged spear and calms the waves; and, calling sea-hued Triton, showing forth above the deep, his shoulders thick o'ergrown with shell-fish, he bids him blow into his loud-resounding conch, and by that signal to recall the flood and streams. He lifts his hollow shell, which twisting from the bottom of a spiral expands into a broad whorl—the shell which, when in mid-sea it has received the Triton's breath, fills with notes the shores that lie beneath the	But Deucalion, floating in the chest over the sea for nine days and as many nights,

(continued)

TABLE 1. A CORRELATION CHART OF FIVE DIFFERENT RECENSIONS OF THE ORIGINALLY MESOPOTAMIAN FLOOD LEGEND *(continued)*

SUMERIAN DELUGE MYTH[1]	AKKADIAN ATRA-HASÎS MYTH[2]	AKKADIAN GILGAMESH MYTH[3]	JUDAIC NOAH MYTH[4]	GRECIAN DEUCALION MYTH[5]	GRECIAN DEUCALION MYTH Alternative account
206 Utu came forth, who sheds light on heaven (and) earth.	*Column v* (First 29 lines of the column missing)	129When the seventh day arrived, **130The flood (-carrying) south-storm subsided in the battle,** **Which it had fought like an army.** **131The sea grew quiet, the tempest was still, the flood ceased.** **132I looked at the weather: stillness had set in,** 133And all of mankind had returned to clay. 134The landscape was as level as a flat roof.	At the end of a hundred and fifty days the waters had abated;		
207Ziusudra opened a window[20] of the huge boat, 208The hero Utu brought his rays into the giant boat.		135I opened a hatch, and light fell upon my face. 136Bowing low, I sat and wept, 137Tears running down onto my face. 138I looked about for coast lines in the expanse of the sea: 139In each of fourteen (regions) There emerged a region (-mountain).		rising and the setting sun. So then, when it had touched the sea-god's lips wet with his dripping beard, and sounded forth the retreat which had been ordered, 'twas heard by all the waters both of land and sea; and all the waters by which 'twas heard it held in check. Now the sea has shores, the rivers, bank full, keep in the channels; the floods subside; and hill tops spring into view; land rises up, the ground increasing as the waves decrease; and now at length, after long burial, the trees show their uncovered tops, whose leaves still hold the slime which the flood has left.	
		140On Mount Nisir the ship came to a halt. 141Mount Nisir held the ship fast, Allowing no motion.	on the sevententh day of the month, the ark came to rest upon the mountains of Ar'arat.		drifted to Parnassus, and there, when the rain ceased, landed

	142One day, a second day, Mount Nisir held the ship fast, Allowing no motion. 143A third day, a fourth day, Mount Nisir held the ship fast, Allowing no motion. 144A fifth, and a sixth (day), Mount Nisir held the ship fast, Allowing no motion.	And the waters continued to abate until the tenth month; in the tenth month, on the first day of the month, the tops of the mountains were seen.		
		At the end of forty days Noah opened the window of the ark which he had made, and sent forth a raven;		Plutach has recorded that Deucalion had sent out pigeons to find out whether land had appeared (see Cuvier, 1827, p. 147, footnote)
	145When the seventh day arrived 146I sent forth and set free a dove. 147The dove went forth, but came back; 148Since no resting-place for it was visible, she turned round.	Then he sent a dove from him, to see if the waters had subsided from the face of the ground; but the dove found no place to set her foot, and she returned to him to the ark, for the waters were still on the face of the whole earth. So he put forth his hand and took her and brought her into the ark with him.		
	149Then I sent forth and set free a swallow. 150The swallow went forth, but came back; 151Since no resting-place for it was visible, she turned round.	He waited another seven days, and again he sent forth the dove out of the ark; and the dove came back to him in the evening, and lo, in her mouth a freshly plucked olive leaf; so Noah knew that the waters had subsided from the earth.		
	152Then I sent forth and set free a raven. 153The raven went forth and, seeing the waters had diminished, 154He easts circles, caws, and turns not round.	Then he waited another seven days, and sent forth the dove; and she did not return to him any more.		

(continued)

TABLE 1. A CORRELATION CHART OF FIVE DIFFERENT RECENSIONS OF THE ORIGINALLY MESOPOTAMIAN FLOOD LEGEND *(continued)*

SUMERIAN DELUGE MYTH[1]	AKKADIAN ATRA-HASÎS MYTH[2]	AKKADIAN GILGAMESH MYTH[3]	JUDAIC NOAH MYTH[4]	GRECIAN DEUCALION MYTH[5]	GRECIAN DEUCALION MYTH Alternative account
			In the six hundred and first year, in the first month, the first day of the month, the waters were dried from off the earth; and Noah removed the covering of the ark, and looked, and behold, the face of the ground was dry. In the second month, on the twenty-seventh day of the month, the earth was dry. Then God said to Noah, "Go forth from the ark, you and your wife, and your sons and your sons' wives with you. Bring forth with you of all flesh-birds and animals and animals and every creeping thing that creeps on the earth —that they may breed abundantly on the earth, and be fruitful and multiply upon the earth." So Noah went forth, and his sons and his wife and his sons' wives with him. And every beast, every creeping thing, and every bird, everything that moves upon the earth, went forth by families out of the ark.		
209Ziusudra, the king, 210Prostrated himself before Utu 211The king kills an ox, slaughters a sheep.	30To the [four] winds 31He put [... 32Providing food [...	Then I let out (all) to the four winds And offered a sacrifice.	Then Noah built an altar to the Lord, and took of every clean animal and every clean bird, and offered burnt offerings on the altar.		and sacrificed to Zeus, the god of Escape.

[1]Preserved only in the lower one-third of the tablet, now in the University Museum, Philadelphia, styled Catalogue of the Babylonian Section (CBS) 10673. It is no older than Late Old Babylonian (ca. 1535–1700 B.C: see Civil, *in* Lambert and Millard, 1970, p. 138). Quoted from Kramer (*in* Pritchard, 1969, p. 42f; also see Civil, in Lambert and Millard, 1970, p. 138–145).

[2]The main recension used by Lambert and Millard (1970), which I follow here, is dated as follows: Tablet II 28th of the eleventh month of 1634 B.C.; Tablet III in the second month of possibly 1635 B.C. The scribe identifies himself as Ku-Aya (see Lambert and Millard, 1970, p. 31f.). For all the manuscripts used here and their provenance, see Lambert and Millard (1970, p. 40ff.)

[3]The text I follow here is that of E.A. Speizer (*in* Pritchard, 1969, p. 93–95). See that article for further references, but I wish to underline here that one of the best English translations of the Gilgamesh Epic with extensive discussion on biblical parallels is the book by Heidel (1949).

[4]The text I have used here is that of the *The New Oxford Annotated Bible with the Apocrypha* (May and Metzger, 1977).

[5]For my main text, I used the most extensive description of Deucalion's flood, which is found in Ovid's *Metamorphoses*, Book I, lines 244–347, as given in the Loeb Series. For some additional information, I also quote, under the "alternative account" heading, excerpts from Apollodorus' *Library* (I, 7, 2) also as published in the Loeb Series, and Cuvier's (1827) references to Lucian and Plutarch. The oldest mention of Deucalion appears in Pindar's *Olympian Odes*, IX, 43 (i.e., early fifth century B.C.), but there is no mention of the flood there. Neither Homer nor Hesiod knew of him.

[6]In one place in the Gilgamesh Epic, Utnapishtim is referred to as Atrahasis (i.e., "exceeding wise"). He is indeed the same person as Atra hasîs of the Atra hasîs myth and as Noah later.

[7]The original deity of the city of Nippur, the head of the Mesopotamian pantheon and the tenant of Anu or An on Earth, Enlil was the Nippurian equivalent of the Storm God (Bottero, 1991, p. 145; see especially Şengör, 1997, p. 33f.), i.e., a Mesopotamian Zeus or Jupiter.

[8]Anu or An, the original god of the city of Uruk, was the sky god. He resided in the heavens and let Enlil rule the world for him (Bottero, 1991, p. 145).

[9]A young warrior god, whom Enki sends against the demonic Anzu, who steals Enlil's tablet of destinies and hence his power to rule over the pantheon (and thus over the world). Ninurta was given the title of the military commander of the Annunaki (assembly of gods) (see Cassin, 1991, p. 173).

[10]"Inspector of canals": see Speizer (*in* Pritchard, 1969, p. 93, note 183).

[11]There may be more than one deity speaking (not quoted here) in the beginning of this story, for the relevant Sumerian verbal forms here seem inconsistent in regard to the use of the singular and plural (Kramer, *in* Pritchard, 1969, p. 42, note 1; also Civil, *in* Lambert and Millard, 1970, p. 168, note 41). The name(s) of the speaker(s) is(are) destroyed. Kramer (*in* Pritchard, 1969, p. 42, note 1 and p. 43, note 7) thinks it is either Enki or Anu and Enlil (perhaps better Anu Enlil, meaning the Enlil to whom the powers of Anu were delegated when he replaced Anu as the head of the Sumerian pantheon; see below). Kramer (*in* Pritchard, 1969, p. 42, note 4) believes that the speaker here quoted is Enki. From the later recensions of this flood story, I agree entirely with this hypothesis. The person addressed here is Ziusudra, a pious, god-fearing king, the counterpart of the biblical Noah. Although in the preserved lower one-third of the only tablet on which the Sumerian flood story had been inscribed, it is not indicated over what city he ruled. We know from the Sumerian king list that he is supposed to have ruled over Sumer from its capital Shuruppak (Kramer, in Pritchard, 1969, p. 42, note 3).

[12]Enki (in Sumerian) or Ea (in Akkadian) was the original god of Eridu and was one of the three great gods of the Mesopotamian pantheon together with Anu and Enlil. In the oldest periods of the Mesopotamian religion, we see a fourth, but female, deity (Ninhursag, "the lady of the mountain") join the three male deities at the head of the pantheon. Enki was the god of the subterranean fresh waters, and in Sumerian his name signified "a jet of water" or a "residence in water" (Bottero, 1991, p. 146).

[13]Apsû, the anti-sky, was the realm of subterranean fresh waters (cf. Bottero, 1991, p. 145). Enki was the ruler of the Apsû.

[14]Lambert and Millard (1970, p. 159, note i. 31) are puzzled as to how the boat could be roofed both above and below. If the "boat" was more of a chest than a boat, it is easy to see how (cf. *The Epic of Gilgamesh* in Pritchard, 1969, p. 93, Tablet XI, line 58, and note 193: "The ship was thus an exact cube;" cf. also Scheuchzer, 1731, plate XL showing the various conceptions about the shape of the ark and he says, on p. 53, that until his time, the most common one was that of a chest, D in his Plate XL). Indeed, in Apollodorus' Greek account, we read: "Δευκαλίων τεκτηνάμαμενος λάρνακα," i.e., "Deukalion made a chest" (The Library, I, 7, 2), λάρναξ meaning a coffer, box, chest, or a coffin, but also an ark. Ovid renders it as a skiff (*ratis*: a raft, a float, poetically, a vessel, a boat, a bark): "Deucalion ... cum consorte tori parva rate vectus adhaesit, ..." ("Deucalion ... with consort in a small skiff carried adhered, ...": *Metamorphoses*, I, 319).

[15]Civil translates this word, *a-ma-ru,* not as flood, but as storm. Oberhuber (1990, p. 547) equates *a.ma.ru* with *abûbu* and interprets both as flood-storm or simply flood.

[16]This word, *abûbu* is variously translated, although the most recent renderings seem to agree that it is the flood or, better, flood-storm (Oberhuber,1990, p. 547) as first suggested by P. Haupt in his commentary to the Deluge in the second edition of *Die Keilschriften und das Alte Testament* (Schrader, 1883). In his letter to Suess, Haupt also admitted, however, that Suess' suggestion of "waterspout" is not impossible, as had already been suggested by Lenormant (1880) as *la trombe diluvienne* in his *Les Origines de l'Histoire.* Other translations include "storm wave" (Haupt, in his *Sumerische Familiengesetze*, 1879, p. 19), "wave raised by the tempest," and "hill of blown sand" (Lotz, in his *Tiglath-pilesar*, 1880, p. 129). The same uncertainty exists in Hebrew. It is not clear what the Hebrew proper noun for the Deluge, *mabbûl,* really means (see Suess, 1883, p. 94f., note 37; in the English edition: Suess, 1904, p. 35, note 1). Professor Walter C. Pitman III tells me that he prefers Suess' interpretation. For all references (except, of course, Oberhuber, 1990) in this footnote, see Suess' flood chapter (1883 or 1904).

[17]In Civil's translation, this line reads: "The storm swept over the capitals." (Civil, *in* Lambert and Millard, 1970, p. 143). The difference lies in the English rendering of two problematic words: *a-ma-ru* and *KAB-dug*$_4$*-ga.* For the diverse interpretations of the latter, ranging from "cult-centre" through "earth's surface" to "city" and "capital," see Civil (*in* Lambert and Millard, 1969, p. 170, note 92).

[18]In Civil's translation, lines 203 and 204 read "After the storm had swept the country for seven days and seven nights" (Civil, *in* Lambert and Millard, 1970, p. 143). The difference again hinges on how one reads *a-ma-ru.*

[19]Civil (*in* Lambert and Millard, 1970, p. 171, note 205) points out that the word ^{gish}*má-gur-gur* is the name of the boat and that the only thing we might infer about this type of boat is its large size.

[20]In the sense of "making an opening" and *not* in the sense of opening a pre-existing window (Civil, *in* Lambert and Millard, 1969, p. 171, note 207).

male god: Enlil in the Mesopotamian versions, Yahweh or El in the Biblical version, and Zeus (or Jove: Latin Iuppiter, Iovis, or Diespiter) in the Greco-Roman version. All these gods are storm gods and, as I have tried to show elsewhere, they were divine representations of *volcanoes* (see especially Şengör, 1997[28]). I therefore think that the flood legends are tied to the storm god and thus, the secondary nature of the flood legends to the storm god myth (i.e., to the volcano god) is demonstrated. The flood, therefore, is a later geological hypothesis than was the bull-god of the storms and of mountains, for in no version is the flood independent of the storm god. Unless a hypothetical, older version existed, I have not yet encountered the slightest trace of the flood story without the storm god. Whoever invented the flood story seems to have already invented, or adopted from elsewhere, the storm-god story (i.e., had experiece with, or knowledge of, a volcano). As shall be demonstrated below, there may also be other apparitions of volcanoes in flood myths.

When the volcano-god was invented is not known. Its first appearance may be recorded in the single preserved wall-painting of Çatal Hüyük, which can be dated to the latest seventh millenium B.C. (see Şengör, 1997 and the literature there). Its physical record is older than that of any flood story, including the Black Sea flood of Ryan and Pitman (1998, especially p. 149–151; also see Ryan et al., 1997). The volcano-god may have been inherited from the Palaeolithic, but I rather agree with Cauvin (1987) that the volcano-god probably post-dated the mother-goddess image and was therefore a Neolithic development. So, it must have been invented when people were already engaged in agriculture in *Anatolia*, for no other place had such grand and violent volcanoes within the primitive planters' environment as did Anatolia and the Turkish-Iranian high plateau forming its easterly continuation (Şengör and Kidd, 1979).

Could the flood myths thus have an element of chthonic contribution to the flood? Since volcanoes could at best inspire an idea of uplift in the minds of unsophisticated observers—though there is no evidence in the storm-god myths that they did—it is unlikely.

In the flood myths tabulated in Table 1, the wind unleashed by the storm-god and associated with the flood was not merely any wind. Evidence preserved in the Gilgamesh epic and in Deucalion's version of the flood story state that it was specifically the *south wind* that brought about the disaster (Table 1; Lisitzin, 1974, p. 258, thinks that a tidal wave, associated with an eruption on the island of Thera {Santorini}, may have been the origin of the myth of Deucalion's flood, but she cites no supporting evidence). In Sumerian mythology, the south (-easterly) wind is the big demon-bird that is vital for the agriculture of Mesopotamia. It is the main rain carrier in the winter (Roux, 1980, p. 107). Both in the Aegean and the northern Black Sea, the *Notas* (the south-westerly; *Lodos* in Turkey) is the main storm-maker and the rain carrier (Erinç, 1969, especially fig. IX-4 and p. 336)[29]. Thus, in both the Mesopotamian and the Greek flood myths, there is meteorological consistency: The wind that accompanies the flood is the main rain carrier *and* the storm-maker.

One aspect of the flood myths, however, that impressed both Suess (1883), the first geologist to undertake a detailed geological analysis of the pre-Biblical flood texts, and Ryan and Pitman (1998), is the slight emphasis placed on the role of rain. In fact, only the Biblical version and Deucalion's version actually mention rain as the main cause of the flood, unless line 87 in the Gilgamesh Epic stating "When he who orders unease at night, will shower down a rain of blight" suggests rain. Most philologists do take it as rain (the role of the rain-carrier southeasterly in the occurrence of the flood corroborates their interpretation) and that is how I coded it in Table 1. Both Suess (1883) and Ryan and Pitman (1998) were impressed that the waters had come from the "deep." Suess (1883, p. 41) interpreted this as "a phenomenon which is a characteristic accompaniment of earthquakes in the alluvial districts of great rivers," giving many examples from delta plains where water spouts formed "in many cases 'fathoms high'" (Suess, 1883, p. 42) during earthquakes.

But it was *not* Enki, the god of the subterranean *fresh* waters, who produced the flood, although he, "the water jet," inevitably contributed to it. The main culprit was Enlil, the storm-god and the head of the pantheon (also referred to as "the great mountain!" by Pritchard {1969, p. 50, col. 2, line 33}). Suess (1883) sees in the very designation of the flood itself, the word *abûbu*[30], the clue to what caused the flood. This word has been variously interpreted (as indicated in the footnotes to Table 1) as "hill of blown sand," "storm wave," or "wave raised by the tempest." Suess therefore thinks that it must have been a cyclonic storm (the south wind!) that hit the coast and, during the course of the storm, an earthquake ocurred. He wrote, "In certain instances, earthquakes have occurred simultaneously with cyclones; this was the case near Calcutta in the fateful night of October 11-12, 1737..." (Suess, 1883, p. 46). But in the epic itself he finds independent evidence for the whirlwind:

> 43 while Nebo and Sarru advance against each other
> 44 the "thronebearers" stride over mountain and plain. (Suess, 1883, p. 40[31])

Suess (1883, p. 41 ff) interprets these lines as follows:

> But what phenomenon of nature do the "thronebearers" [*in the modern interpretation "heralds"*] striding over mountain and plain represent?
>
> Let us glance at Lower Mesopotamia. "Although," writes Schäfli, "true storms are rare, whirlwinds are extremely frequent. Presenting in form the most striking resemblance to a waterspout and differing only in appearance by its whitish colour, the column of sand and dust raised by the wind sweeps majestically and lightly over the desert, losing itself above in the blue cloudless ether ... I remember counting eleven of these pillars of dust in one moment on my journey from Mosul to Bagdad in the middle of June of last year." (1861?).
>
> These pillars certainly sweep along like supports of the sky. But the dust-bearing storm may acquire stupendous power. An example of

this occurred in Bagdad on May 20, 1857, when, with a south-west wind, the sun first became dim and then assumed the appearance of the moon. Then at five o'clock in the afternoon, according to the account of Dr. Dutheuil, a dark cloud of dust appeared; in one moment it had enveloped the whole town, and penetrated into courtyards and rooms. In less than a quarter of a minute the day was turned to the blackest night. The effect was terrifying, everything was in confusion even in the houses. This darkness, deeper than that of the darkest night, lasted five minutes … the terrified inhabitants believed that the end of the world had come. Indeed the noise of the raging winds and the whole phenomenon were such as to inspire the minds of the stoutest with the fear of some great cataclysm.

So while explaining the "heralds," Suess also found an explanation for the disappearance of light reported in the Atra-hasîs legend (line 18 of Table 1, column iii; see Fig. 6) and in the Gilgamesh epic (line 106; see Table 1), stating that consternation over Adad "Who turned to blackness all that had been light" had reached the sky. (Lisitzin, 1974, p. 258, seems to have found Seuss' interpretation of the cause of the Flood acceptable from a modern viewpoint.)

Note that everything Suess counts as a significant observation for the explanation of the cause of the flood—namely the wind, absence of rain, rising of the water table, and the darkening of the sky owing to wind-blown dust—would find a ready explanation also in the hypothesis of Pitman and Ryan (1998). Is there an observation recorded in the preserved myths to decide between the two hypotheses? I think that there may be.

All the myths that I collected in Table 1 agree that the whole earth was covered completely by water, although in Deucalion, it is ambiguous whether Parnassus had remained out of the waters. There is a hint of the survival of Cerambus in addition to Deucalion and Pyrrha, but he was supposedly wafted aloft on the wings by the nymphs over the mountains of Thessaly (presumably to Olympos, the highest mountain and the Paradise of the Greek mythology; Ovid, *Metamorphoses*, VII, 553 ff.). According to Kerényi (1951, p. 223), Zeus' great flood to kill off the "bronze generation" left only a few people behind, those that had taken refuge in the nearest high mountains, plus Deucalion and his wife Pyrrha. In the older versions,

Figure 6. Atra-hasîs (i.e., the Babylonian Noah) in the ark (right). From an early Babylonian cylinder seal as interpreted by George Smith (1876, p. 257).

we see that the flood caused great destruction while moving *in* and that it leveled the landscape, *but that its retreat is not described.* This is how the Gilgamesh epic describes the end of the flood (among the Mesopotamian manuscripts, only it preserves the end of the flood):

129 When the seventh day arrived,
130 The flood (-carrying) south-storm subsided in the battle, Which it had fought like an army.
131 The sea grew quiet, the tempest was still, the flood ceased.
132 I looked at the weather: stillness had set in,
133 And all of mankind had returned to clay.
134 The landscape was as level as a flat roof.

The actual retreat is only implied in the episode of the birds being sent out from the ark (lines 145–154). This, Ryan and Pitman also find significant. Suess, by contrast, points out with emphasis that, at the end of the catastrophe, the *sea retreated into its basin.* He finds the following line significant:

23 became more quiet; the sea subsided and storm and deluge (whirlwind) ceased. (Suess, 1883, p. 47)[32]

Paul Haupt, the Göttingen epigrapher/philologist[33] whom Suess had consulted, thought that a literal rendering would have been: "He made the sea to withdraw into its basin" (Suess, 1883, p. 48). The interpretation of this line is the critical difference between Suess' interpretation (i.e., a flood caused by a whirlwind coinciding in time with a great earthquake) and Ryan and Pitman's interpretation (i.e., a permanent flooding of the Black Sea basin). In the first case, a retreat must have been observed. In the second, there could have been no retreat.

Both the Biblical version and Deucalion's version mention the retreat, but it is only Deucalion's version that describes it in detail and it indeed gives the impression of the sea returning into its former basin. (There seems to be a similar indication in the Akkadian Gilgamesh epic: see Table 1, lines 133 and 134, and below). Interestingly, not only the reappearance of the terrestrial topography (sea shore, mountain tops, rising of land) is mentioned but also "the trees show their uncovered tops, whose leaves still hold the slime which the flood has left" (Ovid, *Metamorphoses*, I, 345 ff.). There seems no evidence for a Mesopotamian counterpart of these descriptions other than the two lines in the Akkadian Gilgamesh epic that I just mentioned. Ryan and Pitman regard this absence as supporting a *permanent* ancient flooding event.

A variant of the Greek flood legend is recorded by Diodorus Siculus in Book V of his *Library of History* from the tiny but rugged island of Samothrace in the northern Aegean:

And the Samothracians have a story that before the floods which befell other peoples, a great one took place among them, in the course of which the outlet [*of the Black Sea*] at the Cyanean Rocks[34] was first rent asunder and then the Hellespont[35]. For the Pontus, which had at that time the form of a lake, was so swollen by the rivers which flow into it, that, because of the great flood which had poured into it, its waters burst forth violently into the Hellespont and flooded a large part of the coast of Asia [*i.e., Asia Minor*] and made no small amount of the level part of the land of Samothrace into a sea; and this is the reason, we are told, why in later times fishermen have now and then brought up in their nets the stone capitals of columns, since even cities were covered by the inundation. The inhabitants who had been caught by the flood, the account continues, ran up to the higher regions of the island; and when the sea kept rising higher and higher, they prayed to the native gods, and since their lives were spared, to commemorate their rescue they set up boundary stones about the entire circuit of the island and dedicated altars upon which they offer sacrifices even to the present day. For these reasons it is potent that they inhabited Samothrace before the flood.

This myth, like so many others, was then turned into a scientific hypothesis on the origin of the Bosphorus and the Dardanelles by Strato[36], who further extended it to the origin of the Gibraltar by the bursting of the Mediterranean into the Atlantic. Eratosthenes[37], as reported by Strabo[38] (I. 3. 4), liked these hypotheses, although Strabo criticized them because he found Strato's reasoning—that the morphology and elevation of the sea bed was the cause of the motion of the water bodies—to be absurd. He rather thought that it was the tectonic and sedimentary changes in the depth of the sea basins that caused these motions (Strabo, I. 3. 4).

Cuvier (1827), in his *Revolutions of the Surface of the Globe,* discussed the Samothracian legend at some length and concluded that it would be impossible to make a flood in the Aegean by bursting the Bosphorus and the Dardanelles open from the Black Sea:

With regard to the blending of traditions and hypotheses, by which it has recently been tried to infer the conclusion[39], that the rupture of the Thracian Bosphorus was the cause of Deucalion's deluge, and even of the opening of the pillars of Hercules[40], by making the waters of the Euxine Sea[41] discharge themselves into the Archipelago[42], supposing them to have been much higher and more extended than they have been since that event, it is not necessary for us to treat in detail, since it has been determined by M. Olivier, that if the Black Sea had been as high as it is imagined to have been, it would have found several passages for its waters, by hills and plains less elevated than the present banks of the Bosphorus; and by those of the Count Andreossy, that had it one day fallen suddenly in the manner of a cascade by this new passage, the small quantity of water that could have flowed at once through so narrow an aperture, would not only be diffused over the immense extent of the Mediterranean, without occasioning a tide of a few fathoms, but that the mere natural inclination necessary for the flowing of the waters, would have reduced to nothing their excess of height above the shores of Attica. (Cuvier, 1827, p. 147 ff., continuation of note ‡ on p. 144[43]).

With regard to the Samothracian legend, it is precisely Cuvier's reasoning that Ryan and Pitman (1998, p. 250 ff.) adopt to argue, however, that the waters did burst the Bosphorus open but by flowing the other way and that the Samothracian legend is an allochthonous one, carried from within the Black Sea basin to Samothrace by the people expelled by the flood. Such secondary flood myths indicating the flooding of lower-lying basins provide additional support for Ryan and Pitman's (1998) contention that after *the* Flood, the waters never receded. Suess' (1883) hypothesis cannot account for this inference.

In the Samothracian deluge, there is another important point (only hinted at in Deucalion's version by the survival of Cerambus): that not all land had disappeared, for most of the Samothracians *were* saved. Another connection with Cerambus, supposedly rescued by the Nymphs, is that their first post-Diluvial king, Saon, was, according to one version, a son of Zeus and Nymphê. This remnant prominence of land, somehow connected to the gods in a paradisiacal setting, is a theme that was inherited from the earliest flood myth (that we know), namely from the Sumerian. That myth ends with King Ziusudra (the Sumerian Noah; Utnapishtim of the Babylonian flood myth) being made immortal by the supreme gods An and Enlil and settling in an oriental land (of the blessed) called *Dilmun* (Pritchard, 1969, p. 44, col. 2; Civil, *in* Lambert and Millard, 1970, p. 145).

The nature and location of Dilmun has long exercised the ingenuity of the Sumerologists and other orientalists interested in the earliest history of the Middle East, although in 1880, Sir Henry Rawlinson had already identified it with the island of Bahrain in the Persian Gulf (cf. Bibby, 1969; Rice, 1994). However, Dilmun was more than just a geographic locality. As Kramer (1981, p. 142) writes, it was a land that was "pure," "clean," and "bright," and it was the "land of the living." Its inhabitants knew neither sickness nor death. It was thought to be the source of the great rivers because the fresh water god Enki endowed it with rich water resources (cf. Kramer, 1981, p. 142–143; Rice, 1994, p. 13).

But Dilmun has still more to it than just being a land of the blessed. In its earliest mention (that we know of), Dilmun is designated with the Sumerian word *kur*, which originally meant *a mountain* and later come to mean *land*, especially a *foreign land* (Civil, *in* Lambert and Millard, 1970, p. 144, line 260; see also Kramer, 1981, p. 154), even *enemy territory* (Wolkstein and Kramer, 1983, p. 157). If Dilmun were the island of Bahrain and identical with the terrestrial paradise, it is strange that the word *kur* should be employed for it, because not only is this island a very flat place (its highest prominence rising to a modest 134 m elevation) but the name this modest elevation carries is most inapposite: *Jebel ad Dukhan* (i.e., the mountain of the smoke!). The word *dukhan* in Arabic comes from the root *ahdak* (i.e., "to burn"). Even more strange, the word *kur* in Sumerian *also denotes* the nether-world, the Great Unknown (Wolkstein and Kramer, 1983, p. 157), that is, hell, and a monstrous dragon holding the waters of the violent primeval sea (the Biblical *Tehom*: cf. Kramer, 1981, p. 154, 167). This dragon, in one tale, carries off the sky-goddess, Ereshkigal, and is punished by the god of the fresh waters, Enki. While trying to defend himself against Enki's onslaught, the myth records that Kur fought by throwing small and large stones at the fresh-water god.

In its various appearances in the Sumerian mythology, Dilmun seems to carry the memory of very different features and environments. It combines in itself the features of a blessed garden and, through the word used to designate it, with those of a violent (?—if equivalent to the dragon *kur*), smoky mountain (?—if the Arabic name of *Jabal ad Dukhan* carries an ancient memory), possibly with the ability of throwing small and big stones (?—if the dragon and the mountain and Dilmun are indeed one)! The record is unfortunately not complete enough to make sense out of the fragments. In a way that remains unknown to us, the flood and the (violent?, smokey?) mountain (volcano?) somehow seem to come together in the myths woven around Dilmun. What is clear, however, is that Dilmun was the place that stood out after the flood, undestroyed and presumbly unaffected, to receive the immortalized king.

The theme of a mountain not covered by the flood will be encountered again in the Iranian version of the deluge; the alleged survival of Methusela for 14 years after the flood as reported in the *Septuagint* may reflect a similar thing[44]; God's statement in *Ezekiel* (22, 24) that Jerusalem is "a land not cleansed, or rained upon in the day of indignation," has been interpreted to imply "mountain not covered by the flood" (Levene, 1951, p. 187); the Samaritans certainly believed that it was true for Mount Gerizim (i.e., Palatinus, which is located about 3 km southwest of 32°13′N, 35°16′E in the middle of Samaria {Levene, 1951, p. 187}); the fourth century monastic master, St. Ephrem the Syrian, sang about it (Brock, 1990, p. 78–79, see below); the Asian traveler Giovanni Marignolli heard about it in Ceylon in the thirteenth century (Yule, 1914[1966], p. 245[45]); and it reappeared in the diluvial hypotheses of the seventeenth and eighteenth centuries to form the basis of a neptunian theory of Asian geological and social evolution (see below), and the latter had its repercussions well into the twentieth century![46]

All the Middle Eastern and the related Greek myths reviewed on the preceeding pages regard the exogenous agents as being responsible for the deluge and the waters as the mobile agent. This basic view remains unchanged in other myths of Asian origin (with a single exception!—see below).

Is there an implication also in any of these myths that the flood leaves a "geological record"? The following lines from the Gilgamesh epic, already quoted above, seems to imply the deposition of silt and clay where the flood raged (cf. Table 1):

133 And all of mankind had returned to clay
134 The landscape was as level as a flat roof.

The same is also implied in Deucalion's version (Table 1): "and now at length, after long burial, the trees show their uncovered tops, whose leaves still hold *the slime* which the flood has left" (italics mine). The flood therefore appears in these myths as an active process of the waters that leaves not only a record of destruction but also of construction (i.e., sedimentation). This is an aspect that has proved vital for the development of lithostratigraphy in the seventeenth century together with all its implications for the later development of the idea of eustatic movements (see below and Şengör and Sakınç, 2001).

Aryan Myths

Other flood myths are also encountered in other Asiatic cultures, and some of these seem to have interfered and/or combined with the Mesopotamian legend to inspire various scientific flood theories in Europe. The most widely known flood myths of Asian origin, apart from the classical Mesopotamian accounts, are those that we encounter in the pre-Islamic religions of Iran and in the Vedic texts of India. (For a recent assessment of direct cultural, including linguistic, connections, even between the Semitic Mesopotamian and the pre-Aryan Harappan cultures, see, for example, Shendge, 1991).

The religious picture in pre-Islamic Iran was dominated by Zoroastrianism and, owing to its pervasive influence and the way it made use of its predecessors, now it is extremely difficult to recover the original Aryan religious substrate, let alone anything that had existed before the Aryan invasions that moved into the present-day Iran and India at the beginning of the second millenium B.C. (Varenne, 1991). From the available evidence, it seems clear that at the head of the original Aryan pantheon was a god whose name appears on the famous inscription of Behistun in Iran, engraved by the great Achaemenid emperor Darius I in the sixth century B.C.[47]: Ahura Mazda[48], the god of the sky, storms, and fire—the Iranian equivalent of the great storm-god, a local Zeus (Diodorus Siculus, II. 13), whose sacred animal was the primeval bull from whose semen the humanity was created (Hansen, 1991, p. 104).

Ahura Mazda created Yima, the first man, whose name in the Veda (*Yama*) means twin, i.e., man *and* woman, who engendered the entire humanity by an incestious union. Yama became king over the humans during the Golden Age of humanity, in which there was neither cold, nor heat, nor age, nor death, nor envy. But under these perfect conditions, humans rapidly multiplied, forcing Ahura Mazda to enlarge the earth three times. But finally, Ahura Mazda thought that a permanent solution to the growth problem had become necessary, and he warned Yima that a Great Winter was about to set in[49]: "First the cloud will snow from the highest mountains to the deepest valleys ... water will flow in great waves and it will be impossible to cross over those places where now the tracks of sheep may be seen!" (*Vidêvdât*, ch. 2, quoted from Varenne, 1991, p. 889).

Gradually the whole world will disappear beneath the waters, except the highest spot, where Yima is ordered to establish a fort in which he is told to keep a pair of each animal—a sort of terrestrial Noah's ark or *Dilmun*. Here Yima was supposed to keep the remnants of the Golden Age until Ahura Mazda was appeased and the flood abated. Then he was to reopen the gates of his fort and let the inmates out to repopulate the world[50]. Yima's fortress is seen by some traditions to be still around and identical with the Paradise. When the world deteriorates through time by the increase of evil, Yima's fort will be enlarged to cover the entire earth and will re-initiate another Golden Age (Varenne, 1991, p. 889).

In the Iranian flood myth, we see some elements of the Greek flood myth as preserved in Samothrace, but more importantly, the idea of *the paradise being the highest spot on earth* appears explicitly here for the first time, an element that is certainly not explicit in the Mesopotamian myths. Although Kramer (in Pritchard, 1969, p. 47, note 3) pointed out that Dilmun may have been located on the mountain Hurrum, it is not in the least certain that this was really so, as discussed in the preceeding pages. Indeed, Dilmun is identified in the Akkadian version as being where the rivers come together (i.e., a lowland). Speizer (in Pritchard, 1969, p. 95, line 196 of Tablet XI) and Bottero (1991, p. 147), following the earlier suggestions of such authors as Huet (1691) and Delitzsch (1881) who spent considerable scholarly efforts on identifying the location of the Paradise (i.e., Dilmun), identified it as being at the head of the Persian Gulf, although the description, "where the rivers come together," seems to signify in one of the Sumerian Tammuz liturgies, *Hasur*, the common point of origin of the rivers, and Dilmun is supposedly the place of their outpouring (Rice, 1994, p. 13).

The Indian mythology has two separate versions of the flood myth: the older one is contained in the *Satapatha Brahmana*[51] (1.8.1.1–10) and is called the Matsya, one of the classical avatars in the form of a fish of the Vishnu, one of the supreme trinity of the Vedic deities (or better, one aspect of *a* deity that has three dominant characters). According to the legend, Matsya is brought (as an oversight?) to Manu in a small bowl of water for his morning ablutions. He asks Manu for protection, lest the larger fish eat him in the ocean, in return for protection from an imminent flood. Manu keeps the little fish in a pot, then in a ditch as he has grown somewhat, and finally returns him to the ocean when he has grown big enough to fend for himself. Matsya tells Manu of the date of the future flood and asks him to call him when he has finished building a boat. When Manu does as he is told, the fish hooks the boat to his horn and draws him to the north slope of the mountains to watch the receding waters. Then Manu must regenerate the humankind. He does this by practicing austerity and offering sacrifices.

In the sacrifice theme we see the appearance of the Vedic creation myth of the Purusa, of the Prajapati, the first cosmic male, dismembered to give rise to the world and humanity (Biardeau, 1991a, 1991b; Scheuer, 1991), and the flood is identified with the night of the eternal cosmic time that constantly goes from night to day and back (Biardeau, 1991c). What is important in this Indian myth is again the importance of the mountain to which Manu was pulled by Matsya. Biardeau (1991c, p. 854) indicated that the majestic Himalaya, rising north of the Indian plains, divided for the Indians the world of the mortals from the world of salvation. Behind them, i.e. in Central Asia, thus was placed the paradise! The Himalaya appears here as the immobile gauge against which the rise and fall of the flood is measured. This relation is not entirely dissimilar to the Iranian flood myth, with which the Indian myth had been once no doubt connected (before the Iranian and Indian Aryans went their separate ways in the beginning of the second millenium B.C.) and indicates an immobile land versus a mobile water body.

Closer parallels to the Mesopotamian story emerge in the later Purânic texts. In the *Matsya Purâna* (1. 12 ff.), for example, Manu is provided with a boat by the gods and is asked to bring aboard all living creatures (Biardeau, 1991c, p. 854). In the *Bhâgavata Purâna* (8. 24. 7–58), the god Vishnu himself is identified with the fish Matsya. In yet another Purânic text, Vishnu, in his incarnation as Krishna, saves mankind from drowning, this time by lifting Mount Govardhana by his fingertip, when the head of the Vedic pantheon—Indra, the storm god—unleashes a deluge for seven days and seven nights. *This is the first instance in all the flood myths that I have reviewed so far that involves differential motion of land independent of the movement of the waters.*

East and Southeast Asian Myths

Farther east in China and in southeast Asia, there are also flood stories, but few of these bear any similarity to the Mesopotamian myth. In southeast Asia, Thierry (1991, p. 916) records a flood myth in which the flood results from an argument amongst the gods and puts an end to an early phase of humanity, much like its Mesopotamian equivalent from which it may have ultimately derived. The Sré of southeast Asia believe that a flood coming from the sea destroyed all mankind except a young man and his sister, who took refuge in a drum. After the flood, they emerged from the drum and engendered the present human population (Dournes, 1991, p. 984). In China, the only similar legend is from the southern part of the country; it is known as the myth of Fuxi and Nüwa, the survivors of the flood and progenitors of mankind (Kaltenmark, 1991, p. 1025).

The flood hypothesis, even in its mythical form, was probably conceived as an explanation of a major geological event (as Suess and Ryan and Pitman assume), or perhaps of a class of events that took, or have kept taking, place in maritime countries. In all of the flood myths, it was clearly believed that it was the *waters that had risen* to cover the land. Only in the Purânic flood story, Vishnu, in his incarnation as Krishna, lifts Mount Govardhana by his fingertip, thereby giving evidence that independent movement of land was thought of by the creators of the story. But this movement was not the cause of the deluge—it was effected to provide a refuge to the inundated mankind.

Mythological Accounts Giving the Rocky Foundation the Active Role in Inundating or Desiccating the Land

It was in a younger family of hybrid Aryan-Middle Eastern mythologies (that of the Hellenes, who probably had begun to identify themselves as such from the beginning of the first millenium B.C. onwards—see, for example, Schachermeyr, 1983, p. 337) that we begin to see statements that may be interpreted as implying the movement of land (i.e., the rocky foundation) as the cause of the recession of the waters when, for example,

Epimenides[52] declares that Rhodes was the daughter of the Ocean.[53] [54] Numerous earlier statements in the Greek literature are, however, difficult to interpret, such as Kerényi's (1951, p. 45) point that one of the daughters at least of Okeanos was *Petraea*, the "rocky" (Hesiod, *Theogony*, 357), or that the island of Delos (from δῆλος meaning visible, manifest, conspicuous) was so called because it had risen out of the waters (see especially Kerényi, 1951, p. 130, for the references to the classical authors to piece this story together). Hesiod further chanted in *Astronomia*[55] that "[*in the Straits of Messina*] the sea was open, but Orion piled up the promontory by Peloris, and founded the close of Poseidon which is especially esteemed by the people thereabouts." In Musaeus[56], we encounter the statements that "Triptolemus was the son of Ocean and Earth" (DK2B10) and "Shooting stars are borne up from Ocean and generated in the Aether" (DK2B17). Could this have been in comparison with incandescent ejecta from sub-sea-level volcanoes?). Pindar's[57] following statement is clearer:

> But the tale is told in ancient story that, when Zeus and the immortals were dividing the earth among them, the isle of Rhodes was not yet to be seen in the open main, but was hidden in the briny depths of the sea; and that, as the Sun-god was absent, no one put forth a lot on his behalf, and so they left him without any allotment of land, though the god himself was pure from blame. But when that god made mention of it, Zeus was about to order a new casting of the lot, but the Sun-god would not suffer it. For, as he said, he could see *a plot of land rising from the bottom of the foaming main*, a plot that was destined to prove rich in substance for men, and kindly for pasture; and he urged that Lachesis of the golden snood should fortwith lift up her hands and take, not in vain, the great oath of the gods, but consent with the Son of Cronus, that that island, when it had risen forth into the light of day, should for ever after be a boon granted to himself alone. (Pindar, *Olympian Odes*, VII, 55–69; italics mine)

In the song *On Delos*, Pindar calls that island "god-made" (also in *Olympian Odes*, VI, 99) and "daughter of the sea." In Pindar's words, there is little question that "*an active rise of land*" within a passive water body is implied. In fact, Kerényi (1951, p. 185) uses the phrase "growing upwards" (*emporwachsen*) for Rhodes' appearance from beneath the waves in rendering Pindar's description into German. It is another interesting observation that almost all the earliest relevant Greek fragments that we possess come from the adherents of Orpheus, thus underlining a strong oriental connection.

Theophrastus'[58] remark about Rhodes and Delos: "The most famous of islands, Rhodes and Delos, formerly invisible and submerged, were covered by the sea, but, later, the level dropped gently, they were elevated little by little and became visible" (fragment no. 784 in Bouillet-Roy, 1976, p. 320–321) has the opposite implication from Pindar's ode. In Theophrastus' passage, the *slow* rise of the islands out of the sea is ascribed *explicitly* to the *dimunition of the sea*[59]! It is unclear to me whether Theophrastus was perpetuating an ancient tradition in ascribing the recession of the waters to an actual dimunition of the waters or whether he was simply exhibiting the influence of the geological theories of Anaximander (see below) on the later generations. In the light of the views of his friend and teacher, Aristotle, on the same question (see below), I rather incline to the latter possibility. If that were so, the Greeks appear to be the first people who have left record of thoughts concerning actual differential movement of various parts of land detected by using sea level as a datum (with the singular exception of the Purânic text cited above, according to which Vishnu saves mankind from drowning by lifting Mount Govardhana by his fingertip, when Indra unleashes a deluge). That the early Greeks did indeed contemplate active rise of parts of the rocky foundation with respect to their surroundings is clearly spelled out in the myth of Mount Helicon's growth toward the sky because of the excitement, caused by the songs of the Muses (Pausanias, IX.29.3).

Anaximander and the Origin of the Natural Sciences in Ionia

Anaximander (*fluorit ca.* 560 B.C.: Fig. 7), co-founder with Thales of the Ionian enlightenment[60] (Popper, 1989), claimed a general desiccation with the consequent enlargement of land[61] in his book, which Heidel (1921) believes was a geographical treat-

Figure 7. Anaximander of Miletus (*fluorit* ca. 560 B.C.), together with Thales of Miletus, were the founders of science-based human civilization as we know it. He developed the first scientific theory of earth history. Figure is from a small portrait relief in the Museo Nazionale Romano, probably early Roman Empire after Hellenistic original (from Kahn, 1960, frontispiece).

ise, although it seems to have been more of a combination of cosmology, historical geology and, perhaps, what we today understand under history proper (see Heidel, 1921, especially p. 287–288). In this sudden reversal[62] of the moving medium—from land to sea[63]—to explain the reciprocal relations of the terra firma and the ocean, we may be witnessing one of two things, or perhaps a mutually reinforcing combination:

1. Allowing the sea to move rather than the land may have been the reaction of the Milesian physicists[64] against the mysterious, the mythical, and the religious.[65] Evaporation was an obvious *physical* mechanism to depress the level of the sea with respect to land. It was not equally obvious what may have raised Rhodes out of the water. Hence, Anaximander opted for depressing the sea level.[66] Heidel (1921, p. 275) argued that it may have been Anaximander's personal acquaintance with the Nile Delta and the conditions there that pushed him to consider fluctuating the sea level: "There fishes spontaneously generated; there existed the ideal conditions for the beginnings of life; there was the cradle of the human race and the fountain head of civilisation. There, we may with reason assume, Anaximander (and perhaps Thales) laid the scene of the early life history of the earth." If that were so, then Anaximander's imagination was fed from the same source that had fed the earlier flood myths. His distinction was to suggest a testable model for the origin of the earth and life, instead of offering a theological tale.

2. The preference to move the sea level and not the land may represent a reflection of the beliefs in the Middle Eastern mythologies, in which ocean and the creative power and/or life itself have always been thought of as antagonistic powers, as Wensinck (1918) showed in his detailed study on the place of the ocean in the thinking of the western Semites. Creation has always confined the waters to beyond barriers erected by the creator so that life would not be endangered. But the relation between the waters and life is not entirely one of hostility. Wensinck (1918, p. 3) points out that St. Ephrem the Syrian (died 373 B.C.) and Jacob of Serugh (451–521 B.C.) thought that, in the beginning, the earth was covered by the waters like an embryo by membranes. God rent this mass of water asunder and made mountains and basins into which he ordered the waters to retreat. Wensinck (1918) says that another frequent comparison, in both the Syriac and the Arabic literature, is that between the earth, surrounded by waters, and the yolk of an egg, surrounded by the glair. Especially this last comparison strongly reminds one of Anaximander's crusty primitive creatures, but the Syrian writers post-date the Ionian by nearly a millennium, and it is unclear as to who influenced whom (owing to the long pedigree of the Syrian writers). The Yima legend bears a strong resemblance to St. Ephrem's *Paradise Song* (quoted below in Chapter VII), but even the latter has a Rabbinic predecessor as I have shown above. Owing to temporal priority, I incline to assign to Anaximander the originality of his views and the thoroughly secular interpretation as also argued by Heidel (1910, 1921) and Burnet (1930).

One of Anaximander's students, the brilliant Xenophanes of Colophon (*fluorit* ca. 530 B.C.)[67], conceived of *repeated* transgressions and regressions—every transgression culminating in a universal deluge (and total destruction of life?) and dissolving the earth and every regression ending in complete desiccation[68]. If he suggested any observational basis or mechanism for this cyclical behavior, they have not been preserved (Lasaulx, 1851, p. 5–6; Schvarcz, 1862, p. 34–35; Zeller, 1919, p. 666–669; Guthrie, 1962, p. 387–390). Xenophanes may have put it forward to reconcile his teacher Anaximander's theories of an eternal universe and the reign of eternal justice, with the record of a progressive drying up of the earth[69].

In this context we should also bear in mind that cyclical regenerations of the world, each ending with a conflagration, is a very old and fundamental Indo-European mythic theme confronting us from Vedic India through Zoroastrian Iran to the Greco-Roman world (Varenne, 1991)[70]. Macrobius, in his *Commentary on the Dream of Scipio* (published sometime before 410 A.D.: Stahl, 1952, p. 5), has offered a natural cause for the alternation of floods and conflagrations (see Stahl, 1952, p. 218):

> While the world goes on, our civilizations often perish almost completely, and they rise again when floods or conflagrations in their turn subside. ... The cause of this alternation is as follows. Natural philosophers have taught us that ethereal fire feeds upon moisture, declaring that directly under the torrid zone of the celestial sphere, which is occupied by the sun's course or zodiac, nature placed Ocean, as is shown by our diagram [*Fig. 8*], in order that the whole broad belt over which the sun, moon, and the five errant planets travel might have the nourishment of moisture from beneath. ... Since heat is nourished by moisture, there is an alternation set up so that now heat, now moisture predominates. The result is that fire, amply fed, reaches huge proportions, and the moisture is drained up. The atmosphere in this changed state lends itself readily to conflagration, and the earth far and wide is ablaze with raging flames; presently their progress is checked and the waters gradually recover their strength since a great part of the fire, allayed by the great conflagration, now consumes less moisture. ... Then, after a great interval of time, the moisture thus increasing prevails far and wide so that a flood covers the lands, and again fire gradually resumes its place; as a result the universe remains, but in the alternation of excessive fire and flood civilisations perish and are born again when temperate conditions return.[71]

From Anaximander to Empedocles: Fire Inside a Porous Earth and the Discovery of Endogenic Geodynamics

Xenophanes' fellow-student, Anaximenes (*fluorit* ca. 546 B.C.), a native of Miletus like his teacher Anaximander[72], thought that the progressive drying up of the earth could be made responsible for the earthquakes. In the *Meteorologica*, Aristotle preserved the essence of this theory: "...when the earth is in the process of becoming wet or dry it breaks, and is shaken by the high ground breaking and falling. Which is why earthquakes occur in droughts and again in heavy rains: for in droughts the earth is dried and so, as just explained, breaks, and when the rains make it excessively wet it falls apart" (Aristotle, *Meteorologica*, 365b). There is nothing remarkable in

Figure 8. The world map of Macrobius Ambrosius Theodosius (*fluorit* ca. 400 A.D.) showing the equatorial ocean (*Alveus Oceani*, ocean trough) and the two continental groups separated by it. The equatorial ocean is flanked by scorched zones (*perusta*) created by the Sun's passage directly overhead from east to west, as explained by Macrobius. The map is from a 1485 printed edition of Macrobius' *Commentary* published in Brescia (from Stahl, 1952, map facing p. 215).

Anaximenes' seismic theory (for its medieval adoption by Ğâbir ibn Haiyân, see Sezgin, 1979, p. 236), nor anything particularly relevant to our topic in this book, except that in objecting to that theory, Aristotle noted that if Anaximenes' hypothesis were right, "the earth ought to be sinking in many places" (Aristotle, *Meteorologica*, 365b). I interpret this statement as saying that, in drying and thus seismically crumbling, *large tracts of land would sink owing to withdrawal of subterranean water*. This simple sentence of Aristotle will gain great significance below in undestanding the Academy's—and of course also the Lyceum's—view of the internal structure and tectonic behavior of our planet and connecting it with the views of the inaugurators of the Ionian natural philosophy.

The Sicilian Empedocles' (*fluorit* ca. 450 B.C.)[73] view of cyclic generation and destruction of the world by alternate domination of Love and Hate[74] clearly owed much to Xenophanes' teaching, in addition to those of Anaximander, Heracleitus, and Parmenides, though it is not the reason I mention him here. In relating Hate's job of disintegrating a spherical and internally placid Cosmos, the Agrigentian describes how fire escapes from the pre-existing mixture:

> As when a man, thinking to make an excursion through a stormy night, prepares a lantern, a flame of burning fire, fitting lantern-plates to keep out every sort of winds, and these plates disperse the breath of the blowing winds; but the light leaps out through them, in so far as it is finer, and shines across the threshold with unwearying beams: so at that time did the aboriginal Fire, confined in membranes and in fine tissues, hide itself in the round pupils; and these [tissues] were pierced throughout with marvellous passages. They kept out the deep reservoir of water surrounding the pupil, but let the Fire through [from within] outwards, since it was so much finer. (DK31B84; Bollack, 1969b, p. 134–135)

Bollack's (1969c, p. 314–329) very detailed and scholarly commentary on this fragment, concentrates (as does the discussion by Guthrie, 1965, p. 234–238) on Empedocles' discussion of the nature of the eye and the vision. However, this quoted fragment is a part of his poem called περὶ φύσεος (*peri phuseos*: on Nature) and seems to belong to that part where Empedocles describes how the earth and its inhabitants develop through the work of Aphrodite (i.e., Love). Although apparently the workings of the eye and the mechanism of vision are described, it is done by using the porous nature of the building materials and the nature of the fire. The eyes were viewed by the Greeks to have been born of the Sun or the sunlight (Kerényi, 1951, p. 186), i.e., they were "like the sun" and themselves issued an inner light. Now in the same poem, Empedocles also stated that "Many fires burn below the surface [of the Earth]" (DK31B52; Bollack, 1969a, p. 88–89) and that "Mightily upwards [rushes fire]"[75] (DK31B51; Bollack, 1969a, p. 62–63). Bollack (1969b, p. 228) quotes the commentary on *Timaeus* by Proclus of Byzantium (fifth century A.D.) to underline that *rivers of fire* flowing beneath the earth had been postulated by Empedocles. Fragment B51 adds that Empedocles believed that this fire may in places vigorously go upwards. How Empedocles imagined this fire to be housed within the earth is given in a testimony by Seneca (Fig. 9)[76]:

Empedocles thinks that the water is heated by fires which the earth covers and conceals in many places, especially if the water lies under earth such as waters can pass through. We commonly construct serpent-shaped containers, cylinders, and vessels of several other designs, in which we arrange thin copper pipes in descending spirals so that water passes round the same fire over and over again, flowing through sufficient space to become hot. So the water enters cold, comes out hot. Empedocles conjectures that the same thing happens under the earth. (*Quaestiones Naturales*, III, 24: DK31A68; see also Bollack, 1969a, p. 88–89; 1969b, p. 227–228)

Figure 9. Lucius Annaeus Seneca (ca. 4 B.C.–65 A.D.), tutor, confidant, and victim of Nero. Seneca's *Quaestiones Naturales* was built upon a Greek foundation and formed one of the chief sources of medieval and Renaissance geology. This third century marble bust, after a first century original, is located in the Staatlichen Museen zu Berlin.

This quote justifies my employment of the fragment DK31B84 to give us a model of the interior of the earth as it was imagined by Empedocles. Now we are ready to inquire in what philosophical context the Agrigentian may have placed the subterranean fires, with a view to asking later what implication this subteranean fire had in his, or in his followers', eyes for the tectonic behavior of the planet, to put it anachronistically.

Schvarcz (1862, p. 2) pointed out that although Aristotle in his *On the Heavens* had written that "The theory of the Italian 'Pythagorian' philosophers is opposed to those who maintain that the Earth is in the middle of the Cosmos. They claim that in the middle is a fire and that the earth is one of the heavenly bodies that produce day and night by moving around the middle in a circle" (DK58B37[77]), Simplicius[78] added (*In de Caelo*, 512. 9: cf. Guthrie, 1962, p. 290): "So does he [i.e., Aristotle] recount the tenets of the Pythagoreans; those however who have a better information place the fire into the interior of the earth as a creative power to vivify thence the whole earth, and to replace the cooling new heat" (see also Sambursky, 1956, p. 67).

Fire in the middle of the earth could not have been an entirely strange concept even for the early Pythagoreans. Pythagoras himself[79] often spoke of the dead going to the nether world and once even interpreted the origin of the earthquakes as "nothing but a concourse of the dead" (DK58C2). But the nether-world where the dead met was also, according to the description of the Goddess Circe, the beautiful daughter of Helios, "... the dank house of Hades. There into Acheron (the river of misery)[80] flow Periphlegeton (one that burns like fire) and Cocytus (wailing; i.e., the River of Wailing)[81], which is a branch of the water of the Styx (the river of hatred)"[82] (Homeros, *Odyssey*, X. 512–514)[83]. So the nether-world, according to the mythology as related by Homer, was a place that contained rivers of water *and rivers of fire*. Although the Pythagoreans were more followers of the rival mythology of Orpheus, even Orpheus himself was believed to have made a journey into Hades to rescue his beloved wife Eurydike (Kerényi, 1958, p. 302 ff.). All of this has its roots way back in the Sumerian mythology, where Kur, as we saw above, also signified Hell. This Hell was thought to be the empty space between the earth's disk and the primeval sea, and to it went all the shades of the dead. A "man-devouring river," the equivalent of the Grecian Styx, had to be crossed to reach it, on a boat steered by a boatman, the Sumerian counterpart of Charon (Kramer, 1981, p. 154). Subterranean caves and galleries were persistent images in all Middle Eastern mythlogies of which we have record.

This fiery Hell image *within the earth* appears as part of a common Indo-European myth in the ancient sacred books of the *Avesta*. When Zoroaster, the god-prophet interrogates the one god Ahura Mazda (Varenne, 1991, p. 879) as to where the earth feels most pain, the divinity responds by saying that "It is the neck of Arezûra, whereon the hosts of fiends rush forth from the burrow of the Drug" (Darmesteter, 1887, p. 24). Drug is interpreted as Hell, and Arezûra or Arezûra grîva could be any one of the active volcanoes crowning the Turkish-Iranian high plateau (cf. Şengör and Kidd, 1979) such as the Demâvend in the Alborz or the Tendürek or the Süphan in Turkey[84].

This view of a volcanic vent as the entrance to a subterranean hell has been a regularly recurring image in literature: Modi (1905–1908) cites al-Mas'ûdi on the Etna, saying: "It is a source of fire from which come out enflamed bodies resembling bodies of men but without head which rise high in the air during the night to fall back afterwards to the sea," and further, Etna throws out "fires accompanied by bodies" (Modi, 1905–1908, p. 329). Modi then compares these passages with Dio Cassius' description of the 79 A.D. eruption of Vesuvius: "Many huge men, surpassing human stature, such as the giants are described to have been, appeared wandering in the air and upon the earth, at one time frequenting the mountain, at another the fields and cities in its neighbourhood. ... Some thought the giants were rising again (for many phanthoms of them were seen in the smoke, and a blast as if of trumpets[85], was heard)" (Modi, 1905–1908, p. 141 ff.).

Modi rightly connects these descriptions with the beliefs that volcanoes were the terrestrial representatives of Hell or its entrances:

> Thus it appears both from an Arab author and a Roman author that people thought that they saw figures of men rising from the volcanoes high into the air. Don [*sic*!] Cassius says that they appeared to hover over cities and fields. Of course, this was due to all the fantastic shapes which the vapours emenating from the craters assumed. But these statements suggest the idea that perhaps it is from the appearance of such phanthoms or fantastical shapes of vapours, added to the terrible sound from within, that the ancients thought that the volcanoes were the localities of Hells where the bodies of the sinful were burnt in suffocating flames and smoke. (Modi, 1905–1908, p. 141 ff)

Indeed, the still-smoking Solfatara in the Campo Flegrei had been thought of as the entrance to Hell; it was the source of inspiration both for Aeneas' trip to the underworld in the IVth book of Virgil's *Aeneid*—which Gibbon believed was the most pleasing and perfect composition of Latin poetry—and Dante's *Inferno* (Bullard, 1976, p. 187). The noise that at times emanated from Mount Hekla in Iceland was thought of as being the moans and groans of the tormented souls, and Mount Etna was believed to be the place where the sinful soul of Anne Boleyn resided (MacDonald, 1972, p. 30)!

Such were probably the pre-scientific foundations of a conviction that *there was fire in the earth*. Empedocles was an admirer of Pythagoras and a follower of at least some of his teaching. Schvarcz (1862) thought that he could see evidence that not only did the Pythagoreans believe in a central fire within a spherical earth, but so did Empedocles *following their footsteps*. This is strongly supported by the evidence cited in Guthrie (1962, p. 289–293) and compatible with what Proclus says (see above), showing that indeed the original Pythagorean cosmography[86] was that of a geocentric universe with the earth having a fiery interior. Guthrie also cited evidence to argue that this Pythagorean view of a terrestrial central fire probably had derived from empirical evidence on volcanoes and hot springs, but that it had also another, "theoretical" reason: Greeks, he says, commonly believed that all life, animal as well as vegetable, originated within the earth, and was being nourished by heat and moisture (see Simplicius' statement above!). Guthrie thinks this belief goes way back to the days of mother-worship in the early Neolithic[87]. Though in Anaximander's exogenic theory the sun is the only source of heat, "both myth and philosophy preserve traces of the idea that the heat as well as the moisture came from inside the earth..." (Guthrie, 1962, p. 292).

That Empedocles believed in a central fire is very commonly thought (Guthrie, 1962, p. 292; 1965, p. 206 ff.) and may have an additional root in Alcmaeon's (*fluorit* ca. 500 B.C.) idea that the eye, a spherical organ, has fire in it to enable it to see (Freeman, 1949, p. 135, 137; Guthrie, 1965, p. 237; Clagett, 1994, p. 39). Plutarch in his *On the Principle of Cold* also mentions Empedocles in the context of possibly the earliest fragment about an *internal geodynamics*: "As for these features that are visible, cliffs and crags and rocks, Empedocles thinks that they have been made to stand and are upheaved by leaning on fire that burns in the depths of the earth."[88] This report, if true,[89] indicates clearly an *endogenic* view of earth behavior, in which pores play a significant role. It was thus distinct from the *exogenic* view of Anaximander and Xenophanes, though in a way connected with theirs via Anaximenes' model of a porous earth. Bollack (1969b, p. 248–249) even brings the following passage from Lucretius' *De Rerum Natura* into connection with Empedocles' endogenic ideas: "... and *so much the more slipped out and flew away those many bodies of heat* and air, and on high far from the earth packed the shining regions of the sky. *The plains settled down, the lofty mountains increased their height*" (Lucretius, V, 489–493, italics mine)[90]. It is perhaps not surprising that Empedocles, writing in the luxury of hot springs (cf. Bollack, 1969b, p. 248, 260{A69}) and under the threat of the towering and smoking Etna (which may have had two major eruptions during his lifetime: 475 and 426 B.C. {Simkin et al., 1981}), should be more aware of the internal heat of the planet than his Milesian predecessors who fluorished on vast alluvial lands quietly encroaching almost by the day onto the Aegean Sea and only occasionally getting shaken by strong earthquakes.

CHAPTER III

TWO KINDS OF MOVEMENT CAUSED BY THE EARTH'S INTERIOR AT ITS SURFACE: PLATO AND HIS STUDENTS

Plato's Picture of the Internal Structure of the Earth and its Sources

Whatever Empedocles may have actually thought and written, by the time Socrates was sentenced to death in 399 B.C., it seems that the importance of subterranean fire and its surficial effects in shaping the landscape were common knowledge in Athens. For example, Antiphon the Sophist[91] stated that earthquakes were caused by internal fire[92]: "(*The fire*) by heating the earth and melting it, makes it corrugated" (DK87B30); "(*Word for earthquake*): Corrugation" (DK87B31)[93]. In Plato's (Fig. 10)[94] *Phaedo*[95], the condemned Socrates explains, upon Simmias'[96] request, his view of the structure of the earth, saying in one place:

> ...and there are everlasting rivers of huge size under the earth, flowing with hot and cold water; and there is much fire, and great rivers of fire, and many streams of mud, some thinner and some thicker, like the rivers of mud that flow before the lava in Sicily, and the lava[97] itself. These fill the various regions as they happen to flow to one or another at any time. (Plato, *Phaedo*, 111E[98])

Figure 10. Plato (428/7–348/7 B.C.). This is a Roman copy, probably from a 4 B.C. Greek original located in the Staatlichen Museen zu Berlin.

So far there is nothing unusual in what Plato has Socrates say here. It is merely a pious picture of the internal structure of the earth taken directly from the mythology as we saw above (see also Pfeiffer, 1963, p. 19–20). It was most likely common knowledge at the time and survived well into the early childhood of modern geology in the seventeenth century (e.g., Kircher, 1665)[99]. The unexpected comes in the next sentence, in which Socrates tells Simmias that:

> ... a kind of oscillation [*Gallop, 1975, p. 66, translates this as "pulsation"*] within the earth moves all these up and down. And the nature of the oscillation is as follows: One of the chasms of the earth is greater than the rest, and it is bored right through the whole earth[100]; this is the one which Homer means when he says "Far off, the lowest abyss benearth the earth;" and which elsewhere he and many other poets called Tartarus. For all the rivers flow together into this chasm and flow out of it again, and they have each the nature of the earth through which they flow. And the reason why all the streams flow in and out here is that this liquid matter has no bottom or foundation. So it oscillates and waves up and down, and the air and wind do the same; for they follow the liquid both when it moves toward the other side of the earth and when it moves toward this side, and just as the breath of those who breathe blows in and out, so the wind there oscillates with the liquid and causes terrible and irresistable blasts as it rushes in and out. (Plato, *Phaedo*, 112A, B, C; for a different translation, refer to See, 1907, p. 238.)

This last quotation *is* surprising because it gives, in the space of a few short sentences in a dialogue devoted to the nature of the soul and to the problem of universals in the form of abstract "forms," an almost complete theory of the internal geodynamics of the earth that reigned supreme almost till the eighteenth century[101], when the first version of a usable theory of an endogenous geodynamics was proposed by the Abbé Antonio-Lazzaro Moro[102] (see the excellent discussion on Plato's ideas on endogenous events of the earth in Serbin, 1893, p. 5–7). It is also surprising, because Socrates talks about surface "oscillations" that are the result of the oscillations of a *fluid* interior (Frank, 1923, who presents a very illuminating discussion of the origin of Plato's earth-hypothesis in *Phaedo*, does not understand the extremely important implication of the *fluid center* and the oscillations it causes like "the breath of those who breathe blows in and out"; see Fig. 11). These oscillations are clearly *independent of volcanic eruptions* and of the mountains that the volcanoes construct. *With Plato, we have thus for the first time a recognition of two independent ways to make topographic prominences*: volcanic mountains, the products of which are of small wavelength (i.e., copeogenic and irreversible), and "oscillations" of the solid part of the globe that have very large wavelengths (i.e., falcogenic) as implied by Socrates' statement in *Phaedo* concerning the large passageways in the earth: "These fill the *various regions* ... Now a kind

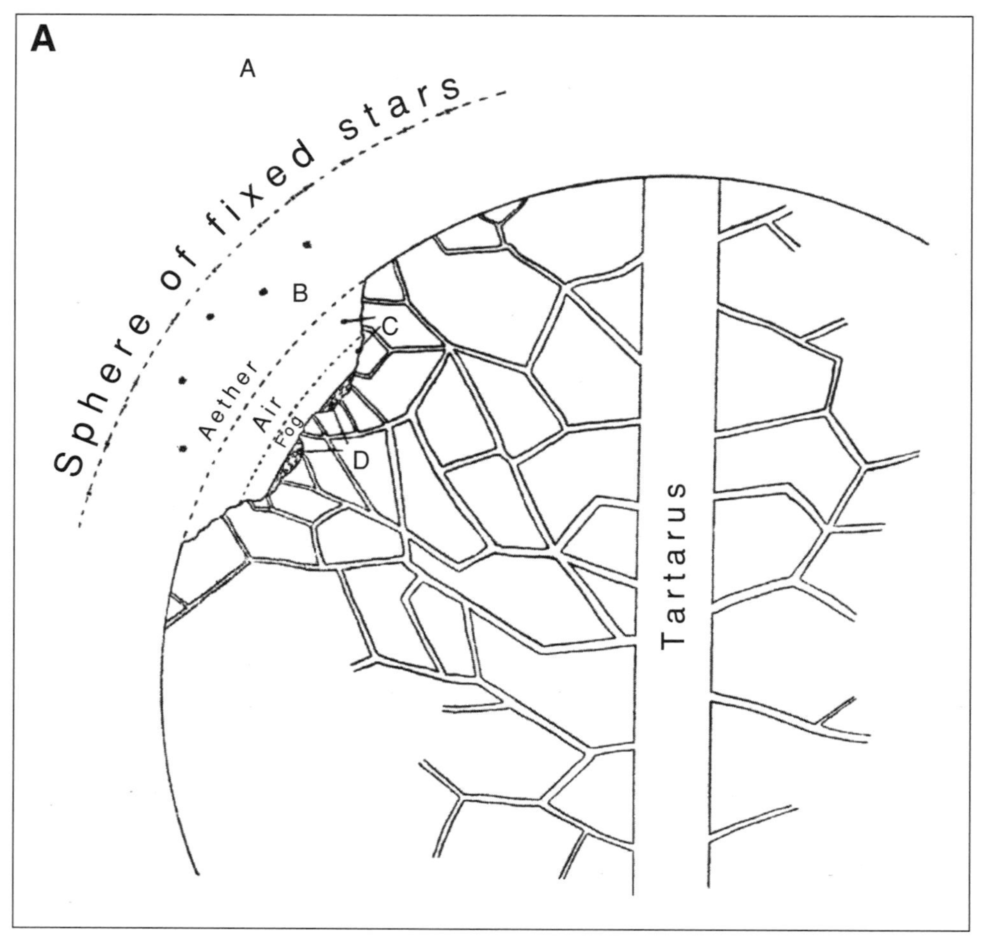

Figure 11. A: Frank's (1923, fig. 2) view of the interior of Plato's earth. Frank's key to letters (for both 11A and my revision in 11B): (A) Extra-universal (*hyper-heavenly*) place of Ideas (ὑπερουράνιος τόπος). (B) Space filled with aether, place of planets (star-gods), the "quantitative world of mathematical astronomy." (C) The world of humans ("qualitative world," air-cave); (D) The world of water (water animals), the apparent world of water, reflexion, etc. Note that the Tartarus (or Hades) is depicted as a cylinder through the globe that intersects the surface of the globe in two places. With this geometry, I find it difficult to visualize how the liquid filling the tubes in Plato's earth "oscillates and waves up and down, and the air and wind do the same; for they follow the liquid both when it moves toward the other side of the earth and when it moves toward this side, and just as the breath of those who breathe blows in and out, so the wind there oscillates with the liquid and causes terrible and irresistible blasts as it rushes in and out" (Plato, *Phaedo*, 112B). *(continued)*

of oscillation in the earth *moves all these* up and down" (Plato, *Phaedo*, 111E; italics mine) and are admittedly reversible[103].

Plato's theory of underground passages most probably had its roots in Empedocles' theory of pores allowing the fire to escape the earth without letting water flow in during the reign of "Hate." Through Empedocles, that theory may be related to the medical ideas that were current among the Pythagoreans in Croton (Zafiropulo, 1953, p. 141), to such thinkers as Alcmaeon, whose Pythagorean connections are not free from suspicion (Freeman, 1949, p. 135, 137; Guthrie, 1965, p. 237), and finally, to the older Middle Eastern mythologies. Plato let his passages into the earth be occupied by water, lava, and winds, the idea of which was then taken over by his disciple Aristotle and used for a theory of earthquakes. Yet another surprising side of Plato's theory is its great similarity to Democritus' ideas, whose books he wished could all be burned! In reviewing previous theories of earthquakes, Aristotle in his *Meteorologica* (365b) also describes Democritus' opinion: "Democritus says that the Earth is full of water and that earthquakes are caused when a large amount of rain water falls besides this; for when there is too much for the existing cavities in the earth to contain it causes an earthquake by forcing its way out. Similarly, when the earth is dried up water is drawn to the empty places from the fuller and causes earthquakes by the impact of the passage" (DK68A97 and 98). Democritus connects Plato with Anaximenes and thus completes the documen-

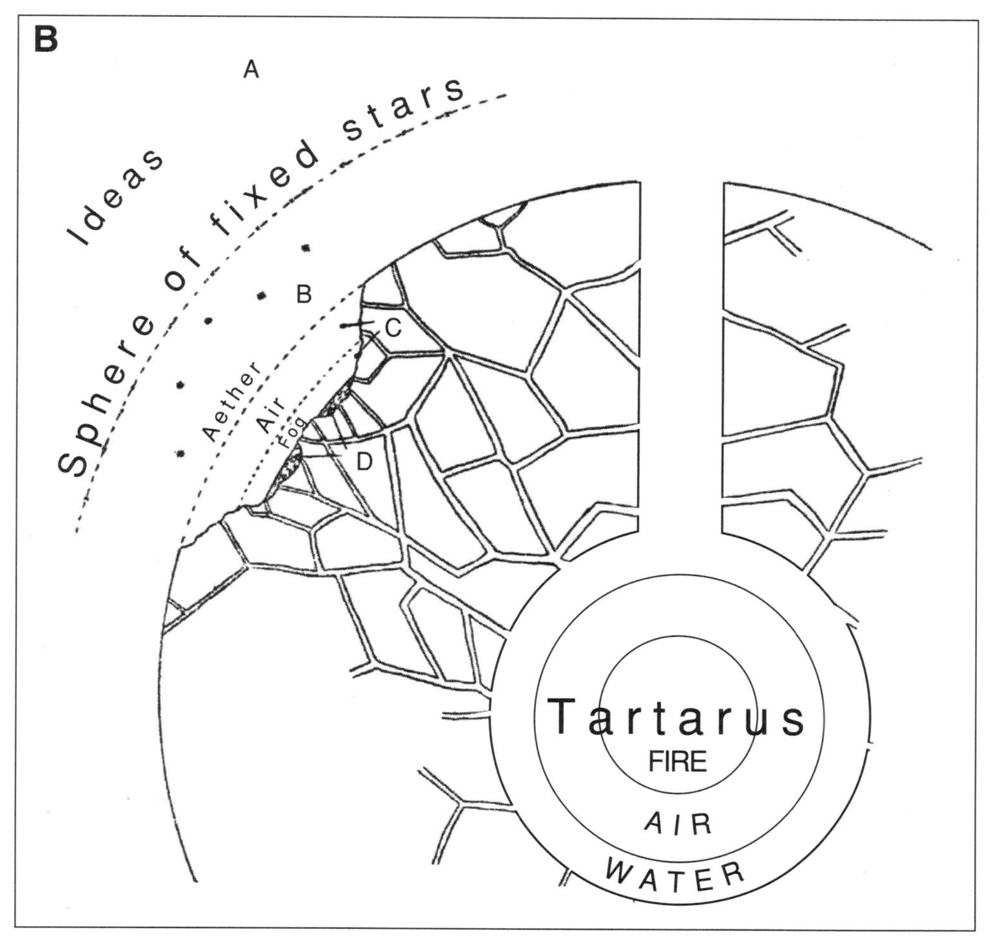

Figure 11 *(continued)*. B: If instead, the cylinder goes only to the center (as argued by Robin, 1941, p. LXXI ff.) and ends in a spherical Tartarus (or Hades) as shown here (which is my interpretation of the internal structure of the Platonic earth), the oscillations become more intelligible and may affect each hemisphere alternately, as Socrates relates, by the motion of water and air in the central cavity of Hades. If Tartarus is not a cul-de-sac in the earth, the breathing analogy becomes unintelligible. What may have misled Frank (1923) was probably the statement that "One of the chasms of the earth is greater than the rest, and it is bored right through the whole earth" (Plato, *Phaedo*, 112B). Following Robin (1941, p. LXXI), I read it to mean through the whole earth *between the surface and Hades* (see endnote 100), i.e., through a thick spherical shell containing all the earthquake-generating channels bound on the outside by the air/aether and on the inside by the fluids filling up Hades (air, water, lava). The following statement by Socrates appears to corroborate my view: "Now it is possible to go down from each side to the center, but not beyond, for there the slope rises upward in front of the streams from either side of the earth" (Plato, *Phaedo*, 112E). I see a further, independent, corroboration of my interpretation in Aristotle's critique of Plato's theory of the hydrological cycle as expounded in *Phaedo* (111C–113D): "Plato's description of rivers and the sea in the *Phaedo* is impossible. He says that they all flow into each other beneath the earth through channels pierced through it, and that their original source is a body of water under the earth called Tartarus, from which all waters running and standing are drawn" (*Meteorologica* 355b–356a; italics are mine). Given Aristotle's belief in a spherical earth (cf., among others, his *De Caelo*, especially book II), I do not see how his words just cited can be interpreted except in terms of Figure 11B here. Figures 11A and 11B show plainly that however one interprets the geometry of Hades, it is clear that the earth is porous and permeable with fire in the center. This will be of great assistance in understanding Aristotle's theory of global tectonics.

tation showing that the roots of Plato's geological ideas were in mythology *and* in pre-Socratic speculative science.

Before I describe the post-Platonic views about endogenous uplifts of the earth's outer rocky envelope, it is necessary to point out that all this talk about underground passages of water and lava had *also* a firm observational basis in the Grecian world of the sixth through the fourth centuries B.C. Not only was the karstic phenomenon of subterranean drainage well-known (cf. Frazer, 1919, p. 203; Pfeiffer, 1963, especially p. 12–23; Eichholz, 1965, p. 26; Bouillet-Roy, 1976, p. 74–79) but also lava tunnels, which in the surroundings of the Mount Etna, for example, had been used as burial sites, storage areas, dwellings, even as temples for cult rituals during the period from late Copper Age to early Bronze Age (~3000–1400 B.C.; see Privitera, 1998; Procelli, 1998). In some cases, the sites remained occupied until about 1250 B.C. Plato himself visited Mount Etna and was inspired by its lava streams to describe the fiery interior of the earth (Guthrie, 1975, p. 336, footnote 3). Given the widespread distribution of karst, especially in the externides of the Alpide chains of the Mediterranean, and the volcanic phenomena in the Kula region of western Anatolia[104], through the Aegean volcanoes to the volcanic regions of Italy (von Hoff, 1824, p. 101–266; 1840, p. 123–168; Serbin, 1893; Ströhle, 1921), it is not surprising that the Greek thinkers developed a theory of subterranean channels containing *both water and lava* to portray the internal structure of the earth. These observations agreed with the mother-earth motif having both moisture and warmth inside and thus nourished the Greek mythology.

Aristotle[105] in Assos: The First Theory of Global Tectonics

Recent re-estimation of the date of Theophrastus'[106] (ca. 372/369–288/285 B.C.) *On Fire* (Gaiser, 1985) opens up new horizons for the history of geological thought, both in the Academy of Plato and in the Lyceum of Aristotle (Fig. 12). Gaiser also claims that the much-debated fourth book of Aristotle's *Meteorologica*[107] was written by Theophrastus and quite early, soon after the completion of *On Fire*, most likely when he was in Assos with Aristotle (347–345 B.C.). These new datings show:

1. That the pore theory of Empedocles was adopted by Theophrastus very early in his scientific career. He may have learned it in Athens from Plato (if he ever was a student of Plato) or acquired it in Assos in the company of Plato's greatest student, Aristotle. This would have pre-disposed Theophrastus to accepting Plato's concept of the earth.

2. That Theophrastus started his own geological studies also very early. He says in the beginning of *On Fire* that he had previously dealt with various, mostly violent, manifestations of fire above, on, and under the earth (Gaiser, 1985, p. 50). This experience very likely happened in a Platonic context and was possibly inspired by such writings as those of Antiphon the Sophist (see above), who seems to have kept alive the views of the Pythagoreans and Empedocles concerning the internal fire of the earth.

As Gaiser repeatedly emphasizes, Theophrastus' relationship to Aristotle must not have been merely as a taker—he also probably gave much, especially an enthusiasm for the observational natural sciences, namely geology and biology in this case. In order to constrain our views of the relationships of the two men and their ideas, it is useful to look at the dating of their relevant books written when they were still in Anatolia or in Lesbos. Table 2 is a reproduction of Gaiser's table (1985, p. 86; indication of the interval in which Aristotle's tectonic theory was formulated added by me).

Table 2 indicates that we should start our inquiry into post-Platonic tectonics with Aristotle's *Meteorologica* (for *Meteorologica*, see Strohm, 1935, 1970; Lee, 1952; Tricot, 1955). We know that Aristotle too, like Plato (and most likely because of Plato), believed in chasms and cavities beneath the earth and that his main observational evidence was the karstic subterranean streams in Greece (Pfeiffer, 1963, p. 21–22); Aristotle mentions those in the Peloponnese and Arcadia in *Meteorologica* (351a; See, 1907, p. 244; Eichholz, 1965, p. 26; Bouillet-Roy, 1976, p. 74–79). But Aristotle also believed the Caspian Sea to be connected to the world ocean by subterranean channels because it had no outlet despite the fact that it received many rivers[108] (*Meteorologica*, 351a)[109]. Aristotle's earthquake theory and the associated uplifts and subsidences were directly related to his views on the geometry and functions of subterranean channels:

Figure 12. Aristotle (384–322 B.C.).This Roman bust, copied from a Greek original dating from the last quarter of the fourth century B.C., is located in the Kunsthistorischen Museum, Vienna.

TABLE 2.

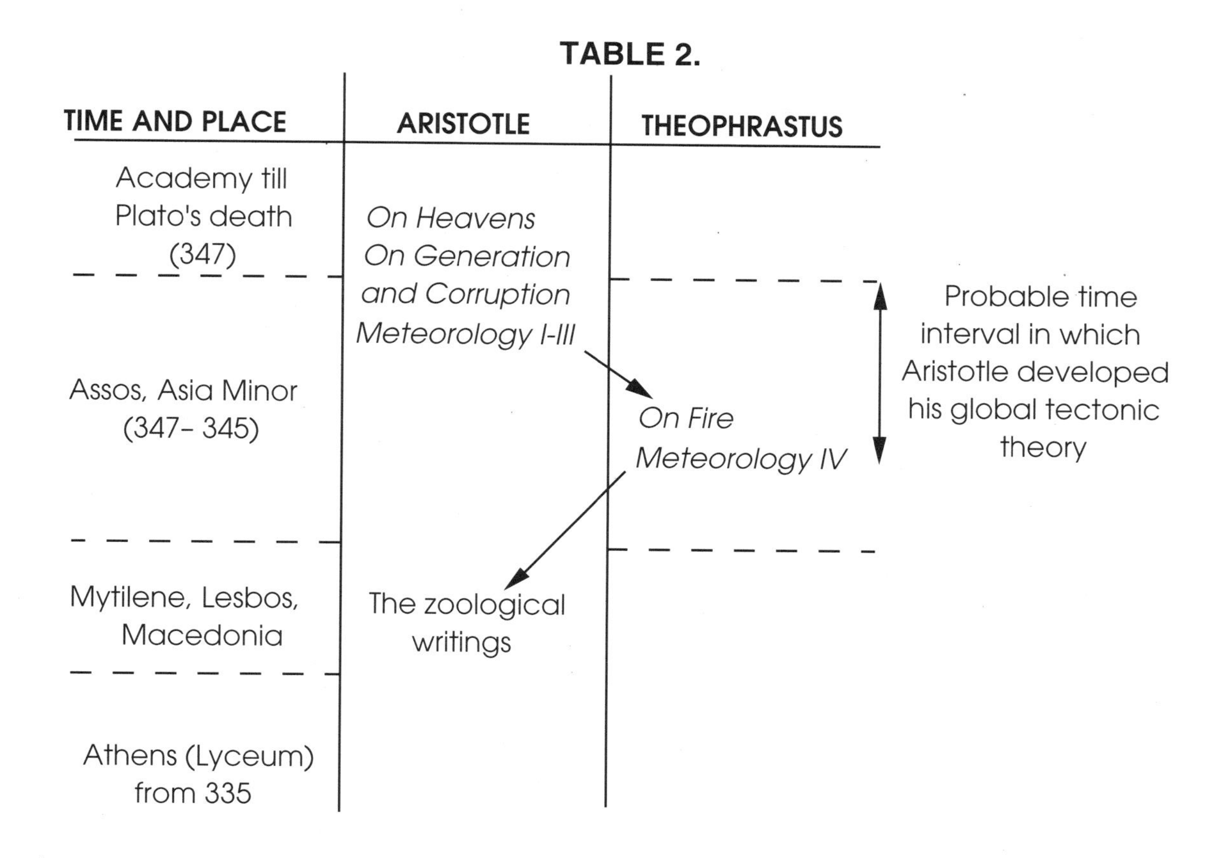

> Now it is clear, as we have already said, that there must be exhalation both from moist and dry, and earthquakes are a necessary result of the existence of these exhalations. For the earth is in itself dry but contains much moisture because of the rain that falls on it; with the result that when it is heated by the sun and its own internal fire, a considerable amount of wind is generated both outside it and inside, and this sometimes all flows out, sometimes all flows in, while sometimes it is split up.
>
> This process is inevitable. Our next step is to consider what substance has the greatest motive power. This must necessarily be the substance whose action is most violent. The substance most violent in action must be that which has the greatest velocity, as its velocity makes its impact most forcible. The farthest mover must be the most penetrating, that is, the finest. If therefore, the natural constitution of wind is of this kind, it must be the substance whose motive power is the greatest. For even fire when conjoined with wind is blown to flame and moves quickly. So the cause of the earth tremors is neither water nor earth but wind, which causes them when the external exhalation flows inwards. (*Meteorologica*, 365b21–366a; also refer to See, 1907, p. 243–244)

This theory presupposes a porous earth in which the sizes of the pores range from vast caverns and tunnels to minute capillaries. Pfeiffer calls this the "sponge theory of the earth" (1963, p. 22). This description makes reference also to earth's own exhalations that are dry and to "its own internal fire." According to Aristotle, "there is in the earth a large amount of fire and heat..." (*Meteorologica*, 360a6–7; also refer to See, 1907, p. 242). Until now, this "internalist" component of Aristotle's geodynamics, based on the dry exhalations and the subterranean lava streams—possibly nourished (at least in part) by them—has not been recognized other than by such remarkable exceptions as the great Arabic polymath genius Ğâbir ibn Haiyân (mainly 8th century AD {730?-820?: Prof. Fuat Sezgin, pers. comm., 2003}: for his life and accmplishments, see Sezgin, 1971, p. 132–268; also see Sezgin, 1979, p. 236), the Persian Syrian physician Alî ibn Rabban al-Tabarî (796 or 801-towards 864, in his *Firdaus al-Hikma*: Sezgin, 1979, p. 239–240) and Agricola (1544[1956], p. 30[114–115]). This has led some historians of earth science to consider Aristotle's geodynamics exclusively exogenic (e.g., Büttner, 1979a, p. 139)[110].

Finally, Aristotle's seismic theory requires the intervention of atmospheric agents, wind and rain, to combine their powers to generate earthquakes. Further investigtion of Aristotle's theory of exhalations would lead us to the roots of the earliest tectonic theories (cf. Şengör, 1997) and to the origin of the concept of the soul[111]—but that would carry us very far away from our present inquiry.

Aristotle connected the earthquakes with volcanism and with the tectonic deformation of the rocky rind of the earth. He clearly recognized that there were two types of earthquake motion: (1) horizontal "like a shudder" (See, 1907, p. 248, translates this as "trembling to and fro"), and (2) vertical "like a throb" (*Meteorologica*, 368b; in See's 1907 translation, p. 248: "as a pulsation, oscillating up and down"). Aristotle thought that the vertical shocks were rarer "since there is many times as much of the exhalation that causes shocks horizontally as of that which causes them from below. But whenever this type of earthquake does occur, large quantities of stones come to the surface, like the chaff in a winnowing sieve" (*Meteorologica*, 386b; See, 1907, p. 248). In the islands far out to the sea, earthquakes were rarer because the mass of the sea water cools the exhalations and diminishes the violence of the subterranean winds (*Meteorologica*, 386b; See, 1907, p. 248):

> [In the Aeolian islands] ...part of the earth swelled up and rose with a noise in a crest-shaped lump; this finally exploded and a large quantity of wind broke out, blowing up cinders and ash which smothered the neighbouring city of Lipara, and even reached as far as some of the cities of Italy. The place where this eruption took place can still be seen. (This too must be regarded as the cause of the fire in the earth; for when the air is broken up into small particles, percussion then causes it to catch fire.) (*Meteorologica*, 367a; See, 1907, p. 245)

Aristotle thus views volcanism as stemming from the same source as the earthquakes connecting all copeogenic events into a single cause, and he ascribes part of the internal fire to a deformation of the air racing through the subterranean galleries of the earth.

The Stagirite was also aware of the larger and slower deformations of the surface of the earth, with which widespread palaeogeographic changes were thought to be associated. Evidence for these thoughts is regrettably scattered in an unsystematic manner in his *Meteorologica*. That is probably why their true significance has not been appreciated until now. In a passage quoted by Geikie (1905, p. 34 ff.), Aristotle speaks of the change of venue of land and sea through time and says that the reason we do not notice these events in our lifetimes is because they proceed so very slowly. He explains the cause of these changes unfortunately in a most circuitous manner, which is probably why, for example, his explanation escaped Geikie's attention. Büttner (1979a, p. 143), who points out the same passage which he considers to be "among the most important" in Aristotle's book, also gives an incomplete account of Aristotle's ideas of the change of place of land and sea, probably for the same reason.

First, Aristotle says that such slow movements occur because "the interior parts of the earth, like the bodies of plants and animals, have their maturity and age. Only whereas the parts of plants and animals are not affected separately but the whole creature must grow to maturity and decay at the same time, the parts of the earth are affected separately, and the cause of the process being cold and heat" (*Meteorologica*, 351a19). After this somewhat unpromising start, he points out that the cold and heat are supplied by the Sun and are thus dependent on the Sun's course (still not terribly clear where he is heading). As some places are scorched by Sun's heat, they dry up; consequently, the rivers dry up; if the rivers disappear, the sea, being unable to draw water, would withdraw from regions of drought and inundate places where rivers have increased their flow. Given a static topography, however, it is difficult to see how this could happen (unless it happens only by evaporation and precipitation with no runoff!). Aristotle does talk here about the growth of alluvial plains into the sea, and thus, about runoff. (*Meteorologica*, 351b), although there is no mention of the *making* of basins into which the sea could retreat. As noted above, we do know, however, that Aristotle mentions large-scale subsidence as a consequence of water withdrawal, when he criticizes Anaximenes' theory of earthquakes a little later. As soon as this connection is made, Aristotle's line of thought suddenly becomes crystal clear: We must read the *Meteorologica*, books I–III as containing a single coherent view of the entire earth-system and put together his palaeogeographic change theory with his critique of Anaximenes' seismic hypothesis. In addition, we must remember that not only his teacher, Plato, but his friend and pupil, Theophrastus, were enthusiastic supporters of the view of a permeable earth, in the channels of which wind, water, and lava allegedly circulated. In fact, Theophrastus ascribed the origin of earthquakes to the motion of water bodies imprisoned in subterranean cavities[112] reminiscent of Plato's description of the subterranean fluid bodies in *Phaedo* cited above. In Aristotle's view, we can now see that abundant supply of atmospheric water would swell up the subterranean channels (like a sponge!), and the topographic surface would rise accordingly (some influence of Democritus? DK68A97), supplying runoff and feeding peripheral alluvial plains. By contrast, drought would cause the underground passages to shrink, upon which the surface would subside and turn into a basin. Because too much liquid in the passages (i.e., all passages completely full

→

Figure 13. Schematic cross-sectional views of Aristotle's meteorological theory of global tectonics according to the interpretation developed in this present book. A: Very schematic display of the most significant aspect of the Aristotelian earth: porous and permeable by means of an interconnected network of subterranean channels. (Aristotle emphasized that the channels can be of various sizes from huge galleries housing immense rivers to mere capillaries. The channels are shown here to be all roughly of the same size for graphic simplicity.) B: Representation of a region of two compartments: The compartment on the right is underlain by swollen channels containing all sorts of interstitial fluid (predominantly water), but the swelling is due to past heavy precipitation in a former wet climate. Because of the swelling of the arteries beneath, the topographic surface has domed up and become a region of falcogenic rise, from which rivers run off to the surrounding depressions. By contrast, the compartment on the left is underlain by shrunken channels indicating a past episode of drought. Most fluid was drained from the subterranean galleries. As a consequence of the arterial shrinkage, the surface has undergone falcogenic subsidence, formed a basin, and is occupied by the sea. The region now receives much precipitation, and the subterranean channels will soon receive fluid and swell up again. By that time, the neighboring high area will have exhausted its fluids (through exhalation, evaporation, and runoff) and will subside. Some of the fluid will pass laterally into the neighboring fluid-poor region (as Democritus originally suggested) and in the process may trigger earthquakes and volcanoes in the border region, where the channels are neither too congested nor too shrunken, but possibly bent and otherwise deformed. C: The reversal of the situation in (B) is now obtained. The former sea-bed falcogenically rose to become a high region (a continent, a plateau), and the former dome falcogenically subsided to become a sea-bottom. Notice also that the climates also shifted.

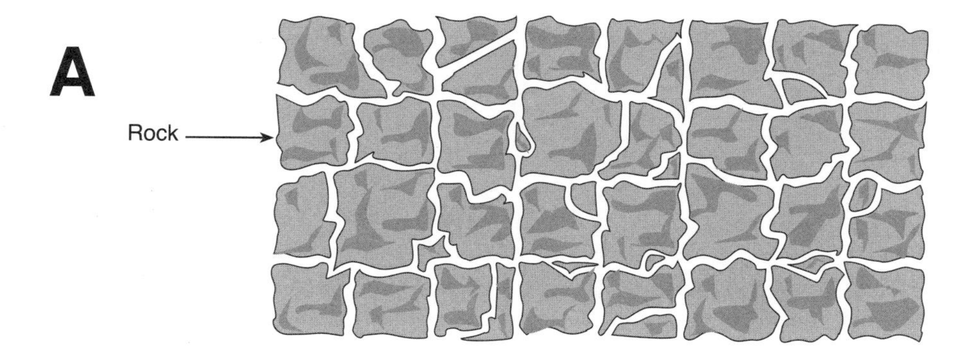
A
Rock

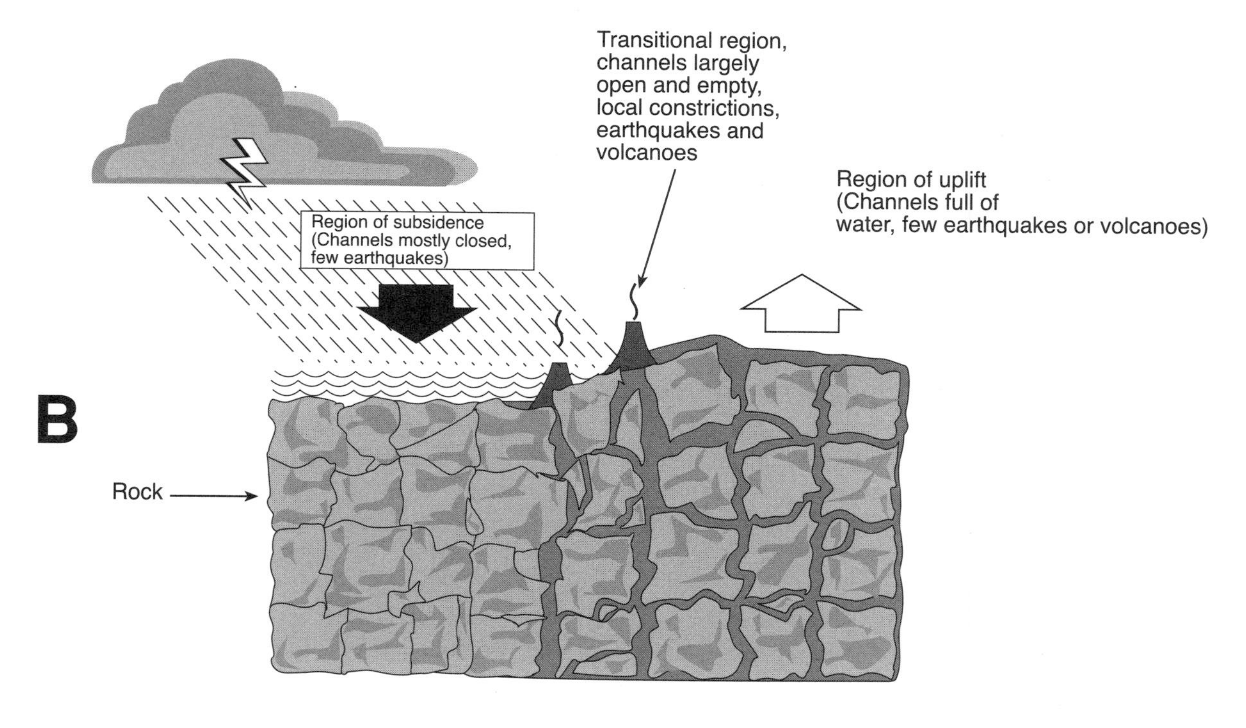
Transitional region, channels largely open and empty, local constrictions, earthquakes and volcanoes
Region of uplift (Channels full of water, few earthquakes or volcanoes)
Region of subsidence (Channels mostly closed, few earthquakes)
B
Rock

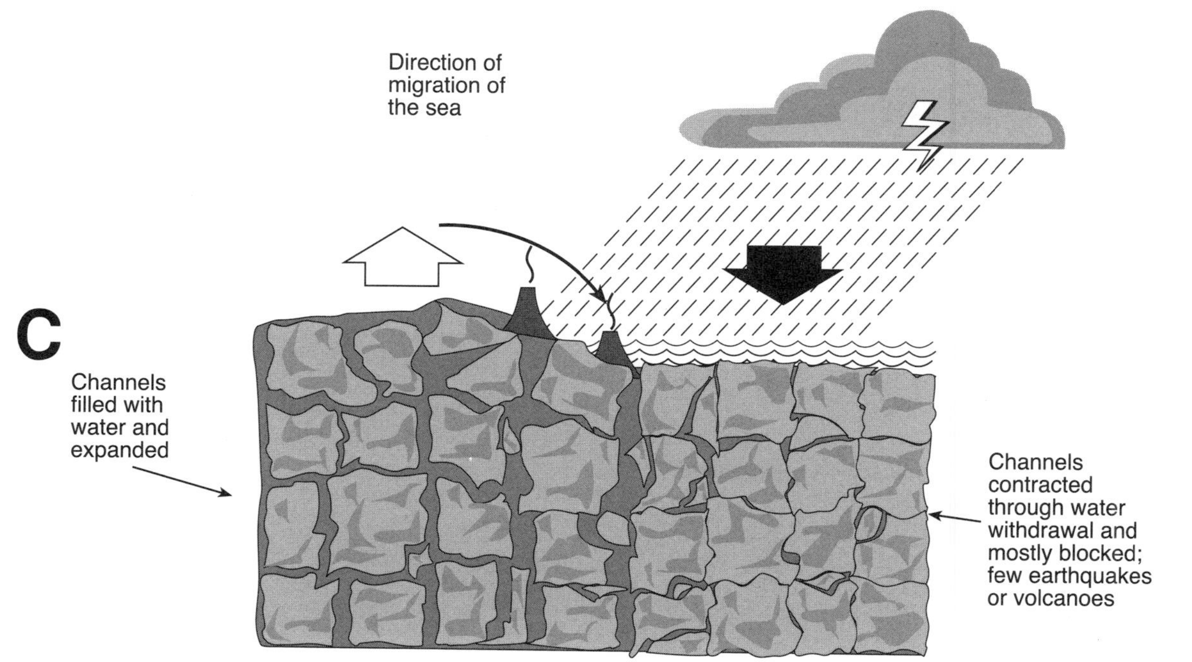
Direction of migration of the sea
C
Channels filled with water and expanded
Channels contracted through water withdrawal and mostly blocked; few earthquakes or volcanoes

of water) and no liquid at all (i.e., completely shrunken passageways) would prevent wind from having free sway in the subterranean galleries, both earthquakes and volcanoes would tend to occur in the transitional zones between the swollen-up areas and the depressed areas, where there might be optimum permeability through the earth's rocky rind and sufficient irregularity to squeeze and inflame the air. Aristotle emphasized meteorological and climatic oscillations as bringing about large-scale palaeogeographical changes. He gave Deucalion's flood as an example of an exceedingly wet episode that affected *only* the old Hellas, the country around Dodona and Acheloüs, the latter being "a river which has frequently changed its course" (*Meteorologica*, 352a).

Aristotle's theory is a complete account of the geological phenomena then known. Like Plato's (and most likely because of Plato's), Aritstotle's account makes a distinction between slow movements that create structures of very large wavelength and essentially free of breaks[113] (falcogenic), and rapid movements that create earthquakes and local intumescences that might blow up into volcanoes and make high mountains (copeogenic movements). Figure 13 gives a schematic account of the structure and evolution of the earth's outer rocky rind according to Aristotle. This theory of Aristotle was certainly a stroke of genius. It explained nearly all available observations at the time, proposed a mechanism that in those days most would have considered perfectly viable, and made some brilliant predictions such as the following: (1) the location of most earthquakes and volcanoes near the margins of the major sea basins, (2) the arid climates of continental interiors as opposed to the wet climates of ocean basins, and (3) the great difference between the rates of movement and sizes of falcogenic and copeogenic structures. Except for its endogenic mechanism, Aristotle's theory greatly resembles the tectonic theories of Haarmann (1930), van Bemmelen (1931a, 1931b; 1932a, 1932b; 1933; 1935; 1949; 1954; 1955), and Beloussov (1948, 1954, 1962). However, Aristotle's theory is superior to these others in its predictive capability, resembling more the views of Babbage and Herschel (*in* Babbage, 1838; see below). Anybody reading Sir Charles Lyell's lucid chapters on the causes and mechanisms of volcanic action and earthquakes and on the rise and fall of land in the ninth edition of *Principles* would be greatly surprised by how little progress had actually occurred between Aristotle and the proponents of the various views then considered modern (Lyell, 1853, p. 533–565; also see endnote 99). Even in the first decade of the twentieth century, Aristotle's tectonics remained modern in some physically and mathematically very capable minds (e.g., See, 1907).

We do not know whether the theory, as reconstructed here, had any other authors than Aristotle. Plato and Theophrastus naturally come to mind, but what we know of the writings, minds, and intellectual tastes of these two men makes it highly unlikely that they helped Aristotle in any significant way.

Theophrastus: Blurring the Falcogenic/Copeogenic Distinction in the Transition to Roman and Medieval Geology

Too many of Theophrastus' writings have perished to enable us to reconstruct completely his theory of earth behavior (see Fortenbaugh et al., 1993a, 1993b). We only have a few fragments, the longest—and the most significant—being in *On the Eternity of the Universe* by Philo the Jew (for a discussion see Duhem, 1958a, p. 241 ff.). In this fragment, Theophrastus argues against those who think that the universe cannot have been eternal because, if it were, the irregularities in the topography would long have disappeared and the progressive dimunition of the sea would have turned the earth into an endless desert. He points out that, although erosion planes down the land,

> ... the fiery element that is enclosed in the earth is driven upwards by the natural force of fire, it moves towards its own proper place, and if it finds any short route by which to escape, it drags up with it a great amount of earthly substance, as much as it can. But this, surrounding the fire from outside, is carried (upwards) more slowly; but being compelled to accompany (the fire) for a great distance it is lifted up to a great height, contracts as it reaches a summit and ends up as a sharp peak which imitates the shape of fire. (Fortenbaugh et al., 1993a, p. 349)

This account, reminiscent of Lucretius' verses cited above and which may have served as inspiration, is entirely compatible with the theory of Aristotle (and corroborates its partly endogenic character), which allows the topography to be rejuvenated at intervals through earthquakes, volcanoes, and alternating subsidences and uplifts as outlined above. Theophrastus, therefore, does not find it surprising that an eternal universe may accommodate a planet that has dynamic topography.

In Theophrastus' paragraph above, we have a description of a theory of *mountain-building*, i.e. one for *copeogenic movements*. We know from Seneca's testimony that both Theophrastus and Aristotle held the same views concerning the cause of earthquakes (*Quaestiones Naturales*, VI, 13: Fortenbaugh et al., 1993a, p. 365), although the testimony of Al-Hassan ibn Baklul (see endnote 112) suggests that the pupil also had ideas of his own. But Theophrastus' opinions on falcogenic movements are unfortunately not known. One may only conjecture that he might have entertained opinions similar to those of Plato and Aristotle on the strength of what Al-Hassan ibn Baklul writes, because the only fragment we have by him describing the retreat and invasion of the sea does not make clear whether any mechanism (other than erosion and mountain-building) was involved. He asks, "How many parts of the mainland have been swallowed up, not only on the coasts but even inland, and how much dry land has become sea and is sailed over by ships of great tonnage?" (Fortenbaugh et al., 1993a, p. 351). This question does read like Aristotle, but he then relates the story of the Messina Strait being opened by the force of wave erosion, during what was believed to have been an unusual tempest. Other islands also broke off the mainland, so he says, in the Peloponnese. Did not Atlantis, "greater than Libya and Asia together," suffer the same fate in a single day and night (as related in Plato's *Timaeus*, 25c6: see Taylor, 1928[1972], p. 5) as a result of extraordinary earthquakes and floods (Fortenbaugh et al., 1993a, p. 351–353; von Humboldt, 1835, p. 425 and footnote*, thought that Plato's Atlantis was conceived in the light of the Indian and Iranian myths!)? Theophrastus describes all these as resulting either from marine erosion or from copeogenic events. Even the description of the emergence of new lands by Pliny the Elder, which has strong Theophrastian elements in it, can be read as remaining entirely within the bounds of copeogeny: "The cause of the birth of new lands is the same, when that same breath although powerful enough to cause an upheaval of the soil has not been able to force an exit" (Pliny, II. 87[114]).

In Theophrastus' extant writings, there is little that is theoretical about the structure of the earth. This is in harmony with what we know of his tastes and inclinations in the sciences: Theophrastus was an empiricist and dealt with theoretical questions only within the framework of empirical data. Copeogenic events such as earthquakes and volcanoes do lend themselves to observation by one individual because they are small and fast. By contrast, falcogenic events, involving almost whole continents and oceans across millenia, require detailed and comprehensive syntheses. That was clearly Aristotle's forte. The available documentation does not indicate that Theophrastus followed his friend and teacher into the realm of tectonic synthesising.

The size and the temporal framework of the falcogenic events have long remained a hindrance to their study and understanding after Aristotle. Following the fall of the Greek enlightenment, it was only the rise of world-wide geology in the nineteenth century that again permitted a realistic assessment of the architecture and the evolution of the falcogenic structures (the falcogens) of our planet. In the writings of such geographers as Eratosthenes (Bernhardy, 1822; Berger, 1880[1964]; Fraser, 1972a, especially p. 525–539; 1972b, p. 756–772), and Strabo (Tozer, 1893, especially p. 1–53; Aujac, 1966) or Pomponius Mela (Silberman, 1988) and Dionysios of Alexandria (Jacob, 1990; Brodersen, 1994), we find nothing that can enlighten us about the opinions on the distinction between falcogenic and copeogenic events in late antiquity. The ideas about the interior of the earth remained entirely Empedoclean/Platonic (e.g., Strabo, V. 4. 9; VI. 2. 9) and, except for one important contrast in the interpretation of the origin of volcanic cones, nothing new was added to the conceptual repertoire of geology concerning tectonic events. The one exception[115] is worth noting because it *seems* like an exact copy of the difference of opinion that divided geologists in the early nineteenth century concerning the origin of volcanic edifices (see below) and involves the first mention (that I know of) that swelling-up of the ground as a consequence of subterranean heat must involve *stretching* of the top of the swelling.

Strabo (I. 3. 18) describes an eruption north of Methana (37°35′N, 23°24′E), Greece, of a volcano within the Methano Peninsula[116] in the northeastern Peloponnese, which may have taken place in the third century B.C.[117]. He writes:

> And about Methone [*sic*!] in the Hermionic Gulf [*actually within the present Saronikos Kolpos*][118] a mountain seven stadia[119] in height was cast up in consequence of a fiery eruption, and this mountain was unapproachable by day on account of the heat and the smell of the sulphur, while at night it shone to a great distance and was so hot that the sea boiled for five stadia and was turbid even for twenty stadia, and was *heaped up with massive broken-off rocks* no smaller than towers." (Strabo. I. 3. 18; italics mine)

In this passage, Strabo describes the origin and growth of a volcanic mountain and clearly spells out that it formed by accumulation of rock debris regurgitated by the volcanic vent (notice the passages I italicized). A near contemporary, not a scientist but a great poet, Publius Ovidius Naso (43 B.C.–18 A.D.) relates the same events in his *Metamorphoses*:

> Near Troezen [*south of Methana*] ruled by Pittheus, there is a hill, high and treeless, which once was a perfectly level plain, but now a hill; for (horrible to relate) the wild forces of the winds [*note the Aristotelian twist!*], shut up in dark regions underground, seeking an outlet for their flowing and striving vainly to obtain a freer space, since there was no chink in all their prison through which their breath could go, *puffed out and stretched the ground, just as when one inflates a bladder with his breath, or the skin of a horned goat*. That swelling in the ground remained, has still the appearance of a high hill, and has hardened as the years went by. (XV, 296–306; italics mine)

In Ovid's poetry, we do not have much of a description but instead an interpretation of how the topographic prominence was created. Strabo's account was based on a synthesis of detailed observation reports undertaken by a scientist. Ovid's, by contrast, based on one or more superficial relations, was a casual interpretation by an educated layman. What is interesting here is that Aristotle's earthquake and volcano theory was so widely known more than two centuries later that even an educated layman could make interpretations on the basis of it.

Alexander von Humboldt (1845, p. 251) compared the origin of the volcanic mountain in Methana—*not* with the elevation craters proposed by his friend, Leopold von Buch (see below), as Neumann and Partsch (1885, p. 308) later did—but with the puncturing of a volcanic blister, similar to the one he himself had earlier described from the Mexican volcano of Jorullo (Gadow, 1930). Von Humboldt thus found in Ovid's verses views "which in a remarkable way agree with those of modern geognosy" (von Humboldt, 1845, p. 251). Despite that, however, Ovid's fantasy was much closer to the ideas of von Buch—who in 1809 conceived the idea that the rock he called *domite* (oligoclase-bearing hornblende- and biotite-trachyte) occurred, in the Puy de Dôme and in other cones, as giant blisters propelled upwards by "internal volcanic power" (von Buch, 1809[1867], p. 483)—than to those ideas developed by such men as Sir Charles Lyell, George Poulett Scrope, and Constant Prévost, who advocated, like Strabo, that volcanic edifices form not by upheaval, but by accumulation. In spite of its resemblance to the craters of elevation controversy, I find the Strabo-Ovid contrast less relevant to that controversy than to an understanding of the enormous influence of Aristotle's theory of the wind-generated earthquakes and volcanoes and of Empedocles' and Plato's theory of porous earth in antiquity. It is these latter theories that dominated the thinking in geology until the seventeenth century, when modern geology began developing through the work of such men as Descartes, Steno, and Hooke.

But some ideas and some observations before then helped to set the scene for the later developments. Among these, the most important post-Aristotelian innovation was the so-called gravitational theories of the relative movements of the hydrosphere and the lithosphere, developed in the Middle Ages.

CHAPTER IV

THE MIDDLE AGES[120]: JEAN BURIDAN AND HIS PSEUDO-ISOSTASY

General

It was Theophrastus' view that remained popular after him in antiquity, as we know from the works of Eratosthenes, Strabo, Seneca, and Pliny the Elder, none of whom make a clear distinction between falcogenic and copeogenic events. Tectonic studies fell into almost total oblivion in the general dissolution of rational inquiry during the early Middle Ages (which the great historian Barthold Georg Niebuhr called the "second night" in the history of mankind, implying that the deep, mythic antiquity was its "first night": Niehbur, 1811, p. 1), despite the spirited but regrettably short-lived attempt of the Muslims to resurrect it (e.g., the treatises of the Brethren of Purity {*Rasâ'il İkhwan aş-Şafâ'*}: Dieterici, 1861[1999]; Duhem, 1958a, *passim*; Sezgin, 1979, p. 284–287; Ellenberger, 1988, p. 77–84[121]; al Hamadânî: Sezgin, 1979, p. 272; Al Biruni: Sachau, 1910, p. 196 ff.; Wiedemann, 1912a, 1912b, 1912c, 1912d; Gardet, 1979; Sezgin, 1979, p. 264, footnote 9; Strohmeier, 1988; Avicenna {Ibn Sînâ}: Holmyard and Mandeville, 1927; Crombie, 1952; Duhem, 1958a, p. 266; Sezgin, 1979, p. 300). I pass in silence the numerous attempts during the Middle Ages to explain the distribution of land and sea (cf. Kretschmer, 1890; Sachau, 1910, especially p. 196 ff.; Holmyard and Mandeville, 1927; Adams, 1938, p. 51–136, 329–342; Duhem, 1958a, p. 79–323; Crombie, 1961, v. I, p. 133–139; Grant, 1971, p. 70–71; Ellenberger, 1988, p. 71–110) because the entire discussion (Christian or Muslim) added essentially nothing to what was inherited from antiquity, except the important pseudo-isostatic model of Jean Buridan (1300–1358) and his disciple Albert of Saxony (1316–1390; see Duhem, 1958a, p. 293–316). The uplift theory of Egidius Romanus (Giles of Rome) may perhaps be viewed as another exception, as I discuss below, but only to the extent that it applies a Theophrastian theory of uplift to an entire continent (Duhem, 1958a, p. 142–146).

Early Middle Ages

In the writings of the early medieval Latin encyclopedists (Stahl, 1959), we see the layered universe with the earth at the center (e.g., Macrobius: Stahl, 1952, p. 104–107). Martianus Capella (first half of the fifth century A.D.) not only speaks of a cavernous earth but even makes an allusion to a centrally located Hades, by quoting Virgil[122] (Stahl et al., 1977, p. 224). We see the reflections of such ideas also in the discussions by Venerable Bede (672/73–735) and the Irish monk Dicuil (*fluorit* 825: Vivien de Saint Martin, 1873, p. 224), the author of *De Mensura Orbis Terrae* (825), regarding whether volcanoes are simply burning mountains or vents for deep-seated fires in the earth (Kimble, 1938, p. 149). Furthermore, Martianus Capella suggests (at the expense of some internal inconsistency of his book) that the position of the earth is not exactly at the center of the universe (Stahl et al., 1977, p. 332). In none of these authors's works is there any expression of worry about why earth and water appeared to lie within the same spheres, which is supposedly contrary to their nature, or why earth and water change position with time, except for some unoriginal tidal theories (e.g., Macrobius: see Stahl, 1952, p. 214). William of Auvergne (Guillermi Alverni, died 1249), the author of the celebrated *De Universo*[123], cut the Gordian Knot by stating that the reciprocal relations of water and land were so simply because the Good Lord ordered it to be so, as stated in *Genesis* (1:9): "*et congregentur aquæ*" (Duhem, 1958a, p. 109 ff.).

In the Muslim world, which was much more rational than the Christian Europe in the few centuries following the advent of Islam, the physician Hunain ibn Ishâq (809-873) came up, in his *Ğawâmi* (Sezgin, 1979, p. 264), with the remarkable view that the hydrosphere could change its volume both by increasing and decreasing it and thereby tried to explain the long-term variations in the palaeogeography mentioned by Aristotle in his *Meteorologica*, discussed above. Hunain pointed out that such volume changes must have been very slow and not perceptible in one or even two lifetimes. Regrettably, the discussion Hunain devotes to the topic of changes of level is far too short to know exactly what led him to this view, but in later Arabic-Muslim writings it is common to ascribe ebb and flow to volume changes of the sea water because of the heating by the Moon's "watery nature" (e.g. Sezgin, 1979, p. 276). Hunain's idea is the first appearance of anything like Suess' eustatic movements in the history of geology.

Later Middle Ages (after the Eleventh Century Church Reform)

St. Albertus Magnus (Albrecht von Bollstdädt, ca. 1193–ca. 1280[124]), the "Universal Doctor" (and since 16 December 1941, the patron saint of all naturalists), in his *Liber de causis et proprietatibus elementorum et planetarum* followed the ideas of Avicenna (at the time incorrectly ascribed to Aristotle) on the building of mountains, and he distinguished an *essential and universal* way from two *particular* ways that were believed to be local and temporally limited. The two particular ways were by means of water erosion to sculpt mountains and by means of wind to accumulte sand and dust into topographic prominences. The essential and universal way, however, was by means of earthquakes:

The essential and universal cause is the following: Mountains are born of earthquakes and in regions where the surface is too solid and too compact to disintegrate. The gas (*ventus, meaning wind*) that forms in abundance in the interior of the earth and which is violently agitated, uplifts the ground and forms mountains. Earthquakes are frequent near the sea or near large bodies of water, because water stops up the pores of the earth and thus imprisons the vapours emitted by the earth in the guts of the earth. Also, the highest mountains are born near the sea, or large water bodies. Under such mountains, large cavities remain that can contain a great quantity of water. Also, very often, mountainous regions are places of abundant springs and, which, by their outflow, create large lakes. (quoted from Duhem, 1958a, p. 273)

In all this, there is nothing that is also not in Aristotle. But then, the Universal Doctor adds his own observations on rocks and fossils in support of this theory of mountain uplift:

Of all this, we find a proof in the debris of acquatic animals and perhaps also in the devices of ships found in the crags of mountains and in the caves hollowed out of the flanks of mounts. The water no doubt carried them there with the loam that glued together whatever it enveloped. The cold and the dryness then prevented them from putrefying completely. One finds a very strong proof of this sort in the rocks of Paris, because very often one encounters there shells, some round, others crescent-shaped, others still bulging like the shell of a tortoise. (quoted from Duhem, 1958a, p. 274)

With St. Albert, we thus have not only the account of a theory of uplift—though not distinguished whether falcogenic or copeogenic—but also first-hand relation of evidence to document it.

However, St. Albertus Magnus also knew that not all of the retreat of the sea had been because of uplift. In Flanders, the sea had retreated, because formation of dunes had cut off access of a lagoon thus formed to the rest of the sea:

One might object that the Sea of England, which is a part of the Ocean, retreated from a city known in the past as Tuag Octavia; we saw with our own eyes that the sea had retreated in a short time from a large area near this city. One could even say that the sea retreats continuously from the city named Burig in Flanders. But we say that this retreat is not continuous. It is not caused by the fixed stars in the heavens and it is purely accidental. ... It has come about because dunes formed at the entrance of the port and that thin marine strips rise continuously. Thus the sea itself cut off its access to these cities and has been retiring little by little. (Duhem, 1909[1984], p. 310)

St. Albertus Magnus' sources are clearly Muslim, mainly Avicenna, and this episode in the history of geology was obviously made possible by the increasing closeness of the relationships between the Muslim and the Christian worlds. But it is remarkable that St. Albertus was not content just to read but actually took the trouble of seeing the first-hand evidence for himself. Gardet (1979) points out that St. Albertus Magnus was in some ways a thinker similar in spirit to the great Turco-Muslim scientist Abû Râihân Muhammad ibn Ahmad al Biruni (973–1051) who had "an opennenss of mind and sympathy" that enabled him to see beyond the circle of vision of his predecessors. This insight and openness could hardly have failed to arouse the curiosity of a kindred soul.

Roger Bacon (*Doctor Mirabilis*, ca.1220–ca.1292)[125] in his *Opus Maius* (which he sent from his Parisian exile to Pope Clement IV[126] upon the Pope's mandate in 1267 or 1268?) confirmed William's observation concerning the distribution of land and water, but chose not to emphasize the ultimate cause (Duhem, 1958a, p. 110 ff.; Suess, 1888, p. 7). But the *Doctor Mirabilis* believed that the water sphere was precisely concentric about the center of the earth and sought to demonstrate the point with the famous thought experiment, most likely following the Muslim scholar al-Khâzinî (Wiedemann, 1890, p. 319; reprinted in Girke, 1984, p. 41), that a goblet would hold more wine in the cellar than in the attic because the radius of curvature of the wine surface is smaller in the cellar than in the attic (Bacon, 1928, part four, ch. XI, p. 179–180).

One of the most remarkable solutions offered for the cause of the distribution of land and water on the terrestrial globe in the thirteenth century came from Giles of Rome (Egidius Romanus, 1247?–1316; for information on his life, see Duhem, 1973, p. 106–108), a student of St. Thomas Aquinas. In his two books, the *Super Secundum Libro Sententiarum* and the *Opus Hexaemeron*, Giles noted that water and earth formed concentric spheres and, ideally, earth ought to lie beneath water. It was clearly the Divine Power that kept the waters at a position lower than some parts of the earth, so that terra firma could emerge and land creatures, including ourselves, could flourish on it. However, Giles found it "useless to resort to miracle, when one could give a natural explanation of the Holy Writ" (quoted from Duhem, 1958a, p. 142). He thus compared the earth sphere with a somewhat shrunken apple. (This is the first earth-apple comparison that I am aware of!). Giles reasoned that, as the surface of such an apple would present irregularities in the form of "high" and "low" areas, so does our earth (is this a distant echo of the idea of Alexander of Aphrodisias that the earth is inhomogeneous, exhibiting different densities in different places and cannot therefore be a perfect sphere? See Duhem, 1958a, p. 159). Giles assumed that a great tumor ("*gibbositas*") of the earth sphere exists in the northern hemisphere forming the inhabited world. As the apple has creases on it, so does this tumor, forming valleys and the bays of the sea. Giles thought that the Mediterranean occupied one such valley. (Interpreting sea basins as valleys will have repercussions as late as the mid-nineteenth century, when Alexander von Humboldt interpreted the Atlantic Ocean as a great valley with corresponding angles (von Humboldt, 1835 {p. 111, 158, 324, 336, 338–339}; 1852 {p. 314–315}; 1858a {p. IV}. For a twentieth century fixist interpretation of the corresponding angles of the Atlantic Ocean likely influenced by von Humboldt, see Stille, 1939b, especially figs. 1–4. Do not let us forget that it was the corresponding angles on both sides of the Atlantic that eventually led to the idea of continental drift.). Giles calculated the uplift of land from its original position as 5/4 of the radius of the original terrestrial sphere on the basis of the relative densities of the earth and water. He assumed that there would be as much earth matter as there would be water matter and that the density of the earth is ten times more than that of water. However, Giles then added that perhaps earth was not so much denser than water after all and that the water sphere would not need to be so voluminous (Duhem, 1958a, p. 144–145).

This conclusion was followed by Andalo di Negro, a Genoese "physicist" of the early fourteenth century. From his *Tractatus Spere Secundum Magnificum Militem et Dominum,* we learn that Giles' mechanism to accomplish the uplift of the earth was to appeal to vapors: solar heat was thought to move the vapors enclosed in the earth, and these vapors had uplifted parts of the earth. One of these uplifted parts reached the surface of the water sphere and became terra firma (Duhem, 1958a, p. 146).

This mechanism (which Andalo disliked) is no different from Theophrastus' or Aristotle's mountain-building mechanism, which they employed to bring about copeogenic deformation. Giles used the same mechanism to create a large falcogenic uplift.

It was considerably after the twelfth century awakening[127] that Jean Buridan[128] and Albert of Saxony[129] began developing what Suess (1888, p. 22) aptly termed *gravitational theories*, i.e., theories that questioned the immutability of the shape of the oceanic waters and that of land, to account for the existence of the continents and ocean basins and their large-scale and slow movements. These ideas considered the earth-sphere to have a center of gravity independent of that of the water-sphere and the two to be able to move with respect to one another. Since a theory of gravity as attraction did not then exist, these theories clashed with the Aristotelian geocentric universe model and were not seriously tested by observation until the nineteenth century (cf. Geikie, 1903, p. 28–29 and the references cited there, and p. 377–388 and the references cited therein).

Jean Buridan's Theory

The first version of the medieval gravitational theories appeared in the two last questions of the first book of Jean Buridan's *Questiones Super Tres Primos Libros Metheororum et Super Majorem Partem Quarti a Magistro* (Duhem, 1958a, p. 293, footnote 1), a book devoted to the topics treated in Aristotle's *Meteorologica*. Of the two last questions of Buridan's first book, question 20, states "Has the dry land been at another time where the sea is today, and, conversely, has the sea been once where the dry land is today and will it return there?" (Duhem, 1958a, p. 293; Gohau, 1987, p. 32). This question had already been answered in the negative by the anonymous author of the *Liber de Elementis*[130] (falsely attributed to Aristotle) and, following him, by St. Albertus Magnus, because it was believed that all movements in the sublunary sphere were under the control of the stars. If, it was argued, one day the sea occupies where there is now human habitation, this would mean that a corresponding change has occurred in the heavens. It was believed that this was impossible. Buridan showed that none of the motions that we know from the heavens were sufficiently slow to cause such changes. Given the rate of motion of the stars and planets, if they had caused any changes in terrestrial geography, we would surely have noticed them. Buridan, however, cites Aristotle (the very same passages of the *Meteorologica* that we discussed above as embodying his global tectonic theory) concerning the occurrence of great floods in the past, i.e., great changes of the place of land and sea. He is aware that Aristotle believed these changes to be periodic, but Buridan seems not to think much of that assertion (possibly because he knows that in the same work, Aristotle also mentioned important floods that are *not* periodic; e.g., *Meteorologica* 368b?). Periodicity never appears in Buridan's own theories (explicitly anyway), although the very mechanism he suggests for changing the places of lands and seas has a sort of periodicity built into it.

Before discussing how sea and land may exchange places, Buridan addresses another related issue, namely the asserted elevation of the surface of the sea above that of land in some places. This is an old problem, mentioned in antiquity by Aristotle[131], Seneca[132], and Olympiodorus the Younger[133] in his commentaries on Aristotle's *Meteorologica* (see Duhem, 1958a, p. 97 ff.) and tackled in the Middle Ages by Sacrobosco (died 1256?) in the first chapter of his *De Sphaera*, the most famous astronomical treatise which the Middle Ages bequeathed to posterity[134]. The problem arose from the Aristotelian proposition that the four elements ought to be concentrically disposed about the center of the universe, with earth in the center, enveloped successively by water, air, and fire. Yet such is manifestly not the case, as land stands above the surface of the sea (i.e., the element water occupies a position lower than the element earth). To circumvent the problem of land lying higher than water, it was suggested that it was only apparent and that the surface of the seas really stands higher than the surface of the land and the reason why the seas do not flood the land is because the water has a natural tendency to form a sphere itself. Hence, the surfaces of the seas are in reality surfaces of partial spheres eccentric with respect to the center of the universe (Duhem, 1958a, p. 126).

Buridan rejected the idea that water stands higher than the land. He said that in Zeeland in Holland, there are indeed tracts of land that are below the surface of the water (see also St. Albertus Magnus as quoted by Duhem, 1909[1984], p. 310), but this condition was brought about "violently" (an Aristotelian expression meaning "not naturally") by means of dykes. If water could stand higher than land, this would be tantamount to admitting the possibilty of universal deluges (Duhem, 1958a, p. 299; note Seneca's influence!), which Buridan also rejects: "It is impossible to produce a universal deluge by natural agencies, i.e. to cover the whole earth by water, although God could do it by supernatural means" (quoted after Duhem, 1958a, p. 298; this is really a response to Seneca's elaborate description of the natural agencies that will bring about a final flood in the chapters of his *Quaestiones Naturales,* as indicated in endnote 132).

There are not only high mountains on earth, but they rise on high continental pedestals. Buridan relates his own observations in southern France, made during his trip to Avignon during the pontificate of Pope John XXII (1316–1334). Sarton

(1947, p. 540) conjectures that he may have climbed Mount Ventoux before Petrarca's famous ascent. Buridan maintains that the high mountains rising from the plains of the Loire and the Allier are four French leagues high (about 20 km!). The plain itself is being drained by rivers, and therefore, it too must have a slope as far as where the rivers empty into the sea. He calculated that this adds another two leagues (about 10 km!) and thus gets a height of six leagues, which is nearly 30 km! Although these height estimates (by the native of the very flat and low Picardy {Artésien}) were way off the mark[135], the lesson he drew from his observations was correct. He thought that these high mountains must be eroded and that their erosion products must ultimately be carried off to the ocean.

Buridan then relates, quoting St. Albertus Magnus, the ways in which mountains can be made. The first way is by "the flux of the sea and the impetuous movement of other waters...; the second is the wind that moves and gathers the dust and the sand...; the third, which I take to be more real than the other two, is by earthquakes which, at times, uplift a great mass of the earth" (quoted from Duhem, 1958a, p. 300). But even earthquakes seem to Buridan not quite capable of creating very high mountains, for which, he thinks a movement of the whole earth is necessary. For this inference, I conjecture that his observation in southern France—that even the largest mountains rest on immensely larger continental platforms—must have been the inspiration. Although I have never come across an explicit statement to that effect in his writings, Buridan's descriptions give me the impression that he must have sensed the difference between the small-wavelenth, supposedly earthquake-generated mountains, and the very much larger wavelength, gentler structures that create the continents.

Although Buridan thinks a movement of the whole earth is necessary, he knows that Aristotle wrote in the second book of his *De Caelo* that the earth is at rest and does not move (*De Caelo*, II. 14: 296a ff.). Buridan overcomes this by following a hint from Aristotle himself (*De Caelo*, II. 14: 297a and b): "If the earth being at the centre and spherical in shape, a weight many times its own were added to one hemisphere, the centre of the Universe would no longer coincide with that of the earth. Either, therefore, it would not remain at the centre, or if it did, it might even as it is be at rest although not occupying the centre, i.e., though in a situation where it is natural for it to be in motion."

Martianus Capella may have based his idea that the earth was not exactly at the center of the universe (cf. Stahl et al., 1977, p. 332) on this passage too. (It is quite amazing to see the very same idea used to explain the falcogenic movements on earth by A.E.H. Love in his presidential address to the British Association in 1907! {see Love, 1907, 1908; and the summary in Hume, 1948, p. 82–88}). Buridan thought that the high continents were high because they were light and that the ocean floors were depressed because they were heavy. He then imagined that ongoing erosion of mountains would further lighten the inhabited world (his world consisted of one major continent occupying one hemisphere, and the ocean the other), and the debris that was carried into the ocean would further load its floor and depress it. If this process continued for a very long time (Buridan had no scruples about thinking in terms of thousands of millions of years! cf. Duhem, 1958a, p. 296; Gohau, 1987, p. 32), one can imagine that the original sea floor would eventually approach the center of the earth and finally emerge from the other hemisphere (though turned inside out) as new mountains.

This extremely bold scheme was complicated because the composition of mountains is not uniform. Buridan noticed that some consisted of materials easily eroded, others much more resistant:

> In different places, the earth presents different aspects: here argillaceous, there sandy, in a different place pebbly. In one place it is more solid and difficult to divide; elsewhere by contrast, more fragile and more divisible. While, therefore, as I have said, the earth is continuously uplifted in its bare part, the parts of the bare surface that are more fragile and more divisible are preferentially carried away by rivers and rains into the lower parts. By contrast, the more solid parts and those less divisible cannot so flow away. They thus stay in place and continue to be uplifted. We also see that there are more boulders and pebbles in the high mountains than on the plains. In this fashion the more a resistant piece of earth is larger in length, width and depth, the larger will be the resulting mountains in length, width and depth. (quoted from Duhem, 1958a, p. 302)

Buridan also knew that if rocks traveled through the center of the earth, this would affect their mineral wealth!

> Let us accept, as many say, that metals, rock crystal, and other stones form by coagulation of a terrestrial or aqueous mixture and that this coagulation would be brought about by a very intense refrigeration or by the failure of heat for a very long time. It is, however, possible that the minerals of these metals are found in the first, second or the third crust of the habitable earth, although one finds there neither an intense cold nor failure of the heat. These minerals may have remained for a long time at or near the centre of the earth. And while their generation requires vapours or hot gases, these minerals nevertheless may reach great depths, although at such depths there are hardly bodies capable of furnishing either the gases or the vapours. Other minerals, it seems, were transported to these places. (quoted from Duhem, 1958a, p. 305)

We thus see a grandiose scheme involving uplifting of an entire continent through a pseudo-isostatic mechanism, carving out mountains (depending on local rock composition and structure), and now and then, creation of local protuberances by earthquakes. While the rocks travel through the earth's body, they become first compacted into sedimentary rocks and then, by infiltration of gases and vapors and by the effects of heat, are induced to produce metals, rock crystal, and other stones. I would not be entirely amiss, I think, if I claim that we see nothing so audacious in conception, so broad and accurate in empirical basis, and so logically sound as Buridan's theory of the earth until we come to Steno's ideas in the seventeenth century, fully three centuries after the time of Master Buridan. What is more, Buridan's theory provides the strongest and the most direct thread that connects Aristotle's ideas with those of the geologists of the Renaissance and the Enlightenment. His theory contains a major falcogenic component (that of the continuous uplift of the inhabited world) and smaller copeogenic structures (namely the earthquake-generated mountains of Albertus Magnus). Buridan's was a dynamic earth of a sort that became popular again only in the nineteenth century.

Buridan's younger colleague in Paris, Nicole Oresme (ca.1320–1382), was ambivalent towards this new theory of the earth. In his last work, entitled *Le livre d'Aristote Appellé du Ciel et du Monde*, Oresme reviewed Buridan's theory and offered some physical objections such as the fact that Buridan had not considered the air resistance (!) while theorizing about the movements of parts of the earth. In the end, however, Oresme confessed that he disliked it essentially because it contradicted the 92nd Psalm (*Dominus regnavit*): "For he hath established the world which shall not be moved" (cf. Duhem, 1958a, p. 306–308).

Albert of Saxony

Buridan's pupil, Albert of Saxony[136], adopted his master's theory with no reservations. This is of great importance because Albert's writings went through a number of editions towards the end of the fifteenth century and into the beginning of the sixteenth century (see especially Duhem, 1906[1984], p. 334–338), whereas Buridan's writings remained unedited until the twentieth century! Thus, the men of science from the Renaissance onwards, Leonardo da Vinci among them (Duhem, 1909[1984], p. 327–331 ff.; 1958, p. 309; Salomon, 1928, especially p. 7 ff.; also see de Lorenzo, 1920, and Weyl, 1958), have learned of Buridan's theory of the earth through Albert's publications.

Although now the concept is often presented as Albert's theory (e.g., Hölder, 1960, p. 19–20; even in Duhem, 1909[1984], p. 328–330; and in Sarton, 1948, p. 1430–1431), Albert has added essentially nothing to Buridan's ideas. The following quotation from Albert pretty much summarizes his views on this topic:

> From the part of the earth not covered by waters numerous earthy masses are carried by rivers to the depths of the sea. Thus, the earth grows in the part covered by the sea, while it is diminished in the part not so covered. Therefore its centre of gravity cannot remain at the same place. As the centre of gravity changes its place, the new centre of gravity strives to be at the centre of the universe. The point that used to be the centre of gravity is thus pushed successively towards the convex surface [of the earth] not covered by waters. By this flow and by this continuous movement, the part of the earth that was once at the centre arrives at its surface and vice versa. (quoted from Duhem, 1958a, p. 310; also see Hölder, 1960, p. 20)

During the Middle Ages, copeogenic structures were believed to have been produced exactly as stated by most of the ancient Greeks (and following them, the Romans): by wind and/or water circulating in subterranean galleries and deforming the surface into bulges when making a failed attempt to exit. This development is how, as we have seen, St. Albertus Magnus imagined mountain-building to happen, and how his contemporary, Ristoro d'Arezzo, believed that mountains can *also* be superimposed on a topography primarily shaped by mimicking the "topography of the starry heaven" (Adams, 1938, p. 335–340).

The tremendous reawakening of rational inquiry and critical discussion towards the end of the Middle Ages and the fifteenth century Renaissance hardly developed anything more, until the seventeenth century, that was in any significant way different from what the Greeks had said almost two millenia before. When something new was finally said, it was entirely on a Greek foundation.

CHAPTER V

THE RENAISSANCE: PERSISTENCE OF THE ANTIQUE AND MEDIEVAL MODELS IN TECTONICS

General

The Ottoman Sultan Bayezid I (1360–1403) began putting pressure on Constantinople immediately after his accession to the throne in 1390 (Shaw, 1976, p. 31)[137]. The Byzantine Emperor Manuel Paleologos II (1350–1425), the "philosopher Emperor" of his friend Demetrius Cydones (ca. 1324–ca. 1398; Norwich, 1997, p. 350), sought the aid of Europe and, to secure it, sent out a number of embassies. The great Constantinopolitan scholar Emanuel Chrysoloras (1353–1415; Fig. 14) was one of his ambassadors (Şengör, 1992). Chrysoloras arrived in Florence in 1391 (thus, immediately after the accession of Manuel) and there met two youths interested in Greek learning: a rich bachelor named Roberto Rossi and the future Papal secretary, Jacopo Angelo da Scarperia (1360–1410)[138]. Desire to know more about the knowledge of the ancients had begun to pick up in the twelfth century as a consequence of the influence of the University of Constantinople and the translations into Latin and Hebrew from the Arabic that had begun in the tenth century and sped up considerably in the eleventh and the twelfth centuries (Sezgin, 2000a, p. XII). The social environment for the thirst for pagan knowledge was largely created by the increasing prosperity of the Lombard cities in northern Italy during the twelfth to the fourteenth centuries (Russell, 1945, p. 431 ff.). Impressed by Chrysoloras' immense learning, Jacopo Angelo went to Constantinople with him both in search of Greek manuscripts and with the hope of persuading Chrysoloras to emigrate. Finally, in 1397, the Constantinopolitan scholar decided, upon the invitation of Coluccio Salutati (1331–1406), the Chancellor of Florence, to move to that city, with his books in his luggage, where he took up teaching at the University of Florence (see Jonathan Harris at http://orb.rhodes.edu/encyclop/late/laterbyz/harris-ren.html. and the references there).

Figure 14. Constantinopolitan scholar and diplomatist, Emanuel Chrysoloras (1353?–1415), whose move from Constantinople to Florence in February 1397 may be taken to mark the beginning of the humanistic movement. The fact that his luggage contained a copy of Ptolemy's *Geographike Uphegesis* has significantly influenced the fortune of the earth sciences after him. From a fifteenth century woodcut by Tobias Stimmer of Reusner (from my private collection).

I think Pagani (1990, p. VII) rightly says that "thus began the Humanistic movement in Italy." A movement such as the Renaissance cannot be dated to a day, or even to a year. For convenience, however, I think it appropriate to take Chrysoloras' move to Florence in February 1397 as a starting date.

The Renaissance in Italy and about a century later the Reformation north of the Alps have been acknowledged to be the events that created modern Europe. The former is generally seen as the rebirth of pagan Greek knowledge, whereas the latter was a revolt against the excesses of what was essentially an alien Church that had considered the pagan knowledge blasphemy for nearly a millenium. The influence of these events on the development of the sciences has been much more complicated, however, than indicated in the few sentences above and, in the history of geology, had initially little to do with intellectual emancipation[139].

South of the Alps, there was very little difference between the theories of geology of the late Middle Ages and the Renaissance. A perusal of such works as Giovanni Boccaccio's (1313–1375) geographical lexicon, *De Montibus, Silvis, Fontibus, Lacubus, Fluminibus, Stagnis seu Paludibus et de Nominibus Maris*[140] (written ca. 1360–1362, printed in Venice in 1473[141]); Pietro Latini's student Dante's (1265–1321) famous oration delivered on 20 January 1320 in Verona, in the chapel of Santa Helena, entitled *Questio de Aqua et Terra* (Dante, 1508[142]); Alessandro Piccolomini's (1508–1578) *Del Trattato della Grandezza della Terra et dell'Aqua* (Piccolomini, 1558); and Camillo Agrippa's (died sometime after 1595) *Sopra la Gener-*

atione de Venti, Baleni, Tuoni, Fulgori, Fiumi, Laghi, Valli & Montagne (Agrippa, 1584) clearly brings out an Aristotelian world, much enriched with Hellenic learning plus Buridan's ingenious speculations, but with little reference to contemporary observation (also see Kelly, 1969).

North of the Alps, the Reformation had an immense influence on the development of geological thinking. Büttner, in numerous publications (see especially 1979c, 1989, 1992), has empasized the importance for the earth sciences of the difference between the Catholic God, who was viewed to be the creator, and the Protestant God, the provider. The job of the Catholic earth scientist in the Renaissance and earlier was to learn about the creation of the Creator. It was assumed that the present-day world offered a true picture of the created world—except for the effects of the deluge—and that was the world one had to know in order to picture what it had been like during the creation. The Catholic approach was a historical view of sorts (especially made so by the idea of original sin and final redemption), but one having a static geographical stage. For the Catholic earth scientists, mathematical geography, the science of *measuring* the earth's surface (i.e., Ptolemy's *Geographike Uphegesis* [Guide to Geography])[143], was thus a convenient place to start.

By contrast, the God of the Reformation was the Providence. In the newly Anglican England, for example, Richard Hakluyt (1552–1616), the immortal chronicler of the English geographical discoveries, thought, according to Lestringant's (1994, p. 6) formulation, that "the beauty of the Cosmos thus resided in its value, and in the profit that a Christian could draw from it." However, on the question of the Providence, the reformers were split among themselves. Luther was interested only with the "now." Accordingly, his lieutenant in educational matters, Philipp Melanchthon (original name: Schwartzert), the *Praeceptor Germaniae* (1497–1560: Elliger, 1961; Büttner, 1979d, 1989), felt himself compelled to incorporate more and more Aristotle, rather than the Bible, into his lectures, to be able to come to grips with the processes now going on around them. Therefore, he had to overstep the boundaries of the (almost) purely mathematical geography of Ptolemy and wander into physical geography and human geography, thus broadening the scope of the earth sciences (Büttner, 1979b, p. 26; 1979d).

By contrast, Calvin's reformed view had a strong historical element. For Calvin, human beings became individually sanctified, and the world was being reconquered for Christ in a stepwise fashion, thus fulfilling the purpose of God. The Calvinist and other "Reformed" earth scientists in the sixteenth century proceeded from the present day earth but projected the present backwards in an attempt to understand the history reported in the Bible. It was the further development of this Reformed mentality that eventually gave rise to an actualistic interpretation of geology in the hands of Hutton (Galbraith, 1974; Dean, 1975, 1992). However, for our present purposes, suffice it to say that during the fifteenth and sixteenth centuries, nothing new was added to our understanding of geology, observationally (except the rapidly expanding geographical horizon) or conceptually[144]. Moreover, everywhere there was a strong revival of Aristotle, more so in Lutheran circles than in Calvinist ones (Büttner, 1979b, p. 16, says that the sixteenth century was the "Aristotle-epoch" in the history of German geography). This revival actually was the continuation of a trend established earlier (Miethke, 1989). Together with Aristotle, one leaned heavily on other naturalists and geographers of antiquity, especially Strabo, Pliny the Elder, Solinus, and Seneca.

South of the Alps, the secularization went on so rapidly that it finally called to life the Counter-Reformation, which did more to re-discipline the Catholic Church than to stem the tide of Reformation north of the Alps. But in the meantime, enough of the Greek knowledge had been revived everywhere to perpetuate the taste for independent rational inquiry.

It is in the light of all this that we must view the developments that took place during the sixteenth and seventeenth centuries in the recognition of the distinction between the falcogenic and copeogenic events and structures. Movements of the earth's outer rocky rind were ascribed mostly to subterranean winds and dry exhalations, following Aristotle and his disciples (both in the west and the east). Fire seen in volcanoes was similarly explained, but many thought that inflammable substances below ground were responsible for fueling it. Following Strabo (V, 4. 6), Pliny the Elder (XXXV. L. 174–177), and especially Seneca (V. 14: Clarke and Geikie, 1910, p. 206), sulphur was seen as the main agent of combustion[145].

One must not view this extraordinary theory with our present spectacles, however. In the sixteenth century, the earth was still Empedoclean/Platonic (enforced by what meager literature there was inherited from the Middle Ages, such as the books of St. Albertus Magnus), i.e., thoroughly perforated by innumerable channels, galleries, and passages (cf. Agricola, 1544[1956], p. 40[128], where he says "As these [*veins*] are distributed through the entire body of the individual, ... so are those [*passages*] through the entire terrestrial sphere, especially in the mountainous regions."), interconnecting larger vacuities that housed large deposits of sulphur and/or bitumen. When one of these deposits caught fire, the channels that may have connected it with the earth's surface became volcanic feeders. (The great chemist Sir Humphry Davy was still an advocate of similar views in the beginning of the nineteenth century as we saw above! {see Siegfried and Dott, 1980, especially lectures 9 and 10}).

Gortani (1963) summarized Alessandro Degli Alessandri's (1461–1523) ingenious hypothesis of shifting the axis of rotation of the earth to solve the Aristotelian problem of land and water occupying different places on the earth's surface at different times. He used fossil shells exposed on the mountains of Calabria as empirical evidence for the land/water exchanges. This hypothesis was the only original (though allied to Buridan's pseudo-isostasy) tectonic hypothesis that the Renaissance produced. A number of other Italian Renaissance authors are listed in Gortani (1963, p. 506) who also made observations on rocks and fossils and inferred upheavals and shifts of water bodies. None (including Leonardo da Vinci, as far as his views on tectonic movements were concerned; see especially Salomon, 1928, p. 7–8), however, showed any greater insight than that of such medieval masters as St. Albert the Great and Jean Buridan. Gortani (1963) rightly insists, though, on the positive influence that the intellectual environment created by all such Italian discussions on geology had on Nicolaus Steno's views about a century later in Toscana.

Georgius Agricola

All the notable figures who devoted their attention to tectonic problems in the sixteenth century lived north of the Alps. The greatest among them, Georgius Agricola (1494–1555; Georg Pawer [Bauer] with his original name: Fig. 15), the father of mineralogy and certainly one of the main heralds of modern geology[146], had an almost thoroughly Theophrastean view of the earth, supplemented by interpretations by later writers of antiquity. Agricola took the basis of his knowledge from Aristotle's books, which he found was largely corroborated by what he could read by later authors in antiquity and a few Muslim authorities, such as Ibn Sina (Avicenna; see Agricola, 1544[1956], p. 38–39[126–127]). Not that he took everything on faith—he was keen to compare the statements he read with "the reality" (Agricola, 1544[1956], p. 34[120]). He adopted the theory that volcanoes were fueled mostly by sulphur, bitumen, and coal, not only because the antique authorities said so, but also because Mount Hecla in Iceland was reported to spew out sulphur at times, while coal mountains in Meißen (not far from where he lived) burned, and the Hephaistian Mountains in Lycia[147] allegedly had fire that burned even in water (Agricola, 1544[1956], p. 36[123]). All these statements had ample support from authors in antiquity. Similarly, he cited observational evidence for the origin of earthquakes through the activity of winds: winds had been noted to pick up just before the earthquake and to last until a few days afterwards (Agricola, 1544[1956], p. 32[117])[148].

Figure 15. Georg Pawer (Bauer; his latinized name: Georgius Agricola [1494–1555]), the founder of mineralogy in the Renaissance. He was a medieval man in his respect for the antique authors, a modern man in his zeal for testing their statements. The portrait is from Joannes Sambucus' *Icones veterum aliquot ac recentium medicorum philosophorumque* ... (Antwerp, 1574).

Agricola's perspective on geology was one of a physician and a miner. He was interested in anatomy and in explaining the origin of that anatomy. He had little knowledge of geography, (i.e., the surface of the earth) and expressed no interest in the great geographical discoveries going on at his time (that, despite the fact that great metal wealth was being reported from the new world). He had more faith in the scholars of classical antiquity than in the adventurers of his own time (Fraustadt and Prescher, 1956, p. 49). Though he was still medieval in that aspect, he was modern in seeking to underpin ideas with reported actual observations. I think he was simply conservative and cautious—like most European miners of his time. He also reflected the tension between the mariner and the scholar that so sharply divided the geography of the Renaissance (for this division, see especially Lestringant, 1994): the scholars were the slaves of their books, the mariners were hypnotized by what they experienced (Gallois, 1890, p. XIV).

The two worlds began converging towards the end of the Renaissance, and Agricola is an outstanding representative of the reconciliation, although he rather believed the experience of a group (the miners), which he thought he could himself test, than the fable-infested reports of the overseas explorers, whose new worlds were beyond his reach. Agricola's limited knowledge of the surface processes, resulting from his disinterest in geography, did not allow him to distinguish copeogenic processes from falcogenic ones. It was simply not possible to become aware of falcogenic structures while being confined to galleries not wider than a few tens of meters at most. He bequeathed this limitation to his friend, Sebastian Münster, although Münster was far more adventurous in using reports from overseas explorers and overland travelers (Oehme *in* Münster, 1550[1968], p. VII).

Sebastian Münster

The great Hebraist scholar and geographer[149], Sebastian Münster (1488–1552: Fig. 16)[150], was a student first of Gregorius Reisch (1470–1525: Hoheisel, 1979a) at Freiburg in Breisgau (sometime between 1507 and 1509) and then of Johannes Stöffler (1452–1531: Hoheisel, 1979b) at the University of Tübingen, between 1515 and probably 1518(?)–1520(?). Reisch enthused Münster for Hebrew and for geography, whereas Stöffler convinced Münster and his fellow student Philipp Melanchthon that natural sciences led one to God. Stöffler was mainly an astrologist, and the emphasis in his teaching was still

Figure 16. The greatest of the German Renaissance geographers, Sebastian Münster (1488–1552), friend and correspondent of Agricola. In *Cosmographei*, Münster's skillful combination of ancient style cosmography, meteorology, and chorography with the observation reports of the age of the great geographical discoveries was one of the first steps in the direction of the creation of modern geography. This portrait of Sebastian Münster by Christoph Amberger is located in the Deutschen Museum, Munich.

Figure 17. Title page of the 1550 German edition of the *Cosmographei* of Sebastian Münster, from a facsimile produced in 1968 by Teatrum Orbis Terrarum Ltd. in Amsterdam.

on mathematical geography, both because of his main interest and because it was the fashion of the day. Reisch and Stöffler together awakened in Münster a desire to combine the dry mathematical geography of Ptolemy with a description of the physical and cultural aspects of the earth's surface in a general book, in vindication of the Bible and in praise of the Lord.

Unlike his schoolmate Melanchthon, Münster became "Reformed" (because he wanted the professorial chair in the Reformed city of Basel). His *Cosmographei* (Münster, 1550[1968]; Fig. 17)[151] was a geographical compendium that he published after his earlier geographical writings (see Gallois, 1890, p. 192–193, 212–213, 222; Büttner and Burmeister, 1979, p. 126) and the commented editions of Solinus[152], of Pomponius Mela[153] (Münster, 1538)[154], and of Ptolemy (Münster, 1540)[155]. Münster introduced his *Cosmographei* with what may be considered a theory of the earth, many elements of which I believe he adopted from his correspondent, Georgius Agricola[156].

Münster's theory of the earth was strictly Biblical[157] and considered two major events as the prime causes of what we see at the surface of the earth today. This was very much like Nicolaus Stenonis' views about a century later (see below) and J.G. Lehmann's two centuries later (Şengör, 1991c, p. 428): first, the creation of the world and, second, the Noachian flood. During the Creation, God ordered the waters to recede from areas destined to become land, making the remaining water-covered areas ("Greeks and Latins call them Ocean;" Münster, 1550[1968], p. ii) twice as deep (this is a sort of mixture of Aristotle's and William of Auvergne's ideas!). Although Münster says that no man can reach the bottom of water-covered areas, the fact that they do have a bottom is stated emphatically. (Remnant of the medieval preoccupation with the "bottom" of the ocean in the framework of Buridan's theory and its successors?) During the flood, the waters *rose* to cover all the highest peaks and, when they receded again, they greatly changed the face of the earth by excavating new avenues and thus creating new mountains (Münster, 1550[1968], p. ii).

The continents were created in the beginning, but the flood altered their shapes by eroding large gulfs (*sinus*) in them, so that "where there was no sea, there now formed new seas, exactly as many mountains and valleys formed in places that had been formerly fields[158] owing to the same reason through the back and forth flowing of marine waters" (Münster, 1550[1968], p. ii). Münster names (as examples of such gulfs) *Sinus Persicus, Sinus Arabicus,* and *Sinus Indicus*. Moreover, the Caspian Sea is implied to be a post-diluvian creation (Münster, 1550[1968], p. ii). Münster's interpretation of the origin of the morphology closely follows not only Agricola, but also Ibn Sina (via St. Albertus Magnus? or Agricola?). Münster pointed out that we do not know in any detail what the earth's face had looked like before the deluge but that Noah and his three sons had known both worlds and that they must have told their descendants what the ante-diluvian world had been like (Münster, 1550[1968], p. xlii). However, he implies that there must be some common points of reference, some measure of "fixism," for he refers his reader to his maps of the present-day world to visualize the theatre of the Biblical story of the deluge[159].

Although numerous islands had been created in the beginning, Münster points out that many (e.g., Delos, Rhodes, Alone, Thera, Sicily, Theresia) were formed in the sea later (Münster, 1550[1968], p. iii[160]). I note here the incredible continuity from the half-mythic sages of Greece, such as Epimenides to the Renaissance (and the implied dearth of original observations)! The movements of the sea in covering and retreating from lands are recognized and ascribed to the will of God, exactly as William of Auvergne had done earlier—providing a thread of continuity between the Middle Ages and the Renaissance. Münster pointed out (p. iiii) that sometimes large inundations occurred in a day, as had happened in the Netherlands, and sometimes such invasions of the sea accompanied earthquakes (as Hooke, *in* Waller, 1705, was to recognize almost exactly a

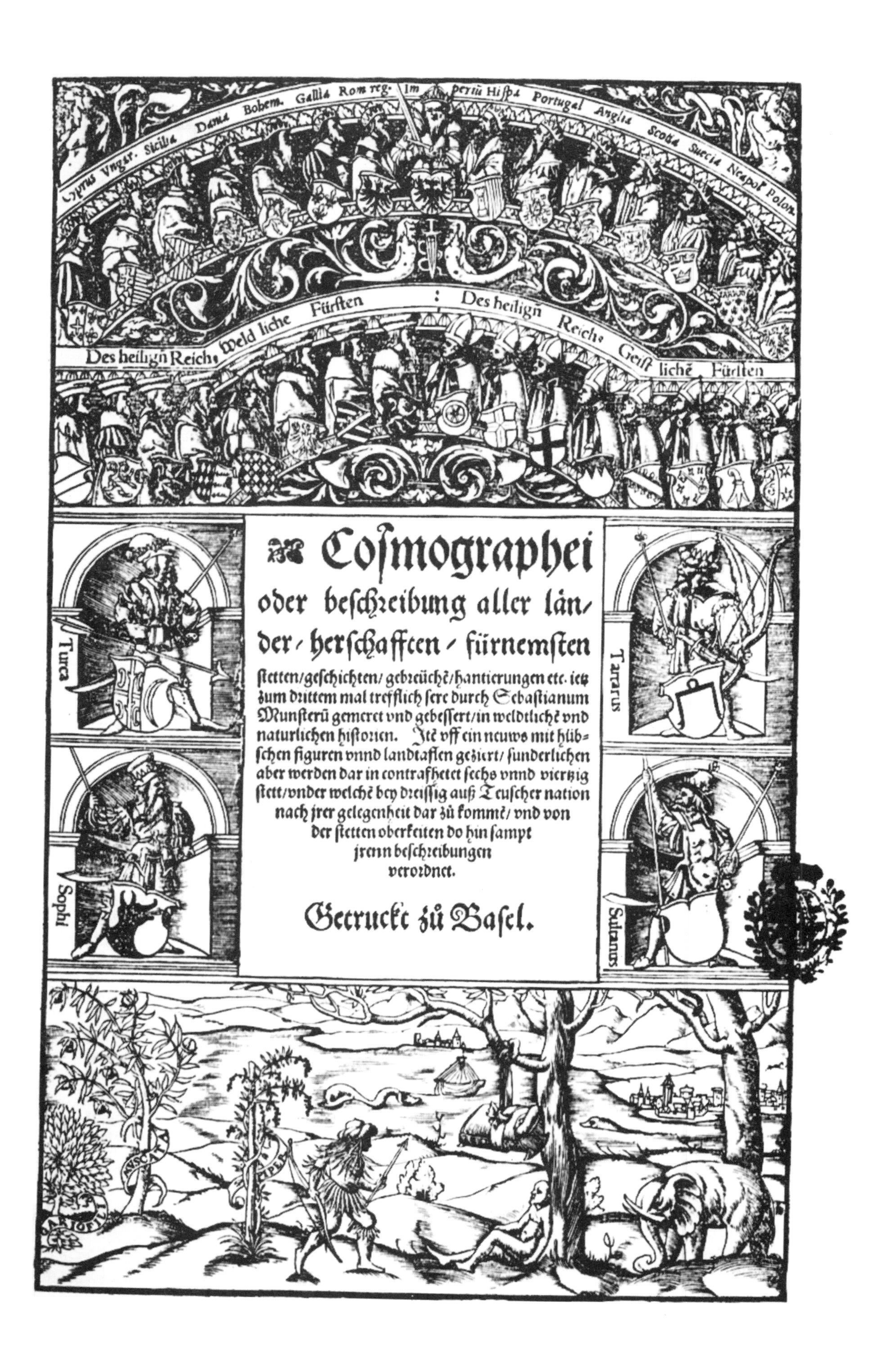

Cosmographei

oder beschreibung aller län-
der/ herschafften/ fürnemsten
stetten/ geschichten/ gebreüchē/ hantierungen etc. ietz
zum drittem mal trefflich sere durch Sebastianum
Munsterū gemeret vnd gebessert/ in weldtlichē vnd
naturlichen historien. Itē vff ein neuws mit hübschen figuren vnnd landtaflen geziert/ sunderlichen
aber werden dar in contrafhetet sechs vnnd viertzig
stett/ vnder welchē bey dreissig auß Teuscher nation
nach jrer gelegenheit dar zü kommē/ vnd von
der stetten oberkeiten do hin sampt
jrenn beschreibungen
verordnet.

Getruckt zü Basel.

century later: see below in Chapter VI). Thus, Münster had an eye for actualistic analogues while seeking to establish the veracity of some of the hypotheses he was entertaining.

Münster seems to have taken particularly what he says of the internal workings of the planet from Agricola, describing those workings under the titles "*Of some Powerful Effects of the Earth*" (ch. v), "*Of Hot Springs Which Well up from the Earth*" (ch. vi), and "*Of the Fire that Burns in the Earth*" (ch. vii). Of the first and last of these chapters, I present below a combined liberal translation of selected passages, with my commentary in square brackets, to give us an idea of the great humanist's view of the behavior of the planet:

> The earth has many dykes and channels hidden in its depths [*porous earth theory of the Greek Mythology and of Anaximenes, Empedocles, Socrates, Plato and others!*] which effect wonderful things in Nature, which partly give rise to violent things. Peculiar things grow on the earth [*in the dykes and channels*] and in the air from its moisture, inclosed air, vapour, fire, and hot exhalations [*Aristotle!*]. A variety of things cook in the earth, for example, all sorts of wonderful soils, congealed juices, gemstones, metals, etc. [*Strabo, Seneca*], which one has to look for in the earth, where they were born through the wonderful workings of the earth. Nature does not have a very peculiar behaviour on [*the surface of the*] earth and in the air, but in the belly of the earth, she has much wonderful metamorphosing effects. Nature dislikes empty places [*Aristotle!*]. But the earth is not fully filled up with earthly materials. In some places it is loose, in others stony or craggy. And between the large crags are many crevasses, loose places, and veins and caves [*porous earth theory of the Greek Mythology and of Anaximenes, Empedocles, Socrates, Plato, and others!*] . Where land and sea come into contact, the earth is made moist and its many holes become filled up [*Democritus, Plato, Aristotle, Theophrastus*]. The moisture forces itself into the earth into its veins and dykes and thus air gets trapped in the earth. When it is heated, it evaporates out of the earth and when it cools, it condenses to become water and does not evaporate out. As the air is warm and moist the cold drives away the heat and necessarily it must turn into water. However, there are large holes and caves in the earth, and one sees here and there written that entire fields and mountains have sunk. Thus large quantities of air can also be trapped in not so small caves and can throw out against the sky large hills, indeed entire mountains, or it can throw up mounds and hills onto flat fields, or give rise to such terrible earthquakes [*so far, only copeogenic structures are discussed in an entirely Aristotelian framework*], whereupon also whole cities are destroyed as it happened a few years ago in Putcolis not far from Naples [*This is reference to the origin of the volcano named Monte Nuovo in 1538: see Sapper, 1917, p. 5–6*]. Sometimes it also happens that the water channels within the earth get ruined and become congested. Such a flow makes itself a new channel or reopens an old vein. This causes the Nature to leave behind in the deep earth peculiar things, exactly as it also happens when some thick juices and coagulated moisture bear them. This happens with metallic materials, especially with copper.
>
> It is known that in the past mountains even fields on earth burned. Indeed even in our days one sees fire coming out of the earth. ... Now where there is a mountain that always burns, it is good, because it means that there the hole is not congested and that is why flame and smoke can have free exit [*Aristotle,* Meteorologica *367a*]. And when it happens that the inner avenue becomes congested, the fire does not burn any less in the inner oven, but in the upper chimney it becomes extinguished for some time, because there it has no material to live on. But when a strong explosion comes from the impetuous wind to the inner oven, the fire often has to break through the formerly congested hole with violence or it has to seek another chimney or any exit and drives with it ashes, sand, suspended materials, pumice, iron chunks, stones and other materials, often not without harm to the surroundings [*This is entirely Aristotelian and refers also to copeogenic events*]. (Münster, 1550[1968], p. v–vii).

Anybody familiar with the writings of Georgius Agricola, especially his *De Ortu et Causis Subterraneorum Libri V* (1544[Fraustadt and Prescher, 1956][161]), cannot fail to see the great resemblance between the opinions of Agricola and Münster, such as the description of earthquakes in Agricola (1544 [Fraustadt and Prescher, 1956], p. 32)[162]. Thus, Münster's *Cosmographei* very much reflects the prevailing opinions on the behavior of the planet during the middle of the sixteenth century. It shows us a theoretical consensus about an Aristotelian world, but—unlike Aristotle's—dominated by the universal flood. Although Agricola ascribed volcanic activity to the burning of subterranean masses of bitumen and sulphur, some of his contemporaries, such as Falloppius (*De Medicatis aquis atque de Fossilibus*, Venice, 1564, p. 102; quoted after Adams, 1938, p. 344), preferred a more genuinely endogenous theory, holding the Aristotelian dry exhalations responsible for mountain uplift, à la Pliny, and criticized the views of Agricola and Münster. But in none of these writers do we see an awareness of falcogenic events. Münster mentions explicitly that the sea level rises and falls at God's command across an immobile landscape, save for the deformation caused by volcanoes and earthquakes.

Gerhard Mercator

The next "geological" book of importance after Münster's *Cosmographei* was *Atlas* by Gerhard Mercator, "the greatest geographer of the 16th century" (Gallois, 1890, p. 240; see especially Averdunk and Müller-Reinhard, 1914; Büttner, 1979a, 1992: Fig. 18). *Atlas* was published in 1595 by Mercator's son, Rumold, a year after his father's death[163]. Though subtitled *"or A Geographicke description of the Regions, Countries and Kingdomes of the world, through Europe, Asia, Africa, and America, represented by new & exact Maps"* in its English translation, the Latin original had a different subtitle, better expressive of the intention of the author: "*Sive Cosmographicæ Meditationes de Fabrica Mundi et Fabricati Figura*" (or "A Cosmographer's Meditations on the Creation of the World and the Shape of the Creation")[164] (Fig. 19). The book consisted of a text and maps, the latter causing the *Atlas* justifiably to become famous. Our attention here is on the text, however. As Büttner (1992, p. 17) pointed out, it is really a Biblical creation exegesis (in the same vein as Münster's and possibly inspired by his)—nowhere is there a description of the "Shape of the Creation," so its title is somewhat misleading. Although Mercator is insistent on the value of observation, he does not always follow his own advice and advances into speculation. *The extremely important thing about this is*

Figure 18. Gerhard Kremer (with his Latinized name, Gerardus Mercator: 1512–1595) at the age of 62 in a portrait engraved by Frans Hogenberg showing the geographer 20 years before his death—and 21 years before the publication of his *Atlas*—when he was busy on his Ptolemy maps intended as a preparation of his *Cosmography* (from Skelton, 1968, p. VII).

that, in his speculations, Mercator was willing to leave the Bible, where the evidence or simple logic demanded it. Büttner (1992, p. 38) showed that Mercator confined the concept of Providence to the human society and considered the natural world outside it. This was the beginning of the emancipation of the earth sciences from theology and, via such authors as Bartholomäus Keckermann (1572–1609), eventually led to the deistic concepts of the eighteenth century (Büttner, 1979e). Along this path, Mercator followed the footsteps of Agricola and Münster.

Mercator's geology contains nothing that we cannot find in Agricola and Münster, except one speculation that leads him to discuss *possible* falcogenic effects:

Here is also to be observed, how great the wisdome of the creator was in making hollow those bayes, and chanels, as receptacles of the waters, for so hath he distributed the Sea throughout the whole world, that all the Kingdomes of the world may have commerce one with an other, and what things forever either nature, or art affordeth, may transport whither they will. And (which is most of all) that the earth, with the waters collected together, making one Sphere, might remaine in one equall balance: for otherwise the earth should not be established upon the waters, but the more heavy weight being collected into one part, should presse downe all the masse of the earth, towards the center of gravitie, and of the world and that depression of the earth having elevated more high, & aloft, the waters lying on the other part, would have caused them to overflow, & possesse the next adjoyning lands. For after that the earth in the same quantitie is heavier than the waters, it is necessary that first the bodie of the earth, consist by it selfe in an equall balance. (Mercator and Hondius, 1636[1968], p. 19)

Figure 19. The title page of the first edition of the *Atlas* published by Mercator's surviving son, Rumold, in Duisburg in 1595 (from an undated facsimile of the maps of the *Atlas* by the Coron Verlag in Lachen, near Zürich).

In the text above, Mercator describes the pseudo-isostatic theories of the later Middle Ages and visualizes wholesale uplift and subsidence of the solid parts of the globe. However, he then says such movements *do not* in fact happen because God has so wisely distributed land and sea upon the globe as to maintain a perfect balance. We shall see that in the seventeenth century (i.e., the century after the one in which Mercator wrote these words), the two great leaders of geological thinking both still considered the medieval pseudo-isostatic theories to be capable, at least in part, to account for the falcogenic movements of the earth's outer rocky rind.

The books by Agricola, Münster, and Mercator give us a good idea of the prevalent tectonic thinking of the sixteenth century. It was dominated by Aristotelian tectonics with winds causing most of surface deformation by circulating in a complex network of subterranean channels and being aided by fire fueled (according to such post-Aristotelian authors as Strabo, Pliny the Elder, Lucretius, and especially Seneca) by burning sulphur, bitumen, and coal. That the Renaissance geologists did not use, and probably did not even know, Aristotle's falcogenic theory was probably because Agricola, with a dominant interest in mining, never worried about the evolution of the surface of the earth. The rapid rise of modern science in the next century, combined with the expanding horizons of the world geography, helped to bring together the subterranean bias of Agricola and the geomorphological bias of Münster and Mercator, from which union the modern geology began to emerge. But, as Stille (1919, p. 165) observed in the beginning of the twentieth century, geologists rooting intellectually into the mining tradition of central Europe still neglected the surface of the earth in their interests and that the falcogenic events remained more in the domain of the geographers. Agricola's influence reached the geologists of the twentieth century through such great central European miner-geologists of the eighteenth century as Füchsel, Lehmann, and Werner (cf. Şengör, 1991c). For the intellectual background of the central European mining tradition in geology, see also Şengör (2001b).

CHAPTER VI

THE DAWN OF MODERN GEOLOGY: DESCARTES, VARENIUS, STENO, HOOKE, AND THE TWO KINDS OF DEFORMATION OF THE EARTH'S ROCKY RIND

Descartes and the Origin of the Modern Ideas on the Evolutionary History of the Earth

In addition to expanding our geographical horizons, "doubling the products of the creation" (to quote Voltaire), destroying the neat geometric conceptions about the earth entertained by the medieval thinkers (e.g., see Behrmann, 1948; Bettex, 1960; Debenham, 1960; Newby, 1975; Fernández-Armesto, 1991), and, in addition to the advancing mining operations, practices and theories, revealing the anatomy of the earth (e.g., Wilsdorf, 1956), an appreciation of time different from the antique concepts of eternity and the medieval preconceptions of Mosaic creation history (cf. Whitrow, 1988) had to emerge to lay the foundations of modern geology. This last element, strangely, came from the developments in astronomy (cf. Gohau, 1983a, p. 277–284; 1983b, p. 6–7; 1990, especially p. 66 ff.).

The Copernican re-invention of a sun-centered heaven and the parallel developments, resuscitating the idea of an endless universe, inspired in the mind of René Descartes (1596–1650: Fig. 20)[165] a model for the evolution of our solar system that developed from a combination of Copernican geometry with an Aristotelian concept of physics (see Descartes, 1644[1842]; Gohau, 1983b and 1990, reproduce the relevant passages on their p. 71–73 and 90–95, respectively). Descartes' universe consisted of agile, yet perfectly fitting, particles creating a continuum. The agility resulted in mutual rubbing, which ended up eroding the original fitting particles and creating round elements. These round elements Descartes strangely called "secondary" and the scraps resulting from their erosion "primary." A third type of element consisted of conglomerations of the primary elements. The primary elements were, according to Descartes, "luminous," the secondary "transparent," and the tertiary "opaque." The future solar system originally consisted of 14 whirls, the nuclei of the future sun, planets, and their satellites. Not all original whirls were of the same size owing to piracy from neighboring whirls and consequent losses in their favor. The sun's whirl was larger than the rest and, thus, eventually engulfed all others.

Descartes thought that the earth originally was as much a star as the sun, as the sun was as much a star as all other stars. The Aristotelian onion-universe was shattered, and privileged locations vanished. Smaller "stars" formed shells around themselves consisting of the third, opaque element, labeled layer *M* by Descartes, enclosing a firey nucleus consisting of the first element (*I;* see Fig. 21A). The interstellar space, called the space *B* by Descartes (Fig. 21A), also consisted of the tertiary element but in a much looser arangement. Its biggest agglomerations soon settled on the layer *M*, forming another opaque layer *C* on the nuclear sphere, whose formation was accompanied by the squeezing out of it a fluid part forming yet another concentric shell outside *C*, called *D* (Fig. 21B). Above *D*, Descartes noted the necessity of yet another shell consisting of layered material, which he called *E* (Fig. 21B). This *E*-shell was what we might consider a sort of equivalent to our earth's crust. Descartes must have considered its formation a kind of by-product of the "evaporation" of the layer *D* because the *D* elements, agitated by the heat of the sun, rose across the pores of the *E*-shell and returned by night owing to the diminished temperature and the resultant loss of agitation. Not all of the *D* elements were able to make it back to *D* but got stuck in the pores of the elemet *E*. Descartes probably thought this is how an original *E*-shell may have formed by the *D*-elements that "evaporated" during the day and "precipitated" at night. That may have been why he emphasized that the *E*-shell was "layered."

Figure 20. René du Perron Descartes (1596–1650), one of the creators of the modern world. While doing so, he also sowed the seeds of modern geological thinking. (From Runes, 1959, p. 175; Runes credits Archives Photographiques, Paris, as the source of picture.)

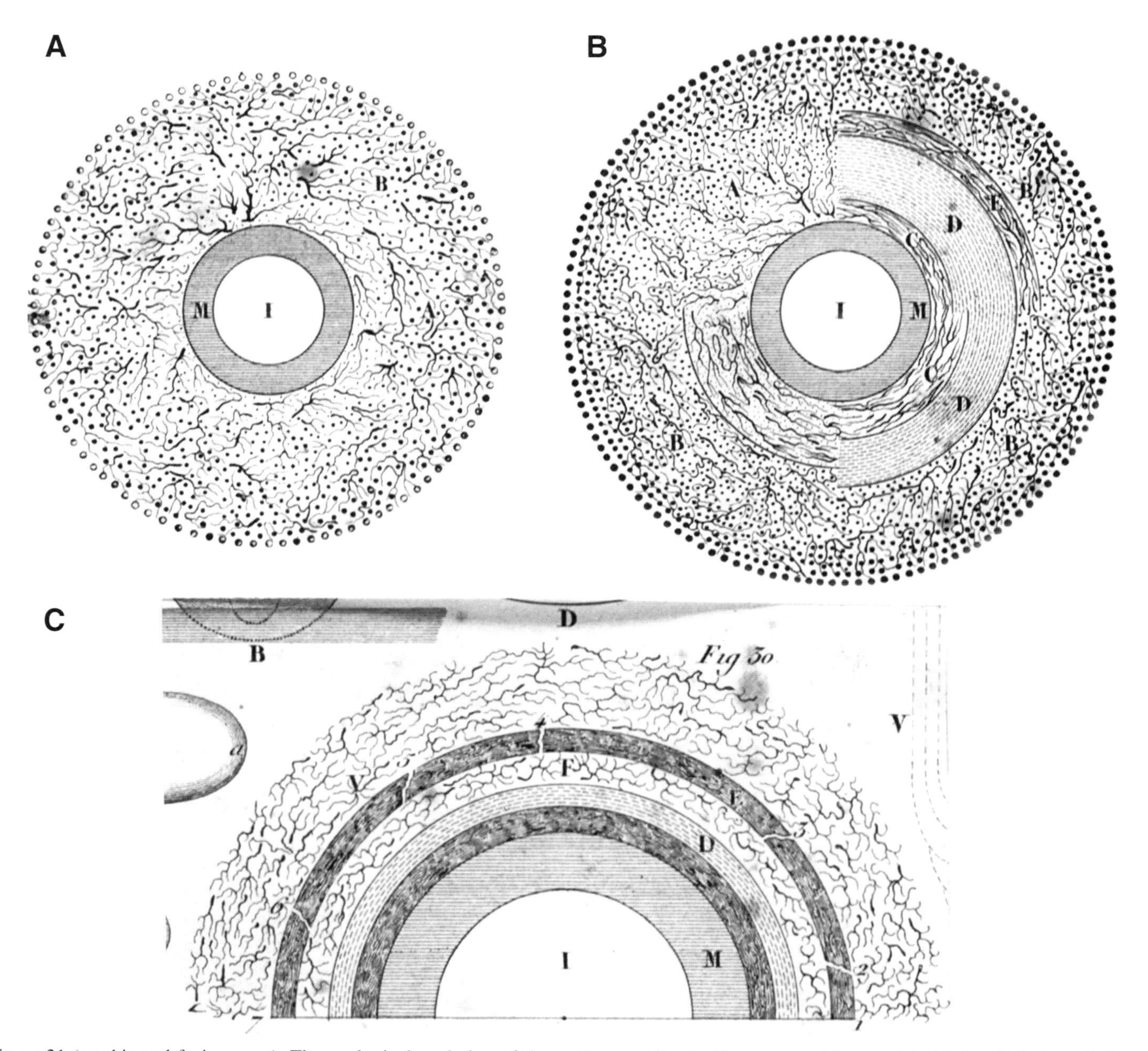

Figure 21 (on this and facing page). The geological evolution of the earth, according to Descartes as illustrated in his *Les Principes de la Philosopie* (figures here were copied from Descartes 1648[1842], figs. 26, 28, 30 and 31). A: Formation of the heavy core. B: The loose tertiary element differentiates into the shells *C, D*, and *E* disposed concentrically with respect to the earlier formed core. C: The formation of the *F*-shell.

In any case, these excess *D* elements caused with time a space to form between *D* and *E*, which Descartes thought was filled by a new shell, called *F* (Fig. 21C), consisting of smaller particles than those making up the *D*-shell (probably because they made it back through the *E*-shell with no trouble, or because some *D* elements became decomposed to form *F* elements?). The *E*-shell had more weight than the *F*-shell. By continuous *D*-element depletion, the *F*-elements attempted to fill in the *D*-pores left behind in the *E*-shell, but their size was insufficient. Vigorous *D*-influx then took place into the *E*-pores to close them (Descartes repeats the old Aristotelian saying that was so popular in the Middle Ages and during the Renaissance: nature abhors the void), but this only ended up enlarging the *E*-pores even more by the intensity of the impacts. Finally, the pores in the *E*-shell became so large that it became incapable of supporting itself above the *F* and the *D* shells. The *E*-shell disintegrated and fell into the underlying fluids (Fig. 21D). But since the radius of the *E*-shell was much larger than that of the *C*-shell onto which it eventually fell, parts of it had to form broken arches to fit into the lesser area below[166]. Thus, Descartes thought, this was the process by which mountains formed (Fig. 21D).

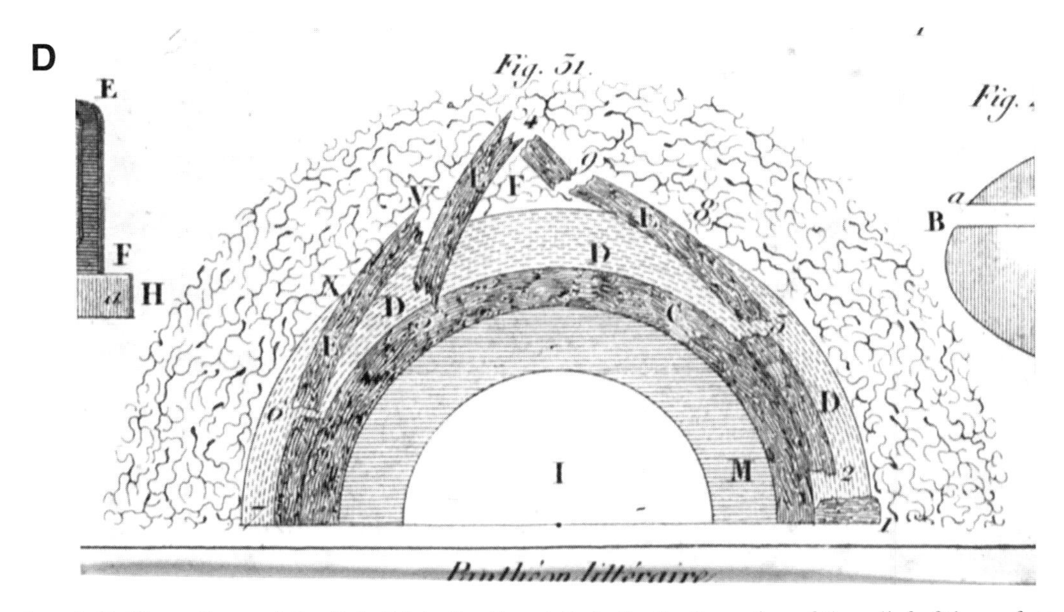

Figure 21 *(continued)*. D: The collapse of the *E*-shell into the *F*- and *D*-shells; the formation of the relief of the surface of the earth. Compare this with Steno's depiction of the tectonic history of Etruria (illustrated in Fig. 23). Note that the hollowing out of the basement of the youngest bed in Figures 23B and 23E is very similar in conception and in terms of its later effects to the formation and evolution of Descartes' *F*-shell.

Note that this is the first theory we have so far encountered that makes possible the creation of copeogenic deformation (as Descartes depicted it: Fig. 21D) and that potentially can also generate falcogenic structures (which apparently Descartes did not think of) without making any use either of exogenic phenomena (wind, water, etc.) or of volcanism. Descartes' concept results from the peculiar evolution of the earth itself as a whole! It is a part of the history of the construction of the planet. He depicts a planet, the large-scale history of which *necessarily* (more so than the gravitational theory of Master Buridan) involves large-scale deformation of its outer rocky rind!

This was a tremendous novelty! Judging only by the great similarity of Descartes' deformational features to those envisioned by Steno in the figure he appended to the *Prodromus* (Stenonis, 1669[1969]; see Fig. 23 herein), I venture to think that it profoundly influenced Steno's tectonic thinking despite the fact that in the year in which the Dane published his first geological book of lasting fame (Stenonis, 1667[1969]) while undergoing a great spiritual turmoil culminating in his conversion to the Catholic faith on 2 November, the Vatican chose to place the great Frenchman's books on the infamous Index! Many writers following Steno simply used Descartes' basic model to account for the deformational surface features of the earth (e.g., Burnet, 1684, especially p. 114, where there is an explicit reference to Descartes; Ray, 1692, especially the chapter entitled "A Digression Concerning the Primitive Chaos and Creation of the World"; Woodward, 1695, especially Part II). Like Descartes, none of these writers, except Ray (see below, Chapter VIII), made a distinction between copeogenic and falcogenic deformations.

Bernhardus Varenius and the Birth of Modern Physical Geography

When the sixteenth century drew to a close, our knowledge about the earth's surface had greatly increased, but the way mankind viewed it was not in any way different from the way the earth and its inhabitants had been regarded in antiquity. Antique geographers described the earth in general and then proceeded to a chorography, in which individual geographical regions were described. This chorography was commonly mixed with history and what might anachronistically be termed ethnography. Geography from Anaximander in the sixth century B.C. to Pomponius Mela in the first century A.D. was regarded as one long description of everything that one encountered on the face of the earth. Specifically "geophysical" topics were usually treated under the heading meteorology, following Aristotle's example (see, for example, Gilbert, 1907[1967]; Strohm, 1935). Both medieval geographers and the great Renaissance geographers, of whom Münster was perhaps the apogee, continued this tradition of "comprehensive and continuous description."

This tradition began to be shaken by the work of what Oskar Peschel called the "Dutch School," many members of which were born Germans (Günther, 1906, p. 5). In the works of these geographers, including Gerard Mercator, Abraham Ortelius (1527–1598), Petrus Plancius (1552–1622), Gemma Frisius (1508–1555), Philipp Clüver (Cluverius) (1580–1622), and Paulus Merula (1558–1607), a "general geography" began to be distinguished from "special" or "particular geography" dedicated to local description. Günther (1906, p. 5–7) argued that Petrus Bertius (1565–1629), Paulus Merula, David Christiani (1610–1688), and Abraham Goellnitz (Golnitzius) (*fluorit* 1631–1642) were the immediate predecessors of Bernhardus Varenius, who is usually credited with having created the modern general geography with his great book *Geographia Generalis*, first published in 1650 in Amsterdam (herein I cite a 1664 reprint by Elzevier that I have in my private library). In addition, Baker (1963, p. 11) cited authorities to the effect that Varenius may have known of *Geographie Delineated in Two Bookes, containing the Sphericall and Topicall Parts thereof* (Oxford, 1625, corrected second edition in 1635) by the English geographer Nathanael Carpenter (1589–1628?). But the evidence is not conclusive. Büttner (1979) documented, however, that Varenius took the concept of "general geography" from his great predecessor Bartholomäus Keckermann (1572–1609) and in fact plagiarized many parts of his *Systema Geographicum* (1611, Hanoviae; possibly written in 1602: Büttner, 1979e, p. 157; for references to Keckermann's writings, see Büttner's paper).

Very little is known of Varenius' life; he is thought to have died when only 28! The following summary is based on the biography published by Günther (1906).

Varenius was born in the small town of Hitzacker near Lüneburg on the Elbe, south of Hamburg. His father was a well-educated court preacher. Varenius received his fundamental education in Hamburg, in the *Akademischen Gymnasium*, under the multi-faceted genius Joachim Jungius (1587–1657), whose wide-ranging interests from pure mathematics through chemistry and botany to linguistics also included geography. Having finished his studies under Jungius in 1643, Varenius went to the university in Königsberg in East Prussia. This proved a disappointment. Varenius left Königsberg and moved to Leiden in Holland in 1645 to study medicine, where the university, founded in 1575 by William the Silent, was one of the best in Europe. But after two years, in 1647, Varenius' deteriorated financial situation forced him to move to Amsterdam, where he became a private manager and, in his free time, a private tutor. However, Varenius was restless. His double vocation left him little spare time for his intellectual pursuits. He finally resigned his managerial position some time in December 1647 to live off his savings while attempting to find a better position. In a year and a half, he finished his studies in medicine and received his M.D. on 22 June 1649. From his letters, Günther (1906) inferred that Varenius was keen to get an academic position in a university, and this he hoped to do by writing a book on some academic subject that interested him. It is to this hope that we owe the first modern geography book that came out the following year. In fact, the last established date in Varenius' short life is 1 August 1650, when he finished writing the dedication for his book (Varenius, 1664, p. 6). After that, as Günther sadly remarks, the author disappears from the sight of the historian. It is believed that he died shortly thereafter, most likely from overwork and exhaustion. In many respects, his short, but fruitful, life and unbending character offer parallels with those of Mozart.

It is not the purpose of this section to discuss Varenius' geographical ideas. Günther (1906, especially p. 44–125) did that in great detail, to which the interested reader is referred (and to Büttner, 1979, as well). Suffice it to say that in his *Epistola Dedicatoria*, Varenius says that he wrote the book because he felt that the general principles of geography had so far been neglected. What interests us is his concept of the tectonics of the globe and specifically his view of falcogenic movements.

Varenius' earth was no different from that of most of his contemporaries. He was not a revolutionary in that regard. He believed volcanoes to be fueled by sulphur and coal (Varenius, 1664, p. 105) and the earth to be porous, in which not only inflammable substances existed, but also rivers circulated. He further considered earthquakes, resulting from collapsing cave roofs, as being among eight good reasons for believing in a porous earth. He believed that volcanoes, islands that emerge from the sea, and mountains had a common origin in the rise and eruption of "spiritus" (Varenius, 1664, p. 314). Günther (1906, p. 204, note 628) interpreted Varenius' "spiritus" as "highly compressed vapour." Varenius divided mountain ranges into two classes after their trends, namely those following the meridians and those following the parallels (Varenius, 1664, p. 92; in fairness to him, let me note that he only said that "those that extend from north to south and those from east to west"). He inherited this idea from antiquity (mainly from Ptolemy) and bequeathed it to posterity. He was certainly not its herald, as Günther, in his enthusiastic biography of Varenius, claims (Günther, 1906, p. 194, note 523). It was popular until almost the end of the eighteenth century. Both Varenius' contemparory, Athanasius Kircher, and Comte de Buffon in the late eighteenth century were among its proponents.

Varenius was aware that not only mountains, islands, and volcanoes have a history of formation, but also entire landscapes. He notes in successive propositions that rivers, swamps, lakes, and oceans now occupy places they previously had not occupied. He argues that the Atlantic Ocean opened by tearing America from Europe ("... *ita Americam à Veteri Orbe, seu sola Europa* avulsam *esse* ...": Varenius, 1664, p. 317, emphasis mine) and thus the American Indians must be also children of Adam.

The question Varenius next answers is whether such changes are sufficient to create complete changes in the present geography. He does not believe they are. He is willing to admit quite significant changes, but cannot believe that whole continents and oceans can exchange their places. He does not think that waters can overstep their bounds, because their surface cannot anywhere lie higher than the surfaces of *entire* continents (he was no advocate of the medieval notion that the center of gravity of the water and land spheres can move with respect to one another); conversely, if one wanted to drain the oceans, Varenius did not believe there were sufficiently large empty reservoirs within the earth to accommodate their waters (Varenius, p. 319–320).

In his small book, Varenius thus appears as an innovator with respect to his view of geography as a science, but not so much with respect to his views as to how the earth works. However, he was not entirely unoriginal. His world was largely Empedoclean/Aristotelian, much like Agricola's and Münster's. He believed it had a perforated structure and that its internal channels contained both inflammable substances and water. This led to local fires and generation of steam. When steam forced an escape to the surface, it either created volcanoes by piercing the earth's crust (the expression "earth's crust" was used by Varenius) or generated mountains or islands if the crust was only uparched. Surface processes created large-scale changes on the surface, but Varenius must not have thought Aristotle's mechanisms discussed in *Meteorologica* sufficient to create wholesale changes in the entire geography of continents. Thus, his earth model was more "fixist" than Aristotle's, but perhaps better in accord with what could be observed or reasonably inferred from the presently active processes. In this important aspect, he was a follower of Theophrastus and a forerunner of the actualists such as Hutton, von Hoff, and Lyell. Günther (1906, p. 123) even sees in him a forerunner of those who believed in the permanence of continents and ocean basins. In his world, there were no sharp divisions between copeogenic and falcogenic events; one group graded into the other as they did in Theophrastus' world and as they were to do in Hutton's and, initially, also Lyell's. But Varenius' greatest contribution was no doubt the circumscription of a science of the earth distinct from the nebulous geography of antiquity. He created physical geography as a science of both space and time, although he had no model that could have determined the direction of time involved in building the structures of the earth. He remained untouched by Descartes' new model of earth evolution, and he did not live to see the great innovation Steno introduced almost two decades after his death. It is a melancholy reflection to think what Varenius himself might have yet done with such stimuli had he lived longer.

Steno

After Descartes and Varenius, two luminaries enlighten the path of the development of geology in the middle of the seventeenth century: the Dane, Niels Stensen (1638–1686: Fig. 22)[167] and the Englishman Robert Hooke (1635–1703)[168]. Both were initially concerned mainly with fossils, and both concluded that the fossils had been enclosed in rocks during deposition and that the enclosed rocks had been subsequently made a part of land by deformation (Rudwick, 1976, p. 53, speculates, with a reference to a 1958 paper by Victor Eyles, that Steno may even have derived some of his ideas from Hooke, whom he may have met in Montpellier in southern France). Both viewed earthquakes as an agent of rock deformation (e.g., Stenonis, 1667[1969], p. 98, 100; Hooke in Waller, 1705, p. 290–291).

Steno was more concerned to document that rock deformation had happened than he was inclined to discuss its causes (see especially Stenonis, 1669[1969], p. 166, 168). However, he did consider two principal ways in which rocks may be deformed:

[1] The first way is the violent upheaval of strata, whether this be due mainly to a sudden flare of subterranean gases or to a violent explosion of air caused by other great subsidences nearby...

[2] The second way is a slipping or subsidence of the upper strata after they have begun to crack because of the withdrawal of the underlying substance or foundation; in consequence the broken strata take up different positions according to the variety of cavities and cracks [*see Fig. 23*]. ... Alteration in the position of strata affords an easy explanation of various fairly difficult problems. Herein may be found a reason for the unevenness of the earth's surface that gives rise to so many controversies, as are found in mountains, valleys, natural reservoirs, elevated plains and low-lying plains. (Stenonis (1669[1969], p. 164–167; italics mine).

The events Steno mentions are clearly Aristotelian/Theophrastian and copeogenic in nature, although subsidence by withdrawal of subterranean "substance" seems influenced by Aristotle's criticism of Anaximenes' seismic theory and also undoubtedly by Descartes' and possibly also Varenius' speculations (see above). However, when it came to *explaining* the geological history of Tuscany, Steno had to deal with at least two major inundations of the sea (Fig. 23A and C), for which he felt that copeogenic phenomena could not account:

Figure 22. Niels Stensen (or Steensen; his Latinized name: Nicolaus Stenonis; 1638–1686) shown in a portrait that was probably made by J. Sustermann as part of a blanket order to paint portraits of the entire Medici court in Florence (see Bierbaum et al., 1989, p. 147 and plate I). The portrait, which is now in the Uffizi in Florence, shows Stensen at the time when he was engaged in his geological studies (1666–1667).

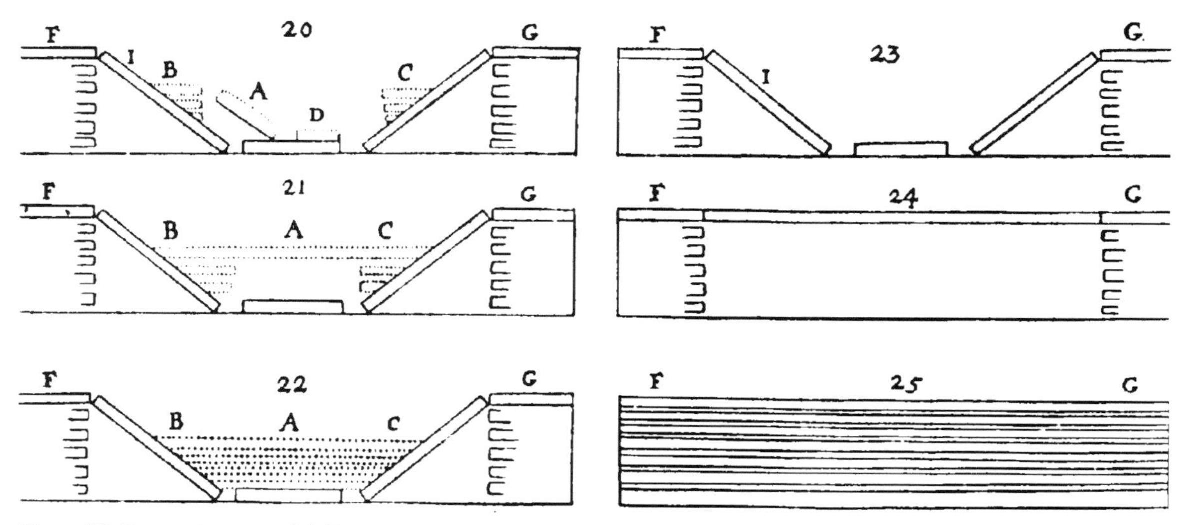

Figure 23. Stensen's sequential diagrams showing the six former states of Etruria (present-day Tuscany in Italy) from the *Prodromus* (Steno, 1669 [1969]). Stensen's caption to this diagram is as follows (translation by John Garrett Winter, 1916, p. 276; italics mine):

The last six figures [*i.e., the ones displayed here*], while they show in what way we infer the six distinct aspects of Tuscany from its present appearance, at the same time serve for the readier comprehension of what we have said about the earth's strata. The dotted lines represent the sandy strata of the earth, so called from the predominant element, although various strata of clay and rock are mixed with them; the rest of the lines represent strata of rock, likewise named from the predominant element, although other strata of a softer substance are sometimes found among them. In the Dissertation itself I have explained the letters of the figures in the order in which the figures follow one another: here I shall briefly review the order of change. [A.] Figure 25 shows the vertical section of Tuscany at the time when the rocky strata were still whole and parallel to the horizon. [B.] Figure 24 shows the huge cavities eaten out by the force of fires or waters while the upper strata remained unbroken [*Note influence of the porous earth model of antiquity and of Kircher and Descartes*]. [C.] Figure 23 shows the mountains and valleys caused by the breaking of the upper strata [*influence of Descartes!*]. [D.] Figure 22 shows new strata, made by the sea, in the valleys. [E.] Figure 21 shows a portion of the lower strata in the new beds destroyed, while the upper strata remain unbroken. [F.] Figure 20 shows the hills and valleys produced there by the breaking of the upper sandy strata.

Steno thought that the stage A corresponded with the creation of the world (Descartes!), and the stage C with the Biblical flood.

Regarding the manner in which the waters rose, we can put forward various agreements with the laws of Nature. If it should be said that the centre of gravity of the earth does not always coincide with the centre of its figure, but sometimes moves away from the side, sometimes from the other, according to the formation of subterranean cavities in different places, it is possible to put forward a ready reason why the fluid that covered everything in the beginning of things left certain places dry, and returned again to occupy them. (Stenonis, 1669[1969], p. 206–207)

This gravitational theory *à la* Buridan, identical with that envisaged by Mercator, somehow did not appeal to Steno (as it had not earlier to Mercator). Perhaps Mercator had inspired Steno to think so because it was not in complete agreement with the Scripture. But Mercator had scrapped the theory because he did not think that large inundations had happened in God's perfect world. Steno documented that large inundations had indeed happened. Consequently, he invented a Platonic-Aristotelian model, in which a greater role was assigned to the central fire in the earth—an idea which, in Steno's day, had long been associated with the generation of metals in the earth (Adams, 1938, p. 279–282; also see the long quotation from Münster above[169]). At the time Steno was writing his *Prodromus*, this idea had been made popular in "geological" circles by the Jesuit father Athanasius Kircher[170] (Fig. 24; see also Kircher, 1657, p. 175, where he talks about the *pyrophylacia*, which are depicted in Fig. 25 and in his ch. IV, especially p. 197 ff. for the origin of metals)[171]. Steno's model had the following features:

1. Passages through which the sea penetrates into hollows in the earth to supply water to the sources of bubbling springs were blocked by the slipping of fragments of certain strata.

2. Water, undoubtedly enclosed by the bowels of the earth, was driven by the force of the well-known subterranean fire partly towards the springs and partly ejected into the atmosphere, through the pores of the earth that were not yet covered with water; then the water, not only that which is always present in the air but also that which was mixed with it by the method described above, fell in the form of rain.

3. *The bottom of the sea was raised up by expansion of subterranean caverns.*

4. The remaining cavities on the earth's surface were filled with earthly material eroded from higher places by the continuous rainfall.

5. The surface of the earth was less uneven since it was nearer in time to its original state.

(Stenonis, 1669[1969], p. 208–209; italics mine)

Figure 24. Jesuit Father Athanasius Kircher (1602–1680), shown at age 62, in an engraved portrait from his *Mundus Subterraneus* (facing the title page).The caption beneath the portrait reads (in Latin): "Painter and poet say in vain, HE IS HERE: His face and name are known to the antipodes," by Jacobus Albanus Ghibbehim, M.D.

Steno's deluge model calls not only for a Platonic-Aristotelian porous earth (which at that time was considered perfectly plausible; see Fig. 25 and 26; Kircher, 1657, p. 64[172] and 175 ff.) but also for the agency of the central fire to uplift the sea bed[173]. However, this uplift was accomplished (in the Aristotelian manner) by enlarging the pores of the earth (which, in Steno's devoutly Catholic mind, probably made his model compatible with the statement "all the fountains of the great deep were broken up ...": *Genesis* 7:11) and was thus falcogenic in character. It also happens to agree with the view of Giles of Rome. In essence, therefore, we are still very much with the Platonic-Aristotelian model of earth behavior featuring two distinct sorts of movement of the rocky rind with a somewhat increased role perhaps being assigned to the central fire. Copeogenic events were obvious at one or few outcrops. Falcogenic events had to be inferred from the study of many outcrops showing the uniform behavior of sea level at many widely separated localities.

Hooke

Writing a year before Steno's *Prodromus* was published, Robert Hooke (in Waller, 1705) agreed in all essentials with Steno concerning the nature of fossils and the former transformations of the earth's surface. Far less original than Steno in matters of detailed outcrop geology, Hooke also conceded:

> That a great part of the Surface of the Earth hath been since Creation transformed and made of another Nature; namely Parts which have been Sea are now Land, and divers other Parts are now Sea which were once a firm Land; Mountains have been turned into Plains, and Plains into Mountains, and the like. ... That most of those Inland Places, where these kinds of Stones are, or have been found, have been heretofore under the Water; and that either by the departing of the Waters to another part or side of the Earth, by the alteration of the Center of Gravity of the whole Bulk [*note Buridan's influence across the ages!*], which is not impossible; or rather by the Eruption of some kind of subterraneous Fires, or Earthquakes, whereby great quantities of earth have then been rais'd above the former Level of those Parts, the Waters have been forc'd away from the Parts they formerly cover'd, and many of those Surfaces are now rais'd above the Level of Water's Surface many scores of Fathoms. (Hooke, 1705, p. 290–291, italics mine)

Hooke pointed out that he knew of eight different kinds of effects of earthquakes, four of uplifts, and four of subsidences. As can be seen in Figure 27, all of these belong in the class of copeogenic events, despite the fact that Hooke envisaged some of them to affect whole plains (but on p. 311, he stressed that mountainous tracts and seashores suffer most from earthquakes; see also p. 421). Hooke did also think of falcogenic events, however, in terms of a shift of the center of gravity of the earth (p. 321–322 and 346 ff.) that might explain a slow shift of past palaeogeographies.

At the dawn of the modern era of geology, its principal theoretical practitioners were thus aware that the earth's upper surface probably had been (and possibly still was) subject to two very different sorts of events. One kind, being of short wavelength and also of short duration, was associated with events that changed the fabric of rocks visible at the outcrop. Earthquakes and volcanic eruptions were readily associated with this class. The other class was more elusive. Clearly it influenced the distribution of land and sea and the changes undergone in a big way, but neither its mechanism nor its more intimate effects (if any) seemed within the easy grasp of the geologist. The major difficulty was to describe and to account for its exact nature. Neither Robert Hooke nor Niels Stensen were any more comfortable with it, in the middle of the seventeenth century A.D., than Anaximander had been at the beginning of the sixth century B.C.!

Figure 25. Kircher's ideal system of fire containers (*Systema ideale pyrophylaciorum*) in the earth (1665, plate located between p. 180 and 181).

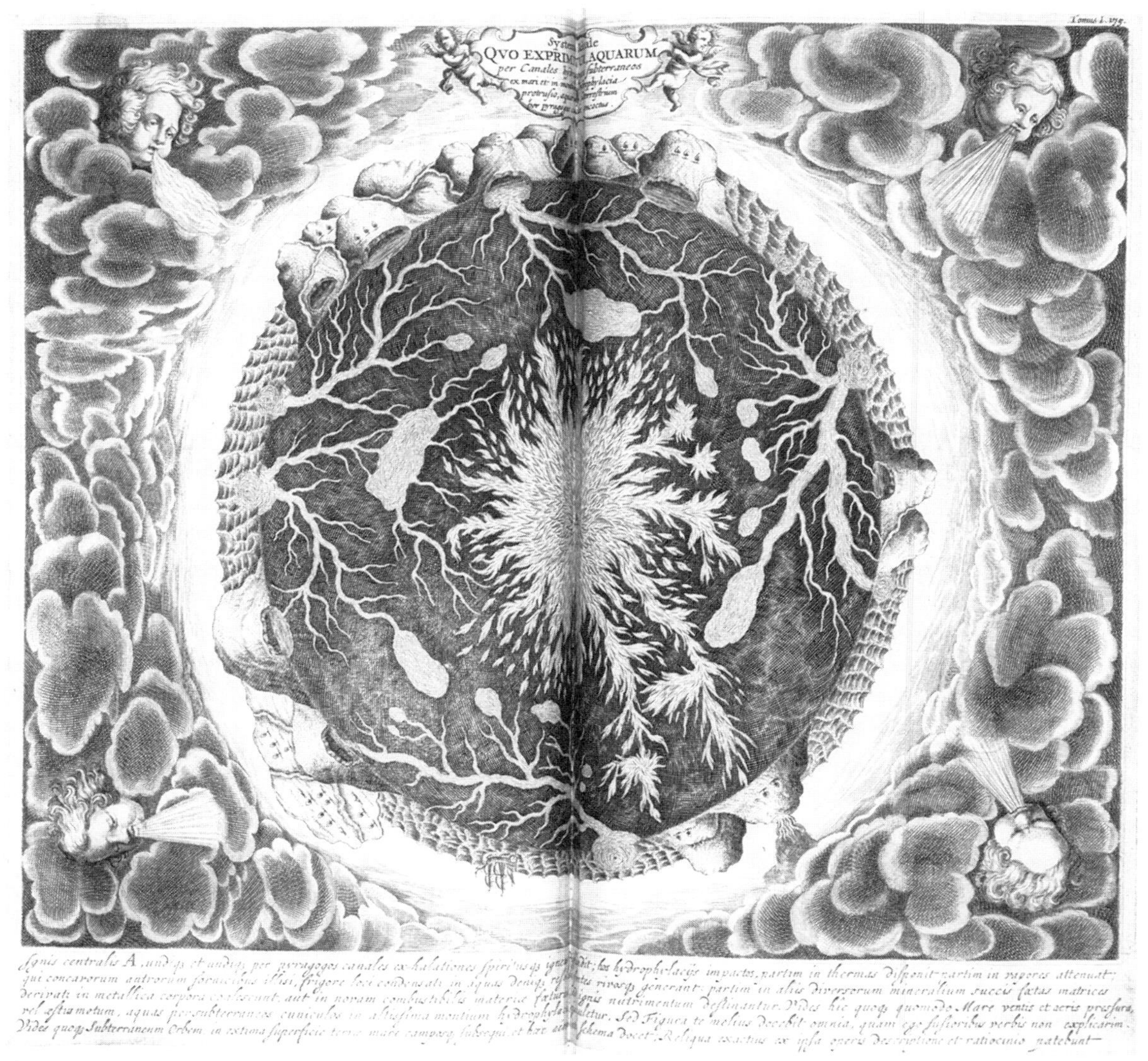

Figure 26. Kircher's ideal water system with subterranean water canals (*Systema ideale quo exprimitur, aquarum per canales hydrogogos subterraneos*) in the earth (1665, plate located between p. 174 and 175).

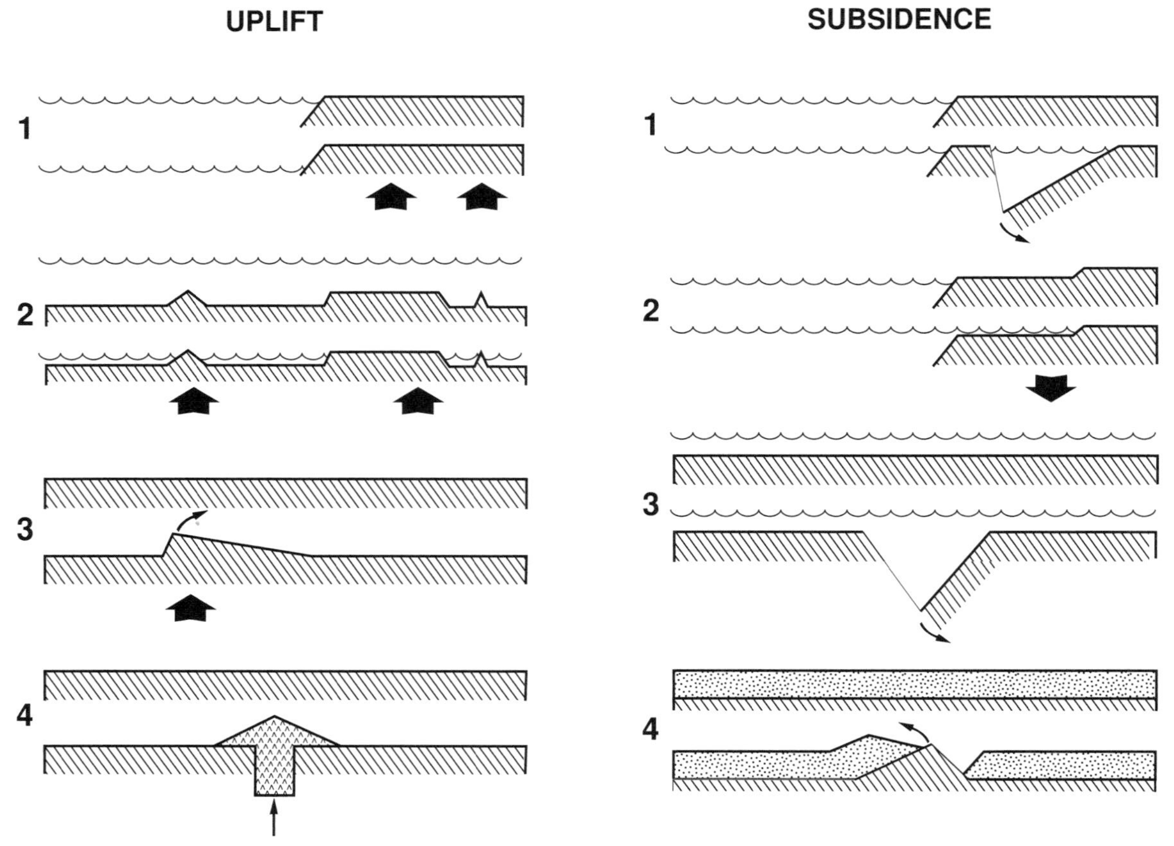

Figure 27. Effects of earthquakes according to Hooke (1705):

Uplifts:

The 1st is the raising of a considerable Part of a Country, which before lay level with the Sea, and making it lye many Feet, nay, sometimes many Fathoms above its former height. A 2nd is the raising of a considerable part of the bottom of the Sea, and making it lye above the Surface of the Water, by which means divers Islands have been generated and produced. A 3rd species is the raising of very considerable Mountains out of a plain and level Country. And a 4th Species is the raising of the Parts of the Earth by throwing on of a great Access of new Earth, and for burying the former Surface under a covering of new Earth many Fathoms thick. (p. 298)

Subsidences:

The *First*, is a sinking of some Part of the Surface of the Earth, lying a good way within the Land, and converting it into a Lake of an almost unmeasurable depth.

The *Second*, is the sinking of a considerable Part of the plain Land, near the Sea, below its former Level, and so suffering the Sea to come in and overflow it, being laid lower than the Surface of the next adjoining Sea.

A Third, is the sinking of the Parts of the bottom of the Sea much lower, and creating therein vast *Vorages* and *Abysses*.

A *Fourth*, is the making bare, or uncovering of divers Parts of the Earth, which were before a good way below the Surface; and this either by suddenly throwing away these upper Parts by some subterraneous Motion, or else by washing them away by some kind of Eruption of Waters from unusual Places, vomited out by some Earthquakes. (p. 298–299).

To these effects Hooke added (not illustrated here): "A Third sort of Effects produced by earthquakes, are the Subversions, Conversions, and Transpositions of the Parts of the Earth. A Fourth sort of *Effects*, are *Liquefaction, Baking, Calcining, Petrifaction, Transformation, Sublimation, Distillation*, &c." (p. 299; italics his).

CHAPTER VII

SCANDINAVIA: FALCOGENY IN ACTION?

Urban Hiärne

The Vikings must have been aware that some of their harbors were steadily shallowing and rendering them useless and that some were even in the process of emerging out of the sea, because the rate of uplift around Scandinavia is truly phenomenal, reaching nearly a centimeter a year in the northern part of the Gulf of Bothnia (Mörner, 1979; 1980, Part 2B, p. 251–354; Ekman, 1989, Fjeldskaar and Cathles, 1991). Figure 28 shows the increase of land in the Vasa (or Vaasa) region in Finland from 300 A.D. to 1900 A.D., and there is little question that it is impressive even for non-geologists.

It is possible that in the *Edda*, the mythology of the Nordic peoples, the mention of a retreating flood and the rise of land from the flood in the poem entitled "The Face of the Seer" may actually reflect the awareness of this regression (Genzmer and Schier, 1997, p. 36, 40).[174] Accordingly, the Swedish polymath Urban Hiärne (1641–1724; see von Beskow, 1857; also Frängsmyr, 1990; Königsson, 1990; and the delightful historical novel by Cederborg, 1946) included two questions on a possible increase of land in a questionnaire on natural science problems that he circulated in 1694, publishing the results in 1702 and 1706 (Mörner, 1979). Hiärne questioned a number of people and obtained observation reports stating that previously underwater areas in the outermost rocks of the Swedish archipelago were rising and coming out of water. In such areas, ship's keels had begun hitting rocks in places where such a thing had not happened earlier. The architect Abraham Swanskiöld (1644–1709) told Hiärne that mostly around Gotland and Öland but also around Carl's Islands (Karlskrona), the water had been 12–15 fathoms higher in the past as seen on what we today would call recessional terraces "as projected from AB to CC" (Hiärne, 1702, p. 99: see Fig. 29). Hiärne asked Swanskiöld whether he thought it was the water being taken out or land waxing or rising. The architect said that *if* it was the land rising, the amount of uplift was very considerable indeed: 15 fathoms! Hiärne claimed that the rise had not been noticed in the North Sea (Arctic shores) or in the West Sea (present-day North Sea). Indeed in many places flooding had been reported, as in north Germany or in the Netherlands (also see the discussion on water areas becoming land and vice versa in Hiärne, 1706, p. 282–291). Hiärne thus concluded that (1) the land could not be rising (the amount seemed unreasonable), and (2) the recession of the waters was not universal. He formulated the novel hypothesis that the Baltic Sea had been a lake that had an outlet too small to compensate for all the incoming waters from around it. Hence, its level had been higher. But the rigorous outflow had since enlarged the outlet and that is why the sea level in the Baltic Sea was rapidly dropping (Hiärne, 1702, p. 100). As we shall see in Chapter XIII, this is similar to old hypotheses concerning the origin of the Bosphorus (as we saw in Chapter II) and also what essentially Eduard Suess concluded in 1888 about the cause of the recession in the Baltic Sea (see Chapter XIII, below).

Emanuel Swedenborg

Although Emanuel Swedenborg[175] (Fig. 30) at first played with the idea of changes in the speed of rotation of the earth to alter the shape of the water envelope of the globe (Suess, 1888, p. 10), he later noted that this was impossible because "the seas toward the equator are but little elevated, or retain their horizontal altitute [*sic*!]" (Swedenborg, 1847, p. 31). Swedenborg then agreed with Hiärne that only a draining of the Baltic Sea could explain the observations, and he cited in support the allegedly different base levels of the rivers draining into the North Sea and into the Baltic.

As in Ionia more than two millenia earlier, the first attempts at a scientific explanation of the sea-level change around the Baltic Sea did not view land uplift as a serious option. However, similarities between the Scandinavian and the Ionian cases were to be increased by the creation of a neptunistic global theory associated with the former.

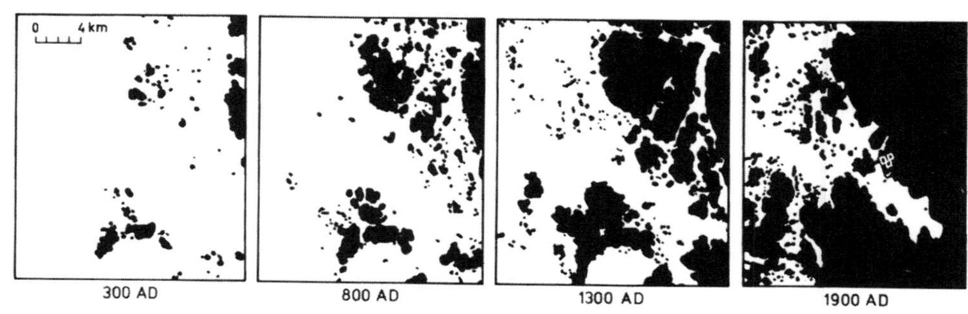

Figure 28. Uplift reflected in the increase of land in the Vasa (Vaasa) region of Finland. This area is where the Gulf of Bothnia is narrowest. After Renquist, from Mörner (1979, fig. 1).

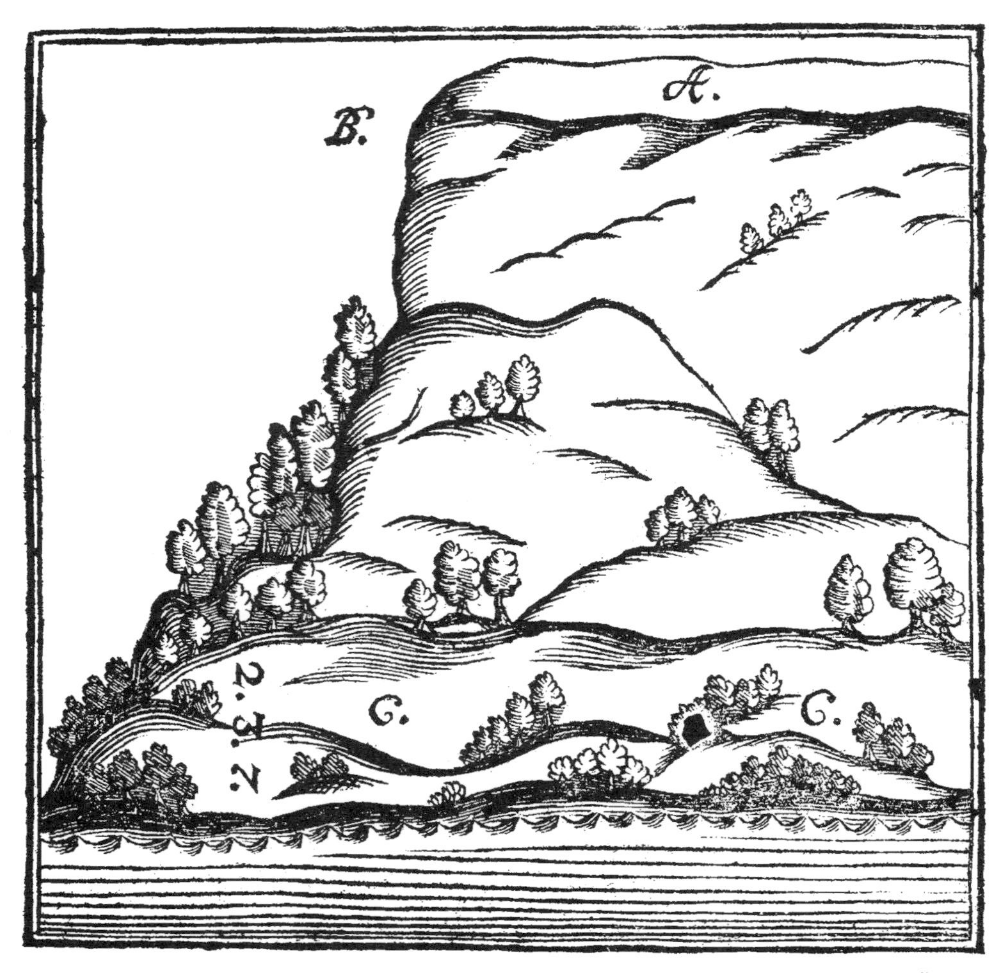

Figure 29. Urban Hiärne's depiction of marine terraces in the Swedish Archipelago (Gotland, Öland, and Carl's Islands {Karlskrona}; from Hiärne, 1702, p. 99). AB is the soil horizon. The sea has retreated from AB to CC (some 15 fathoms).

Celsius and Linnaeus

Anders Celsius (1701–1744: Fig. 31) and Carl von Linné (1707–1778): Fig. 32)[176] agreed with Hiärne's observation but thought that the recession of the strand was a global, ongoing affair much along the lines of Anaximander's original theory (Celsius, 1744; von Linné, 1744; for details on sources and influence, see Nathorst, 1908 and Frängsmyr, 1994b). Von Linné's elaborate earth history—starting with the Biblical flood, which allegedly had only spared the Paradise located atop the highest mountain on earth—had been especially influenced by the Christian mythology, through the Bible itself (*Ezekiel* 22, 24) and the Rabbinic tradition and early Syrian fathers (Levene, 1951, p. 187), and also perhaps via such medieval schoolmen as William of Auvergne and Renaissance geographers as Gregorius Reisch (cf. Hoheisel, 1979a, p. 63)[177]. For instance, in his *Hymns on Paradise*, St. Ephrem the Syrian (d. 373 A.D.)[178] sang:

> With the eye of my mind
> I gazed upon Paradise;
> the summit of every mountain
> is lower than its summit,
> the crest of the Flood
> reached only its foothills;
> these it kissed with reverence
> before turning back
> to rise above and subdue every peak
> of every hill and mountain.
> The foothills of Paradise it kisses,
> while every summit it buffets.
> (Brock, 1990, p. 78–79)

But, as we have seen above, the influence could easily have come also from the classical mythology (see endnote 27), or possibly even from the extra-Biblical Asia through such travelers as Nils Matsson Kiöping, whom von Linné read and quoted on Ceylon (Frängsmyr, 1994b, p. 120). If we remember that earlier in the thirteenth century Giovanni Marignolli had heard about the Paradise mountain and its having escaped the Flood in Ceylon (Yule, 1914[1966], p. 245), we might be able to weave a fairly robust thread leading to von Linné's Paradise hypothesis consisting of both Biblical and other south Asiatic material and of fragments of the classical mythology, which he knew so well.

This is, however, not the place to discuss von Linné's neptunistic theory of the earth, which was also the first scientific theory of biological dispersion. Suffice it to note that its influence was immense (see Nathorst, 1908 and Browne, 1983), not only because of the great authority of its author and the sources of its inspiration, but also because ideas such as this were in the air at the time (e.g., de Maillet, 1748[1968]; Leibniz, 1749, 1949; Oldroyd and Howes, 1978; Waschkies, 1989). Although both Jessen in 1763 (in a work entitled *Kongeriget Norge, Fremsittelt Efter dets Naturlige og Borgerlige Tilstand* {see Naumann, 1850, p. 269, note *}), and E.D. Rüneberg in 1765 (von Zittel, 1899) ascribed the negative movement of the strand in Scandinavia to uplift owing to earthquakes, it was only after Hutton shattered the neptunian earth theory that such opinions gained any currency.

Figure 30. Emanuel Swedenborg (1688–1772), portrait by Per Krafft the Elder (now in the Grisholm Castle in Sweden), depicted in his "mystical" years (from Benz, 1948, facing p. 256).

Figure 31. Anders Celsius (1701–1744) from a portrait by an unknown eighteenth century artist.

Figure 32. Carl von Linné (Latinized name: Carolus Linnaeus: 1707–1778) from the frontispiece in Stoever (1794).

CHAPTER VIII

KINDS OF UPLIFT IN A HUTTONIAN WORLD AND THE FOREPLAY TO THE CRATERS OF ELEVATION THEORY

General

Descartes gave the seventeenth century a reason to expect a greater mobility of the earth's rocky rind than hitherto thought possible. His successors mixed his ideas with those inherited from antiquity concerning a porous earth to generate models of continent-, ocean-, and mountain-building. In chapter VI, we saw the efforts of Varenius, Steno, and Hooke in this direction. In this chapter, we shall see how these ideas led to those of the nineteenth century, during which understanding of an explicit separation of large-wavelength, slowly evolving structures from short-wavelength, fast-evolving structures became commonplace and embraced the "strange" observations made earlier in Scandinavia as outlined in chapter VII.

John Ray: Synthesis of an Empedoclean Earth with a Cartesian Earth

The English naturalist John Ray (1627–1705; for his life, see Webster, 1981; Baldwin, 1986, with a fine bibliography including Raven's great biography of Ray) stands at the end of this style of thinking and provides a connecting link between Descartes' successors and Antonio-Lazzaro Moro, who fully returned to the Pythagoran/Empedoclean idea of a central fire in the earth, which made Hutton's theory possible.

Ray's consideration of the motions of the rocky rind of the earth is occasioned by his account of the creation. Starting with the primeval chaos, Ray assumes that God created the earth and "the solid and more ponderous naturally subsided, the fluid and more watery, as being more light, got above them" (Ray, 1692, p. 151). He finds support for this supposition in *Genesis* (ch. I, verses 2 and 9). Finally, it pleased God to separate the waters from the land. Ray confesses ignorance as to how the Good Lord did all this, whether by fiat or by invoking natural agencies. If the latter, "It might possibly be effected by the same Causes that Earthquakes are, *viz.* Subterraneous Fires and *Flatuses*. We see," he wrote with enthusiasm, "what incredible effects the Accension of Gunpowder hath: It rends Rocks, and blows up the most ponderous and solid Walls, Towers, and Edifices, so that its force is almost irresistible. Why then might not such a proportionable quantity of such Materials set on fire together raise up Mountains themselves, how great and ponderous soever they be, yea the whole Superficies of the dry Land (for it must all be elevated) above the waters?" (Ray, 1692, p. 153).

In the above quotation, we see Ray making a distinction between the rise of the mountains and the rise of the entire surface of land. This is not dissimilar to Buridan's distinction and, like his, points in the direction of a vague appreciation that mountain uplift and continental uplift are two different things.

To support his fire-driven uplift theory, Ray quotes Ovid's third century B.C. description of the rise of the volcano near Troezen (*Metamorphoses*, XV, 296–306; see p. 47) and the 1538 A.D. origin of Monte Nuovo near Pozzuoli. Having described the origin of these small volcanoes, Ray then asks the pertinent question: "If such Hills, I say, as these may be, and have been elevated by subterranous Wild-fire; *flatus* or Earthquakes: ... if we may compare great things with small, why might not the greatest and highest Mountains in the World be raised up in like manner by a subterraneous *Flatus* or Wild-fire, of quantity and force sufficient to work such an effect, that is, that bears as great a proportion to the superincumbent weight and bulk to be elevated, as those under these smaller Hills did theirs?" (Ray, 1692, p. 155–156; italics his). This is a reasoning most likely borrowed from Strabo—"For it cannot be that burning masses may be raised aloft, and small islands, but not large islands; not yet that islands may thus appear, but not continents." (Strabo, I, 3. 10)—although Ray does not cite him.

Ray then goes on to develop arguments to show that on the earth's surface the seas and the land are of equal proportions and, despite changes in their shapes and locations, they maintain this proportion through time. He seems to think that the sea bed had been gradually depressed, for he asks: "How the Sea comes to be gradually depressed, and deepest about the middle part; whereas the bottom of it was in all likelihood equal while the Waters covered the whole Earth?" (Ray, 1692, p. 159). It is in answering this question that Ray talks about the origin and evolution of large-wavelength structures of the outer rocky rind of our planet:

> ...the same Cause that raised up the Earth, whether a subterraneous Fire or *flatus*, raised up also the skirts of the Sea, the ascent gradually decreasing to the middle part, where, by reason of the solidity of the Earth, or gravity of the incumbent Water, the bottom was not elevated at all. For the enclosed Fire in those parts where its first accension or greatest strength was, raised up the Earth first, and cast off the Waters, and thence spreading by degrees, still elevated the Land, and drove the Waters further and further; till at length the weight of them was too great to be raised, and then the Fire broke forth at the tops of the Mountains, where it found the least resistance, and disperst it self into the open Air. (Ray, 1692, p. 159)

Ray thus imagined that a very large swelling developed a negative dent under the weight of the superincumbent water or because at that point the earth was too stiff to be swelled, as it was being inflated. The dent was eccentric with respect to the center of application of the "first accension or greatest strength" of the subterranean fire (Ray, 1692, p. 159). This dent formed the ocean basins. Owing to continued inflation, the sides of the dent were further raised and finally burst in places and created volcanoes. The initial phase would correspond in our present terminology to a falcogenic event; the origin of volcanoes and associated mountains correspond to a copeogenic event.

Clearly, the Pacific Ocean served as a model of ocean formation for Ray, for he showed the Andes as evidence for his model: "But we cannot doubt that this may be done. When we are well assured that the like hath been done. For the greatest and highest Ridge of Mountains in the World, the *Andes of Peru*, have been for some hundreds of Leagues in length violently shaken, and many alterations made therein by an earthquake that happened in the year 1646. [*sic*] mentioned by *Kircher* in his *Arca Noæ*, from the Letters of the Jesuits" (Ray, 1692, p. 156, italics his).

Ray naturally did not know the depth conditions of the oceans. He simply deduced, from his earlier assumption that land and sea must maintain equality, that the depth conditions in the world ocean must resemble those on land:

> It hath been observed by some, That where there are high Cliffs or Downs along the Shore, there the Sea adjoining is deep; and where there are low and level Grounds, it is shallow: the depth of the Sea answering to the Elevation of the Earth above it: and as the Earth from the Shores is gradually higher and higher, to the middle and parts most remote from the Sea, as is evident by the descent of the Rivers, they requiring a constant declivity to carry them down; so the Sea likewise is proportionably deeper and deeper from the Shores to the Middle. So that the rising of the earth from the Shores to the Mid-land is answerable to the descent or declivity of the bottom of the sea from the same Shores to the Mid-Sea. (Ray, 1692, p. 160)

Ray considered these enormous, gentle slopes to lead to elevations (and, by implication, to depressions) much higher than mountains: "This rising of the Earth from the Shores gradually to the Midland, is so considerable, that it is very likely the Altitude of the earth in those Mid land parts above the *Superficies* of the Sea, is greater than that of the Mountains above the level of the adjacent Lands" (Ray, 1692, p. 160, italics his). Thus, in retrospect, Ray seems to have sensed that the falcogenic structures were much larger than copeogenic ones and, like Élie de Beaumont 150 years and Grove Karl Gilbert 200 years later, he ascribed continent and ocean-basin generation essentially to falcogenic movements.

In Ray's theory, interestingly, falcogeny *necessarily* eventually led to copeogeny. Of course, this is a retrospective interpretation of his ideas, but he did separate large-wavelength, slowly evolving structures of the earth's surface from those of short wavelength and much faster evolution. This was the first explicit separation of these two kinds of structures since the Middle Ages. As we shall see below, this separation did not become explicit again until the nineteenth century.

Central Fire Before Hutton: Moro

It was the Abbé Antonio-Lazzaro Moro (1687–1740: see Thomasian, 1981) who made uplift caused by igneous activity a subject of central geological debate (Moro, 1740). This conflict happened even though giants such as de Saussure, who certainly thought internal fire significant, slighted his contribution largely because his pyric enthusiasm was fueled by sedimentary naivété (de Saussure, 1779, p. XIV; also in Élie de Beaumont, 1852, p. 1325; for other pros and cons to Moro's views and his influence, see Thomasian, 1981). In Moro's book, it seems fairly clear that he himself made no distinction between what I call falcogenic and copeogenic structures—in contradistinction to Aristotle, let us say. He was far too concerned to persuade his reader that the internal fire was capable of bringing about most of what we see on the earth's surface today. While doing so, he made use of an ingenious concept put forward by the remarkable Count Luigi Ferdinando Marsili (1658–1730; see Rodolico, 1981. The name is variously spelled as Marsili, Marsilli, or Marsigli). Marsili distinguished an *essential floor of the sea* (*essenziale fondo del mare*) that formed the primeval surface on which the first waters had accumulated, from an *accidental* (*accidentale*) floor of the sea made up of rocks later deposited on the *essential* floor (Moro, 1740, p. 284; see Marsilli, 1725, p. 15: "one can see … how these accidental floors cover the essential."). [179] Moro describes in great detail the growth of volcanic mountains by accumulation of material ejected from craters. But he also points out that the previously formed layers, those that lie between the "essential floor" and the "accidental floor" may be again thrown up by subterranean fire fed from material within the layers and form mountains: "And from this, mountains form, which we call *secondary*, and which consist entirely of strata" (Moro, 1740, p. 274[180], *emphasis Moro's*)[181]. While explaining his Plate VIII (Fig. 33 here), Moro depicts wholesale uplifts (e.g., R, M, P *in* Fig. 33) and depressions (H, N, S) that deform the "essential floor" as well as the "accidental floor." It is not possible to discern from his book whether Moro consciously makes a distinction between these large-scale bendings of the crust and the deformations caused by the piercing of the "essential floor" by primeval volcanoes. Although one can read out of his narrative and figure such a distinction *in retrospect*, I do not think Moro himself would have considered it significant.

James Hutton

Some 50 years later, James Hutton (1726–1797: Fig. 34) published his epoch-making paper essentially resurrecting Theophrastus' view of the evolution of the surface of the earth with an internal energy (much like Moro's concept), but with an understanding of the surface processes that was far in advance of Moro's (Hutton, 1788; see especially Dean, 1992; McIntyre and McKirdy, 1997[182]). Neither in that epoch-making paper nor in his classic book that succeeded it (Hutton, 1795a, 1795b; 1899[1997]; Hutton, *in* Dean, 1997) did the ingenious doctor make a distinction between falcogenic and copeogenic structures. Just as in the case of Moro, one can read in retrospect such a distinction out of his statements, but I think Hutton himself was unaware of, or at least uninterested in, that distinction.

Views of geologists on uplifts after Hutton can be largely connected with his views. They developed mostly within the

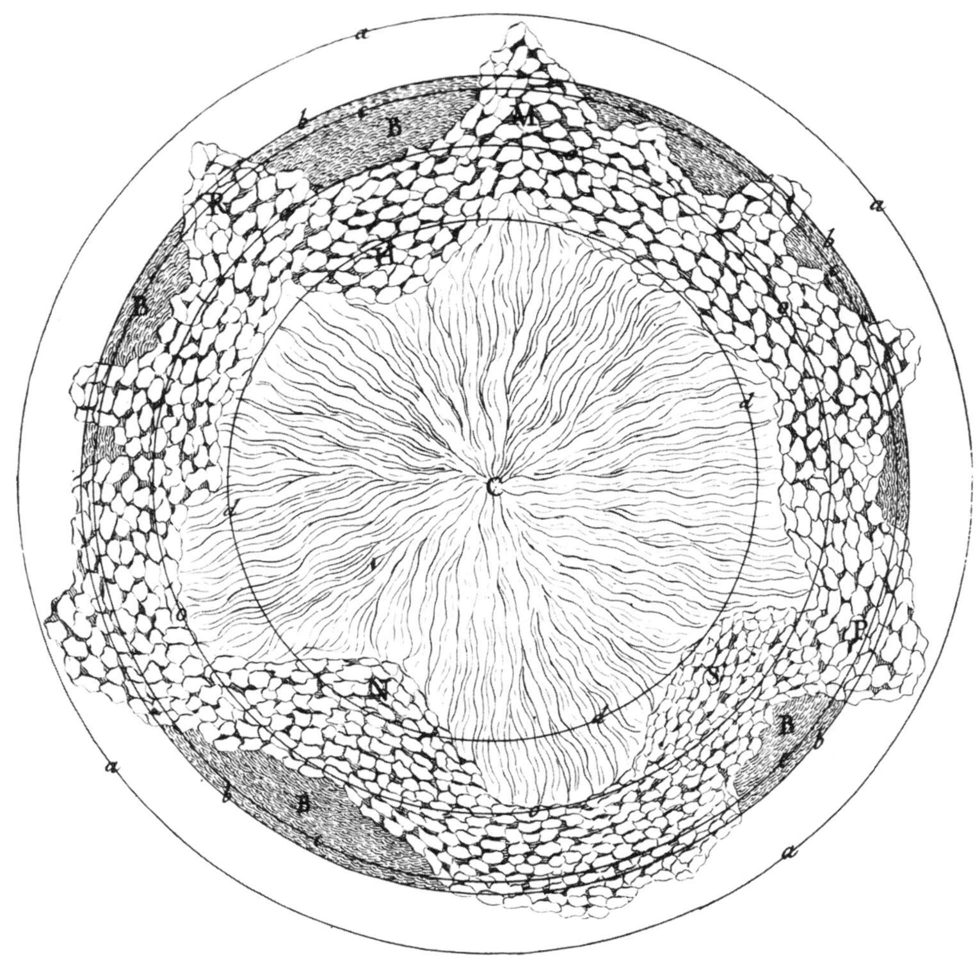

Figure 33. Interior of the earth according to Antonio-Lazzaro Moro's "second hypothesis," which he says is similar to that of Empedocles.

framework of the vulcanist-neptunist debate in the early nineteenth century. We need to examine these views in some detail, owing to their direct bearing not only on our present opinions on the vertical motions of the lithosphere and their consequences, but also on some of the particulars of the resulting structures.

John Playfair

In Hutton's great champion John Playfair (1748–1819: Fig. 35) [183], we find a reflection of the ideas that had developed from Moro to Hutton himself. Playfair (1802) considers that both what he calls *angular elevation* (i.e., elevation of strata to remove it from its original relation to the horizontal) and *absolute elevation* (i.e., translation to a greater distance from the center of the earth) were brought about by "an expanding power, which has acted on the strata with incredible energy, and has been directed from the centre toward the circumference" (Playfair, 1802, p. 53). It is perhaps easier to visualize this if we say that Playfair's "angular elevation" changes the "fabric" of what we see at an outcrop, whereas his "absolute elevation" moves the rock bodily, and unless having a datum such as sea level, we would be unconscious of its operation at an outcrop. The absolute elevation, Playfair understandably finds more difficult to deal with, because it is always difficult to determine whether the land or the sea (or both?) moved with respect to the center of the earth. About Scandinavia, however, he has little doubt. He reviews a number of cases of change in sea level and concludes that it must have been the land that moved because (1) depressing or uplifting the sea level would involve volume changes of the ocean on such a stupendous scale as to make the whole thing unlikely, and (2) sites of ongoing, or very recent, rise and fall of sea level are so distributed on the face of the earth as to exclude any global mechanism, such as acceleration of the diurnal rotation (Playfair, 1802, p. 441–457). Exactly as presented in the book of Moro or in the writings of Hutton, Playfair gives the impression that although he is aware of deformations of short wavelength and deformations of very long wavelength, he attaches no significance to the difference.

Figure 34. James Hutton (1726–1797) became the father of modern geology through his balanced treatment of internal and external agencies sculpting the face of the earth. This portrait by Sir Henry Raeburn is located in the Scottish National Portrait Gallery in Edinburgh.

Figure 35. John Playfair (1749-1819). Portrait by Sir Henry Raeburn, showing him ca. 1814, is now in the University of Edinburgh.

CHAPTER IX

LEOPOLD VON BUCH AND THE DEVELOPMENT OF THE THEORY OF CRATERS OF ELEVATION: ELEVATION AND SUBSIDENCE IN THE POST-HUTTONIAN WORLD

Leopold von Buch

After having called the negative movement of the strand in the Baltic "a most peculiar, odd, striking phenomenon" (von Buch, 1810[1870], p. 503), the most influential geologist of the first half of the nineteenth century, Christian Leopold von Buch, Baron of Gelmersdorf (1774–1853: Fig. 36)[184], stated categorically, "It is certain that the sea-level cannot sink; the balance of the seas will not allow it. But as the phenomenon of reduction cannot be doubted, as far as we can now see, there remains only one way out and that is the conviction that the whole of Sweden is slowly rising" (von Buch, 1810[1870], p. 504). He pointed out that, according to the information he was able to gather during his 1807 trip, this rising was not confined to the Baltic but was also felt along the North Sea coasts. This was not the first encounter of the former student of Werner with the consequences of an earth interior not entirely rigid, but it was one that left a profound impression on him (cf. Gohau, 1987, p. 161).

Figure 36. Christian Leopold von Buch, Baron of Gelmersdorf (1774–1853), as he appeared to a portrait artist in 1823, at the time when his elaboration of the theory of the elevation craters reached its zenith. Engraved by Ambroise Tardieu.

In his tour to the volcanic districts of central France in April 1802, while in Neuchâtel on an official visit from the Prussian government, Leopold von Buch, who was at the time still a neptunist, not only had become convinced that basalt could *also* be a volcanic rock, but at the same time conceived the idea that the rock that he called *domite* (oligoclase-bearing hornblende- and biotite-trachyte) occurred in the Puy de Dôme and in other cones as giant blisters propelled upwards by "internal volcanic power" (von Buch, 1809[1867], p. 483). Mont Dore (Fig. 37) was a further surprise for him because he could see on this "volcano" neither a crater nor lava flows resembling those familiar to him from his observations of Mount Vesuvius. At Mont Dore, a uniform cover of basalt lay over multifarious porphyries (for a modern description of Mont Dore and its rocks, see Peterlongo, 1972, p. 94–116; for a popular account with a fine colored geological map that may be compared with that in Fig. 37 here, see Brulé-Peyronie and Lécuyer, 1998). Now that von Buch knew basalt had been once molten, he developed the peculiar idea that basalts first must have formed in molten lakes, then solidified into flat layers, and were only then uplifted—clearly a neptunist hang-up of a recent convert to volcanism![185] The "circus" on top of Mont Dore was interpreted by von Buch as an *extensional* collapse structure, *not* as a crater (von Buch, 1809[1867], p. 513 ff.). His visit to the Canaries in 1815 convinced von Buch completely that basaltic islands were *not* volcanoes—not in the ordinary sense anyway. He applied the idea he had conceived in Auvergne to the individual edifices there and concluded that they each consisted of originally flat-lying basalt layers, later uplifted at their center by a volcanic force acting from below; this volcanic force also assisted in fracturing the top of the upblown blister and formed a large caldera. Such "craters" were called by von Buch "craters of elevation" to distinguish them from "craters of eruption" (von Buch, 1820[1877]). The uplifted edifices had to extend in their middle, and von Buch interpreted the numerous gullies (the "*barrancos*")[186] radiating away from central calderas, as the expression of as many fissures that opened up during the expansion of the surface of the basalt flows[187] (Fig. 38). Von Buch's interpretation of the large basaltic shield volcanoes not only dominated the theory of craters of elevation but also set the tone for the tectonic study of *all* axisymmetric uplifts for the next 150 years.

George Poulett Scrope

Although von Buch's peculiar interpretation relating to the *craters of elevation*, betraying his neptunistic roots, was severely criticized by Scrope (1825) and Lyell (1830, p. 386

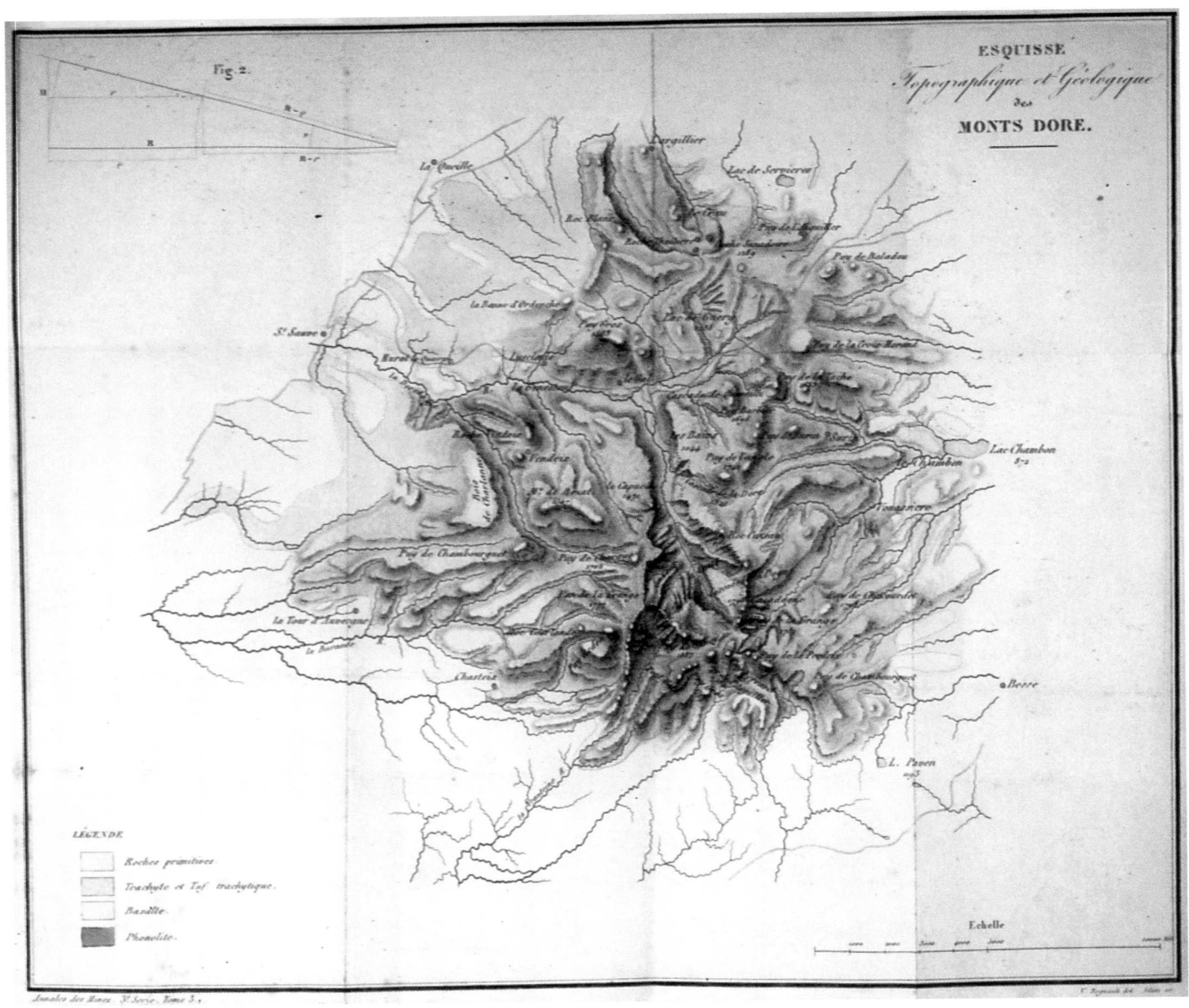

Figure 37. Topographical and geological sketch of Monts Dore (from Dufrénoy and Élie de Beaumont, 1834, Plate XI). Note the concentric arrangement of the outcrop of successive units.

ff.), Scrope's theory of tectonism developed for the earth's crust was very much along the lines von Buch had earlier depicted for the supposed elevation of lava layers. In the following long quotation, we have Scrope's (1797–1876: Fig. 39)[188] view of vulcanicity and terrestrial tectonism:

…a continual supply of caloric passes off from the interior of the globe towards its circumference. …

If … the phenomena of volcanos, … together with their accompaniments of earthquakes, &c. &c. and perhaps many of the more ambiguous and obscure indications of congenerous causes visible in the constitution of the globe's surface, can be accounted for in the simplest and most satisfactory manner, according to well-known principles of physics, by this single assumption of the exposure of subterranean masses of crystalline rocks, which we know to exist, to a continual accession of caloric from below, which we have the strongest reasons for presuming a priori—in this event we shall be bound by common sense and the simplest rules of induction to accept this hypothesis with the utmost confidence, and it would be the height of irrationality and scepticism to refuse our acquiescence in it. (Scrope, 1825, p. 30–31)

Since the accession of caloric takes place by our assumption from below, the temperature of the mass will be unequal throughout, diminishing more or less gradually from below upwards. …

Whenever the overlying rocks yield in any degree to the general expansive force of the mass, the consequent ebullition will take place *first*, and *with the greatest violence* where the expansive force is highest, that is, in the *lower* strata. The sudden dilatation of these inferior strata must forcibly elevate the upper *solidified* parts, as it were, *en masse*;

the pressure they sustain between the expansion of the lower part, and the weight and cohesion of the solid rocks above them, suffering them to preserve the water they enclose unvaporized, notwithstanding their intense heat.

This forcible elevation of solid rocks by a violent expansion, taking effect at a considerable depth, cannot occur without considerable rupture and dislocation. ...

... Fissures must be created by these disruptions, both in the superficial rocks of low temperature which overlie the mass of lava, and in the upper part of this crystalline mass itself, solidified, as we have seen above, by the superior expansive force of the lower parts. Of these crevices some must be supposed to open downwards, towards the confined mass of lava, and others outwardly; the accompanying figure will best illustrate this position. (Scrope, 1825, p. 32–33, fig. 1 and 2 {Fig. 40A, B herein}).

Scrope (1825) makes the important distinction between *primary elevations* and *secondary volcanic phenomena*: "The volcanic phenomena are then only *secondary* and attendant circumstances on the more immediate and *primary* results of subterranean expansion, *viz.* the partial elevations of the solid crust of the globe" (p. 199; compare this with Clarence Dutton's views of the causes of uplift of the plateau country in the western United States and the vulcanicity there {discussed below, Chapter XII}; compare also with Burke and Whiteman {1973, p. 735} on African uplifts: "Uplift has normally been followed by alkaline vulcanicity..." Neither is it any different in principle from what Moro and Michell had earlier said {see Chapter VIII above}).

Here we have a much more general theory than that of von Buch, but like his, it failed to explain why in certain places the internal heat made volcanoes or other sorts of uplifts and in others did not. Scrope did not make a distinction between the generation of volcanoes, the uplift of individual mountain ranges, and the elevation of entire continents. All uplift was, for him, due to the rise of heat from the interior of the earth, much as it had been for Moro, Michell, Hutton, Dolomieu, and Playfair before him and was to be for Lyell and Darwin after him. He made a half-hearted attempt to explain the reason for the diversity of the surficial manifestations of the central heat by appealing to imperfections in the structure of the crust. A full-fledged theory of mountain and/or highland origin was never developed by Scrope.

Charles Lyell and the Inevitability of Subsidence in a Uniformitarian World

Ever since the ancient Greeks, the making of high ground by vertical uplift, propelled through fiery, or igneous, agencies (to use the contemporary jargon of Moro, 1740, and especially of Hutton, 1788) had been a respectable view, remaining so well into the first half of the nineteenth century. As we have just seen, most respectable geologists of the day accepted it, however they may have differed among themselves as to its details.

Vertical uplift by igneous agencies was also the mechanism Sir Charles Lyell (1797–1875: Fig. 41)[189] thought was responsible for creating the mountain ranges of the globe. However, he noted that the prevalent opinion of the eighteenth and the early nineteenth centuries, mainly owing to Hutton's influence, was that while internal forces uplifted land, the external forces alone reduced it back. This, Sir Charles perceptively observed, could not be, unless the volume of the planet was increasing:

We have said in a former chapter* that the aqueous and igneous agents may be regarded as antagonist forces, the aqueous labouring incessantly to reduce the inequalities of the earth's surface to a level, while the igneous are equally active in restoring the unevenness of the crust of the globe. But an erroneous theory appears to have been entertained by many geologists, and is indeed as old as the time of Lazzaro Moro, that the levelling power of running water was opposed rather to the *elevating* force of earthquakes than to their action generally. To such an opinion the numerous well-attested facts of subsidences must always have appeared a serious objection, but the same hypothesis would lead to other assumptions of a very arbitrary and improbable kind, inasmuch as it would be necessary to imagine the magnitude of our planet to be always on the increase if the elevation of the earth's surface by subterranean movements exceeded the depression. The sediment carried into the depths of the sea by rivers, tides, and currents, tends to diminish the height of the land; but, on the other hand, it tends, in a degree, to augment the height of the ocean, since water, equal in volume to the matter carried in, is displaced. The mean distance therefore, of the surface, whether occupied by land or water from the centre of the earth, remains unchanged by the action of rivers, tides, and currents. Now suppose that while these agents are destroying islands and continents, the restoration of land should take place solely by the forcing out of the earth's envelope—it will be seen that this would imply a continual distension of the whole mass of the earth.

*Chap. x, p. 167

(Lyell, 1830, p. 474–475)

An expanding earth did not appeal to Lyell's uniformitarian mind, so to avoid expanding the earth, he thought that tectonic subsidence must equal tectonic elevation. This concept, however, appeared to him still unsatisfactory because, as he pointed out, the earth's interior regularly divulges material onto the surface, which did not exist before, such as lavas and deposits of mineral springs. These materials must leave some vacuities down below in the earth, and these empty spaces eventually must be eliminated by the collapse of the superjacent crust into them. Thus, there must not be subsidence only to compensate for tectonic uplifts, but also additional subsidence to compensate for material extracted from the interior of the earth by various means. Lyell therefore concluded that tectonic subsidence at any one time must exceed tectonic uplift. He drew stratigraphic consequences from this deduction:

If we find, therefore, ancient deposits full of fresh-water remains which evidently originated in a delta or shallow estuary, covered subsequently by purely marine formations of vast thickness, we shall not be surprised; for we must expect that a greater number of existing deltas and estuary formations will sink below, than those which will rise above their present level. (Lyell, 1830, p. 477–478)

A

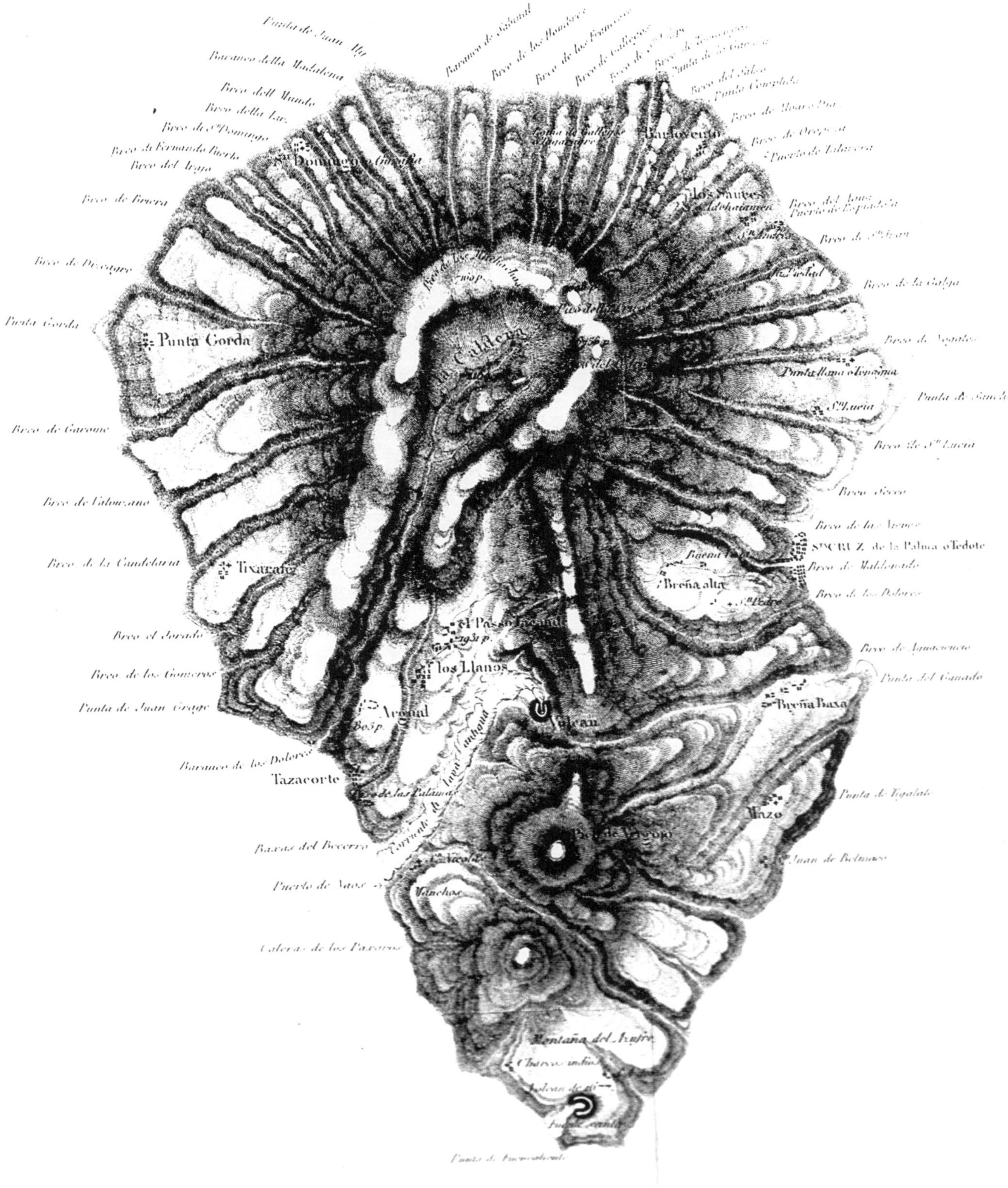

Punta de Juan Ady
Baranco della Madalena
Brco dell Mundo
Brco della Luz
Brco de Sto Domingo
Brco di Fernando Porto
Brco del Agua
Brco de Briera
Brco de Dezcagre
Punta Gorda
Brco de Garome
Brco de Valoezano
Brco de la Candelaria
Brco el Jorado
Brco de los Gomeros
Punta de Juan Grage
Baranco de los Dolores
Baxas del Becerro
Puerto de Naos
Caleras de los Pazaros
Punta de Fuencaliente
Baranco de Sabinal
Brco de los Hombres
Brco de los Franceses
Brco de Gallegos
Brco del Sabco
Punta Cumplida
Brco de Oropesa
Puerto de Talavera
Brco del Juan
Puerto de Espindola
Brco de la Galga
Punta de Sancha
Brco de Sta Lucia
Brco Seco
Brco de las Nieves
Brco de Maldonado
Brco de los Dolores
Brco de Aguacencio
Punta del Ganado
Punta de Tigalate
Sto Domingo
Barlovento
los Sauces
Punta Gorda
La Caldera
Punta llana o Tenagua
Sta Lucia
Sta CRUZ de la Palma o Tedote
Breña alta
S. Pedro
Tixarafe
El Passo
los Llanos
Volcan
Breña Baxa
Argual
Tazacorte
Mazo
Sn Juan de Belmaco
Montaña del Azufre
Volcan de 1677

Figure 38 (here and on facing page). A: Map of the Island of La Palma in the Canaries, showing the alleged "crater of elevation" with the numerous *barrancos* radiating away from it as depicted by Leopold von Buch in 1825. Very characteristic is the northern part of the island centered around the Taburiente Caldera, which von Buch called "La Caldera" and which since has become the type caldera in volcanological literature (e.g., MacDonald, 1972, p. 321–322; Bullard, 1976, p. 79). For modern descriptions of the geology and the evolution of La Palma and the relevant literature, see Abdel-Monem et al. (1972) and Middlemost (1972). B: La Palma seen from the west-southwest, as it appeared to Leopold von Buch in 1815 (von Buch, 1825[1877]).

Figure 39. George Poulett Scrope (1797–1876) at about the time when he was fighting the theory of craters of elevation. Lithograph is from a portrait by E.U. Eddis (*in* Wilson, 1972, fig. 26).

Lyell ended up by affirming that, although the number of observations then were inadequate to provide empirical support for his deduction, the number of instances of subsidence on record was higher than those of uplift.

The nature of the compensating uplifts and subsidences was left vague, except that they were brought into connection with earthquakes. Lyell wrote: "This cause so often the source of death and terror to the inhabitants of the globe, which visits, in succession, every zone, and fills the earth with monuments of ruin and disorder, is nevertheless, a conservative principle in the highest degree, and, above all others, essential to the stability of the system" (Lyell, 1830, p. 479).

Lyell's view of compensation of uplift and subsidence later played a great role in many tectonic theories and was particularly applied to falcogenic movements. Its immediate effect, we shall see in the development of Darwin's ideas concerning the mobility of the earth's crust as expressed in his theory of the growth of the coral reefs and the origin of the three principal types of coral islands. But, first, let us review the ideas on compensatory uplift and subsidence of another earth scentist who also had a great influence on the tectonic ideas of his successors.

Alexander von Humboldt on Compensatory Uplift and Subsidence

Ideas on compensatory subsidences and elevations were also published by Alexander von Humboldt (1769–1859: Fig. 42)[190] only a year after the appearance of Lyell's book and most likely independently of it. These concepts were inspired by his travels in Asia in 1829. Von Humboldt had long wished to go to Asia to study the Himalaya, not only to look at their tectonics but also to check the elevation of the snow line and to compare it with his Andean observations. But the Napoleonic wars, his deteriorating financial situation, and the repeated delays in the publication of his mammoth, multi-volume work on America had not permitted him to realize this dream.

A first opportunity presented itself in 1811, when Russia organized an expedition to Tibet via Kashgar. The Russian Prime Minister, Count Romanzov, knew von Humboldt personally and instructed von Rennenkampf to invite him. Von Humboldt accepted enthusiastically, but Napoleon's invasion of Russia made the trip impractical.

A second opportunity was created by the Prussian King Friedrich Wilhelm in 1816, but could not come to fruition presumably because of the British East India Company's fears that a visit from a man of von Humboldt's great fame and liberal convictions regarding colonialism might not be in the best interests of the British possessions in Asia. (This attitude of the East India Company so frustrated von Humboldt that, for a time, he considered leaving Europe permanently and settling in Mexico City: Meyer-Abich, 1969, pp. 66, 78–80 and note 100 on p. 160.)

These frustrated attempts at visiting Asia did at least some good, however: von Humboldt kept his interest in Asian geology fresh, maintained close contact with noted oriental linguists (such as Abbé Gregoire, Abel-Rémusat, Letronne, Hase, Freytag, Klaproth, Villoisin, and Champollion) and travelers such as Andrea de Nerciat (Persia) and the great orientalist Sylvestrès de Sacy, studied some Persian (mainly with de Sacy), and collected documents.

Especially fruitful was his contacts with his countryman, the polyglot scholar Heinrich Julius Klaproth, who at that time was working on a map of central Asia upon the commission of the Prussian government. Klaproth had diligently studied all the Chinese and other Asian sources he could find in the rich libraries of Paris and out of these he constructed various maps of Asia (see Klaproth, 1826a, 1826b; 1828. especially map facing p. 416; 1831[191]), including the famous four-sheet map of central Asia (Klaproth, 1836).

Through these preliminary studies, von Humboldt first realized that the idea of a vast central Asian highland, the immense, allegedly monolithic and homogeneous high plateau of "High Tartary," was untenable. The celebrated author of the *Essai sur la Géographie des Plantes* (von Humboldt, 1805) could not accept the assertion that places where cotton, grapes, and pomegranates were raised were located on a plateau having an

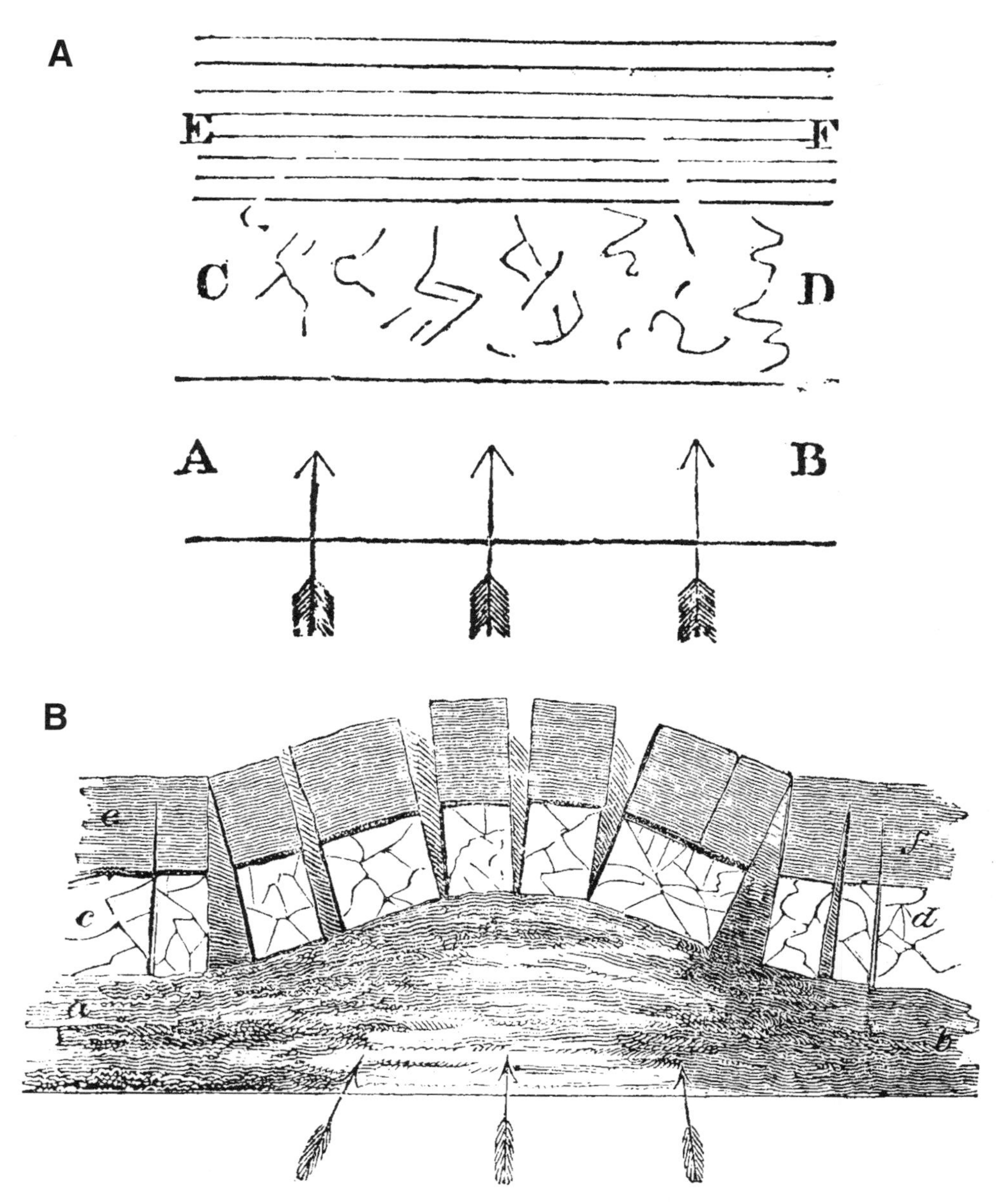

Figure 40. George Poulett Scrope's two figures illustrating crustal uplift by transfer of heat from the earth's interior:

A: If ABCD represent the subterranean mass of lava, confined to the overlying strata EF, the accession of caloric to the lower part of the mass, viz. AB, so far increases its expansive force as to consolidate the upper mass CD (by condensing its enclosed vapour;) and by the continuation of this process, the general expansive force, acting from below upon the overlying rocks EF, at last becomes superior to their powers of resistance,...

B: ...and they yield more or less to the dilation of the lower strata of lava ab ...; the fissures broken towards the centre or convex part of the space elevated will open outwardly, those towards the limits of this space will open downwards.

Such in fact is the natural and constant effect of any forcible elevation of a solid crust by an impulse from below. (Scrope, 1825, p. 33)

Figure 41 Charles Lyell (1797–1875), in the years when the craters of elevation controversy was raging. Drawing is by J.M. Wright in 1836 (*in* Wilson, 1972, fig. 45).

Figure 42. Baron Friedrich Wilhelm Heinrich Alexander von Humboldt (1769–1859), the last man to claim the entire natural sciences as his province and whose writings helped the recognition of the significance of falcogenic movements in tectonic evolution. From an original portrait painted in 1813 in Paris by Carl von Steuben, engraved by Ambroise Tardieu.

alleged average height of 5 km![192] One of the early products of von Humboldt's preliminary Asian studies was his publication *Sur l'elévation des montagnes de l'Inde* ("On the Elevation of the Mountains of India"; 1816) in which he noted that

> In Central Asia, the mountains appear at first to form an immense massif, whose surface area equals that of Australia[193]. This Dauria[194], up to Berlour-tagh, is from east to west 47° longitude long and from the Altay to the Himalaya, from north to south, 20° latitude wide. This is the massif which is called, if somewhat vaguely, *the plateau of Tartary*, which, however, presents in its western extremity, grand inequalities as indicated by the produce and the climate of Songaria [*Junggar Basin*], of Little Bukharia[195], of Turfan and of Hami (Chamul or Chamil) famous for its raisins. One could assume with much likelihood that this plateau in no way forms a continuous massif, but that in more than a third of its extent its elevation above the level of the Ocean is modest. (de Humboldt, 1816, p. 307–308; italics von Humboldt's)

In a second paper, entitled *Sur la limite inférieure des neiges perpétuelles dans les montagnes de l'Himâlaya et les régiones equatoriales* ("On the snowline in the Himalayan mountains and the equatorial regions"), von Humboldt re-emphasized that "these massifs [*of Central Asia*] without doubt do not form a continuous plateau" (von Humboldt, 1820, p. 54). This dislike of the idea of a vast central Asian plateau was no doubt nourished by von Humboldt's bias for individual, continuous, and fairly straight mountain chains propelled upwards by internal magmatic energy (despite his knowledge of the high plateau of New Spain {i.e., present-day Mexico} and the southwestern United States: see Chapter XII below), and much strengthened by Klaproth's and Abel-Rémusat's philological detective work on the orography of Asia.

His chance to check personally the tectonics of Asia finally appeared on 15 August 1827, when the conservative Russian finance minister, Count Egor Frantsevich Kankrin, wrote to von Humboldt to inquire whether platinum might be used in coinage and about the would-be relative value of platinum coins with respect to the gold and silver coins. The Minister added in a second letter dated 22 October 1827, answering von Humboldt's questions on the size of the platinum reserves and the projected yearly production, that "the Ural would be well worth a visit for naturalistic purposes" (von Humboldt and von Cancrin, 1869, p. 8). Von Humboldt answered on 19 November that platinum would be unlikely to maintain a stable value as against gold and silver as a monetary standard. He closed his letter by saying that it was his wish to pay a personal visit to the Minister in Russia: "I picture to myself the Ural, the Ararat, which is soon to become Russian, and even the Lake Baykal, as lovely vignettes"[196] (von Humboldt and von Cancrin, 1869, p.18; Bruhns, 1872a, p. 435).

The Count grabbed the opportunity to invite the great naturalist to Russia[197] and the resulting trip lasted from 12 April to 28 December 1929. Von Humboldt was accompanied by two professors from the University of Berlin, the mineralogist Gustav Rose and the zoologist Christian Gottfried Ehrenberg. The main purpose of the Russian government was "to further science," and as far as compatible with scientific interests, to help the Russian mining interests (Bruhns, 1872a, p. 438). By river transport on the Volga, the travelers went from Nijni Novgorod to Kasan to see the Tartar ruins of Bulgari, and then via Perm onto Yekaterinenburg on the Asiatic side of the Ural. In one month, von Humboldt visited the central and the northern parts of the range and studied its rich minerals. From Yekaterinenburg, they went via Tumen to Tobolsk on the Irtysh and then through the Barabansk steppes to Barnaul north of the Altay. After studying the southwestern slopes of the Altay, they continued to Ust-Kamenogorsk and via Bukhtarminsk they reached the Sino-Russian frontier in the Junggar basin. They even received permission to cross the frontier and visited the Mongolian station of Baty.

On the way back to Ust-Kamenogorsk, along the desolate shores of the Irtysh, the travelers saw the locality first described and interpreted by B.F. Hermann (1801, p. 108–113), where for a distance of more than 5 km., granite seemed to overlay schists. Von Humboldt noted in his notebook that "this superposition of the slates by the granite, already noted by Hermann, is indubitable" (von Humboldt, 1843, p. 306). The great plutonist interpreted the contact, however, as transgressive bedding in an igneous relationship (von Humboldt, 1843, p. 307). He thought that the hot granite had flowed atop the schists (it is in reality a slightly overturned strike-slip contact).

From this remarkable locality, the travelers went via Semipalatinsk and Omsk to the Ishim River and thence to the southern Ural Mountains. They traversed the chain near Orsk where pretty green jasper quarries operated and went to Astrakhan to collect water samples from the Caspian Sea (undertaken by Gustav Rose). From Astrakhan, the travelers went back to Moscow and then to St. Petersburg (Cuvier in von Humboldt, 1832, p. 1–5).

Von Humboldt published two books on Asia. One came out shortly after he returned home and the other some 14 years later, which von Humboldt considered not so much a revised edition of the first, "but rather an entirely different book" (von Humboldt to Charles Darwin, 18 September 1839: in Burkhardt and Smith, 1986, p. 221). Both were much enriched by his study of the available literature and especially by what he learned from his linguist colleagues in Paris.

The first of the two books was his *Fragmens de Géologie et de Climatologie Asiatiques*, published in two volumes in Paris (de Humboldt, 1831) with one map at the end of volume one (in Şengör, 1998, fig. 18, the German translation of this map, which is identical with the French except for the lettering, has been reproduced). In this first volume, von Humboldt summarized his views of the structure of Asia and added a number of geologically and climatologically important observations that he and others had made. He argued that four major mountain ranges of *east-west* trend dominate the structure of Asia (e.g., von Humboldt, 1831, p. 85–86): namely, from north to south, the Altay, the Tien Shan, the Kuen-Lun, and the Himalaya.

Another major structure in Asia that captured von Humboldt's attention was the vast West Siberian Depression, which he compared with the giant craters on the Moon (von Humboldt, 1831, p. 137 ff.). Von Humboldt interpreted the structure as a *counterpart* of the equally impressive highland that stretched from Iran via Tibet to Mongolia, and he regarded the depression to be coeval and cogenetic with the uplift of the immense and topographically much differentiated plateau. The Ural chain, however, he thought had to be younger because, had it been older, the subsidence would have long erased any expression of that low mountain range. As we shall see below, von Humboldt's observations and inferences concerning the vast lowlands of western Siberia and the North Caspian region prompted both Darwin and Élie de Beaumont to speculate about the nature and the cause of large-scale subsidences of the earth's crust.

In his second volume of the *Principles*, which came out a year after von Humboldt's book, Lyell returned to the question of subsidence of large areas while trying to account for the origin of atolls (Lyell, 1832, ch. XVIII, especially p. 286, 296; also see Stoddart, 1976, especially p. 203–204). He repeatedly stressed that not only incremental subsidence was brought about by earthquakes, but also there were uplifted atolls (such as Elizabeth or Henderson's Island: see Lyell, 1832, fig. 8 and 9; Fig. 43 herein). He believed, in parallel with his earlier arguments presented in the first volume of his book (see above), that while subsidence predominated in the Pacific Ocean, uplifts may also create, from the linear atoll chains, "large continents, mountain chains ... capped and flanked by calcareous strata of great thickness..." (Lyell, 1832, p. 298). Lyell was careful always to stress that incidence of uplifts and subsidences must be so adjusted as to allow only a small margin of predominance of subsidences to accommodate the extra material extravasated by volcanoes and mineral springs. He also stressed that alternating episodes of uplift and subsidence must affect the same areas so as not to deviate from the uniformitarian picture of earth evolution. As we saw above, his main objection to von Buch's ideas of uplift, both of elevation craters and of Scandinavia, was because they did not allow the sort of alternation of up and down motions in time.

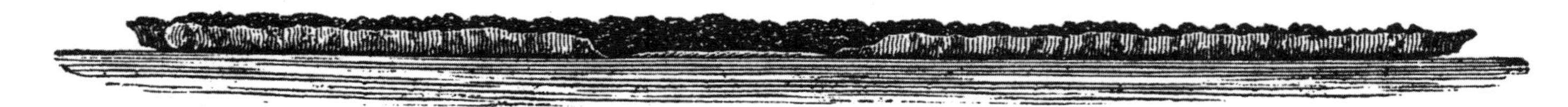

Elizabeth or Henderson's Island.

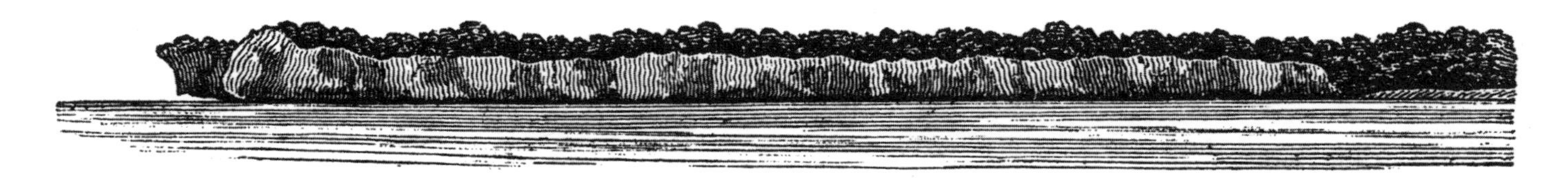

Enlarged view of part of Elizabeth or Henderson's Island.

Figure 43. Uplifted coral islands (from Lyell, 1832, figs. 8 and 9).

In the first three decades of the nineteenth century, the idea that the earth's crust was subject to large-wavelength and compensatory uplifts and subsidences was developed both by the uniformitarian and catastrophist camps. Neither side worked out a detailed mechanism nor the anatomy of the resulting structures, although Lyell repeatedly stressed in the *Principles* that such movements were always accompanied by earthquakes. This alleged association influenced Darwin's thinking. Lyell himself was, however, to be forced to alter his opinion soon, as we shall see below.

CHAPTER X

TIME OF TRANSITION FROM "RADIAL THEORIES" TO "TANGENTIAL THEORIES"

Élie de Beaumont: A Central Figure in Nineteenth Century Tectonics

The first three decades of the nineteenth century were a time of rapid development of the theory of mountain-uplift by central heat, when both Baron Leopold von Buch and his life-long friend Baron Alexander von Humboldt worked out an elaborate hypothesis of rapid, catastrophic uplift along fault lines running the length of mountain chains. (For von Humboldt's conversion to vulcanist views, see the detailed study of Hoppe, 1994; also von Engelhardt, in press). Leopold von Buch even showed the effects of successive, non-coaxial mountain uplift in his famous 1824 paper on the geognostic systems of Germany (von Buch, 1824a). It was this paper that placed the necessary emphasis on the *narrow* and *elongate* aspect of the mountain belts[198]. Through Élie de Beaumont's studies on the mountain ranges of the globe, von Buch's paper helped to separate mountain uplift from larger wavelength uplifts.

The great French tectonician Léonce Élie de Beaumont (1798–1874: Fig. 44)[199] proposed, in 1829, that mountain ranges (or *mountain systems* as Leopold von Buch circumscribed them) might be due to *lateral shortening* resulting from the thermal contraction of a cooling earth (Élie de Beaumont, 1829, 1831). He, Leopold von Buch, and Robert Bakewell (1768–1843)[200] earlier, recognized that different mountain chains were raised at different times *and that chains thus raised had internal structures different from areas of broad uplift*. However, before we discuss this important distinction, it is necessary to cast a glance at the work that Élie de Beaumont did in the French Alps and later in the Auvergne together with his colleague and friend Ours-Pierre-Armand Petit-Dufrénoy (1792–1857)[201]. These two pieces of work established the foundations of what was later called the theory of geosynclines, developed further the structural geological aspects of Leopold von Buch's theory of elevation craters, and led eventually to a general theory of large-wavelength, crustal or lithospheric domal uplifts, the geometrical aspects of which still remain useful (as discussed in another publication: Şengör, 2001a).

Figure 44. Jean-Baptiste-Armand-Louis-Léonce Élie de Beaumont (1798–1874). Courtesy of the Académie des Sciences (Paris).

Élie de Beaumont and the Harbingers of the Theory of Geosynclines

In the first decade of the nineteenth century, the great French geologist and a former Werner student A.-J.-M. Brochant de Villiers (1772–1840) discovered a passage of sedimentary rocks into crystalline rocks in the Tarentaise Alps in France (eastern margin of the Belledonne Massif: Brochant de Villiers, 1808). Using arguments similar to Hutton's, he put forward the hypothesis that the crystalline, micaceous, talciferous limestones and the micaceous, talcifreous, and amphibolitic schists in this region must have been originally sedimentary and probably a part of the so-called "transition rocks" of Werner (and not of his "primitive rocks" as until then supposed) owing to their content of plant fossils and their similarity with alleged correlatives in Germany (Daubrée, 1860, p. 19).

Brochant's student and later associate Élie de Beaumont returned to his master's old mapping area in the late 1820s to take a closer look at the crystalline rocks of the area of the Belledonne Massif and its sedimentary frame. He mapped the region and investigated in detail the partly metasedimentary terrain southeast of the Belledonne Massif, now known to be occupied by a series of slices belonging to the Ultradauphinois tectonic unit (e.g., Gwinner, 1978, p. 181, 297, 351, 352; Debelmas, 1982, fig. 7). Élie de Beaumont first studied those in the vicinity of Petit Cœur (Élie de Beaumont, 1828a) and then those exposed at the Chardonet Pass farther to the southwest (Élie de Beaumont, 1828b). He confirmed Brochant's observations and disagreed with Bakewell (1823a, especially ch. VIII), who had earlier interpreted the intercalation of dark schists, crystalline limestones, and quartzites as secondary mixing owing to the Liassic belemnites in the limestones and Carboniferous plants in the schists. Élie de Beaumont interpreted the entire section as representing a stratigraphic sequence. He ascribed the whole of it to the Lias on the strength of the belemnites found in the limestones and argued that the plant fossils of alleged "coal age" (i.e., Carboniferous, as correctly identified by Bakewell, 1823a, ch. VIII; 1823b, p. 410–411) also belonged to the same epoch (i.e., to the Lias). This assignment was the beginning of the controversy known as "the affair of Petit-Cœur,"[202] between those who denied the plant fossils their age-diagnostic characteristic (Élie de Beaumont and his party) and those who denied that there was at Petit-Cœur an undisturbed sequence (Bakewell and his supporters). For a history of this affair, see Gaudry (1855), Favre (1867, p. 358–382), and Ellenberger (1958, p. 19–20). As late as 1850, Élie de Beaumont's interpretation was presented as the "current interpretation" by some continental Europen authors (e.g., von Cotta, 1850, p. 22).

In this region, Élie de Beaumont noticed an increase in the "modification" of rocks towards the crystallines expressed by the progressive coalification of the plant material, passing of black slates into green- and wine-colored schists, and the limestones becoming [*sic*] gypsum. On the first geological map of France (scale 1:500,000; Dufrénoy and Élie de Beaumont, 1840, sheets Lyon and Marseillée), this transition is marked by a sharp line passing through the Chardonet Pass (4°10′E and 44°55′N on the Marseillée sheet) and following the arc of the western Alps. The line of transition separates the "unmodified" rocks from the "modified" rocks. Élie de Beaumont likened the transition to a piece of wood that is half-coalified: in one half, the plant tissue would be recognizable; in the other, all plant tissue would have disappeared, coalification would be complete, and even graphite could be generated (Élie de Beaumont, 1828b, p. 362). The implication of a metamorphic *front* is clear.

Élie de Beaumont then pointed out the enormous thickness of the deposits that he studied: "I regard as obvious that this thickness (composed of the sum of the thicknesses of all the intermediary beds measured perpendicular to the plane of stratification) is not less that two thousand metres" (Élie de Beaumont, 1828b, p. 376). He emphasized that nowhere else in Europe, where one so far had searched for the type areas for various formations (he really means *stages* in our present terminology), had such an enormous thickness been encountered, although it was also clear that the Alpine formations (*stages*) and the others in Europe did not correspond with one another exactly from the viewpont of their mineralogy (Élie de Beaumont, 1828b, p. 376–377). But he thought that the fossil content of the formations he was looking at made it clear that they were the equivalents of the much thinner formations in the "non-dislocated parts of Europe" (Élie de Beaumont, 1828b, p. 377). He also thought that the sections he studied in the Tarentaise indicated that they were deposited "at the depths of a very deep sea, when the most intensively studied parts of the Jurassic deposits were laid down along the shores, crowned at intervals by great coral reefs. The central part of the Alps seem to offer to our regards *pelagic deposition*; [*whereas*] hills in the vicinity of Bath and Oxford present to us *littoral deposition*" (Élie de Beaumont, 1828b, p. 377, italics Élie de Beaumont's). The "modification" (i.e., metamorphism), then, affected these very deep-sea deposits. How the intrusions came to affect these, we can only understand if we understand the distinction Élie de Beaumont made between *normal* metamorphism and *accidental* metamorphism.[203]

Élie de Beaumont thought that at a time when there was already life on earth, sufficient temperatures may have obtained at depths of 1000 m or so from the surface of the earth to keep most rocks at those depths in a liquid state. In other words, the thickness of the earth's crust at those times may have been only about a kilometer. Now, Élie de Beaumont further believed that the accumulation of coal beds, corals, and mussel-banks showed that most Paleozoic seas had had a shallow depth. Yet the entire thickness of the Paleozoic strata reaches several thousand meters. The weight of even a small basin would thus have been enough to "fold in" its basement. Every newly laid down bed would push the basin bottom closer to the red-hot interior. Élie de Beaumont thought that this would heat up the lower parts of the basin sufficiently to change the texture, even the structure, of the original sediments. The thicker the sediment package in a basin, the greater would be the effects of metamorphism (Vogt, 1846, p. 247–248).[204]

Such conditions, however, would only obtain if there is an extraordinarily high geotherm. Élie de Beaumont believed that this must have been the case in the Paleozoic. In the later eras, however, the earth would have cooled down sufficiently to tolerate a much thicker crust (estimated to be less than 50 km thick: Élie de Beaumont, 1852, p. 1237) and the "normal metamorphism" would not take place as easily and as generally as it did during the Paleozoic. This was one reason, he further thought, why metamorphism was more widespread in Cambrian rocks than in those of later times (Vogt, 1846, p. 165). In later times, metamorphism became confined to the vicinity of large intrusions. Until 1875, the external massifs of the Alps were believed by most geologists to be intrusive bodies, and Élie de Beaumont thought their intrusion had been responsible for the metamorphism of the Jurassic rocks, not only in the Tarentaise but in all regions of the Alps, where the sedimentary rocks came close to these massifs (Vogt, 1846, p. 293). Metamorphism that was a consequence of heating owing to the proximity of an intrusive body was termed "accidental metamorphism" by Élie de Beaumont who thought that accidental metamorphism became relatively more widespread with respect to normal metamorphism in the younger periods of earth history.

The idea that the grade of metamorphism is an indication of the antiquity of a rock body, which remained prevalent well into the second half of the twentieth century, is a leftover of this anti-uniformitarian interpretation of Élie de Beaumont. In 1860, his student, Auguste Daubrée, was to call *accidental metamorphism* "metamorphism of juxtaposition" (corresponding with the currently used term "contact metamorphism": Daubrée, 1860, p. 54–59) and *normal metamorphism* "regional metamorphism" (Daubrée, 1860, p. 59–65), a term still in use.

In Élie de Beaumont's 1828 papers, there is no intimation what may have caused the change from a realm of thick accumulation of sediments in an open sea environment in the deformed regions making up the Alpine chain, to a shallow sea, to a littoral environment outside it in regions not dislocated. A year later, Élie de Beaumont published his memoir on the mountains of Oisans (in the region of the Pelvoux Massif in the western Alps: Élie de Beaumont, 1829). In a long footnote added to the memoir (p. 15–19), he introduced the idea that thermal contraction may be the cause of the deformations of the earth's crust. Because Élie de Beaumont hid this suggestion in a footnote appended to a regional paper, the original form in which he first propounded the idea of contraction has remained little known and has led to much misquotation. Many thought that he proposed it in his long paper on the revolutions of the surface of the globe giving rise to mountain ranges (Élie de Beaumont, 1829–1830). I myself have been guilty of a similar mistake, thinking that he first introduced the idea in the English summary of his revolutions paper published in 1831, because I could find it in none of the French versions of the revolutions paper (Şengör, 1990, p. 17, note 7). I later discovered the footnote in the 1829 Oisans paper while I was reading it for an entirely different purpose and asked my late lamented friend Professor François Ellenberger whether this was the first publication of Élie de Beaumont's contraction idea. He confirmed that it was. Because it is so little known, because it had such an immense influence on the evolution of tectonic ideas, and because it bears such a resemblance to Dana's first global tectonic paper of 1846 that was one of the vehicles that spread the influence of Élie de Beaumont's ideas (and about which I shall talk below), I give the following full translation of the part of the footnote in question in which the contraction idea is introduced.

After having compared, with Leopold von Buch's elevation craters, the circular shape of the Pelvoux Massif and the arc-like disposition of the Alpine external massifs in the Mont Blanc-Aiguilles Rouges area with no metamorphism outside the arc, Élie de Beaumont continued as follows:

> This position of a high mountain in the middle of a depression reminds one, up to a certain point, of the pitons that rise in the middle of the cirques that are present in such large numbers on the surface of the Moon.
>
> For an observer placed in a balloon at a great height above the mountains surrounding the Bérarde, and who could in imagination remove the secondary masses masking a part of their mass as well as the effects of daily degradations to which these mountains have been, and continue to be, subjected through the action of atmospheric precipitation and torrents, these mountains certainly present a great similarity, in general form, to certain circular mountains, but without central pitons, which a good binocular allows us to see towards the centre of the Lunar disc, when this satellite presents to us its half illuminated by the sun. The very detailed maps accompanying the topography of the visible part of the surface of the Moon, published in 1824 by M.W.G. Lohrmann in Dresden, show that mountains with a more-or-less complete circular ridge surrounding a less elevated area are numerous there. One sees there these kinds of cirques which have a diameter of 20 myriameters [=*200 km*], that is to say nearly equal to that of Bohemia, or Bulgaria, or Wallachia between the Balkan mountains and the frontiers of Transylvania, or the extent of the sea separating Santiago (Cuba) from Cayes (Haiti), and all the sizes below these. These complete or incomplete crown-shaped protuberances are often combined with other protuberances which are more or less irregularly disposed, and which the former seem to cut. I have copied from the maps of Mr. Lohrmann four of these systems (V, plate I, figs. 2, 3, 4 and 5 [*Fig. 45 herein*]) and I drew next to them and, on the same scale, the shapes of the mountains of Oisans and the elevation crater of the island of Palma, so that one can judge the similarities of the size and general disposition of the forms that I seek to compare. It seems to me to result from this comparison that certain mountains of the earth, if transported to a map of the Moon would not look strange, especially if one ignores the degradation they are daily subjected to by atmospheric precipitation and by torrents.
>
> If one thing emerges with certainty from an inspection of the surface of the Moon, it is that this surface has none of the convergent valleys that always form a prominent trait of an exact representation of an elevated part of the terrestrial surface and which are due to the action of waters that play the final role in the configuration of our continents. Even the precise observations on the Moon have not detected the phenomena of reflection and refraction that would have been the necessary consequence of the presence of an ocean and an atmosphere. From this it seems permitted to conclude the only fact that I wish to talk about, that no liquid of considerable quantity exists there.
>
> The surface of our satellite, on which, it seems, on the basis of the preceding, impossible to conceive the formation of any sedimentary

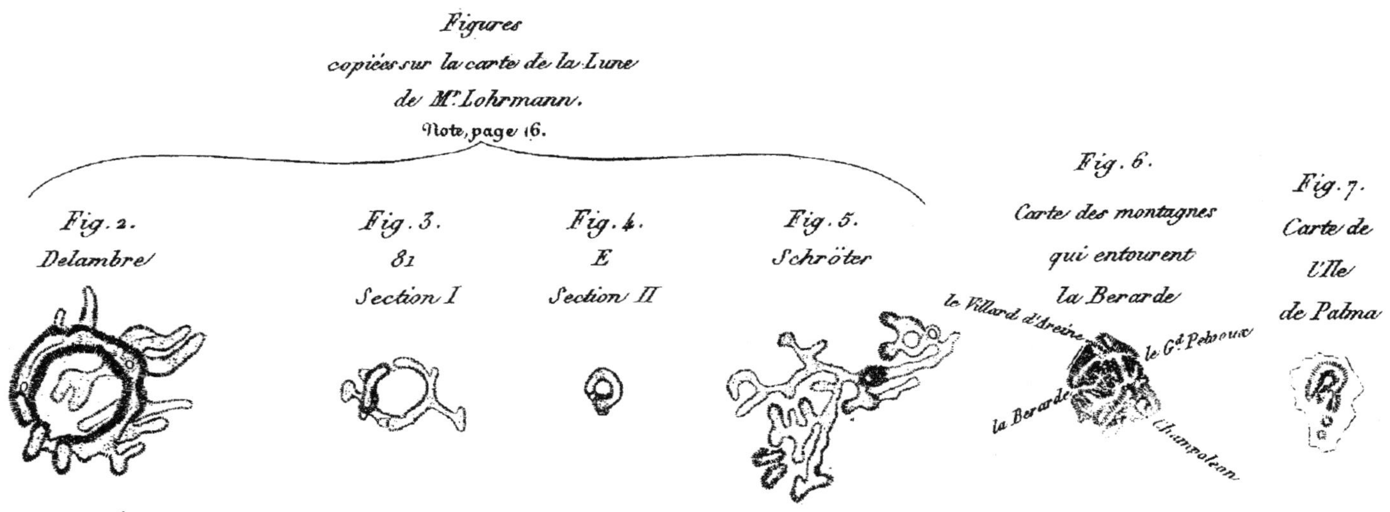

Figure 45. Élie de Beaumont's comparison of lunar crater geometries with terrestrial "elevation craters" (from Élie de Beaumont, 1829, Plate 1).

deposit, appears at the same time similar to those parts of the surface of the earth that are most remote from the seas and great lakes and on which there is no volcano in activity.

Without doubt, we do not know of which material, whether metallic, oxidised, or other, the exterior crust of the Moon is composed. As to its form, beyond the relationships that I shall indicate between the grand masses of a certain regularity, the external configuration of the surface of the Moon in general reminds one very much that presented by certain parts of the surface of the earth—if one ignores the erosional valleys—where the primitive masses have received in very ancient times the exterior configuration they show us, such as Auvergne, Bohemia, Bretagne and especially that which these same terrains would have shown us if they could have presented to us their primitive rocks, forming their basement, stripped of the sedimentary deposits covering them.

It seems therefore that among the mechanical causes of different natures which have contributed to give the mountains of the earth their forms which they present to us, those, that have primarily influenced the exterior configuration of the mountains composed of the so-called primitive rocks, have been similar to the phenomena of which the surface of the Moon seems to have been the theatre.

No doubt, some of the similarities I shall indicate would be somewhat vague, one must nevertheless note their agreement with the observations leading us to view the phenomena, to which our primitive mountains owe their forms, as being of a more elevated order than those producing the sediments and even than the actions that are entirely volcanic. They tend, it seems to me, to distance themselves more and more from the idea that the general form of the masses of primitive rock could have resulted from their crystallisation from a liquid standing higher than their present peaks, and thus they furnish a new argument in favour of the hypothesis of uplift. They would perhaps also lead us to ask whether the emanations of gases that seem to have taken place at the time of this uplift would not rather be an effect than one of the essential causes. Thermal springs, like the earthquakes, being most common in regions where the sedimentary beds are dislocated, it is almost necessary to derive from one and the same hypothesis the explanation of the gradual increase of the temperature as we go deeper into the earth and that of the uplift of the mountains. If a pursued examination would show that there really is a compatibility between the asperities of the crust of the Moon and a certain class of the asperities of the crust of the earth, it would become at the same time necessary that the hypothesis being sought for should be equally applicable naturally to the two celestial bodies, whose surfaces are found in such disparate conditions as the earth and the Moon. And, perhaps, of all the hypotheses proposed so far, the only one whose difficulties are not augmented by this consideration would be that which seeks in a secular refrigeration of the planets the primary cause of the production of mountains making their surfaces rugged.

Secular refrigeration seems to me to contain an element of a nature that can be used in the explanation of geological phenomena and M. Fénéon has for some time been thinking of using it for the same purpose. That element is the relationship that an advanced refrigeration of the planetary masses establishes between their solid envelope and the volume of their internal mass. In a given time, the temperature of the interior of planets diminishes much more than that of their surface, whose refrigeration can no longer be felt. Without doubt, we are ignorant of the materials of which the interiors of these bodies are composed. But the most natural analogies lead us to think that the inequality of the cooling, of which I shall speak, must necessitate for their envelopes to continue to diminish their volume continually, despite the rigorous constance of their temperature, in order not to stop embracing exactly their internal masses, whose temperatures would diminish appreciably. As a result, they must somewhat depart from their spheroidal figure and the tendency to return to it, be it by this sole agent, be it in combination with other internal causes of change that the planet may contain, could provide a cause for the formation of the ridges and diverse tumescences that are produced episodically in the solid exterior crust of these spheroids.

It is concluded from geological observations, on the publication of some of which I am currently engaged, that the ridges of the mineral crust of the terrestrial globe do not date back to the same instant, but, by contrast, they have been produced successively at intervals of considerable time and the examination of the Selenography of Cassini and Mr. Lohrmann leads one to presume that the same kind of distinctions between the inequalities of the surface of the Moon can be established.

It would seem very natural to think that the phenomena, the traces of which are observed by geologists on the surface of the terrestrial globe are not essentially sublunar things, things entirely peculiar to the surface of our planet. By contrast, everything leads to presume that those phenomena governed by the internal causes must be common to all planetary bodies. (Élie de Beaumont, 1829, p. 16–19, note 1)

Thus, by 1829, Élie de Beaumont already knew and had made known that: (1) in the Alps, strata are thicker and much more deformed than are the correlative strata of undislocated Europe; (2) the thicker strata in the Alps have a pelagic character as opposed to the littoral facies of the undislocated extra-Alpine areas; (3) the thickness of the strata in the Alps is probably a result of the downbuckling of the earth's crust; (4) this downbuckling may be a result of the progressive refrigeration and consequent shrinking of the planet and of the weight of the sediment accumulating in the downbuckle; (5) metamorphism and granite plutonism are consequences of the melting of the bottom of the downbuckled part of the crust in the mountain ranges; and (6) early in the earth's history, the earth's crust was thinner (because the earth was hotter), and consequently downbuckling in future mountain areas was easier. This resulted in (a) thicker sedimentary successions, and (b) more widespread metamorphism. Élie de Beaumont thought that (b) was the reason geologists encountered more metamorphic rocks in older terrains. It is mainly the reason for the bias implanted into the heads of geologists that metamorphic grade may be an indication of age!

I hardly need to emphasize that the 1828 and 1829 papers discussed above had immense influence on the thinking of geologists for more than a century to come and that they contain a complete theory of what later was to be called geosynclines (i.e., large-wavelenth, low-amplitude downbends of the earth's outer rind). However, Élie de Beaumont did not leave it there. He developed his "geosynclinal" theory further in the second volume of the explanatory text of the first edition of the geological map of France (Dufrénoy and Élie de Beaumont, 1848) and in his *Notice sur les Systèmes de Montagnes* (Élie de Beaumont, 1852) which I discuss below. I believe his publications were a major influence on both Hall's and especially Dana's thinking in America and, clearly, the American version of the geosynclinal theory did not develop in isolation. In other words, the geosyncline was not a major concept "made in America" (Dott, 1979). Long before James Hall wrote anything on the subject, European geologists emphasized that mountains had been deeply-subsident basins before mountain-building commenced (e.g., von Cotta, 1850, p. III, where it is stated in the Contents, second letter, "Alps, Once a Basin"; also see p. 25–26 for a discussion; for a discussion specifically of Élie de Beaumont's idea of the Alpine pre-orogenic basin, see Burat, 1858, esp. p. 279–280 and Suess, 1858, p. 21). This idea was imported to America and, when it was re-exported back to Europe, it was mainly Élie de Beaumont's deep-water version (his "Alpine version") that the European geologists used.

But before we go that far, we must discuss Élie de Beaumont's contribution to the theory of elevation craters and its relevance to the understanding of the large-wavelenth structures of the lithosphere.

Further Development of the Theory of Elevation Craters by Élie de Beaumont

In a memoir published in 1834 with Dufrénoy, Élie de Beaumont developed von Buch's idea on the *barrancos* into a complete theory of fracturing of an inflated layer of rock having a circular circumference. Below, I give first von Buch's theory (from his own pen) to understand the foundation on which Élie de Beaumont was to build (refer to Fig. 40A, B, for the description that follows):

With the overview of this remarkable, outstretching island [*he means La Palma or, more accurately, San Miguel de la Palma, in the Canaries*], with the view of the size and the depth of the central cauldron, with the thought that here not lava flows, but beds rise uniformly from the sea to the highest point, one can almost see the whole island rising from the bottom of the sea. The beds were raised by the uplifting agent, the elastic powers of the interior and in the middle these vapours broke out and opened the interior [*Note the great similarity to Scrope's description quoted above. Von Buch's paper was read seven years before Scrope's book and published five years before; Scrope did not refer to von Buch*]. This crater would then be a result of the elevation of the island and that is why I call it a *crater of elevation* in order never to confuse it with the *craters of eruption*, via which the volcanoes communicate with the atmosphere. Even the wonderful barancos [*sic*!], which dissect the slopes in such an incredible number, appear to be a direct result of this elevation. *They are true cracks across the outer periphery of the beds*; ... Water flows in them only in those short intervals when there is snow on the mountains. One cannot ascribe to such waters the origin of these valleys, as even the strongest stream cannot dissect firm rocks like a knife. *The beds were elevated towards the middle. So, they must tear and leave cracks behind, because the same inextensible material must distribute itself across a larger area on the surface of a cone.*[205] We see the same effect, when we suddenly and strongly push upwards a firm piece of clay[206]. (von Buch, 1820[1877], p. 9–10, italics mine)

Élie de Beaumont developed von Buch's theory of the formation of radial fissures about the center of elevation in a conical protuberance, in terms of what he termed the "starring" or better, "star-making" (*étoilement*) of an originally flat-lying layer by raising its center (i.e., developing star-shaped, radiating fissures from a common center; for excellent contemporary theoretical illustrations, see Prévost, 1935, plate VI, fig. 11a–d). Élie de Beaumont assumed for his calculations that a circular layer of radius **R** be uplifted by height **H** into a conical edifice (Fig. 46A, B). His question was: what would be the total area of the uplift-generated extension? Because the area of the base of the cone (i.e., the area of the surface to be deformed into a conical tumescence before the uplift) is $\pi\mathbf{R}^2$ and the surface of the cone (without the base) is given by:

$$\pi\mathbf{R}\sqrt{\mathbf{R}^2+\mathbf{H}^2},$$

then the area generated during uplift is simply

$$\pi\mathbf{R}\sqrt{\mathbf{R}^2+\mathbf{H}^2}-\pi\mathbf{R}^2.$$

This can be rewritten as:

$$\pi\mathbf{R}^2\left(\sqrt{\mathbf{1}+\frac{\mathbf{H}^2}{\mathbf{R}^2}}-\mathbf{1}\right).$$

If we solve this expression, it opens into an infinite series. However, $\mathbf{H} << \mathbf{R}$ obtains for most shield volcanoes in which Élie de Beaumont was interested. For such cases, the series can be very closely approximated by $\mathbf{1/2\pi H^2}$. Note here that the radius of the uplift has fallen out of the solution and only the magnitude of the uplift appears to matter.

We need to note here that, two years after the paper by Dufrénoy and Élie de Beaumont, Hopkins[207] (1836, p. 47 ff.), in an independent study, pointed out that in a conical elevation of a layer of rock, first the center of elevation will yield forming an orifice. This would release the tensional stresses set up parallel with the slope of the cone; otherwise ring fissures would originate, forming inverted cones with apices at the center of the base of the cone of elevation, similar to the ring dykes and cone sheets in Scotland (Richey et al., 1975). Once the orifice is formed at the apex, the slope-parallel stresses are released, and extension that is tangential to the circumference of the cone will dominate at all elevations, with a minimum near the base to a maximum near the apex.

Élie de Beaumont showed that, if we wish to have the total area of the radial fissures at an arbitrary cross-section of the cone, we could obtain it in the following manner and derive from it a number of corollaries. Consider Figure 46C. If $\Sigma\mathbf{f}$ represents the total area of the extensional fissures formed by the rise of the truncated cone as shown, then

$$\Sigma\mathbf{f} = \mathbf{2\pi r} - \mathbf{2\pi}\rho, \text{ where } \rho = \mathbf{R(R - r/\cos\Theta)}.$$

Substituting ρ, we get

$$\Sigma\mathbf{f} = \mathbf{2\pi(R - r)(1 - \cos\Theta/\cos\Theta)}.$$

This exact value may be approximated by the following equation that eliminates the Θ:

$$\Sigma\mathbf{f} = \pi\mathbf{(R - r)(H^2/R^2)}.$$

As the conical tumescence becomes enlarged without changing its height, we get:

$$\mathbf{dSf/dR} = \pi\mathbf{(H^2/R^3)(2r - R)}.$$

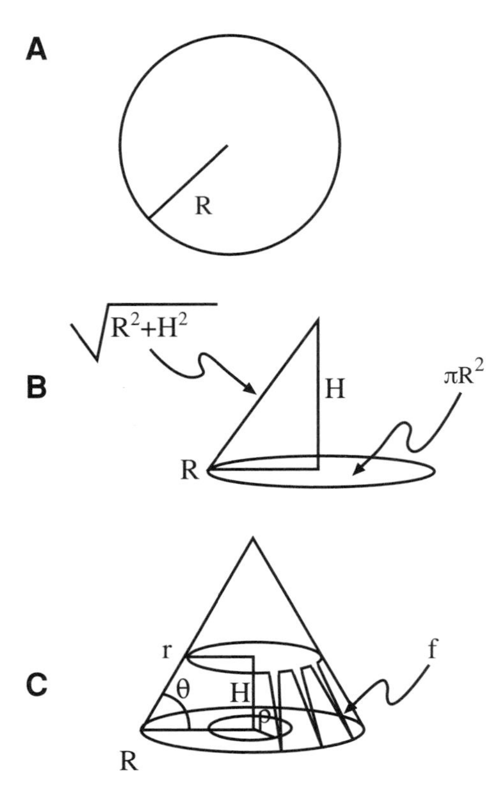

Figure 46. A: Base of a cone of elevation with radius **R**. B: A cone of elevation of radius **R** and height **H**. C: A truncated cone of elevation with *barranco*-like radial fissures emanating from a central opening. **R** is the basal radius, whereas **r** is the radius of the summital opening. Θ is the angle of slope dip, ρ is the radius of a circle obtained by rotating the side of the truncated cone down by Θ (see Dufrénoy and Elie de Beaumont, 1834, Plate XI, fig. 2), **f**—*barranco*-type extensional fissure.

This value is negative if **r** < **R/2**, positive if **r** > **R/2**; it means that, as **R** becomes larger, fractures near the center become narrower and near the periphery, wider. Consequently, given the same height, broad domes will be less fractured in the center than small domes, a conclusion amply corroborated by experience (cf. Şengör, 2001a).

Élie de Beaumont's equations, which are the earliest ones that I know of in the geological literature, provide a rigorous method of calculating uplift-related extension. Given the knowledge of the global geology of his time, it was as yet not straightforward to deduce firm tectonic consequences from them. The equations were generated for a study of the elevation craters and the *barrancos* that form around them. Although Élie de Beaumont used the same theory of extensional fissure formation for what he thought to be proper (or, better, "hybrid") volcanoes such as Mount Etna, it was not applied to any continental elevation of the kind about which Scrope (1825) had earlier written. However, as von Humboldt remarked in the *Kosmos* (1845, p. 312), the idea of continental upheaval was then being supported not only by direct observations on the continents themselves, but also by analogy with "volcanic phenomena," (by which he clearly meant the theory of elevation craters[208]). Today, with our incomparably larger and sounder database, Élie de Beaumont's equations can be widely used to estimate the total extension atop crustal domes (but see Şengör, 2001a).

Charles Lyell and the Pitfalls of Uniformitarianism

In the first volume of the first edition of the *Principles of Geology*, Lyell (1830) mounted a well-coordinated attack against von Buch's theory of elevation craters. He used both the negative evidence of there not being any such craters made up of marine or lacustrine sedimentary strata (although nobody knew it then, he was wrong; see endnote 208) and the positive evidence that all the well-studied volcanic edifices being a result of accumulation and not of elevation (Lyell, 1830, p. 386 ff.; also see 1835b, p. 205 ff.). However, those closely familiar with Lyell's way of thinking also recognize among the causes of this attack the antagonist's Anaximandrian dislike of ascribing any activity to the earth that is not *immediately perceptible to the senses now*[209]. This tendency was noted by Eduard Suess, who not only knew Lyell personally but also had a great veneration and perhaps an even greater affection for him (see Suess, 1904, p. iv). Suess called Lyell's attitude "quietism" and rightly complained that this reduced everything to the scale of man in a most Protagorian manner; he reminded his readers that whereas the planet is measured *by* man, it is not to be measured *according* to man (Suess, 1883, p. 25, emphasis his).

When confronted with von Buch's interpretation of the rise of land in Sweden, Lyell judged in 1830 that "the phenomena do not lend the slightest support to the Celsian hypothesis, nor to that extraordinary notion proposed in our times by Von Buch, who imagines that the whole of the land along the northern and western shores of the Baltic is slowly and insensibly rising!" (Lyell, 1830, p. 231 ff.). Here was more prejudice than impartial reflection speaking.[210] The evidence was well-known, repeatedly tested, and never seriously disputed, except with regard to what was moving—sea or land. Von Buch's authoritarian attitude, and his advocation of elevation craters that Lyell believed he had thoroughly demolished, lent von Buch's views on Scandinavia little credibility in Lyell's eyes. Nevertheless, the opinion on the "retreat of the sea" in Scandinavia was disturbingly widespread (see Wilson, 1972, p. 389). Lyell finally decided to go and see the evidence for himself with a view of putting the whole affair to rest, "because [*he*] suspected that it might be explained by reference to more ordinary causes ... and ... because it appeared to [*him*] improbable that such great effects of subterranean expansion should take place in countries which, like Sweden or Norway, have been remarkably free within the times of history from violent earthquakes" (Lyell, 1835a, p. 2). So, in May 1834, he got under way. He was in for the shock of his life (for the details of his journey, see Wilson, 1972, p. 391–408)!

We read, in his journal, in the entry dated "*Oregrund: July 1.* ... It seems true, as Galileo said in a different sense, 'that the earth moves'" (Lyell, 1881a, p. 433). By October 1834, Lyell was writing to Gideon Mantell that "In Sweden I satisfied myself that both on the Baltic and Ocean side, part of that country is really undergoing a gradual and insensibly slow rise" (Lyell, 1881a, p. 442). On 27 November and 18 December 1834, Lyell presented the Bakerian Lecture to the Royal Society of London in two installments (Wilson, 1972, p. 410). In his lecture, he was "willing to confess, after reviewing all the statements published previously to my late tour for and against the reality of the change of level in Sweden, that my scepticism appears to have been unwarrantable" (Lyell, 1835a, p. 2 ff.). However, a part of Lyell's incredulity was due to the enormous difference he perceived between the "intermittent" manifestations of volcanoes and earthquakes that included "sudden rising and falling" of land and "the slow, constant, and insensible elevation of a large tract of land" (Lyell, 1835a, p. 2), which he had originally thought inconceivable. Lyell clearly perceived the difference between what I herein call *copeogenic structures* (his earthquake uplifts and volcanoes) and *falcogenic structures* (his slow, constant, insensible elevation of land). Unfortunately, his "geological quietism" (Suess, 1883, p. 26) and "his allergy to tectonics" (Ellenberger, 1994, p. 311) did not allow him to pursue its implications (also see Lawrence, 1978). Lyell's long paper is a remarkably careful, detailed, and conscientious account of the evidence he saw or heard. In that paper, there is neither an attempt to test von Buch's statement that the north was rising faster than the south in Sweden, nor the slightest intimation as to what the cause of the observed upheaval might be.

It was only in the fourth edition of the *Principles* that Lyell suggested some possible causes:

> The foundations of the country, thus gradually uplifted in Sweden, must be undergoing important modifications. Whether we ascribe these to the expansion of solid matter by continually increasing heat, or to the liquefaction of rock, or to the crystallization of a dense fluid, or the accumulation of pent-up gases, in whatever conjecture we may indulge, we can never doubt for a moment, that at some unknown depth the structure of the globe is in our own times becoming changed from day to day, throughout a space probably more than a thousand miles in length, and several hundred in breadth. (Lyell, 1835b, p. 349)

Lyell's preferred mechanism was heat, generated by chemical reactions (cf. Lawrence, 1978), expanding, and heaving up a crust about 200 mi. (~322 km) in thickness. He believed that such a process might explain his observations regarding the uplift of land in Scandinavia (Lyell, 1835b, p. 384). Though Lyell's least successful attempts at geological theorizing comprised those pertaining to tectonics, his eventual conversion to the slow, continuous, and aseismic upheaval hypothesis helped the ideas on continental uplift, as distinct from mountain uplift, to gain wide currency in the middle of the nineteenth century.

Lyell's attempts at theorizing spurred some of his friends, who were more apt than he at understanding the physical processes, to generate some extremely interesting and fruitful speculations. Sir Henry T. de la Beche (1796–1855)[211], in the first edition of his *Researches in Theoretical Geology*, adopted the contraction hypothesis of Élie de Beaumont (with whom he had been in contact earlier and helped to ventilate his ideas in Britain: see Şengör, 1990). He pointed out that contraction "would not only appear to raise large areas, composing continents, bodily out of the water, by producing great depressions, but would squeeze the principal surface fractures into mountain ranges" (de la Beche, 1837[212], p. 139).

The views of both Charles Babbage (1792–1871: Gridgeman, 1981; Hyman, 1987; Babbage, 1994; Fig. 47) and of Sir John Herschel (1792–1871: Evans, 1981; Fig. 48), published in the appendices to the *Ninth Bridgewater Treatise* (Babbage, 1838, notes F through I, p. 204–247), are concerned primarily with the means of generating uplifts and depressions through the internal heat of the earth. Both contend that the lines of equal temperature must mimic the topography grossly, subaerial or subaqueous. While erosion depresses (with respect to the center of the earth) the geotherm below a given point near the original surface, deposition raises it. This may cause metamorphism or even melting under thick sedimentary piles and might liberate water vapor and other gases, causing volcanic eruptions. Herschel, in his letter to Lyell (*in* Babbage, 1838, p. 225–236), pointed out that since a fluid substratum must exist

Figure 47. Charles Babbage (1792–1871) from a drawing by William Brockedon in 1842, only two years after the publication of the second edition of the *Ninth Bridgewater Treatise* (from Burkhardt and Smith, 1986, facing p. 89).

Figure 48. Sir John Frederick William Herschel (1792–1871), oil painting, dated 1843, by Christian A. Jensen in the Royal Society of London.

beneath the crust, sedimentation would load any basin floor and depress the crust underneath into the substratum. By contrast, erosion would occasion uplift.

This is an early form of the theory of isostasy (Longwell, 1928) and is identical with that of Élie de Beaumont's earlier idea (although Herschel seems to have conceived it independently). Herschel's concept remained well known throughout the nineteenth century, influencing the ideas of the American James Hall on what eventually became the theory of geosynclines (see Şengör, 1998 and below) and those of Captain Dutton on isostasy proper (Greene, 1982, p. 108; Oreskes, 1999, p. 31; also see below, Chapter XII). Both Babbage and Herschel were mainly concerned about explaining the motor of the uplifts. Their theories satisfied them as far as the causes of volcanoes and of broad uplifts and subsidences were concerned. Their arguments must have also pleased Lyell, especially Herschel's assurance that a central heat (in the sense of Cordier, 1827, with which Lyell disagreed: see Wilson, 1972, p. 386–387; Lawrence, 1978; Rudwick, 1990) was not a necessary condition for their theories to be true (in Babbage, 1838, p. 246)—although Herschel did emphasize the "*frightfully* rapid progression" of temperature downwards into the earth (in Babbage, 1838, p. 246, emphasis Herschel's). Neither Lyell nor anybody else made use of these geophysical speculations until much later, despite the fact that, within a decade, gravity observations began to make it possible to constrain the thickness of the crust (cf. Petit, 1849; Pratt, 1855; Airy, 1855; see Daly, 1940, p. 36–64, for a well-informed and concise history of gravity observations; also Oreskes, 1999).

Return of the Emphasis on the Distinction Between Slow Deformations of Long Wavelength and Fast Deformations of Short Wavelength

As early as 1836, Ami Boué (1794–1881; Fig. 49)[213], in a paper on Élie de Beaumont's views on mountain uplifts, had remarked that mountain uplift and continental elevation were different things both in terms of *mechanism* and in terms of *rate*. Although numerous books published around that time emphasized the same inference (e.g., Reboul, 1835, p. 65, 68–69), Robert Bakewell gave the clearest expression to this idea in his popular textbook (Bakewell, 1839), in which he noted, resonating with reminiscences of Playfair (1802), that

> The emergence of large islands and continents from the ocean, was not effected by the same operation as that which tilted up the beds of primary rocks in many mountain ranges. The lower or primary rocks, after they were tilted up [*Playfair's* angular elevation, *see above*], were still beneath the level of the ocean, when they were covered by the secondary strata unconformably, as many of these strata are marine formations. ...
>
> The elevation of the uptilted beds was a distinct operation from that which raised them, together with the rocks that cover them, above the ocean, and which converted the former bed of the sea into dry land [absolute elevation *of Playfair; Reboul says essentially the same*].
>
> I consider it probable that all large tracts of country or continents, emerged slowly from the ocean, forming at first mountainous islands, before the lower countries were raised above the level of the sea. The power which could upheave a continent, or, in other words, occasion a large portion of the crust of the globe to swell out, must be very different from the force which acted along certain lines, and elevated mountain ranges. This power may be dependent on a more general law of subterranean motion, with which we are at present unacquainted." (Bakewell, 1839, p. 408–409)

Charles Darwin on Falcogenic Subsidence in a Uniformitarian World

This was the time when Charles Darwin (1809–1882: Fig. 50)[214] announced to the world the slow, continuous rise and fall of large tracts of the earth's crust in the Pacific, on the basis of his interpretation of the evolution of coral islands (Darwin, 1838[1977], 1842,1899[1987])[215]. Darwin confessed his inability to understand how the secular shrinking of the earth's crust—of the kind advocated by Élie de Beamont and advertised in Great Britain by Sir Henry T. de la Beche (1837, p. 139)—could be considered a sufficient cause for "the slow *elevation*, not only of linear spaces, but of great continents" (Darwin, 1840[1977], p. 82, emphasis Darwin's[216]). Like so many others interested in questions of uplift at his time, Darwin had likewise started with volcanoes. In fact, it was his suspicion that the "perfectly horizontal white rock" he observed in St. Iago in the Cape Verde Archipelago, sandwiched between two successive volcanic formations (Darwin, [1876], p. 5 ff.; see Fig. 51), indicated first a subsidence and "then the whole island has been upheaved" (Darwin, *in* Barlow, 1958, p. 81[217]), which gave him his first piece of self-confidence to write about geology.

Figure 49. Ami Boué (1794–1881) was one of those geologists whose writings helped to establish the view that mountain-building and continental-scale uplift belonged to two different types of phenomena. Reproduced from the frontispiece of the *Livre Jubilaire* of the Geological Society of France (de Margerie, 1930).

Figure 50. Charles Robert Darwin (1809–1882) at age 43, about a decade after the publication of his atoll theory, holding his eldest child William. From a reproduced daguerrotype (*in* Barlow, 1958, facing p. 111).

While the *Beagle* was skirting the South American coasts, Darwin was eagerly collecting observations on continental uplift. As the *Beagle* was sailing towards Callao along the western coast of Peru, Darwin jotted down the following notes into his notebook: "On the Atlantic side my proofs of recent rise become more abundant at the very point where on the other side they fail. Collect all data concerning recent rise of Continent" (Barlow, 1946, p. 245).

It is clear that Darwin was interested in continental uplift already when the *Beagle* was still in the Atlantic Ocean. But he was also looking at relative movements of parts of the continent. His remarks, quoted above, may be read to indicate some embryonic thoughts concerning compensatory movements. All of this was clearly Lyell's influence. Darwin's mentor, the "sagacious" Henslow, had advised Darwin before they set sail from England that he should read Lyell's newly published first volume of the *Principles of Geology* but on no account accept the views expressed in it (Barlow, 1958, p. 101). Captain FitzRoy had indeed just given Darwin the first volume (Stoddart, 1976, p. 204). The second volume reached him in Montevideo on 26 October 1832, sent by Henslow (Barlow, 1933, p. 435, note 22). Darwin learned Lyell's ideas on compensatory uplifts and subsidences of the earth's crust from the first volume and the role of subsidence in forming atolls from the second. Stoddart (1976, p. 203) noted that the paragraph in the last chapter of the second volume in which "subsidence by earthquakes" is advocated to account for the form of the atolls (Lyell, 1832, p. 294, §3) had been scored by Darwin[218]. Darwin also knew from Lyell's second volume that J.R. Quoy and J.P. Gaimard, of the *Uranie* and *Physicienne* expedition (1817–1820), under the command of de Freycinet, had noted that below a depth of 25–30 ft. (~10 m), reef-building corals did not grow (Stoddart, 1976, p. 200 and the references to his note 19). He was further aware of Forster's (1778, p. 14; also see p. 148 ff.) classification of the islands of Polynesia into (1) high islands without coral reefs, (2) islands encircled by coral reefs "as a picture is by a frame" (Darwin, 1962, p. 6), and (3) "the low, half-drowned islands composed entirely of coral" (Darwin, 1962, p. 6). Darwin noted that a

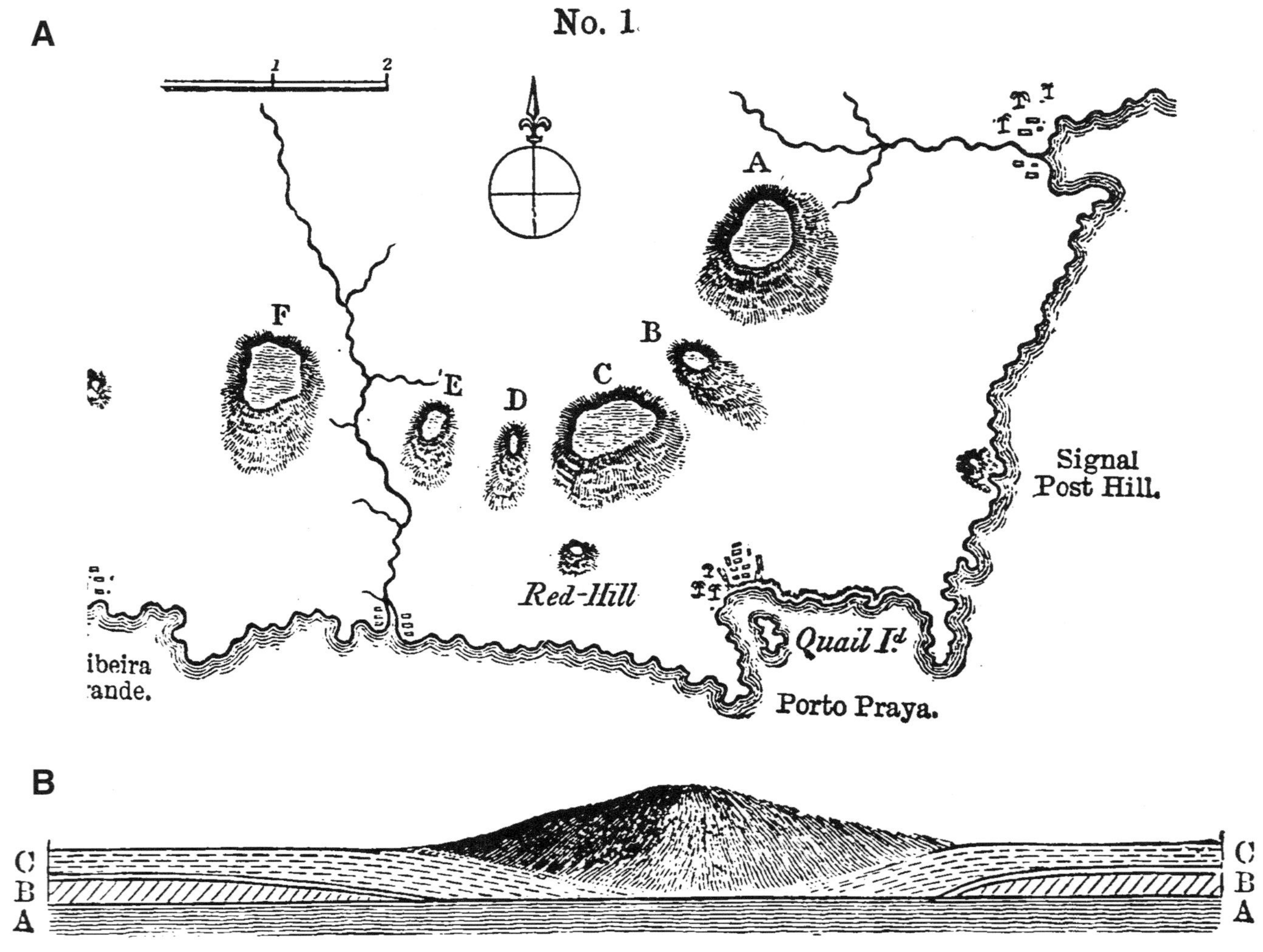

Figure 51. A: Port of St. Jago, one of the Cape Verde Islands' (reproduced from Darwin, [1876], p. 4). B: Signal Post Hill, eastern St. Jago (see Fig. 51A). Calcareous stratum (B) sandwiched between two volcanic formations (reproduced from Darwin, [1876], p. 12).

fourth variety existed, described by Captain Beechey, consisting of uplifted coral reefs. He emphasized that Captain Beechy had observed that members of this last class were rare.

These are the only elements that, in retrospect, Darwin needed to be able to formulate his theory of the origin and evolution of coral islands. They had to form on platforms close to sea level because below 30 ft. (~10 m) they could not live (according to Quoy and Gaimard). As corals grew vertically upward, the various classes of islands described by Forster could best be accounted for if a central, roughly conical island gradually subsided to disappear from sight eventually, leaving only its coralline crown at the surface. That there were also uplifted coral islands (Beechey's "fourth" class) clearly pointed to the tectonic instability of the foundation on which the islands rose. Lyell's point about subsidence predominating in the Pacific Ocean was perfectly suitable to explaning the abundance of coral islands in this part of the world. As South America as a whole appeared to be rising by means of earthquakes, it was only reasonable (in Lyell's view of the world) that a correspondingly large area should subside. Darwin thought that that area was the Pacific and the coral islands its tide gauges.

Darwin's earliest notes about the origin of coral islands appears in a field notebook entitled "Santiago Book" (Down House no. 1.18: Burkhardt and Smith, 1985, p. 567). As Burkhardt and Smith (1985, p. 568) emphasize, the coral formations are treated therein as evidence for subsidence, rather than as formations whose origin and structure had to be explained. It was thus the tectonic problem of subsidence, rather than the general problem of coral islands, that was foremost in Darwin's mind at this time:

> As in Pacific a Corall bed. forming as land sunk. would abound with. those genera which live near the surface. (mixed with those of deep water) & what would more easily be told the Lamelliform: Corall forming Coralls.—I should perceive in Pacific. wear & tear of Reefs must form strata of mixed. broken sorts & perfect deep-water shells (& Milleporæ).—

Parts of reefs themselves would remain amidst these deposits, & filled up with infiltrated calcareous matter.— Does such appearance correspond to any of the great Calcareous formations of Europe.—

Is there a **large** proportion of those Coralls which only live near surface.—If so we may suppose the land sinking: I believe much conglomerate on the other hand is an index of bottom coming near the surface. If so Red Sandstone Epoch of England. will point out this: Mountain Limestone the epoch of depression.— Do the numerous alternations of these two grand classes of rock point out a corresponding opposite & repeated motion of the surface of that part of the Globe. (Notebook 1.18, p. 6–18: Burkhardt and Smith, 1985, p. 568)

Darwin was thus led immediately to the great question of the oscillatory motions of the earth's crust and the facies that might allow the geologist to decipher their history. He re-emphasized this a few pages later in the same notebook: "May we not imagine each band of conglomerates marks an epoch when that part of the ocean's bottom was near to a continent or shoal water; & that having again been depressed. Calcareous fine sediments were deposited. (if under circumstances to allow of corall reefs, such would be abundant)—" (Notebook 1.18, p. 12: Burkhardt and Smith, 1985, p. 568). On p. 15 of his notebook, Darwin further observed that "The Test of depression in strata is where great thickness has. shallow. coralls growing in situ: this could only happen. when bottom of ocean subsiding" (Burkhardt and Smith, 1985, p. 568).

All of this Darwin thought out without laying eyes on a single atoll. He said as much himself years later in his autobiography in a much-quoted passage:

No other work of mine was begun in so deductive a spirit as this; for the whole theory was thought out on the west coast of S. America before I had seen a true coral reef. I had therefore only to verify and extend my views by a careful examination of living reefs. But it should be observed that I had during the two previous years been incessantly attending to the effects on the shores of S. America of the intermittent elevation of the land, together with the denudation and the deposition of sediment. This necessarily led me to reflect much on the effects of subsidence, and it was easy to replace in imagination the continued deposition of sediment by the upward growth of the coral [*Note here the great influence of Lyell's reasoning quoted above on p. 85*]. To do this was to form my theory of the formation of barrier-reefs and atolls. (Darwin, *in* Barlow, 1958, p. 98–99)[219].

Darwin first saw an atoll in the Tuamotus ("Low or Dangerous Archipelago") on 13 November 1835 (Barlow, 1933, p. 344), and then he saw the island of Eimeo (Moorea), which is a volcanic island surounded by a barrier reef, from a height 2000 and 3000 ft. (~600–1000 m) on Tahiti on 17 November 1835 (Barlow, 1933, p. 348; Yonge, 1958, p. 246–247; Stoddart, 1976, p. 204). It is interesting to note here that in Tahiti, Darwin thought the geomorphology and the structure of the island suggested that it had been recently uplifted and "cut by numerous profound ravines, which all diverge from a common center" (see Barlow, 1933, p. 347). Such observations, reminding us of those of Leopold von Buch on the Canaries, have long influenced Darwin, and he always remained sympathetic to the idea of elevation craters, if not a proponent of it. These observations and inferences also further strengthened his belief in the vertical mobility of the floor of the Pacific Ocean.

After these very sparse and preliminary observations, Darwin jotted down the earliest outline we have of his theory (Darwin, 1962). Stoddart (*in* Darwin, 1962, p. 2) dates the writing of this first outline (entitled "Coral Islands") between 3 and 21 December 1835, during *Beagle*'s passage from Tahiti to New Zealand.

Darwin opens the outline by pointing out how little he has seen of the coral islands of the Pacific Ocean. He then observes that the island chains rimming the ocean to the west have trends fundamentally different from those within it: "All the Islands [*of Polynesia*] ought rather to be considered as so many short parallel lines, than the continuation of the great volcanic band which sweeps round the eastern shores of Asia.—I have pointed out this fact, as showing a degree of physical connection in the Islands of Polynesia" (Darwin, 1962, p. 5).

After discussing the island chains, Darwin focuses on the individual islands and presents Forster's grouping them into three classes as indicated above (plus the uplifted class added by Captain Beechey). Darwin then concludes that "it appears to me that the distinction between the II and III division, or the high islands with reefs and the lagoon ones, is artificial" (Darwin, 1962, p. 6). He lists the observations that led him to this conclusion: (1) the form and the size of the reefs in islands with central peaks are the same as those in which only a lagoon is left; (2) their structure appears similar, and (3) islands that have central heights are arranged in lines parallel with those along which islands with only lagoons are arranged (for example, the Tuamotu and the Society Archipelagos). Finally, Darwin went one step farther and claimed that he could see no difference between island-encircling reefs and those extending along the northeastern shores of Australia and, so he believed, along the northern shore of Brazil. He was encouraged to make this last statement by the fact that "in Tahiti M. Hoffman found Granite. M. Ellis states that in several of the Society Is^{ds}. Granite, Hornblendic rock, Limestone & rock with Garnets is found" (Darwin, 1962, p. 8).[220] This mental leap was possible because Darwin had earlier contemplated how the Andes must have risen from beneath the waters and what they must have looked like before their complete clearance of the waters:

I have certain proof that the S. part of the continent of S. America has been elevated from 4 to 500 feet [~120–150 m] within the epoch of the existence of such shells as are now found on the coasts. It may possibly have been much more on the sea coast & probably more in the Cordilleras. If the Andes were lowered till they formed (perhaps 3-4000 ft.) [~1000–1200 m] a mere peninsula with outlying Islands, would not the climate probably be more like that of the S. Sea Islands [*i.e., the Pacific Islands*], than its present parched nature? At a remote Geological æra, I can show that this grand chain consisted of Volcanic Islands, covered with luxuriant forests; some of the trees, one of <which> 15 feet [~5 m] in circumference, I have seen silicified & imbedded in marine strata. If the mountains rose slowly, the change of climate would also deteriorate slowly; I know no reason for denying that a large part of this may have taken place since S. America was peopled. (Barlow, 1933, p. 303)

If the same origin for the fringing and barrier reefs is admitted, Darwin reasoned, then there would be no necessity to have the atolls nucleate on craters. He said that some of the atolls were far too large and far too irregular to be mimicking submerged craters anyway.

For much of the the remainder of this outline, Darwin discussed the details of coral growth, the role of the freshwater springs or streams in inhibiting coral growth, and the reasons for the rigorous growth of corals on the outer edges of the coral islands where the surf is strong. He concludes by presenting his subsidence hypothesis:

Better to explain my views, I will take the case of an Island situated in a part of the ocean. which we will suppose at last becomes favourable to the growth of Corall.—The circumstances which determine the presence or absence of the Saxigenous Polypi are sufficiently obscure, but they do not enter into this discussion.—Let AB represent the slope of an Island so circumstanced & CD the level of the ocean [*see Fig. 52 herein*] Then Corall would immediately commence to grow on the shore (D) & would extend Sea-ward as far as the depth of water would permit its rising from the bottom.—Let this point be (H).- The breadth of the reef (HD) would then depend, on the angle of inclination of the bottom.- This space might either be converted into a piece of Alluvial ground, or even, from the Corall springing up vertically from E & so protecting the inner space, might exist as a Lagoon.-

__________________________(page break in Darwin's manuscript)

This reef would however essentially differ from those in the South Sea, in the depth of the water. (I exclude any few exceptions) beyond the Wall not suddenly becoming excessive.- if the level of this Island should remain stationary. I cannot imagine any change.- But if land should be raised. (or sea sink): the outline would be as represented by the dotted line.- And on the shores. a fringe of Dry Coral rock would be left: This circumstance is known to happen in the East & West Indian Isds [221].— … Now if we suppose the land gradually to subside (See Fig. II [*Fig. 52 herein*]. I have represented the water rising; the effect of course is the same) the level of the sea will stand at Cl instead of CD.- The Coral of the outer wall favoured by the heavy surf. will soon recover its former level.(a) - If this process

(a) or the whole may be supposed to have the same tendency to grow up & recovers its former level: but that the sediment &c from the land checks its growth.

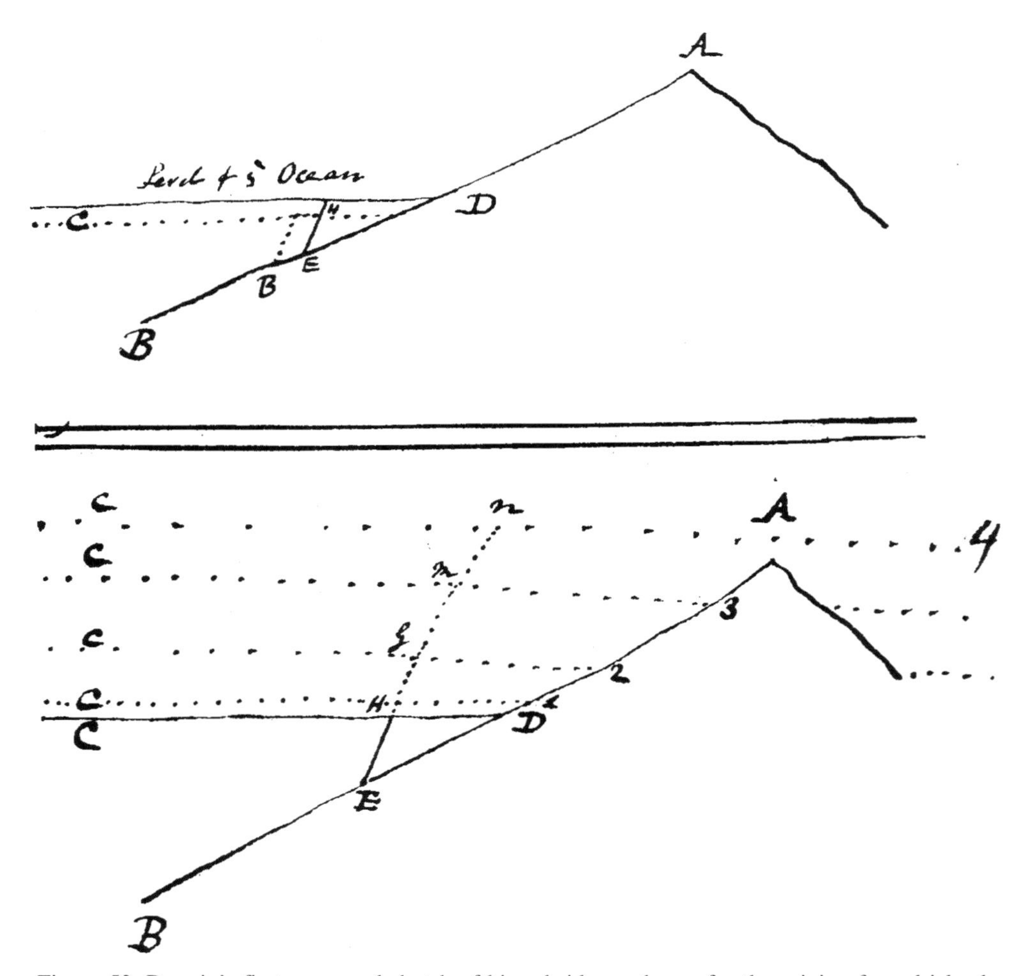

Figure 52. Darwin's first preserved sketch of his subsidence theory for the origin of coral islands (From Darwin, 1962, before p. 15). Upper sketch: Level of ocean sinks to CD (or the island is uplifted) and the original reef EH is exposed. New reef forms above point B. Lower sketch: the sea level rises continuously C1, C2, C3 … (or the island subsides). The reef (originally EH) grows to keep up with rising sea level through j, m, n to form first a barrier reef and then an atoll.

____________________________(page break in Darwin's manuscript) is repeated each time the sea will gain on the land while. the reef rises, nearly vertically on its first foundation.- I say nearly vertically, because any & every small portion removed in front of the lower part & the building being continued upwards before its repair, this must throw backwards the whole of the superstructure. When the level stands at (C3), the space between the reefs & the land, will be more, than twice as broad as at first. This space will probably be occupied by a lake of water. such still water. not being favourable to the growth of the most efficient species of Coral. (Darwin, 1962, p. 14–15; {p. 16–18 of the manuscript}).

Darwin then explored some of the consequences of his hypothesis. He deduced that in regions of slow elevation, atolls must not occur, while they would be abundant in regions of slow subsidence. His manuscript ends on a tectonic note:

Before finally concluding this subject, I may remark that the general horizontal uplifting which I have proved has & is raising upwards the greater part of S. America & as it would appear likewise of N. America, would of necessity be compensated by an equal subsidence in some other part of the world.—Does not the great extent of the Northern & Southern Pacifick include this corresponding Area?—Humboldt carrys a similar idea still further; In the Fragmens Asiatiques, P. 95. he says. "Par consequent l'epoque de l'affaisement de l'Asie occidentale coincide plutot avec celle de l'exhaussement du plateau de l'Iran, du plateau de l'Asie centrale; de l'Himalaya, du Kuen Lun, du Thian shan & et tous les anciens systemes de montages [*sic*] diriges de l'est a l'ouest; peut etre aussi celle de l'exhaussement du Caucau [*sic*], & du noeud de montagnes de l'Armenie & de l'Erzeroum."[222] (Darwin, 1962, p. 17 {p. 22a of his manuscript}).

Presumably after Darwin completed the manuscript discussed above, the *Beagle* arrived on 1 April 1836 at the isolated atoll of the Cocos (Keeling) Island, which is located some 1000 km south-southwest of Sumatra. (For the geography and geology of this atoll, which consists of five main and many smaller coral islands, and for the controversy that developed regarding it between Darwin and his antagonists, see Woodroffe et al., 1990a, 1990b; Woodroffe and McLean, 1994; Woodroffe and Falkland, 1997). This is the only atoll that Darwin could study in any detail (Yonge, 1958, p. 247–249; Stoddart, 1976, p. 204–205; for Darwin's description of his visit, see Barlow, 1933, p. 394–400). The *Beagle* stayed in the atoll's lagoon until 12 April, and Darwin landed on the coral island to look at its geology, fauna, and flora. The only observation he seems to have made in connection with the tectonic part of his theory was that the lagoon was shallow and lay in stark contrast to the precipitous depth immediately outside it (cf. Yonge, 1958, p. 247 and 249). Darwin noted the following in his diary on 12 April, as they were leaving the Cocos (Keeling) Islands:

If the opinion that the rock-making Polypi continue to build upwards as the foundation of the Is^{d} from volcanic energy, after intervals, gradually subsides, is granted to be true; then probably the Coral limestone must be of great thickness. We see certain Is^{ds} in the Pacifick, such as Tahiti & Eimeo, mentioned in this journal, which are encircled by a Coral reef separated from the shore by channels & basins of still water. Various causes tend to check the growth of the most efficient kinds of Corals in these situations. Hence if we imagine such an Island, after long successive intervals to subside a few feet, in a manner similar, but with a movement opposite to the continent of S. America; the coral would be continued upwards, rising from the foundation of the encircling reef. In time the central land would sink beneath the level of the sea & disappear, but the coral would have completed its circular wall. Should we then not have a lagoon island?—Under this view, we must look at a Lagoon Is^{d} as a monument raised by myriads of tiny architects, to mark the spot where a former land lies buried in the depths of the ocean. (Barlow, 1933, p. 400; also see Burkhardt and Smith, 1985, p. 570)

Darwin's last encounter with coral islands was in Mauritius, where the *Beagle* stayed from 29 April to 9 May 1836 (Yonge, 1958, p. 249). There, Darwin studied the zoology and the depth conditions of the coral reefs, in part in the company of Captain Lloyd, the surveyor-general, on 5 May (Barlow, 1933, p. 403). He noted again that, near the mouths of rivers, corals did not grow and that the sheltered side was low (he could not study the eastern, windward side first-hand, but heard reports that it was higher and better fortified by reefs).

This is how far his observations on the coral reefs reached. The rest of his work on coral islands was essentially devoted to collecting more material from the literature to test his theory and to elaborate its consequences. In October 1838, Darwin started writing his book on *The Structure and Distribution of Coral Reefs*. The writing was interrupted by his marriage in January 1839 and the following period of illness. He took up writing again in 1841 and handed the publisher his manuscript in January 1842 (Stoddart, 1976, p. 206). He corrected the last set of proofs on 6 May 1842, and the book was out soon thereafter (Yonge, 1958, p. 250).

What is of interest to us here is Darwin's tectonic conception of the nature and progress of the subsidences occasioning the growth of the coral islands and of the correlative uplifts. His preconception (inherited from Lyell) would have been that subsidences were somehow jerky and local, related to earthquakes. His observations in South America would have supported that concept. However, both Lyell's reasoning that uplifts and subsidences must somehow be compensatory (with tectonic subsidence being a trifle more preponderant than tectonic uplift) and von Humboldt's grand vision that subsidences and uplifts embrace large parts of entire continents must have inspired Darwin to think in terms of large areas. We know that he did so indeed by considering nearly the whole of the Pacific basin a region of subsidence and its compensatory uplifts being in the double continent of America. In fact, Burkhardt and Smith (1985, p. 568–569) wrote that "Paradoxically, C[*harles*] D[*arwin*]'s adoption of the compensatory crustal changes led him to depart from Lyell's own view of the geology of the Pacific. In his chapter on coral reefs in the second volume of the Principles of Geology, Lyell had adopted the prevailing view of the time that the Pacific was a region of general volcanic uplift and that the reefs had been formed by corals building on volcanic mountains that had subsided as a result of local earthquake action." This is not entirely true because at the time, neither Lyell nor Darwin had well-formulated ideas on the nature and different types of on crustal motions. They were both convinced that the crustal motions were slow and that subsidences had to compensate for uplifts lest the earth expands. They both thought that earthquakes accompanied crustal movements. Lyell's opinion ascribed more jerkyness to crustal motions (by the time Darwin returned home, Lyell had changed his opinion in view of the gentle up-arching of Scandinavia, as described above), while Darwin saw a more continuous and more tranquil subsidence recorded in the coral reefs.

Darwin also noted other tectonic characteristics of areas of subsidence and uplift. During the journey of the *Beagle*, his reading and observations of the coral islands and his inference that coral islands must display their greatest variety of types on a subsiding floor showed him that not only here was a subsident counterpart of the slow continental elevation of the kind known from Scandinavia, *but also that active volcanoes were found only atop rising parts of the crust* (Darwin, 1889[1987], p. 186–189): "The absence of active volcanos throughout the great areas of subsidence on our map ... is a very striking fact. So is the presence of active volcanic vents and chains on or near many of the shores coloured red on our map, and which are fringed with reefs; for as we have just seen, these fringed coasts have been recently upheaved in a large number of cases" (Darwin, 1889[1987], p. 186–187).[223]

What was the nature of the uplifts? Were they block uplifts bounded by faults (as the hypothesis that they were the results of jerky motions related to earthquakes would lead one to surmise), or perhaps they were dome-shaped large intumescences of the crust? In 1836, on his way home aboard *H.M.S. Beagle* in the Atlantic Ocean, Darwin wrote into his notebook: "Try on globe, with slip paper a gradually curved enlargement, see its increased length which will represent the dilation which dilated cracks must be filled up by dikes and mountain chains" (Barlow, 1946, p. 261). Here we see Darwin thinking along lines similar to those of Leopold von Buch, Élie de Beaumont, Alexander von Humboldt, and Carl Ritter (see especially Şengör, 1998, p. 32–34) but partly independently of them. We have no record that he had access to Élie de Beaumont's or Carl Ritter's writings on board the *Beagle* (see the list of books Darwin had along or had access to in Stoddart's appendix in Darwin, 1962, p. 18–20), concerning the relationship of uplift and stretching of the uplifted surface to create cracks and fissures, which magma could use to reach the surface. Darwin was also thinking not of jerky increments of uplift but of gradual uplifts, bending the crust into "curved enlargements." Could he have previously seen Scrope's (1825) figure, reproduced herein as Figure 40? Most likely: we know that it was among the books that he requested from his sister, Catherine, in a letter he wrote on 22 May 1833 from Maldonado, Rio Plata (Burkhardt and Smith, 1985, p. 314).

The upheaved and subsided areas had immense dimensions. Whole continents, thought Darwin, were witnesses to wholesale elevation. In particular, along the western coast of South America, he believed to have demonstrated the continuing action of such upheaval, whereas one had good reasons to suspect the same of the western shores of the Indian Ocean (Darwin, 1889[1987], p. 190). He distinguished the lofty tablelands, which "prove[...] that large surfaces have been upraised in mass to a great height..." (Darwin, 1889[1987], p. 192), from the "highest points consist[ing] of upturned strata..." (Darwin, 1889[1987], p. 192), but did not exploit the significance of this distinction. He also noticed that ocean islands tend to occur along lines and pointed out in the case of the Hawaiian line that while the northwestern end was subsiding, the southeastern end was rising. Again, he did not pursue any possible implications of this observation. He was more impressed with the alternation of adjacent lines of uplift and subsidence (because, although he sensed that they were somehow different, he could not appreciate the nature of the difference between what we today know to be mid-plate islands and subduction zone islands, trenches, and marginal basins[224]).

In the corals Darwin found a faithful recorder of the slow but continuous subsidence or rise of the earth's crust. But this led him to a far more significant inference, as I pointed out above: "...volcanos are often present in the areas which have lately risen or are still rising, and are invariably absent in those which have lately subsided or are still subsiding; and this, I think, is the most important generalization to which the study of coral-reefs have indirectly led me" (Darwin, 1889[1987], p. 190). Like Scrope before him, he viewed the up and down movements of the crust affecting whole oceans or whole continents, as resulting from the one and only igneous agency: "The argument may be finally thus put:—mountain-chains are the effects of continental elevations; continental elevations and the eruptive force of volcanoes are due to one great motive power, now in progressive action; therefore the formation of mountain chains is likewise in progress, and at a rate which may be judged of by either phenomenon, but most nearly by the growth of volcanoes" (Darwin, 1840[1977], p. 80).

So Darwin seems to have disagreed with almost everybody, except he seemed in broad agreement with Hutton, Lyell, and a few followers such as Babbage and Herschel, by considering not only the up and down motions of regions of the size of a continent or an ocean, but also the rise of mountain ranges and even single volcanoes, slow events. These events were not only slow, but all exhibited motion rates of a similar order of magnitude. His views that did not distinguish mountain uplift from continent uplift must have seemed out of date to his continental colleagues (except for the employment of coral reefs as indicators of vertical motions of the crust with respect to sea level) even when he was writing (although the question of rate of movement of crustal parts with respect to the center of the earth and to one another was to occupy geologists for more than a century to come).

Constant Prévost as Embryonic-Suess: Logical Conclusions of the Contraction Theory

Although the issue of elevation craters remained contentious, there was general agreement among geologists, before the first half of the nineteenth century closed, that most of the relief of the earth was due to uplifts and subsidences governed by the internal processes of the planet. One dissonant voice, that of Constant Prévost (1787–1856)[225] (Fig. 53), disturbed this complacent picture and prepared not only the ground for the theory of eustatic movements (to be developed by Eduard Suess in 1888) but, by profoundly influencing Suess' overall thinking in tectonics, became a critical catalyst in the generation of the grandest tectonic synthesis of the earth before the rise of plate tectonics in the sixties of the twentieth century.

Prévost was a student of Cuvier and Brongniart and, as such, started his geological career with an excellent foundation in palaeontology and stratigraphy much along the lines he had learned from his masters. His study of the Tertiary deposits of the Vienna basin (Prévost, 1820), which was highly praised by his former teacher, Brongniart (1820), and undertaken while he was the director of a spinning mill in Hirtenberg (near Vienna) between 1815 and 1818 (von Zittel, 1899, p. 288), shows not only his competence as a palaeontologist and stratigrapher, but his inclination to apply a global approach to geological problems.

Doubts began to emerge in Prévost's critical mind concerning what he had learned from Cuvier and Brongniart about the Tertiary stratigraphy of the Paris Basin, when he found in 1821 mixed faunas of marine and freshwater environments. These separate faunas had been earlier used by Cuvier and Brongniart (1811) as proofs for successive revolutions of the surface of the globe separating distinct time periods in the history of the earth (Prévost, 1821; cf. Gohau, 1995). Prévost's continuing studies revealed in the subsequent few weeks that the mixture was much greater than could be dismissed as a local accident (Prévost, 1822). Prévost became convinced that there had been no repeated deluges and retreats of the sea. He thought that the sea had retired only once, and the mixture of faunas had been a consequence of subaerial erosion, transport, and sedimentation. He thus became the first to emphasize the great dangers of "reworked fossils" for biostratigraphy, to use an anachronistic label.

Figure 53. Louis-Constant Prévost (1787–1856), one of the most influential geologists of the nineteenth century. From de Margerie (1930, plate XIX).

His conviction of "the retreat of the sea once" gradually led him to an Anaximandrian view of the historical geology of the earth, assuming that it had been characterized by a progressive regression of the seas from the continents. It was mainly this idea that eventually led him to deny the possibility of any primary uplifts of the earth's crust. But, for that idea to reach fruition, his involvement in the debates on the elevation craters had to intervene.

When the so-called Islet of Julia (Graham Island, as commonly known in the English-language literature) suddenly made its appearance in the strait of Sicily in September 1831, Prévost was charged with the mission of studying the phenomenon on the spot. He went on board ship on 16 September in Toulon. Although in his first letter to the president of the Academy he stated that "around the island of Julia there must be a belt of uplifted rocks, which would be the rim of an elevation crater," (Prévost, 1831; cf. Gohau, 1995), he later claimed that, in his first report, he had been circumspect and said that nothing in the structure of the new island indicated that it was created by an uplift of the ground (cf. Gohau, 1995).

Prévost's major memoir summarizing his findings came out in 1835, and it embodied an all out attack on the theory of craters of elevation. He pointed out that not only the new Islet of Julia, but neither Vesuvius, nor Etna, nor indeed Stromboli and Vulcano (which he had opportunity to see first-hand: Prévost, 1835, p. 120), showed any evidence of uplift. They were cones of accumulation, and the radial valleys emanating from the crater rims were products of fluvial erosion. In this he was in complete agreement with Scrope and Lyell (see, Prévost, 1835, p. 124). He later visited the groups of Cantal and Mont-Dore, which Dufrénoy and Élie de Beaumont (1834) had already interpreted as elevation craters, to take a look for himself. He found them perfectly comparable with the other volcanoes he had come to know and summarized his conclusions in eight items in a letter to the president of the Academy of Sciences, appended to his report on the Islet of Julia (Prévost, 1835, p. 121–124), concluding:

> The most attentive and impartial examination has led me to see, in the groups of Mont-Dore, Cantal and Mézenc, only three volcanoes formed like the Vesuvius, and even better, like the Etna, by the successive accumulation of volcanic materials, erupted from numerous orifices in the form of flows, or pulverised and fragmentary projectiles.

My trip in Auvergne has gone to confirm the ideas that had been formed in my mind by my study of the volcanic terrains of Sicily and Italy and convinced me especially that products of volcanism only locally, nay, even rarely, dislocate the ground across which they reach the surface. The Tertiary terrains of the Limagne and of the surroundings of Clermont, those of the basin of the Puy, the granites which surround the red rock, furnish proof that the most violent eruptions of scoriae and cinders, the most abundant eruptions of trachytes, of basalts and lavas could take place, amidst terrains of diverse nature, without producing any noticeable disruption. (Prévost, 1835, p. 123–124)

With these words, Prévost not only denied the existence of elevation craters but further affirmed that the most violent magmatism may take place without creating any noticeable deformation of the ground. The latter observation was a direct negation of the claim by von Buch (1824b) that mountain chains were raised by the upwelling of augite porphry and quite catastrophically[226] (von Buch, 1827[1877]).

Were mountain chains really "raised"? Was *anything* in the rocky rind of the earth really "raised"? In the thirties of the nineteenth century, Constant Prévost came to believe that geomorphological and/or structural features that were commonly thought to have been raised, were in fact products of subsidence. In a lengthy reply to defend this opinion against Rozet, Prévost began defining what he meant by upheaval:

Upheaval of the ground means, according to me, the raising of this ground above its original level by an *uplifting* force, that is to say, applied under it and working from the inside towards the outside of the terrestrial shere.

The theory of upheavals is therefore that, which consists in creating reliefs on the surface of the earth such as volcanoes, mountain chains, plateaus, inclination and verticality of strata, faults, etc. by the *uplift* of the masses forming the ground by means of an *agent* placed under the consolidated exterior of the earth which pushes out this resistant part, *uplifts*, *deforms* and *splits* the dislocated panels. (Prévost, 1840, p. 184, italics his)

Uplift as expressed by the theory of elevation was impossible, thought Prévost, simply because his stratigraphic studies had shown him that earth history had been characterized by a continuous, irreversible regression. He rightly pointed out that if elevations occurred within the existing ocean basins, this would diminish their capacity, leading to overflowing and therefore to sea-level rise (influence of Lyell? {1830, p. 474–475}; see the quotation above, p. 85). Since he denied that any sea-level rise had occurred, he was inescapably driven to assume that only subsidences were possible. (Note here that Prévost's argument loses power on a spherical earth; moreover, accumulations affect the sea level precisely in the same way as uplifts do, as Seuss showed half a century later; see below, Chapter XIII.)

At this point, his experience with volcanoes and the so-called elevation craters came to his aid. Had they not shown him that all claimed uplifts had in fact turned out not to have been uplifts but accumulations, that magmatic rocks had been seen not to have caused any appreciable deformation of the basement they had traversed? Prévost concluded that subsidence alone was sufficient to give rise, by "rebound" (he says "*par contre-coup*"[227]), to local absolute elevations, lateral shortenings, bendings, foldings, ruptures, squeezings, faults, etc. (Prévost, 1840, p. 186). He viewed magmatic rocks only as passive, rising wherever they could find an opening or some other opportune condition; this view prepared the way for Suess (1875) and Heim (1878a, 1878b, 1878c), who finally laid to rest the hypothesis of the active role of magmatic intrusions in mountain-building nearly four decades later.

All of this kind of deformaton was a consequence of the thermal contraction of the globe:

I believe very simply, as do nearly all geologists of our day, that the terrestrial spheroid is a cooling body. The consolidated outer crust floats on a still fluid or soft interior. The body loses volume in such a way that the exterior, while trying to keep pace with this centripetal motion of the interior tends to get folded, ondulated, broken, engulfed etc. By analogy, I think that the same cause produces in numerous phases the folds, the ondulations, the ruptures, the depressions which constitute the present surface features of the earth. (Prévost, 1840, p. 201)

Ever since I became familiar with the ideas of Prévost and those of Élie de Beaumont (i.e., about 30 years ago), I have never been able to understand why they regarded each other as opponents. I was delighted to discover that Gohau (1995, p. 81), one of the most knowledgeable students of the history of French geology, is similarly baffled. Their only difference consisted in the interpretation of the craters of elevation. True, Élie de Beaumont also thought that the main mountain ranges of the globe may have been assisted in their rise by igneous material filling their cores, but, like Prévost, he thought that the big mountain ranges mainly resulted from lateral shortening ("*ridement!*"). Perhaps Gohau (1995, p. 81) is correct in thinking that here personalities rather than ideas were clashing. In any case, Prévost's ideas had to incubate for another 35 years to hatch and be married with those of his great adversary, Élie de Beaumont, in the grand event marked by the publication of Eduard Suess' *Die Entstehung der Alpen*. The offspring of this happy union was the greatest tectonic synthesis of the globe the world had seen prior to the rise of the theory of plate tectonics: Suess' *Das Antlitz der Erde*.

James Dwight Dana: Darwinian Explorer and Beaumontian Theoretician

James Dwight Dana (1813–1895; Fig. 54)[228] was the best-known and the most influential American geologist of the nineteenth century. In his review of the history of geology to the end of the nineteenth century, von Zittel said of him,"he was incontestably the first geologist of North America" (von Zittel, 1899, p. 459; also see Schuchert, 1915, p. vii). Dana's concept of the tectonic behavior of the earth was taken completely from Élie de Beaumont and Constant Prévost but applied with a religious conviction to the tectonics of North America, which Dana viewed to be the "type" continent (see especially Dott, 1997). It was

Figure 54. Portrait of James Dwight Dana (1813–1895) at the time when he was most actively engaged in the coral reef research (from Davis, 1928, fig. 1).

Dana's version of the contraction theory that had the greatest influence in the late nineteenth century and in much of the twentieth century until the rise of the theory of plate tectonics. Dana was more often cited by European tectonicists than were Élie de Beaumont and Prévost, although he owed his entire conceptual framework to them.

By the time Dana published his first tectonic paper, he had already gone out with the United States Exploring Expedition in 1838–1842 and had seen a greater part of the Pacific Ocean than Darwin had.[229] After he found out about Darwin's theory on the origin and growth of the coral islands, Dana's main tectonic interest during the expedition centered on the subsidence question. In fact, his first tectonic paper is devoted to it (Dana, 1843), which is really an advertisement for the full report that had to be a part of the government publication to be published later.

Dana pointed out in his 1843 paper first that Darwin's theory "has been fully confirmed by the investigations of the Exploring Expedition" (Dana, 1843, p. 131; also see Dana, [1849], especially p. 123–134; 1853, p. 88–89; Davis, 1913, 1928, p. 1 and 45 ff.), although he felt that Darwin's own observations were inadequate to support many of his specific assertions, especially about the ongoing motions and the distribution of the areas of subsidence and uplift in the Pacific.[230] Dana rightly pointed out that the instantaneous motions could only be ascertained by observations over a number of years, whereas the cumulative coral thicknesses only inform us about the past during which those thicknesses have accumulated. He then set out to map the areas of subsidence in the Pacific (see also, Dana, [1849], p. 394; 1853, ch. V).

Dana observed that "On examining a map of the Pacific, we find a large area just north of the equator with scarcely an island [*Fig. 55 herein*]. To the south, the islands increase in number, and off Tahiti, to the northward and eastward, they become so numerous, and so crowded together, as to form a true archipelago. They are all, too, coral islands, throughout this interval. This then is a remarkable fact in the distribution of these islands. But let us look farther" (Dana, 1843, p. 133; [1849], p. 394–401).

When he did look farther, he noted that between the Sandwich Islands (i.e., the Hawaiian Islands) and the Society Group, a large area north of the equator had scarcely an island in it. To the south of the equator, the number of islands increased: north of a line running east-southeast from New Ireland, Wallis' Island, Samoa or the Navigators, the Society Islands, and thence bending a little southward to the Gambiers, all the islands (with two or three exceptions) were purely of coral, whereas those to the south of that line were high basaltic volcanoes. Dana noted that the basaltic islands were all bordered by reefs and that the reefs were most extensive near the line demarcating pure coral islands from those that had central high volcanoes. As one went north within the coral group, the lagoon islands gradually gave way to individual points of reef.

Dana concluded from his observations that, in the south, subsidence was either not rapid or had not been going on for long enough a time to produce atolls; whereas in the north, subsidence was either very rapid or long-lived, so as to drown even the atolls—to reduce the remaining few to mere points or pinnacles of reef (Dana, 1853, p. 120–121). From the shape of the subsiding area, Dana assumed that subsidence would be greater still to the west, and he remarkably predicted that drowned atolls must exist in the west Pacific, a prediction verified almost exactly a century later by Harry Hess (1946; Hamilton, 1956; Natland, 1997). He wrote the following about the timing of subsidence:

> The time of these changes we cannot definitely ascertain; neither when the subsidence ceased, for it appears to be no longer in progress. The latter part of the tertiary and the succeeding ages may have witnessed it. Although I am by no means confident of any connection, yet for those who find a balance motion in the changes, I would suggest that the tertiary rocks of the Andes and North America, indicate great elevation since their deposition; and possibly during this great Pacific subsidence, America, the other scale of the balance, was in part undergoing as great or greater elevation. (Dana, 1843, p. 135; also [1849], p. 401; 1853, p. 125–126).

Dana disagreed with Darwin that the subsidence was continuing. He also clearly disliked Lyell's notion of balancing any uplift with subsidence, but he was as yet not ready to challenge Lyell's theory so he offered observations that might have been taken as supporting Darwin's postulate that the rise of the double continent of America may have been the counterpart of the Pacific subsidence. Why he disliked Lyell's notion of keeping

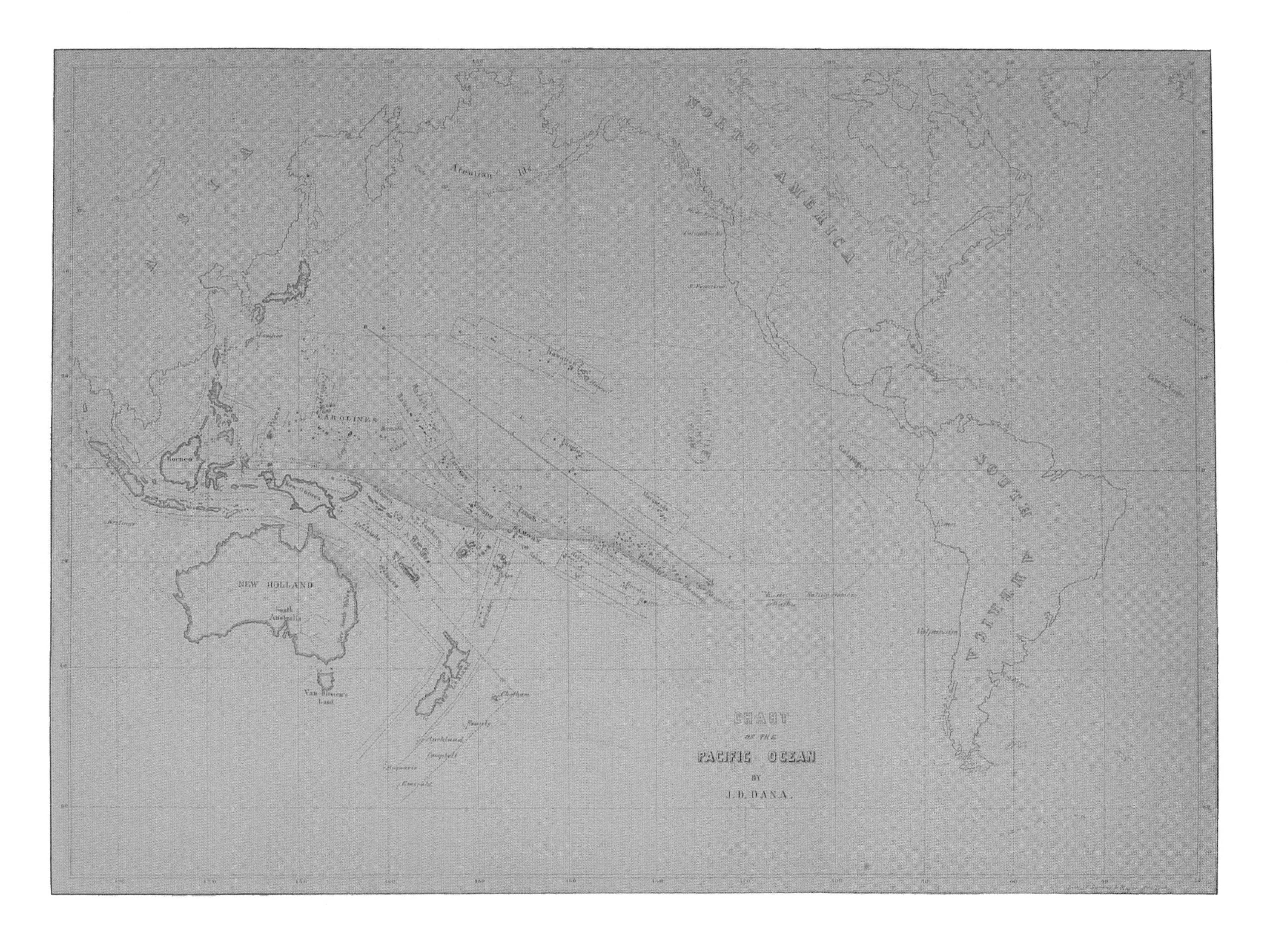
CHART
OF THE
PACIFIC OCEAN
BY
J. D. DANA.
NORTH AMERICA
SOUTH AMERICA
ASIA
NEW HOLLAND
South Australia
Van Diemen's Land
Borneo
New Guinea
CAROLINES
Aleutian Ids.
Galapagos
Chatham
Auckland
Campbell
Macquarie
Emerald
Lima
Valparaiso
Azores
Canaries
Cape de Verde
Viti
SAMOAN
Radack
Ralick
Kingsmill

Figure 55. Dana's map of the Pacific Ocean showing the distribution of coral islands (from Dana, [1849], frontispiece).

subsidences and uplifts in balance, his readers were to find out three years later.

Significantly, in a paper on the volcanoes on the Moon, Dana (1846) chose to present his preferred theory of global tectonics. His choice of the Moon as the best introduction to terrestrial tectonics was significant because that was exactly how, as we saw above, Élie de Beaumont had nearly two decades earlier chosen to introduce contraction as the overarching cause of terrestrial tectonism. Dana twice refers to Élie de Beaumont in his 1846 paper, both in reference to Élie de Beaumont's 1829 Oisans paper, without explicitly citing the paper: once on p. 336, where he says Élie de Beaumont used the recent Moon maps to support "certain geological theories." Here "certain geological theories" clearly refer to von Buch's elevation craters and the theory of thermal contraction of the earth. Dana's second reference to the same paper by Élie de Beaumont is on p. 345, footnote *: "The elevation theory of Von Buch has been supported from facts in the moon. We offer nothing here on that subject." Dana further acknowledged Constant Prévost's ideas on terrestrial contraction and its effects on global tectonics. So, he was fully current with the French thinking of his day on global tectonics. It is very curious, however, that he rarely acknowledged Élie de Beaumont's obvious influence on him.

The 1846 paper is ostensibly on the volcanoes of the Moon. Dana thought that the lunar craters were all volcanic. This idea was not new and indeed constituted the conventional wisdom of the day (cf. Koeberl, 2001). Dana's novelty consisted in comparing those craters with the pit craters of Hawaii, which he had come to know first-hand during the U.S. Exploring Expedition (see Dana, [1849], p. 155–284, 353–456; 1891, p. 39; Appleman, 1985; Natland, 1997). Having established to his satisfaction that the lunar surface was once a boiling inferno, he noted that there are areas on the Moon that are smoother and that have relatively few craters. He concluded that these areas must have cooled earlier than the others and hence the volcanic activity in them had turned off earlier.

Turning his attention to the earth, Dana noted that since the "early Silurian epoch" (he means since the beginning of the Paleozoic because the Cambrian as the first period of the Paleozoic era did not become universally acknowledged until much later), vast areas in the central parts of North America and Russia were free both of "eruptive fires" (1846, p. 353) and deformation. Dana thought these areas (the continents) were the first areas to have cooled, and consequently, they switched off their igneous activity and further contraction. The oceans, by contrast, were still volcanic. He compared the earth with "a melted globe of lead or iron, when cooling unequally, becomes depressed by contraction on the side which cools last" (Dana, 1946, p. 353). The oceans were low with respect to the continents because they were still in the process of cooling.

Most of these ideas were derivative. Dana defended himself by pointing out that he had arrived at these independently of Prévost (Dana, 1846, p. 355), but a similar claim with respect to Élie de Beaumont's ideas would have been hardly credible. That is probably one reason why Dana was so quiet about him.

In the 1846 paper, we do not learn anything about the *structural* nature of the oceanic depressions: were they simple downbends, as Élie de Beaumont assumed, or downfaulted cauldrons, as Prévost thought and Suess later agreed (see below)? The answer came in the second of four papers on tectonics, which Dana published in 1847. All four of these papers are mere abstracts and outlines of his ideas, except that the second one (Dana, 1847a) presents, in graphic form, how Dana thought oceans basins formed. Figure 56 herein is a reproduction of Dana's (1947a) figures.

Dana's figure 1 (in Fig. 56) represents his concept of the earlier times of the earth, sometime in the very beginning of the Paleozoic or a bit earlier. There were no continents and no ocean basins, and the hydrosphere formed a sheet of nearly uniform thickness on the lithosphere: oo' over ct. In figure 2 (in Fig. 56 herein), the globe has contracted from the dotted line ct to $c't'$. In this state, the parts $c'o$ and $o't'$ are the portions that are free from volcanic action. In other words, they are the coagulating continental nuclei, and they clearly form saddles in a folded crust, where the oceans (oo') represent a trough in the same fold train. p represents an area upon the continent ($o't''$) and seems to occupy a smaller trough in an area of a larger saddle. In figure 3 (in Fig. 56) is simply a further accentuation of this situation whereby the oceans have deepened sufficiently to drain the continents of all the epeiric seas. Dana thought that the subsidence of the oceans would exercise a pressure against the continent (because the earth's crust is now being crowded into a smaller area) and further saddle and trough systems would form: m, n, r and s (shown on fig. 3 in Fig. 56), which are the saddle hinges of such upfolds. The troughs of these smaller continental folds will be called *geoclinals* by Dana in 1862[231] (Dana, 1863, p. 722: Fig. 57) and *geosynclinals* in 1873, after the New York palaeontologist and stratigrapher, James Hall, had drawn attention to the enormous thicknesses of the Paleozoic strata in the Appalachians and interpreted them as having been accumulated in a synclinal trough, as we shall see below. Dana thus thought that his geoclines (later geosynclines) and oceans were fundamentally the same sorts of structures differing only in size, a view to be followed three quarters of a century later by his European disciple, Émile Haug.[232]

Dana thought that contraction and the formation of low-amplitude and very long-wavelenth folds, in essence forming the continents and the oceans, continued uninterrupted for long periods of time. These relatively quiescent periods were then episodically interrupted when the resistance of the crust was overcome and rapid mountain-building catastrophes occurred. Dana thought that such catastrophes marked the ends of geological eras (Dana, 1847a, p. 187). He pointed out that large areas of low-amplitude, long-wavelenth uplifts on the continents were also due to the same mechanism: "If the material subjected to lateral pressure be not capable of folding, or *only partially so*, the region operated upon instead of rising into a series of elevations, would be raised into one or more ridges of much greater height. Has not this last been the case on the Pacific side of the continent?" (Dana, 1847a, p. 185, italics Dana's). In his first of the 1847 papers, Dana described the Pacific side of America in the following words, on the basis of the great 1845 publication of John Charles Frémont, "The Pathfinder":

> On the *Pacific* side of the continent, we observe the Rocky Mountain range rising with a gentle swell from the coast. From the mouth of the Kansas to the top, and on the opposite or western side, the average slope is hardly twelve feet to the mile [*0.13*°]. The summit is about eight thousand feet high. But there are ridges which add five or six thousand feet to the chain: these form a crest to parts of the range, but are not properly of the range itself, though often so recognized. The Rocky Mountains appear, then, to be another effect of the contraction, viz. a gradual swelling of the surface, accompanied by fissures and dislocations over its area. (Dana, 1847b, p. 98; see Figs. 85 and 86 herein)

The four 1847 papers set out clearly and concisely how Dana thought the earth behaved tectonically. I can perhaps itemize his view of the terrestrial tectonics as follows:

(1) Terrestrial tectonics ultimately results from the thermal contraction of the globe.

(2) Continents represent earlier cooled parts that have ceased contracting.

(3) Oceans are still losing heat and are contracting as shown by their active volcanicity and lower elevation than the continents.

(4) Oceans and continents are parts of large folds of the earth's crust. Continents are the gentle arches, oceans the gentle depressions of these folds (see especially Le Conte, 1872a, figure on p. 346).

(5) This large-scale folding is a continuous process that progressively deepens the oceans and elevates the continents with respect to one another. A consequence of this action is that oceanic waters progressively recede from the continents and become concentrated in the ocean basins.

(6) Now and then the contraction-related tangential stresses increase to a point so that the ocean/continent boundary catastrophically fails, creating long, narrow, linear to curvilinear mountain ranges parallelling the continental margin. These ranges are characterized by much closer folding that involves overturning toward the continent (compared with the earlier broader undations). Such catastrophic mountain building paroxysms terminate the geological eras. Dana's favorite example was the Appalachian revolution that supposedly ended the Paleozoic era.

Dana thus clearly recognized that there were two classes of structures that resulted from the thermal contraction of the

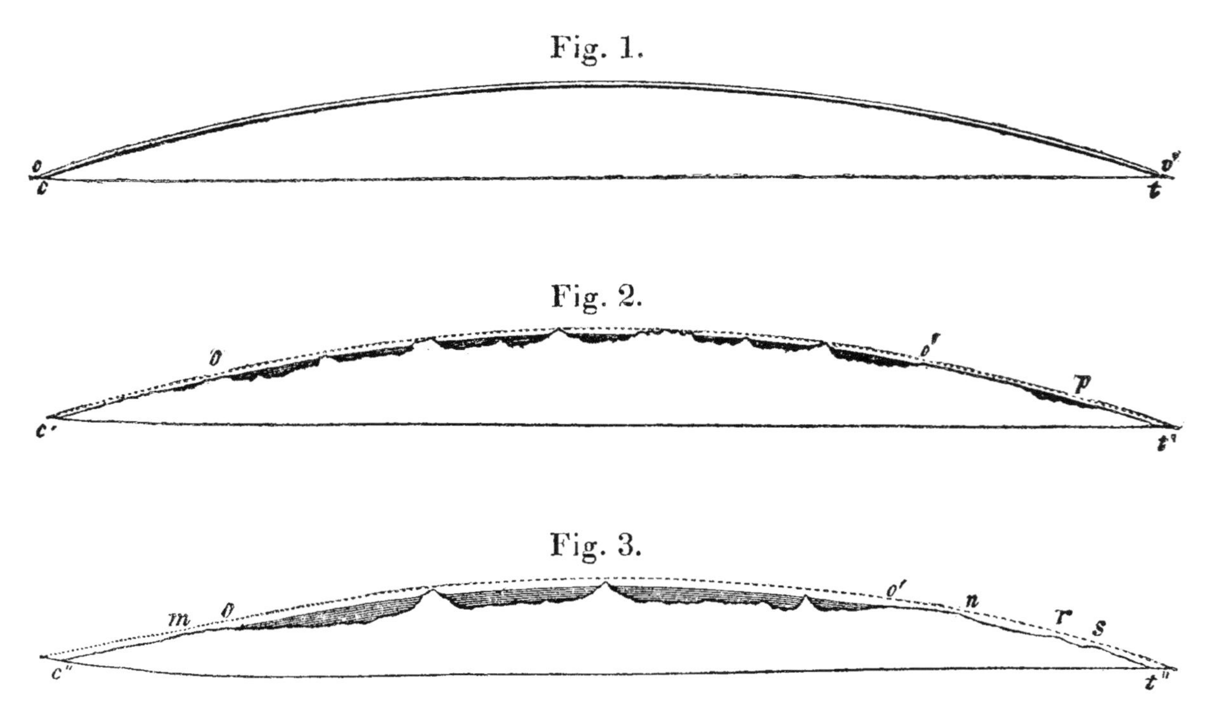

Figure 56. Dana's sequential cross sections illustrating topographical results of the thermal contraction of the earth, which he alleged to have happened since the origin of the planet (from Dana, 1847a, figs. 1, 2, 3). Fig. 1: The crust (*ct*) is represented as covered with water (*oo'*). Fig. 2: The globe has contracted from the dotted line to *c't'*; *c'o*, *o't'*, which are the portions free from volcanic action (i.e., contracted earlier). *p* is an area of water upon *o't'*. *oo'* represents an incipient oceanic depression. It was considered still contracting. Fig. 3: As the crust below the oceanic depression becomes thicker by cooling, the contraction, not now causing fractures over its own area alone, would produce a tension laterally against the non-contracting area and occasion pressure, fissures, and upheavals; and thus elevations *m*, *n*, *r*, *s*.

planet: (1) the low-amplitude, long-wavelenth (falcogenic) structures; and (2) the high-amplitude, short-wavelenth (copeogenic) structures. The distinction Dana made was not original and was mostly borrowed from Élie de Beaumont without acknowledgment. However, Dana's influence became not only more widespread, but also more enduring than that of Élie de Beaumont. The reason for this was, I believe, that he wrote in simple terms that anybody—even non-geologists—could clearly understand. This allowed his handling of the subject to be far less rigorous than Élie de Beaumont's rigorous, but because of that, commonly opaque, quantitative treatment. Dana also wrote clearly and concisely, and in numerous iterations (he being the editor of the most widely read American scientific journal for many years made this easy for him), generally without encumbering his short papers with scholarly apparatus. This frequently gave the impression that the ideas he was presenting were his own. Finally, he became the author of one of the most sucessful geology textbooks ever in the history of the earth sciences, namely his *Manual of Geology*, in which his ideas were systematized, summarized, and illustrated mainly with American examples. The *Manual* was first published in the December 1862, reprinted in 1863 with corrections, and then ran through four more editions until 1895.

Dana's choice of North America as his illustrative material and the way he presented it to his readers also contributed to the seductiveness of his material. He presented North America as a perfect illustration of the principles of the sort of tectonics he was advocating. He believed this continent to have a simple structure with a central stable region and two mountain ranges on both sides bordering oceans, the subsidences of which were the causes of the crumpling of the marginal mountain belts. The bigger of the mountain ranges faced the larger ocean. The mountain ranges verged towards the central stable region, and their more metamorphic and intruded parts were nearer the ocean. Volcanism was confined to the oceans or to the mountain ranges actively being built.

It is astonishing that Dana's views survived, especially in North America, into the 1960s! In 1966, for example, B.C. Burchfiel and G.A. Davis submitted to *Science* a paper entitled "Two-sided nature of the Cordilleran Orogen and its tectonic implications" with a view to combatting the then still prevalent opinion that the entire North American Cordillera was an

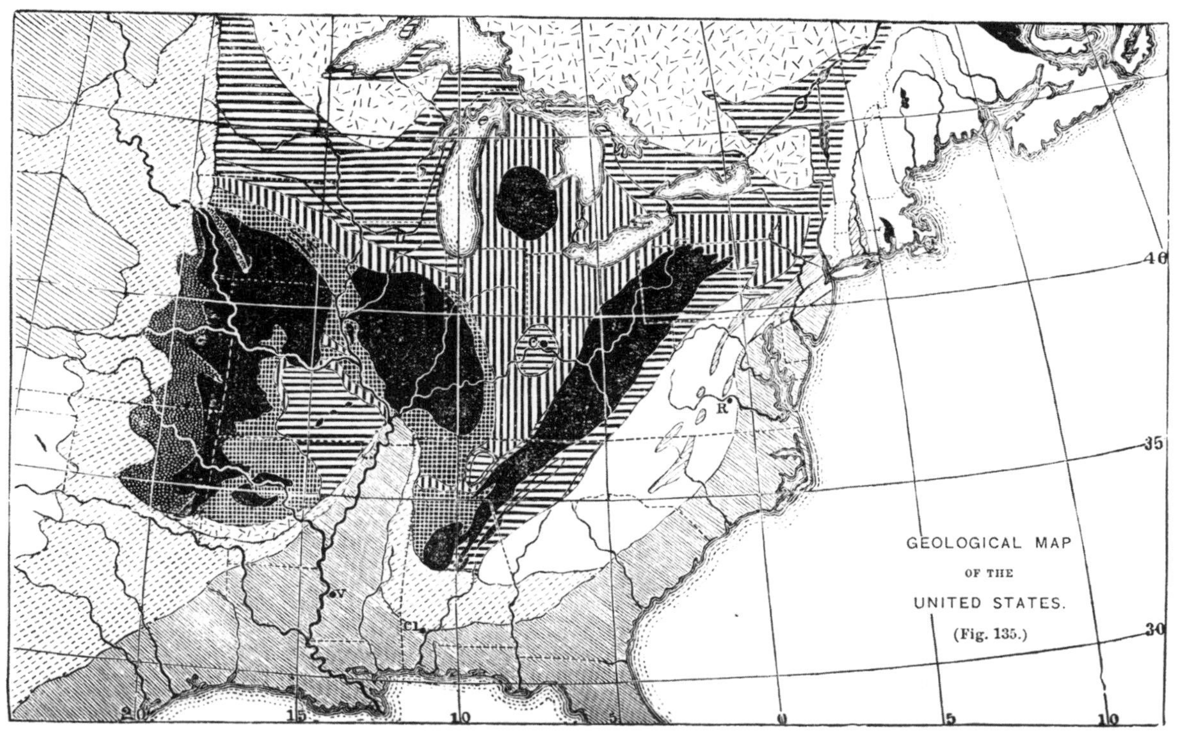

Figure 57. Geological map of the United States east of the Rocky Mountains (from Dana, 1863, fig. 135), showing the location of "geoclinals" (i.e., large, crustal scale down-bendings creating valleys). Legend: Heavy horizontal lining indicates Silurian (i.e., Cambrian through Silurian in the present terminology). Heavy vertical lines indicate Devonian; Carboniferous is divided into three subdivisions: (1) Sub-carboniferous is light cross-lines on a black background; (2) the Coal formation is black surface; and (3) Permian is dots on a black background. Diagonal lines sloping from top right to lower left indicate *Reptilian* (Mesozoic), including the Triassic, Jurassic and the Cretaceous (the Cretaceous being distinguished by broken lines). Lines sloping from top left to lower right indicate Tertiary. Irregular line dottings indicate Azoic (Precambrian). The surface without markings is occupied by rocks of undetermined age that, on the east, is mostly crystalline. Note on this map the great "geocline" of the Mississippi and the smaller "geoclines" represented by narrow areas of Mesozoic signature on unpatterned crystallines of the Appalachians.

asymmetric, east-vergent orogen, exactly as Dana had portrayed it more than a century earlier. Their paper was rejected twice with such typical comments as: "The idea proposed does not merit publication *except as an unfounded speculation*" (from an unpublished referee's comment; italics mine: Şengör, 1999b, p. 35). Such a comment would have surprised any European geologist already in 1909, after Suess' meticulous documentation of the double-sided structure of the North American Cordillera (cf. Şengör, 1999b, p. 35). But Dana's influence was also very great in Europe. Even Suess, who ended up deviating from most of Dana's interpretations, derived much of the initial inspiration for his tectonic studies from Dana's writings. What Şengör (1982a, 1982b, 1991c, 1998, 2000) called the Kober-Stille school of tectonics in the twentieth century was entirely a continuation of Dana's way of thinking. That is why it is a serious mistake to consider the twentieth century fixist schools in Europe a continuation of Suess' way of thought or to characterize Suess' ideas as "*the* European ideas" on tectonics, as I have pointed out repeatedly since 1979 (Şengör, 1979; for the most recent iteration, see Şengör, 2003).

Vertical Uplift Theory in the Mid-Century: Alexander von Humboldt and the Textbook Writers Before 1850

In his epoch-making *Asie Centrale*, which demolished the long-standing belief in an immensely high and uniform plateau of High Tartary in the middle of Asia (Schmaler, 1904, p. 106 ff.), Alexander von Humboldt held onto the vertical uplift theory of the mountain ranges and continental plateaux (despite Prévost), which he had developed together with his friend Leopold von Buch (see especially Şengör, 1998, p. 23–34, with rich, in part color, illustrations). A number of kinds of uplifts were recognized, which von Humboldt classified according to *speed of upheaval* and *continuity* of action: "To constrain better the dependent geological causes of a change of level of the ground, we distinguish among partial, brusque, and instantaneous uplifts, such as those frequently brought about by earthquakes in the southern part of the New World or the mud volcanoes of the Apsheron Peninsula on the western shore of the Caspian Sea, and the slow and continuous uplifts of Scandinavia" (de Humboldt, 1843, p. 287). In his *magnum opus*, the *Kosmos*, von Humboldt warned the reader not to confuse the slow, continuous movement of the Swedish coasts, documented by historical records, with the older elevations of the strand at Nordcap[233] or in Spirtzbergen, neither should they be confused with the instantaneous jump of shorelines during earthquakes. He indicated that the continuous movements now and then take the aspect of folding and, while some regions ascend, others descend (von Humboldt, 1845, p. 313–314; this may have influenced Suess, 1875, p. 151: see below, Chapter XIII). It is clear that von Humboldt did not consider Élie de Beaumonts "mountain systems" as belonging to this class of slow and continuous vertical movements.

The Swiss geologist Bernhard Studer (1794–1887)[234], in his influential textbook of 1847, not only distinguished *mountain-building uplifts* from *continent-building uplifts*, but he also further divided uplifts into three distinct groups as follows:

> In the activity of the interior of the earth, called "heat" after its most prominent character, lies a much more fruitful and general principle for the explanation of the irregularities of the surface of the earth and the origin of mountains. The pressure, applied from the depths upwards to the outer crust of the Earth, can, so it is believed, annul the effects of gravity and work against a general levelling. … According to where the impulse is applied, whether at a point, along a crack, or across a large surface, a *central*, a *linear*, or an *areal* uplift will occur. The uplifted mass will accordingly take the form of a *dome mountain*, a *mountain chain*, or a *plateau*. At a higher level of development, these three types are found repeated in the form of *central mountain systems*, *zones of mountains*, and *continents*. (Studer, 1847, p. 178)

Thus, the idea of continental elevation by internal agencies, popular almost since the beginning of the century, was finally married with the ideas of Leopold von Buch, developed for volcanic phenomena and applied by himself otherwise only to mountain ranges. Studer held vertical uplift responsible for all three classes of uplift and used this idea extensively in interpreting the tectonic evolution of his native Alps. His mention of the heat impulse from within the earth, being applied at a point to create a dome mountain, is, to my knowledge, the earliest harbinger of the present theory of mantle plumes underlying uplifted hot spots[235]. Studer also drew attention to the fact that "the uplift and the delimitation of land areas occurred independent of the configuration of their surface and of their internal structure" (Studer, 1847, p. 242). In other words, he realized that falcogenic structures did not visibly affect the internal fabric of rock masses, and neither were falcogenic structures much influenced by them.

Only a year later than Studer's textbook came the comprehensive study—by the notorious author[236] of the *Vestiges of the Natural History of Creation*—of the ancient strand marks (Chambers, 1848). Already in the *Vestiges*, Chambers (Fig. 58) had followed the conventional wisdom of his day in describing two types of deformation of the earth's crust. Describing the events at the end of the "Era of the Primary Rocks" (i.e., end of the Devonian), Chambers writes:

> It was only now that the central granitic masses of the great mountain ranges were thrown up, carrying up with them broken edges of the primary strata; a process which seems to have had this difference from the other ["*volcanic disturbances and protrusions of trap rock*" (p. 73)], that it was the effect of more tremendous force exerted at a lower depth in the earth, and generally acting in lines pervading a considerable portion of the earth's surface. … There is no part of geological science more clear than that which refers to the ages of mountains. It is as certain that the Grampian mountains of Scotland are older than the Alps and Apennines, as it is that civilisation had visited Italy, and had enabled her to subdue the world, while Scotland was the residence of "roving barbarians." The Pyrenees, Carpathians, and other ranges of continental Europe, are all younger than the Grampians, or even the insignificant Mendip Hills southern England. Stratification tells this tale as plainly as Livy tells us the history of the Roman republic. (Chambers, 1844, p. 73–74)

Figure 58. Robert Chambers (1802–1871), whose *Ancient Sea Margins*, together with the opinions of Constant Prévost, exercised a strong influence on Eduard Suess' ideas on the causes of the movement of the strand.

This is a perfectly Beaumontian picture of mountain-building written in a language of naïve self-confidence enough to annoy even a man of Lyell's placid character. (One wonders how this book was ever ascribed to Lyell?) One thing is clear, however: Chambers sees disruption of structural fabric visible to the eye and, with the aid of this disruption, the movement is dated. When describing the age of the disturbances at the end of the Carboniferous, he writes: "That these disturbances took place about the close of the formation, and not later, is shewn in the fact of the next higher group of strata being comparatively undisturbed" (Chambers, 1844, p. 92). He thus uses unconformities exactly in the sense recommended by Élie de Beaumont (1829–1830, 1830, 1831, 1833, 1849). But Chambers also saw another kind of movement, expressed in very different structures:

> These consist of *terraces*, which have been detected near, and at some distance inland from, the coast lines of Scandinavia, Britain, America, and other regions; being evidently ancient beaches, or platforms, on which the margin of the sea at one time rested. ... Taking a particular beach, it is generally observed that the level continues the same along a considerable number of miles, and nothing like breaks or hitches has as yet been detected in any case. A second and a third beach are also observed to be exactly parallel to the first. These facts would seem to indicate quiet elevating movements, uniform over a large tract. (Chambers, 1844, p. 140–142, italics Chambers')

Thus on one hand, he recognized the "effect of more tremendous force exerted at a lower depth in the earth, and generally acting in lines pervading a considerable portion of the earth's surface," and on the other "quiet elevating movements, uniform over a large tract." The latter class was much less studied and certainly much less understood than the former. That is why Chambers devoted his one book, published in 1848, concerning geology only to that latter class of structures. However, he did not very much like them. He was entirely under the spell of the diluvial hypothesis and was convinced that the northern countries at least had been subjected to "one last long submersion" before the Recent commenced (Chambers, 1844, p. 140). It was the re-emersion, he thought, that had created the terraces. They were far too extensive and far too regular to have resulted from local tectonism of any sort, however long-waved. Chambers consulted Darwin, who could also "adduce no recent uprise preserving equality over a wide surface, but on the contrary, believes that such uprises as those of South America which he has described, must present inequality" (Chambers, 1848, p. 319).

However, Chambers found a ready aid in Darwin's hypothesis of oceanic subsidence to account for the apparent rise of the land:

> Perhaps we should be in a more hopeful course, if we were to turn our eyes to Mr Darwin's views regarding the subsidences of great oceanic basins, as implied by the phenomena of coral islands. The undoubted effect of such extensive subsidences must be the lowering of the sea round all the shores of the world. The sinking of an area measuring the twentieth of the aqueous surface of the globe, to the extent of half a mile, would cause a sinking of the entire sea to a depth embracing several of the intervals of our British terraces—about 130 feet. Such may have been the history of the changes of relative level in our region, while, in other districts, both risings and fallings of the land may have taken place. There are, it will be remembered, clear proofs that the sea, after falling, had risen again in our island. (Chambers, 1848, p. 319–320)

Clearly, Chambers showed amazing insight (as he had done earlier on the question of the evolution of life in the wildly controversial *Vestiges*: Williams, 1981; Secord, 1994) and formulated a model for what Suess was to call exactly 40 years later "eustatic movements." Chambers' model, perhaps inspired by Prévost's (1840), was substantially the same as Suess' own. Suess quoted him in detail (cf. Dott, 1992) and wrote with characteristic generosity: "This is, so far as I know, the first attempt to bring into causal connection the formation of atolls in the tropics and that of terraces in the higher latitudes" (Suess, 1888, p. 21). But Chambers' model was superior to Prévost's (and to Suess' own: see below) because the Scotsman allowed also for some vertical oscillations of the lithosphere (as much as he disliked them), which, a decade before the Frenchman, and four decades later the Austrian, completely denied. Thus, the plate tectonics mechanism excepted, Chambers' view of the causes of the changes of level are much the same as our modern views and involves slow, secular rises of stretches of land without altering their mesoscopic structural fabric and also slow subsidences of ocean floors (cf. Dewey and Pitman, 1998).

The most influential textbook of the mid-century was the *Lehrbuch der Geognosie*[237] of Carl Friedrich Naumann (1797–1873: Fig. 59)[238], to whom among others, we owe the word *tectonics* in geology (in the form *Geotektonik*). Naumann's book (Fig. 60) went through two editions, the first two-volume edition (with an additional atlas of fossils in loose sheets) in 1850 and 1854, and the second, uncompleted three-volume edition between 1858 and 1872. In both editions, Naumann pointed out that the fast movements (i.e., mountain-building), even instantaneous (earthquakes), were to be distinguished from the slow, *secular* movements. The secular movements, Naumann thought, created the sea basins and the continents (see especially Naumann, 1850, p. 397). By contrast, the fast movements gave rise to uplifts of much more restricted horizontal dimensions. The secular movements, Naumann conjectured, must be associated with very large stresses. (Emile Argand was to agree with him on this issue three quarters of a century later.) Here and there these stresses would tear the crust "which would result in a higher rise of the individual parts of the uplift field and thus create plateaus and stepped lands. Uplifts near the fracture margins could have been raised episodically; as then, the slow secular movements could be interrupted frequently by stronger instantaneous movements and be supported in their effects by them" (Naumann, 1850, p. 398). Although similar to those of Dana's, Naumann's ideas were derived independently of Dana and were based on those of Élie de Beaumont (with frequent acknowledgments), which further underlines the unoriginality of Dana's ideas in global tectonics.

Figure 59. Carl Friedrich Naumann (1797–1873), whose textbook *Lehrbuch der Geognosie* exercised a strong influence on tectonic studies in the third quarter of the nineteenth century. From the *Leipziger Illustrierte Zeitung*, v. 53, no. 1368 (18 September 1869), p. 221.

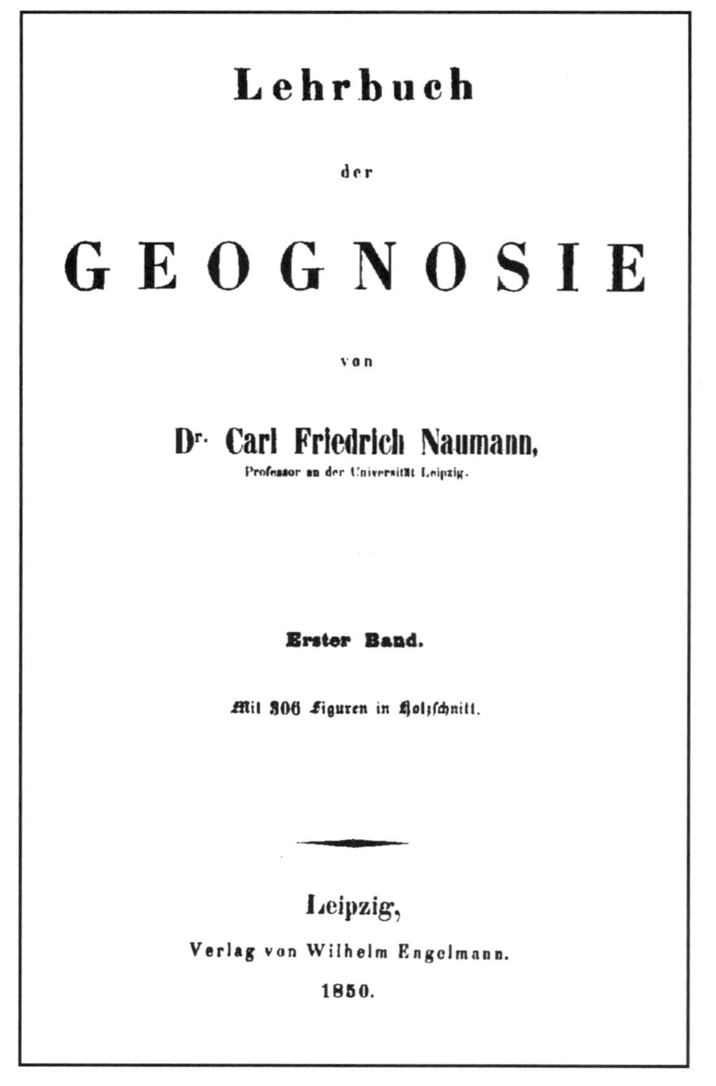

Lehrbuch

der

GEOGNOSIE

von

Dr. Carl Friedrich Naumann,

Professor an der Universität Leipzig.

Erster Band.

Mit 306 Figuren in Holzschnitt.

Leipzig,

Verlag von Wilhelm Engelmann.

1850.

Figure 60. Title page of the first edition of Naumann's influential *Lehrbuch der Geognosie*, volume I.

When the first half of the nineteenth century closed, and when Sir Charles Lyell read to the Geological Society of London his presidential address outlining the incredible success geology had in establishing all the (currently used) world-wide stratigraphic systems of the Phanerozoic Eon (with the singular exception of the Ordovician) and the victory of an actualistic view of the geological past of our planet, global tectonics still retained a touch of catastrophism despite the best efforts of the illustrious speaker. It was generally admitted (despite Élie de Beaumont) that all tectonism was a consequence of vertical uplift caused by igneous intrusion. However, the rise of mountain belts, the growth of volcanoes, and the occurrence of earthquakes were believed to be fast and episodic affairs affecting limited areas (or at least those areas with a limited width), expressed in terms of tens to hundreds of kilometers. By contrast, slow and continuous motions, called secular, affected large, commonly equant areas of horizontal dimensions measurable by thousands of kilometers. Whereas the first group generated structural amplitudes commonly in excess of several kilometers, the secular motions generally formed structures whose amplitudes were usually about a kilometer or two. The first group of structures were irreversible. Secular movements were generally believed to create reversible structures. Uplifts in one place were believed to be compensated by subsidences elsewhere, and this picture itself was thought to be reversible. By 1850, many geologists had come to acknowledge the influence of the secular cooling of the globe on its tectonic life, but most denied that this alone could explain all the phenomena (e.g., Naumann, 1850, p. 399). Three tiny volumes published only two years into the second half of the century by Élie de Beaumont did not change this recognized repertory of structures, but these publications did change the causal interpretation of these structures in a way that was to affect tectonics for more than a century to come—possibly longer.

Notice sur les Systèmes de Montagnes and the Transition to Modern Tectonics

There is no need here to review the controversy brought about by the clashing ideas of Lyell and Darwin on one side and Élie de Beaumont on the other in the mid-century (see Greene, 1982; Şengör, 1982a, 1982b, 1991c). The idea that vertical uplift motivated by the internal heat of the planet was responsible for all deformation became popular—despite Élie de Beaumont's cogent arguments for, and Dana's seductive portrayal in his *Manual of Geology* of, transversal shortening across mountain ranges—and entered textbooks everywhere, adorning many a colorful "ideal section" across the earth's crust published around that time (Fig. 61). Élie de Beaumont (1852) showed, in his most influential yet least accessible (cf. Suess, 1904, p. iv) publication, the much-quoted but very-seldom-read *Notice sur les Systèmes de Montagnes* (Fig. 62)[239], how terribly misleading all such idealized cross sections could be.

The key-note of the *Notice* is that the earth is thermally contracting and that all its major surface features, such as continents and oceans or mountain ranges and basins, are a result of this contraction. The contraction expresses itself in two main structure types. One is what Élie de Beaumont has called *bosselement* [240], a sort of tumescence of the crust (which Élie de Beaumont {1852, p. 1237} thought was less than 50 km thick) but used by its inventor in the sense both of an outcurvity and an incurvity. These *bosselements* form when the outer, already refrigerated crust becomes too large for the still contracting hotter inside of the earth. Élie de Beamont rightly pointed out that the crust was far too weak to support itself as a dome above a void between the spherical interior and an up-arched crustal segment. (Later, many believed that he had thought exactly the opposite and criticized him for it, showing how rarely the *Notice* was actually read). He argued that the sphere turned into a spheroid and adapted itself to the *bosselements*. However, the very formation of the *bosselements* threw certain parts of the crust into compression and others into extension. The structures, to which these stresses might give rise, were Élie de Beaumont's principal concern. Nobody has yet given a concise and accurate account of his ideas, and neither can I do so here without deviating from my purpose.[241] All I present below is a *résumé* of his ideas on the nature of the *bosselements* representing the falcogenic elements in his theory.

Élie de Beaumont had presented a brief account of his ideas on *bosselements* already in the second volume of the explanatory text of the geological map of France (Dufrénoy and Élie de Beaumont, 1848, p. 605–621). While describing the Jurassic deposits of the Paris Basin, Élie de Beaumont noted that the deposits all showed a remarkable continuity across the basin, as revealed by the peripheral outcrop belt and the few available boreholes. The sediments thickened towards the interior of the basin from a feather-edge around it. The fossils (such as oysters, *Gryphea*, and *Exogyra*) showed that the sediments were not deposited in water depths exceeding 100 m. The Jurassic sediments were all laid down on surfaces convex upwards owing to the spherical shape of the earth (Fig. 63). The sub-Jurassic surface itself within the basin, which Élie de Beaumont estimated to lie "at many hundreds of metres deep,"[242] must have been convex upwards throughout the deposition of the Jurassic sediments. He appealed to the weight of the sediments as a mechanism for depressing the crust (Dufrénoy and Élie de Beaumont, 1848, p. 611) but left it unclear in this particular text how the pre-Jurassic sedimentary rocks had been deposited.

The depression of a convex piece of the crust must lead to shortening; Élie de Beaumont pointed out that this shortening was generally taken up by the uplift of the basin margins, not by shortening within it. He argued that these "up-bends" would exploit old structures in the crust, while their orientations would deviate from those dictated purely by elastic bending. The whole structure would then fall prey to further mountain-building compressions. He had hoped to discuss this question more fully when describing the Cretaceous sedimentary rocks of the Paris

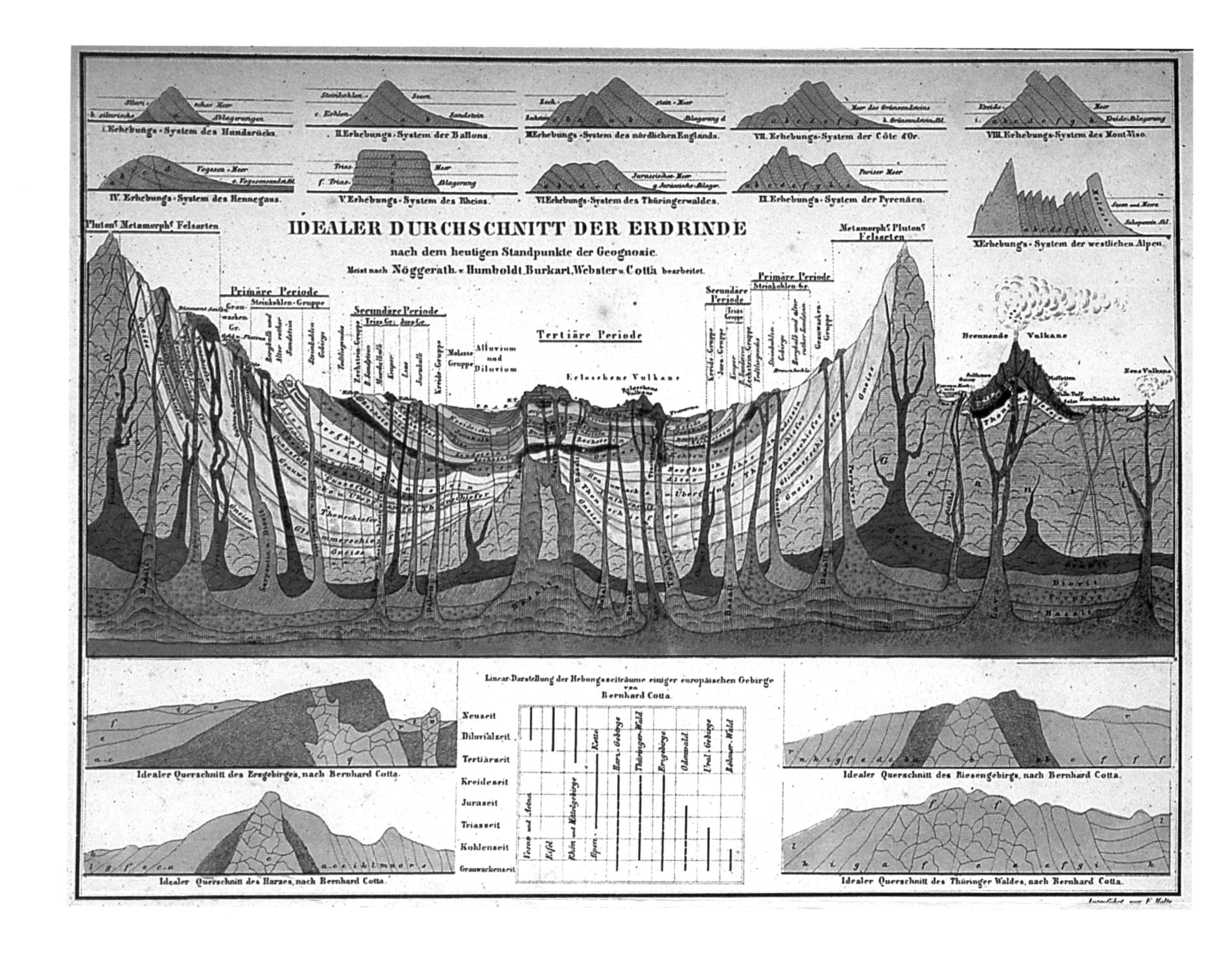
I. Erhebungs-System des Hundsrücks.
II. Erhebungs-System der Ballons.
III. Erhebungs-System des nördlichen Englands.
VII. Erhebungs-System der Côte d'Or.
VIII. Erhebungs-System des Mont-Viso.
IV. Erhebungs-System des Hennegaus.
V. Erhebungs-System des Rheins.
VI. Erhebungs-System des Thüringerwaldes.
IX. Erhebungs-System der Pyrenäen.
X. Erhebungs-System der westlichen Alpen.
IDEALER DURCHSCHNITT DER ERDRINDE
nach dem heutigen Standpunkte der Geognosie.
Meist nach Nöggerath, v. Humboldt, Burkart, Webster u. Cotta bearbeitet.
Plutonˢ Metamorphˢ Felsarten
Primäre Periode
Steinkohlen-Gruppe
Secundäre Periode
Tertiäre Periode
Alluvium und Diluvium
Erloschene Vulkane
Metamorphˢ Plutonˢ Felsarten
Brennende Vulkane
Neue Vulkane
Granit
Diorit
Porphyr
Basalt
Gneiss
Glimmerschiefer
Thonschiefer
Linear-Darstellung der Hebungszeiträume einiger europäischen Gebirge von Bernhard Cotta.
Neuzeit
Diluvialzeit
Tertiärzeit
Kreidezeit
Jurazeit
Triaszeit
Kohlenzeit
Grauwackenzeit
Vesuv und Aetna
Eifel
Rhön und Mittelgebirge
Alpen
Kette
Harz-Gebirge
Thüringer-Wald
Erzgebirge
Odenwald
Ural-Gebirge
Böhmer-Wald
Idealer Querschnitt des Erzgebirges, nach Bernhard Cotta.
Idealer Querschnitt des Harzes, nach Bernhard Cotta.
Idealer Querschnitt des Riesengebirgs, nach Bernhard Cotta.
Idealer Querschnitt des Thüringer Waldes, nach Bernhard Cotta.

Figure. 61. Ideal cross section of the earth's crust as depicted in the *Atlas* (published by Bromme, 1851, plate 8) to accompany von Humboldt's *Kosmos*. The cross section is based mostly on the work of such mid-century workers as Nöggerath, von Humboldt, Burkart, Webster, and von Cotta. Note the extreme exaggeration making all structures look as if their amplitudes are on the same order of magnitude as their wavelengths. This is what most upset Élie de Beaumont.

Basin (Dufrénoy and Élie de Beaumont, 1848, p. 621), but, regrettably, only a tiny descriptive summary of the Cretaceous rocks of France appeared in the third volume of the *Notice Explicative* from the pen of Dufrénoy. It was published in 1873, 16 years after Dufrénoy's death and only one year before the demise of his great friend Élie de Beaumont himself! In the *Avertissement* of the second volume of the *Notice Explicative*, Élie de Beaumont and Dufrénoy pointed out, repeating what Élie de Beaumont had already said in 1828 concerning the Alps, that both in the French Alps and in the Pyrenees, sedimentary sequences became thicker and more complete towards the interior of the mountains, exactly as they did in the Paris Basin towards the interior of the basin (Dufrénoy and Élie de Beaumont, 1848, p. X). As I said above and also elsewhere, this was as good a geosynclinal theory as any (Şengör, 1998).[243]

However, Élie de Beaumont did even better in the *Notice sur les Systèmes de Montagnes*. He pointed out that, on the face of the earth, small mountain chains commonly alternate with basins within the area of semicircular "slices" having a maximum width of some 2000 km. (resembling melon slices, which he called *fuseau* {spindle} on account of their map shape; see Fig. 64. The term *fuseau* continued to be used in tectonics for the same purpose long after Élie de Beaumont's ideas had become obsolete: see Charles Jacob's translation of Suess' "*grosser meridionaler Ausschnitt*" as "*grand fuseau méridien*": Suess, 1911a, p. 694). Some of these basins were truly concave upwards (in Dufrénoy and Élie de Beaumont, 1848, p. 616, he had shown that basins with diameters not larger than 100 km could be concave upwards), but the big ones had to be convex upwards. Seas, gulfs, lakes, and river basins commonly occupy such basins (Élie de Beumont, 1852, p. 1258)[244]. To illustrate their overall character, he took one of the largest—and at the time most popular, owing to von Humboldt's visit—the Pre-Caspian depression (de Humboldt, 1843, p. 311). He noted that the Pre-Caspian depression had a diameter of 841,044 m, and near Astrakhan (i.e., near its geographical center), had a depression of -24.75277 m. (Élie de Beumont was much given to mathematical precision, apparently without much regard to its significance.) This depression, he calculated, caused a shortening along the diameters of one in two million! However, shortening was not uniform over the entire surface of the basin. Near the center it was zero; it rose to a maximum towards the margins[245]. Élie de Beaumont argued that this meant much of the shortening was concentrated in small areas and, in such areas, the ability of rocks to absorb the shortening without yielding would be overcome. So "when the limit of yielding is reached in one place owing to the unequal distribution of the dimunition of the surface imposed by the *bosselement*, the yielding material surges to the surface in the form of mountains" (Élie de Beaumont, 1852, p. 1271). Here we have a description, much like the one he had given earlier of the Paris Basin, of first a depression forming, which then fails in a certain pre-destined place—where there is a maximum of shortening—to create a mountain chain. If we remember that Dana had at least known about Élie de Beaumont's book at the time he wrote his famous 1873 paper[246], we could perhaps better appreciate the roots of the idea of geosynclines. Further, Élie de Beaumont was careful to point out that the *bosselements* left intact the visible fabric of the rocks affected: "Such a feeble inflexion cannot dislocate, disrupt, or even tilt in a way perceptible to the eye, the sedimentary beds

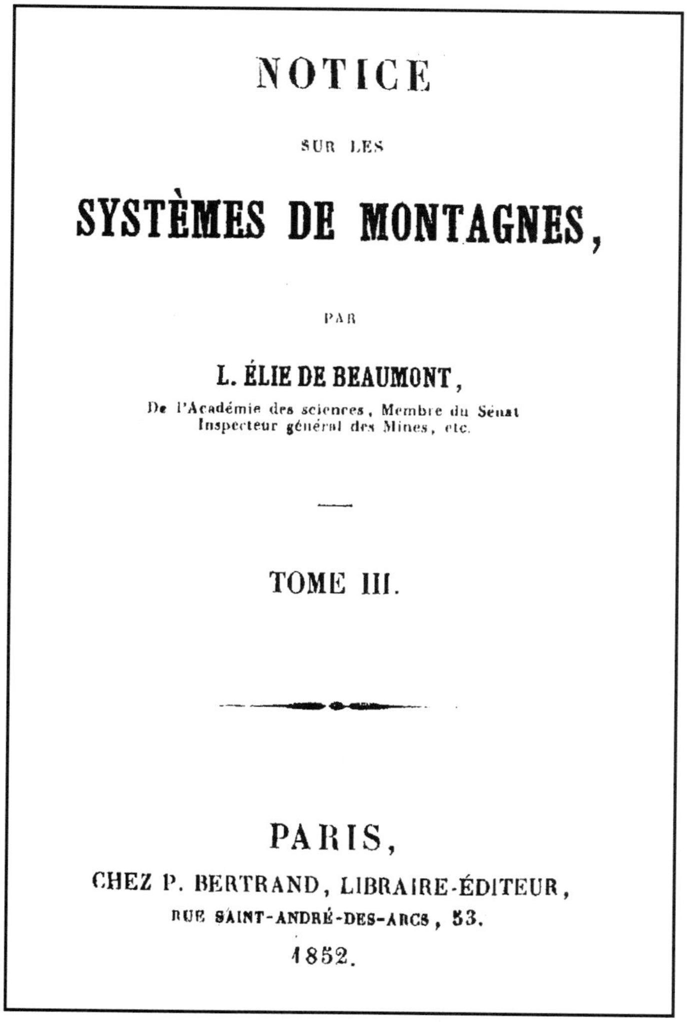

NOTICE

SUR LES

SYSTÈMES DE MONTAGNES,

PAR

L. ÉLIE DE BEAUMONT,

De l'Académie des sciences, Membre du Sénat
Inspecteur général des Mines, etc.

TOME III.

PARIS,

CHEZ P. BERTRAND, LIBRAIRE-ÉDITEUR,
RUE SAINT-ANDRÉ-DES-ARCS, 53.
1852.

Figure 62. Title page of the third volume of Élie de Beaumont's (1852) *Notice sur les Systèmes de Montanges*, in which he discussed the difference between large-wavelength/low-amplitude and small-wavelength/large-amplitude structures.

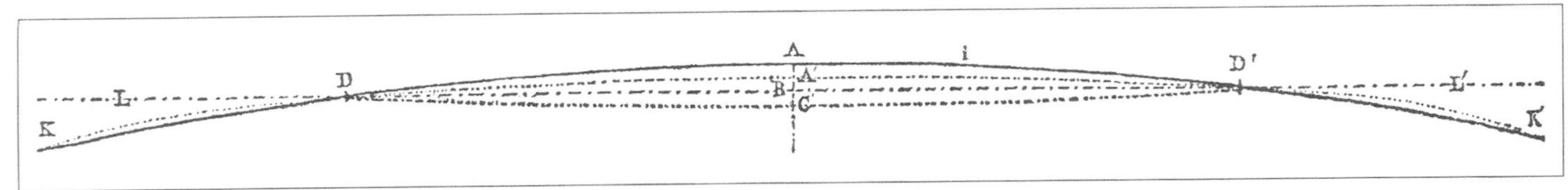

Figure 63. Origin of a concave basin on the convex surface of the earth according to Élie de Beaumont, as illustrated on the example of the Jurassic Paris Basin (from Dufrénoy and Élie de Beaumont, 1848, p. 614). In order to create a true concave surface (DCD′), the amount of subsidence must exceed AB. The arc (KDAD′K) of a circle represents the surface of the earth. When the middle point of this arc is depressed (from A to A′), its margins are raised elastically (KLDA′D′L′K′). In sharp contrast to James Hall a decade later, Élie de Beaumont stressed that the length of the line segment KLDA′D′L′K′ could not be much longer than that of KDAD′K′. The amount of uplift from L to L′ is dependent on the amount of depression to A′. Élie de Beaumont emphasized that the uplifts forming the margins of the basin "would naturally be located where the crust of the earth offers the least resistance to uparching, for example, along lines which had been axes of uplifts before the deposition of the Jurassic" (Dufrénoy and Élie de Beaumont, 1848, p. 619–620).

extending on the surface" (Élie de Beaumont, 1852, p. 1280). He criticized the common topographical and geological sections, vertically exaggerated many times, for giving a false impression of the extremely gentle structure of the *bosselements*; every basin drawn in such sections appeared to have steep margins, which gave the viewer the wrong impression that they were fault-bounded (Fig. 61). Such structures, if depicted at proper scale, would hardly be visible even on a model globe (Élie de Beaumont, 1852, p. 1314). That was one reason why it was so hard to recognize and to study them. Despite that, such structures were of immense importance. They showed the *general mobility of the crust of the earth*: "These phenomena, large and little pronounced, less clear, less easy to grasp and to study than an unconformity of stratification or the structure of a mountain chain, but despite that appear very worthy of attention, as offering a proof of the general mobility of the crust of the earth and as an almost certain index to its thickness[247] and to its flexibility" (Élie de Beaumont, 1852, p. 1289–1290).

Élie de Beaumont clearly separated the two major structural families of the earth's crust:

> Two different phenomena in the inflexions of the earth's crust are superposed and their effects mixed up, similar to the small ondulations on the surface of a liquid. On the one hand the general *bosselemets* owing to the excess area of the crust, which are the causes of the new ridges into which the thin crust now and then contracts. On the other, the more or less pronounced curvatures of these ridges themselves, the formation of which accompanies the formation of the mountains. (Élie de Beaumont, 1852, p. 1296).

With hindsight, we can see that Élie de Beaumont separated the large-wavelength/small-amplitude, reversible structures that generated no visible deformation on the outcrop and that represented the general mobility of the crust, from the small-wavelength, irreversible structures that crushed and folded the rocks and that formed not continuously but only now and then, when the yield strength of the affected rocks was attained. The amazing thing about this summary of Élie de Beaumont's views is that it would have been equally applicable to Hans Stille's views in 1960! The form that views on falco-

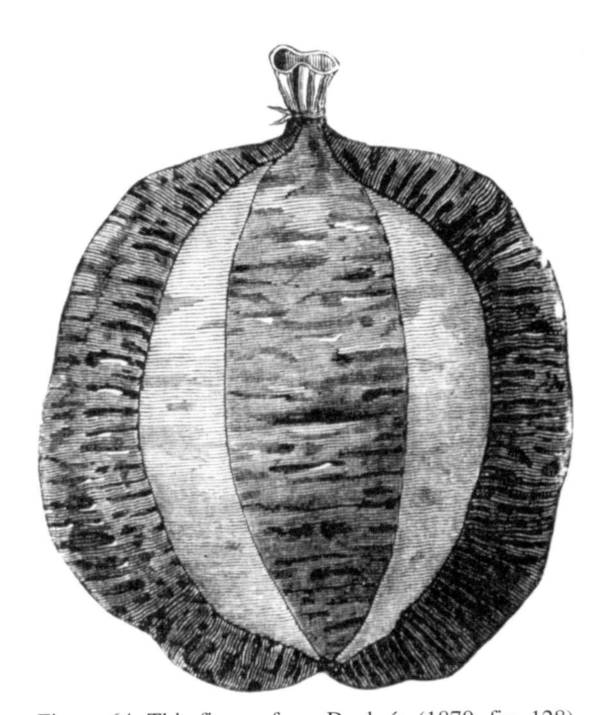

Figure 64. This figure, from Daubrée (1879, fig. 128), illustrates an experimental production of *fuseaus* using a shrinking baloon, on which contractable and non-contractable materials were made to alternate along meridional lines. These semicircular "slices" allegedly correspond with those of a maximum width of some 2000 km on the face of the earth, along which (according to Élie de Beaumont) small mountain chains commonly alternate with basins. Élie de Beaumont thought that the *fuseau* had to form as a consequence of the shrinking of the planet along great circles *and* because displacement could not be transferred from one great circle to another (because at the time, strike-slip faults were neither hypothesized nor yet recognized—thus, Élie de Beaumont's *fuseaus* had to form because he could not conceive what is in essence transform faults in a contractionist framework!). Different episodes of contraction employed differently orientated *fuseaus* on the face of the earth, the collective traces of which formed his famous "pentagonal network" (*réseau pentagonal*: see Élie de Beaumont, 1852 *passim*).

genic and copeogenic events were to take towards the end of the nineteenth century and in the first half of the twentieth century was thus given to them by Élie de Beaumont. Moreover, these views were based on still older ones that generally took shape in the 1830s (e.g., Reboul, 1835, p. 69–70):

> The consolidation of the earth's crust, the establishment of the waters that descended on this cooled crust, the rise of isles and continents above the surface of this universal aqueous envelope are the grandest events of the history of the earth. Everything functioned in terms of what its logic prescribed. Everything required and occupied a long time. The revolutions that resulted were slow. Only local accidents were sudden.

These ideas found ready acceptance among Élie de Beaumont's contemporaries. Zimmermann, for example, in his popular (and otherwise very anti-French!) book on historical geology, wrote "That plateaux originated by slow uplift is so little doubtful as that mountains are built by violent eruptions. For both of these we have sufficient facts before us in the present. The coasts of Scandinavia and Chile rise continuously. For the rapid uplift of mountains we shall provide most convincing proofs later while dealing with volcanism" (Zimmermann, 1861, p. 391–392).

Eduard Suess effectively bypassed these ideas because he denied the existence of any primary vertical uplift of the lithosphere. The pre-Suess ideas were resurrected especially by Émile Haug and Hans Stille at the beginning of the twentieth century and remained dominant until the 1950s. It is surprising how the role Élie de Beaumont played in the recognition and delineation of falcogenic structures has been so little appreciated by geologists and historians of geology.

CHAPTER XI

THE REINVENTION AND CHRISTENING OF THE CONCEPT OF GEOSYNCLINE IN AMERICA

The European Birth of the Concept of Geosyncline: Recapitulation

As the discussion of the work of Élie de Beaumont in the preceding pages has clearly shown, the idea of a preparatory trough, filled with sediments, mostly pelagic in environment, and coincident with the future lie of a mountain-range has been explicit in his writings since 1828. Already in 1828, he forcefully stated that the extra-mountain parts of a continent have thinner sedimentary covers that are not dislocated and that differ from the much thicker and highly dislocated sediments, generally having different, more pelagic, facies within mountain ranges. He elaborated, in succesive iterations, on the mechanism of the formation of such troughs under the influence of tangential stresses derived from the secular thermal contraction of the earth. Those iterations were in 1829, in his Oisans paper, then in 1848, in the first edition of the geological map of France (second volume of the explanatory text, which he wrote in collaboration with his friend Dufrénoy), and finally in 1852, in great detail, in his *magnum opus*, *Notice sur les Systèmes de Montagnes*. Élie de Beaumont initially had written only about the weight of the accumulating sediments in the trough as being responsible for the subsidence, but already in 1829 he implied, and in 1848 explicitly stated, the role of thermal contraction of the globe in creating the pre-orogenic troughs. He thought that these long-wavelength/small-amplitude undations of the crust formed slowly but were destroyed swiftly during a catastrophic collapse of the trough owing to accumulated contraction-related stresses. He gave examples of both now-destroyed (e.g., Alps, Pyrenees) and extant and active (Paris Basin, North Caspian Depression) examples of such troughs destined to develop into mountain ranges.

We saw further that all of these ideas were known to the leading American geologists of the mid-nineteenth century who were influenced by them—especially James Dwight Dana. Tidings of the related ideas of Herschel and Babbage also reached the New World and no doubt amplified (and, in part, complemented) the influence of Élie de Beaumont's much more comprehensive ideas.

James Hall

It is in the light of this background knowledge that one must approach the rise of the concept of geosynclines in America.[248] The man who precipitated it, James Hall (1811–1898; Fig. 65) of the New York Geological Survey, was a palaeontologist and stratigrapher[249]. He was familiar with the European ideas but did not seem to have the necessary background in the physical sciences or in mathematics to understand those that pertained to tectonics. His involvement in the tectonic questions relating to mountain belts came strictly as a consequence of his important stratigraphic discoveries in the State of New York and subsequently in the American midwest in the forties and fifties of the nineteenth century.

Already in the spring of 1841, Hall toured the states of Ohio, Indiana, Illinois, a part of Michigan, Kentucky, and Missouri, and the then still-territories of Iowa and Wisconsin with the purpose of comparing the Paleozoic succession there with that which he had become familiar with in the east, in New York (Hall, 1842; 1843, p. 500–515; also see the table of correlation on p. 519; Schuchert, 1918[1973], p. 87). For the first time, Hall became aware that many of the formations that he knew from New York greatly thinned out, and some even entirely disappeared westward (Hall, 1843, p. 500; he had been alerted to some of the similarities and equivalences and informed of the localities where these could be seen in the west by Lardner Vanuxem {1792–1848}; Hall, 1843, p. 500; Merrill, 1904, p. 292, footnote *a*; 1924, p. 123, footnote 25). Concomitantly, clastic material also diminished, and some clastic formations of the east became replaced by limestone equivalents in the midwest U.S. (Hall, 1842, especially p. 62; 1843, p. 515). We know from a letter Hall wrote to Alexander Dallas Bache (the superintendent of the U.S. Coast and Geodetic Survey and the great-grandson of Benjamin Franklin) on 14 January 1852, that he was interested not only in establishing the stratigraphy of the Paleozoic rocks in New York and farther west, but he was also intent on understanding the message of the rocks he was mapping and correlating, and, of the the fossils they contained, con-

Figure 65. James Hall (1811–1898) of the New York Geological Survey in 1856, at the time when he was developing his version of the geosynclinal hypothesis.

cerning the physical conditions of the past. He was interested in the geography of the past and in the forces responsible for changing that geography from what it had been to what it is now. He wished to know:

> positively what changes were taking place in the bed of the ocean during these apparently consecutive and continuous formations. This point, taken in connexion with the fauna of the successive periods, would be a subject of very great interest. We wish also to know, whether the force that has apparently uplifted all our formations, has acted equally and simultaneously over the whole, or whether some parts of our sedimentary deposits may have, in the course of uplifting, suffered undulatory movements in the time of strike. This knowledge would I am sure, give us a clue to the explanation of many physical phenomena which have occurred since the elevation of our strata, and which in the absence of knowledge, we fail to understand and properly appreciate. (from a letter in the Rhees collection, Huntington Library; quoted from Daniels, 1968[1994], p. 24)

So, in his further studies, Hall kept such palaeogeographic and tectonic questions in mind, as well as his main palaeontological and stratigraphic interests. (For a graphic summary of Hall's studies of the geological formations of New York State and of the mid-western region in 1841, see his colored geological map and cross-sections in Hall, 1843).

Hall was called in 1855 to head the Iowa Survey, to employ his skills as a palaeontologist to help establish the Paleozoic stratigraphy in the Midwest (Dott, 1985, p. 162–164). While working in Iowa, he corroborated his earlier observation concerning the greatly reduced thickness and the incompleteness of the Paleozoic section compared with that in New York. What he knew to be kilometer-thick, nearly continuous sequences in New York appeared in Iowa as at most a few hundred meter-thick sections having numerous gaps in the succession: "In tracing westward the geological formations as known in New York and Pennsylvania, we find them, with one or two exceptions, gradually becoming thinner, until at last several of them are scarcely recognisable, or are so attenuated as to be overlooked in a country deeply covered by modern deposits" (Hall, 1858, p. 38).

From this simple, but important observation, Hall derived an extraordinary conclusion:

> This remarkable fact of the thinning out westwardly of all the sedimentary formations, points to a cause in the conditions of the ancient ocean, and the currents which transported the great mass of materials along certain lines which became the lines of greatest accumulation of sediments, and consequently present the greatest thickness of strata at the present time. It is this great thickness of strata, whether disturbed and inclined as in the Green and White mountains and the Appalachians generally, or lying horizontally as in the Catskill mountains, that gives the strong features to the hilly and mountainous country of the east, and which gradually dies out as we go westward, just as in proportion as the strata becomes attenuated.
>
> The subdued features of the West are therefore due, not alone to the absence of great disturbing forces, but to the absence or the great tenuity of the formations, or paucity of materials or strata to be disturbed. The thickness of the entire series of sedimentary rocks, no matter how much disturbed or denuded, is not here great enough to produce mountain features; and the most elevated portions of the region are those where no disturbing force essentially affecting the horizontality of the strata has acted.
>
> Thus it would appear that the height of these mountains is not due to upheaval from beneath, or to the folding and plication of beds; but that the dislocations of the strata and consequent denudation render the elevation always less than it would have been made by the actual thickness of the strata, had they remained undisturbed and piled upon each other in their horizontal condition, and these subjected to the same denuding agencies. (Hall, 1858, p. 41–43)

In 1857, Hall addressed the American Association for the Advancement of Science as its President at its 31st meeting in Montreal. Two brief summaries of his address (published by others anonymously, but cited as Hall {1857a} and Hall {1857b}, below) appeared in the same year, and they both reported that Hall had impressed upon his audience that previously little attention had been paid to the material making up the mountains. He showed that the sedimentary strata making up the Appalachian chain was thickest in the mountains and thinned out westwards where mountainous topography also diminished gradually. He attributed the great thickness of sediments along the Appalachian line of trend to the action of currents bringing material from the northeast. Hall's emphasis on the role of major northeast- to southwest-directed currents in carrying sediment and influencing the physical geography of the sea bottom was probably influenced by his countryman Horace Henry Hayden's[250] book on the influence of marine currents especially in creating the great alluvial plain skirting the North American continent to the southeast and south (Hayden, 1820, especially his Preface, and ch. I and VIII). Hall denied that any folding or upheaval could be responsible for the origin of the high topography, which he believed was solely caused by the thick accumulation of sedimentary rocks (Hall, 1857a, 1857b).

The full text of his address was published, essentially unaltered, 26 years later (Hall, 1883). Its contents were much expanded in his more famous writing (Hall, 1859) published as an introduction to the third volume of the *Palæontology of New York*, and that is why I do not discuss the 1883 publication here. Hall took the opportunity of the later publication, however, to respond to what he called a "facetious" criticism by Dana (Hall, 1857, p. 68). I shall discuss that response after I present Dana's criticism below.

The most detailed statement that Hall ever published concerning his tectonic views about what we might term anachronistically "the geosyncline question" is in the Introduction to the third volume of his monumental *Palæontology* volumes of the *Natural History of New York*. In that Introduction, he reviewed all the sedimentary formations of New York, and almost one-by-one he compared them with their continuations and/or equivalents in the midwestern states (Illinois, Iowa, Missouri, Michigan, Kentucky, and Tennessee) and Canada. His comparisons were based in part on his personal observations (in Illinois, Wisconsin, and Iowa), in part on his extensive communications with the workers in these states (some of whom had sent him fossil material to be described), and also in part on the basis of the literature (see Dott, 1985, for a list of the people with whom Hall was in contact).

Hall's basic philosophy of geology was much influenced by his Troy, New York, teacher, Amos Eaton (Clarke, 1921, especially p. 24–40; also see Friedman, 1979, especially p. 4) and the reading he later did professionally. Hall specifically cites as having influenced his thinking Sedgwick and Murchison in stratigraphy and Sir James Hall, Sir Henry de la Beche, Sir Charles Lyell, Babbage, Herschel, Hopkins, Martin, Fitton, Weaver, Dumont, Studer, and "others" (Hall, 1859, p. 81). Among the "others" was Élie de Beaumont, as we know from Hall's discussion of Élie de Beaumont's ideas on the directions of mountain chains in note A to the Introduction in the third volume of the *Palæontology of New York* (Hall, 1859, p. 86–87).

Hall commenced the discussion of the Paleozoic rocks of New York with the Potsdam sandstone (now known to be Upper Cambrian[251]). He immediately emphasized two points: (1) the sediment is of shallow-water origin, indeed littoral in facies; and (2) we know this by actualistic analogs: "… the shells are broken and comminuted, and are drifted together precisely in the same manner as we find seashells upon a modern beach" (Hall, 1859, p. 2). Hall kept these two emphases as he continued his descriptions up the section into younger deposits. Up to the time of the deposition of the Trenton limestone (Upper Ordovician), the units between New York and the midwest U.S. (Iowa, Wisconsin, Minnesota) correlate almost one-to-one (with the exception of the Galena Limestone of the midwest U.S. which does not show up in New York), and there are no serious thickness changes (Hall, 1859, p. 8–9). Beginning with the Trenton time, Hall claimed, accumulations became thicker in the east than in the west (p. 50). But earlier, (e.g., Hall, 1859, p. 12), he had emphasized that already the Trenton thinned westward, as did the other limestone units beneath it. In his 1857 address, Hall had emphasized that only the Potsdam sandstone was deposited under completely equable conditions: "At no period [*as during the Potsdam time*] has deposition been so uniformly diffused or animal life so widely distributed in the same forms over so huge a proportion of our continent" (Hall, 1883, p. 41). Hall also noted that westward from the Green Mountains, both the deformation and the metamorphism diminished and eventually disappeared (Hall, 1859, p. 16). The Hudson-river group of Hall (Upper Ordovician clastic rocks, mainly shales and sandstones of the Utica, Pilaski, Oswego and Queenston formations) also showed dramatic thinning westward and "accumulated the immense amount of its materials" towards the east. (In his 1857 address, Hall had emphasized that the "form and outline of our present continent were determined" at the Hudson-river group time: Hall, 1883, p. 41). He also noted that its clastics were fed from sources that were to the northeast and east:

> We have been accustomed to look to the northeast for the source of the sedimentary materials of this group; and to regard some part of the present North Atlantic ocean bed as having been occupied by land, the destruction of which furnished the sedimentary materials for this formation. We are scarcely prepared, therefore, for the information which comes as a result of investigations in the Canada Survey, that while this source may have been to the northeast from us in New York, and far beyond the limits of our explorations, it lies in a direction more to the east than we have been accustomed to believe. (Hall, 1859, p. 21[252])

It was already recognized by some in the thirties of the twentieth century that what Hall was talking about as the preparatory troughs of the mountain ranges were nothing more than foredeeps that developed in front of moving Taconic thrust sheets (i.e., were part of the mountain-building process itself; cf., Suess, 1937, p. VI). In fact, Hall's Hudson-river group is nothing more than the distal equivalents of the clastic wedges consisting of the Normanskill and the Illinois Mountain Formations shed from the advancing nappes (cf. Bird and Dewey, 1970, fig. 5; Colton, 1970, fig. 2).

Hall interpreted the abundance of clastics as being carried from the northeast by a southeastwardly flowing current and as defining the edge of the continent to the west. Farther west, the deposition of very fine-grained clastics ("finer mud": Hall, 1859, p. 20) suggested to him that the effects of the current diminished westward. The clastics became so sparse as to allow the incipient growth of coral reefs. "Thus from the St. Lawrence on the north, through the Appalachian chain, the coarse sandstones and conglomerates indicate the close of this period; while the same geognostic line, from the northern side of Lake Huron, by the course of the Cincinnati axis, quite to the center of Tennessee and still farther to the south, is marked by bands of coral limestone" (Hall, 1859, p. 20–21). So, by the close of the Ordovician Period (using our present-day terminology), Hall had observed the formation and filling of a trough that paralleled the future trend of the Appalachian orogen. Demarcating this clastic trough from the continent to the west was a string of coral reefs that adorned the Cincinnati axis.

Hall further assumed that since the materials came from the northeast and east, the present-day North Atlantic ocean must have been a continent (Hall, 1859, p. 21). This assumption gave him the vision of a trough lying between two continents (Hall, 1859, p. 22).

The Hudson River group was succeeded by the Medina Sandstone and the Clinton group (dolomites, shales, and sandstones), but Hall noted no significant changes in thickness westwards in them in his report. (In his 1857 address, he had pointed out that they "die out in the direction of the Great Lakes on the westward and scarcely have a vestige in the Ohio valley; while they are, like the Hudson-river group, much more persistent in the direction of the Appalachian chain": Hall, 1883, p. 45; see Colton, fig. 2 for the current interpretation: Hall was essentially right.) He did note, however, before the deposition of the next package of rocks, that for the Helderberg group of calcareous and argillaceous mudstones, the strata deposited previously had become deformed and an unconformity formed between the Helderberg group and the rocks underlying it (Hall, 1859, p. 39). This furnished Hall with one piece of evidence to lead him to believe that the sinking in his sedimentary trough was episodic and not uniformly continuous (see Hall, 1859, p. 70, last paragraph). In his 1857 address, he had used this as evidence to show that deformation and deposition were essentially continuous.

Hall noted that the Upper Helderberg group (Lower Devonian) also thinned westward. So did the succeeding Hamilton group (Middle Devonian). Hall estimated (rightly: Colton, 1970, fig. 2) that the Hamilton and the higher Chemung groups have an aggregate thickness of 3000 ft.(~900 m) in eastern New York, where they had been defined, and probably "much more" (Hall, 1859, p. 48). Their equivalents in Indiana, Illinois, and Michigan had barely 200 ft. (~60 m) total thickness (cf. Colton, 1970, fig. 2). In the Devonian, Hall found the greatest facies changes between eastern and western New York. The eastern facies was coarser clastic, whereas the western facies had mainly shales and limestones (cf. Boucot, 1968, fig. 6-4, 6-5). The fossil content also betrayed a difference in the eastern and the western parts of the depositional environment (Hall, 1859, p. 45 ff.).

Now Hall turned his attention to the metamorphic series making up the Green and White Mountains and concluded that they also represented altered sediments of lower and middle Paleozoic age. The Green Mountains he thought to consist of Silurian rocks (Hall, 1859, p. 50; we now know them to be Grenville basement rock but containing metamorphosed rift and continental margin deposits not entirely dissimilar to some of the early Paleozoic sections with which Hall was familiar from the unmetamorphosed sections of the Appalachian orogen in the northeastern U.S.: see Osberg et al., 1989, p. 218). The White Mountains Hall thought were younger and at least in part correctly guessed that they must have contained mostly metamorphosed Devonian and Carboniferous rocks (Hall, 1859, p. 50; they are now known to be mostly Devonian: Rodgers, 1970, p. 106 ff.). He thus considered that the sediments of Paleozoic age "must everywhere contribute largely to the matter forming the metamorphic portion of the Appalachian chain, as well as the non-metamorphic zone immediately west of it" (Hall, 1859, p. 50). From this conclusion, Hall moved to a generalization that formed the core of his ideas concerning tectonics in general:

From the facts here stated, the student is prepared to appreciate the conclusion, that all the sedimentary formations above the Trenton limestone have had a line of greater accumulation; and that it is demonstrable, from the combined investigations of geologists, that this line was along the course of the Appalachian range. In the second place, all the observations carried on through New-York, Ohio, Indiana, Michigan, Wisconsin, Illinois, Iowa, and Missouri, show a thinning of these sediments in a westerly direction, until, in the Mississippi valley, they have greatly attenuated or entirely disappeared. (Hall, 1859, p. 50–51)

Hall then sought a connection between the thickness of the sediments deposited in any one place and the topography:

The accumulations of the Coal period were the last that have given form and contour to the eastern side of our continent, from the Gulf of St. Lawrence to the Gulf of Mexico. And as we have shown that the great sedimentary deposits of successive periods have followed essentially the same course, parallel to the mountain ranges, we very naturally inquire: What influence has this accumulation had upon the topography of our country? And is the present line of mountain elevation, from northeast to southwest, in any manner connected with this original accumulation of sediments? (Hall, 1859, p. 66)

The answer Hall gave to this question is one of the most peculiar and difficult to understand statements in the history of geology[253]. Neither his friends nor geologists elsewhere were able to understand what he really meant and yet, as a result of his statement (without acknowledgment to his predecessors) that mountain-building and thick sedimentary deposits had a genetic relationship, his name has become forever linked with one of the longest-lived generalizations in the history of geology. Few have realized that what had become accepted as the geosynclinal theory of mountain-building was (1) Dana's interpretation of Hall's observation on sedimentary thickness changes in and around the Appalachians on the basis of Élie de Beaumont's theoretical views, and (2) the "geosynclinal theory" (as it has become known) had nothing to do with Hall except the inspiration his observations and statements gave to Dana to rehash Élie de Beaumont's ideas. Dana's ability to communicate his ideas clearly and concisely in terms of a happy neologism and the great difficulty of reconstructing the pre-deformation geometries of rock packages in mountain belts until the middle of the twentieth century were the factors that won the day for the geosyncline idea.

Having enunciated the question as to the relationship between the sedimentary thickness and topographic elevation, Hall proceeded to answer it:

We are accustomed to believe that mountains are produced by upheaval, folding and plication of strata; and that from some unexplained cause, these lines of elevation extend along certain directions, gradually dying out on either side, and subsiding at one or each extremity [*so far Hall is following Élie de Beaumont's* Notice sur les Systèmes de Montagnes; *see his note A*]. In these pages, I believe I have shown conclusively that the line of accumulation of sediments has been along the direction of the Appalachian chain; and, with slight variations at different epochs, the course of the current has been essentially the same throughout. The line of our mountain chain, and of the oceanic current which deposited these sediments, is therefore coincident and parallel; or the line of the greatest accumulation is the line of the mountain chain. (Hall, 1859, p. 68)

So far, Hall's reasoning is clear and easy to understand. What makes his "theory" thoroughly unintelligible is the sentence that immediately follows the quoted sentences above: "In other words, the great Appalachian barrier is due to original deposition of materials, and not to any subsequent action or influence breaking up and dislocating the strata of which it is composed" (Hall, 1859, p. 68).

To prove his point, Hall refers to the relief in the Appalachians and to that in the valley of the Mississippi. In the Appalachians, where he thinks the total sediment thickness is some 40,000 ft. (~12 km), the mountain ranges rise to several thousand feet on either side of valleys cutting down to Potsdam sandstone. However, in the valley of the Mississippi, where he estimates the total thickness to be some 4000 ft. (~1200 m):

The same denuding action has produced low cliffs or sloping banks of one or two hundred feet in height. Therefore had the country been evenly elevated without metamorphism or folding of the strata, making the lowest palæozoic [*sic*] rocks the base line, in the States bordering the Atlantic we should have had higher mountains and deeper valleys, wherever the series was complete. At the same time, the great plateau on either side of the Mississippi river would have presented the feature it now does, of valleys extending to the Lower Palæozoic beds, with cliffs of the height represented by the actual thickness of the beds which there constitute the entire series.

The gradual declination of the country westward is due primarily to the thinning out of all the formations which have accumulated with such great force in the Appalachian region. It is also susceptible of proof, that no beds of older date have contributed to elevate the later ones, or to form a part of the mountain chain. (Hall, 1859, p. 68–69)

But Hall knew that the difference in elevation between the midwest U.S. and the Appalachians was not 36,000 ft. (~11 km). Even if it were so, as the Appalachian strata accumulated, the midwestern sea would have progressively deepened because, at any given time, the midwest accumulated a much thinner section than the Appalachian realm of deposition. But Hall repeatedly emphasized the shallow-water nature of most of the sedimentary deposits both in the Appalachians and in the midwest U.S.

But what caused the folding in the Appalachians? What influence did this folding have on the generation of topography in the Appalachians? What effects were due to metamorphism that accompanied the mountain chain?

As these questions presented themselves to him (Hall, 1859, p. 69), Hall referred to the European authorities (as he tells us later, mainly Babbage, Herschel, and Lyell: Hall, 1859, p. 81 and note E; also 1883, p. 69) to argue that thick sediment accumulation would cause subsidence. He believed that this subsidence was the answer to all of his questions concerning tectonics:

We have evidence for this subsidence in the great amount of material accumulated; for we cannot suppose that the sea has been originally as deep as the thickness of these accumulations. On the contrary, the evidence from ripplemarks, marine plants, and other conditions, prove that the sea in which these deposits have been successively made was at all times shallow, or of moderate depth. The accumulation, therefore, could only have been made by a gradual subsidence of the ocean bed; and we may then enquire, what would be the result of such subsidence upon the accumulated stratified sediments spread over the sea bottom. (Hall, 1859, p. 69–70)

Hall clearly misunderstood Herschel's physics in that he did not realize that a basin was necessary to begin accumulating enough sediments to depress the crust. But he is hardly to blame because precisely the same assumptions had been made by others much better schooled than he in the physical sciences, both before (e.g., Élie de Beaumont, 1928b) and after him (e.g., Dutton, 1882, p. 60, footnote *).

Hall further assumed that the subsidence would take the simple form of a synclinal down-sagging[254]:

This sinking down of the mass produces a great synclinal axis; and within this axis, whether on a large or small scale, will be produced numerous smaller synclinal and anticlinal axes. And the same is true of every synclinal axis, where the condition of the beds is such as to admit of a careful examination [*Hall points out in a footnote here that this idea was suggested to him by Sir William Logan*]. I hold, therefore that it is impossible to have any subsidence along a certain line of the earth's crust, from the accumulation of sediments, without producing the phenomena which are observed in the Appalachian and other mountain ranges. (Hall, 1859, p. 70)

Hall assumed a flexural slip folding mechanism for the formation of his big syncline. In a footnote, he describes it thus: "To have an idea of this folding, it is only necessary to take a package of flat sheets of paper, and hold the edges firmly in the same position and relation they had when in a horizontal position, depressing the center, and as the lower sheets assume the curved direction the upper ones will curve upwards or wrinkle" (Hall, 1859, p. 70–71, note †). Here Hall seems not to have realized that his folding paper stack analogy did not hold because layer after layer was *added* to his growing syncline as it sagged and shortened horizontally in a corresponding amount (see Fig. 66 herein). The added layers would simply sag down and stretch as more and ever shorter layers are added on top of them, thus obviating the necessity of creating any secondary folds (as Le Conte {1872a, p. 461} clearly recognized; see Fig. 67). Hall's entire orogenic picture is based on this simple misunderstanding:

This is an illustration after a different manner of the old elementary process of producing foldings in sheets of paper, as illustrative of folded strata by lateral pressure. Now, as a set of strata one or two hundred miles in width cannot slide over each other, as sheets of paper do if left to themselves during the process of depression, the beds on the lower side must either become extremely broken, or the higher portions become folded and plicated. That some fractures will take place below there can be no doubt, and these are probably such as we see filled with trappean [*i.e., basaltic and doleritic*] matter. But the greater movement would undoubtedly take place in the higher beds, which necessarily assume positions and relations as have been pointed out. This condition and movement offers, moreover, an explanation of the form of trapdykes, which are often narrower above in synclinals and on synclinal

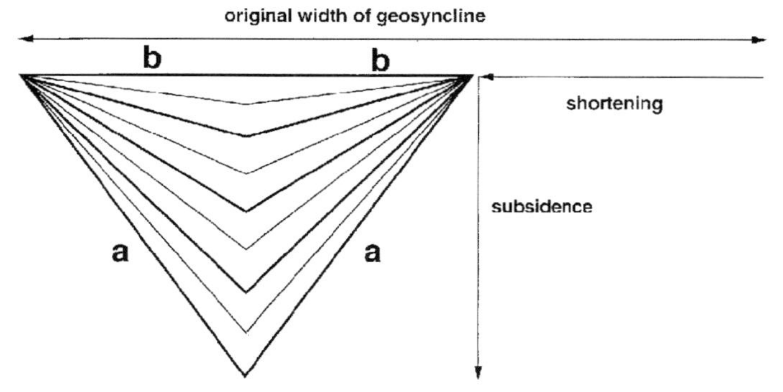

Figure 66. Geosynclinal shortening and subsidence under the weight of accumulating sediments. Every layer newly added has a shorter cross-geosynclinal width than that of the preceding one. So 2b is *originally* much shorter than 2a. Thus, geosynclinal folding owing to subsidence is not flexural-slip folding as erroneously thought by Hall. See also Figure 67.

slopes, the matter filling a fracture opened from below; while in the case of such matter penetrating an anticlinal, it would necessarily widen above from the reversed conditions attending the fracture. (Hall, 1859, p. 71, continuation of footnote † from p. 70)

Thus, Hall believed that he explained in one simple model both the formation of the great synclinal basin and the folding affecting its contents. The model was based on two misunderstandings: First, to account for the subsidence, Hall tried to use Herschel's "isostatic" model, but did not understand its physics and thus derived invalid conclusions from it. Second, Hall tried to use a flexural slip folding model to account for the folding of the contents of his subsiding synclinal basin, but he failed to see that the growing syncline as he described it could not be a flexual slip fold, but must be a generative one and would thus not generate intrados shortening. Hall evidently never bothered to draw accurately his model to see whether it would work. Based on these two misunderstandings, he triumphantly announced his conclusions:

This successive accumulation, and the consequent depression of the crust along this line, serves only to make more conspicuous the feature which appears to be the great characteristic, that the range of mountains is the great synclinal axis, and the anticlinals within it are due to the same cause which produced the synclinal; and as a consequence, these smaller anticlinals, and their corresponding synclinals, gradually decline towards the margin of the great synclinal axis, or towards the margin of the zone of depression which corresponds to the zone of greatest accumulation. (Hall, 1859, p. 71)

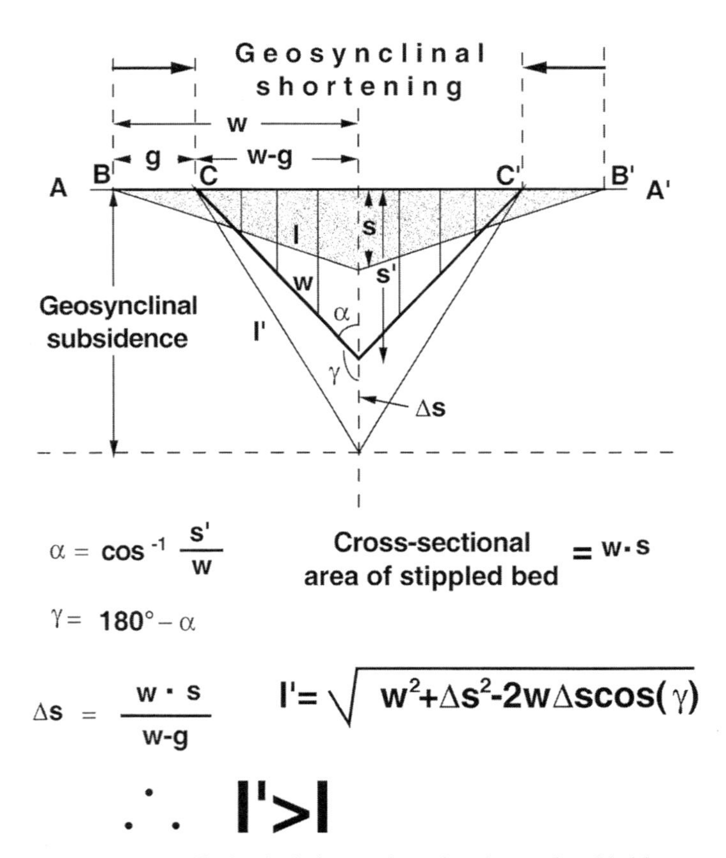

Figure 67. Simplified calculation to show that the newly added layers would be shorter than the previous one in a geosyncline that subsides and shortens as it is being loaded. (I am much indebted to my friend, Professor Michael C. Gurnis of California Institute of Technology, for his help with the calculations.)

But this was not all. Hall thought he could explain both the uplift and the metamorphism on the basis of his simple model. He flatly denied that any elevating agency had ever acted on the mountain belt. (He had to retract this statement implicitly in 1883 {Hall, 1883}.) The uplift, he believed, was one of continent-wide influence, and the Appalachians were high simply because they had thicker sediment packages in them:

It is possible that the suggestion may be made, that if the folding and plication be the result of a sinking or depression of the mass, then these wrinkles would be removed on the subsequent elevation; and the beds might assume, in a degree at least, their original position. *But this is not the mode of elevation. The elevation has been one of continental, and not of local origin*; and there is no more evidence of local elevation along the Appalachian chain, than there is along the plateau in the west. (Hall, 1859, p. 72, italics mine)

Hall believed that uplift was nevertheless caused along the margins of the Appalachian line of subsidence because of Herchel's isostatic model. He ascribed the unconformity between the Lower Helderberg group and the inclined Hudson-river group to this uplifting process (thus deviating from his 1857 interpretation that had ascribed the unconformity to folding due to intrados shortening of the down-sagging synclinal trough). Hall explained:

This process of subsidence of the sea-bottom when loaded by accumulating sediments, is clearly recognised by Herschel in his explanation of the rising of Scandinavia, which he says may be caused by accumulation of sediments on the adjacent ocean bed; which giving way beneath the pressure, will drive a portion of the yielding matter beneath the adjacent continent, thus causing the elevation.

This process of depression at one point and elevation at another by the yielding mass beneath, doubtless offers an explanation of many phenomena both of recent and more ancient geologic times. I have shown in the preceding pages that the strata composing the lower Helderberg group, and to a great extent the Oriskany sandstone also, follow a line parallel to the Appalachian chain, and do not extend far to the westward: at the same time it is shown that there had been a movement in the accumulated sediments before that date, and these beds lie unconformably above the inclined beds of the Hudson-river rocks below. The depression of the accumulated matter along the axis of the Appalachians, displacing the yielding mass beneath, would cause an elevation or bulging of the ocean bed on the western side, which, at the distance of hundred miles, might have risen so near to the surface as to prevent the accumulation of sediments; while the slope of gradually deepening waters towards the present mountain range would allow the formation of just such a set of strata as we now find, having their thickening edges towards the east, while they gradually thin out on the west. (Hall, 1859, p. 88, note C)

Hall's acknowledgment and application to the Appalachians of Herschel's isostatic model both for subsidence and uplift is important because it influenced, in less than two decades, another American geologist, Clarence Edward Dutton, to start thinking of the implications of vertical movements under the influence of increasing or decreasing crustal loads (see Dutton, 1874, 1876, 1880, 1892; Barrell, 1918[1973], p. 183–185).

Hall further believed that the metamorphism observed along the Appalachians was also a result of the sediment accumulation. According to Hall's concept, the metamorphism was due more to pressure because of burial than to increase in temperature. Volcanism, as seen in such structures as the Palisades Sill or the basalt flows in the Connecticut Valley, he believed was also because of the great sediment accumulations. He thought that all volcanoes were Tertiary or younger and that no volcanoes existed earlier (this sounds like a Wernerian legacy, possibly inherited via Amos Eaton):

> In the comparatively slow accumulation over large areas along the course of the Laurentian and Appalachian mountains, the depression would be slowly accomplished, and, as I suppose, comparatively few extensive rents or fractures would be produced. These would be filled, as we find them in the dykes, with rarely overflows of the same matter. On the contrary, we may readily conceive that where very rapid accumulation has taken place over certain areas of limited extent, the crust below might give way, from the overload, and the whole be plunged into the semi-fluid mass beneath, causing it to overflow. Whether this reasoning be correct or otherwise, I believe trappean matter are always coincident with rapid acccumulation of sedimentary materials.
>
> …
>
> Following the evidences from the oldest geological times, we find in the later periods a greater accumulation of trappean or volcanic products, which in many instances have added largely to the mass of the sedimentary deposits with which they are associated, or of themselves have produced extensive masses.
>
> Volcanoes proper, and their products, are of modern date [*this is so Wernerian that it could have been almost a quote from Werner himself*]; and it has been shown by the observations of numerous geologists, that these phenomena are always associated with the tertiary or more modern geological formations. I believe that these phenomena have been produced in regions of rapid accumulation of other deposits, and can never occur except as a result of such conditions. These igneous outflows, therefore, I regard as produced by and dependent upon other agencies, and are but the manifestations of rapid accumulations of sedimentary matter. (Hall, 1859, p. 79–80)

Dott (1985) presented an excellent summary of Hall's studies of the cratonic interior of the United States. However, I must strongly dissent from his statement that Hall discovered the craton. Craton is a word introduced by Stille in 1932 (*Kraton* in German {Stille, 1936b}; anglicized by Kay, 1944, 1947, 1951) in the place of Kober's (1921, p. 21) *Kratogen*. Kober derived it from the Greek κράτος (meaning strength, might, power) and γένεσις (meaning manner of birth). The defining characteristic of a craton is that it is strong, able to withstand forces that create tectonic deformations. This characteristic is usually inferred from the fact that the sedimentary rocks lying above a craton are so gently deformed as not to be noticeable on the outcrop or not deformed at all. We should first note that Hall was by no means the first person to note that there were large plains in the world where sedimentary rocks (since the beginning of the Paleozoic) lay flat, in contrast to mountainous areas where they were deformed. As far as I have been able to ascertain, Sir Roderick I. Murchison was the first to state this concept explicitly. He noted in his great *The Geology of Russia in Europe and the Ural Mountains* that in Russia east of the Ural Mountains, all rocks since the Silurian (he meant since the beginning of the Paleozoic) lay in a horizontal position and that this was in sharp contrast to the situation in Great Britain and in the Ural Mountains (Murchison et al., 1845, p. 24–26 and 583–586; for an historical–critical assessment of Murchison's trip to Russia, see especially Shatskiy, 1941, ch. entitled "Murchison's trips to European Russia and the Ural;" see especially p. 52, where Shatskiy points out Murchison's influence on the definition of the "Russian plate" {*Russische Tafel*} by Suess).

Moreover, Hall believed that mountain regions represented the strong areas of the crust and that the plains having flat-lying sedimentary cover were the weaklings:

> The original idea that the dislocations, fractures, or mountain elevations have taken place along the weaker lines of the earth's crust, is shown to be fallacious, from the accumulations known to exist, not only along the Appalachian chain, but also in the Rocky mountains and in other mountain chains. So far, therefore, as thickness of accumulated deposits have any influence in strengthening the crust of the earth, these lines should be the stronger ones; while the really weaker lines would lie in the great plains where the strata are thinner, and as a consequence we might suppose weaker. (Hall, 1859, note A, p. 86–87)

If there was ever an anti-craton statement, this must be it. Hall not only thoroughly misunderstood the tectonics of the mountain belt he was examining, he also misunderstood the message the craton was trying to give him.

It is important to place Hall's errors into his own time. Almost all of his contemporaries realized that his tectonic interpretations were nonsense. His biographer pointed out that geologists went away from the 1857 Montreal meeting of the American Association for the Advancement of Science "shaking their heads" (Clarke, 1921, p. 327). The earliest and perhaps the most frank reaction came from his friend, Joseph Henry, the great physicist, who had started life as a geologist under the tutelage of Amos Eaton in Troy, New York[255] (Clarke, 1921, p. 27). In friendly words, Henry tried to tell Hall that what he had said in Montreal made no sense:

> I should be pleased to have an opportunity to discuss with you your new views of geology. They are, as I understand them from your remarks at Montreal, of such a remarkable character that did they not come from you I would suppose there would be nothing in them. Your opinions, however, are entitled to my attention and respect though they may be considered at variance with what have long been regarded as established principles. If after having brought your views to the test of widest collection of facts you still are assured they are correct, then give them to the world. But I beg that you will be cautious and not commit yourself prematurely.
>
> Forgive the freedom of my remarks—they are dictated by a regard for your reputation which belongs to science of the country and is now powerful in the advance of truth or in the propagation of error. (Joseph Henry in Clarke, 1921, p. 327–328)

Henry was a keen intellect with broad interests and no doubt saw immediately the errors in Hall's ideas resulting from ill-digested knowledge of certain physical principles and plain internal inconsistencies arising from his geometrical ineptitude. Hall learned nothing from his friend's cautionary note, however, and replied, repeating his same arguments reviewed above. Below, I quote his long reply to Henry, dated 26 December 1857, in full, at the risk of boring the reader, because that letter is the best summary statement Hall ever gave of his views that leaves no room for doubt as to what he meant and thus constitutes a precious document:

> I very much regret that it has not been in my power to discuss fully with you the points which I have brought forward in my address, and which appear to some of my friends so strange and hazard, or rather as the expression is, to "compromise" my scientific reputation. I agree with you that no one should advance new views or theories till well considered, and I should be extremely sorry to advance anything which was not founded on the manifestations of Nature. I can say that thus far I have exercised the most scrupulous care that all I have advanced should bear the test of the most careful reexamination—and I would sooner commit a moral falsehood than a scientific one, if I could deliberately do either.
>
> My views are the most simple and natural conclusions from the observed facts, and so simple that I am surprised that the same idea should not have occurred to every observer. In the first place geological accumulations are spread over an ocean bed; towards the source of this material and along the line of the stronger current there will be the greatest accumulation. It is quite impossible from the nature of the material and of the forces in operation that you can have deposits of uniform thickness over wide areas. The lines of greatest accumulation have been necessarily the lines or areas of subsidence, for the sea has not been deep originally, but the bed has gradually subsided to admit the accumulation of thousands of feet. Simple subsidence of the crust may account for the plications of the formation. When these accumulations subsequently emerge, it is or has been on this continent a continental emergence, and not an emergence along certain lines of fracture or uplifting, as we have been taught to believe. If we take as an example the Appalachian chain, we find that it is composed of numerous parallel ranges, as has been well demonstrated by Rogers and others, but the greatest height of the mountain chain scarcely exceeds half of the original thickness of the deposits of which they are composed. The highest rock of the Green mountains, say 4000 feet [*~1200 m*] above tide water, is the upper member of the Hudson river group; now the entire thickness of the sediment, from the base of the Potsdam to the top of the Hudson river is scarcely less than 10,000 feet [*~3 km*]. You will see that there is much below the sea level as there is above it, and this I believe to be true in all similar mountain chains. It is not therefore elevation or uplifting, if you please to call it so, that has given geographical height, but the original thickness of the deposit, and no disturbance or uplifting of strata, that is uplifting of beds, can ever give you as great an elevation as the original pile in its horizontal and unaltered condition. As an example we have the Catskill Mts., nearly 4000 feet [*~1200 m*] above the level of the sea, composed of nearly horizontal beds, while on the east side of the Hudson the disturbed region consisting of Lower Silurian formations altogether at least 10,000 feet [*~3 km*] in thickness give no mountains of 4000 feet [*~1200 m*] high. There is another point for consideration also. All theoretical sections give you the elevation, as if produced by the bulging up of the granite or some part of the central primary nucleus; on the contrary nearly all worked or actual sections show nothing of this or only insignificant effects from some local outbreak of volcanic matter. Geologists are pretty well agreed to abandon the term *primary*, but they have not at the same time dropped the theoretical views connected with it and we still reason as if we had proved the existence of an unstratified primary mass, which in truth exists in theory only; though doubtless existing, it nowhere comes to the surface. The foldings and plications of strata which give elevation seem generally to involve nothing beyond that set of strata, as may be shown in numerous sections made in this country and Europe. Nor do elevations thus produced remain elevations, for so soon as strata are bent upwards they are weakened by cracks & otherwise, and subjected to erosion, so that we never or almost never find the exhibition which we might suppose would result from a folding and plication of strata. If we show a set of strata thus wrinkled, we shall find that the anticlinal axes are all eroded so that instead of being mountains, these parts are really valleys, while the original valley, the synclinal, is the mountain, the erosion having gone on so as to remove all that part above the red lines, while the line of sea level is about midway between the base and the top of the group of beds; or there may be often a much larger proportion of the material beneath the sea level. If you will examine some sections in the first volume of the memoirs of the Geol. Survey of Great Britain you will see that the representation of the amount which has eroded, the proportion above the sea level and that below. Had these beds continued unbroken, we should have had high hills where there are now valleys. The valleys are lines of greater disturbance, while the mountains and higher grounds are those parts where there has been least disturbance. See also, if you will, any set of really worked geological sections and you will find essentially these features. (Hall in Clarke, 1921, p. 328–331, italics Hall's)

Seven years later, Hall reiterated some of these same views in a letter he wrote to the engineer Vose:

> If I can sustain the great principle which I advocate viz., that mountains are not produced by upheaval but by accumulation and continental elevation I shall feel that I have done something to advance the Science of Geology in true principles. I feel quite sure that it is the only true explanation, the only mode of making mountain ranges, for they cannot be made without material and no imaginary upheavals will ever explain their existence. (Hall in Merrill, 1924, p. 688)

On 10 January 1876, Hall was still holding onto his own theory of mountain-making in a letter he wrote to Clarence King (Clarke, 1921, p. 332–333), although admitting that at the time he formulated it, it had not occurred to him to use the thermal contraction argument. He was perhaps grudgingly coming around that he had left something significant out—but he was not quite as yet ready to admit it. He did so somewhat more openly in 1883 while responding to Dana's criticism, as I shall show below.

James Dwight Dana and the Falcogenic Deformations of the Earth

Vose's (1866, p. 47–55) favorable treatment of the subsidence and metamorphism aspects of Hall's ideas pushed Dana to protest. In a short discussion of Hall's theory as in part favored by Vose, Dana pointed out that (1) the foot-per-foot subsidence assumed by Hall could not be brought about because the earth's crust, even if it were only 5 mi. thick, would be too strong to bend down under the weight of a single bed of sediment; (2) Hall's theory required continuous folding and metamorphism as sediments accumulated. Dana pointed out that there were distinct episodes of great amounts of shortening and associated metamorphism, as, for example at the end of the Carboniferous Period; (3) Dana could not see how Hall's folding allegedly resulting from intrados shortening would generate mountains. He famously wrote, referring to Hall's hypothesis of mountain buiding, that "It is a theory for the origin of the mountains, with the origin of the mountains left out" (Dana, 1866, p. 210). In 1873, Dana criticized Hall's views again, but more severely as a consequence of a defense published by Hunt (1873) in their favor, declaring that the foot-per-foot subsidence under the weight of the accumulating sediments and the folding allegedly resulting from the subsidence were "physical impossibilities" (Dana, 1873a, p. 349). Le Conte, in a paper in the same issue of *The American Journal of Science and Arts* as Dana's paper, pointed out that Hall and Hunt "leave the sediments just after the whole preparation had been made, but before the actual mountain formation had taken place" (Le Conte, 1873, p. 450). To Hunt's charge that his and Hall's statements had been misunderstood, Le Conte replied that "neither he [*Hunt*], nor Hall ever produced any theory of mountain formations at all, but only a return to the views of Buffon and Montlosier, that 'mountains are fragments of denuded continents'" (Le Conte, 1873, p. 450). Hall took advantage of the late publication of his 1857 presidential address to the American Association for the Advancement of Science in Montreal to respond to Dana. It is in that response we see the first reluctant admission that there may have indeed been uplift along the Appalachians! I quote his response in full:

The Address has been facetiously criticized as proposing a system of mountain making with the mountains left out. The address was not intended to propose any system of mountain making, but to show that mountain ranges were coincident with lines of great sedimentary accumulation. That this accumulation of sediments with its subsidence and consequent folding and plication, and the subsequent elevation of the mass and erosion of the anticlinals, had shaped the mountains; that no mountain elevations could take place where the sediments composing the area were thin; and that the mountain elevations were never equal to the vertical thickness of the strata composing them. I intended to imply that mountain elevation was due to sedimentary accumulation and subsequent continental elevation—trusting to the intelligence of my hearers and readers to interpret my suggestions.[256]

From various sources giving the thickness, I stated that the maximum of the palæozoic sediments, entering into the formation of the Appalachian chain, was 40,000 feet [~*12 km*]. Perhaps it would be more prudent, as a basis of argument to accept a medium and place the thickness at 25,000 feet [~*7.5 km*] (though the aggregate is much greater), out of which have come mountains of 5000 feet [~*1.5 km*] in height. It may not be easy to account for the manner in which the enormous erosion has been accomplished, for this could not have taken place beneath the sea, and the most natural explanation is that the eastern part of the continent has at some time been greatly elevated to allow such erosion. I did not pretend to offer any new theory of elevation, nor to propound any principle as involved beyond what had been suggested by Babbage and Herschel. I did not propose to discuss the theory of the contraction of the globe from cooling, or of the crumpling of the earth's crust from the gradual cooling and shrinking of the interior mass, because such arguments are not always philosophical for want of a basis in facts, and are always unsatisfactory as giving a very inadequate solution to the problem. This question cannot be properly discussed in a note.

I am satisfied that a region where the ocean bed, during palæozoic time, subsided so as to permit a deposition, under water, of more than 25,000 feet [~*7.5 km*] of strata, is sufficiently unstable in character to come up again to an elevation required for the erosion of the anticlinal valleys. This great subsidence alone may indicate that the area was one of weakness and liable to elevation or depression according to the action of the forces. [*This is a complete reversal of Hall's earlier view that the mountain areas were the strong parts of the crust! It also implies localized elevation along the mountain chain, which he had also vehemently denied earlier.*]

During the long palæozoic time the area of subsidence was in the Appalachian region, though clearly enough, during some portion of that time great uplifting occurred on the northeast, to be succeedded by subsidence which may have been equal to the elevation. Why could not the area of subsidence be changed from the Appalachian region to the ocean on the east? Subsidence in one locality means a corresponding, but not necessarily equal, elevation elsewhere; so while the ocean bed subsiding may not the Appalachians have risen? (Hall, 1883, p. 68–69)

This was a complete *volte-face*! Dana had clearly hit the target, and Hall may have been beaten down by the nearly universal condemnation of his theory of mountain-building. Even the gentlemanly Suess, who never let an injuring remark about a colleague or his ideas escape his lips or his pen, had been driven to confess publicly that he could not understand how, by the sinking and softening of the floor of a wide marine basin, mountains could be made (Suess, 1875, p. 97). But by the time Hall published his response to Dana, the world of tectonics had long by-passed him, and his *volte-face* aroused little attention.

For all his dislike of Hall's tectonic views, Dana was ultimately responsible for building and perpetuating Hall's fame in the literature of tectonics almost exactly for a century to come. In the first edition of his *Manual of Geology*, Dana already had emphasized that the margins of continents were unstable and subject to much greater oscillations than the interior, and this was reflected in the "wonderful contrast in the thickness of the strata" (Dana, 1863, p. 198). In a famous paper published in four installments in 1873 (Dana, 1873b, c, d, e), Dana reiterated his conviction that the continents and the oceans were permanent features of the face of the earth and that the continents had almost completed their thermal contraction, whereas the oceans were still vigorously subject to it. He pointed out that the principal mountain ranges of the globe were created by lateral pressure generated by the thermal contraction. He then noted that

Owing to the lateral pressure from contraction over both the continental and oceanic areas, and to the fact that the latter are the regions of greatest contraction and subsidence, and that their subsidence pushed, like the ends of an arch, against the borders of the continents, therefore, along these borders, within 300 to 1000 miles of the coast [~*500 to 1500 km*], a continent experienced its profoundest oscillations of level, had accumulated its thickest deposits of rocks, underwent the most numerous uplifts, fractures and plications, had raised its highest and longest mountain chains, and became the scene of the most extensive metamorphic operations, and the most abundant outflows of liquid rock. (Dana, 1873b, p. 424)

Dana then asked the pertinent question of whether any subsidences were created by the lateral pressure. He reminded his readers that he had shown elsewhere (e.g., in Dana {1866 and 1873a}), as had others, that Hall's interpretation of subsidence under the weight of sedimentary deposits alone was "wholly at variance with physical law" (Dana, 1873b, p. 426). He then reviewed Le Conte's (1872a, 1872b) idea of the heat induration of rocks which supposedly would lead to a density increase and hence to subsidence. But Dana pointed out that if any contraction had occurred, this would have been in the basement of the 40,000-ft. (~12-km) thick sedimentary pile described by Hall, whereas Le Conte's theory considered the very pile itself to undergo the density increase. Dana thus thought Le Conte's idea not applicable. As no other cause had been put forward to explain the subsidence, Dana thought that lateral pressure was the only mechanism left to be considered.

What exact shape did this subsidence induced by lateral pressure take? Dana reviewed the history of the individual parts (what he called the *monogenetic ranges*) of the Appalachians, into which he divided them; namely, the Highland range (including the Bule Ridge and the Adirondacks), the Green Mountain range, and the Alleghany [*sic*!] range, making up the polygenetic mountains of the Appalachians. He argued that both the Alleghany range and the Green Mountain range had been created first by a long-continued subsidence. This subsident feature, the cradle of the future mountain range, Dana called a "geosynclinal" (Dana, 1873b, p. 430, but he did *not* call the resulting range as geosynclinal, as implied in Kay, 1967, p. 311). Dana thought that Hall's statement "may be made right" if it is assumed that the long-continued subsidence occured first as a preparation to the mountain-making: "Regions of monogenetic mountains were, previous, and preparatory, to the making of the mountains, areas each of a slowly progressing geosynclinal, and *consequently*, of thick accumulations of sediment" (Dana, 1873b, p. 431). Dana proposed to call the mountains born in geosynclines "synclinoria" and contrasted them with "anticlinoria," formed from progressing "geanticlines," (i.e., "upward bendings in the oscillations of the earth's crust," Dana, 1873b, p. 432). He did not distinguish between a geanticline and an anticlinorium and defined them to be equivalent terms. He gave the Cincinnati uplift as an example of "a geoanticline or an anticlinorium" (Dana, 1873b, p. 432).

Dana pointed out that geosynclinal ranges (or synclinoria) have experienced in almost all cases since their completion, true elevation by means of geanticlinal movements. He emphasized that a new generation geanticline was always of a wider wavelength than an older generation synclinorium as it embraced the older feature as a whole and uplifted it. Already, in his paper on the origin of continents (Dana, 1947b), Dana had described the gentle swell nature of the western one-third of the U.S. territory (see the quotation, p. 113). He now gave that feature as an example of a geanticlinal: "The great uplift of the Rocky Mountain region of more than 8000 feet [~*2.5 km*], which began after the Cretaceous, had nothing to do, as I have said, with crushing or plication, although there was disturbance of the beds in certain local Cretaceous and Tertiary areas; it appears to have been a true geanticlinal elevation of the Rocky Mountain mass, itself mainly, if not wholly, a combination of synclinoria" (Dana, 1873b, p. 432–433).

Dana considered that the formation and progress of geosynclines and geanticlines lasted a long time but that the collapse of geosynclines to create synclinoria (what we would today call orogenic belts) was a short-lived, indeed a catastrophic event. Normally contraction subjected the earth's crust to large oscillations in the form of geosynclines and geanticlines: "If a geanticlinal were in progress over the middle of the Atlantic crust, as a result of the lateral thrust in the continental and oceanic crusts, there might also be a reverse movement or general sinking along the continental borders,[257] as well as a rise of water about the continents from the dimunition in the ocean's depth; and when the oceanic geanticlinal flattened out again through subsidence, the subsiding crust would naturally produce a reverse movement along one or both continental borders" (Dana, 1873b, p. 443). In the conclusion of his 1873 essay, Dana emphasized that such oscillations were made possible by a fluid rock substratum under the crust (Dana, 1873e, p. 170). Now and then, one or more of the geosynclines would give way, and its collapse would create a mountain belt consisting of highly folded and broken strata:

> ... in the work of mountain-making in eastern North America, there was first the commencing and progressing geanticlinal on the seaborder; and, as a concomitant effect of the lateral pressure, a parallel geosynclinal farther west, along the border of the continent. Concurrently, the deepening trough of the geosynclinal was kept filled to the water level, or nearly, by sedimentary accumulations, until these had become seven miles [~*11 km*] in thickness; and, as a consequence, the lines of equal temperature (isogeotherms) in the crust beneath gradually rose upward seven miles; and further, the geosynclinal crust, owing to this rising of the heat from below, lost part of its strength up to a higher level by the softening action of the heat, while it received, as the only compensation for the loss of thickness, the addition of half-consolidated sediments above. Finally, the geosynclinal region, owing to its position against the more stable continental mass beyond it, and to the weakness produced in its crust in the manner explained, became, under the continued lateral pressure and the gravity of the geanticlinal, a scene of catastrophe and mountain-making after the manner described. (Dana, 1873c, p. 12–13).

Dana thought that granitic and trachytic (i.e., what we now would call andesitic) magmatism and metamorphism in the continents would occur at these times of disturbance principally because of frictional heating of the strata being deformed (Dana, 1873c, p. 14). However, he emphasized that the major source of volcanism was the "plastic layer situated beneath the crust, or local fire-seas derived from that layer" (Dana, 1873e, p. 172).

Dana thus assured Hall's name a lasting fame by citing his observations on the thickness of strata in the mountain belts and marrying that observation with his own contractionist scheme. Nothing in his scheme—not even the observation that mountain belts are loci of thicker sediment packages than those outside them which he repeatedly credited to Hall—was new and had been already elaborated, using essentially the same theoretical framework (minus the permanence of ocean basins and continents), by Élie de Beaumont, as I documented above. Nor did Dana lack American predecessors: the veteran New England geologist, Edward Hitchcock, spoke of large anticlines and synclines encompassing whole mountain ranges and continental plains. In Hitchcock's (1841) "First Anniversary Address" before the American Association of Geologists[258] at its second annual meeting, he had announced:

> There is no small reason to believe, indeed, that on the western side of this continent, from Cape Horn to the northern Arctic Ocean, one vast anticlinal axis exists, along the crest of the Rocky Mountains. Subordinate and perhaps intersecting systems of strata will undoubtedly be found along this extended line; but this appears to be the great controlling and probably the most recent uplift on the continent. The occurrence of volcanic vents along the whole line, while they do not exist in the eastern part of the continent, renders it probable that the former has been upheaved at a later epoch than the latter.
>
> . . .
>
> The Appalachian range of mountains forms another anticlinal ridge, extending northeasterly through New England, and not improbably to Labrador. The rise of this chain elevated the Cretaceous and Tertiary rocks on the Atlantic slope, as well as the new red sandstone, and tilted up the southeastern margin of the transition rocks in the valley of the Mississippi. The uplift of the Rocky Mountains raised the western side of the same rocks, and produced the easterly slope of the strata extending to the Mississippi. That river, therefore, flows through a synclinal valley, and it was the existence of that valley which determined its course. (Hitchock, 1841, p. 264–265)

Deformations of small wavelength happening fast (even catastrophically) at given sharp episodes and creating orogenic belts (the *ridements* of Élie de Beaumont) and those of large wavelength progressing slowly over many geological periods and creating the large plateaus and depressions (essentially Élie de Beaumont's *bosselements*) became a firm part of American tectonic world-view. Dana even recognized that the magmatism associated with what we now call orogeny was of a different nature from that happening atop regions of what he called "anticlinorial" or "geanticlinal" uplift. This was the first glimmer of the recognition that what we today recognize as subduction magmatism is of a fundamentally different nature from taphrogenic and intra-plate magmatism. The great geological surveys of the United States, sent to explore the area west of the 100th meridian, contributed large amounts of critical observations concerning the large vertical motions and associated magmatism to the geologist's databank in the second half of the nineteenth century. Some of the geologists of these surveys were well-trained officers who had more of an engineering background than many of the geology professors active at the time. They combined their knowledge of physical sciences with their vast store of novel observations and came up with concepts pertaining to the nature and causes of the great vertical movements seen in the territory of the western United States. Most of their models are still among the fundamental concepts of tectonics and stand as a testimony to their incredible industry and utter devotion to science under hostile terrain conditions, penetrating observational skills, tremendous store of knowledge, brilliant insights, and boundless creativity. The history of the advancement of geology on the basis of the work done in the American West constitutes one of the golden pages of the annals of our science. Its fascinating details cannot be told here. In the next section, I summarize briefly what the geologists learned concerning the falcogenic structures and events on the basis of that work.

CHAPTER XII

THE EXPLORATION OF THE AMERICAN WEST: FALCOGENY IN THE PLATEAU COUNTRY

A Strange Landscape

> Looking southward from the brink of the Markágunt the eye is attracted to the features of a broad middle terrace upon its southward flank, named The Colob. It is a veritable wonderland. It lies beyond the Cretaceous belt and is far enough away to be obscure in details, yet exciting curiosity. If we descend to it we shall perceive numberless rock forms of nameless shapes, but often grotesque and ludicrous, starting up from the earth as isolated freaks of carving or standing in clusters and rows along the white walls of sandstone. They bear little likeness to anything we can think of, and yet they tease the imagination to find something whereonto they may be likened. Yet the forms are in a certain sense very definite, and many of them look merry and farcical. The land here is full of comedy. It is a singular display of Nature's art mingled with nonsense.

With these words, Clarence Edward Dutton (1882, p. 78), one of the most elegant and powerful prose writers in the history of geology, described the grand view southward onto the Colorado Plateau, a vast table of only gently tilted and sparsely broken layer-cake stratigraphy, standing at the incredible elevation of 2500–3000 m. To the north, these elevations continue and embrace the Uinta Mountains. Beginning with the Uintas and farther north, the high elevations become less surprising to the geologist because of the more intense and conspicuous dislocations of the rocks forming the U.S. Rocky Mountains (Rockies), called the "backbone of the continent" by Bernardo de Miero y Pacheco, the cartographer of the Domínguez-Escalante expedition of 1776.[259] Yet the appearance is deceptive. The whole country seems still much higher than it ought to be. When we descend from the mountains onto their flanking plains, onto the cratonic hinterland of the Cordillera, we still stand more than 1500 m above sea level in Denver (the "mile-high city"), and from there it is only with extreme gentleness that the topography drops eastward into what the early Spanish explorers had called the "Great Valley" of the mighty Mississippi, half way to the Atlantic coast. By contrast, Frémont's arid Great Basin (Frémont, 1845, p. 275) west of the barrier of the Rockies is chopped up by Gilbert's basin ranges (Gilbert, 1875, p. 22), bounded by numerous generally north-striking normal faults that, along with their flatter ancestors, we now know to have extended the Great Basin by more than by 100% (see, for example, Snow and Wernicke, 2000; Dickinson, 2002; for a popular geology of the Great Basin, see Fiero, 1986). Despite the tremendous stretching, the Great Basin still stands at an average elevation of about 1700 to 1800 m, in places more, as already closely estimated by Frémont (1845, foldout map; also see Goetzmann, 1993, p. 313). Regions (such as the Aegean Sea or the North Sea) that extended in a similar style with a similar present-day crustal thickness elsewhere are now below sea level. In fact, the earliest geological characterizations of the high region of the western United States as a broad anticlinal uparching of the crust occupying almost one-third of the width of the continent (Hitchcock, 1841, p. 264), or as a gentle swell along the eastern flank of which the average slope is only 0.13° (Dana, 1847b, p. 98; actually the average slope is even less, between 0.04° and 0.08°: see below) seem very apt. This high region is one of the most puzzling regions of the globe and, since the end of the nineteenth century, has been a focus of intense research interest.

The grotesque and ludicrous morphology characterizing large areas in this high country that enlivened Dutton's humor is unique in the world, its nearest analogs appearing in the Roraima region of the Guyana Shield in South America (Gansser, 1973; Gibbs and Barron, 1993, ch. 15.3 and 15.4), in the tablelands of South Africa (King, 1967b), and the Sahara (e.g., in Tibesti {Vincent, 1963} and in the northern part of the basin of Taudenni {Villemur, 1967}, where the "farcical" landforms of Dutton reappear). Yet in the twentieth century, this morphology of the plateau country of the western U.S. has become well-known internationally thanks to the numerous western movies shot amidst its multitudinous morphological features.[260] As late as the first quarter of the nineteenth century, however, this vast and extraordinary region had remained a geographical terra incognita.[261]

In the following pages, I present a summary review of the growth of knowledge on the geography and geology of the high plateaus of the western United States from the sixteenth century to 1880, when Clarence Dutton published his great book, *Geology of the High Plateaus of Utah*, which contained the first comprehensive modern synthesis of their tectonics. Although the following is only a skeleton summary, the reader will notice that it is more detailed than most other parts of this book. This is because, despite the existence of excellent scholarly accounts on the history of exploration of the western United States, the history of the growth of geological knowledge of the plateau country has never been told in a satisfactory manner. We have delightful accounts of the field operations of the numerous expeditions and surveys; we know a good deal of the lives and adventures of their leaders and some of the participants; and we are well aware of the social contexts in which the expeditions and the surveys were conceived and executed. Yet we have no single narrative telling the story of the geological ideas that were inspired, tested, refuted, or corroborated by their work. The references mentioned in the narrative below are given in the hope of aiding those who might be enticed to pursue the extraordinarily rich and instructive history of those ideas, with their implications not only for the history of tectonics or geology in general, but also for the philosophy of science. In few other places can one see with the same clarity as in the western United States how knowledge advances by creating bold conjectures in the face of ignorance and by falsifying them through most daring attempts at checking their predictions. The history of the growth of geological knowledge in the plateau country is one of great heroism by titanic men dedicated to eradicating ignorance.

The Spanish Exploration of the Plateau Country of the Western United States and Its Legacy[262]

Although an accurate geographical picture of the western parts of the present-day conterminous United States was not available as late as the beginning of the nineteenth century, parts of it had been already explored in some detail by Spanish conquistadores and Catholic, especially Franciscan, priests (Fig. 68). The knowledge they gathered formed the basis of the early nineteenth century geographical and tectonic syntheses of the central parts of the North American Cordillera, showing the presence of a major swelling of the ground in the western one-third of the continent and provided a springboard for new explorations.

Cabeza de Vaca

The presence of an integrated high western margin of North America remained unknown as late as 1700 and until sometime later. Glimpses of it had been caught much earlier though. Alvar Nuñez Cabeza de Vaca (1490?–1556?), the great defender of human rights in the "Indies," probably saw the southeasternmost end of the topographic Rockies in the Davis and Guadalupe Mountains of western Texas (DeVoto, 1952, p. 18; Pupo-Walker and López-Morillas, 1993, p. 89) and near the present-day town of Ures in Mexico, on the so-called "shell trail" (cf. Wood, 2000, p. 251 ff.), where he heard the Indians speak of "some very high mountains toward the north"

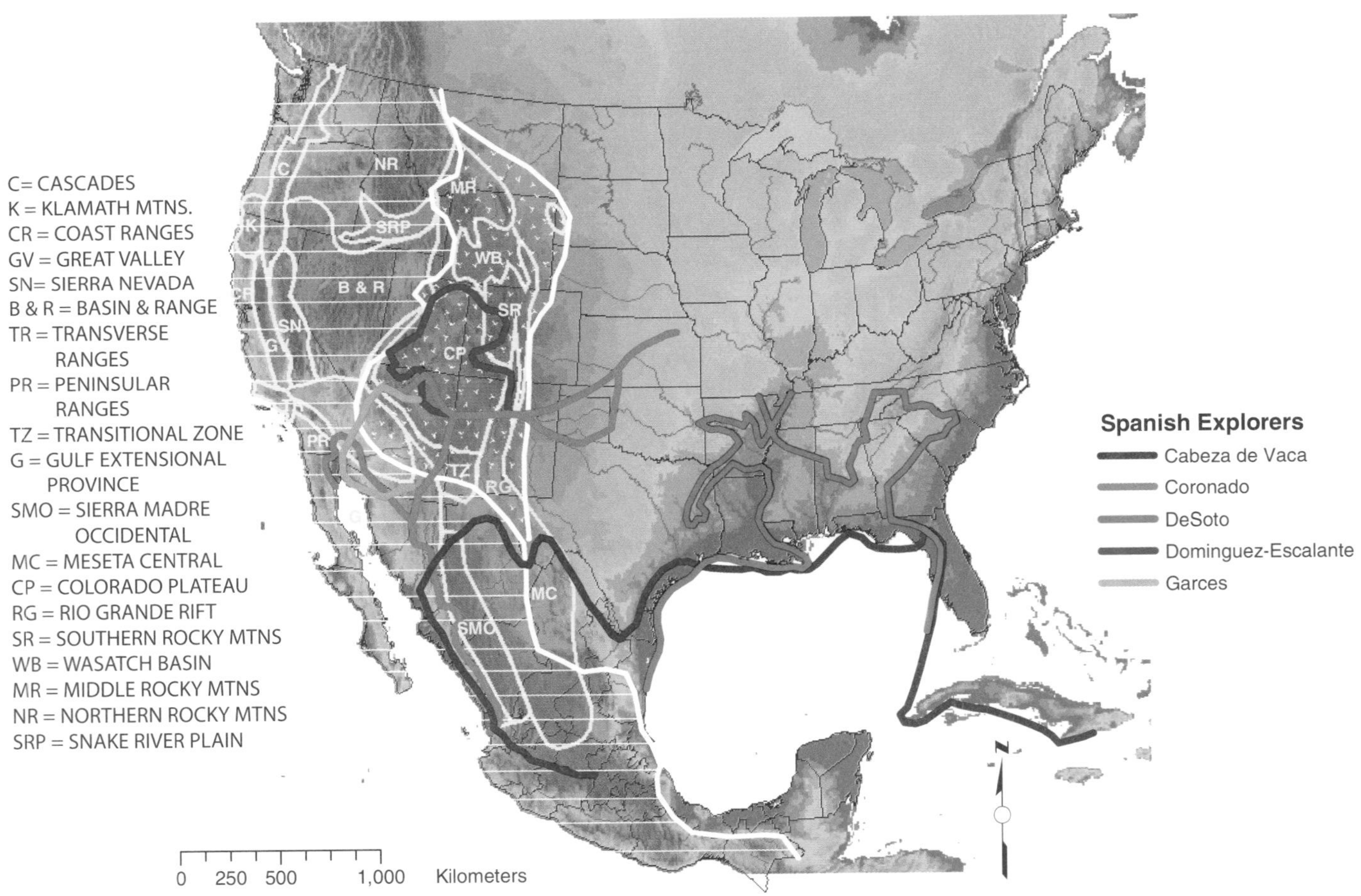

Figure 68. Spanish explorers' routes discussed in this book. The topography is from the ESRI-Data and Maps plotted by Arc Map in Arc GIS. The palaeotectonic features (white) are from Burchfiel et al. (1992a: for the United States) and Ortega-Gutiérrez and Guerrero-García (1982: for Mexico). The geomorphological provinces (which correspond largely with neotectonic provinces, yellow) are from Burchfiel et al. (1992b) and Dickinson (2002) for the United States, and Dickinson (2002) for Mexico. I have plotted the exploration routes from Goetzmann and Williams (1992), except for Coronado's march through the Llano Estacado (and the consequent changes that it imposed onto his route), which I adopted from Morris (1997).

(DeVoto, 1952, p. 10; Pupo-Walker and López-Morillas, 1993, p. 104).[263] These were the very first reports that the Europeans ever received of the North American Cordillera.

Coronado

We get a faint feeling (and only a faint, indirect one) concerning the flatness of the topography in the high country, later to be called the "Plateau province," comprising major parts of the present states of Utah, Colorado, Arizona, and New Mexico, from the various reports, letters, and testimonies that resulted from Francisco Vázquez de Coronado's (Fig. 69) expedition in the years 1540–1542, which was sent out by the remarkable first Viceroy of New Spain, Antonio de Mendoza [1490–1552] in search of the legendary seven cities.[264] The Spaniards were extremely good observers, and their commanders generally included men of great courage, considerable curiosity, and fine education. To quote Davidson (1886, p. 155): "There were giants in the earth in those days."

When expeditions were sent out, either from Spain directly or from the Viceroyalty of New Spain, they were given detailed instructions regarding what especially to observe and to record. Contrary to what is commonly emphasized by most of the historians of our times and caricaturized by the popular press (e.g., Newby, 1975, p. 94, one of numerous examples), they were not instructed only to look for gold and silver and slaves, despite the fact that these were, like all other explorations at the time, mainly for commerce and conquest (Wheeler, 1889, p. 485). In his letter to the Viceroy, Coronado says, for example, that he asked the Indians "to have a cloth painted for me, with all the animals they know in that country, and, although they are poor painters, they quickly painted two for us, one of the animals and the other of the birds and fishes" (Winship, 1896, p. 561; Hammond and Rey, 1940, p. 173)—in essence, requesting a natural history report of the country, which he sent on to the Viceroy (Winship, 1896, p. 562; Hammond and Rey, 1940, p. 176).[265]

The explorers were expressly ordered to be friendly to the natives, and those who were not were later tried and persecuted, as happened, for example, to the discoverer of the Grand Canyon, García López de Cárdenas[266]. The instructions by the Viceroy given to Friar Marcos de Niza (who preceded Coronado into what is today the U.S. states of Arizona and New Mexico and whose reports to the Viceroy encouraged him to send out the Coronado expedition) do not sound in spirit and content all that different from those given by the President of the Royal Society, Lord Morton, to Captain Cook, or from those given by Thomas Jefferson, the scholarly President of the United States, to Captain Meriwether Lewis more than two centuries later (compare Mendoza's orders to Friar Marcos {in Bandelier, 1886[1981], p. 71; and in Day, 1940, p. 34} with Morton's instructions to Cook {in Branagan, 1994} or with Jefferson's instructions to Lewis {in Jackson, 1962, p. 61–66}).

In order to be able to assess how much the Spanish conquistadores were able to understand the geography of the country they were exploring, it is useful to inquire to what degree they realized what the overall shape and surface configuration of North America was like[267]. In his narrative of the Coronado expedition, Pedro de Castañeda de Nájera[268] estimated that the distance from coast to coast, at about the latitude of southern California, was "more than six hundred leagues [*i.e., 2512 km*]." Given the fact that he had only the surveys of the Hernando DeSoto expedition (1539–1543; for the area covered, see

Figure 69. Spanish general, administrator, explorer, and gentleman—Francisco Vázquez de Coronado (1510?-1554) as conceived by the artist, Bill Ahrendt, appropriately grasping a map (from *Arizona Highways Magazine*, 1984, v. 60, no. 4, p. 4). The civilized world owes to Coronado and his heroic men its first pieces of accurate information concerning the plateau country of the present-day southwestern United States.

Goetzmann and Williams, 1992, p. 34–35) and those of Alarcón and Coronado (Goetzmann and Williams, 1992, p. 36–37), his estimate is remarkable (the real distance is 3300 km), as Bolton (1949, p. 397) also emphasized. Castañeda wished

> ... to give a detailed account of the inhabited region seen and discovered by this [*i.e., Coronado's*] expedition, and some of their ceremonies and habits [*sic*], in accordance with what we came to know about them, and the limits within which each province falls, so that hereafter it may be possible to understand in what direction Florida[269] lies and in what direction Greater India [*i.e., south Asia*]; and this land of New Spain [*i.e., southern Mexico*] is part of the mainland with Peru, and with Greater India or China as well, there not being any strait between to separate them[270]. On the other hand, the country is so wide that there is room for these vast deserts which lie between the two seas, for the coast of the North sea [*i.e., the Atlantic Ocean*] beyond Florida stretches toward the Bacallaos[271] and then turns toward Norway, while that of the South sea [*i.e., the Pacific Ocean*] turns toward the west, making another bend down toward the south almost like a bow and stretches away toward India, leaving room for the lands that border on the mountains on both sides to stretch out in such a way as to have between them these great plains[272] which are full of cattle and many other animals of different sorts, since they are not inhabited.... (Winship, 1896, p. 513, {Spanish text on p. 447}; Hammond and Rey, 1940, p. 247).

The plains were indeed very large. Castañeda carefully noted that the eastern mountain ranges (the Appalachians) could not be seen from the western ones (the Rocky Mountains). He imagined that in the endless plains between the eastern and the western mountain ranges of North America, he could see evidence for the "rotundity" of the earth, which he likened to a ball because even the smallest obstacle such as a buffalo would obscure the horizon which completely surrounded the observer (Winship, 1896, p. 527 {Spanish text: p. 456}, also p. 542 {Spanish text: p. 467}; Hammond and Rey, 1940, p. 261, 280; see also the discussion of Castañeda's great sense of place in describing this sensation in Morris, 1997, p. 119). Despite sparse habitation and enormous extent of the plains, Bandelier (1886[1981], p. 77) pointed out that the Mississippi River had been known to the Indians of the Rio Grande valley in the sixteenth century owing to trade relations that existed across the plains, although the knowledge was entirely oral. The Spanish explorers collected this information to combine it with their own observations and concluded that a "great valley" (DeVoto, 1952, *passim*) existed between the vast plains of the west and the eastern mountains. The lower parts of this great valley, the river (i.e., the Mississippi) of which was named Espíritu Santo by De Soto[273], was swampland, which Castañeda thought was "the very worst country that is warmed by the sun" (Winship,1896, p. 545 {Spanish text: p. 468–469}; Hammond and Rey, 1940, p. 282).

The Spaniards were aware that a great mountain range, having many intervening longitudinal valleys, trended northward from the province of New Galicia (nearly coincident with the present federal state of Sinaloa in Mexico, from which all northward expeditions of the Spaniards started[274]), because they had to cross it in their northward marches (e.g., Winship, 1896, p. 553; Hammond and Rey, 1940, p. 164). From there, the country gradually became higher towards Cíbola (the Zuñi Pueblo at the Arizona/New Mexico border; Winship, 1896, p. 517 {Spanish text: p. 450}; Hammond and Rey, 1940, p. 252). Parallel with the rise of the land, the climate became colder. Melchior Díaz, who had been sent ahead of the main expeditionary force by Coronado, reported that the country was mountainous and sent a map of his route to his general, who passed it on to the Viceroy. The farther north Díaz traveled (towards present Arizona), the colder he found the country. Finally, some of the Indians he had brought along froze to death, and two Spaniards became critically cold-stricken (Hammond and Rey, 1940, p. 157, 210; also Winship, 1896, p. 485 {Spanish text, p. 426} also see p. 550).

Díaz got to hear about Cíbola and reported back what he had heard: that Cíbola had fine mountains[275], that people raised maize and beans (although the area was arid), and that they had no fruit trees (Winship, 1896, p. 550; Hammond and Rey, 1940, p. 159).

Coronado himself finally reached Cíbola on 7 July 1540, at the present-day Hawikuh Pueblo (Goetzmann and Williams, 1992, p. 36), always keeping the north to their left, (Fig. 68; Winship, 1896, p. 517 {Spanish text, p. 450}; Hammond and Rey, 1940, p. 252; also see Goetzmann and Williams, 1992, p. 36 and 37 for Cíbola's location; the Dellenbaugh, {1897} deviant interpretation has not been vindicated). Coranado found Cíbola to be "all level and is nowhere shut in by mountains, although there are some hills and rough passages" (Winship, 1896, p. 559; Hammond and Rey, 1940, p. 172; for an early twentieth century picture of Hawikuh, see Sedgwick, 1926, photograph facing p. 56). The country was cold, and no cotton could be raised. He noted that there were not many birds "because of the cold and because there were no mountains near" (Winship, 1896, p. 559; Hammond and Rey, 1940, p. 172). Neither were there any trees fit for firewood in the immediate vicinity. The natives had to haul it from four leagues away (about 17 km). By contrast, the grass was good, and Coronado noted with pleasure that they could use it for pasturage for the horses and for mowing it to make hay.

All first-hand reporters of the Coronado expedition emphasized the extreme cold and flatness of the country around Cíbola and its wider surroundings between the Grand Canyon and the Rio Grande valley. For example, Castañeda wrote that it was so cold that the snow would not wet the surfaces on which it fell (Winship, 1896, p. 494 {Spanish text, p. 433}; Hammond and Rey, 1940, p. 222), and the keen observer Juan de Jaramillo emphasized that Cíbola "is a cold country" (Winship, 1896, p. 586; Hammond and Rey, 1940, p. 298). They also mentioned that it was high. When Coronado sent Pedro de Tovar and Cárdenas to explore the reports of more villages farther westward and to find a great river reported by the Indians, they traveled on flat country (for slightly different interpretations of their possible routes, see Bartlett, 1940, and Goetzmann and Williams, 1992, p. 37). It is most regrettable that Cárdenas' report to Coronado, written by the young chronicler of the army, Pedro Méndez de Sotomayor, about their discovery of the

Grand Canyon, is lost (Dellenbaugh, 1987, p. 417; Day, 1940, p. 142; Hammond and Rey, 1940, p. 217, footnote *), but we have Castañeda's relation of the event (Fig. 70):

> … when they marched for twenty days they came to the gorges of the river, from the edge of which it looked as if the opposite side must have been more than three or four leagues away [*i.e., about 12.5 to 16.5 km; an amazingly accurate assessment of the actual distance of about 16 km*]. This region was high and covered with low and twisted pine trees; it was extremely cold, being open to the north, so that, although this was the warm season [*late August; Coues, 1900a, p. 144, thinks that the Grand Canyon was discovered on or around 15 September, but does not tell us on what the precise date is based*], no one could live in this country because of the cold.[276] (Hammond and Rey, 1940, p. 215, also p. 18; Winship, 1896, p. 489 {Spanish text: p. 429} also p. 390; Dellenbaugh, 1903,[277] p. 34–35; DeVoto, 1952, p. 38; Pyne, 1998, p. 6)

The Spaniards also noted the peculiar morphology of the hills and mountains that rose above the flat plateau of Cíbola. These features were also flat-topped and steep-sided[278] and were as difficult to climb[279] as the Grand Canyon had proved difficult to descend into.[280] The best impression we get of the appreciation of this kind of morphology is from the descriptions Coronado's men gave of the great rock of Acoma (see Sedgwick, 1926; for location of Acoma Rock, see Sedgwick, 1926 endpapers; also Baars, 1995, map on p. VI–VII). Acoma Rock was discovered by Captain Hernando Alvarado's advance party as they moved east from Cíbola in 1540 (Hammond and Rey, 1940, p. 19; for a discussion of the route taken, see Bolton, 1949, p. 182–183). Alvarado admired the rock's strategic position and noted the difficulty of climbing it.[281] Castañeda observed that at the top of the rock was place for planting and growing "a large amount of maize" (Winship, 1896, p. 491 {Spanish text: p. 431}; Hammond and Rey, 1940, p. 218). The *Relacion Postrera de Cíbola* adds that there were about 200 houses in addition (Winship, 1896, p. 569 {Spanish text: p. 566–567}; Hammond and Rey, 1940, p. 309; cf. Sedgwick, 1926, photographs facing p. 20 and 34: "General view of Acoma pueblo.").

Figure 70. Captain García López de Cárdenas and his men discovering the Grand Canyon, as imagined by Bill Ahrendt (from *Plateau*, 1991, v. 62, no. 3, p. 6). This memorable event took place sometime during the last days of the summer of 1540. Because the report of the discovery by the chronicler, Pedro de Sotomayor, has not yet been found, the date of the event cannot be fixed more precisely.

The word Acoma is derived from the *Keres* (or *Queres*; Castañeda's *Quirix*: see Sedgwick, 1926, p. 294–295), consisting of the Indian words "ako" meaning *white rock* and "mi" meaning *people* (Sedgwick, 1926, p. 291[282]; also see Baars, 1995, p. 189). The pueblo sits atop the white Zuni Sandstone capped by the Dakota Sandstone of Albian to Cenomanian age (~112–90 Ma ago; Baars, 1995, p. 189–190; also see fig. 14 on p. 72). It is a kind of North American Masada, of which King Herod would have been proud. The Spaniards used the extremely apposite word *mesa*, meaning table, to describe this kind of flat-topped prominence rising from the flat surface of the high plateau.

This apposite description found its way into the later accounts as "tableland" (e.g., Murchison, 1849, p. 228, footnote *, where there is reference to the tableland of Mexico; also see below) and was extensively used by the American geologists who studied the geology of the flat-lying highlands of the western United States. From there it entered the vocabulary of the European geologists who worked in the United States (e.g., Marcou, 1856, p. 151, where he uses the English term *tableland* directly in his French text) and from both sources into the terminology of Eduard Suess, who used the terms table (*Tafel*) or tableland (*Tafelland*) to describe extensive areas of flat-lying sedimentary rocks and to contrast them with mountainous areas with their highly deformed and contorted rocks and jagged morphology. Some use the term tableland for only a flat surface without regard to the attitude of the strata underlying it (e.g., Süssmilch, 1909); I think such a usage is inappropriate. That contrast was finally formalized by Kober's definition of kratogens and orogens (Kober, 1921, p. 21), which became the craton-orogen distinction in Stille's writings in 1935 (cf. Şengör, 1999a).

Coronado's expeditionary force wintered in the Rift Valley of the Rio Grande, where they discovered the high-lying Tiguex and Taos Pueblos (the latter was named Valladolid by the Spaniards), which were "very high and extremely cold" (Winship, 1896, p. 511 {Spanish text: p. 445}; Hammond and Rey,

1940, p. 244). Here they were told by an Indian whom they called "the Turk" (owing to his looks) that a rich center of culture, called Quivira, existed to the northeast, on the plains. Coronado thus decided to extend his explorations farther east than originally planned. This excursion, accompanied by much hardship, took him as far northeast as present-day Kansas in the summer of 1541. Finally, they discovered that the Indian had lied to them, and they turned back.

The expedition to Quivira (for the details of the route taken, see Morris, 1997, ch. 3–7) acquainted the Spaniards with the prairie and with the morphology of the North American craton. In his account of the expedition, the widely-traveled Captain Juan de Jaramillo noted that "This country has a fine appearance, the like of which I have never seen anywhere in our Spain, Italy, or part of France, nor indeed in other lands where I have traveled in the service of His Majesty" (Winship, 1896, p. 591; Hammond and Rey, 1940, p. 305). He noted that it was not rough, but contained hillocks, low ridges (*lomas* in the Spanish original: Winship, 1896, p. 591, footnote 1) and charming rivers with fine waters. He observed prophetically that "it will be very productive for all sorts of commodities"[283] (Winship, 1896, p. 591; Hammond and Rey, 1940, p. 305). In his letter to the King, Coronado himself chose to emphasize the monotonous flatness of these immense plains:

> After traveling nine days [*from Tiguex*], I came to some plains, so vast that in my travels I did not reach their end, although I marched over them for more than three hundred leagues [*about 1255 km*].
> ...
> For five days [*more: Winship, 1896, p. 581*] I went wherever they led me, until we reached some plains so bare of landmarks as if we were surrounded by the sea. Here the guides lost their bearings because there is nowhere a stone, hill, tree, bush, or anything of the sort. There are many excellent pastures with fine grass. (Winship, 1896, p. 580–581; Hammond and Rey, 1940, p. 186)

Castañeda remarked that "...since the land is so level, when they had wandered aimlessly until noon, following the game, they had to remain by their kill, without straying, until the Sun began to go down in order to learn which direction they then had to take to get back to their starting point." (Winship, 1896, p. 509 {Spanish text: p. 443}; Hammond and Rey, 1940, p. 241[284]). In the *Relación del Suceso*, Indians' employment of dogs as pack animals is ascribed to the flatness of the land, such that they were able to drag the A-frames (Winship, 1896, p. 578; Hammond and Rey, 1940, p. 293). Around Quivira, the *Relacíon* records "Traveling in these plains is like traveling at sea, since there are no roads other than the cattle trails. Since the land is so level, without a mountain or a hill, it was dangerous to travel alone or become separated from the army, for, on losing sight of it, one was lost" (Winship, 1896, p. 578; Hammond and Ray, 1940, p. 292).

Coronado's likening of the flatness of the prairie to the surface of the sea was not only repeated by his men and also in the *Relacíon Postrera de Cíbola* (Winship, 1896, p. 570 {Spanish text: p. 567}; Hammond and Rey, 1940, p. 310; cf. Morris, 1997, p. 48–49), but three centuries after him, the great American geologist John Strong Newberry (1822–1892)[285] expressed precisely the same sentiment about the morphology of the high prairie around the valley of the Arkansas: "In this plateau the tributaries of the Arkansas have excavated valleys of greater or less breadth, but they are generally narrow and are separated by 'divides' of the high prairie, which to the eye are as level as the surface of still water, and are everywhere covered with velvety carpet of buffalo grass" (Newberry, 1876, p. 22). John Charles Frémont likened the climate of "the vast prairie" to the ocean (Frémont, 1845, p. 122). Topographer John Lambert of the Stevens expedition along the northernmost Pacific Railroad survey gave a detailed description of the level prairie in terms of its resemblance to the sea *(in* Stevens, 1855, p. 160; see the quotation from Lambert on p. 174). It is amusing to note that when a mountain man, a denizen of the prairies and the mountains, saw the Pacific Ocean for the first time, he opened his arms and exclaimed, "Lord! There is a great prairie without a tree" (Emory, 1848, p. 112–113; Goetzmann, 1993, p. 255). Indeed, the great historian Arnold Toynbee also drew attention to the similarity of the effects the sea and the steppe had on the human life dependent on them, because of the very flatness and emptiness of their surfaces:

> ... in its relationship to man, the Steppe, with its surface of grass and gravel, actually bears a greater resemblance to "the unharvested sea" (as Homer so often calls it) than it bears to *terra firma* that is amenable to hoe and plough. Steppe-surface and water surface have this in common, that they are both accessible to man only as pilgrim and sojourner. Neither offers him anywhere on its broad surface, apart from islands and oases, a place where he can settle down to a sedentary existence. Both provide strikingly greater facilities for travel and transport than those parts of the Earth's surface on which human communities are accustomed to make their permanent homes, but both exact, as a penalty for trespassing on them, the necessity of constantly moving on, or else moving off their surface altogether on to the coasts of *terra firma* which surround them. (Toynbee, 1947, p. 166; also see pp. 185–186)

From the reports of the Coronado expedition, we thus learn the Spanish expeditionary force's discovery of a high but flat-lying region occupying much of the area of the present-day U.S. states of Arizona and New Mexico. We also learn that a short distance east of the Rio Grande valley, the aspect of the country changes dramatically: instead of high plateau ornamented by numerous higher-lying mesas, we have here lower, featureless plains that stretch as if into infinity. The high plateau was deeply dissected by canyons, the largest of which belonged to the Tizon (or the "Firebrand" River, the present-day Colorado River of the West), discovered by Captain García López de Cárdenas. The high plateau was recognized to be different from the "cordilleras" and the "sierras" on both sides of the continent: it was a highland, but not a mountain-land. The Spaniards recognized that the continental water divide lies in this highland.[286]

After Coronado

It is commonly written that Coronado's gains were lost for the next three centuries and historians later dug them up. That is emphatically not true. Not only his geographical discoveries found their way into the work of contemporary cartographers (see endnote 264 and 267), but his expedition spurred others to follow him. In 1583, Antonio de Espejo visited many of the localities seen by Coronado's expedition, including Acoma, on his rescue expedition in search of the two priests (Francisco Lopez and Friar Santa Maria) and a lay brother (Augustin Ruis), who had left for the north country in 1581 (Harris, 1909, p. 50–51). It was Espejo who gave New Mexico its name (but then its boundaries were very different from those of the present day U.S. state that bears that name). He was followed in 1596 by Juan de Oñate, the founder of Santa Fé (in 1606). In 1604, Oñate explored the Colorado Plateau on his way from San Juan to the South Sea (in this instance, the Gulf of California). He passed through Cíbola and named the present-day Little Colorado River (or Colorado Chiquito) the Rio Colorado, the first time this name was ever applied to any part of the great river. He next came upon the two branches of the Rio Verde, thus revisiting Espejo's localities 23 years later. From there he descended down the great plateau (for a summary of these expeditions, see Coues, 1900a, p. 394–395, 476–479; Harris, 1909, p. 33–34, 53–54).

These expeditions opened the way to ardent missionary activity by the Catholic Church in the present southwestern United States. The zealous fathers fearlessly roamed the terrain now covered by northern Mexico and Baja California, and the U.S. states of California, Arizona, New Mexico, Utah, and portions of Colorado. These indefatigable men bequeathed to us a rich library containing their accounts of the physical geography, botany, zoology, geology, and ethnology of the areas they visited (Harris, 1909, p. 36–37; also see Priestley, 1946). From the viewpoint of the history of appreciation of the geomorphology of the plateau country, the diaries of two of the fathers are of prime importance: Father Francisco Garcés[287] and Father Francisco Silvestre Vélez de Escalante.[288] Both were Franciscan priests, and both were trying to find a road of communication between Santa Fé and the Pacific Ocean.

Garcés and Escalante

Father Garcés made five trips, but it was on his last, which he undertook in the years 1775 and 1776, that he traveled onto and atop the Colorado Plateau (for his route, see Fig. 68; also Bolton, 1930a, map entitled "Map of Western New Spain in the Later Eighteenth Century"; and Galvin, 1965, foldout map entitled "Father Garcés' Travels 1775–1776"; Briggs, 1976, map on p. 7). What impressed Father Garcés was that north of the Jamajab Indian nation (where the present Mohave Mountains are located just northeast of Lake Havasu City in Arizona at 34°28′ N and 114°20′ W), the "Rio Colorado comes thorough profound caxones [*sic*]" (Coues, 1900b, p. 443, 472). Indeed, north of that point the mighty river flows between the Black Mountains to the east and the Dead and Eldorado Mountains to the west. Only after it takes the great easterly bend at Lake Mead, the Grand Canyon proper starts between the Shivwits Plateau to the north and the Grand Wash Cliffs to the south. Father Garcés thus correctly identified the southwestern boundary of the plateau country along what is now generally called the Mogollon Rim or Mogollon Hingeline.

Father Garcés noted the depth and narrowness of the canyon near where the Havasupai Indians lived in Cataract Canyon by pointing out that "it is ten o'clock in the day when the sun begins to shine" (Coues, 1900b, p. 345; DeVoto, 1952, p. 290, wrote, "That he [*Father Garcés*] could reach the Cataract Canyon where they lived and descend its vertical wall is against reason but he did so."). Because he traveled from the southwest, Father Garcés gradually incorporated sharp mountain ridges into his descriptions of the Colorado Plateau. For this reason, he used the term *sierra* both for independent mountain ridges and for the cliffs of the canyons. This is confusing to those readers not familiar with the terrain, but once one follows him with a good physical map at hand, it becomes apparent how awestruck he was by the sheer canyon walls. He referred to these precipitous walls as *peña viva* (live rock) and to the canyon itself as a *foso* (a veritable trench or trough: Coues, 1900b, p. 355). So the Grand Canyon was a trough cut into live rock.

Once he was out of the canyons, in which he felt himself buried alive (Coues, 1900b, p. 408), Father Garcés noticed the dominance of a flat topography, here and there studded by flat-topped mesas (Coues, 1900b, p. 357, 358, 361, 382, 392). His account is illustrated by Pedro Font's map (1777: in Wheat, 1957, p. 91–92; also see endnote 287), of which a true tracing is reproduced as a foldout frontispiece in Coues (1900a)[289]. This map, in the versions reproduced in Coues (1900a) and in Galvin (1965), reveals nothing unusual in terms of the depiction of landform types in the areas visited by Father Garcés, where all elevations, mountains, highlands, and mesas were uniformly drawn by the characteristic hummocky patterns of the time. The situation is dramatically different in the case of the map that resulted from the Domínguez-Escalante journey recorded by Father Escalante.

The Domínguez-Escalante journey took place almost entirely within the confines of the Colorado Plateau[290]. Together with his superior, Father Francisco Atanasio Domínguez, Superior of the New Mexico Franciscans and Commissary Visitor of the Custody of the Conversion of St. Paul[291], and with the engineer and retired militia captain Bernardo Miera y Pacheco (plus seven others[292], some of whom could speak Ute), he left Santa Fé on 29 July 1776 and returned there on 2 January 1777. Together they traveled almost 2500 km. Escalante's previous journey to the Hopi pueblos on the Colorado Plateau had persuaded him that a road to connect Monterey with Santa Fé could not pass that way owing to harsh terrain conditions, aridity of the climate, lack of pasturage, and the hostility of the natives (Adams, 1976, p. 48). So, their route instead passed through the Rio Grande Rift, continued north into the central Rocky Mountains (which Miera called "the backbone of North America" on his map: see endnote 259), along the Wasatch, and then swung east to the shores of Utah Lake (Fig. 71). Then they went back

Figure 71. "Father Escalante Discovers the Utah Valley": an oil painting by Keith Eddington, 1950 (from Bolton, 1950). The Domínguez-Escalante expedition reached the Valley of Lake Utah, on the western edge of the plateau country, on 23 September 1776.

south, down to the present Cedar City, and thence southeastwards to the Hurricane Cliffs, to the Paria Plateau across the Grand Canyon, and then to the mesas of Oraibi, Hopi Buttes, Zuñi (the old Cíbola!) via Acoma and back to Santa Fé (Fig. 68). Earth scientists are grateful that they had Captain Miera y Pacheco[293] with them. He was a skilled cartographer and had his surveying instruments with him.

The map compiled by Captain Miera was long thought lost. Its "discovery" was reported in 1941 by J. Cecil Alter. Wheat (1957, p. 99 ff.) counted six distinct manuscript copies of this map in existence. Goetzmann (1995, p. 109) indicated the existence of a seventh copy in the Bienecke Library of the Yale University.[294] Wheat (1957) divided the existing manuscripts into three basic types and designated them as *A* (the "undecorated" type), *B* (the "Tree and Serpent" type), and *C* (the "Bearded Indian" type). It is not known whether any of these copies were actually produced by Miera's own hand or if all of them are copyists' works. Wheat believes that the type designated *A* and the earliest of the *C*-type maps may have come from Miera's own hand.

The *A*-type map is represented by a single manuscript now in the British Museum, with a call number *Additional Manuscripts No. 17,661-C* (Wheat, 1957, p. 100). This map is dated 1777 and from the dedication it carries to the Viceroy Antonio Maria Bucareli y Ursua,[295] it is believed to be the earliest version of the Miera map. Photostats of this map are filed in the Library of Congress, the Newberry Library in Chicago (Wheat, 1957, p. 227), and the Berkeley campus of the University of California. This is the version that is reproduced here in Figure 72A.

The original of the *B*-type, the so-called "Tree and Serpent" type (so named after a serpent figure wrapped around a tree), made most likely in 1777 and dedicated to the Caballero de Croix, the Comandante General of the *Provincias Internas* of the Viceroyalty of New Spain[296], is in the *Deposito de la Guerra* at Madrid, Spain (*No. LM 8a-1a-a.40*; cf. Alter, 1941, p. 64[297]). Alter (1941) presents a photographic reproduction of this map in his paper. Figure 72B is a reproduction of this map.

There is yet another version of the Miera map dated 1778, the *C*-type or the "Bearded Indian-type" (after the bearded Indians depicted in it). Wheat (1957, p. 107) records four distinct copies of this type: one in the British Museum (call number *Additional Manuscripts No. 17,661-D*), another in the *Archivo General* of Mexico City, hand-copied in Bolton (1950; see Wheat, 1957, p. 112, footnote 37). The other two copies were made by different scribes for the Kohl collection[298] in the Library of Congress. Figure 72C herein is a reproduction of Bolton's facsimile. (I chose Bolton's facsimile owing to the sharpness of the topographic features.)

What is most interesting about the landform depiction on these so-called Miera maps (with the exception of the unremarkable Yale University copy, herein reproduced as Fig. 72D) is that Captain Miera employed two distinct symbols to draw topographic prominences[299]: the ordinary conical mountain signature to depict what Father Escalante in his diary calls *sierras*, and

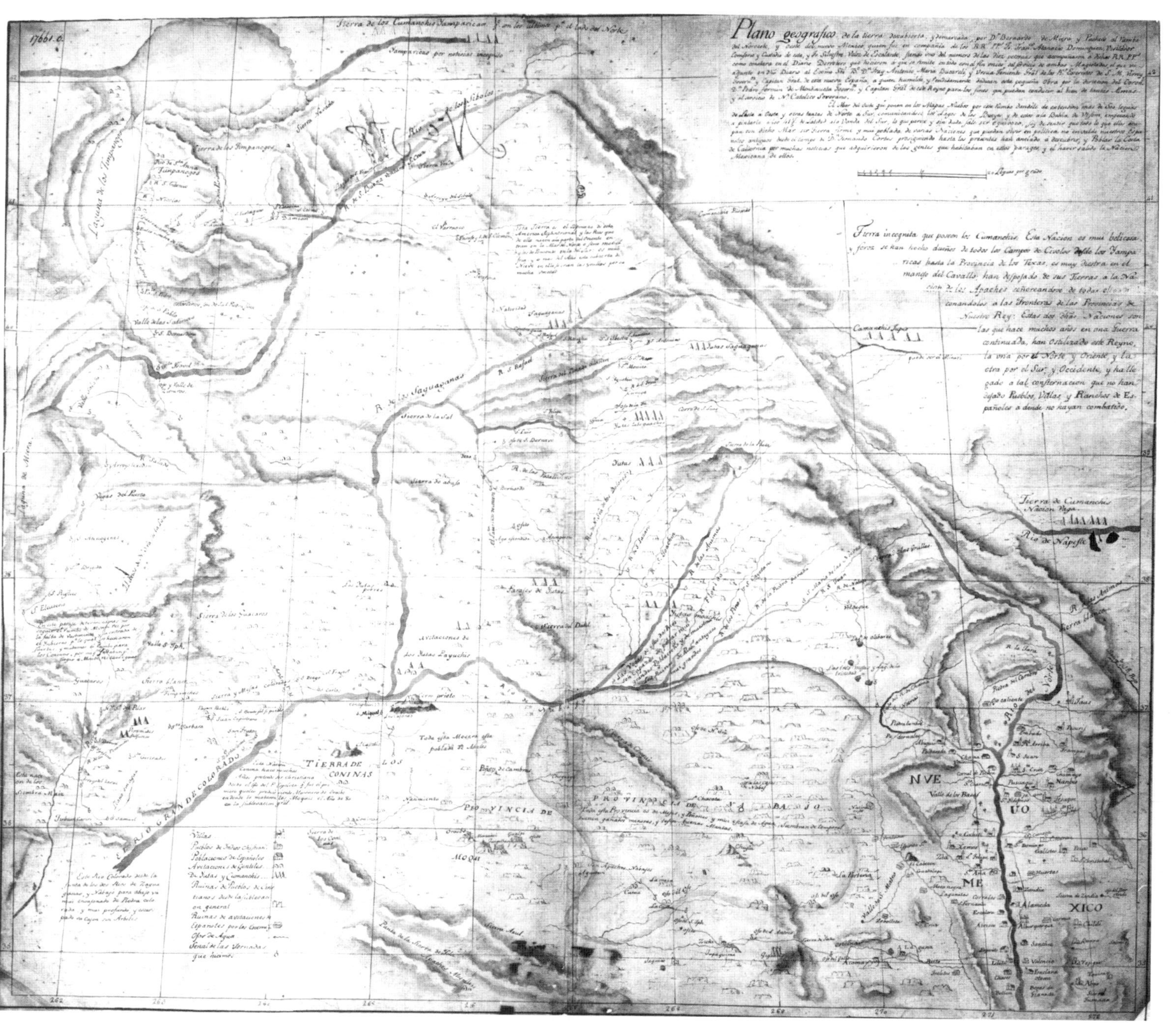

Figure 72 (on this and following three pages). A: The "undecorated" or the "A-type" of the Miera maps. (Courtesy of the Library of the University of California at Berkeley.) In this map and the following two versions, note the richness of the topographic detail.

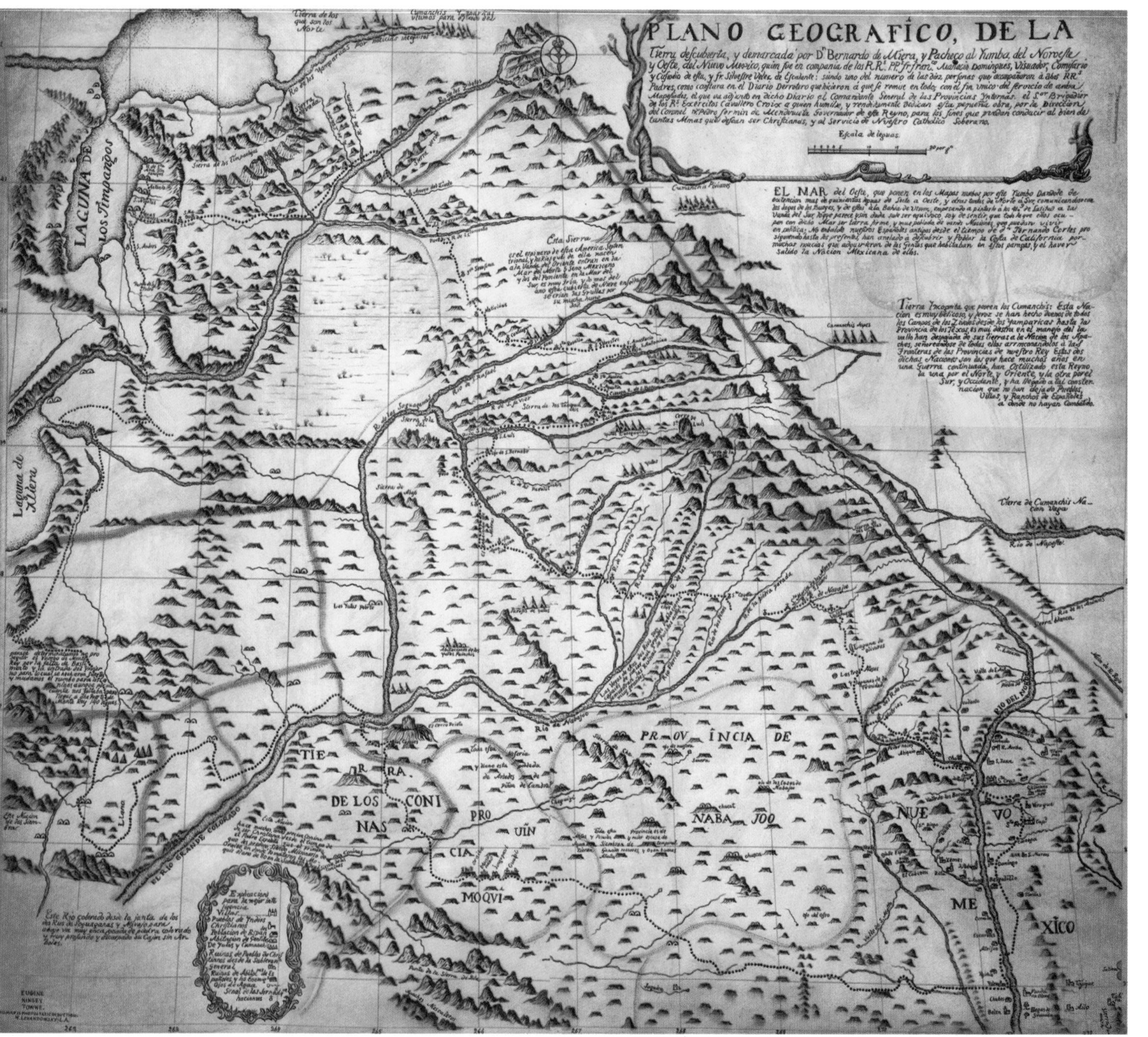

Figure 72 *(continued)*. B: The "Tree and Serpent" or the "B-type" of Miera maps. (Courtesy of the Library of the University of California at Berkeley.)

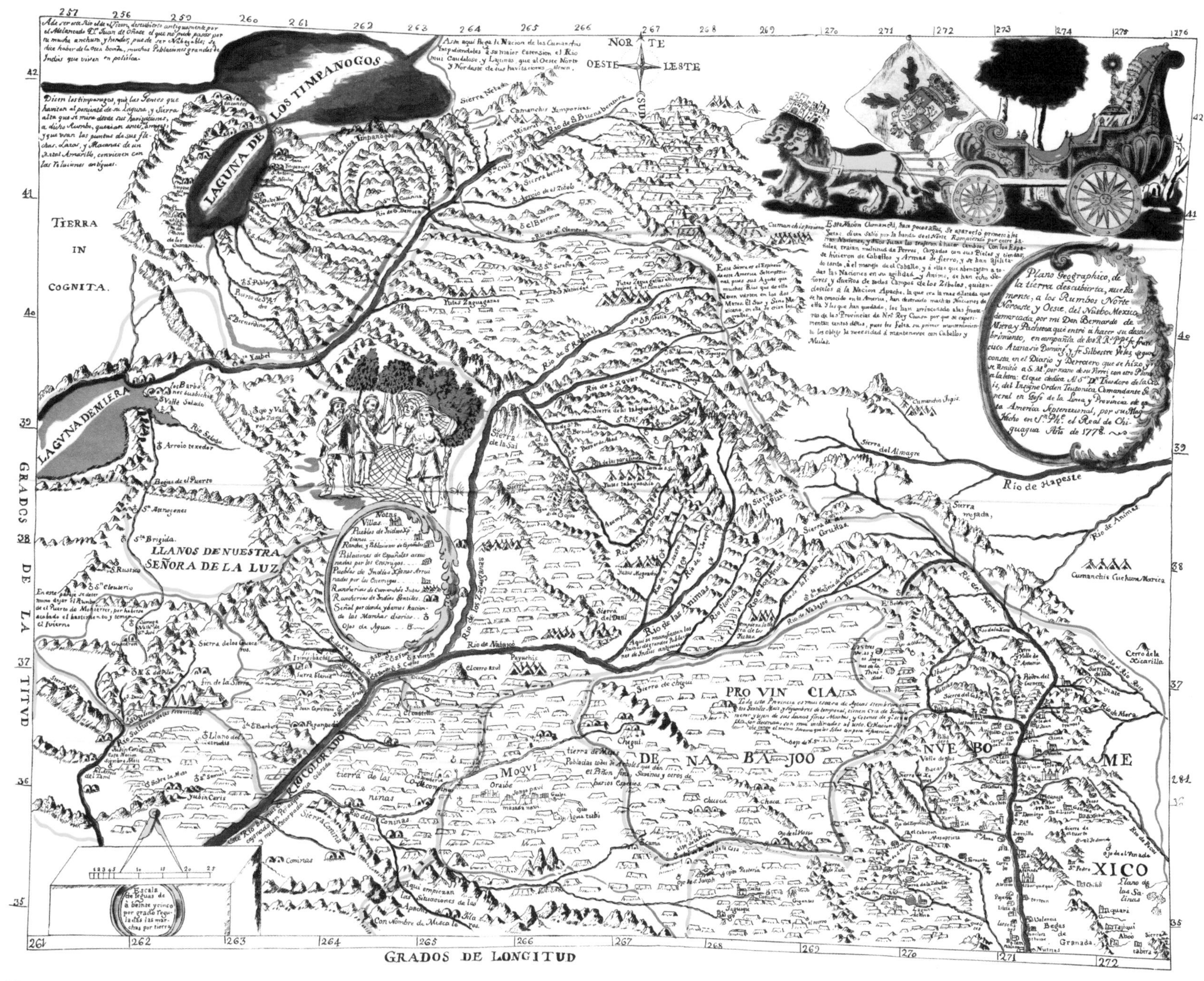

Figure 72 *(continued)*. C: The "Bearded Indian" or the "C-type" of the Miera maps. This is Bolton's hand-copied and "corrected" version. (Copied from Bolton, 1950, with permission).

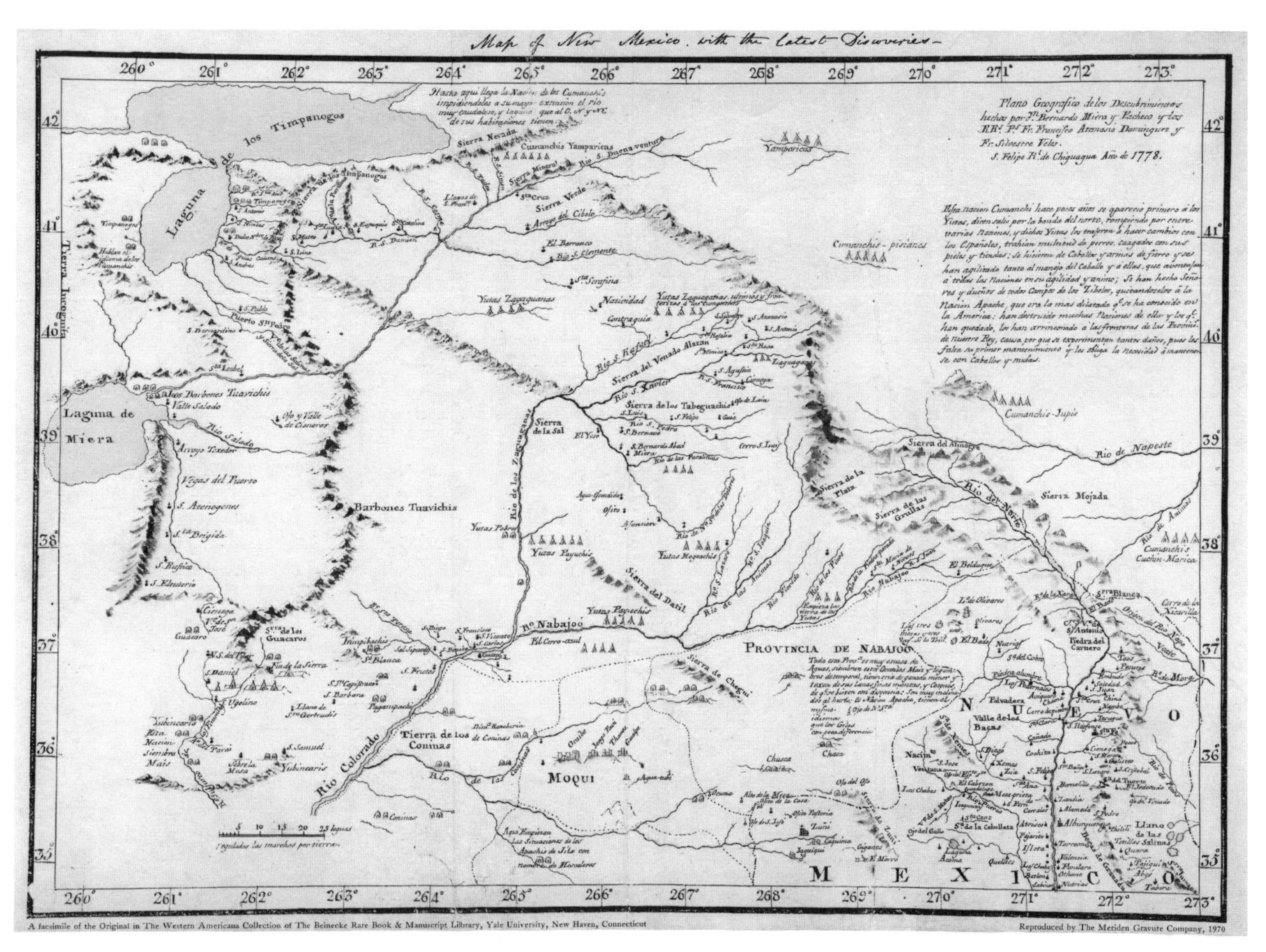

Figure 72 *(continued)*. D: The Yale copy of the Miera map. From the facsimile produced by the Meriden Gravure Company and sold by the Bienecke Library of the Yale University. Note the sparsity of topographic detail.

trapezoids to indicate what Farther Escalante called *mesas*. (The Yale copy is without the abundant trapezoids depicting mesas, except for a few where the mesas were inhabited and consequently not nearly as significant as those copies in the British Museum, Madrid, or Mexico City from the viewpoint of the history of the appreciation of the physical geography of the plateau country). To my knowledge, this is the first time a separate signature is used to depict flat-topped plateaus on a physical map.

As a result, we get from Miera's map a very good idea of the extent of the flat-topped plateau country. It is clear from Figures 72A, 72B, and 72C that the flat-topped plateau has a triangular shape with an apex pointing northward. The triangular shape results from the east-west foreshortening due to the impossibility for Miera to fix longitudes (Alter, 1941, p. 66). When allowance is made for this error, the Colorado Plateau acquires more-or-less its present trapezoidal outlines in the Miera map. Even some of its individual topographic features can be recognized: To the east-northeast, the San Juan Mountains and the Uncompahgre Uplift are represented by a series of northwest-trending sierras in the Miera map. They merge with the Monument Upwarp westwards, depicted in the Miera map by a mixture of sierra and mesa signatures. Farther to the northwest, Powell's (1875, especially p. 169–181) various cuesta cliffs (see Fig. 73 showing a bird's eye view of the "Terrace Cañons" of Powell) are represented again by northwest-trending sierra signs. Southwards, the Black Mesa Basin is represented entirely by Mesa signs. It is delimited in the Miera map to the east by the Defiance Uplift, which Miera depicted as a northwest-trending long sierra connected with the Zuni Uplift (roughly concident with Kelley's {1955, fig. 9} Zuni Lineament, on which the Zuni, Defiance, Monument and Circle Uplifts are located; coincident with Marcou's {1856} Sierra Madre; also see Blake, 1856) amidst a large mesa country, essentially formed from the Black Mesa and the San Juan Basins. The Escalante party was thus fully aware of the peculiarity of the highland on which they were traveling. Escalante's diary provides additional support for this claim[300].

Where the Colorado River cuts across Battlement Mesa and Book Cliffs in west-central Colorado near the present town of Grand Valley, Father Escalante noted north of the river that "On this side there is a chain of high mesas, whose upper half is of white earth and the lower half evenly streaked with yellow, white, and not very dark coloured red earth" (Bolton, 1950, p. 163; Vélez de Escalante, 1995, p. 45). In this remarkable passage, the good Father not only describes the morphology but also the stratigraphy and the geological structure of the mesas: The "white earth" he refers to is the white shales and sands of the Green River Formation, whereas the reddish earth is the Wasatch Formation (Tweto, 1979). The Green River Formation consists of shale, sandstone, and beds of oolitic rock, among which the shaly beds that predominate are very compact and firmly-bedded. Escarpments exist along mesa walls and high bluffs, where the weathered beds have a characteristic chalky-white color. In the lower parts, massive white sandstones are seen in places (Willis, 1912, p. 759). Below the Green River Formation is the Wasatch Formation of mainly clays and shales, in which various shades of red and drab predominate (Willis, 1912, p. 759). That Escalante noted that these colors divide the mesas into a higher and lower parts shows that he was aware of the horizontality of the layers parallel with the mesa tops. Miera recorded the topographic observation in the form of a series of mesas around Natividad on the present day Roan Plateau (refer to Fig. 72A, 72B, 72C; for locality identification, compare the facsimile of the Miera map and the map of "The Escalante Trail 1776" in Bolton {1950} or the reproduction of its relevant part in Vélez de Escalante, {1995, p. 37}).

On the way, Escalante continuously paid attention to the details of the topography, carefully distinguishing mesa land from the sierras and rock type, mentioning gypsum, salt, and gold occurrences; the multifarious colors of rocks; and the mineralogical nature of various hot springs. Around the Grand Canyon, he again emphasized the great height and the flatness of the country, both on the basis of his own observations and of what he heard from his guides (Bolton, 1950, p. 207; Vélez de Escalante, 1995, p. 98). On the Kanab Plateau, the flatness of the terrain reached after a fatiguing climb impressed him (Bolton, 1950, p. 208, 210; Vélez de Escalante, 1995, p. 98, 101). Like many others before him, Escalante noted the extreme cold of the weather atop steep cliffs (Bolton, 1950, p. 229–230; Vélez de Escalante, 1995, p. 126–127). On the Kaibab Plateau, his descriptions (Bolton, 1950, p. 218; Vélez de Escalante, 1995, p. 111) are so precise that one recognizes at once that he was probably walking over the Harrisburg Member (Sorauf, 1962) of the Kaibab Formation of McKee (1938; see Hopkins, 1990). While describing their traverse across the volcanic rocks of the Uinkaret Plateau, Escalante wrote: "We swung south now over stony malpais (which is like slag, although heavier and less porous) ..." (Bolton, 1950, p. 204; Vélez de Escalante, 1995, p. 95: Father Chavez translates "volcanic slag" instead of just slag as in Bolton) and described the hot and sulphurous springs (Bolton, 1950, p. 205; Vélez de Escalante, 1995, p. 95). He distinguished mesa-studded highlands from flat-topped plateaus without many mesas (Bolton, 1950, p. 231–232; Vélez de Escalante, 1995, p. 128, 130).

Although their primary purpose was to reach Monterey and thus to establish a direct route between Santa Fé and the coastal missions while converting as many heathens as possible to Catholicism, Escalante was aware of the value of exploring unknown lands and informing the civilized world about them (e.g., Bolton, 1950, p. 198; Vélez de Escalante, 1995, p. 87). It is otherwise inexplicable why he paid such minute attention to the details of the local physical geography (and one might anachronistically say, geology) and recorded his observations so carefully. His observations were still the basis, more than a quarter of a century later, of Alexander von Humboldt's view of the physical geography and tectonics of the area covered by what are today the U.S. states of Utah, Colorado and Arizona.

With the Domínguez-Escalante expedition, we come to the end of the great entrada period (1540–1777), during which the

Figure 73. Bird's eye view of the Terrace Canyons. (From Powell, 1875, fig. 61).

Spanish explorers established the basic outlines of the physical geography of what we today call the *plateau country* of the western United States. They recognized the presence of a massive, high-lying plateau between the "Backbone Mountains of North America," which von Humboldt was to include in his Cordillera and which we today know under the name of the Rocky Mountains, and the much lesser "sierras" to the west, the basin ranges of Gilbert (1875). They noted its flat top, and Escalante even noticed that the strata composing it was at least in places flat-lying, parallel with its flat surface. I do not think they had any inkling that they were looking at a peculiar earth feature, though. Their service was to record what they saw faithfully so that those who would be interested in the landforms in that part of the world and who had a better understanding of earth processes than they had, could use their results. The fact that the Spanish explorers did not understand (although they saw and recorded) the distinction between large plateaus and mountain ranges is probably why the earliest general maps of the American West by Antoine Soulard (1795, *in* Wheat, 1957, p. 157–158, map 235a), Victor Collot (1796, *in* Wheat, 1957, p. 160–161, map 236), and Juan Pedro Walker[301] (pre-1810?: Fig. 74) show the North American Cordillera as one or more narrow and long ranges and do not honor the earlier Spanish observations concerning the great plateau land. It is only natural that none of the maps based on these earlier ones (many derived from Soulard's map: see Wheat, 1958, ch. XI) show any plateau land either. Only four years later did Alexander von Humboldt show in his famous map (von Humboldt, 1812) large, flat-topped areas next to the main Cordilleran axis in the present-day areas of Arizona and Utah (Fig. 75) and compared them with the vast plateau lands of the Old World.

Alexander von Humboldt and the Tectonics of the Plateau Country

Alexander von Humboldt came to Mexico (then still the Viceroyalty of New Spain) on 22 March 1803 after having completed his explorations in South America with his friend and scientific companion Aimé Bonpland. On the 28 March 1803, Von Humboldt sent a letter to the Viceroy requesting passports and help to be able to consult the archives of Mexico, and reached the capital towards the end of April. Von Humboldt stayed in the capital until 20 January 1804 (Bruhns, 1872a, p. 386–392; for von Humboldt's life and accomplishments, see the references given in endnote 190). During this time interval, he undertook a number of field investigations in the southern and central parts of the country to establish the hypsometric and geological conditions and to visit some of the mines. Between 20 January and 7 March 1804, von Humboldt and his friend Bonpland went from the capital to Veracruz, from where he traveled to Cuba. On this last excursion in Mexico, the travelers made barometric height measurements and studied the volcanoes of Popocatapetl and Orizaba (Bruhns, 1872a, p. 392–393).

Von Humboldt was given free access to the archives of Mexico by the Viceroy Don Vincente José de Iturrigaray (1742–1815) (Bruhns, 1872a, p. 386–387; Beck, 1966; Chevalier, 1997). With his habitual enthusiasm and industry, he went through the mountain of material with astonishing speed and produced a geographical synthesis of what was then comprehended under the designation of New Spain (which today covers Mexico and the U.S. states of California, Arizona, New Mexico, and Texas, and parts of Utah and Colorado). He submitted a Spanish manuscript to the Viceroy before he left. On his way back to Europe, he stopped in Washington, D.C., and visited President Thomas Jefferson, a fellow scholar. It was during this visit that he deposited a manuscript copy of his "Carte du Mexique et des Pays Limitrophes Situés au Nord et à l'Est" ("Map of Mexico and Bordering Countries to the North and to the East") with the Secretary of State, Mr. James Madison, together with a report (see von Humboldt's letter to Jefferson, dated 20 December 1811 in Jackson, 1966b, p. 377–378; Wheat, 1958, p. 24)[302]. After he returned to Europe, he worked over his manuscript and published it in 1811 in two quarto volumes under the title *L'Essai Politique sur le Royaume de la Nouvelle-Espagne* (Political Essay of the Kingdom of New Spain) (de Humboldt, 1811a)[303], plus a folio atlas with 20 plates (de Humboldt (1812)[304].

Unlike many who considered the geography and the structure (what the French geographer Philippe Buache had called the "charpente" of the continent in 1761) of western North America before him, von Humboldt had learned the distinction between mountain ranges and plateaus. This distinction was old, going as far back as Strabo, and had been in common use among the European geographers and geologists in the eighteenth century (see Schulten, 1914, although I disagree with his belittling the contributions of the pre-Humboldtian geographers regarding the concept of plateau). As von Humboldt later pointed out (1843, p. 58, note 1), it was Strabo's important contribution to have distinguished "mountains" from "plateaus"; for the latter, he introduced the technical term *oropedia*[305]. Strabo noted that Eratosthenes' Taurus System (see Şengör, 1998, p. 10–12, and fig. 2 and 3) "has in many places as great a breadth as three thousand stadia"[306] (Strabo, XI. 1. 3). Strabo also knew that "the Taurus has numerous branches toward the north" (Strabo, XI. 12. 4), which he described under the names Antitaurus (XI. 12. 4), Scydises (XI. 2. 15), Moschici (XI. 2. 4), and Pariadres (XII. 3. 18; some topographic details at XII. 3. 28). East of these branches, he noted a number of parallel chains in present-day eastern Turkey, which, according to Strabo, "comprise many mountains [*ore*], many plateaux [*oropedia*]" (XI. 12. 4), all of which comprising the main trunk of the Taurus System of Eratosthenes here. In the eighteenth century, Buache (1761) developed the hypothesis that all continents possessed a central plateau, from which mountain chains radiated away and continued beneath the oceans, thus forming the geological structure of the globe (Fig. 76). This concept was nothing more than an extension of the idea of a highland hypothesized to exist in the center of North America by Sir Humfrey Gilbert (1539?–1583: see De Voto, 1952, p. 61) and of the high Asia hypothesis of Leibniz (1749, 1949) and von Linné (1744). It was warmly adopted by Pallas (1779: see Schmidt-Thomé, 1960), who gave it great currency. Although von Humboldt was not a follower of the "central plateau" hypothesis of Buache, he was certainly aware of the difference between mountain ranges and vast plauteau highlands when he went to America in 1799. This knowledge came in handy when he considered the structure of the Cordillera in Mexico and in the regions to the north. I let von Humboldt describe the high plateau of Mexico in his own words:

> There is hardly a point on the globe whose mountains present such an extraordinary construction as those of New Spain. In Europe, Switzerland, Savoy, the Tyrol are regarded as very elevated countries. But this opinion is based only on the aspect consisting of a grouping of a large number of peaks perpetually covered by snow and disposed in chains parallel with the great central chain.
>
> The peaks of the Alps reach 3900 even 4700 metres of height, although the plains near the canton of Berne rise only to 400 to 600 metres. The former elevation could be considered as that of most of the

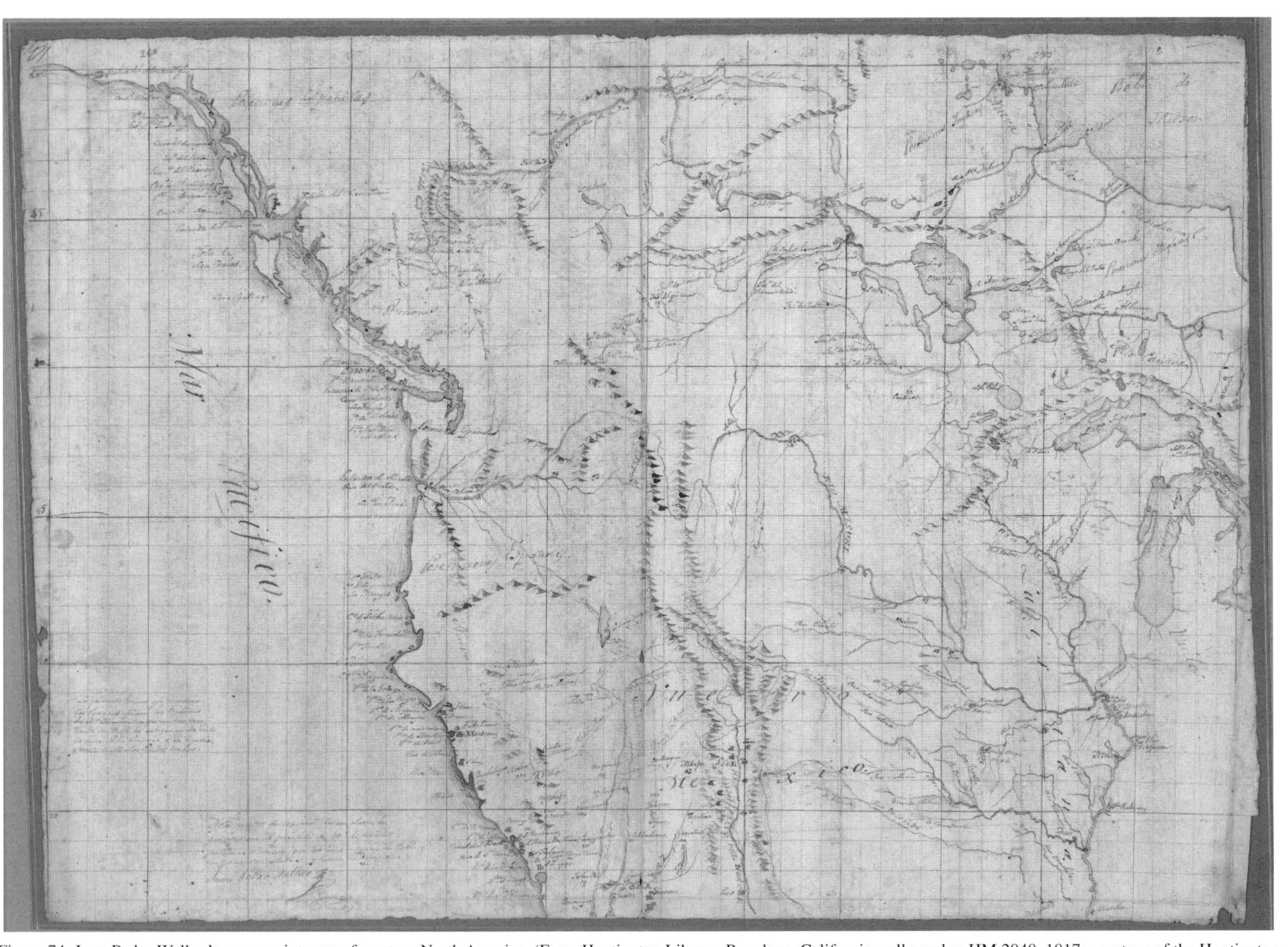

Figure 74. Juan Pedro Walker's manuscript map of western North America. (From Huntington Library, Pasadena, California, call number HM 2048, 1817; courtesy of the Huntington Library.) Notice the absence of any indication of the plateau country and the chain aspect of the North American Cordillera.

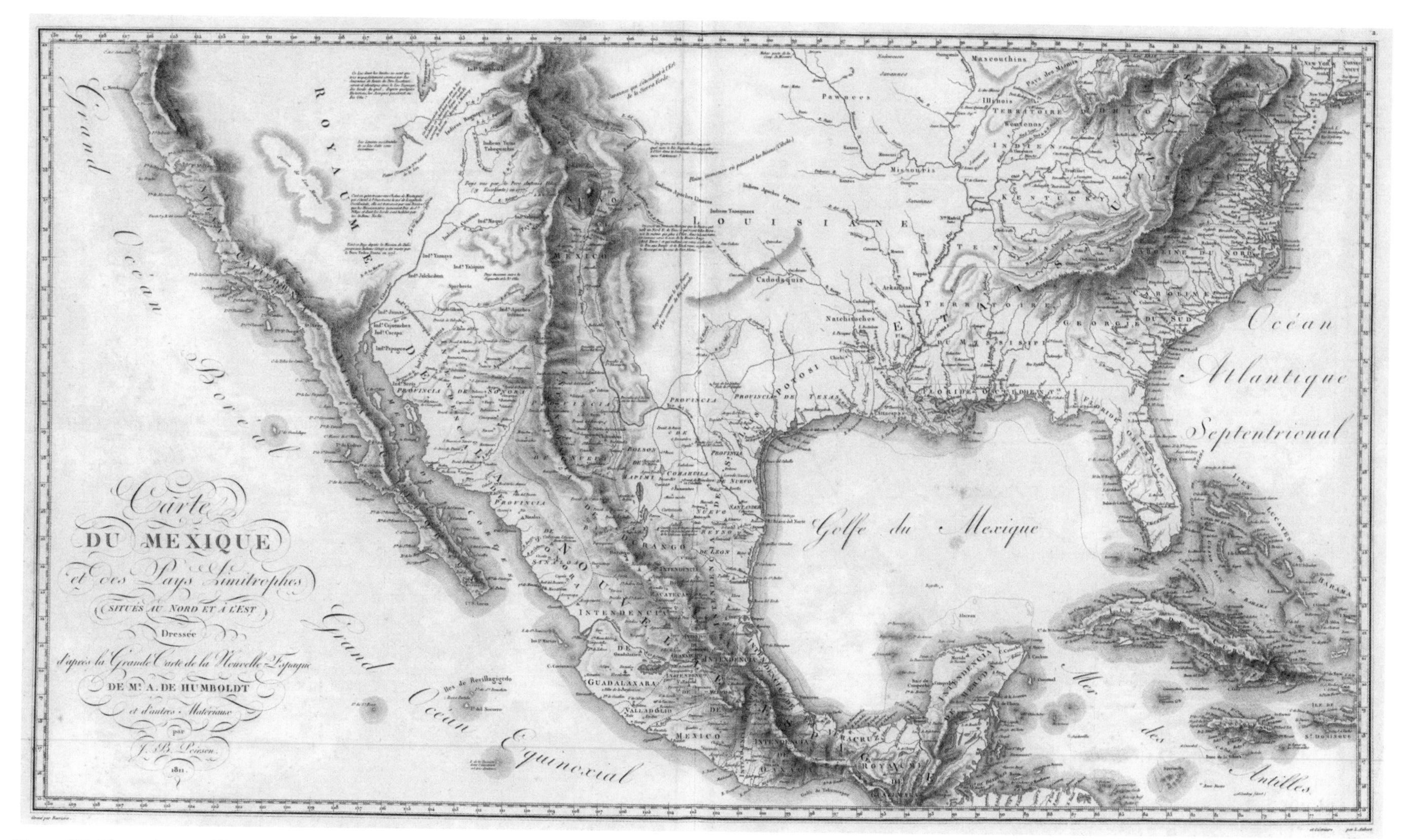

Figure 75. Alexander von Humboldt's great map of "Mexico and Bordering Countries Situated to the North and to the East," published in the Atlas volume to his *L'Essai Politique sur le Royaume de la Nouvelle-Espagne.* Although the Cordillera looks like a narrow range at first sight, when carefully examined, one sees the various highlands sketched in the plateau country. Compare this map with those of Captain Miera y Pacheco shown in Figure 72. It is clear that all of von Humboldt's information concerning the present-day Utah, Arizona, and parts of New Mexico came from Miera's map. (Courtesy of David Rumsey Collection, Cartography Associates.)

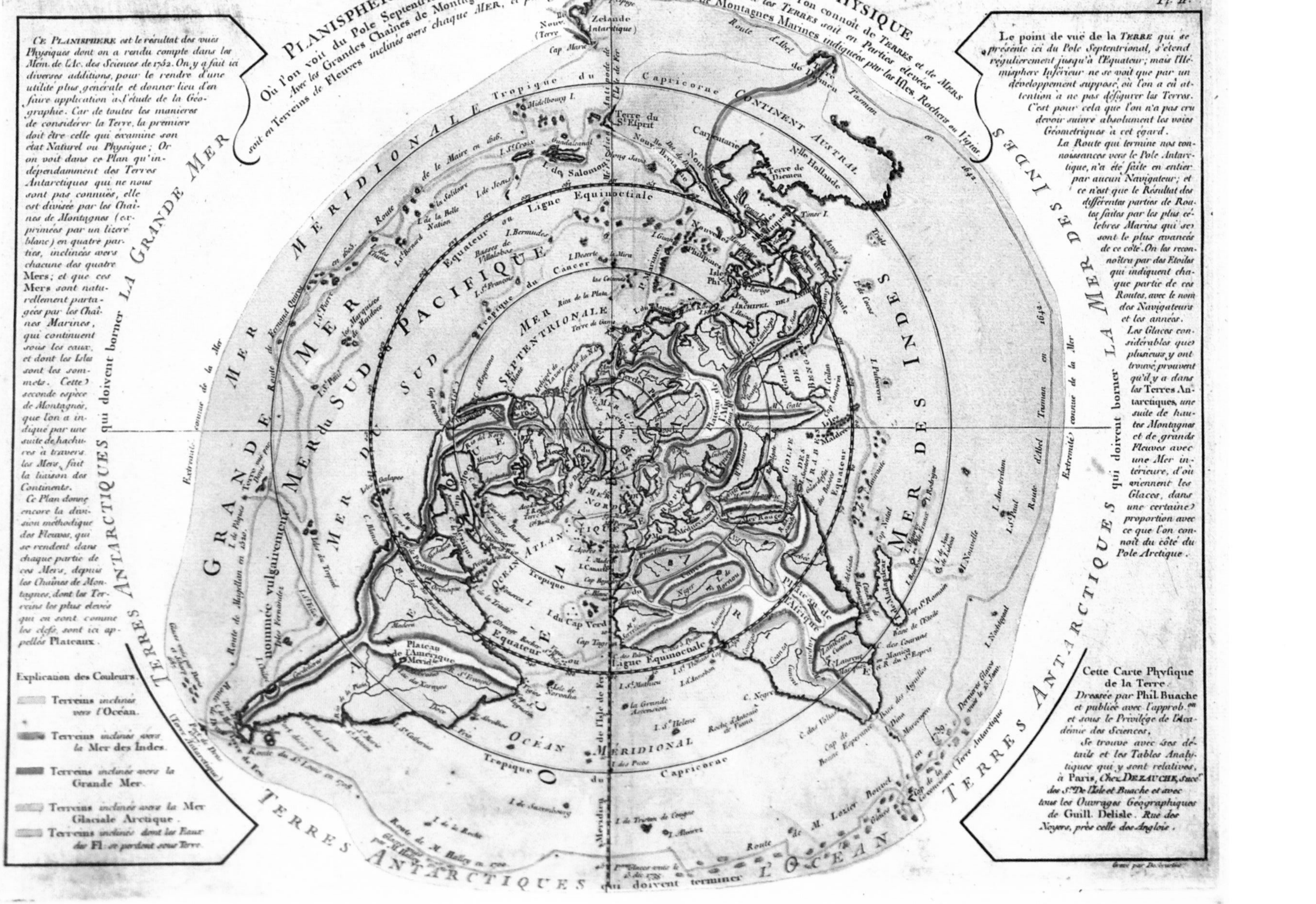

PLANISPHERE PHYSIQUE

Où l'on voit du Pole Septentrional ce que l'on connoît de Terres et de Mers
Avec les Grandes Chaînes de Montagnes, qui, traversant le Globe, divisent naturellement les Terres soit en Parties élevées
soit en Terreins de Fleuves inclinés vers chaque Mer, et partagent les Mers par une suite de Montagnes Marines indiquées par les Isles, Rochers ou Vigias.

Ce Planisphere est le résultat des vuës Physiques dont on a rendu compte dans les Mém. de l'Ac. des Sciences de 1752. On y a fait ici diverses additions, pour le rendre d'une utilité plus générale et donner lieu d'en faire application à l'étude de la Géographie. Car de toutes les manieres de considérer la Terre, la premiere doit être celle qui examine son état Naturel ou Physique; Or on voit dans ce Plan qu'indépendamment des Terres Antarctiques qui ne nous sont pas connuës, elle est divisée par les Chaînes de Montagnes (exprimées par un lizeré blanc) en quatre parties, inclinées vers chacune des quatre Mers; *et que ces* Mers *sont naturellement partagées par les Chaînes Marines, qui continuent sous les eaux, et dont les Isles sont les sommets. Cette seconde espèce de Montagnes, que l'on a indiqué par une suite de hachures à travers les Mers, fait la liaison des Continents.*

Ce Plan donne encore la division méthodique des Fleuves, qui se rendent dans chaque partie de ces Mers, depuis les Chaînes de Montagnes, dont les Terreins les plus élevés qui en sont comme les clefs, sont ici appellés Plateaux.

Explication des Couleurs.

- Terreins *inclinés vers* l'Océan.
- Terreins *inclinés vers* la Mer des Indes.
- Terreins *inclinés vers* la Grande Mer.
- Terreins *inclinés vers* la Mer Glaciale Arctique.
- Terreins *inclinés dont les Eaux des* Fl. *se perdent sous Terre.*

Le point de vuë de la Terre *qui se présente ici du Pole Septentrional, s'étend régulierement jusqu'à l'Equateur; mais l'Hémisphere Inférieur ne se voit que par un développement supposé, où l'on a eû attention à ne pas défigurer les Terres. C'est pour cela que l'on n'a pas cru devoir suivre absolument les voies Géometriques à cet égard.*

La Route qui termine nos connoissances vers le Pole Antarctique, n'a été faite en entier par aucun Navigateur; et ce n'est que le Résultat des différentes parties de Routes faites par les plus célebres Marins qui se sont le plus avancés de ce côté. On les reconnoîtra par des Etoiles qui indiquent chaque partie de ces Routes, avec le nom des Navigateurs et les années.

Les Glaces considérables que plusieurs y ont trouvé, prouvent qu'il y a dans les Terres Antarctiques, *une suite de hautes Montagnes et de grands Fleuves avec une Mer intérieure, d'où viennent les Glaces, dans une certaine proportion avec ce que l'on connoît du côté du Pole Arctique.*

Cette Carte Physique de la Terre. *Dressée par* Phil. Buache *et publiée avec l'approb.*[on] *et sous le Privilége de l'Académie des Sciences, Se trouve avec ses détails et les Tables Analytiques qui y sont relatives,* à Paris, *Chez* Dezauche, *Succ.*[r] *des S.*[rs] *De l'Isle et Buache et avec tous les Ouvrages Géographiques de* Guill. Delisle. *Rue des Noyers, près celle des Anglois.*

Gravé par Dezauche

TERRES ANTARCTIQUES qui doivent borner LA GRANDE MER
TERRES ANTARCTIQUES qui doivent borner LA MER DES INDES
TERRES ANTARCTIQUES qui doivent terminer L'OCEAN

plateaux that extend in Swabia, Bavaria and in New Silesia near the sources of the Warta and the Pilica. In Spain, the ground of the two Castilles has an elevation a little above 580 metres (300 toises). In France, the highest plateau is that of the Auvergne on which Mont-d'Or, Cantal and the Puy-de-Dôme sit. According to Mr. von Buch's observations, this plateau has an elevation of 720 metres (370 toises). These examples show that in Europe elevated terrains having the character of plains have elevations hardly more than 400 to 800 metres (200 to 400 toises) above the level of the ocean.

Perhaps in Africa, towards the sources of the Nile, and in Asia, between the 34° and 37° northern latitudes, one finds plateaux analogous to those of Mexico[307]. But the voyagers who traversed these latter regions have left us in perfect ignorance of the elevation of Tibet. According to the work of Father Du Halde, the elevation of the great desert of the Gobi, in northwest China, is above 1400 metres. Colonel Gordon has assured Mr. Labillardière that from the Cape of Good Hope to the 21st degree of southern latitude, the ground in Africa rises imperceptibly to 2000 metres (1000 toises) of elevation. This fact, new and surprising, had not been noted by other physicists. (de Humboldt, 1811[1997]a, p. 66–67)

With this introduction, von Humboldt tells us that the mountainous tracts of Mexico are to be seen as plateaus and not as individual mountain ranges (see also his hypsometric cross-sections in de Humboldt, 1812, plates 12, 13, and 14, which show the topographic character of the Mexican plateau[308]; also see the two-page frontispiece in de Humbodt, 1811[1997]b showing a complete hypsometric cross-section from Acapulco to Veracruz). We also learn that von Humboldt makes no distinction between what we today know to be the mainly copeogenic plateaux of Tibet and the falcogenic plateaux of Africa.

Yet von Humboldt sees Mexico as nothing but a continuation of the Andes. He does emphasize the difference in the structure, however, and he uses Buache's word, *charpente*, to express the idea of the structure of a mountain range: "The mountain chain forming the vast plateau of Mexico is the same as that, which, under the name of the Andes, traverses all of South America. However, the construction, I dare say the structure ["*charpente*"] of this chain is different on both sides of the equator" (de Humboldt, 1811 [1997]a, p. 67).

The difference von Humboldt saw was that the Andes were cleft by long valleys (he called them "crevasses") that run parallel with the chain and divided it into longitudinal ranges. He likened these valleys to gaping fissures[309] not filled in by "heterogeneous" extra material. Although plains existed with elevations of ~2700–3000 toises (~1400–1500 m), as those around Quito or in the province of Los Pastos, von Humboldt did not think that their exent was comparable with the highlands of Mexico. The high plains around Quito and farther north, north of the equator around Pastos, were nothing more than intermontane valleys limited on both sides by the main ridges of the "grand Cordillera of the Andes." In Mexico, by contrast, von Humboldt thought that the "...backs of the mountains themselves form the plateau. It is the direction of the plateau, so to say, that determines the direction of the entire chain" (de Humboldt, 1811[1997]a, p. 67). "In Peru, the highest peaks constitute the crest of the Andes. In Mexico, these same peaks, less colossal indeed, but nevertheless with elevations of 4900 to 5400 metres (2500 to 2700 toises), are more dispersed on the plateau or aligned along lines that have no relation to the direction of the Cordillera" (de Humboldt, 1811[1997]a, p. 67). Von Humboldt noted that, in the Andes, the existence of the deep valleys inhibited transportation; goods or persons that could normally be carried by horses had to be carried either on foot or on the backs of native porters (called *cargadores*). In New Spain, by contrast, he observed that a carriage could "roll on" from Mexico City to Santa Fe in the province of New Mexico (de Humboldt, 1811[1997]a, p. 67).

Von Humboldt thought that the plateau of Mexico extended from 18° N latitude to 40° N (well into the present-day U.S. states of Utah and Colorado: de Humboldt, 1811[1997]a, p. 69). He thought that the surface of this plateau dipped imperceptibly northwards. He regretted that no altitude measurements were available from regions to the north of Durango (24°01′ N latitude), but the voyagers he consulted had pointed out that the elevations dipped towards New Mexico and towards the sources of the Rio Colorado (the Colorado River of the West). North of the 19° N latitude, von Humboldt noted that the Cordillera assumed the name of Sierra Madre. North of Guanajuato (21° N, 101°16′ W), he observed that the Cordillera attained an immense breadth (de Humboldt, 1811[1997]a, p. 74) and soon afterwards became divided into three branches: The easternmost one followed pretty much the course of the present-day Pan-American Highway and disappeared around the city of Santander Jiménez. The eastern branch followed the coast and rapidly lost altitude, but did reach the Gila River in Arizona via Culiacán and Sonora.

It was in the main central branch of the Sierra Madre that von Humboldt recognized the true northerly continuation of the Andes. He believed that he was able to follow this branch (see

←

Figure 76. Philippe Buache's (1761) map of the world showing its main morphological features that express its structure (what Buache called the "charpente"). The left-hand legend that explains Buache's morpho-tectonic view is as follows:

This planisphere is the result of the views, of which an account was given in the Memoirs of the Academy of Sciences of 1752. Here diverse additions have been made to give it a more general utility and make it applicable to the study of geography, because, of all the ways of considering the earth, the first must be to examine its natural or physical state. It is seen here, with the exception of the Antarctic lands, which remain unknown, that it is divided by chains of mountains (expressed by a white signature) into four parts, inclined towards each of the four seas and that these seas are naturally divided by marine chains that continue under the waters and the islands form their summits. This second type of mountains, which is indicated by a series of hachures across the seas make up the liaison among the continents. This plan also gives a methodical division of the rivers which flow into each part of these seas from the chains of mountains, whose highest summits are like keystones and are here called plateaus.

It is interesting that the central plateaus of the continents are what Buache compared with keystones forming the summits of arches. The second legend to the right simply explains how the projection used was construed.

Fig. 75) from Durango all the way along the Sierra de las Grullas (*Crest of the Cranes*[310], the main ridge of the Rocky Mountains) to the Sierra Verde (*Green Crest*, the Yampa Plateau in northeastern Colorado: Crampton, 1958). This northern part of his great plateau (those parts now in the U.S. states of Arizona, Utah, Colorado, and New Mexico) von Humboldt learned from the maps of Miera and Font and the diary of Father Escalante (de Humboldt, 1811[1997]a, p. 74). He cites the latter two by name, and Father Escalante is further mentioned in the notes on his map (see Fig. 75). A comparison of Figures 72 and 75 will make the degree of von Humboldt's indebtedness to the Spanish explorers clear. However, Wheat (1957, p. 135, footnote 21) and Crampton (1958, p. 271) pointed out that von Humboldt's information of the cartographic fruits of the Escalante expedition and Font's map was most likely drawn from a secondary source, namely the maps of Miguel Costansó and Manuel Mascarò that were prepared from 1779 to 1782 to show the area between the Mississippi and the Pacific Ocean and from the Great Salt Lake to approximately Lake Chiapas (i.e., to about 16° N latitude; for the details of this group of maps, see Wheat, 1957, p. 121–124).

A remarkable feature on von Humboldt's map is the great valley of the Rio Grande. On both its sides, north-south–trending mountain ranges are seen. The one to the west, the Sierra de los Mimbres (*Crest of Wickers*) and its northerly continuation into the Sierra de los Grullas are broader. Moreover, its western slope is shorter than its eastern slope, indicating that in the west the mountain range borders a highland. This highland is studded with mesas and isolated plateaus, as reported by the Miera map and in Father Escalante's journal (Fig. 75). The larger of the plateaus, such as the unnamed ones on both sides of the Gila River, or the Sierra de la Florida to the south, or the Sierra de los Cosninas, or the Sierra de Chegui, all strike east-west or deviate from that orientation to the southwest or to the northwest. Almost all are at high angles to the main ridge of the Sierra de los Mimbres and the Sierra de los Grullas. All in all one gets the impression of a broad and long main trunk being joined in a sort of fish-bone fashion by auxiliary shorter and narrower branches (not dissimilar to the picture painted by Strabo's descriptions of the mountains of the Middle East).

We saw above that von Humboldt thought that this was also the case in the south. All of these topographic axes were thought of as independent axes of upheaval in line with the thinking von Humboldt was developing with his friend Leopold von Buch on the origin of mountain ranges (cf. Şengör, 1982a, 1998). Plateaus formed where a number of main axes came together and were joined by cross-axes. Von Humboldt used miner's terminology to describe the relationships of the individual axes of upheaval to one another. Such axes either crossed each other, or they crowded together into what a later generation was to call *virgations* and *syntaxes*. In the plateau country, the cross-axes drew von Humboldt's special attention ever since the origin of the volcano of Jorullo in the Trans-Mexican volcanic axis on 29 September 1759 (Gadow, 1930).[311] The event had shaken his confidence in his master Werner's neptunistic theory of the earth, and von Humboldt began to turn his allegiance to the views of Hutton (de Humboldt, 1805, p. 130; 1811[1997]a, p. 82 and 266–267; 1845, p. 251). That was why he later urged John Strong Newberry to search for axes of elevation coincident with centers of volcanism before Newberry embarked on his explorations of the Colorado Plateau. Von Humboldt knew, at least since the publication of the reports of the Pacific Railway Surveys (especially the Whipple Survey, the route near the 35th parallel, from the Mississippi River to the Pacific Ocean: see below), that a series of recent to subrecent volcanic centers extended from southern California to New Mexico. These volcanic centers, which at that time were believed to form an east-west "volcanic lineament," had also been reported by von Humboldt's friend, Balduin Möllhausen in the description of his voyage under the command of Lieutenant Whipple, based on the descriptions of geologist Jules Marcou (Möllhausen, 1858, p. 323–335 and notes 23, 24, 25 on p. 492–493; also see the wonderful colored plate showing the San Francisco Mountain {facing p. 324}). Möllhausen brought back rock samples from that voyage, and von Humboldt described them in the fourth volume of his great work *Kosmos* (von Humboldt, 1858b, p. 470–471) in the light of the publications by Marcou (see von Humboldt, 1858b, p. 471 for references to Marcou's papers; also see Möllhausen, 1861, v. II, p. 398–399, note 17).

Referring to the volcanics south of the Moqui (now Hopi) villages, in the area of the gorgeous San Francisco Mountain near Flagstaff, Arizona, Newberry wrote in response to von Humboldt's query:

> As I have previously remarked, the eruption of the material composing and surrounding these volcanoes has produced comparativly little disturbance of the sedimentary rocks upon which they rest. We have as yet no evidence that they form part of any well-marked line of upheaval, nor is there any obvious connection between these mountains and any of the chains which surround them, or with any of the other volcanic vents which are scattered over the great central plateau. My attention was particularly directed to this point by Baron Humboldt, through Mr. Möllhausen, before visiting New Mexico; and I was particularly requested to examine the country between the San Francisco Mountain[312] and San Mateo[313], to detect, if possible, some connecting link, such as lines of upheaval or volcanic vents. I was, however, able to discover no proof whatever of any relationship between the two other than perfect correspondence with their local phenomena present. It is true that east of the Little Colorado, just south of the Moqui villages, is a series of buttes composed of comparatively recent volcanic matter [*see also Möllhausen, 1858, p. 493, note 23*], which made its exit from a number of vents in that vicinity; but the sedimentary rocks are scarcely at all disturbed, even in the intervals between these trap buttes, and the country on either side is entirely free from dikes, faults, or displacements of any kind which would indicate the action of disturbing forces along the line connecting San Francisco Mountain and San Mateo. (Newberry, 1876, p. 61)

The expedition on which these observations were made took place in 1859, in the year von Humboldt died, so the aged Baron never received Newberry's answer to his queries. But from these queries we know that the great geographer was thinking in terms of igneous forces uplifting the plateaus of western North America. In his mind, all these plateaus thus belonged to the fast, abrupt movements of the earth's crust that created mountain ranges and, by grouping and crowding together, plateaus. This is hardly surprising, as von Humboldt had essentially no data to be able to assess the details of the hypsometry of the entire western third of the continent. The requisite data started to accrue as Americans began venturing westward under diverse pressures ranging from frontier troubles with the Spanish Empire to competition with the British in the extremely lucrative fur trade. [314] The culminating events of this post-von Humboldt period were the grand expeditions of John Charles Frémont that provided the first pieces of information on which American geologists could begin to speculate seriously concerning the tectonics of the western part of their newly acquired trans-continental empire in the 1840s.

Figure 77. Brigadier-General Zebulon Montgomery Pike (1779–1813), explorer, patriot, and an "acknoledged hero" in the words of Thomas Jefferson. (From Harris, 1990, p. 27, with permission.)

Mapping the Great Western Continental Swell: American Surveys to Frémont

Two main motives have contributed to the American exploration activity in the post-von Humboldtian period until the great explorations of Frémont: (1) gaining sufficient intelligence on the lay of the land between the recently purchased Louisiana territory and the Spanish Empire for the purpose of negotiating a boundary with the Spanish; and (2) to gain an upper hand in the fur trade and, in the process, to expand westwards at the expense of the British. Neither of these motives was specifically geological, but both contributed to the gathering of much topographic data and the inital geological data from the west of the Mississippi.[315]

Zebulon Montgomery Pike

After the Louisiana purchase, General James Wilkinson (1757–1825), the sinister governor of the newly acquired territory, sent the infantry lieutenant Zebulon Montgomery Pike (1779–1813; Fig. 77)[316] in 1805 to explore the Mississippi to its sources. Pike was only 26 years of age at the time, but peformed brilliantly. Consequently, immediately upon his return to St. Louis, General Wilkinson this time sent him west. The ostensible purposes of the expedition were to convey home some native dignitaries of the Osage Nation who had been to Washington, D.C., to visit the "Great White Father"; to return some 51 captives belonging to the same Nation who were liberated from their enemies, the Potawatomi (Viola, 1987, p. 30); to try to establish a permanent peace between the Osages and the Kansas; to establish friendly relations with the Comanches; and to find out as much about the southwestern boundary regions of Louisiana as feasible. During this journey, Pike has been accused of spying for General Wilkinson, with treacherous intent with respect to both the United States and the Spanish Empire (e.g., Coues, 1895a, p. lv, footnote 13). Others have fervently denied this accusation (e.g., Hart and Hulbert, 1932, especially p. lxiii–lxxvii). As we are only interested in the geographical-geological results of Pike's expedition, I shall not dwell on this topic any longer, except to say that Pike was singularly ill-suited to lead an expedition to collect geographic intelligence on account of his modest education in the natural sciences.

The expedition left St. Louis on Tuesday 15 July 1806. Pike's route took him gradually upslope on the craton towards the Rocky Mountains along the Missouri and the Osage Rivers (Fig. 78). On 11 August, about a week before reaching 95° W meridian, Pike noted the first trout encountered west of the Allegheny Mountains (Coues, 1895b, p. 378), a clear indication of the cooling of the stream waters as a result of the modest rise in elevation. On August 13th, he noted: "This day, for the first time, we have prairie hills" (Coues, 1895b, p. 380). They continued into September, skirting the Ozark Dome to the north and admiring "The prairie rising and falling in regular swells, as far as the sight can extend, [*which*] produces a very beautiful appearance" (Coues, 1895b, p. 397). By September 11th, Pike was already speaking of "high, hilly prairie" (p. 400). He was by no means oblivious to the geology through which they were traveling. Within the permission of his limited knowledge, he noted rock types (e.g., "This ridge was covered with a layer of stone, which was strongly impregnated with iron ore": Coues, 1895b, p. 402) and such features as spa springs, always with an eye on possible economic importance.

In eastern Kansas, where average elevation begins to rise above 1000 m, Pike noted, on 10 November, that the "face of the country considerably changed, being hilly, with springs" (Coues, 1895b, p. 440). The next day, he wrote in his diary that the hills were increasing (p. 441). Finally, on 15 November, Pike sighted, from eastern Kansas, the main ridge of the Rocky

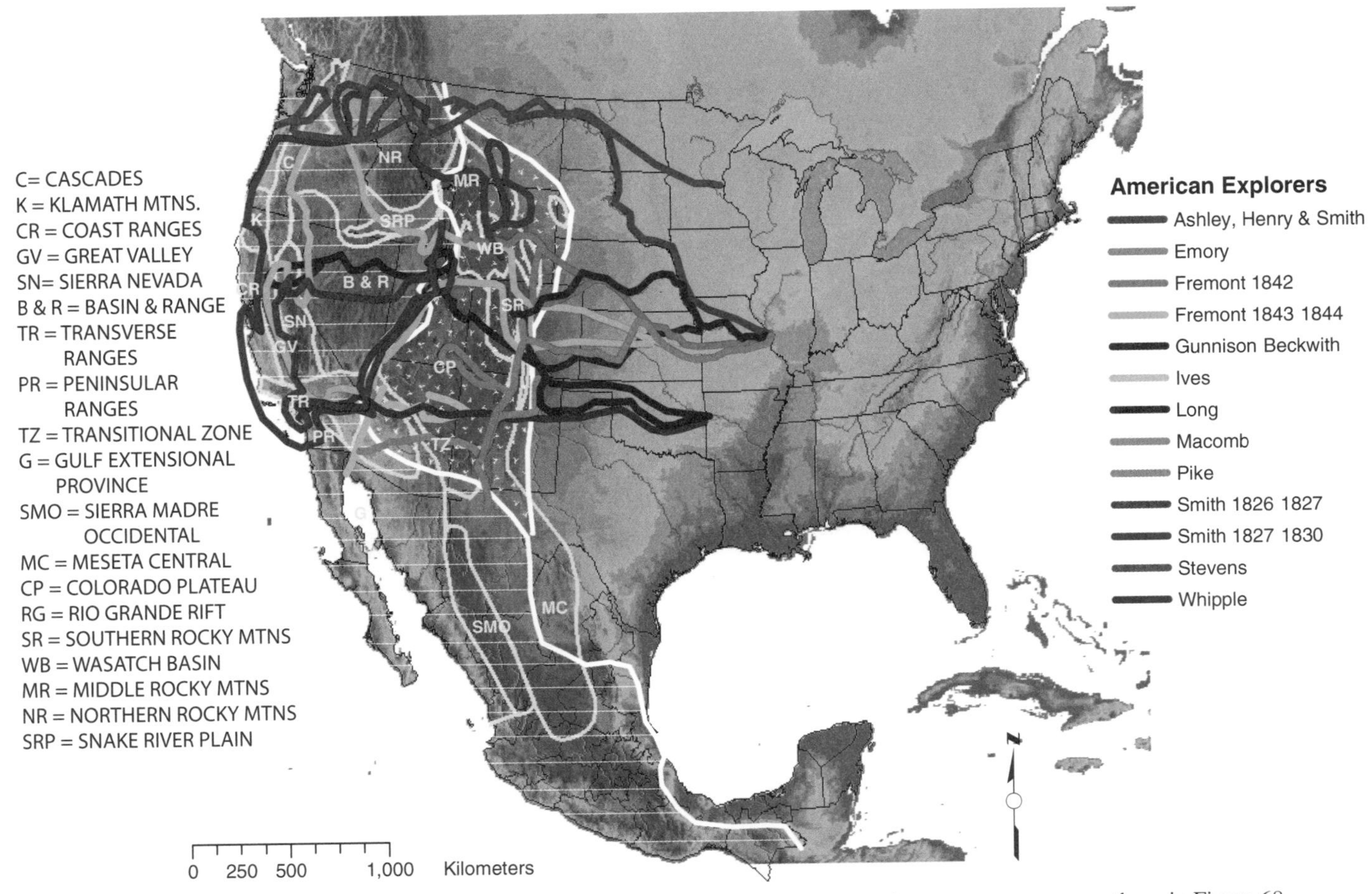

Figure 78. Routes of the American expeditions discussed in this book. Base map and route sources same as those in Figure 68.

Mountains (p. 444, see also Coues' explanatory comment in footnote 36 on the same page). On 3 December, they had reached the main ridge, and Pike measured the height of the highest peak in view (now called Pike's Peak) above the surrounding plains to be 10,581 ft. (~3200 m). He assumed that the plains they were standing on were 8000 ft. high (i.e., nearly 2500 m, whereas the actual elevation is ~1600 m). This would have made the height of the peak some 5700 m above sea level, only about 800 m lower than the then-known height of the volcano Chimborazo in the Andes, which was then thought to be the highest mountain in the world. Chimborazo's height was thought to be 6544 m, measured barometrically by von Humboldt on 23 June 1802 (von Humboldt, 1837a, 1837b; 1853, p. 132–174; the most recent measurements provide 6310 m as the height of Chimborazo).

On Christmas Eve, Pike compared his own exploration results along the Red River and in the Ozark Mountains with those by Dr. George Hunter, William Dunbar, and Thomas Freeman[317]. He noted that he had seen the sources of the Little Osage and White (Neosho) Rivers, he had been around the head of the Kans River (i.e., above the confluence of the Smoky Hill and Republican forks) and had seen the headwaters of the South Platte River (Coues, 1895b, p. 473). He was keen to explore also the upper course of the Red River which could not be reached by the Hunter et al. party owing to the hostility of the Osage Indians (Goetzmann, 1993, p. 42).

They were in the Rockies in the winter of 1806–1807, and as all previous explorers, Pike and his men suffered from the extreme cold. They had left the warm lowland of the Mississippi and reached the cold highlands of the Rockies. In the process, they recorded, for the first time, that the land was rising gently and continuously westward. This corroborated the old Spanish idea of a Great Valley in which the Mississippi was flowing, but Pike's party recorded for the first time, by daily observations, the extremely gradual rise of the land westward. Pike summarized this as follows:

As you approach the Arkansaw [*sic*] on this route within 15 or 20 miles [*~25–30 km*] the country appears to be low and swampy; or the land is covered with ponds extending out from the river some distance. The river at the place where I struck it is nearly 500 yards [*450 m*] wide, from bank to bank, those banks not more than four feet [*~1.32 m*] high, thinly covered with cottonwood. The north side is a swampy low prairie and the south a sandy sterile desert. Thence, about halfway to the mountains, the country continued with low prairie hills, and scarcely any streams putting into the river; and on the bottom are many bare spots on which, when the sun is in the meridian, is congealed a species of salt sufficiently thick to be accumulated, but so strongly impregnated with nitritic qualities as to render it unfit for use until purified. The grass in this district, on the river bottoms, has a great appearance of the grass in our salt marshes. From the first fork the borders of the river have more wood, and the hills are higher, until you arrive at its entrance into the mountains. (Coues, 1895b, p.517)

Pike also noticed that as one nears the mountains along the prairie rivers, the river beds become gravelly, deep and narrow; by contrast, as one moves in the direction of the descending prairie, they become sandier, wider, and shallower (Coues, 1895b, p. 521–522). He thought that the sources of all the great east-, west-, and south-flowing rivers was to be found in a small area in the main ridge of the Rocky Mountains:

The source of La Platte is situated in the same chain of mountains with the Arkansaw [*sic*] (see chart), and comes from that grand reservoir of snows and fountains which gives birth on its northern side to the Red river of the Missouri (the yellow stone river of Lewis), its great southwestern branch, and La Platte; on its southwestern side it produces the Rio Colorado of California; on its east the Arkansaw; and on its south the Rio del Norte of North Mexico. I have no hesitation in asserting that I can take a position in the mountains, whence I can visit the source of any of those rivers in one day. (Coues, 1895b, p. 523–524; see Pike's map[318] entitled "A Map of the Internal Provinces of New Spain" in Coues, 1895c; also reproduced in Jackson, 1966b, map 5.)

Pike's map shows that he conceived of the mountain ranges of western North America much in the spirit of the eighteenth century geographers, as narrow and long ridges. His contact with Juan Pedro Walker in Mexico may have further strengthened his views in this regard. On Pike's map we see none of Miera's or von Humboldt's flat-topped plateaus, and neither is it possible to get the impression of a vast, flat-topped highland around the Colorado Plateau from his cartographic depiction.

At the time Pike was writing the lines I just quoted, the central Rockies had not yet been explored, and New Mexico was believed to be neighboring the terrain traversed by the expedition of Lewis and Clark. This ignorance of the geography led Pike to assume a central highland somewhere in the Rockies close to where he was located, from which all the grand rivers of western North America supposedly derived their waters and the main mountain ranges radiated away. This notion was dispelled only following the explorations by the fur trapping mountain men organized by General Ashley in 1822 (Goetzmann, 1993, p. 105). To this group belonged the Mozart of western North American exploration history, the great explorer Jedediah Strong Smith. But before we summarize the geographical and geological legacy of the mountain men in relation to the long-wavelenth lithospheric structures in western North America, we must take a brief look at the expedition of Major Stephen Harriman Long (1784–1864), which produced the first graphic display of the gradual upwards slope from the prairies to the foot of the Rocky Mountains.

Stephen Harriman Long

Major Long (Fig. 79) proposed to the U.S. government to investigate the sources of the great east-flowing rivers of the Platta (i.e., Platte), Arkansas, and the Red Rivers (Wood, 1966, p. 89 ff.). The Secretary of War, John C. Calhoun, liked the idea and encouraged Major Long that "the farther you can extend your route to the West with safety, the more interesting and important it will be, as it will take you into a portion of our Country heretofore less explored" (quoted *in* Wood, 1966, p. 92). President Monroe finally approved the venture, and the expedition set off on 6 June 1820 from the "Engineer Cantonment" located about 8 km south of Council Bluffs (Warren, 1855, p. 24; Wood, 1966, p. 95; Goetzmann, 1993, p. 60). Long's party progressed along the Platte River. They were in the open plains country, but, regrettably, all their barometers were broken by the time they reached the forks of the Platte River (i.e., around 100° W meridian: Warren, 1855, p. 25), which meant that they were reduced to estimating altitude above sea level by noting the boiling temperature of water. Despite that, they noted the gently rising elevation, and when they reached the foot of the Rocky Mountains, the nights had become significantly cooler.

Unlike his predecessor Pike, Major Long was well-versed in the natural sciences, including geology (Wood, 1966, p. 42) and in addition, he had Edwin James along, who was responsible for botany, geology, and surgery when needed (Wood, 1966, p. 94). Smith (1918[1973], p. 195) considers the Long report authored by James to be the beginning of the federal work on

Figure 79. Major Stephen Harriman Long (from Wood, 1966, frontispiece).

the geology of the West. Thus, the resulting report is much more informative with respect to geology than was Pike's. Figure 80 is a reproduction of the cross section Long and James drew along the 38° N latitude (see James, 1823, for the narrative of the journey). The most remarkable aspect of this cross section is the depicted extremely gentle slope rising towards the Rocky Mountains, underlain by "Extensive Plains with insulated tracts of high table land." As one approached the Ozark Dome, the country became swelly and even mountainous. Long and James correctly show this being a result of rigorous downcutting in a country where the geological structure was not that of tablelands. The tablelands also disappeared once they were in the Rockies. Long and James took a neptunian view of geological theory (as was then prevalent in America: see especially Ospovat, 1960, p. 211–213) and depicted the "supposed level of the primitive ocean" (Fig. 80) where the granite-gneiss cores poked through the sedimentary rock of the Rockies, thus submerging the entire prairie country below its waves.

A year before Long's results were published by James, the great Philadelphia cartographer Henry S. Tanner (1786–1858)[319] brought out his Map of North America (Fig. 81), which Wheat regards as "monumental" and says that Tanner had made use of Long's results, as well as the results of the Kotzebue and Vancouver expeditions with respect to the northwest, McKenzie and Harmon's Journal for the British possessions, and Hearne's map for the Coppermine River (Wheat, 1958, p. 82–83). Tanner's map is the first one I know of showing remarkably well the swelling of the ground in the western United States. Although he depicts continuous ridges along the Rockies, the fact that the headwaters of rivers interfinger in a broad area (between the western longitudes of 100° and 110°) gives the impression of the presence of a broad swell rather than a sharp water divide. The publication of Long's expedition report in less than a year only strengthened this impression. It was substantiated by the geographical reconnaissance of the mountain men related to the fur trade in the next two decades.

Summary of Knowledge Obtained by Von Humboldt and His Successors

If we sum up what was learned by the early twenties of the nineteenth century concerning the general morphology and the tectonics of the western United States, we see that the concept of a broad ground swelling between the Mississippi Valley and the Pacific Ocean was well-established. Especially the eastern flank of that swelling had become well-known as a result of the expeditions of Pike and Long (and to a lesser degree by that of Lewis and Clark). Von Humboldt was still the only one who had a comprehensive theoretical view of what all that meant, and he maintained correspondence with his American friends to probe deeper into the nature of the great swelling. He viewed the whole structure as a (copeogenic) high plateau, no different from that which he had personally investigated in southern and central Mexico and from those he knew of only second-hand in central

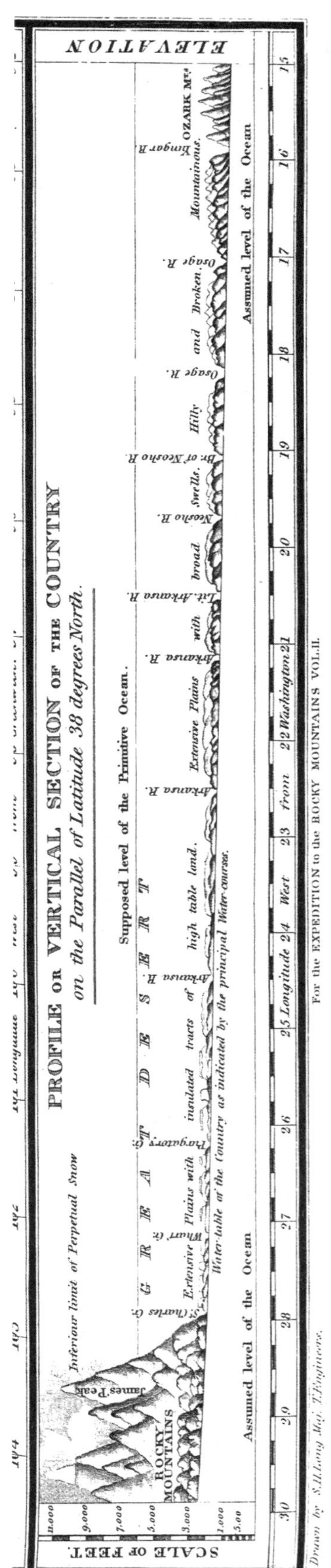

Figure 80. Topographical-geological cross section from the Ozark Mountains to the Rocky Mountains along the 38th north parallel (cf. Fig. 78; from James, 1823, foldout map in v. 1; copied from Wheat {1858, reproduction 353}, with permission). Note the gently rising topography westward, giving rise to the high tablelands before the Rockies are reached. Note also the depiction of the "Supposed level of the Primitive Ocean," betraying the explorers' Wernerian background.

Figure 81. A part of the great North America map by Henry S. Tanner, published in 1822, showing the topography of the United States. Note that the hydrographic net clearly implies the presence of the great western swell between the Mississippi River valley and the Pacific Ocean and that the broad crestal region of the swell is indicated by the interfingering of the headwaters of the rivers. (Courtesy of the David Rumsey Collection, Cartography Associates.)

Asia and central-eastern Africa. He thought that such plateaus were made no differently from mountain ranges and that they were essentially "broad mountain ranges" with numerous axes of elevation (as opposed to one dominant axis). That was why Von Humboldt wrote to Newberry and urged him to keep an eye on any possible axes of elevation on the Colorado Plateau.

The attitude of strata was too poorly known to attempt a synthesis. Even if it could have been done in the early 1820s, the dominantly flat attitude of the strata in the plateau country of the southwestern United States would hardly have aroused surprise, as all elevations were believed to be generated by vertical uplift as a result of magmatic injection at depth. In the 1820s and 1830s, two important developments happened to change this belief. In 1829, Élie de Beaumont showed that mountain belts are made by lateral shortening and folding of strata, and he re-inroduced the thermal contraction theory in a modern context, as we saw above. In the 1830s and 1840s, knowledge of the topography and geology of the West in America became sufficient to realize that the swelling of the ground between the Mississippi River valley and the Pacific Ocean was indeed very gentle and that it was a *deformational feature* of very long wavelength. Two groups contributed to this knowledge: the fur-trapping mountain men and Frémont's explorations.

Contributions of the Mountain Men to the Growth of Knowledge Concerning the Physical Geography of the Western United States

I follow Goetzmann (1993) in grouping under "mountain men" the fur trappers who had been organized in 1822 as a central company by General William Henry Ashley (1778–1838)[320] and Andrew Henry (born between 1773 and 1778, died 1832)[321]. General Ashley put an advertisement in the St. Louis *Gazette and Public Advertiser* for "Enterprising Young men … to ascend the Missouri to its source, there to be employed for one, two, or three years" (Goetzmann, 1993, p. 105)[322]. The area covered by Ashley's "Enterprising young men" in the 1820s and 1830s far exceeded the headwaters of the Missouri and eventually reached the shores of the Pacific Ocean. They discovered new routes across the Rocky Mountains, crossed what was to be called the Great Basin, blazed paths across the Sierra Nevada, and descended into what was then known as the Alta (Upper) or New California. In so doing, they paved the way for new settlers from the east and helped greatly the western expansion of the empire of the United States of America. They also mapped vast areas of the west and greatly clarified the geographical knowledge of the country west of the Rockies[323]. It was that knowledge that showed, for the first time, the gentleness of the great, north-south–trending topographic swell making up nearly the western half of the continent. By the time the General retired from the active fur-trapping business in 1826, his men had rediscovered the South Pass, originally discovered on 23 October 1812 by Robert Stuart[324] during the last expedition sent inland from Fort Astoria, founded and operated by John (Johannes) Jacob Astor's (1763–1848) Pacific Fur Company (Goetzmann, 1993, p. 34). They had also discovered the Great Salt Lake and thus made a first foray into the Great Basin, in addition to opening up the area of the future Yellowstone Park (Goetzmann, 1993, p. 129).

Ashley sold his interest to Jedediah Smith (1799–1831)[325], who continued trapping and exploring. After the 1826 rendezvous in Cache Valley, Utah, Jedediah Smith made his grand overland journey across the Great Basin to California and back. During that journey, he became the first civilized man to see the terrains described by Father Escalante and Father Garcés. In 1827, he again went to California. This time he swung way north, reached Vancouver, contacted the British authorities, and then headed back. In addition to charting much hitherto untrodden terrain, Smith wrote to the Secretary of War, John H. Eaton, on his return on 29 October 1830, to emphasize the relative ease with which not only wagons but also herds of cattle and milk cows may be taken across the South Pass all the way to California (Goetzmann, 1993, p. 103, 139). This was the first public announcement of the gentleness of the topography across the great western swell, nothwithstandig the great mountains—the Rockies and the Sierra Nevada, plus what were later to be called the basin ranges in the Great Basin—that ornament its surface in places. By 1832, all the important trails in the west had been explored, and by the end of the 1830s, the Oregon Trail was well-established (Goetzmann, 1993, p. 77, 169). During the period 1833–1834, Joseph Walker's (1798–1876) parties also established the great Emigrant Trail to California (Warren, 1855, p. 33–34; also see Gilbert, 1983, p. 119–152)[326].

The results of the efforts of the mountain men were made public mostly by stray reports in local and more rarely in national newspapers in the United States. The great *Map of the United States of North America* of David H. Burr, Geographer to the House of Representatives, published in 1839 (Fig. 82), incorporated all the then available information including detailed indications of Jedediah Smith from a manuscript map Burr is believed to have had open before him while compiling his own map (Morgan and Wheat, 1954, p. 20–23). Smith's manuscript map was most likely provided to Burr by Smith's friend and one-time partner, General Ashley, who was then a Member of the House.

The most remarkable aspect of the Burr map is the Humboldtian nature of the central part of its western half (shown in Fig. 82). Burr's addition of much detail concerning the hydrographic network of the West greatly enhanced the swell appearance of the plateau country, and the broad-topped plateaus stand in sharp contrast to the narrow mountain depictions of most previous maps and with similar depictions of mountains elsewhere (including the narrow and long ridges of the Valley and Ridge province of Appalachians, which Burr depicted appropriately as narrow ridges). His map shows that by 1840, the American West was no longer a terra incognita: its vast dimensions had been revealed and the main routes within it had been charted (see Goetzmann, 1993, p. 179).

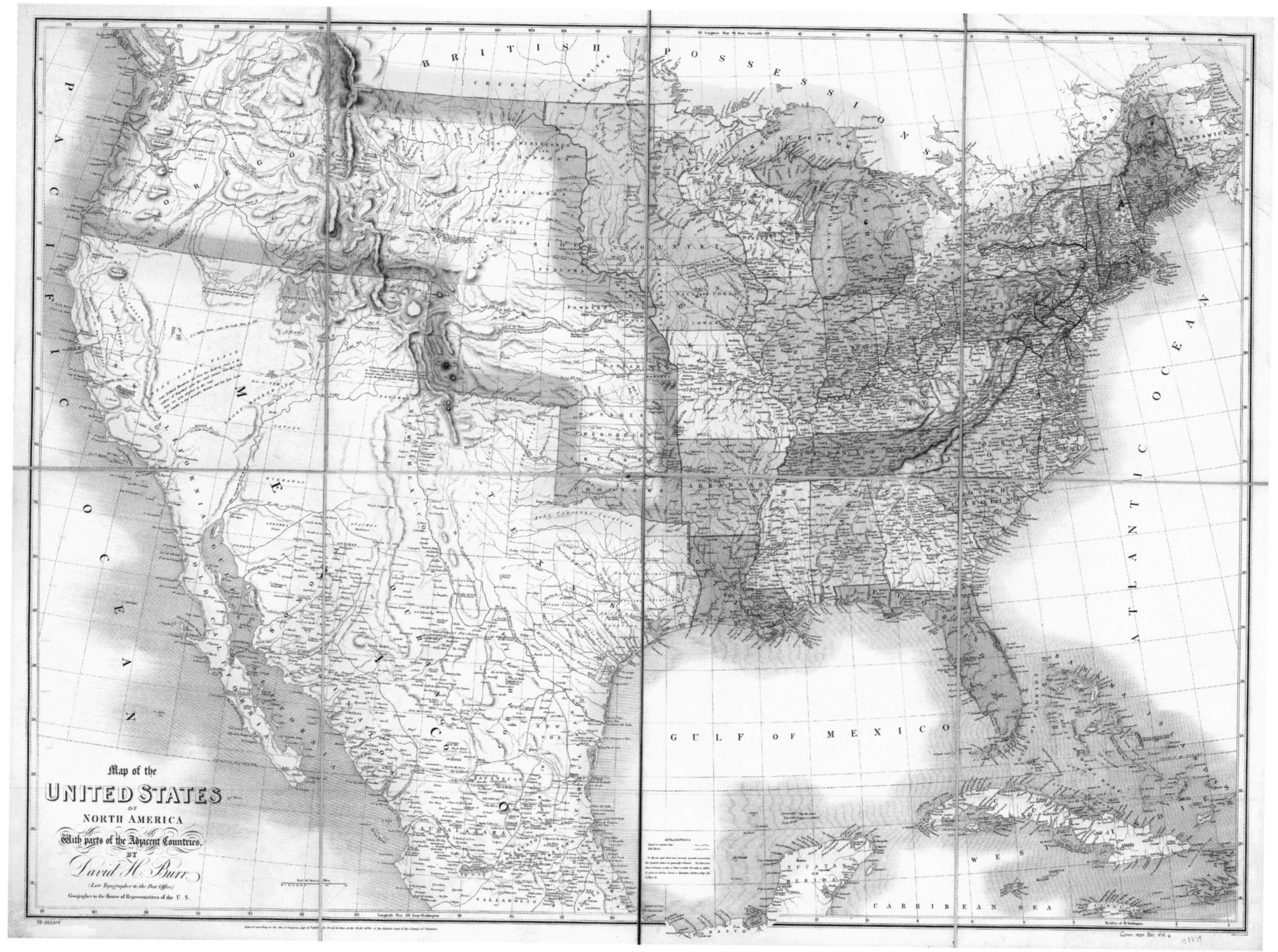

Figure 82. The *Map of the United States of North America* by David H. Burr, Geographer to the House of Representatives, published in 1839. (Courtesy of the Library of Congress). Note the "Humboldtian" appearance of the West. Contrast the great breadth of the plateaus of the West with the narrow folded ridges of the Valley and Ridge province of the Appalachians. Although Burr's flat-topped ranges are not as broad as von Humboldt's, nor are they nearly as broad as they ought to be when compared with the real terrain, his map shows that the appreciation of the flat-topped highlands in the West had started growing. Compare this map with Walker's (shown in Fig. 74), and von Humboldt's (Fig. 75), and with Tanner's (Fig. 81).

Even conjectures began forthcoming now from American geologists as to its tectonic nature. Above, I have quoted Edward Hitchcock's hypothesis that the entire North America consisted essentially of two large crustal anticlines forming the Appalachians and the Rockies and a vast, shallow syncline housing the Mississippi Valley (Hitchock, 1841, p. 264–265). In his textbook, Hitchcock pointed out that the Rockies had a "primitive core," successively overlain by "extensive secondary deposits ... among which are clay slate, greywacke, and limestone" (Hitchcock, 1847, p. 342). Although Hitchcock learned new facts concerning the western geology from Nicollet's observations beween 1839 and 1847 and no doubt from Frémont's two expeditions (see below), I quote his 1847 book to show that his overall picture of a mountain range was very largely that of the eighteenth century and that of Élie de Beaumont in the early nineteenth century. This picture was to undergo a significant change with Dana's work.

Dana's picture of his home continent was greatly influenced not only by his predecessors (such as Hitchcock) but also by the geographical explorations in the western United States that took place in the early part of the 1840s. It was that picture that dominated Dana's vision of terrestrial tectonics to the end of his life in 1895 and that decisively influenced tectonic thinking in Europe into the second half of the twentieth century! The more modern concepts concerning the nature and origin of long-wavelength structures of the lithosphere on the basis of observations in the western United States were developed in the latter half of the nineteenth century by criticizing Dana's tectonic world-picture. That is why Dana's picture of the tectonics of the United States, and especially of the large topographic swell in its western one-third of the country, is of the greatest historical importance. Without understanding its origins, observational basis, and subsequent influence, it would be impossible to understand the developments in tectonics in the latter half of the nineteenth and the first half of the twentieth centuries.

Frémont and Dana

As the mist covering the West was gradually lifting to allow the bare outlines of its physical geography to emerge, the U.S. government was becoming progressively more interested in exploring the region on a more solidly scientific basis as a prelude to its expansion westward following the "Manifest Destiny" of the young republic. Among the staunchest advocates of this policy was the senior U.S. Senator from Missouri—then one of only three states west of the Mississippi—Thomas Hart Benton (1782–1858), generally known as the "Old Bullion"[327]. Benton believed that the West could only be occupied in a sensible and enduring way if the government was properly informed of its potential. This meant exploration, but no longer of the kind undertaken by the fur trappers and the mountain men. The government needed to know, by means of properly documented reports, the detailed topography, the climate, the soil, the fauna and flora, the prospective mines, and the present-day inhabitants, if all which were to be embraced by the civilization to be expanded westward by the United States.

In Washington, Benton and Lieutenant-Colonel John James Abert (1788–1863), the chief of the newly-created Army Bureau of Topographical Engineers[328], used to visit frequently the home of the eccentric Swiss scientist Ferdinand R. Hassler (1770–1843) in 1839 and 1840. At the time, as the head of the U.S. Coast Survey, Hassler was a close friend of Joseph Nicolas Nicollet (1786–1843), the immigrant French mathematician and geographer, who had explored the upper reaches of the Mississippi and the terrain between the Mississippi and the Missouri in addition to the Alleghenys and the Red and the Arkansas Rivers since his arrival in the United States in 1832[329]. In his last two expeditions up the Mississippi and the Missouri Rivers, Nicollet had been provided as an assistant a young civilian member of the Bureau, John Charles Frémont (1813–1890: Fig. 83), a very young protegée of the Secretary of War, Joel R. Poinsett. A man of remarkable abilities[330], Frémont had been already familiar with the rudiments of surveying, but it was under Nicollet's enthusiastic tutoring that he had ripened into a first-rate explorer. His apprenticeship under Nicollet had brought him into contact with another man of great western fame, fur trapper and the first white man ever to see the Great Salt Lake in Utah (Chittenden, 1902[1954], p. 280;

Figure 83. The great American explorer and geographer, John Charles Frémont (1813–1890) during the years of his early expeditions.

Goetzmann, 1993, p. 120), the French-Canadian Étienne Provot (1785–1850)[331]. From Provot, the young Frémont had learned the skills and the delights of western wild life.

During their visits, Senator Benton instilled in young Frémont a desire to explore the West. Nicollet would have been the ideal leader for such an expedition, but his health was rapidly failing. Instead, his able assistant Frémont, now commissioned a lieutenant in the Bureau, was chosen to lead an expedition "to explore and to report upon the country between the frontiers of the Missouri and the South Pass in the Rocky mountains, and on the line of the Kansas and Great Platte rivers" (Frémont, 1845, p. 9). The expedition set off from Cyprian Chouteau's landing at the junction of the Missouri River with the Kansas River on Friday, 10 June 1842. Frémont had with him the Prussian cartographer and landscape artist, Carl Preuss, and the party was guided by the legendary mountain man Christopher "Kit" Carson (1809–1868)[332].

Frémont's expedition went from the confluence of the Kansas and the Missouri Rivers to South Pass (located in present-day Wyoming), south of the Wind River Range of the U.S. Rockies, mainly following the courses of the Republican and the North Platte Rivers (see Fig. 78). Frémont measured the elevation of their starting point (94°25′46″ W and 39°5′57″ N) to be 700 ft. (~212 m) above sea level (Frémont, 1845, p. 9). By the time they reached 98°45′49″ W and 40°41′06″ N, Frémont observed that they had reached an altitude of about 2000 ft. (~600 m) and were traveling on "lime and sandstone[333], covered by the ... erratic deposit of sand and gravel which forms the surface rock of the prairies between the Missouri and Mississippi rivers. Except in occasional limestone boulders, I had met with no fossils" (Frémont, 1845, p. 16). As they continued their westerly march, the prairie continued gaining in elevation (Frémont, 1845, 1 July 1842 entry, p. 19). On 7 July, Frémont noted a change in rock type from sandy soil to hard marly clay at 103°07′00″ W and 40°51′17″ N (p. 25; he was in the Tertiary continental sedimentary rocks: King and Beikman, 1974), but the next day they were traveling atop an alternation of sandstones and shales. The barometer observations indicated that the elevation was continuing its gentle rise (Frémont, 1845, p. 27). On 9 July 1842, they caught the first glimpses of the Rocky Mountains. The next day, they reached St. Vrain's Fort (~6 km south of Milliken, Colorado, which is at 40°21′ N, 104°53′ W). Here Frémont observed that their elevation was 5400 ft. above sea level (~1630 m), and he noticed that the region was free of "the limestones and marls which give the lower Platte its yellow and dirty color" (Frémont, 1845, p. 31). He was right: They had come out of the Tertiary deposits and were back in the Upper Cretaceous (King and Beikman, 1974; also see Hall, 1845, p. 296) at the foot of the Rockies.

Until 4 August 1842, Frémont skirted the Laramie Mountains forming the northern Front Range of the Rockies and led his party to South Pass across what is now known as the Granite Mountains, a south-vergent basement-cored thrust mass between the Wind River Mountains in the northwest and the Laramie Range and the Medicine Bow Mountains in the southeast and east. It was during this trek that we see him reporting, for the first time, dip measurements on sedimentary rocks: on 30 July 1842, at the foot of the Rattlesnake Range near the future Pathfinder Dam (constructed in 1909 and named after him), forming a part of the Granite Mountains. He reported "pudding-stones... the pebbles in the numerous strata increasing in size from the top to the bottom, where they are as large as a man's head. So far as I was able to determine, these strata incline to the northeast, with a dip of about 15°" (Frémont, 1845, p. 55)[334]. Up to this point, Frémont had been annoyingly uninformative about the attitude of the strata he was observing. It was probably because he did not find it necessary to report horizontal attitudes. Only where significant deformation began to command his attention did he commence reporting attitudes. Wherever he encountered horizontal strata, amidst deformed rocks, he did report their attitude also (e.g., Frémont, 1845, p. 57).

As they marched farther west along the Sweet Water River, a fine-grained granite[335] was encountered, cut by "trap rocks,"[336] on 2 August 1842. On 4 August, they camped at the foot of granite mountains (literally within what is today called the Granite Mountains) in view of the Wind River Mountains: "a low and dark mountainous ridge," Frémont noted (1845, p. 58: Fig. 84). Before reaching South Pass, they entered a canyon along the Sweet Water River on 6 August, at the immediate entrance of which, "superimposed directly upon the granite, are strata of compact calcareous sandstone and chert, alternating with fine white and reddish white, and fine gray and red sandstones. These strata dip to the eastward at an angle of about 18°, and form the western limit of the sandstone and limestone formations on the line of our route. Here we entered among the primitive rocks" (Frémont, 1845, p. 58). Up this narrow valley they later noticed boulders of "gneiss, mica slate, and a white granite" where "on both sides granite rocks rose precipitously to heights of three hundred and five hundred feet, terminating in jagged and broken pointed peaks" (Frémont, 1845, p. 59). On 10 August 1842, they proceeded farther up the Sweet Water River and encountered 45°-dipping (direction unspecified!) compact mica slate, "alternating with a light-colored granite...; the beds varying in thickness from two or three feet to six or eight hundred" (p. 59).

When they were observing the dipping mica-slates, they were already within a few miles of South Pass. It was so inconspicuous that:

> with all the intimate knowledge possessed by Carson, who had made this country his home for seventeen years, we were obliged to watch very closely to find the place at which we had reached the culminating point. This was between two low hills, rising on either hand fifty or sixty feet [*i.e., barely 20 m*]. When I looked back at them, from the foot of the immediate slope on the western plain, their summits appeared to be about one hundred and twenty feet above. From the impression on my mind at this time, and subsequently on our return, I should compare the elevation which we surmounted immediately at the Pass, to the ascent of the Capitol hill from the avenue, at Washington. (Frémont, 1845, p. 60).

Figure 84. View of the Wind River Mountains as beheld by Frémont on 4 August 1842 (from Frémont, 1845, facing p. 66).

Frémont thought this place was hardly worthy of the name "Pass":

> It will be seen that it in no manner resembles the places to which the term is commonly applied—nothing of the gorge-like character and winding ascents of the Allegheny passes in America: nothing of the Great St. Bernard and Simplon passes in Europe. Approaching it from the mouth of the Sweet Water, a sandy plain, one hundred and twenty miles long, conducts, by a gradual and regular ascent, to the summit, about seven thousand feet above the sea [*~195 km*]; and the traveller, without being reminded of any change by toilsome ascents, suddenly finds himself on the waters which flow to the Pacific Ocean. (Frémont, 1845, p. 60)

He was also careful to point out that even the surrounding mountains were "not the Alps" (p. 61).

From the above descriptions by Frémont, his readers in the 1840s obtained the following picture: From the confluence of the Kansas and the Missouri Rivers to the foot of the Rocky Mountains, the ground rises very gently, about an average of 0.094°. All along this route are rocks that, for all practical purposes, are flat-lying. When South Pass is reached, there are some mountains around, which are granite-cored and overlain by schists, slates, and clastic rocks. Dips increase towards the cores of the mountains that are dominated by "primitive" formations. Despite all this, South Pass itself is a very gentle inflexion, barely recognizable on the ground as such. It is clear that there is a gentle swelling of the ground in the western half of the continent, the culminating point (or line) of which is pierced with granites creating local protuberances appearing as middle-size mountains.

As we have seen above, this is exactly the picture Edward Hitchcock formed for himself of the structure of the Rockies (Hitchcock, 1847, p. 342). Von Humboldt (1849, p. 59) summarized his impressions as follows:

> Under the middle latitudes of 37° to 43°, the Rocky Mountains display, apart from snow-covered peaks, whose elevations are comparable with that of the Pic of Tenerife, high plains of such extent that they can be seen nowhere else in the world. In east-west breadth, they are more than twice as wide as the Mexican highland. From the mountain mass, which begins somewhat to the west of Fort Laramie, to the other side of the Wahsatch Mountains, a swelling of the ground uninterruptedly maintains an elevation of five to seven thousand feet [*~1620–2268 m*] above sea level. Indeed, it fills the entire space, from 34° to 45° longitude, between the Rocky Mountains *sensu stricto* and the snowy range of the Californian coast. This area, a sort of broad longitudinal valley like that of Lake Titicaca, is called **the Great Basin** by Joseph Walker, an experienced western traveler, and Captain Frémont. It is a terra incognita of at least 8,000 geographical square miles [*~440,418 km²*] barren, almost free of humans, and full of salt lakes, the largest of which lies 3,940 Paris feet [*~1277 m*] above sea level and is connected with the narrow Lake Utah.
>
> …
>
> Although the water divide reaches an elevation, which comes close to those of the passes of Simplon (6,170 feet [*~1999 m*]), Gotthard (6,440 feet [*~2087 m*]), and Great St. Bernhard (7,476 feet [*~2422 m*]), the ascent is so gentle and gradual that nothing hinders the traffic on wheels and wagons between the Missouri and the Oregon region, between the Atlantic states and the new settlements along the Oregon or Columbia rivers, between the shores facing Europe and China. The distance from Boston to the old Astoria on the South Sea [*i.e., the Pacific Ocean*] is 550 geographic miles [*~4081 km*], as the crow flies, on the basis of the longitude difference. With such a gentle ascent of the highland, leading from the Missouri to California and to the Oregon region (from Fort and River Laramie on the northern fork of the Platte River to Fort Hall on the Lewis Fork of the Columbia River all measured camp sites were five to seven thousand feet [*~1620–2268 m*],

in Old Park 9,760 feet [*~3162 m*], high), the culmination point, the *divorta aquarum*, was established not without trouble. It is found south of the Wind-River Mountains, fairly exactly in the middle of the way from the Mississippi to the littoral of the South Sea, at an elevation of 7,027 feet [*~2277 m*]), thus only 450 feet [*~146 m*]), lower than the Great St. Bernhard Pass. The locals call the culmination point the South Pass. (von Humboldt, 1849, p. 59–62)

Frémont returned to Cyprian Chouteau's place on 10 October 1842. He was back in St. Louis on October 17 and arrived back in Washington, D.C., on October 29. He finished writing his report to Colonel Abert on 1 March 1843, and by March 10, he had already received new orders to proceed back to the West to complete a profile from the interior to Oregon, in order to connect the observations in the interior of the continent with those of the Wilkes expedition along the Oregon coast.

Frémont's descriptions of the geology along the route of his second expedition are far scantier than the first, but to compensate for it, there is a report by James Hall of the New York Survey based on the observations and on rock and fossil speciments brought back by Frémont. Frémont noted with habitual care the rise of elevations towards the Rockies on his way out. The routes of his two expeditions were almost coincident, except this time the explorers followed the South Platte River instead of the North Platte River. On 14 June 1843, Frémont noted that some 265 mi. (~425 km) from their starting point (Kansas City), they were already at an elevation of 1520 ft. (~460 m). They had been "gradually and regularly ascending" (Frémont, 1845, p. 108). On June 22, they had reached 2130 ft. (~645 m) elevation, on June 23, they were at 2350 ft. (~712 m), and on June 25 they were already traveling on a road 3100 ft. (~940 m) above sea level. On 8 July 1843, they were 21 mi. (~33.5 km) from St. Vrain's Fort, and Frémont recorded their elevation as 5500 ft. (~1667 m). This time they diagonally cut across the northernmost part of the Laramie Range, which displayed a "red, feldspathic granite, overlying a decomposing mass of the same rock, forming the soil of this region, which everywhere is red and gravelly" (Frémont, 1845, p. 123). After having crossed the Medicine "Butte" (where Frémont reported an elevation of 8300 ft. {~2515 m}: p. 125), the party reached South Pass on 13 August 1843. Frémont measured the elevation at South Pass "with a good barometer" and recorded it at 7490 ft. (~2270 m; p. 128; see the quotation above on p. 164 in which Frémont's estimate of 7000 ft. was recorded during the first expedition).

Having passed South Pass and with it the watershed, the elevations began dropping again. At Bear Springs in present-day Idaho (which they recorded at 42°39′57″ N and 111°46′00″ W), their elevation was 5840 ft. (~1770 m: Frémont, 1845, p. 138). As they came down the valley of the Bear River (which we now recognize to be one of the innumerable rift valleys of the Basin and Range province) to the Great Salt Lake, the elevations further dropped to 5100 ft. (~1545 m). On 9 September 1843, a group of five, consisting of the cartographer Preuss, the *voyageurs* Bernier and Basil Lajeunesse, Kit Carson, and headed by Frémont, was afloat in a little vessel on the Lake, the surface of which they determined to stand at an elevation of 4200 ft. (~1273 m) above sea-level.

Having finished their exploration of Salt Lake, they skirted the "northern boundary of the Great Basin" (Frémont, 1845, p. 170) and by the time they reached the divide between the Burnt and the Powder Rivers (where present-day Baker City, Oregon, is located, i.e., 44°46′ N, 117°50′ W), they had descended to 3300 ft. (~1000 m). On October 18, they were at 117°28′26″ W and 45°26′47″ N and at an elevation of only 2600 ft. (~787 m). They thus reached the great basaltic plateau of the Columbia River. Frémont described the basalts as being uniform in the areas where he got to see them and "very compact, with a few round cavities" (Frémont, 1845, p. 187).

The rest of Frémont's route is of little interest from the viewpoint of this book except his establishment of the elevation of the Great Basin, his surprise in finding the Sierra Nevada higher than the Rockies, and the connection he assumed to exist between the Cascades and the Sierra Nevada ranges. All three had important repercussions in the development of theoretical tectonics in the mid-nineteenth century.

Frémont took his exploring party from Oregon, where he confirmed the volcanic character of the Cascades Range (Frémont, 1845, p. 193), south in search of the legendary San Buenaventura River (p. 196), which supposedly connected the Great Salt Lake with the ocean and drained the western part of the continent at that latitude (for the history of the idea of a San Buenaventura River, see Crampton and Griffin, 1956). He marched south and crossed the Sierra Nevada just south of Lake Tahoe. Before this, he named and measured the elevation of Pyramid Lake and noted that it lay 700 ft. (~212 m) higher than the Great Salt Lake: "The position and elevation of this lake make it an object of geographical interest. It is the nearest lake to the western rim, as the Great Salt Lake is to the eastern rim, of the the Great Basin which lies between the base of the Rocky mountains and the Sierra Nevada; and the extent and character of which, its whole circumference and contents, it is desirable to know" (Frémont, 1845, p. 217–218).

By the time Frémont crossed the Sierra Nevada, he had decided that the San Buenaventura River was a myth (p. 226; see also Crampton and Griffen, 1956, especially p. 170–171), and while crossing the Sierra Nevada, he was greatly surprised to find it to be higher in elevation than the Rocky Mountains. On 19 February 1844, he estimated the height of his camp atop the Sierra Nevada at 38°44′ N and 120°28′ W and found it 9330 ft. above sea level (~2827 m).

This was 2,000 feet [*~606 m*] higher than the South Pass in the Rocky mountains, and several peaks in view rose several thousand feet still higher. Thus, at the extremity of the continent, and near the coast, the phenomenon was seen of a range of mountains still higher than the great Rocky mountains themselves. This extraordinary fact accounts for the Great Basin, and shows that there must be a system of small lakes and rivers here scattered over a flat country, and which the extended and lofty range of the Sierra Nevada prevents from escaping to the Pacific Ocean. (Frémont, 1845, p. 235)

Elsewhere, he noted that the fact that the Sierra Nevada, at the margin of the continent, was higher than the Rocky Mountains, which are located much nearer the center of the continent, appeared "so contrary to the natural order of such formations" (p. 274). This surprise must have been because of Frémont's biases acquired by a reading of the contemporary geographical and geological literature. That literature was almost exclusively European and dominated by ideas derived from contemplation of the conditions of the Old World, especially Asia. In the early nineteenth century, everybody "knew" that Central Asia was the highest place in the world, dominated by immense mountain ranges, both real (for example, the Altay, the Tien Shan, the Kuen Lun, and the Himalaya) and imaginary (such as the Bolor)! Although von Humboldt had started showing the untenability of its factual basis, a vast central plateau in Central Asia still exercised the imaginations of many geographers and geologists. Buache's (1761) hypothesis that each continent possessed one central plateau from which both its main mountain ranges and major rivers radiated (Fig. 76) had not yet been entirely forgotten owing to Peter Simon Pallas' great authority. However, Frémont's explorations had painted an entirely different continental picture for him, in which the high ranges were not central but marginal. His findings endorsed the opinions that the Spanish geographers had long held regarding the geography and architecture of North America, as we noted above, and influenced the course of the development of tectonic ideas in North America for a century-and-a-half to come!

On 22 April 1844, Frémont estimated the altitude in the eastern Mojave Desert to be 2250 ft. (~682 m). After May 5, they reached the Virgin River valley, where Frémont noted the elevation to have risen to 4060 ft. (~1230 m) above sea level. Shortly afterwards, they arrived at the Utah Lake and thus "completed an immense circuit of twelve degrees diameter north and south, and ten degrees east and west" (Frémont, 1845, p. 274). In the few places he was able to measure the elevations in the Great Basin, those recordings were around 4000–5000 ft. (~1330–1500 m), except that they declined in a southwesterly direction and were less than 700 m in its southwestern extremity. In the Great Basin too, elevations gently rose from west to east to culminate in the Rocky Mountains ridge. Only near the westernmost edge was this situation reversed, and they rose again, but this time abruptly, towards the Sierra Nevada.

Frémont noted that the core of the Sierra Nevada was granitic ("white granite," he wrote: Frémont, 1845, p. 214). He connected that range with the volcanic Cascades to the north. His forced foray into the San Francisco Bay area "made me well acquainted with the great range of the Sierra Nevada of the Alta California, and showed that this broad and elevated snowy ridge was a continuation of the Cascade Range of Oregon, between which and the ocean there is another and a lower range, parallel to the former and to the coast, and which may be called the Coast Range" (p. 255)[337].

Thus, Frémont had established that an immense topographic swell, averaging ~1500 m in crestal elevation, occupied the western part of North America between the Mississippi Valley and the Pacific Ocean. The mountains on top of this swell had a structure different from the swell itself and appeared only as small protuberances above the overall level of the swell (Fig. 85; see loose insert accompanying this volume). The existence of the swell could not be noticed locally by the naked eye. Rocks making up its architucture everywhere appeared flat-lying to the local observer. Only the detailed topographic profile Frémont drew (following a method introduced by von Humboldt: see von Humboldt, 1849, p. 59–60) in the combined map and section foldout of his 1845 book (which was a combination of the reports of the 1842 and 1843–1844 expeditions) made the swell visible (Fig. 85). This swell incorporated the high plateau country so meticulously explored earlier by the Spaniards and made their observations on the flat topography and flat-lying sedimentary strata comprehensible. The anonymous reviewer (probably J.S. Dana) of Frémont's book in the *American Journal of Science and Arts,* emphasized that:

> It was before known that the slopes of the Rocky Mountains were very gradual in inclination; but the fact is brought out with greater definiteness and more distinctly to the eye, in the section presented by Lieut. Frémont, of which the following is a reduced copy [*see Fig. 86 herein*] It corrects at once a common impression that these mountains are a narrow line of heights [*this wrong impression was common in those days owing to such maps as those displayed in Fig. 74, 75, and 81 herein*], showing that they stretch over a breadth of twelve or fifteen hundred miles [*~2000–2200 km*], and that the ridges at summit like those on some part of the declivities, are properly ranges of heights upon the great Rocky Mountain elevation. To be fully apprehended, it should be observed that, although the inclination in the above view is so gradual, the height as related to the breadth is actually on the scale adopted, exaggerated *thirty times.* (Anonymous, 1847, p. 196, italics from the original)

James Hall's geological report (Hall, 1845) did little to improve the picture presented by the topography and by Frémont's own observations, because essentially nothing was added in terms of attitudes and rock types. Only in a few places Hall thought he could say something about the ages of the rocks on the basis of scanty and badly preserved fossils. Hall thought that he could perhaps place the three specimens found between 96°15′ and 105° W into the Cretaceous. If we consider that these three specimens had been collected at elevations of approximately 212 m, 606 m, and 1636 m, it might have helped to corroborate the idea that the topographic slope up to the Rockies also corresponded with a geological slope.

The other flank of the topographic swell was impossible to confirm as a geological slope (except during the time of the volcanic products that covered large parts of the swell in Oregon). Along the coastal strip, the rock types appeared too numerous to paint a simple picture, but Frémont's idea of a mountain range cored by granites and maintined active by ongoing volcanism seemed unrefuted; the somber positivist that he was, Hall (1845), however, did not at all go into the tectonic consequences of Frémont's observations and specimens.

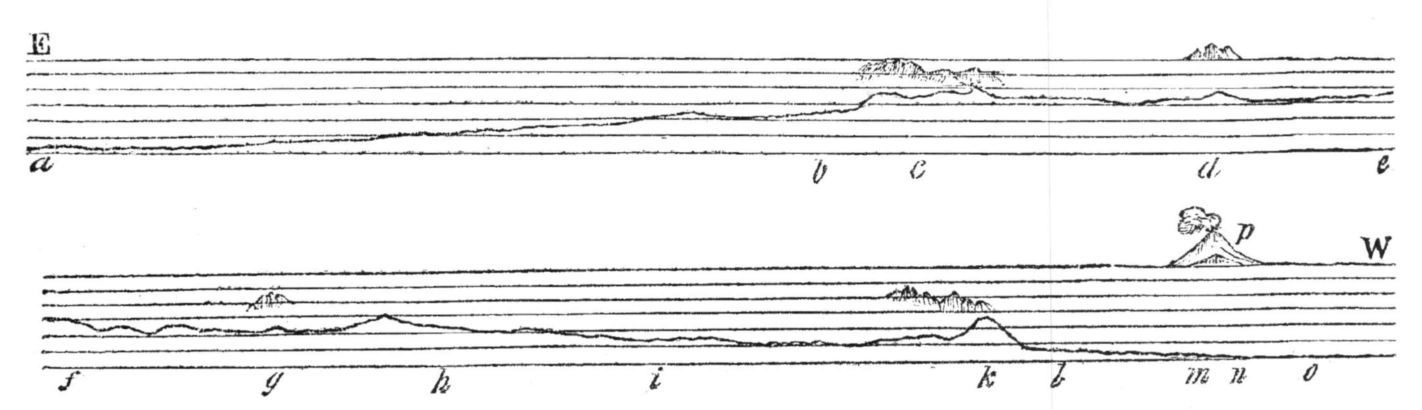

Figure 86. The anonymous reviewer's rendering of Frémont's east-west topographic profile (reproduced herein in Fig. 85; see loose insert accompanying this volume) in the May 1847 issue of the *American Journal of Science and Arts*. It was in this form that the majority of the geographers and geologists in the world found out about the great falcogenic swelling in the American West.

As we saw above, Dana, on the other hand, was quick to seize on Frémont's results as a basis for tectonic model-building. In the very same May issue, in which Frémont's book was reviewed, Dana published the first of his 1847 papers on tectonics, in which he announced that the great Rocky Mountain swell occupying more than one-third of the western part of the continent was an anticlinal upfold formed as a consequence of the thermal contraction of the planet, and he implied that the swell was a different sort of structure from the fractures that here and there disrupted its smooth surface and formed mountain ridges (Dana, 1847b).

From the consideration of the great western swell, Dana returned to the consideration of the structure of the entire continent, and in doing so almost echoed, contractile cause apart, Castañeda from nearly three centuries earlier: "Thus each great oceanic depression, the Atlantic and Pacific, has its border range of heights thrown up by the very contraction which occasioned the depression; and between lies a vast plain, scarcely affected at all by these changes, the great central area of the continent" (Dana, 1847b, p. 98). This was also in line with Hitchcock's (1841) interpretation.

As I have at length discussed Dana's views on falcogenic movements above, I shall not repeat them here. The purpose of this long section was in part to show how observations and ideas since the sixteenth century finally culminated in Dana's theory of continental deformation and the great influence Frémont's publications had on him. However, observations on the falcogenic structures in the western United States did not stop with Frémont. They continued in the framework of government surveys associated with the Mexican War, the construction of the transcontinental railroad, and the Mormon War. These surveys prepared the ground for the much more extensive studies of the four great western surveys after the Civil War, namely those of King, Hayden, Powell, and Wheeler. The work of these great surveys connects up with the work of Suess and with other tectonicians living at the end of the nineteenth century. It influenced them to such an extent that for more than the next half century, American ideas (i.e., mostly Élie de Beaumont's ideas and conceptual framework massively enlarged by the American observations in the western United States) dominated the thought on the origin and evolution of falcogenic structures. The only non-Beaumontian component in the American models came from Cambridge, England, on the basis of the amazingly accurate geodetic work conducted in India by the Great Trigonometrical Survey (Phillimore, 1945, 1950, 1954, 1958; Keay, 2000).

But below, I must take up again the story of the growth of the observations and the evolution of ideas on falcogenic structures in the western United States from where Frémont and Dana had left them in 1847.

Major William H. Emory's Grand Profile (Fig. 87; see loose insert accompanying this volume)

When Mexico refused to recognize the independence of Texas and its subsequent annexation by the United States in 1845, and the dispute over the territory between the Rio Grande and Nueces Rivers continued, the seeds of war between the two countries were sewn. The United States recognized how little the geography of the regions bordering Mexico were known, making war planning difficult. Accordingly, three expeditions were sent out. First, the cavalry Colonel Stephen Watts Kearny retraced with his five companies of dragoons the route of Frémont's first expedition with the purpose of scaring the Indians into pacification. He had with him Lieutenant William B. Franklin of the Topographical Engineers, who drew a map and wrote a report containing a single page of geology (Goetzmann, 1991, p. 112–116). This military expedition corrobrated Frémont's earlier results without adding to them anything that would have been of importance from a general tectonic viewpoint.

The second and third expeditions initially started under Frémont's command. The Pathfinder crossed the Rockies and the Great Basin after sending a part of his force, under the command of Lieutenant James W. Abert (Colonel Abert's son), to reconnoiter the country along the Canadian River. The two expeditions parted company on 9 August 1845, and Frémont commenced his westerly march a week later, on 16 August. He reached California in late autumn, where he became instrumental in launching the Bear Flag revolt against Mexico (Goetzmann, 1991, p. 116–122). The events in California led to the collection of another critical piece of topographical and geological information related to the long-wavelength lithospheric structures in the western United States.

At the suggestion of the "Old Bullion" Senator Benton, Frémont's father-in-law, President James K. Polk, ordered Colonel Kearny to take a force of some 1,700 mounted men (grandly styled "The Army of the West") to California to aid the establishment of American hegemony there. On 5 June 1846, First Lieutenant William Hemsley Emory (1811–1887) of the Topographical Engineers, a Maryland aristocrat who had a strong scientific bent, received his orders from Colonel Abert to join Kearny's force, together with First Lieutenant William H. Warner, Second Lieutenant James W. Abert, and Second Lieutenant William G. Peck. Notwithstanding its small numbers, this was a formidable survey team; it had not only experienced surveyors in it such as Abert and Peck, but both Peck and Emory were men of remarkable scientific ability (Goetzmann, 1991, p. 127–144; 1993, p. 253–257). In addition, they had the assistance of two civilians: Norman Bestor, a statistician, and John Mix Stanley, a landscape artist (Goetzmann, 1991, p. 131).

The team started at Fort Leavenworth on 27 June 1846. For their march to the Rockies, Emory divided the trip into three segments for the purposes of narrative: Fort Leavenworth (39°21′14″ N and 94°44′00″ W) to Pawnee Fork (38°10′10″ N and <99° W), from Pawnee Fork to Bent's Fort (~38° N, 103° W), and from Bent's Fort to Santa Fé (Emory, 1848, p. 10). Between Fort Leavenworth and the Pawnee Fork, the country was a high, rolling prairie, traversed by streams, the banks of which were almost vertical, exposing fossiliferous limestones having crinoids (Emory, 1848, p. 11). At their camp near Pawnee Fork (38°10′10″ N and 98°55′22″ W), they found their elevation to be 1932 ft. (~585 m) above sea level by barometric observation. By the time they reached Bent's Fort (their camp was at 38°02′53″ N and 103°01′34″ W), the elevation had risen to 3958 ft. (~1199 m). The aspect of the country had changed, but very gradually, becoming drier (Emory, 1848, p. 12). Emory emphasized that in traversing 311 mi. (~500 km), they had risen 2300 ft. (~697 m). This gave a slope of "seven feet and four-tenths per mile" (Emory, 1848, p. 12; in other words ~0.08°). They had moved along the Arkansas River, whose bed consisted of sand and in places of pebbles of "primitive rock." Near Bent's Fort, they saw river-cut banks where strata were exposed to view. Emory summarized his geological observations as follows:

> On the lower part of the river, it is a conglomerate of pebbles, sometimes shells cemented by lime and clay overlying a stratum of soft sandstone, which in turn, overlays a blue shale, and sometimes the richest description of marl.
>
> Higher up the river, we find the same formation, but in addition argillaceous limestone, containing ammonites and other impressions of shells in great variety, and in more than one instance distinct impressions of oyster shells. The dip in both cases is about 6°, and a little north of east.[338] (Emory, 1848, p. 12–13)

The scientific crew joined "The Army of the West" on 2 August 1846. On that day they traveled 45 km along the Timpa River southwestwards from Bent's Fort in the foreland of the Rockies and rose to an elevation of 4523 ft. (~1371 m) following a considerable ascent (Emory, 1848, p. 15). The next day they marched along the same river, encountering "argillaceous limestone, containing, now and then, the impression of oyster shells very distinctly. The valley in which we encamped presented the appearance of a crater, being surrounded with buttes, capped with stunted cedar. The stratification, however, appeared regular, and to correspond on different sides of the valley" (Emory, 1848, p. 16). They thus remained entirely in the essentially flat-lying same Upper Cretaceous deposits that they had been traveling along in the upper course of the Arkansas River. However, today's (i.e., 3 August 1846) march took them farther, up to 4761 ft. (~1443 m) above sea level, which Emory noted in his report with due regularity (Emory, 1848, p. 16).

After 5 August, the elevations began to rise rapidly above 2000 m. Their camp on August 5th had an elevation of 5896 ft. (~1787 m). The next day, they marched along the Purgatoire River (the Los Animos of the Spaniards) and compared the beautiful scenery with that of Palestine: they were surrounded by light-colored sandstones (now known to be Paleocene continental deposits: King and Beikman, 1974) that were essentially horizontal, and their road was covered with fragments of volcanic rock (Pliocene: King and Beikman, 1974). On 7 August, they encountered gently east-dipping sandstones at an elevation of 7500 ft. (2272 m). The valley in which they were traveling was "strewn with pebbles and fragments of trap rock, and the fusible rock described yesterday, cellular lava, and some pumice" (Emory, 1848, p. 19).

On 11 August 1846, they were still in the plains east of the Rocky Mountains. They had moved farther to the southwest, to the Ocaté River, and found the hills on the way to be "composed principally of basalt and porous volcanic stone; very hard, with metallic fracture and lustre, traversed by dykes of trap. The lava is underlayed [*sic*] by sandstone. From the uniform height of these hills, one would think they originally formed the table land, and that the valleys had been formed by some denuding process, and their limits determine the alternate existence or non-existence of the hard crust of volcanic rocks" (Emory, 1848, p. 22–23). Lieutenant Emory correctly judged the shapes of the tablelands to be a result of denudation. Although he did not specify water as the denuding agent, I cannot see any reason to doubt, as does Goetzmann (1991, p 131), that Emory thought it was anything else but water effecting the denuding. The diluvialist/fluvialist debate in America had already died down when Emory was writing, and the glacial theory had not yet become common knowledge. It is true that Lyell continued to push his marine denudation idea vigorously during his American visits, but even that idea involves only water to do the denudation (cf. Chorley et al., 1964, p. 253–279). In the highlands of the West, Lyell's idea was not easy to entertain, and Lieutenant Emory was probably thinking of fluvial erosion only. Elsewhere in his report there are references to "stones, rounded by attrition of the water" (e.g., p. 64). In any case, his denudation idea was later taken up by his countrymen who investigated landscape evolution in the West and provided the foundations and the principal terminology of modern fluvial geomorphology.

On August 15th, 16th, and 17th, the explorers marched across the Sangre de Cristo Range, peacefully occupying the Mexican towns on the route. On August 16th, the troop was traveling in a canyon that had 2000-ft. (~600-m) vertical walls with flat tops and that exposed "an immense stratum of red earth" (Emory, 1848, p. 29). When they began descending towards Santa Fé, they encountered sandstone and granite in the mountains. The elevation of Santa Fé itself, which they occupied without a shot being fired, owing to the monumental incompetence of the Mexican General Armijo, was measured barometrically to be 6846 ft. (~2075 m), although it is located in a broad valley.

As the explorers descended down the Rio Grande, Emory's thoughts went to "the hardships, trials, and perseverance of the gallant Pike" (Emory, 1848, p. 38). On 3 September 1846, between San Felipe and Angosturas, 6 mi. (~9.6 km) downstream, the valley of the river was seen to be very narrow. On the west, tablelands capped by basalt stretched before the eye (Fig. 88). On the east were rolling sand hills, rising gradually to the base of the mountains, and covered with large round pebbles (Emory, 1848, p. 39). Emory measured the elevation of the valley floor to be 5000 ft. (1,515 m: Emory, 1848, p. 39). South of Albuquerque, Emory explored the tablelands to the west. He found them to be a succession of rolling sand hills. He came down them through a ravine, "where the lava, in a seam of about six feet, overlaid soft sand-stone. At the point of junction, the sand was but slightly colored. The lava was cellular, and the holes so large that the hawks were building nests in them" (p. 47).

They continued their descent downstream along the Rio Grande and on October 8th, some 15 km north of latitude 33°22′02″, Emory noted the continued presence of the tablelands: "The table lands, reaching to the base of the mountains to the west, are of sand and large round pebbles, terminating in steep hills from a quarter to a half mile from the river, capped with seams of basalt. Some curious specimen of soft sandstone were seen today, of all shapes and forms, from a batch of rolls to a boned turkey" (Emory, 1848, p. 54). The curious landforms of the plateau country had already started to exercise the humor of the naturalists.

By 10 October 1846, the troop, in its journey along the Rio Grande, had descended in elevation some 1800 ft. (~545 m) over a distance of about 203 mi. (~327 km), giving an average slope of descent of 0.1°. On 17 October, they were already

Figure 88. Flat-topped buttes in San Felipe, New Mexico, as they appeared to Major William H. Emory (from Emory, 1848, facing p. 38).

marching westward along the Mogollon Rim of the Colorado Plateau. Elevations were above 1800 m, and Emory noted that "The mountains appeared to be formed chiefly of a reddish amygdaloid and a brown altered sandstone, with chalcedonic coating. In places, immense piles of conglomerate protruded; disposed in regular strata, dipping to the south at an angle of 45°" (Emory, 1848, p. 58): He was looking at the faulted and tilted Pliocene volcanics and continental sandstones of the Piman subtaphrogen (King and Beikman, 1974; Şengör and Natal'in, 2001; Dickinson, 2002).

On 21 October 1846, they had crossed the 109° W meridian and were still pretty much following the 32°30′ N parallel along the Gila River when, at an elevation already less than 4500 ft. (~1364 m), they noticed what we today call the basin-and-range topography:

> After going a few miles, crossing and recrossing the river [*i.e., the Gila*] a dozen times, it was necessary to leave its bed to avoid a cañon. This led us over very broken country, traversed by huge dykes of trap and walls of basalt. The ground was literally covered with the angular fragments of these hard rocks.
>
> From one of these peaks we had an extended view of the country in all directions. The mountains run from northwest to southeast, and rise abruptly from the plains in long narrow ridges, resembling trap dykes on a great scale. These chains seem to terminate at a certain distance to the south, leaving a level road, from the Del Norte [*i.e, the Rio Grande River*] about the 32nd parallel of latitude, westward to the Gila. …
>
> The mountains were of volcanic rock, rock of various colors, feldspathic granite, and red sandstone, with a dip to the northwest, huge hills of a conglomerate and angular and rounded fragments of quartz, basalt, and trap, cemented by a substance that agrees well with the description I have read of the puzzolana of Rome. (Emory, 1848, p. 62)

As they marched along the Gila River, the learned Emory thought of Father Marcos of Nizza and hoped that they might find the fabulous Indian settlements he had discussed (Emory, 1848, p. 64). As they continued their westerly march, the average elevations continued to diminish. By 6 November 1846, when they reached the point where the Gila River crosses the 111° W meridian, their elevation had already descended down to 2115 ft. (~641 m). From camp 76 (on October 18th), they had come ~273.7 km, as the crow flies, and descended 128 m. This gives an average slope of descent of 0.25°. On 8 November, they were at an elevation of only 1751 ft. (~531 m). The elevations continued to descend and reached practically sea level at the confuluence of the Gila River with the Colorado River. Elevations then rose again to what Emory called the Cordilleras of California. Above San Diego, the highest point they estimated was not more than 3000 ft. (~900 m).

I presented a detailed account of Emory's topographical and geological log to give an idea of the state of geological knowledge that existed by the time the great railroad surveys began. It consisted of occasional geological notes taken by well-educated army officers whose main duty was topographical surveying. When combined with their detailed maps and topographic sections, these notes yielded enough of a geological picture to consider at least the western half of the United States as a very broad and very flat anticline, here and there disrupted by faulting and bled by volcanism. With the sort of geological reconnaissance Frémont, Emory, or Abert were able to make, no finer picture was possible to paint. What Hitchcock (1841, 1847) and Dana (1847b) did represented the maximum of what could be done in terms of a tectonic synthesis of the existing knowledge. Only the railroad surveys brought in an abundance of newer data. The railroad reports, for the first time in the American West, contained detailed geological descriptions, by professional geologists, of the country the survey lines traversed. Together with the reports of John Strong Newberry as part of the Ives and Macomb surveys during the Mormon War, geology reported by the railroad surveys formed the basis on which the operations of the post-Civil War Great Surveys were planned.

But before we begin a discussion of the contributions of the railroad surveys to our knowledge of the long-wavelength lithospheric structures in the western United States, we need to take a fleeting look at the remarkable discovery and interpretation of the terraces of Great Salt Lake in Utah by Captain Howard Stansbury of the U.S. Army Corps of Topographical Engineers in 1849–1850.

Captain Howard Stansbury and the Discovery of the Falcogenic Movements Around the Great Salt Lake Area

In the middle of the nineteenth century, the United States had already become a transcontinental empire, and the routes to the western settlements in California and Oregon, both to carry emigrants and to serve as economic arteries, had gained importance. It was clear that a Pacific railroad was soon to be built, and various interest groups had already started a race to win the greatest favor for a possible railroad through the areas they represented. As the U.S. Army Corps of Topographical Engineers was the federal agency responsible for exploration and development beyond the Mississippi, it found itself amidst the political skirmishes arising from this race. The head of the Corps, Colonel Abert, was of the opinion that a southwestern route was the most advantageous to the entire nation. Despite his own conviction, he had to satisfy the other contenders and explore other possible or "desired" routes. Captain Howard Stansbury, an experienced member of his Corps, was thus ordered on 11 April 1849 to join Colonel William Wing Loring's regiment of Mounted Rifles along the Oregon Trail as far as Fort Hall in southern Idaho (43°03′ N, 112°26′ W), from where he was to begin a detailed survey for a military post to help emigrants get prepared for their desert crossing on the way to California (see Stansbury, 1852, p. 13). Colonel Abert gave further orders for the entire Great Salt Lake area to be surveyed. Captain Stansbury was to collect topographical, geological, zoological, botanical, and anthropological material (Goetzmann, 1991, p. 219).

The non-geological details of Stansbury's wonderful explorations do not concern us here. I begin the discussion of his geological results by quoting the three remarkable passages in which he announced the discovery of the magnificent terraces around the basin of the lake and his speculations concerning their origin and implications for tectonic processes:

In regarding the extensive, low-lying clay and sand flats forming the northern shores of the Great Salt Lake (Locomotive Springs Water Fowl Management Area, Salt Wells Flat, and Rozel Flat: see Currey, 1980, plate I), Stansbury inferred that:

> This extensive flat appears to have formed, at one time, the northern portion of the lake, for it is now but slightly above its present level. Upon the slope of a ridge connected with this plain, thirteen distinct successive benches, or water marks, were counted which had evidently, at one time, been washed by the lake, and must have been the result of its action continued for some time at each level. The highest of these is now about two hundred feet above the valley [*now called the Stansbury shoreline in his honor: see Gilbert, 1890, p. 134 and Currey, 1980, p. 75*], which has itself been left by the lake, owing probably to gradual elevation occasioned by subterraneous causes. If this supposition is correct, and all appearances conspire to support it, there must have been here at some former period a vast *inland* sea, extending for hundreds of miles; and the isolated mountains which now tower from the flats, forming its western and south-western shores, where doubtless huge islands, similar to those which now rise from the diminished waters of the lake. (Stansbury, 1852, p. 105, italics his)

On Antelope Island (see Currey, 1980, plate I), they found similar signs of former high levels of the lake, which they sketched (see Fig. 89). It appeared to them that the lowest levels were vacated by the waters very recently, for they found on them driftwood:

> Drift-wood is scattered along the shores at an elevation of four or five feet [*~1.20 or 1.50 m*] above the present lake, which must have maintained that height for a considerable period, since in numerous spots along the drift line unmistakable evidences of a well-defined beach are still to be traced with perfect precision. The wood is small and generally sound, but very dry, and must, from the appearance, have been deposited there for many years. It came doubtless, from Bear River, the Weber, and the Jordan. (i.e., from the east: see Currey, 1980, plate I; Stansbury, 1852, p. 158)

Observations of the raised beaches were also made on Deception Island of Frémont. Captain Stansbury decided to call it, in honor of the great explorer, Frémont Island (Stansbury, 1852, p. 160), a name that it happily still bears (see Currey, 1980, plate I): "In approaching the island from the water, it presented the appearance of regular beaches, bounded by what seemed to have been well-defined and perfectly horizontal water-lines, at different heights above each other, as if the water had settled at intervals to a lower level, leaving the marks of its former elevation distinctly upon the hillside. This continued nearly to the summit, and was most apparent on the north-eastern side of the island" (Stansbury, 1852, p. 160).

Captain Stansbury thus not only recognized that the level of the Great Salt Lake had been much higher in a recent geological past, but that its volume had diminished in numerous successive phases, each of which was interrupted by a period of stability during which the beaches were cut into the basin margins. He further inferred that the volume loss probably occurred

Figure 89. Lacustrine recessional terraces on Antelope Island in the Great Salt Lake (middle ground to the left), as sketched by Captain Howard Stansbury (from Stansbury, 1852, facing p. 102).

by drainage as a result of upheaval of the lake floor in response to subterranean phenomena. This last inference had a direct bearing on the inferences James Hall made from Stansbury's geological notes and collections between the Mississippi and Great Salt Lake.

Stansbury collected specimens and made notes on the geology of the route he traversed all the way from Fort Leavenworth to the surroundings of Great Salt Lake, and he submitted them to James Hall of the New York Survey for scientific evaluation. Hall wrote an extended letter to Stansbury, which amounted to a small treatise on the geology of the areas visited and the fossils collected by the Stansbury expedition (Hall, *in* Stansbury, 1852). Hall largely corroborated the observations he earlier made on Frémont's material: From the vicinity of Fort Leavenworth to the Big Blue River (about 96°45′ W), Carboniferous limestones were encountered (now known to be largely Upper Pennsylvanian {i.e., uppermost Carboniferous} and Wolfcampian {i.e., lowermost Permian}: King and Beikman, 1974). At the Big Blue River, the late Paleozoic was found to disappear under "strata of Cretaceous age" (Hall, *in* Stansbury, 1852, p. 402; now known to be Upper Cretaceous: King and Beikman, 1974). Hall guessed that "these beds extend much farther" (in Stansbury, 1852, p. 402; presumably *much farther west* is here meant), but owing to extensive drift cover, he could not be sure. No mention of any geological formation other than the drift was made until the forks of the Platte River were approached. There, some clays were collected containing what Hall believed to be small marine shells. Hall could not determine their nature but guessed the age of the formation to be Tertiary (1852, p. 402; Hall's guess as to the age was correct but not as to the environment of deposition; we now know these to be continental deposits of Pliocene age[339]: King and Beikman, 1974).

Material brought back from the surroundings of Fort Laramie once more indicated the presence of the Carboniferous limestones. Hall very significantly observed that "Some of the fossils are identical with species collected between the Missouri and the Big Blue, and we can only suppose, from the great similarity of the specimens, that it is a continuation of the same formation. From the dates marked upon the specimens, it is evident that this limestone extends to some distance on the east and west of Fort Laramie" (Hall., 1852, p. 403). When Hall examined the limestones capping the metamorphic rocks of what he believed to be of early Paleozoic age around the Great Salt Lake, he found them all to be of Carboniferous age. Thus, the Carboniferous limestones passed over the water divide of the Rockies, which, according to the notes and specimens submitted to him by Captain Stansbury, Hall found to consist of

> feldspathic granite, with little quartz or mica. The rocks of this locality are doubtless of metamorphic origin, probably rocks of silurian [*sic*] age. The specimens collected three days' march in advance of this place, on the North Fork of the Platte River, are shaly sandstone and thinly laminted sandstones containig fossils. The fossils are some brachiopods, with others similar to *Monotis*, and we may presume from the described position of the beds, and from the characters of the fossils, that these beds are of devonian [*sic*] age. In the journal these beds are recorded as dipping at the rate of 15° to the northeast. (Hall, 1852, p. 403)

What Hall was describing was the northeastern foothills of the Laramie Mountains.

Hall thus clearly recognized that the Carboniferous limestone rose together with the imperceptibly rising topography from the Mississippi to the Rocky Mountains. Deformation in the Rockies, as recorded by the dips of 15°, had exposed the underbelly of the Carboniferous. The same Carboniferous then capped the mountains around the Great Salt Lake. In these mountains, Hall inferred "two lines of elevation, corresponding with the divisions of the lake" (Hall, *in* Stansbury, 1852, p. 405) and interpreted them as metamorphic-cored anticlines. From his descriptions, his readers clearly obtained the image of a very gentle and very broad anticlinal slope from the Mississippi to the Rockies. The crest of this gentle anticline was complicated by axes of elevation throwing the local beds into much narrower and steeper anticlines cored by metamorphic rocks. From Hall's mention of marine Tertiary rocks in present-day Nebraska, it would be clear to his readers that the large anticline spanning the distance from the Mississippi to at least the Great Salt Lake would have had to have risen after some time in the Tertiary. This was entirely compatible with Captain Stansbury's inference of the very recent rise of the Great Salt Lake area, a "gradual elevation occasioned by subterraneous causes" (Stansbury, 1852, p. 105). This inference was to be corroborated by the geologists who worked along the routes of the railroad surveys, especially that which followed the 35th north parallel (see below).

Hall's interpretations involving local axes of elevation is interesting from the viewpoint of his vehement denial of them only a few years later, when he came to believe that thick sediments made mountains with the aid of a wholesale continental elevation without any local uplifting events as discussed above. It seems that Hall must have changed his ideas concerning the nature and causes of mountain uplift after February 1852, when he submitted his report to Captain Stansbury, and before August 1857, when he read his famous presidential address to the American Association for the Advancement of Science at its 11th meeting in Montreal.

The Railroad Surveys

For the history of the ideas concerning long-wavelenth structures of the lithosphere in America, the railroad surveys of the mid-century are important only in so far as the additional data they contributed to understanding the local topography and geology along the routes. Being railroad surveys, their members were particularly careful about regional slopes, their steepness, and their nature. As such, they greatly refined the topographic database that hitherto existed. Because railroads require both construction materials and fuel in addition to knowledge about the engineering properties of the grounds on which they were to be built, the members of these surveys had to keep their eyes open about the geological constitution of the regions traversed. These observations massively expanded the previously existing database about the regional geology of the West—much more in the south than in the north for the reasons I discuss below.

The railroad surveys were organized by the U.S. Army Corps of Topographical Engineers upon the decree of the Pacific Railroad Survey Bill passed by the U.S. Congress on 2 March 1853 (Goetzmann, 1991, p. 274). Their job was to perform a reconnaissance survey along all the feasible routes, all of which were hotly defended by various interest groups in the American capital[340]. Only five surveys spanned the entire width of the Trans-Mississippi West and of these only one, the Whipple Survey that surveyed the 35° N latitude route, ended up presenting a detailed geological report with some general statements concerning the geological history of the traverse. This inclusion of detailed geology was because only that survey had a geologist (who had a frankly global interest) in the person of the Frenchman Jules Marcou (1824–1898: Fig. 90)[341], a student of Élie de Beaumont. As Goetzmann (1991, p. 311) pointed out, Marcou was the first professional geologist to run a survey across the whole North American continent.

In the following, I briefly review the contributions of the three main east-west surveys that spanned the distance between the Mississippi and the Pacific Ocean. The last two partial surveys that together spanned a fourth traverse—those of Lieutenant John G. Parke and John Pope (both in 1854)—worked along the southernmost route (for their itineraries, see Goetzmann, 1991, p. 276, map 10; Goetzmann and Williams, 1992, p. 166–167. For Parke's survey, see Goetzmann, 1991, p. 290–291. For Pope's, see Goetzmann, 1991, p. 291–292). Those two surveys did not traverse an ideal profile to see the large-scale swelling of the western part of the continent; therefore, I do no discuss their results here, except the interpretations of Thomas Antisell, the geologist who went with Parke, concerning the nature of tectonic phenomena in the western United States. Antisell's discussion encompasses a much larger area than the responsibility of the survey to which he was attached.

Figure 90. Jules Marcou (1824–1898), whose obstinate and quarrelsome character greatly annoyed his American colleagues. His contributions to American geology were nevertheless great, and he ended up spending the latter years of his life in the United States (from de Margerie, 1952, p. 118).

Stevens Survey (Route near the 47th and 49th Parallels from St. Paul to Puget Sound)

The main route surveyed by Governor Isaac Ingalls Stevens' (1818–1862) elaborate party was located west of meridian 105° W, which largely coincided with the terrain explored earlier by Lewis and Clark, and between 105° W and 100° W, which coincided with that explored by General Ashley's party, which included Smith and Henry, in 1822–1823 (Goetzmann and Williams, 1992, p. 136–137, 150–151, 166–167). Neither of the earlier parties had observed and recorded any geology to speak of, and Governor Stevens' party had detailed instructions from John Evans as to what and how to observe concerning the geology of the route (Evans, *in* Stevens, 1855, p. 11–13). Although these instructions are very detailed about the local geology, Evans did not emphasize the large-scale tectonics of the terrain to be traversed, although he might have done so in his final report. It is so much more regrettable that the final report he sent was lost *en route* (see Evans, *in* Stevens, 1855, p. 177), and the only geological account that accompanied the Steven's report was that submitted by George Gibbs[342]: "Upon the Geology of the Central Portion of Washington Territory" (Gibbs, *in* Stevens, p. 473–486.). Goetzmann (1991, p. 317) rightly complains, however, that "his contributions to the Pacific surveys in geology were lamentably slight" being largely a superficial geomorphological narrative, the main rock types encountered being given a passing mention. Geological relationships, attitude of strata, and structure were hardly addressed. The topographical descriptions by others often touched upon the geology seen, especially with respect to coal, building materials, and the magnificent volcanic morphology of the Columbia Plateau, thus slightly offsetting the disadvantage created by the flimsiness of the geological contributions in the Stevens report.

John Lambert's report on the topography, from the Mississippi River to the Columbia River, largely corroborated what the earlier explorers had found (see Fig. 91; see loose insert accompanying this volume). In the eastern part of this traverse, the vast prairie was broken into level and undulating segments. Where the level prairie dominated, the horizon is as unbroken as that of a calm sea. Nor are other points of resemblance wanting: the long grass, which in such places is unusually rank, bending gracefully to the passing breeze as it sweeps along the plain, gives the idea of waves, (as indeed they are such;) and the solitary horseman on the horizon is so indistinctly seen as to complete the picture by the suggestion of a sail, raising the first feeling of novelty to a character of wonder and delight. (Lambert, *in* Stevens, 1855, p. 160)

The country near the source of the Shayenne River (now spelled *Sheyenne*: at about 47°45′ N and 100°30′ W), some 366.5 mi. (~590 km) west from their starting point at the Little Falls of the Mississippi River, assumed a bolder character and began to rise:

... the swelling surface takes the forms of terraces and ridges; ponds and marshes occur more frequently; timber disappears from the uplands; the prairie becomes gravelly and abounds in granite boulders; and the river itself, moderately fringed with wood of different kinds, flows through a deep intervale enclosed by sand and clay bluffs from one hundred and fifty to two hundred feet [*~50–75 m*] or more of elevation, which are again surmounted by occasional hills sufficiently conspicuous to serve as landmarks to the hunters ... (Lambert, *in* Stevens, 1855, p. 161)

As they marched farther west for a number of miles, the hilliness increased and the country acquired a wilder character.

Between the Mouse River and the Missouri River, the country was found to be high prairie and tableland, but the elevations had become still higher:

The plateau between Missouri and Mouse rivers cannot be called simply a rolling prairie, though in detail resembling the hilly prairies noticed, but in a very exaggerted degree: a general similarity of outlines; the absence of wood and rocks in place; boulders plentiful; ponds and marshes if possible more frequent; but the elevations so much greater as to be almost considered mountainous, and becoming still more rugged on approach to Fort Union, where it ends abruptly on the level intervale of the Missouri. (Lambert, in Stevens, 1855, p. 162)

Farther west, they encountered the famous Badlands of Dakota, the "Mauvaises Terres" of the French hunters and trappers. Although "the curiosity of the mere tourist is soon sated in such arid and gloomy wilds ... The drooping spirits of the geologist are not, however, permitted to flag. The fossil treasures of the way well repay its dullness and fatigue" (Lambert, *in* Stevens, 1855, p. 164; for a geological and palaeontological treatment of the Badlands of Dakota, with review of earlier explorations, see O'Hara, 1920[1976]).

From Fort Union (i.e., the 104° W meridian; in other words from the point where the Missouri and the Yellowstone Rivers meet) westward, they found the country almost uniformly wild and barren. Lambert noted that the great mountains to the west provided a rain shadow and were the reason for the dryness of the high plains:

The eye grows weary traveling over the naked outlines of the successive plateaus, which, divided and bounded by the various rivers noticed, form but subdivisions of the great tract of country stretching from Missouri and Milk rivers on the south, to the Saskachawan [*sic*] on the north—this tract itself is a subdivision of the Great Plains—an extent embracing every variety of surface, from large and level plains to abrupt bluffs and ranges of summit hills that might be considered mountains. Let it be remembered that a few minutes' reading embraces sections which require tedious weeks to traverse; and that even traveling over and observing them with the patient labor of months, leaves but a *feeling* of their vastness, which baffles the effort to express it. (Lambert, *in* Stevens, 1855, p. 166; italics Lambert's)

After their starting point on the Mississippi River, 1039 mi. (~1675 km), the expeditionists reached the Lewis and Clark's Pass at the Rocky Mountain water divide above 6000 ft. (~1800 m), from which they continuously descended to the Spokane Plateau on the Pacific side. From the topographic profile, their average rate of ascent to the water divide was 0.05° (here I took the average elevation of the Rockies as 5000 ft. {~1500 m} as shown on Figure 91 (see loose insert accompanying this volume), instead of the extreme height of the Lewis and Clark's Pass). Their rate of descent was higher: from the Pass to the Columbia River Valley, they traversed 505 mi. (~813 km) and descended to an elevation of 630 ft. (~190 m), down a slope of 0.1°. The geology on both sides of this asymmetric swell was not dissimilar to what others had seen farther south, with the exception of the vast extent of the Columbia River basalt plateau. When Gibbs climbed up the Methow River on the northern boundary of the basalt plateau, he encountered "horizontal gneisses" displaced only by "the intrusion of trap" (Gibbs, *in* Stevens, 1855, p. 483). Farther west, in the valley of the Okinakane (present Okanogan), they noted that the gneisses were "often contorted" (p. 484). All of these rocks were counted as making up the Cascade Chain (Gibbs, *in* Stevens, 1855, p. 484). The rest of what Gibbs reported was just continuous volcanic geomorphology with volcanic rock-types.

Gunnison-Beckwith Survey (Route near the 38th and 39th Parallels from Kansas City to Salt Lake City; near the 41st Parallel from Salt Lake City to the Sacramento Valley)

This route very grossly paralleled Pike's and Frémont's treks to the Rockies and, beyond that, Frémont's 1845 expedition to California across the Great Basin. Thus, topographically, this survey did not add anything new to the grand picture already established, but greatly refined the details (see Beckwith, 1854a, 1854b). Figure 92 (see loose insert accompanying this volume) shows the grand profile, consisting of profiles No. 1 and No. 2 in the profile plate prepared for Lieutenant E.G. Beckwith's second report[343] in the second volume of the railroad surveys (Beckwith, 1854b; profiles in Anonymous, 1859). Here we see the same picture that both Frémont's and Emory's surveys had already drawn, showing a vast topographic swell of extremely gentle slopes, the culminating point of its general outline reaching above 1500 m along the Rockies and variously accentuated by numerous mountain ranges of diverse sizes, the highest being the Sierra Nevada.

Between Westport (just south of Kansas City) and the Little Arkansas River (± 98° W meridian), the exploration party wandered over gently northeasterly dipping limestones full of late Paleozoic fossils, which German physician James Schiel (who was both the surgeon and the geologist of the expedition) identified as those of the Carboniferous (Schiel, *in* Beckwith, 1854b, p. 96; these strata are now known to be Carboniferous and Lower Permian).[344]

On the Little Arkansas River, the surveyors encountered a horizontal white, fine-grained, non-fossiliferous limestone, a red ferrugineous sandstone, and conglomerates. Schiel thought that these belonged to the "Chalk" formation (i.e., Cretaceous). Farther westward, they came across different sorts of limestones, all flat-lying, and Schiel believed all to be Cretaceous.

On these Cretaceous hills, Schiel thought he could identify "lines which mark the banks of an ancient sea; they lie in one and the same horizontal plane, in whatever direction these hills may run" (Schiel, *in* Beckwith, 1854b p. 97). Immediately after this sentence in the Beckwith report is placed Schiel's figure (which is reproduced here as Fig. 93) and below that to the end of p. 97 in the report is a remarkable note by Lieutenant Beckwith. That note is reproduced here in its entirety owing to its great historical importance: It quotes with approval and enlarges on the first interpretation concerning a large former extent of the present-day Great Salt Lake and the existence of other similar ancient lakes in the Great Basin on the basis of the Great Salt Lake terraces and their significance for our understanding of crustal movements.

**Note by Lieut. Beckwith.*

The old shore-lines existing in the vicinity of the Great Salt Lake present an interesting study. Some of them are elevated but a few feet (from five to twenty [*~1.5–6 m*]) above the present level of the lake, and are distinct and as well-defined and preserved as its present beaches; and Stansbury speaks, in the Report of his Explorations, pages 158-160[[345]], of drift-wood still existing upon those having an elevation of five feet above the lake, which unmistakably indicates the remarkably recent recession of waters which formed them, whilst their magnitude and smoothly-worn forms as unmistakably indicate the levels which the waters maintained, at their respective formations, for very considerable periods.

In the Tuilla valley, at the south end of the lake, they are so remarkably distinct and peculiar in form and position, that one of them, on which we traveled in crossing that valley on the 7th of May, attracted the observation of the least informed teamsters of our party—to whom it appeared artificial. Its elevation we judged to be twenty feet [*~6 m*] above the present level of the lake. It is also twelve or fifteen feet [*~3.6–4.5 m*] above the plain to the south of it, and is several miles long; but it is narrow, only affording a fine roadway, and is crescent-formed, and terminates to the west as though it had once formed a cape, projecting into the lake from the mountains on the east—in miniature, perhaps, not unlike the strip of land dividing the sea of Azoff [*sic*] from the Putrid sea[[346]]. From this beach the Tuilla valley ascends gradually towards the south, and in a few miles becomes partly blocked up by a cross-range of mountains, with passages at either end, however, leading over quite as remarkable beaches into what is known, to the Mormons, as Rush valley, in which there are still small lakes or ponds, once, doubtless, forming a part of the Great Salt Lake.

The recessions of the waters of the *lake* from the beaches at these comparatively slight elevations, took place, beyond all doubt, within a very modern geological period; and the volume of water of the lake at each subsidence—by whatever cause produced, and whether by gradual or spasmodic action—seems as plainly to have been diminished; for its present volume is not sufficient to form a lake of even two or three feet in depth, over the area indicated by these shores, and, if existing, would be annually dried up during the summer.

These banks—which so clearly seem to have been formed and left dry within a period so recent that it would seem impossible for the waters which formed them to have escaped into the sea, either by great convulsions, opening passages for them, or by the gradual breaking of the distant shore (rim of the Basin) and drainig them off, without having left abundant records of the escaping waters, as legible at least as the old shores they formed—are not peculiar to the vicinity of this lake of the Basin, but were observed near the shores of the lakes in Franklin valley, and will probably be found near other lakes, and in the numerous small basins which, united, form the Great Basin[[347]].

But high above these dimunitive banks of recent date, on the mountains to the east, south, and west, and on the islands of the Great Salt Lake, formations are seen, preserving, apparently, a uniform elevation as far as the eye can extend—formations on a magnificent scale, which, hastily examined, seem no less unmistakably than the former to indicate their shore origin. They are elevated from two or three hundred to six or eight hundred feet [*~60 or 90 to 200 or 250 m*] above the present lake; and if upon a thorough examination they prove to be ancient shores, they will perhaps afford (being essentially traced on the numerous mountains of the Basin) the means of determining the character of the sea by which they were formed, whether an internal one, subsequently drained off by the breaking or wearing away of the *rim* of

Figure 93. View of the limestone hills of the valley of the Upper Arkansas River. The dashed line, a------a, shows the position of ancient shores according to the interpretation of the Gunnison-Beckwith survey (from Schiel, *in* Beckwith, 1854b, p. 97).

the Basin—of the existence of which at any time, in the form of continuous elevated mountain chains, there seems at present but little ground for believing—or an arm of the main sea, which, with the continent, has been elevated to its present position, and drained by the successive stages indicated by these shores. (Beckwith, 1854b, p. 97, italics his).

Beckwith's inference is based on local evidence (albeit derived from a false interpretation of the high lake terraces as marine terraces) of the great uplift of the western United States in a comparatively recent time. It seems to go beyond Captain Stansbury's interpretation by assuming that the higher terraces were the terraces of the "main sea." From what Schiel wrote of his inferences regarding the Great Plains, it is clear that evidence for uplift above the level of the world ocean was a topic of discussion during the Gunnison-Beckwith party explorations.

When they passed Bent's Fort (~38° N, 103° W), for a few miles they encountered a chain of high and steep bluffs made up of two kinds of horizontally layered limestone (now known to be Upper Cretaceous). A few kilometers to the east of where they entered the mountains proper, the dip of the strata was reported to be 8° to 9° to the northeast. Schiel was puzzled by seeing an isolated "foliating shale" butte (Fig. 94) on the same level as the horizontal limestones (Schiel, in Beckwith, 1854b, p. 98).

In the mountains, strata became steeply dipping, even vertical: "On entering the mountains we find a white, fine-grained, very hard sandstone, torn, fractured, and upheaved to nearly a vertical position by plutonic rock" (Schiel, *in* Beckwith, 1854b, p. 99). The uplifting agent, they thought, was a "trachytic porphyry, which seems to have given to those mountains [*i.e., the Sangre de Cristo Range of the Rockies*] their peculiar shape and elevation" (p. 99). But the predominating rock of the Sangre de Cristo Range they found to be "a feldspathic granite, passing gradually into gneiss on the right bank of the creek[[348]], the gneiss supporting a hard shale, sandstone, and a bluish brittle limestone. The latter belong perhaps to that class of non-fossiliferous transition rocks lying under the silurian system, [[349]] and the existence of which on this continent has been recognised by several distinguished geologists" (Schiel, *in* Beckwith, 1854b, p. 99).

West of the Coochetopa Pass, the group encountered further "porphyries" (now recognized as Lower Tertiary volcanics: King and Beikman, 1974). Schiel found these to be puzzling and thought their contacts tectonic "for it has polished surfaces which could only be produced by its sliding over some other rock" (Schiel, *in* Beckwith, 1854b, p. 101). When describing a conglomerate made up of igneous clasts, Schiel makes reference to the "*gradual* upheaval of the mountains" (p. 1010, italics mine) in agreement with the gradual elevation around the Great Salt Lake as inferred by Captain Stansbury (see the quotation above on p. 171). All the sandstones dipping slightly to the northeast making up the Elk Mountains, Schiel noted to be concordant. Near the Grand River (i.e., the Colorado River), they encountered numerous blocks of Cretaceous dark gray limestone, but could not see it at outcrop.

Here, Schiel, the European, was impressed by the amazing power of erosion so conspicuous in an arid climate of a terrain

Figure 94. Isolated shale butte, standing adjacent to the limestone hills in the valley of the Upper Arkansas (from Schiel, *in* Beckwith, 1854b, p. 98).

of moderately deformed rocks, which his home continent never displays (except in restricted areas on the Iberian Meseta). In a rare passage, in which he allowed himself to generalize, he recorded his impression of the erosion:

> It is a remarkable feature in the character of the country between the Rocky mountains and the Sierra Nevada, that whole formations disappear, as it were, before our eyes. The wearing and washing away of mountains takes place here on an immense scale, and is the more easily observed, as no vegetation of any account covers the country, hiding the destruction from the eye. Nature here seems only to demolish, without showing any compensating creative activity. (Schiel, *in* Beckwith, 1854b, p. 102).

The scattered inferences of broad uplifts, mountain uplifts within these broader upheavals, and the immense power of erosion carving out of them the "ludicrous landcapes" of Dutton—making the mainly flat-lying formations disappear one by one as if they were pages in a book—were to become the chief themes to be pondered by the geologists of the Great Surveys after the Civil War. It was such considerations, to be emphasized again by Newberry in his wanderings over the southern parts of the same terrain, that eventually led geologists such as Powell, Dutton, and Gilbert to begin thinking in terms of broad, *en bloc* vertical uplifts without any significant folding or thrusting in the plateau country.

Around the Green River and the Wasatch Range, heavy autumn snows hindered geological exploration (Schiel, *in* Beckwith, 1854b, p. 103). When they descended into the Great Basin, Schiel was impressed by what he called "island mountains" rising from the floor of the Great Basin. The stratigraphy of these "island mountains" he found to be similar to the regions to the east of the Great Salt Lake, but he gives no account of the structure (Schiel, *in* Beckwith, 1854b, p. 104). Reading his account, it is thus easy to form a picture of a very broad anticlinal structure dominating the western United States along the traverse that Schiel describes. The pre-existing account by James Hall appended to the Stansbury survey (Hall, *in* Stansbury, 1852) and Jules Marcou's geological report and cross-section along the 35th parallel which the Whipple team surveyed (see below) only accentuated this impression in the mid-1850s.

Whipple Survey (Route near the 35th Parallel from the Mississippi River to the Pacific Ocean)

The entire route of this survey was confined to the area betwen the 34° and 36° north parallels and extended from Fort Smith, Arkansas, to essentially Los Angeles, California, for a distance of 1892 mi. (~3050 km). From the Mississippi River, the explorers reached Fort Smith by boat on the Arkansas River, where the operations of the survey commenced on 14 July 1853. They then followed the Canadian River. The Rockies were crossed at San Antonio Pass south of the Sandia Mountains. They crossed the Rio Grande valley and ascended the San Jose Mountains to Mount San Mateo (now Mount Taylor). Two separate groups of the party crossed the Zuni Mountains on the Colorado Plateau at two passes: Campbell's (north) and Camino del Obispo (south; Camino del Obispo is "Bishop's Road" in Spanish) and reached the Zuñi ruins. They then followed the Little Colorado River (*Colorado Chiquito*) and reached San Francisco Mountain. They crossed the Colorado River at the intersection by Bill Williams' Fork and then crossed the Great Basin at its southernmost section to reach the Transverse Ranges, which William P. Blake had called the Bernardino Sierra (Fig. 95; see loose insert accompanying this volume).

Their route on the Great Plains continuously ascended towards the Rockies, but the ascent was so gentle as to be imperceptible to the traveler (see the topographic profile in Fig. 95):

> The approach to the Santa Fé mountains from the Mississippi is by a gentle ascent from an elevation of 460 feet [~*140 m*] at Fort Smith to near 6,500 [~*2000 m*] at the base of the mountains. To reach this elevation a horizontal distance of about eight hundred miles [~*1300 km*] is traversed; and the rate of ascent [350] being nearly uniform, or but slightly increasing with altitude[351], the slope becomes imperceptible to the traveler, and the country over which he passes has the aspect of a wide plain. As the elevation is increased and the streams have worn deep valleys, the slope becomes a vast table-land. This is the character of the slope traversed by the expedition, and its uniformity is shown by the barometric profile taken along the valley of the Canadian and Washita rivers. (Blake, 1856, p. 7)

They started their march in the foreland of the Ouachita and the Wichita Uplifts and encountered deformed Carboniferous rocks, which the geologist of the expedition, Jules Marcou, correctly identified (Fig. 96A and B; see loose insert accompanying this volume) in part on the basis of his previous knowledge (Marcou, 1853, map; also see Möllhausen, 1858, p. 485–487, endnotes 2–5). Between camp 19 and camp 20 (just west of latitude 97° W, to the south of Old Fort Arbuckle), they came upon "red and blue clays, pretty hard, with a brecciated sandstone. Some fragments of dolomite …; which indicate that the gypsum[352] is not far from us" (Marcou, 1856, p. 128). Near camp 23 (just to the east of 98° W longitude, close to the south bank of the Canadian River), they encountered flat-lying red-colored marls and sandstones (Fig. 97) with enclosed salt crystals. Finally, between camps 25 and 26, in the terrain astride the 98° W longitude, the predicted gypsum finally showed up. Marcou thought this to be the "New Sandstone" (i.e., Triassic) because of its evaporite content and red color (Marcou, 1856, p. 130; 1858, p. 10–16; 1888, p. 31–32; Möllhausen, 1858, p. 487–488, endnotes 7 and 8)[353]. He traced this formation all the way to the Rockies[354]. This led to a disagreement with William P. Blake[355], who ended up writing the final geological report for the Whipple survey because Marcou had hastily left for Europe after submitting only a short report, much to the annoyance of the Secretary of War, Jefferson Davis (Goetzmann, 1991, p. 323–324). Blake, a geologist trained by James D. Dana and James Hall of New York at the height of the American Baconian empiricist tradition (cf. Daniels, 1968[1994]), was unwilling to accept an age assignment based entirely on the rock type and dubious analogies with far-flung regions (in some, such as the Lake Superior region, Marcou earlier had misidentified some of the Mesoproterozoic Keweenawan red clastics and basalt flows as Triassic, contrary to the opinion of American geologists who regarded them as "Silurian": see Marcou, 1853, p. 74–75; on this matter, he was under Élie de Beaumont's influence: see p. 76; Marcou later tenaciously held onto his erroneous opinion, despite the fact that he acknowledged the lack of palaeontological evidence: 1888, especially p. 34). The uniformitarianist Blake rightly pointed out that similar gypsiferous and salt-bearing red sandstones and shales occur at all ages.

Neither could Blake agree with Marcou's estimate of thickness of what Marcou believed to be Triassic. In the east (for example, near camp 44, near 102° W longitude), Marcou estimated the upper part of his Triassic to be "not less than three hundred feet [~*90 m*] above the gypsum" (Marcou, 1856, p. 135). Farther west, near camp 54 (some 10–20′ west of 105° W longitude), there is talk of 500 ft. (~150 m) of the upper part of Triassic (p. 139). Finally, just to the east of Albuquerque, he writes of total Triassic thicknesses of 4000–5000 ft. (~1200–1500 m)! Figure 98 illustrates Marcou's rough sketch where he thinks he saw 1200–1500 m of Triassic. The mountains shown on his sketch are ~900 m above the valley floor, so his estimate is not unreasonable if what he thought was Triassic really was Triassic. We now know, however, that in his "Triassic," there is much Lower Permian evaporite and clastic rock (Leonardian dolomitic limestones, gypsum beds, sandstones and siltstones of the Yeso Formation {cf. King and Beikman, 1974; Frenzel et al., 1988}).

In any case, on the basis of what they knew, Blake's attack on Marcou's estimates of thickness was not justified. Given what was seen in the field and the state of theoretical geology then, Marcou's age assignments and thickness estimates constituted a perfectly sensible working hypothesis. Moreover, when Marcou plotted the estimated thicknesses on a cross section (Fig. 96B), they made much sense within the framework of the mountain-building theories of his teacher Élie de Beaumont. Figure 96B illustrates Marcou's cross section across the southern United States, between the Mississippi River and the Pacific Ocean, along the 35th north parallel. Note on this figure how Marcou drew the Triassic: it thickens from the Mississippi River towards the mountains and thins again (this representation of the Triassic is neither supported nor opposed by his

COLUMNS OF SANDSTONE, SOUTH BANK OF THE CANADIAN RIVER.

Figure 97. Earth pillars and monuments in flat-lying "Carboniferous" (now known to be Permian) rocks in the Great Plains, encountered by the Whipple survey along the south bank of the Canadian River (Blake, 1856, p. 18).

observations!) before reaching the San Antonio Pass south of the Sandia Mountains. (Marcou had drawn the same geometry in his first cross section of this region, when he thought the rocks were all "Silurian" without having seen them himself: see Marcou, 1853, the top cross section on the map sheet.) The Triassic is shown as very thick again between Albuquerque and the Sierra Madre (i.e., the Zuni Mountains of modern nomenclature) and thins once more towards the Sierra Madre. It thickens yet again between the Sierra Madre and San Francisco Mountain. If we retro-deform the cross section (as would have been done by Marcou), we obtain the picture shown in Figure 99. This figure shows a large negative *bosselement* (i.e., a downbend similar to a geosyncline) to use an anachronistic term, in Élie de Beaumont's sense, that already had created a *ridement* (i.e., localized orogenic deformation) in the future Sandia Mountains area during the time between the late Carboniferous and the Triassic (note the Upper Carboniferous being cut out along an angular unconformity at the San Antonio Pass: see Fig. 96A). The Triassic red clastics and evaporites would have been laid down on the geography prepared by the negative *bosselement* and the *ridements* that had formed along its western edge.

Blake notes that Marcou had claimed in Whipple's preliminary report (Whipple et al., 1855), that the Rocky Mountains were of terminal Jurassic age (Blake, 1856, p. 75). The basis for this concept was that Marcou believed that the Upper Cretaceous was unconformable on all older sedimentary rocks (Marcou, [1854], p. 46; 1856, p. 169). He also thought that what he called the Sierra de Mogoyon (the Mogollon Rim of our present-

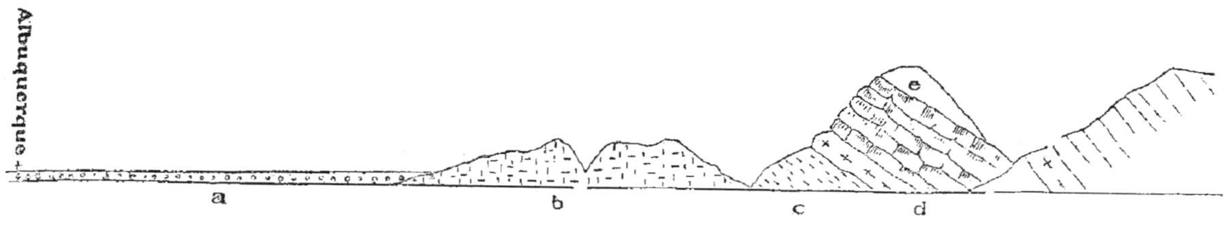

Figure 98. Marcou's cross section across the Albuquerque (Sandia) Mountains (Marcou, 1856, p. 143).

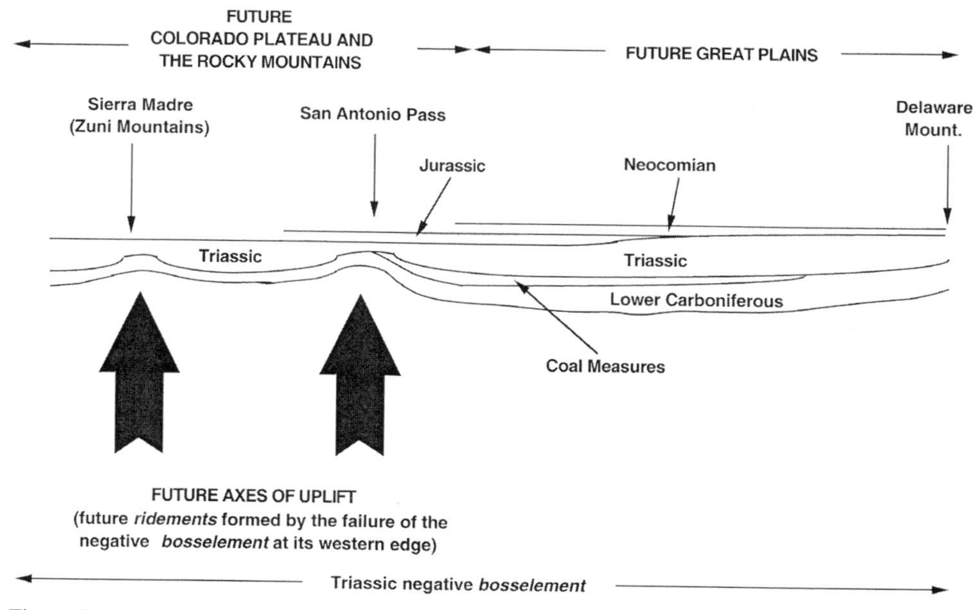

Figure 99. Marcou's cross section shown in Figure 96B, retro-deformed by reversing the uplifts until the top of his "Neocomian" reaches a horizontal position. This yields a major negative *bosselement* (in Élie de Beaumont's sense) that existed already in the early Carboniferous. The uplifts clearly were active before the deposition of the Triassic, as the Triassic sits across an angular unconformity atop the Carboniferous. Neocomian in turn sits across an angular unconformity atop the Triassic. Marcou was of the opinion that the boundary between the Triassic and the Jurassic was also an angular unconformity. The way that he drew the Triassic thickness changes (for which he had no direct evidence) is consistent with this interpretation.

day nomenclature) had been created at the end of the Triassic, before the deposition of the Jurassic (Marcou, [1854], p. 47; 1856, p. 170). The thinning of the Triassic under the Jurassic near the mountains[356], as Marcou depicted on his cross section (Fig. 96B; see also his colored geological cross section in Marcou, 1858), may have further strengthened his belief in a post-Triassic phase of deformation. I was, however, unable to find evidence for this supporting observation in Marcou's own notes, except that is how he depicted the Triassic and the Jurassic strata on his cross section (Marcou, 1856). He says that the Tertiary conglomerates on the Colorado Plateau are "raised up"; in his French original, Marcou says "redressés," (*deformed* only: Marcou, 1856, p. 159), so he must have also thought of a post-Tertiary phase of upheaval. His cross section is equivocal in this respect (see also Goetzmann, 1991, p. 324–325), but in the explanatory text to his first map of the United States, he had certainly spoken of numerous episodes of uplifting of the Rockies extending till after the Miocene: "As to the Rocky Mountains, the chains which form them are certainly to be ascribed to several systems of dislocation, and perhaps even little chains may be found there as ancient as the Silurian epoch. At last, it may be said now that several chains date from the epoch of the systems of the Green Mountains and the Alleghanies, and that the cretaceous, eocene and Miocene [*sic*] formations, even, are very much upheaved, and consequently have been subjected to dislocations after their deposit" (Marcou, 1853, p. 75).

In contrast to Marcou, Blake interpreted the relationships of the Mesozoic strata to the Rocky Mountains as one of abuttment: He thought that the Rocky Mountains had already formed islands in the Mesozoic seas and that the Triassic to Cretaceous deposits had been laid down around them. *Then, he believed, the entire western part of the continent had been upheaved*:

> We may for the most part regard the strata as horizontal, and undisturbed by the uplift of either of the great granitic ranges [*i.e., the Albuquerque (or Sandia) and the Gold Ranges east and northeast of Albuquerque, New Mexico, making up the eastern boundary of the Rio Grande Rift*], which are more recent than the Carboniferous. We may conclude from the notes [*i.e., Marcou's*], and the observations of others, and from the topographical indications, that the formation extends continuously in nearly horizontal beds from one side of the central chain of mountains to the other, occupying the wide depressions, or passes, between the ranges. It occurs, according to Mr. Marcou, along the valley of the Galisteo river, and a short distance north of Camp 56 (Galisteo.) This place is directly in the line of the Santa Fé mountains, and between their south end and the north end of the Gold mountains. The white sandstone and calcareous strata of the Llano are also found to extend through this break in the mountains, and are cut by a trap dyke four hundred yards [*~390 m*] north of Camp 56. (See notes, October 2 [*Marcou, 1856, p. 140*]). Mr. Marcou also records passing from Camp D (probably at Galisteo) to Camp E, or the Pecos village, through "cañons in the Trias, as far as old Pecos, the top of the cañons being of Jurassic sandstone." From this I conclude that the "Trias" and the "Jurassic" were horizontal at the east base of the Santa Fé mountain, and this conclusion is in accordance with the topography, as given by Abert and Peck, in their map of 1846–7. I have been thus particular to present the evidence of continuity of these deposits from one side of the mountains to the other, as Mr. Marcou presents them in his Resumé, as upraised and dislocated by the "Rocky mountains," the dislocation of which, he states, took place at the end of the Jurassic period.[357]
>
> Having thus shown that the gypsum formation extends from one side of the mountains to the other, undisturbed or dislocated only by local intrusions of trap, we may conclude that the principal uplifts of the central chain took place before its deposition, and that a grand continental elevation of over seven thousand feet [*~2100 m*] has taken place since that time. (Blake, 1856, p. 75)

It is instructive to look at the writings of two geologists nearly a century-and-a-half later and see that both were only partly correct. Blake was certainly correct in seeing the continuity of the Mesozoic strata across the Rockies into the Colorado Plateau. Marcou also saw this. Blake was clearly wrong in interpreting the relationship of the Mesozoic strata to the Rocky Mountain ranges as one of abuttment. He seems to have ignored the previously accumulated evidence that the strata do get more deformed near these uplifts. This evidence had already been observed and recorded by Frémont in the present-day Wyoming, by Stansbury (Hall, *in* Stansbury, 1852; see above), and by Schiel in Colorado and Utah, all the way down into New Mexico and Arizona by Emory and Marcou (see especially Marcou, 1853, top cross section on map sheet). As we saw above, Hitchcock (1847) had already used part of that information to intrepret the Rockies as granite-cored ranges surrounded by dislocated superjacent strata. Marcou clearly knew all this. His error lay in his over-enthusiastic application of the lithostratigraphy—to use an anachronistic term—to correlate the sedimentary sequences in the western part of North America with those in western Europe (for his even wider ranging correlations of the Permo-Triassic, see Marcou, 1859), and in that, he was justly rebuked by Blake (all this despite Marcou's clear appreciation of the "much less value" of lithostratigraphy than biostratigraphy for time correlation: Marcou, 1853, p. 58). Blake was also correct in seeing that a grand continental elevation (a falcogenic event) had taken place since the end of the Mesozoic era to uplift the Mesozoic section on the route of the Whipple Survey for over 2000 m. Blake may very well have been influenced in this by his teacher Dana's ideas and the inferences made farther north by James Hall. The terminology he uses ("continental elevation"), however, unmistakably reminds one the ideas of Hall, *the denier of local axes of uplift in 1857,* and not of Hall, *the advocate of local axes of uplift in 1852*. The former set of ideas were to be presented by Hall to the American Association for the Advancement of Science in Montréal only a year later! Hall may very well have changed his opinion concerning local uplifts before Blake took to the field, or the two may have corresponded before Blake wrote his report sometime before 26 September 1856 (Blake, 1856, title page).

By contrast, the way Marcou drew his cross section across the continental swell (Fig 96B; see loose plate accompanying this volume; also see Marcou, 1853, 1858) betrays Marcou's thinking in terms of a copeogenic process only. He appears to have interpreted the entire uplift simply as orogeny, but one of large width with multiple centers of upheaval and igneous intrusion. That Marcou was comparing the directions of the

Rockies and the Sierra Nevada and expressing surprise that despite their parallel trends the Sierra Nevada appeared to be "much younger" than the Rockies clearly show that he had in mind his teacher Élie de Beaumont's theory of mountain-building and his pentagonal network (see Marcou, 1853, p. 58, and his careful review of Élie de Beaumont's theoretical ideas on p. 64–67; also [1854], p. 48). We would not be entirely amiss if we saw in this interpretation also the influence of Alexander von Humboldt's great book on New Spain (de Humboldt, 1811a, 1811b; 1812) and his subsequent publications on mountain-building in Asia (von Humboldt, 1831, 1843). Perhaps because of that influence, Marcou seems not to have considered that the entire uplift between the Mississippi River and California had the character of a positive *bosselement* in terms of his teacher Élie de Beaumont's theory, that is, a broad gentle anticline in the sense of Hitchcock (1841) and Dana (1847b). That is understandable because it was the narrow mountain ranges, the *ridements* (i.e., the copeogenic structures) that had been very popular among Élie de Beaumont's ideas (most likey owing to the influence of the ideas of Alexander von Humboldt and Leopold von Buch and the new and popular method of dating mountain ranges by means of angular unconformities), not the hard-to-recognize falcogenic *bosselements* (see Marcou, 1853, p. 58).

The Sierra Madre (Zuni Mountains) formed the water divide, and the explorers duly noted this fact (also see Möllhausen, 1858, p. 265). The summit of Campbell's Pass was 7750 ft. (~2350 m) in elevation, whereas that of Camino del Obispo was measured to be 7949 ft. (~2400 m). The explorers estimated that the highest points of the range could not be much lower than 12,000 ft. (~3600 m). They further noted that the mountains on the eastern side of the Rio Grande had their gentle slopes facing east, whereas to the west, it was the other way around. West of the Zuni Mountains, they marched on flat highlands around the Little Colorado River, which Marcou correctly colored largely as Triassic. Farther east, they stepped down onto the Carboniferous. (As I mentioned in endnote 344, Permian was not yet internationally recognized; in Dana's (1863) *Manual of Geology,* for example, we see that the Permian was still regarded as a subdivision of the Carboniferous: Dana, 1863, p. 378, fig. 618.)

West of the Sierra Madre, the topography was one of an immense tableland. Here Blake gave a wonderful description of what we today call the Colorado Plateau on the basis of Marcou's notes, in complete agreement with what can be seen on Miera's map and read in the numerous Spanish descriptions from Coronado to Escalante:

> After passing the range of the Sierra Madre, an immense expanse of table-land is spread out before the explorer. From this range to the volcanic cone of San Francisco, two hundred and fifty miles [*~400 km*] distant, there is not a single mountain ridge or sudden swell of the surface to break its monotony. It is a region of horizontally stratified rocks, cut and eroded by streams as on the eastern or Mississippi slope. The descent here, however, is towards the west and very gentle. The same or similar rock-formations are found on this side of the great dividing range, and in the same horizontal position: the topography is consequently similar. The erosions and bluffs produced by the head waters and tributaries of the Canadian, on the elevated plateau of the Llano, find their counterpart on this side of the mountains in the Colorado Chiquito and its tributaries. As in approaching the mountains on the east, the survey followed the pathway thus cut out by the streams; so in descending on the west, the expedition followed similar paths, cut by the streams which flow into the Pacific. The extent of this wide area of table-land is not yet accurtely known. It was seen stretching out indefinitely towards the north, and doubtless is continuous, and of the same character, as far as the head-waters of the Great Colorado, and Grand and Green rivers. The great cañon of the Colorado is also in this plain, and on the south it appears to extend to the base of the Mogoyon mountains. (Blake, 1856, p. 8)

When they reached the Aztec Mountains, Marcou noted that they too had a "table-like summit" (Blake, 1856, p. 3; see Möllhausen, 1858, p. 351, for local topography along the route of the expedition).

West of the Aztec Ranges, they reached the boundary of the great tableland (whose mean altitude was estimated to be 6000 ft. {~1800 m}: Blake, 1856, p. 9), with the Cerbat Range corresponding with the present-day Hualapai Mountains (the Cerbat Range is now considered to form the north-northeasterly continuation of the Hualapai Mountains). West of there, Marcou consistently noted highly deformed Tertiary deposits. Towards the Mojave River, they found horizontally bedded, brecciated limestones. The elevations had already dropped to 4500 ft. (~1300 m). Here Blake reports that:

> It has generally been considered that the surface of the Great Basin was of nearly uniform elevation, and that it was like a plain or table-land. The point, however, reached by the survey, or the bed of the Soda lake [*this refers to Baker Soda Lake in California, at about 35° N and 35°30′ N and 116*], which is probably the end of the Mojave river, is very low, being only 1,116 feet [*~340 m*] above the sea, and very much lower than the average elevation of the surface of the Basin. It is indeed the lowest point of the Great Basin now known. From the Soda Lake the valley or dry bed of the Mojave river furnished a gradual ascent until within twenty miles of the crest of the Bernardino Sierra, when the road leaves the river and commences the ascent of a gently rising slope, which terminates at the summit of the Cajon Pass. This slope may be regarded as a fair type of those which make up the wide surface of the basin, which is, in fact, but a combination of slopes flanking the ridges, producing by their intersection a series of basin-like depressions. (Blake, 1856, p. 9–10)

The structure of the Great Basin was also very different from that of the high plateau to the east:

> The surface of the Great Basin, unlike that of the great plain between the Sierra Madre and the San Francisco volcano, is not formed of horizontal strata, which leave table-like areas where cut by rivers or exposed to denudation. The materials composing the surface appear to be derived from the adjoining ridges and mountains, and are laid down around them with inclined surfaces, the coarser parts being nearest the elevations, while the finer materials are transported further out, and the sloping character of the surface is thus produced. (Blake, 1856, p. 10; to form a mental picture of what Blake here means, see Antisell, 1856, especially plate VI, the "Section from Mojave River to Soda Lake.")

The Whipple survey along the 35th parallel thus found essentially the same basic picture as did the Stevens Survey along the 47th and 49th parallels, and as did the Gunnison-Beckwith Survey along the 38th, 39th, and the 41st parallels: a broad swell spanning the entire width of the continent from the Mississippi River to the western end of the Great Basin, whose slopes were a few hundredths of a degree. The slopes were smooth in the east, but from the Rocky Mountains westwards, local, short-wavelenth deformations interrupted the continuity of the gentle structural slopes and produced topographic irregularities. Despite that, between the Rio Grande and the Great Basin, the terrain still had the appearance of a high tableland with an average elevation nearing 2000 m. Westwards it dipped down, and in the Great Basin its surface became much corrugated.

Summary of the Results of the Railroad Surveys in Regard to the Falcogenic Structures in the Western United States

Hall's (1857: Fig. 100 herein) and Marcou's (1858) geological maps summarized the attainments of the railroad surveys. All three surveys had found that west of the Rockies the geology and the morphology became much more complicated than it was in the Great Plains. In the north, the complication was the huge Columbia Basalt Plateau and the Cascades; in the center and the south, the Great Basin with its mostly north-south–trending "island-like mountains" and the Sierra Nevada created a colorful and varied geological picture. However, in all cases, only the Cascades and the Sierra Nevada presented anomalies to the westerly decreasing elevations west of the Rockies. Especially in the southernmost route, it was recognized that the stratigraphy of the Great Plains could be traced as far west as the Great Basin and that the immense continental swelling had the aspect of a flat anticline, as already surmised by Hitchcock and Dana.

Almost everbody agreed that this anticlinal swell had begun rising in the Cenozoic. Its relations to the Rockies were controversial. The European "old school," adhering to the ideas of von Humboldt and Élie de Beaumont, considered the individual elevations (such as the main range of the Rocky Mountains, the various uplifts on the Colorado Plateau, and the Sierra Nevada) as individual axes of elevation cored by intrusions. This school interpreted the entire swell in the western half of the North American continent vaguely as being a result of repeated episodes of orogenesis (i.e., formation of *ridements* in the sense of Élie de Beaumont) related to the intrusions, as perhaps best illustrated by Marcou's controversial *Geology of North America* (Marcou, 1858), the first book that presented a comprehensive synthesis of the geology of the areas west of the Mississippi River. His geological map best conveyed the image of an immense and complex anticlinal structure encompassing the region extending from the Great Plains to the Great Basin, with its axial region, consisting of a bundle of parallel axes, located along the Rocky Mountains, the Alvarado Ridge of Eaton (1987).

Thomas Antisell[358], in his geological report of the Parke expeditions, pointed out that in the "middle Tertiary," both the Pacific and the Atlantic Oceans must have extended continentward considerably and were perhaps only separated by land in the present-day areas of the Rockies. He inferred from the fossil contents of their deposits that the waters of the two oceans did not intermingle at that time (Antisell, 1856, p. 21; contrary to what Goetzmann, 1991, p. 323, claimed him to have stated).

The Americans took a broader view conditioned by the rapidly expanding knowledge concerning the western part of their home continent. The eastern coast theoreticians saw in the immense swelling an anticlinal structure, essentially a *bosselement* of Élie de Beaumont. This was echoed by their disciples who went west and wandered over the structure itself. William Blake, for instance, implied that the entire Rocky Mountains had been uplifted in a passive way by the continental elevation.

The European Antisell basically agreed with von Humboldt and Marcou but had a more complicated picture in mind, taking into account the immense areas affected by the gentle up and down movements. He first argued that the rise of a great chain of mountains such as the Sierra Nevada could not have happened without affecting the country on both its sides. He then noted that the Great Basin east of the Sierra Nevada was considerably higher than the Great Valley to the west of the Sierra. He thought that the uplift of the Sierra Nevada had raised the Great Valley and cut it off from the sea and uplifted the Great Basin further:

> Inasmuch as the elevation of the Sierra in the north of California and that of the Cascade mountains of Oregon is much greater than that of the southern portion of the Sierra, it might be supposed that the elevation of the contiguous crust would be in proportion. That if the elevation of the Sierra up to 7,000 feet [*~2100m*] was sufficient to lift the Colorado desert [*the Yuma Desert in the United States and the Gran Desierto in Mexico*] up to the sea level—as it now stands—then an elevation of the same Sierra to the north to an altitude of 12,000 or even 17,000 feet [*~3600 or 5100 m*], might suffice to raise the Great Basin to the level of Salt Lake valley [*which Antisell quoted as being 4000 ft., i.e. ~1200 m, above sea level*]. (Antisell, 1856, p. 21)

Antisell thus saw not only a west-to-east increase in elevations towards the Rockies, but also one from south to north, from the mouth of the Colorado River to the Great Salt Lake area (in contradiction to von Humboldt's earlier statement that the plateau elevations were decreasing from Mexico to Utah continuously). He pointed out that entire mountain axes rose and fell along this uplift direction. To my knowledge, Antisell was the first to mention axial culminations and depressions in the Cordillera of North America (except, of course, von Humboldt's implicit statement that the high plateaus making up the central branch of the Sierra Madre in Mexico lost elevation gradually as they extended northwards into the present-day Utah and Colorado; see above, p. 153).

What was the agent of upheaval creating these axial level fluctuations? Antisell does not say directly, except on p. 103,

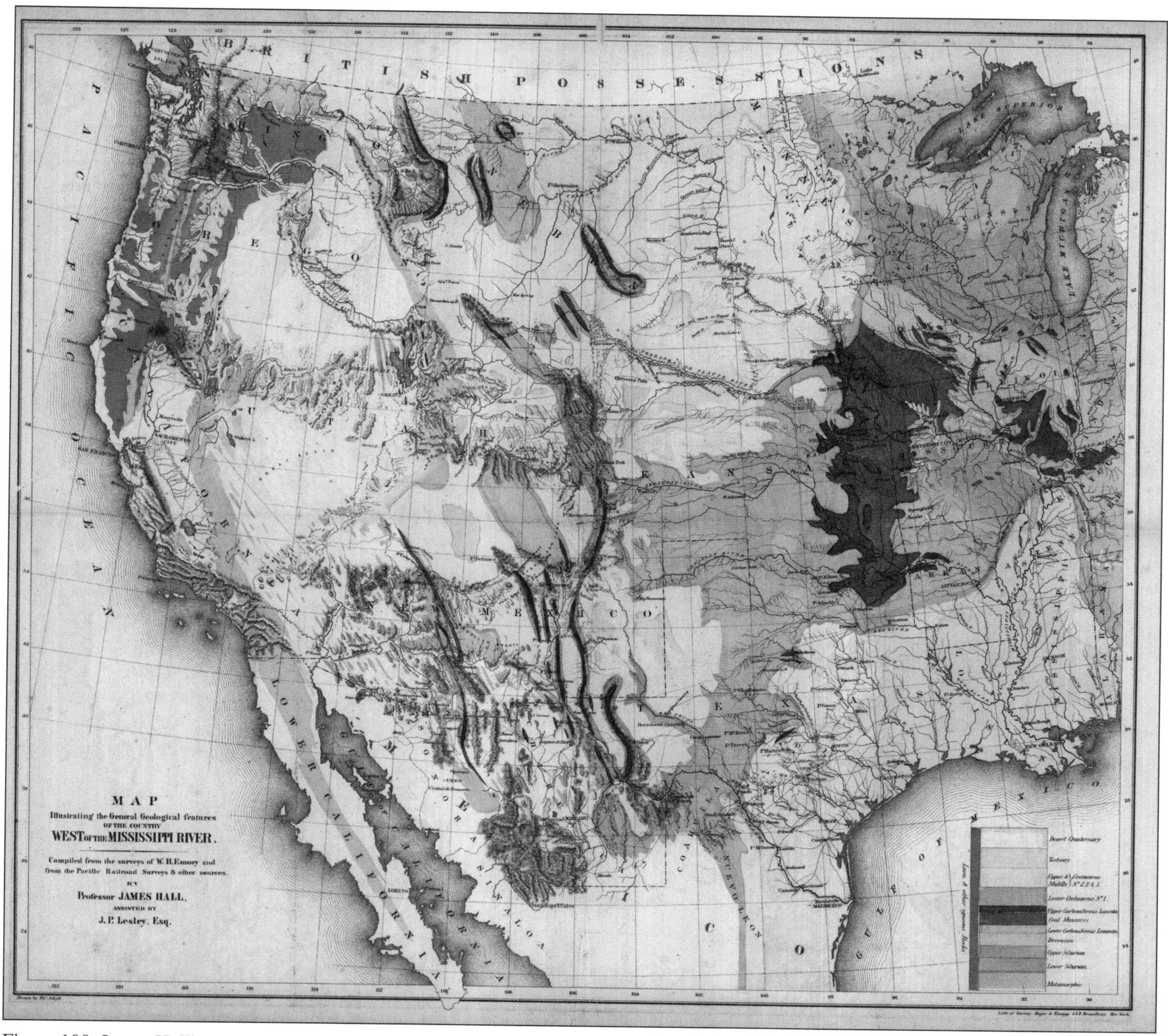

Figure 100. James Hall's compilation geological map of the United States west of the Mississippi River (from Hall, 1857c). This map incorporates the results of all expeditions, beginning with Frémont's, the geological data of which had been submitted to Hall. (Courtesy of the Wichita State University Libraries, Department of Special Collections.)

where there is talk of "upheaving plutonic rock" (Antisell, 1856). On p. 16, he talks about the longitudinal ranges delimiting the valleys of Napa, Petaluma, and Sonoma. He maintains that after their uplift, "a great depression or chasm has been produced across the strike of these ranges, by the exertion of *volcanic forces* …; and in the depressed valley running east and west, thus produced, the waters of the ocean have advanced to meet the Sacramento and Joachim [*i.e,. the San Joaquin*] rivers, which roll down their several valleys from opposite points" (Antisell, 1856, p. 16, italics mine). This chasm was produced, according to Antisell, in the following manner:

> The elevation of the Coast Ranges must have taken place from two points, one in the north and one in the south; the latter force commencing in the southern part of San Luis Obispo and the eastern of Santa Barbara counties, and thence extending north; as the upheaving force passed northward, its power become spent, and unable to lift the imposed strata; a similar action from the north, acting in a southerly direction with less vigour, produced an uplift, whose action ceased between latitude 37° and 38°. So that while the consolidated crust of the State was uplifted at each end, it was quiescent, or nearly so, in the middle; and the two forces acting against each other may have produced a rupture of the superficial strata, and even a depression of the surface below the sea level, in which the waters of San Pablo, Suisun, and San Francisco, have taken their resting place. (Antisell, 1856, p. 20)

He later remarked that

Depressions of the strata and fissures from east to west across the line of mountain ranges are common along the Pacific, north of this point [*i.e., San Francisco Bay*], latitude 38°, and extend inland even east of the Sierra Nevada. In the course of these depressions rivers run. The Klamath and the Columbia are examples; which rivers might possibly never have emptied their waters into the Pacific but for this fracturing effect produced by opposing volcanic forces. (Antisell, 1856, p. 20)

Thus, Antisell's tectonic picture was similar to Marcou's in that he imagined individual axes of upheaval motored by igneous forces. He imagined centers and axes of plutonic or volcanic upheaval, and the areas between these were left behind and sagged down; sometimes ruptured to create transverse fissures (compare this description with Marcou's east-west cross section in Fig 94B). This tectonic world view also had similarities to von Humboldt's in that Antisell imagined both longitudinal (such as the Sierra Nevada) and transverse uplifts (such, indeed, as the Transverse Ranges), similar in kind to that which von Humboldt thought he had discovered in the south Mexican transverse volcanic axis between Puerto Vallarta and Veracruz and to that hypothetical one he asked John Newberry to look for between San Francisco Mountain and San Mateo (Mount Taylor) on the Colorado Plateau (see his letter to Newberry quoted on p. 154). Like von Humboldt, Antisell compared the high interior plateaus and the Great Basin with Tibet, the Sierra Nevada with the Himalayas, and the rest of California with India south of the Himalayas:

In travelling west, across these upper plateaus, the Sierra Nevada is not the lofty mountains known in California, on account of the basin level being so much above that of the Sacramento valley. Several thousand feet of altitude are lost to the mountains when viewed from the basin [*i.e., the Great Basin*]. Something like this occurs in the steppes of Thibet [*sic*] and Tartary, where, travelling south, the Himalaya mountains are apt to be under estimated, because the plateau of the steppe country is so elevated; but, on crossing these mountains into India, the traveller descends several thousand feet, and attains a much lower level of land on the Hindostan side. So it happens in travelling through any of the northern passes of California, at Noble's or Carson's passes, the ascent is comparatively small until the summit is reached, when the descent is more sudden and much greater until the valley is descended. (Antisell, 1856, p. 21)

In the tectonic interpretation of the great western swell of North America by the last great expedition geologist before the Civil War, namely John Strong Newberry, we see the reflections of both the American and the European views of mountain- and plateau-making, without achieving a satisfactory and lasting synthesis. By contrast, Newberry established a fine stratigraphic basis on the Colorado Plateau that became a beacon for Powell's survey after the war. A sound tectonic synthesis came mainly as a result of the post-war work of the Great Surveys, especially that of the Powell survey, building mainly on the work done by Newberry on the Colorado Plateau.

John Strong Newberry with the Ives and Macomb Expeditions on the Colorado Plateau

The Ives Expedition

The September 1857 to June 1858 Mormon War, precipitated mainly by the Friday, 11 September 1857 Mountain Meadow massacre of the emigrants by the Mormons and their Indian allies, prompted the United States government to explore direct supply routes to the heart of the Deseret. Even before the war, the federal government had shown interest in expanding the geographical knowledge of the new territories acquired during the Mexican War and to consolidate the physical coherence of the union. Already in 1852, for example, Lieutenant George Horatio Derby was sent up the Colorado River to map its course between the Gila River and the Gulf of California. The main purpose of this reconnaissance was to see whether Fort Yuma could be supplied by the river (for Derby's expedition, see Miller, 1970, p. 144). It was realized that the river could supply Fort Yuma, and the Colorado Steam Navigation Company was thus born, presided over by Alonso Johnson. This company held the monopoly of supplying the fort and derived its subsistence entirely from it. However, Johnson was interested in expanding the company's business. Consequently, he approached Jefferson Davis, then the U.S. Secretary of War, proposing to explore the Colorado River with a view to establishing whether a riverine suppy route to the Great Basin was feasible. His proposal was backed by U.S. Senator John B. Weller, and some \$70,000–\$75,000[359] were included in the Army appropriation for 1856–1857 for western geographic exploration. Although Johnson had hoped that much of this funding had been created for his project, a delay in the approval of the U.S. Senate and the appointment of John B. Floyd to the position of the Secretary of War in the meantime, frustrated his hopes. The planned expedition was placed in the charge of First Lieutenant Joseph Christmas Ives. Although the indignant Johnson ascribed this appointment to nepotism in Washington (Ives was the husband of Cora Semmes, the niece of the new Secretary of War: Miller, 1970, p. 147, footnote 9), the choice had been wisely made (for Johnson, his initiative, and sources about them, see Miller, 1970, p. 144–148). Ives was a well-educated man who had previous exploration experience on the same ground: He had been educated at Yale and West Point and assisted Lieutenant Amiel Whipple on the Pacific Railroad survey (Goetzmann, 1991, p. 375–380; 1993, p. 306). He was extremely enthusiastic and proved to be a fine leader during the course of the expedition.

The exploring party included the experienced geologist of the Williamson expedition for the railroad exploration between northern California and Washington state, John Strong Newberry (Fig. 101; cf. Newberry, 1856) and two Germans: Baron Wilhelm von Egloffstein, who was responsible for topographic mapping and landscape sketches, and Heinrich Balduin Möllhausen, official artist of the expedition but who also functioned as naturalist. Möllhausen had joined the expedition with

Figure 101. John Strong Newberry (1822–1892), who first erected a usable stratigraphic column for the Grand Canyon country and illuminated the path of his successors who workd on the geology of the high plateau (from the *Geological Society of America Bulletin*, v. 4, frontispiece.)

Alexander von Humboldt's recommendation and had brought queries from the great geographer to Newberry pertaining to geological problems of the area to be traversed[360] (see p. 154). P.H. Taylor served as astronomical assistant, and C.K. Booker served as meteorological assistant.

The main purpose being the discovery of easily navigable routes, the explorers were to proceed on the Colorado River aboard the *Explorer,* a shallow-draft, 54-ft. (~16.5 m) steamboat specifically constructed for the purpose by A.J. Carroll on the East coast and brought to the mouth of the Colorado in prefabricated pieces to be assembled. Carroll himself joined the expedition as the engineer of his creation, which he reconstructed with the aid of his fellow explorers before the astonished eyes of the native Cocopa Indians and the local steamboat company people at Robinson's Landing at the mouth of the Colorado River (see Ives, 1861, plate I).

On 30 December 1857, the explorers launched their boat amidst feelings of "much admiration and complacency" (Ives, 1861, p. 37), and the expedition commenced at midnight to make maximum use of the tide. The next day, a steamer coming down the river from Fort Yuma brought the ominous news of the outbreak of hostilities between the federal troops and the Mormons. Captain Cadmus Wilcox brought, along with the bad news, new orders to explore the possibility of conveying large bodies of troops from Fort Yuma to Great Salt Lake along the Colorado and the Virgin Rivers as soon as possible (Ives, 1861, p. 39; Möllhausen, 1861, v. I, p. 146–147; Goetzmann, 1991, p. 382).

Newberry had been unable to join the expedition as it started owing to illness that detained him in Fort Yuma (Ives, 1861, p. 36). On 6 January, Newberry discovered the occurrence of gold, iron, and lead there (Ives, 1861, p. 48), but his geological narrative does not start until the explorers passed the mouth of Bill Williams' Fork. He was able to join the rest of the party when they reached Fort Yuma on 9 January 1858. Before the explorers set out upstream again, he and Möllhausen reconnoitered the country around the Fort and made natural history collections (Ives, 1861, p. 43). On the evening of 10 January, a party was held in the halls of the Fort to celebrate the departure (Möllhausen, 1861, v. I, p. 153–154; Miller, 1972, p. 1). They departed the next day, and on 1 February 1858, the explorers reached Bill Williams' Fork, about two-fifths of the way to where the head of navigation ultimately proved to be. They noticed the gradually rising country, and Newberry found much to do, in contrast to Möllhausen whom the season prevented from making a rich zoological collection (Ives, 1861, p. 52). They steamed upstream passing the awe-inspiring geological beauties of Mojave Canyon, Mojave Valley, and finally on 8 March they reached the foot of Black Canyon. Typical, high, flat-topped mesa country had already made its appearance, and Möllhausen recorded it in a woodcut (Fig. 102). It was farther up in the forbidding Black Canyon (Fig. 103), that Lieutenant Ives decided on March 12th, after a careful reconnaissance, that they had reached the head of navigation and turned the steamer back (Ives, 1861, p. 87).

Newberry noticed that the Black Mountains were made up of volcanic rocks (now known to be mostly Lower Tertiary: King and Beikman, 1974). After Ives divided his expedition party and and sent some downstream with the steamer, he took the remaining men (including Newberry, Baron von Egloffstein, and Möllhausen) and decided to go eastward in search of another route to the Mormon country (Miller, 1972b; Goetzmann, 1991, p. 388). This march carried them over the Colorado Plateau and eventually to Fort Defiance near the present Arizona-New Mexico border (Fig. 78). It was during this overland journey that Newberry established a stratigraphic column for the great plateau and then compared it with that known from the Great Plains. This showed that since the beginning of the Paleozoic, the Colorado Plateau was an integral part of the North American continent; indeed, it had the same stratigraphy as what Dana had called the stable interior of the continent. Its sedimentary pile was disposed in mostly flat-lying strata. Yet the region was significantly higher than the Great Plains and had reached that elevation only recently. Even though the height and the flatness of both the topography and the underlying sedimentary layers were no novelties, that the flat layers extended as far down in the Paleozoic as the earliest rocks of that era reached was news and eventually posed a grand problem to be solved in tectonics that led Powell and Dutton to novel insights

Figure 102. The beginning of the high table-land country: gravel bluffs south of Black Mountains. Woodcut by Möllhausen (Ives, 1861, fig. 24).

concerning the nature of tectonic processes affecting the outer rocky rind of the planet.

As he was riding over the great plateau, Newberry could afford the luxury of taking a grand view of it within the context of the tectonics of the entire western United States. Generations of observers—since the heroic Spaniards, and including the diligent geographers and geologists of both America and Europe, who traversed the forbidding highlands and the fearful deserts in numerous expeditions—had already accumulated enough for him to know roughly what he was traveling on. He thus started his geological account of the Colorado Plateau with a sweeping generalization, which, to the readers of this book, is already closely familiar:

> The geology of the country traversed by our party east of the great bend of the Colorado [*where the Lake Mead reservoir is today located*] may be conveniently considered in several distinct sections, as there are embraced in this vast region a number of well-marked geographical districts, of which the geological features are, in some respects, peculiar, and are not repeated. And yet these different districts form but parts of the great central plateau of the continent [361], and the relations which the structure of each part sustains to that of other portions of the geological arch—if I may use the simile—which spans the interval between the lower Colorado and the Mississippi are such, that it is quite as important it should be studied as a part of a great whole as in its local and minor details. (Newberry, 1861, p. 41)

The topographic details collected by generations of travelers, and especially the geological data gathered by the railroad surveys, had enabled this grand picture to be painted. Hitchcock's and Dana's speculations and the great geological map of the despised Marcou looked to Newberry to have an appearance of fact. The plateau he was regarding with the eyes of a competent geologist, seemed to his well-informed and intelligent mind to be a part of a great anticlinal up-arching of the earth's crust spanning the entire trans-Mississippi west. However much he may have disagreed with the details of Marcou's stratigraphy, the structure the Frenchman depicted presented itself as eminently plausible.

Newberry then proceded to give a somewhat more detailed view of the geomorphology of the great arch. It is impressive to read not only in view of how much had been learned in such a short time by a handful of dedicated people working with the most primitive means and under frightfully difficult conditions, but also in view of the fact that that body of knowledge stands today intact, in all its details, as a monument to those wonderful men who created its observational basis:

> I shall take the liberty, therefore, of anticipating in some degree my geological narrative, and give very briefly here, as the most convenient and appropriate place, the results of a line of observation carried

Figure 103. The Black Canyon as sketched by Lieutenant Ives and drawn by Baron von Egloffstein (Ives, 1861, plate V).

quite across the great plateau, of which the geological structure is so clearly revealed in the magnificent sections of the banks of the Colorado, not very far distant from the point we had reached at the close of the last chapter [*at the Black Canyon*].

The Colorado rises in a thousand sources, at an elevation of from ten to twelve thousand feet [*~3000–3600 m*] above the sea, on the western side of the Rocky mountains. Descending from their fountainheads its tributaries fall upon a high plateau of sedimentary rocks, which forms the western base of these mountains and occupies all the interval between them and the great bend of the Colorado, where the river enters the volcanic district already described [*reference here is to the Black Mountains: see Newberry, 1861, ch. IV*]. From that point its course trends northeasterly into Utah, where its outline has not been traced. Southward it follows the trend of the Black and Cerbat mountains, which bound it on the southwest, and extends far into Mexico [*note von Humboldt's influence*]. In the intervals between the ranges of the Rocky mountain system portions of the same "mesa" are seen [*Spanish geomorphic terminology dating from the sixteenth century!*], often much disturbed, and flanking the axes of the comparatively modern lines of elevation.

East of the mountains it still continues, forming the high prairies which everywhere skirt their bases. Cut into somewhat detached plateaus by the streams flowing from the mountains, a belt of country in that region has been designated by the name of the "high tablelands" [*Major Steven Harriman Long's terminlogy; see Fig. 80 herein*]; but there is no well-defined geographical area to which the name is strictly applicable, as the most remarkable unity, both of topographical and geological structure, prevails over the entire area of the "plains," which reach the mountains from the Mississippi. The geological elements which compose the great table-lands of the Colorado here reappear, exhibiting the same harmonious stratification. The strata all dip very greatly [*here Newberry probably intended to write "gently" rather than "greatly"*] eastward, and from the western slope of the Mississippi valley. (Newberry, 1861, p. 41)

Newberry then pointed out that the flank of the arch lying west of the Rockies had an average elevation of about 6000 ft. (~1800 m) along their route. The Colorado River had formerly flowed on the surface of this tableland, but now it had cut a deep gorge with walls rising in many places vertically 3000–6000 ft. (~900–1800 m) in height. "This is the 'Great Cañon of the Colorado,' the most magnificent gorge, as well as the grandest geological section, of which we have any knowledge" (Newberry, 1861, p. 42).

Newberry was able to determine, on the basis of a section he measured a few miles to the east of the mouth of Diamond Creek, where the plateau surface rises to an elevation of 7000 ft. (~2100 m) and the cliff height is >5280 ft. (>1600 m), a stratigraphic sequence sitting nonconformably on a granite[362] (Fig. 104) and going from Silurian to Carboniferous (Fig. 105 and 106). He found that: "Silurian and Devonian strata are entirely conformable among themselves and with the Carboniferous rocks. They lie nearly horizontally upon the granite, forming a series of sandstones, limestones, and shales, about 2,000 feet [*~600 m*] in thickness. The Carboniferous series consists of

Figure 104. Granite pinnacles in Diamond Creek Canyon (Newberry, 1861, figure on p. 57).

over 2,000 feet [*~600 m*] of limestones, sandstones, and gypsum, apparently all marine, and often highly fossiliferous" (Newberry, 1861, p. 42–43). Figure 107 illustrates the present-day stratigraphy of the Grand Canyon, which may be compared with Newberry's section. In making that comparison, one should bear in mind that in the mid-nineteenth century Cambrian had not yet been accepted as a general system term, and Silurian was generally accepted to be the lowest system of the Paleozoic. Newberry also criticized Marcou's ascription of the thick red sandstones and the overlying limestones to the Devonian and to the Lower Carboniferous. He showed that rocks which Marcou had described as middle Carboniferous at Partridge Creek, immediately to the west-northwest of Bill Williams Mountain, Marcou himself had ascribed to the Lower Carboniferous and to the Devonian near the Whipple Survey camps 103 and 104, a distance of ~55 km southwest of the Partridge Creek locality. Newberry noted that Marcou's observations had been made under snowstorm conditions, and he felt sure that the Partridge Creek and the camp 103 and 104 limestones both belonged to the middle Carboniferous limestones. From thickness considerations and correlation with his own much better controled section "within a distance of only fifty miles" (Newberry, 1861, p. 70; the actual distance is more like 100 km), Newberry rightly concluded that the schematic section Marcou had drawn near camp 103 could not possibly be correct and that all the red sandstones and the overlying limestones could not be older than the middle Carboniferous (Newberry, 1861, p. 69–70). We now know that Newberry was right. He was looking at the Upper Carboniferous to lowermost Permian Supai Group, plus the Permian Hermit Shale and the overlying sandstones as the "red sandstone" and the next overlying Permian Kaibab Formation as the "magnesian limestone" of Marcou (i.e., the Permian, which in those days was still viewed as part of the Upper Carboniferous[363].

Newberry misjudged the nature of the base of his "Silurian," which led him to some erroneous tectonic considerations. These considerations, however, seemed reasonable in the light of the state of theoretical tectonics then. He thought that the lowermost Paleozoic west of the great plateau was "seen dipping eastward, resting on the flanks of mountain chains which I have described as bounding the plateau in that direction. They here present bold escarpments toward the west, oftener the result of erosion than fracture. They have evidently been elevated by the upheaval of the plutonic rocks upon which they rest, but as they are usually quite unchanged, the igneous rocks could not then have been in a state of fusion" (Newberry, 1861, p. 43). It is clear that here (i.e., most likely in the Middle Granite Gorge near the Yampais village, representing the northernmost point that the

Figure 105. Head of the Diamond Creek Canyon showing the sedimentary rocks lying on the granite farther below the creek (Newberry, 1861, General Report Plate I).

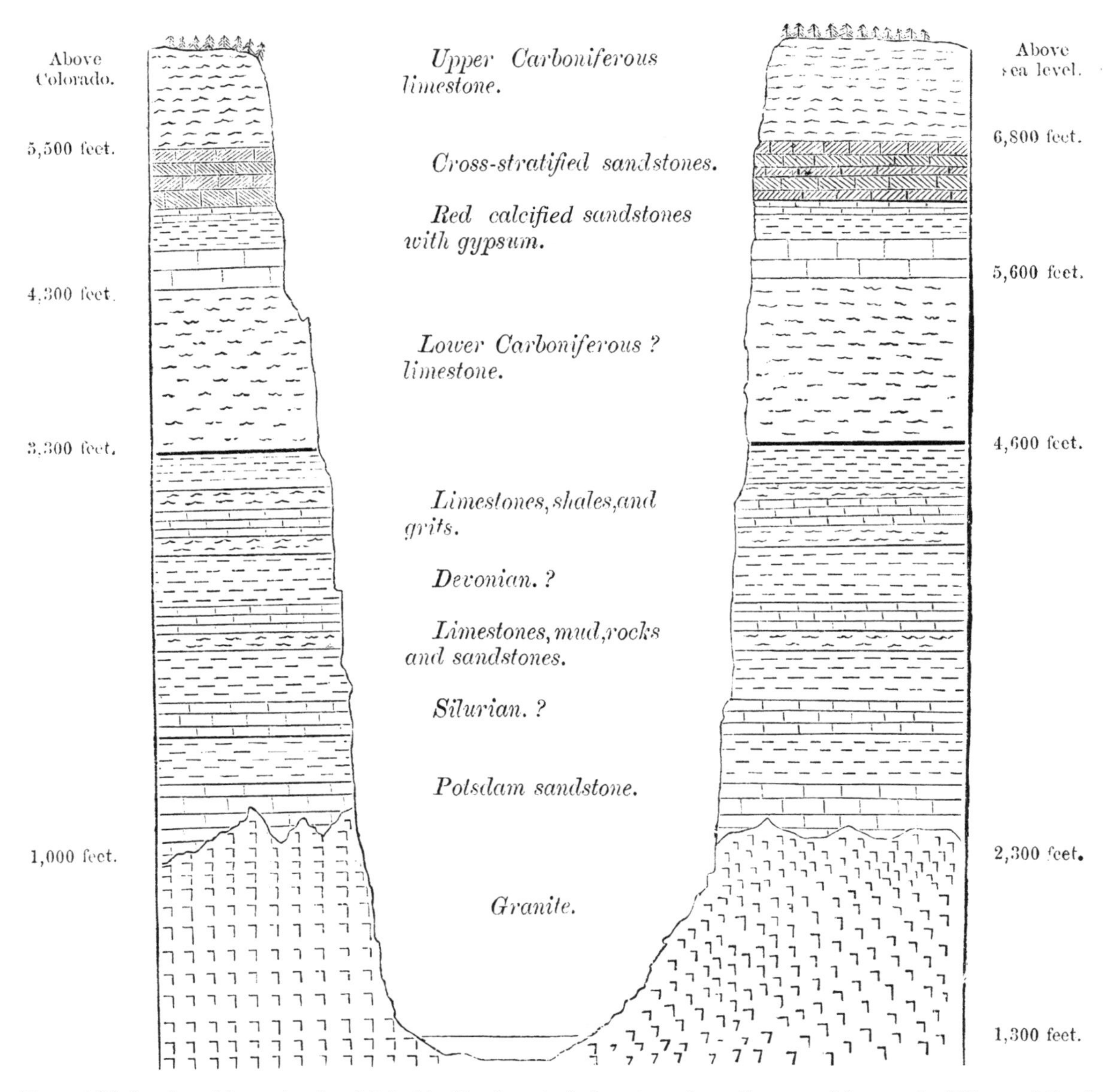

Figure 106. Stratigraphic section (established by Newberry) of a location a few miles east of the mouth of Diamond Creek along the high canyon walls (Newberry, 1861, fig. 12; also reproduced in Goetzmann, 1993, p. 342, lower figure).

explorers visited) Newberry was looking at the Precambrian Unkar Group of the Grand Canyon Supergroup (Powell's {1876, p. 70} Grand Canyon Group), which is composed of unmetamorphosed sedimentary rocks (cf. Hendricks and Stevenson, 1990, especially fig. 1), that he mistakenly took as Paleozoic. Newberry's observation that the igneous rocks had been already solidified at the time of the upheaval was of great importance in retrospect because less that 15 years later, Suess (1872, 1875) and 18 years later Heim (1878) used similar observations to show that the Alpine "central massifs" were not igneous intrusions that had upheaved the mountains, but represented basement that had deformed together with the folded and thrust cover under the influence of other—in the case of the Alps—horizontal, forces. However, Newberry could not take that step, because evidence for horizontal motions in his area was thought to be non-existent. He wrote: "Aside from the slight local disturbance of the sedimentary rocks about the San Francisco mountain, from the spurs of the Rocky mountains, near Fort Defiance, to those off the Cerbat and Aztec mountains on the west, the strata of the table-lands are as entirely unbroken as when first deposited" (Newberry, 1861, p. 46; today the shortening across the plateau is estimated to be less than 1%: McQuarrie and Chase, 2000). Because the upheaval appeared to be purely vertical, Newberry was thrown back to igneous agencies as causes of uplift. Regarding San Francisco Mountain, he wrote: "The great volcanic vent of the last-mentioned mountain has been opened up through this mesa, and has doubtless been an important agent in its elevation. Apparently little disruption has been caused by it..." (Newberry,

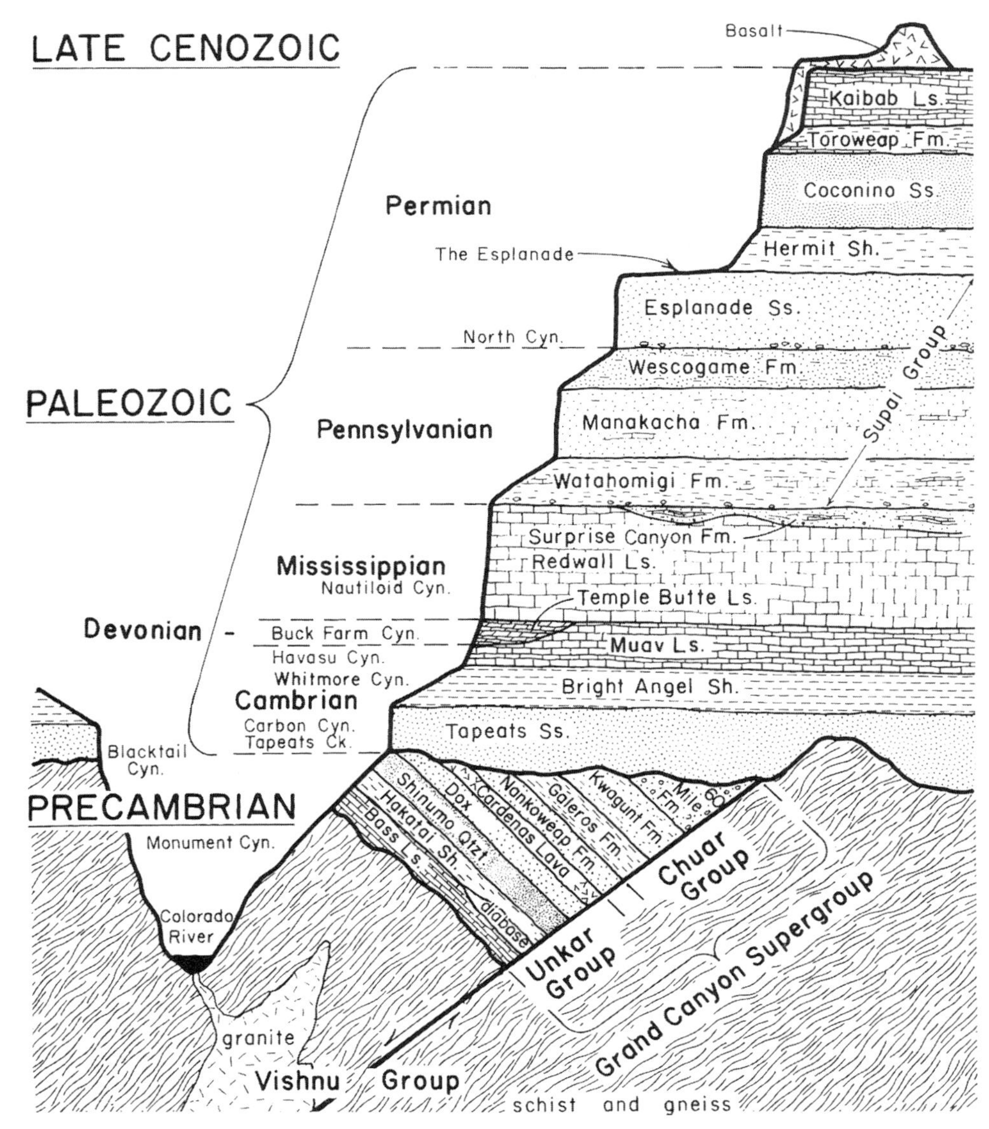

Figure 107. Generalized stratigraphic profile of the Grand Canyon as conceived by present-day geologists. After Potochnik and Reynolds (1990). Also reproduced in Baars (2000, fig. on p. 12, and 2002, fig. 69).

1861, p. 44). He reiterated this view in a later chapter, significantly pointing out that he knew that lateral shortening was in part responsible for mountain-building:

> Little disruption of the stratified rocks attended this grand exhibition of volcanic force [*i.e., in San Francisco Mountain*]; and the formation of the mountain mass seems to have been effected entirely by the ejection of matter in a state of complete fusion, through narrow orifices of unfathomable depth.
>
> Comparatively few mountains have been wholly formed in this manner; probably none but those having the same isolated character with that under consideration. All the mountain *chains* which have come under my observation have been composed, in a great measure, of upheaved strata of a decided sedimentary character. Some of them more or less metamorphosed[364].
>
> *Lines* of upheaval seem to mark features in the earth's crust, and bear evidence of the action of lateral pressure as well as elevatory force. In solitary mountains, on the contrary, I have observed a marked abundance of disrupted and metamorphosed sedimentary rocks, and the prevalence of masses of a purely plutonic character. The conspicuous summits which so generally mark the prominent angles in important mountain chains should generally be included in the same category with isolated cones, as they are also, as far as my observation has extended, are principally composed of ejected materials. (Newberry, 1861, p. 66, italics his)

It is significant that, in contrast to Marcou, Newberry realized that the volcanism had not caused any great disruption of the strata, but that it was still somehow connected with the rise of the plateau. This inference was already leading the way to

Dutton, who was to emphasize the great uplift without small-wavelength deformation as being caused by "plutonic forces," as we shall see below.

The normal faulting that succeeded the deposition of the Grand Canyon Supergroup created a jagged basement top in the Grand Canyon area (see for example, fig. 3 *in* Hendricks and Stevenson, 1990; and Sears, 1990). Newberry mistook the palaeotopography created by normal faulting as original topography and claimed them to be mountain chains "in embryo" that later gave rise to the Cerbat and Aztec Mountains[365] and those bounding the plateau to the east. Although Newberry had noted that the granites forming the jagged mountains (now buried by the "Paleozoic" sedimentary rocks) had been already fully crystalized at the time of the deposition of the superjacent sedimentaries, he still could not but ascribe these mountains, both then as embryos and now as two uplifted ends of the plateau, to elevation "by upheaval of the plutonic rocks upon which they rest" (Newberry, 1861, p. 43).

Newberry further showed that a continent already existed as far southwest as the Colorado Plateau and "Hence the theory generally received [*i.e., Dana's*] that the formation of the continent began in a nucleus about Lake Superior, and that places of the Rocky and California mountains were, until the Tertiary period, occupied by an open sea is proved untenable" (Newberry, 1861, p. 48). This "continent formation" was brought about by vertical uplifts and without any movements that we would today call copeogenic. Herein probably lies Gilbert's future choice of the term "epeirogenic" to designate the falcogenic "continent-making" movements that involve only broad flexing and warping.

Newberry thought that the entire continent west of the Mississippi had been uplifted in the Cenozoic in such a way that the western end of the uplift had a much steeper fall to the sea level than the eastern parts. This asymmetry, in part, is how he sought to explain the deep dissection of the surface in the west by the Colorado River canyons that stood in such great contrast to the broad and shallow valleys of the lazy rivers of the great prairies.

When they exited the mountains onto the plains on their way back east, Newberry noted that the "geological structure of this section of our route is scarcely in any respect different from that of the country immediately west of the Rio Grande" (Newberry, 1861, p. 102). He was thus forced to find an explanation for the great difference in the geomorphology of the two flanks of his great continental arch. This, he thought "may be referred to the combined action of several distinct causes, mainly geological, but in part atmospheric" (Newberry, 1861, p. 103). One factor was the very uniformly eastern dip of the beds in the east; in the west, the structure was more complicated. The next factor was the aridity of the West, which prevented the enlargement of river valleys sideways and led to the formation of the deep canyons. Newberry pointed out that his comparison applied only to that part of the eastern flank of his great arch "bordering the Santa Fé road" (Newberry, 1861, p. 104). The Llano Estacado, for example, "seems to produce, on a smaller scale, the scenery of the table-land of the Colorado" (p. 104).

The Macomb Expedition

In the summer of 1859, yet another army expedition was sent out in search of a direct route to the Great Salt Lake from the southwestern United States. It was placed under the command of Captain John N. Macomb, who was authorized to take along the geologist John S. Newberry plus four assistants (F.P. Fisher for astronomical observations, C.H. Dimmock for topography of the route, and Dorsey and Vail for the meteorological observations. Dorsey also assisted Newberry for the geological collections). The explorers were escorted by a detachment of the Eighth Infantry Regiment under the command of Lieutenant Milton Cogswell to guard them against hostile Indians, and by the Ute sub-agent Albert H. Pfeiffer, with his interpreter Neponocina Valdez, who went along to deal with the friendly ones (Goetzmann, 1991, p. 394).

The explorers set out from Santa Fe "about the middle of July, 1859" (Macomb, 1876, p. 5) pursuing a northwesterly course along the Old Spanish Trail. They followed the valley of the Chama (Fig. 108) and reached the continental divide (Fig. 78; with slight variations, also in Goetzmann, 1991, map 13; and 1993, map facing p. 306; Goetzmann and Williams, 1992, p. 165). They crossed the San Juan River at 37°14′48″ N latitude and 107°02′47″ W longitude (close to the present-day Navajo Resevoir) and proceeded to Mesa Verde, where they found Indian ruins (which have since led to the creation of the Mesa Verde National Park). At Ojo Verde (38°14′50″ N, 109°26′40″ W), they abandoned the Old Spanish Trail and turned west. They reached the junction of the Green and the Grand (now called Colorado) Rivers and became the first civilized men to set eyes on this important geographical locality. They found the country to the junction dangerous, owing to the harsh topography, and worthless, owing to its aridity (Macomb, 1876, p. 6). From the junction, they struck south-southwest and re-crossed the San Juan River, which they then followed on their way back to Santa Fe in the autumn of 1859.

The Macomb expedition established that an easy route to the Great Salt Lake from Santa Fe did not exist. The geological harvest was, by contrast, more positive: Newberry extended the Colorado Plateau stratigraphy northward and established it up to the San Juan Mountains; the area was underlain by a large plateau with layer-cake stratigraphy. He noted that lower and middle Paleozoic were missing in the Rocky Mountain foreland and concluded that the regions lacking these deposits must have been above sea level at the time of their deposition or uplifted after they were laid down to allow erosion to strip them off again. This uplift, he concluded, must have occurred along the axis of the present-day Rocky Mountains:

> The facts revealed in the cañon of the Colorado show plainly that the granitic basis of this country was consolidated previous to the deposition of the Paleozoic strata, and that over many of the minor irregularities of the sea-bottom the older sedimentary rocks were quietly and horizontally laid down, surrounding and abutting against granitic pinnacles, which rose above the shallow waters in which they were

Figure 108. The Chama valley displaying flat-lying Upper Cretaceous beds forming mesas on the high table-land (from Newberry, 1876, plate II).

deposited. These inferences, if confirmed by future observations, will considerably modify the hitherto accepted ideas in regard to the age of the ranges of the Rocky Mountains. We are at least warranted in the conclusion that these great lines of fracture in the earth's crust are few of them wholly of modern date, and it even seems probable that through all the geological ages they have served as hinges upon which the great plates of the earth's crust turned, as, in repeated elevations and depressions, the angles of which they inclose have been ever varying. (Newberry, 1876, p. 42)

Newberry's conclusion was not surprising. Marcou (1853, p. 75) had already claimed that the initial formation of the U.S. Rockies was to be dated back to the "Silurian" (i.e., to the beginning of the Paleozoic). Newberry also had before him the report of William Blake referring to the same country in which Blake had tried to make the Rockies older than the Mesozoic, claiming that the Triassic and the Cretaceous sedimentary layers abutted against them. He also had Dana's preferred directions of dislocation on earth and Élie de Beaumont's *reseau pentagonal* to consider.

In the light of such pre-existing ideas, it was only natural to think of the earth's crust as being compartmentalized along major, long-lasting fracture lines that are repeatedly re-used. His own misinterpretation of the relationship of the lower Paleozoic sedimentary rocks to the pre-Paleozoic topography in the Grand Canyon had already led him to believe that the Aztec and Cerbat Mountains had been in existence in embryo since pre-Paleozoic times. All these concepts inevitably conveyed him to the conclusion that the straight trend-lines of the Rockies were nothing but expressions of old lines of fracture along which repeated movements had taken place to delimit large areas—which Newberry, by a happy coincidence, called "plates"—of sedimentation. This conclusion is identical in principle to what Cloos (1948, p. 133) later called "a conservative earth-picture," consisting of repeatedly reactivated lines of fracture and relatively undeformed blocks surrounded by them, and has also had adherents until very recently (e.g., Brock, 1972). One of Cloos' examples of an old block delimited by lines of repeatedly reactivated fracture was in fact what he called the "Rocky Mountain Block" (*Rocky.-Mtn.-Scholle*: Cloos, 1948, fig. 10[366]). Many works on "lineament tectonics," from the nineteenth century to our own day, are based on a similar concept, and most of their authors have been most likely unaware of Newberry's ideas. Cloos, for example, clearly took his inspiration from generations of work in the central European block-faulting region (the *Schollengebiet* of the German geologists: see Cloos, 1948, fig. 1a; Şengör, 1995, p. 103–104 and fig. 2.10 for a current review). As we shall see below, however, Newberry may indeed have influenced Clarence King's view of the tectonic history of the western United States.

As the explorers wandered around Santa Fe, the erosional removal of almost everything above the Carboniferous drew their attention. Only when they were on what is now called the

Colorado Plateau did they see large tracts of Upper Cretaceous sedimentary rocks (Fig. 108). This gave occasion to Newberry to reiterate the structure and the geomorphological history of what he called "the great central plateau of our continent":

> The Upper Cretaceous rocks are also soft, and it is now necessary to go a long way from Santa Fé before anything like a fair representation of the upper portion of this series can be found. Indeed, east of the mountains [*i.e., the Rocky Mountains*] the extreme Upper Cretaceous strata are only seen in place near the Mississippi and the Gulf of Mexico. In the valley of the Rio Grande none remain, and it is only after crossing the main divide, between the waters of the Atlantic and Pacific, and seeing the magnificent exposures of the Cretaceous series in the valley of the San Juan, that we can form a just conception of the grand scale on which the Chalk formation was orginally built up in New Mexico, or of the enormous denudation which this region has suffered since it was raised above the surface of the ocean. The attention of every traveler over the great central plateau of our continent is attracted to the cañons which give character to the scenery, and when he learns that they are simply the effects of surface erosion, they become sources of unending wonder and interest. (Newberry, 1876, p. 50)

The high plains of the Colorado Plateau presented the morphology of a high tableland, a veritable plateau, to John Newberry. Yet, geologically, it seemed a basin with high Carboniferous tracts on the east and west and the Cretaceous in the middle. The eastern Carboniferous arch was narrow, but the western one was broad, reaching from the valley of the Little Colorado River to the western margin of the Plateau. The Grand Canyon was largely sunk into this broad arch. On the east side of the Little Colorado River, the Triassic was found to underlie the Plateau. As the traveler went northward towards the Wasatch Range, the Cretaceous occupied ever larger tracts: first, outliers of the Lower Cretaceous were encountered; then, an unbroken sheet of Cretaceous covered the landscape.

The tablelands between the Little Colorado River and the San Juan River were seen to have elevations of 8000 ft. (~2400 m), forming the northern extension of the high plateau. Newberry, like Willam Blake before him, found the northern margin not well-defined (Newberry, 1876, p. 62–64). The Sage Plain, between the San Juan Mountains and the Colorado valley, stretched "out nearly horizontal, unmarked by any prominent feature, to the distance of a hundred miles [*~161 km*]" (Newberry, 1876, p. 85). Looking from north to south, Newberry obtained the impression of looking over a sea-like flat surface, into which the southerly spurs of the San Juan Mountains extended like peninsulas (Newberry, 1876, p. 76). The spur forming the dividing range between the Rio Los Pinos and the Rio Piedra seemed to him to form an axis of upheaval: "It is composed of Cretaceous rocks irregularly broken up, generally inclined at a high angle. It has an altitude of about sixteen hundred feet [*~480 m*] above our camp on the Piedra. The view from its summit is peculiarly grand and interesting" (Newberry, 1876, p. 78; they were in reality on the southwestern flank of a southeasterly plunging anticline). To the west of the Rio Los Pinos, the broad Tertiary valley was interpreted as a fault-controlled depression: "...in the breaking-up of the tablelands, a basin like depression was left, into which the Animas flowed and which it partially filled with gravel and bowlders [*sic*] brought down from the mountains above" (Newberry, 1876, p. 81).

This image of a gently warped, vast highland, here and there broken down into tectonic depressions along high-angle faults, had a lasting effect on subsequent generations of researchers of the Colorado Plateau and had a profound influence on the tectonic world picture of Eduard Suess on the other side of the Atlantic, leading him to imagine the radial effects of the thermal contraction of the globe to take the shape of what he called "cauldron subsidences" (i.e., fault-bounded, roughly equant, round depressions). As we shall see below, Suess came to think that the world's ocean basins were nothing more than giant cauldron subsidences, and the continents giant upstanding plateaus similar to the Colorado Plateau.

Newberry repeatedly emphasized the power of fluvial erosion in sculpting the landforms on the high tableland. Looking south into the entrance of the valley of the Colorado proper, from what the explorers called Labyrinth Canyon (Newberry, 1876, p. 95), he gave a vivid description of the landscape that inevitably reminds us of Dutton's "ludicrous and farcical" landforms quoted below (Fig. 109). Newberry followed his military predecessors in using terms freely borrowed from architecture to describe the fantastic, castellated shapes sculpted out from flat-lying rocks and which incessantly exercised his imagination. In this, too, he was to be followed by his great successors, Powell and Dutton:

> A great basin or sunken plain stretched out before us as on a map. Not a particle of vegetation was anywhere discernible; nothing but bare and barren rocks of rich and varied colors shimmering in the sunlight. Scattered over the plain were thousands of the fantastically formed buttes to which I have so often referred in my notes; pyramids, domes, towers, columns, spires of every conceivable form and size. Among these by far the most remarkable was the forest of Gothic spires, first and imperfectly seen as we issued from the mouth of the Cañon Colorado [*Not the Grand Canyon! This appellation refers to a side canyon coming from the east and issuing near the junction of the Green and the Grand Rivers*]. Nothing I can say will give an adequate idea of the singular and surprising appearance which they presented from this new and advantageous point of view. Singley, or in groups, they extend like a belt of timber for a distance of several miles. Nothing in nature or in art offers a parallel to these singular objects, but some idea of their appearance may be gained by imagining the island of New York thickly set with spires like that of Trinity church, but many of them full twice its height. (Newberry, 1876, p. 97)

Eastward, the Rio Grande trough appeared to Newberry to have a synclinal structure (p. 65). This downwarp was between anticlinal upwarps that perturbed the gentle bow of the giant continental arch that spanned the distance between the Mississippi River and the Pacific Ocean.

An important feature of the high tableland was recent and active volcanism. Newberry pointed out that the occurrence of volcanoes in the plateau country contradicted the idea that volcanoes only occur near continental margins or in the oceans

Figure 109. Head of Labyrinth Creek, looking southwestwards (from Newberry, 1876, plate VII). "Scattered over the plain were thousands of the fantastically formed buttes to which I have so often referred in my notes; pyramids, domes, towers, columns, spires of every conceivable form and size. Among these by far the most remarkable was the forest of Gothic spires ... Singley, or in groups, they extend like a belt of timber for a distance of several miles. Nothing in nature or in art offers a parallel to these singular objects, but some idea of their appearance may be gained by imagining the island of New York thickly set with spires like that of Trinity church, but many of them full twice its height" (Newberry, 1876, p. 97).

(Newberry, 1876, p. 62). Neither did volcanism cause any local upheaval (see quotation from Newberry above, p. 154).

With Newberry, the mist surrounding the actual geology of the Colorado Plateau and its stratigraphic relationships to its surroundings began to dissipate. From the Spanish explorers to the railroad surveys, the civilized world had obtained a fairly clear image of the topography of not only the Colorado Plateau and its surronding plateau country in general, but mainly owing to the efforts of the fur trappers and of the Frémont and Emory expeditions, one had gained a fairly clear view of the great continental arch spanning the distance from the Mississippi River to the Pacific Ocean. Both the expeditions of Frémont and Emory, but mainly the railroad surveys, began to put geological flesh onto the topographic bones of the previous explorations. With Newberry, the stratigraphical picture cleared sufficiently to be compared in some detail with the stratigraphy of the prairies. Yet, the picture was still out of focus. Structural details were lacking. Stratigraphic units were enumerated but not cast into their actual setting. There were guesses as to the relationships, but most concepts were not tested. What is more, theoretical tectonics had not yet settled into a sufficiently stable model of continental structure and evolution to guide further research.

The Era of the Great Surveys

As American geologists looked hopefully into the future at the close of the 1850s, the antislavery Republican candidate Abraham Lincoln was elected President on 6 November 1860, with a clear majority in the electoral college, and the slaveholding southern states (from which not one vote had gone to him) began seceding from the Union. On the fateful date of 12 April 1861, Confederate artillery opened fire on Fort Sumter, South Carolina, and the American Civil War began. Hundreds of thousands of Americans died in the bloodiest war history had until then seen. In 1865, the war ended in the victory of reason and human rights, and general reconciliation followed in unprecedented and exemplary rapidity. As the 90-year old transatlantic republic was nursing its wounds, it once more turned its eyes westward. The coming decade in American geology was to be known as the era of the Great Surveys. Four groups of geologists, geographers, and other naturalists were organized under four great leaders—namely, Ferdinand Vandeveer Hayden[367] (Fig. 110A), Clarence Rivers King[368] (Fig. 110B), John Wesley Powell[369] (Fig. 110C), and George Montague Wheeler[370] (Fig. 110D)—to explore and map the geography, geology, and natural history of vast areas of the United States west of the 100th west meridian with a view to planning for new settlements. The story of the Great Surveys forms one of the most glorious episodes in the history of geology. Their members have engraved their names in gold into the annals of our science, not only by the new and wonderful observations with which they enriched geology in an unprecedented way, but also by the new ideas that stimulated the science to new heights of understanding of the architecture and behavior of our planet. But this present book is not the place to recount this grand saga. Although it has been told and retold in numerous recensions[371], a detailed history of their great accomplishments in geology is yet to be written.

What I shall do in the following section is simply outline the contributions of the Great Surveys to our understanding of the long-wavelenth structures of the lithosphere. To do that, I shall depart from the format I have so far followed in tracing the history of ideas in what is now the western United States and no longer summarize the expeditions. They are both too well-known to necessitate retelling and too long and involved to be compressed into short summaries. Instead, I shall directly discuss the ideas generated in the period between 1869 and 1882. This will enable us to connect the concepts formed in the American West with those of Eduard Suess and his contemporaries on the other side of the Atlantic Ocean. From the 1880s onward, Americans largely caught up with Europeans in theoretical geology, and the world geological community became much more coherent, making it no longer necessary to divide the narrative geographically.

From the viewpoint of the subject-matter of the present book, it is best to narrate the contributions the Great Surveys made to our understanding of the long-wavelength structures of the lithosphere around an axis identified by the work of the Powell Survey. Not only were the most enduring conceptual models and terminology generated by Powell and his geologists, but also the piece of country they worked in happens to be the most critical terrain for the discussion of the problem of long-wavelength structures in the western United States.

When Powell first ventured out West in 1867 with his class from the Illinois Wesleyan University "with the purpose of studying geology" of the mountains of Colorado (Watson, 1954, p. 1), Hayden and his men had already corroborated the earlier inferences that the Cretaceous had lain flat right across the Rockies all the way to the Wasatch Mountains when the Cenozoic era dawned (see especially the geological maps *in* Hayden, 1857; 1858; 1869, especially p. 10). This flat surface was at or very near sea level, as the immediately succeeding giant lakes started out as lagoons communicating with the sea (around which, Hayden believed, most of the coal deposits of the West had formed). As these water bodies became progressively cut-off from the sea, their waters freshened and their sizes diminished. Their shores, containing lush tropical flora, became the factories of the vast Tertiary lignite deposits, about which Hayden and his men wrote so extensively (e.g., Hayden, 1871, especially ch. XIII; Newberry, 1871). It was this low, flat surface that Powell and his men found at elevations exceeding 11,000 ft. (~3330 m), for example, in the high plateaus of Utah (Dutton, 1880, p. 8), but still flat-lying!

Powell's two epic journeys down the Grand Canyon and the associated surveys (Powell, 1875, 1895; Dellenbaugh, 1908[undated]; Dolnick, 2001) greatly refined John Newberry's stratigraphy of these vast tablelands. Powell laid before the astonished eyes of the geological community a succession of sedimentary rocks exceeding 60,000 ft. (~18 km) in thickness (Powell, 1876, p. 37). Something quite extraordinary had happened here! The top of the continent had plunged down into the bowels of the earth for many kilometers throughout the Paleozoic and Mesozoic eras, ever keeping the top of its accumulating load of sediments not far from the surface of the ocean. Then, this motion was reversed and it rose, majestically eroding all of the Mesozoic section (some 3.5 km) from its top in the region of the Grand Canyon. And all of this had happened without any remarkable evidence of folding or any other sort of shortening, as Newberry had already noted with some surprise.

Dutton developed Élie de Beaumont's and Herschel's old idea that, as sedimentary loading depresses the crust, material must wander off to neighboring regions to uplift them (Dutton, 1876); however, the massive scale of the uplift seen in the Plateau province and the absence of reciprocally sinking neighboring regions pushed Dutton to desperation:

> We cannot, therefore, attribute the faulting and monoclinal flexing of the plateaus to erosion of the uplifts and the deposition of the *débris* at their flanks, for no such (relatively greater) amount of erosion is found upon the uplifts, and no such deposits take place upon their flanks. The Kaibabs have been enormously denuded, but not much more upon the highest than upon the lowest portions. The High Plateaus have, compared with the Kaibabs, suffered but little from erosion. In neither district can we look for the same causation of faults and flexures as we might at first feel inclined to employ to explain those of Colorado and the Uintas. (Dutton, 1880, p. 52–53, italics Dutton's)

What was the geology of this strange region like? At the opening of this chapter, I have quoted Dutton's humorous description of a part of this region's morphology. A part of Dutton's humor clearly was stimulated by the utter amazement at the novelty of the rock types, their structures, and the landforms to which they gave rise. It was Powell who had brought him into this fairyland of geology in 1875, in which Dutton found the forms and modes of occurrence of geological structures to be "somewhat peculiar, especially when brought into comparison with displacements found in other regions" (Dutton, 1880, p. xvi).

Powell himself had recognized three great tectonic/morphologic provinces in the western United States east of the Sierra Nevada. He called these the "The Park Province, the Plateau Province, and the Basin Province—in order from east to west"

A

Figure 110. A: Dr. Ferdinand Vandeveer Hayden (1828–1887), head of the Geological and Geographical Survey of the Territories, during the years of the activity of his survey. (From the archive of the U.S. National Academy of Sciences). B: Clarence King (1842–1901), geologist-in-charge of the U.S. Exploration of the Fortieth Parallel, at age 27, after having completed three field seasons of the Fortieth Parallel Survey. (From Wilkins, 1988, with permission.) C: John Wesley Powell (1834–1902), head of the Geographical and Geological Survey of the Rocky Mountain Region, at the time of his writing of the *Exploration of the Colorado River of the West and its Tributaries Explored in 1869, 1870, 1871, and 1872*. (From Worster, 2001, with permission.) D: George Montague Wheeler (1842–1905), head of the U.S. Geographical Surveys West of the One Hundredth Meridian. (From Karrow, 1986, with permission.)

(Powell, 1876, p. 7). The Park Province is coincident with what is today known as the U.S. Rockies. Powell thought that the name "Rocky Mountains" ought to be reserved for the entire North American Cordillera, indeed for the entire high region west of the 100th west meridian, very much in the sense of the older geographers and James Dwight Dana[372] (Dana, 1847b, p. 98; see above, p. 167).

That portion of the United States west of the one hundredth meridian lies at a great altitude above the sea. The exceptions to this, as immediately along the Pacific coast and the narrow valleys of some of the principal streams, are but trivial. The rivers descend so rapidly from the upper regions that few of them are of value as highways of commerce; the valleys proper are narrow; treeless plains, cold, arid tablelands, and desolate mountains are the principal topographic features. The more conspicuous of these are the mountains; lone mountains, single ranges and great groups of ranges or systems of mountains prevail. Owing to great and widely spread aridity, the mountains are scantily clothed with vegetation, and the indurated lithologic formations are rarely masked with soils, and the rocks, as they are popularly called, are everywhere exposed; hence all these mountains are popularly known as the Rocky Mountains. But there is more than one system of mountains, and later writers wishing to be more definite speak of the Cascade Mountains, the Coast ranges, the Sierra Nevada, the Wasatch Mountains, &c. But in an important sense the region is a unit; it is the generally elevated region of the United States; it is the principal region of the precious metals; it is the region without important navigable streams; it is the arid land of our country where irrigation is necessary to successful agriculture. But above all it is the rocky region; rocks are strewn along the valleys, over the plains and plateaus; the cañon walls are of naked rock; long escarpments of cliffs of rock stand athwart the country, and everywhre are mountains of rock. It is the Rocky Mountain region. (Powell, 1876, p. 4–5)

Powell's tripartite distinction was not original with him but had been borrowed from Gilbert (1875), who had distinguished the Basin Range System (his p. 22) from the Colorado Plateau System (the term *Colorado Plateau* had been introduced by Wheeler in 1868: see Wheeler, 1889, p. 13; see Fig. 111 herein). Gilbert noted that the Colorado Plateau System lay between the Basin Range System in the west and the Rocky Mountain System (in its restricted U.S. sense) in the east and stretched northward to the Uintah [*sic*] Range (Gilbert, 1875, p. 43). Thus, Gilbert's Colorado Plateau System is identical with Powell's plateau country.

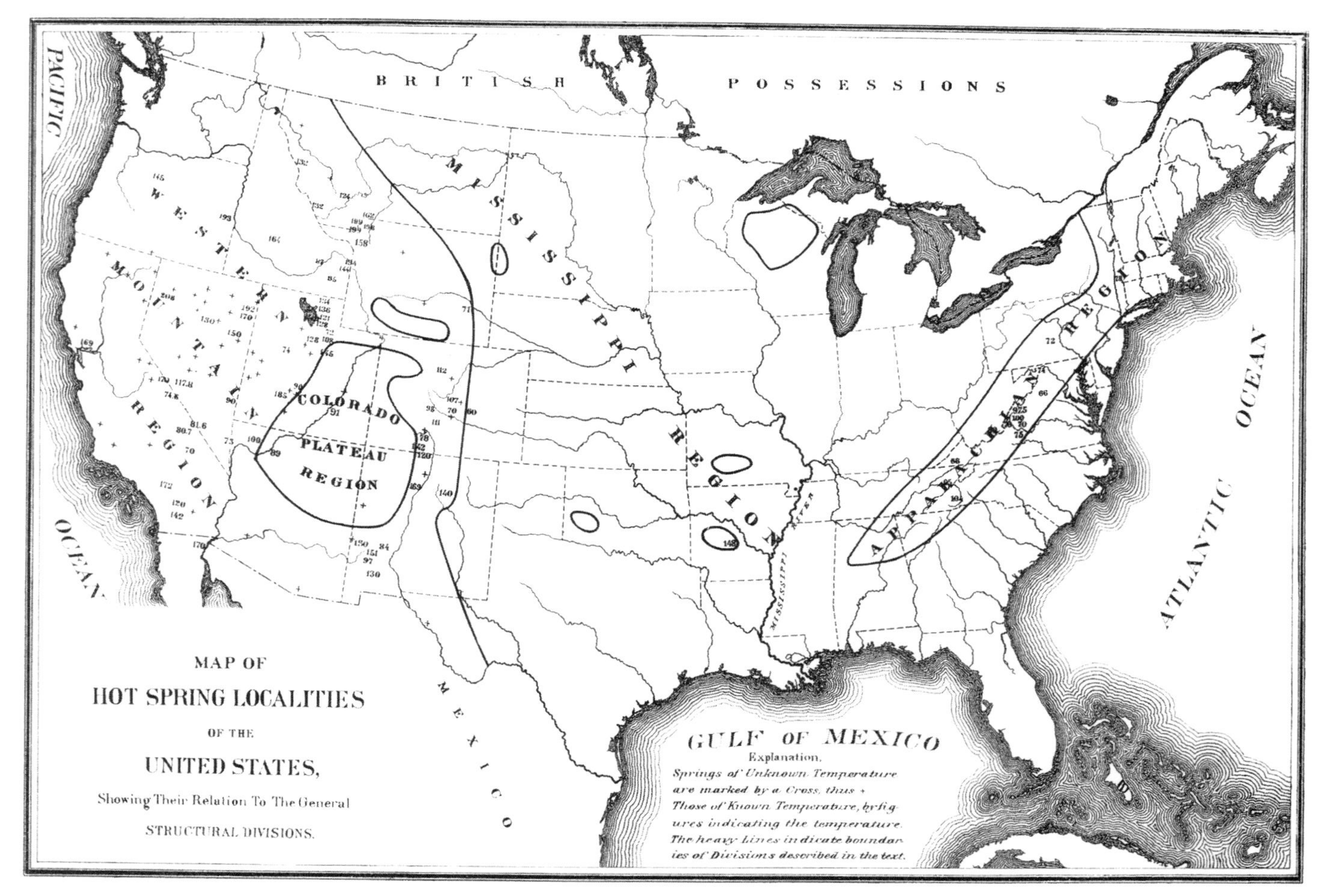

Figure 111. Outlines of the Colorado Plateau and its setting within the morphotectonic regions of the United States, as understood by the members of the Wheeler Survey (from Gilbert, 1875, plate III).

Gilbert undertook a careful, albeit necessarily restricted, study of the Basin Ranges and the Colorado Plateau during the rapid mapping exercises of the Wheeler Survey. He noticed that the Basin Ranges were formed by faulting that was parallel with the ranges and also noticed that many of these faults were dipping basinward (see, for example, Fig. 112A–E). He summarized their characteristics (refer to Fig. 112F) as follows:

> The sections accumulated by our geological observers admit of the following classifications:
> 1. Faulted monoclines occur, in which the strata on one side of the fault have been lifted, while those on the opposite side either do not appear (A), or (less frequently) have been elevated a less amount (B). Two thirds of the mountain ridges can be referred to this class.
> 2. Other ridges are uplifts limited by parallel faults (C), and to these may be assigned a few instances of isolated synclinals (D), occurring under circumstances that produce the idea that they are remnants omitted by denudation.
> 3. True anticlinals (E) are very rare, except as local, subsidiary features, but many ranges are built of faulted and dislocated rock masses (F), with an imperfect anticlinal arrangement.
>
> Not only is it impossible to formulate these features, by the aid of any hypothetical denudation, in such a system of undulations and foldings as the Messrs. Rogers have so thoroughly demonstrated in Pennsylvania and Virginia, but the structure of the Basin Range system stands in strong contrast to that of the Appalachians. In the latter, corrugation has been produced commonly by folding, exceptionally by faulting; in the former, commonly by faulting, exceptionally by flexure. In the latter, few eruptive rocks occur; in the former volcanic phenomena abound and intimately associated with ridges of upheaval. The regular alternations of curved anticlinals and synclinals of the Appalachians demand the assumption of great horizontal diminution of space covered by the disturbed strata, and suggest lateral pressure as the immediate force concerned; while in the Basin Ranges, the displacement of comparatively rigid bodies of strata by vertical or nearly vertical faults involves little horizontal diminution, and suggests the application of vertical pressure from below. (Gilbert, 1874, as quoted *in* Powell, 1876, p. 24)

Gilbert believed “that the forces which have been concerned in the upheaval of the basin ranges have been uniform in kind over large areas; that whatever may have been their ultimate sources and directions, they have manifested themselves at the surface as simple agents of uplift, acting in vertical, or nearly vertical, planes and that their loci are below the immediate surface of the earth’s crust” (Gilbert, 1875, p. 42).

Gilbert thought that the structures observed on the Colorado Plateau were of the same type as those of the Basin Ranges, but fewer in number and inferior in the displacement achieved. Figure 113 shows his general east-west cross section across the Plateau, and the contrast with his cross sections (Fig 112A–E) of the Basin Ranges is clear. This inferred similarity in structures led Gilbert to assume that both provinces were the products of vertically acting forces in contrast to the folded belts that were considered to be the products of horizontal forces. He summarized these ideas in the concluding section of his considerations on the structural geology of the Basin Range System and the Colorado Plateau System. In the following I quote him in full because his ideas had a lasting influence on the models developed not only by his compatriots concerning the tectonics of the western United States, but internationally on models generated on the tectonic behavior of our planet:

> We have already been led to conclude that the forces which have produced the Basin Ranges were uniform in character over large areas, and in horizontal direction over minor, but still considerable areas; that they produced parallel ranges by nearly vertical upheaval; and that they were deep-seated. We have reached the same conclusions in regard to the forces which have produced the conjoint system of faults and ridges in the Colorado Plateau. We have also seen that the loci of the latter forces are in part coincident with those of the former. And a single short step brings us to the important conclusion that the forces were identical, (except in time and distribution;) that the whole phenomena belong to one great system of mountain formation, of which the ranges exemplify advanced, and the plateau faults the inital, stages. If this be granted, as I think it must, then it is impossible to overestimate the advantages of this field for the study of what may be called the embryology of mountain-building [*note the medical man Newberry’s influence! See endnote 365*]. In it can be found differentiated the simplest initiatory phenomena, not obscured, but rather exposed, by denudation, and the process can be followed from step to step, until the complicated results of successive dislocations and erosions baffle analysis. The field is a broad one and its study has but begun; but with its progress I conceive there will accrue to the science of orographic geology a more valuable body of geological data than has been added since the Messrs. Rogers developed the structure of the Appalachians. Of late years the most important contributions have come from physicists, and in their scales have been weighed the old theories of geologists. Here will be an opportunity to compare the speculations of the physicists with new geological data.
>
> The Appalachian mountain system, as the best studied great system—at least of those which exhibit unity of structure—has formed the geological basis of many theoretical structures, although, as Professor Whitney has pointed out, it is rather exceptional than typical in character. The system we have described resembles it in the absence of any great central axis and in the general tendency to uniformity throughout, but differs widely in other respects. In the Appalachians corrugation has been produced commonly by folding, exceptionally by faulting; in the basin Ranges, commonly by faulting, exceptionally by flexure. The regular alternation of curved synclinals and anticlinals is contrasted with rigid bodies of inclined strata, bounded by parallel faults. The former demand the assumption of great horizontal diminution of the space covered by the disturbed strata, and suggest lateral pressure as the immediate force concerned; the latter involve little horizontal diminution, and suggest the application of vertical pressure from below. Almost no eruptive rocks occur in the former; massive eruptions and volcanoes abound among the latter, and intimately associated with them.
>
> To attempt a reconciliation of these antithetical phenomena is premature, before the character of the basin ranges shall have received more thorough study than has been possible for us; I do not desire to undertake here a discussion of theoretical orology, but I cannot forbear a brief suggestion before leaving the subject. It is, that in the case of the Appalachians the primary phenomena are superficial; and in that of the Basin Ranges they are deep-seated, the superficial being secondary; that such a force has crowded together the strata of the Appalachians—whatever may have been its source—has acted in the Ranges on some portion of the earth’s crust beneath the immediate surface; and the upper strata, by continually adapting themselves, under gravity, to the inequalities of the lower, have assumed the forms we see. Such a hypothesis, assigning to subterranean determination the position and direction of lines of uplift in the Range System, and leaving the character of the superficial phenomena to depend on the character and condition of the superficial materials, accords well with

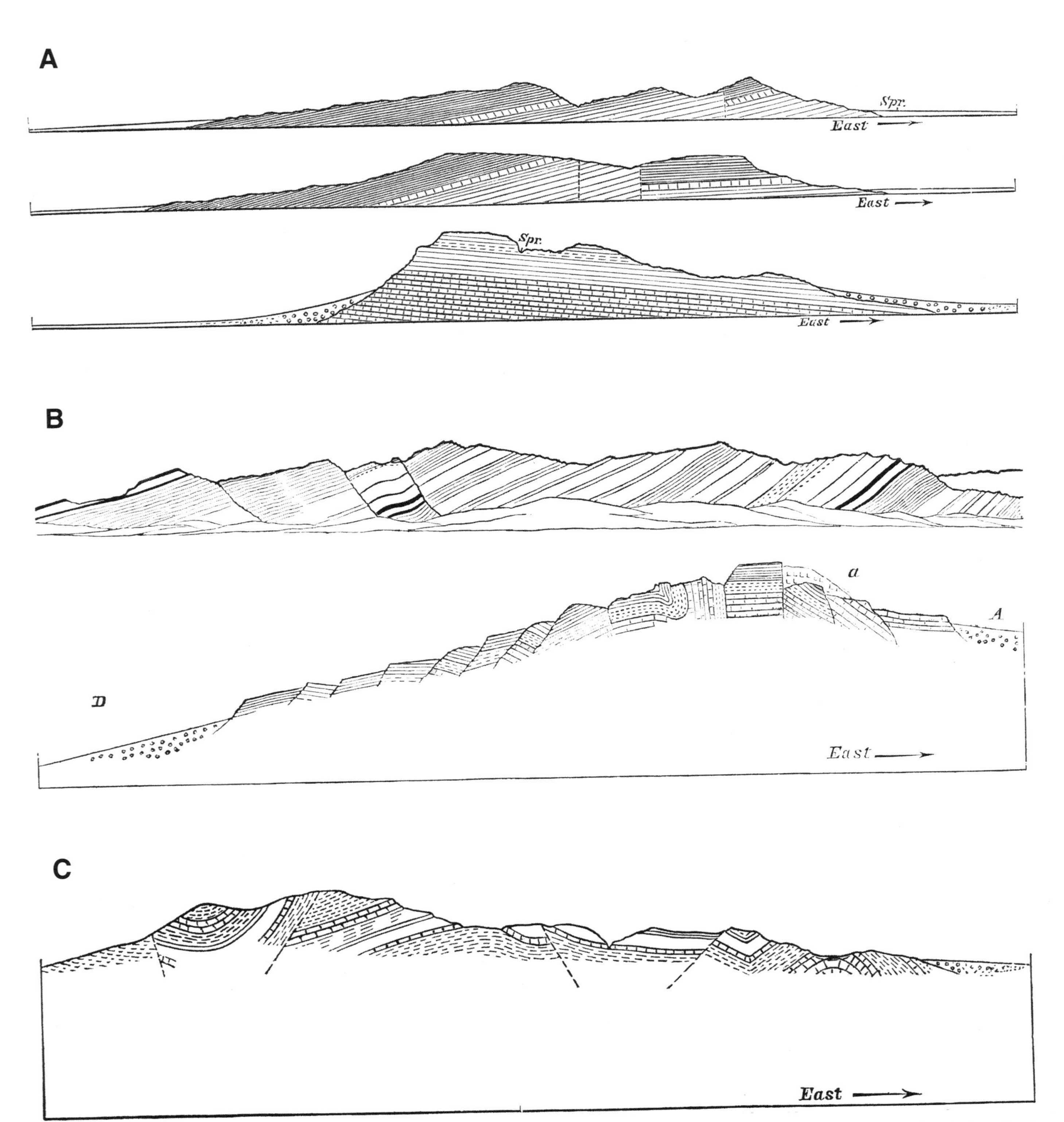

Figure 112 (on this and facing page). Gilbert's cross sections across the Basin Ranges. A: Sections of the House Range at Fish Spring, 2 mi. (3.2 km) south of Fish Spring, and at Dome Canyon (from Gilbert, 1875, figs 3, 4, 5). Note the monoclinally dipping beds and the vertical faults! B: Sections of the Amargosa Range at Boundary Canyon. Base line is sea level. (a—Rhyolite, A—Amargosa Desert, D—Death Valley; from Gilbert, 1875, fig. 12). Note the faults dipping towards the depressions and the antithetically and synthetically rotated blocks. This is the classical area, where in 1941 Levi Noble was to recognize what he termed the "chaos structure" (cf. Şengör and Sakınç, 2001, p. 26). C: Cross section of the Inyo Range near Deep Springs Valley. Base line is sea level (from Gilbert, 1875, fig. 14).

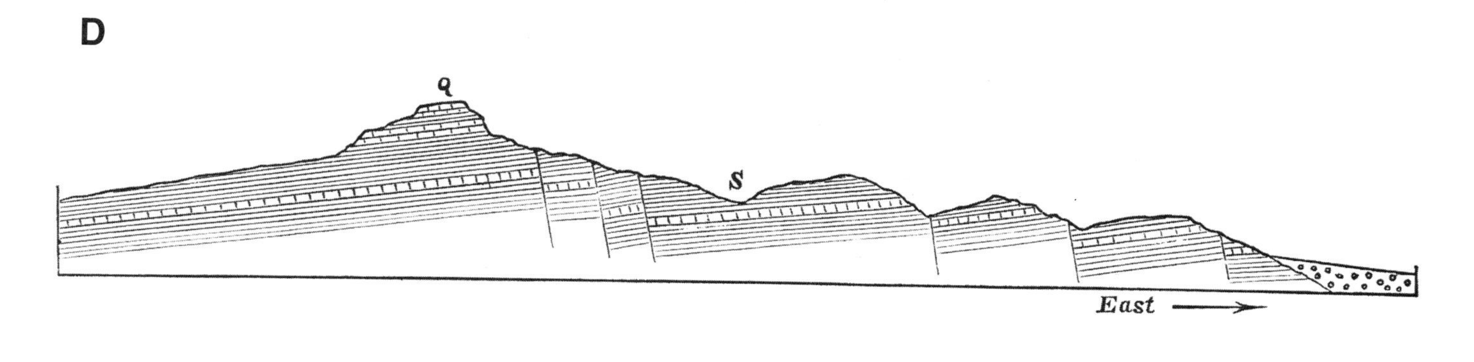

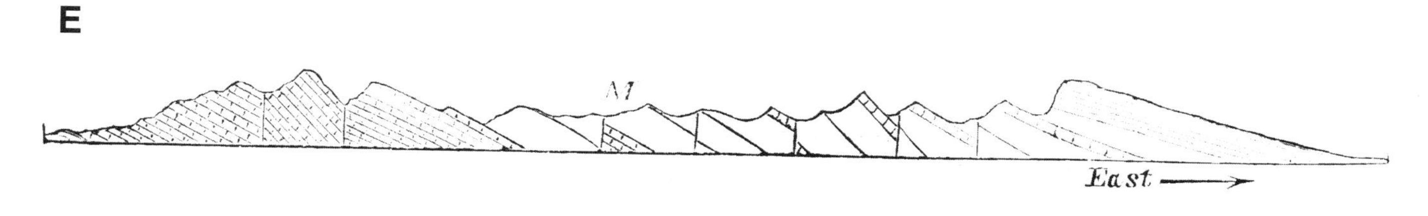

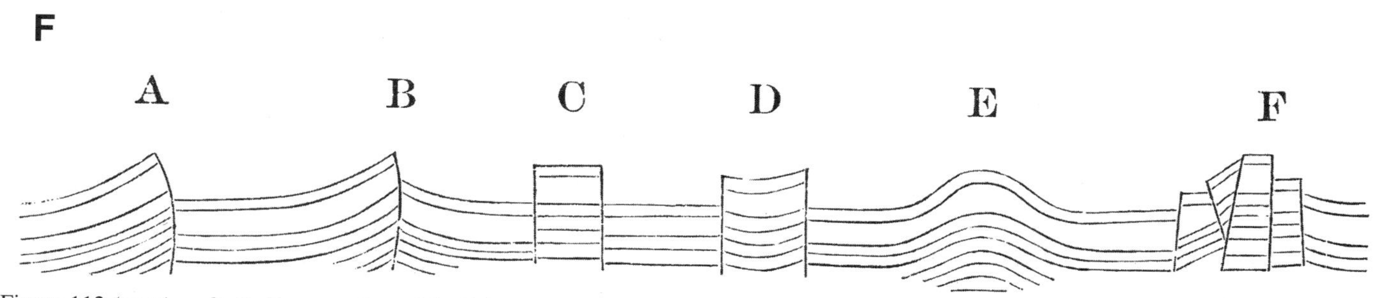

Figure 112 *(continued)*. D: Cross section of the Pahranagat Range at Silver Canyon. Base level is the level of the Great Salt Lake. (Q—Quartz Peak, S—Sanders Canyon; from Gilbert, 1875, fig. 18). E: Cross section across the Timpahute Range at the Groom mining camp. Base level is the level of the Great Salt Lake. (M—location of the mines; from Gilbert, 1875, fig. 19). F: Gilbert's conceptual section across the Basin Ranges (from Powell, 1876, fig. 7).

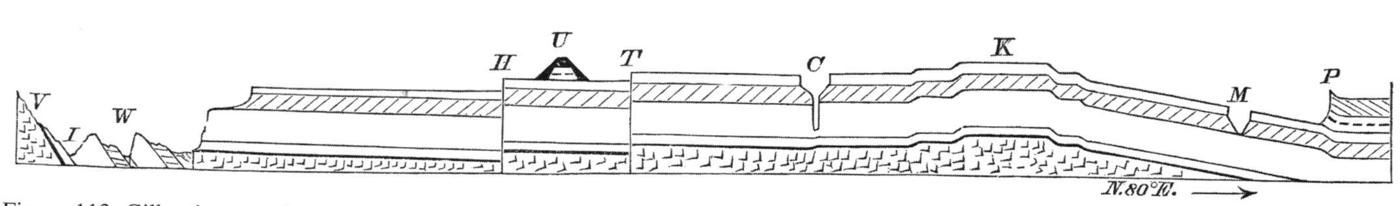

Figure 113. Gilbert's general east-west cross section across the Colorado Plateau from the Virgin Range to the Paria Fold. Vertical scale is about eight times as the horizontal. Base line is sea level. (V—Virgin Range, I—Colorado River in Iceberg Canyon, W—Colorado River at Grand Wash, H—Hurricane fault, U—Uinkaret Mountains, T—Toroweap fault, C—Kanab Creek, K—Kaibab Plateau, M—Colorado River in Marble Canyon, P—Paria Fold; from Gilbert, 1875, fig. 26).

many of the observed facts, and especially with the persistence of ridges where structures are changed. It supposes that a ridge, created below, and slowly upheaving the superposed strata, would find them at one point coherent and flexible, and there produce an anticlinal; at another hard and rigid, and there uplift a fractured monoclinal; at a third seamed and incoherent, and there produce a pseudo-anticlinal, like that of the Amargosa Range. (Gilbert, 1875, p. 60–62)

There is a great similarity between the structures recognized by Powell on the Colorado Plateau and those by Gilbert. From Powell's frequent references to Gilbert, we infer that Gilbert had the priority of recognizing most, if not all, of them. Powell, however, combined the structural geology with erosion and generated a method of studying geomorphology to infer structural history, which was later much improved by Gilbert, Dutton, and especially William Morris Davis. On the Colorado Plateau, he recognized a variety of faults and monoclines (Fig. 114). Powell further noticed, while studying the tectonic morphology of the Uinta Range, that an antecedent drainage (Powell, 1875, p. 163) in the north, across the Uintas, gave way to a superimposed drainage in the plateau country, which developed irrespective of the underlying structure on which it is now perched (Powell, 1875, p. 166). This development created diverse types of valleys, which Powell classified according to their relation to the structures on which they appear superimposed (Fig. 115).

So, when Powell looked at the Colorado Plateau (Fig. 116), he saw, exactly as Gilbert had seen, a high-lying tableland sparsely and diversely deformed along high-angle faults and monoclines (Fig. 117). Although his friend and later close associate, Clarence Edward Dutton, had come out strongly against the thermal contraction hypothesis to explain the terrestrial tectonism (Dutton, 1874), Powell initally was a contractionist and implied that the Uintas as well as the highlands had been uplifted by the tangential compression caused by the "contracting or shrivelling of the earth" (Powell, 1975, p. 153). But he was quick to retreat (probably under pressure from the scholarly Dutton, who was much better versed in the physical sciences in general than was Powell). Already a year later Powell was writing with respect to faults that "in blocks which are bounded by faults and tilted, I shall speak of such portions as are at a higher level as having been uplifted, and portions occupying a lower level as thrown. In such cases I do not wish to commit myself to any theory of upheaval or collapse in the change of the relation of the several parts of these beds to the centre of the earth" (Powell, 1876, p. 9).

He then proceeded to classify the types of mountain structure. The first type he considered was what he called the *Appalachian structure*. It was one "with closely appressed folds and axial planes tipped back from the sea, the modifications of these folds by faults" (Powell, 1876, p. 9). This type he did not discuss in any detail because it did not occur in the Uintas or in the plateau country. What Powell called *simple anticlinal structure* did occur in his area of consideration and when exaggerated, a side of the anticline would be torn by a steep fault to give rise to what Powell called the *Uinta type structure*. He thought most of his Park Ranges in Colorado, Wyoming, Idaho, and Utah displayed Uinta-type structure. A tableland bounded by monoclines or homologous faults Powell called the *Kaibab-type structure*. His classical block diagram, herein reproduced as Figure 114, is considered as illustrative of the type examples of the Kaibab-type structures. When Archibald Marvine of the Hayden Survey (Marvine, 1874, p. 188–190) wrote that the Kaibab type structures were similar to those of the Park Ranges (i.e., the U.S. Rockies), Powell objected and pointed out that the Kaibab type structure was distinct from the Uinta-type structure, which was the common structure type of the Park Ranges, because the Uinta-type structure displayed "many more complexities than the faults and monoclinal flexures usually found in the Plateau Province" (Powell, 1876, p. 28). To these structure types Powell then added Gilbert's *Basin Range-type structure* and what he called "zones of diverse displacement" usually showing hybrid structures.

From Gilbert's and Powell's descriptions and inferences, one inevitably was driven to conclude that the entire Colorado Plateau Province and the neighboring Great Basin were areas of vertical uplift, and all the deformation one saw in them was related to vertical motions. Both Gilbert and Powell emphasized the contrast these regions displayed to areas of tangential compression (such as the Appalachians). The causative agent of the upilft of the Great Basin and the Colorado Plateau Province was believed by Gilbert to be located directly beneath them. Powell initially toyed with the idea of accommodating these uplifts within the framework of the contraction theory as perhaps Dana would have advised him to do, but was evidently quickly scared away from it by his formidable younger friends.

Now that they knew the stratigraphy, structural geology, and the geomorphology of the area between the Great Plains in the east and the Sierra Nevada in the west in some detail, Powell felt that the time had come to sketch a geological history of it. This he did in his 1876 book on the Uintas. He pointed out that in the latter part of the Mesozoic Era, the greater part of the Great Basin, or what he called the Basin Province, was dry land (refer to Fig. 118A–C to compare Powell's inference with our present state of knowledge of the geological history of the western United States in the Cretaceous and the Cenozoic). The Plateau Province was an open but shallow sea. In the Park Province a series of islands dotted the sea surface. Powell wrote:

The Cenozoic time was inaugurated by a series of movements, which, continued to the present time, have produced the topographic features now observed. This part of the crust of the earth, and I mean by the term "crust" simply that portion of the earth which we are able to study by actual observation in truncated folds and eroded faults—this portion of the crust, then, was gradually broken and contorted. The Plateau and Park Provinces were cut off from the sea, and great bodies of fresh water accumulated in the basins, while to the east in the region of the Great Plains, in earlier Tertiary times at least, there was an open sea. Slowly through Cenozoic times the outlines of these lakes were changed, doubtless in two ways: first, by the gradual displacement of

A

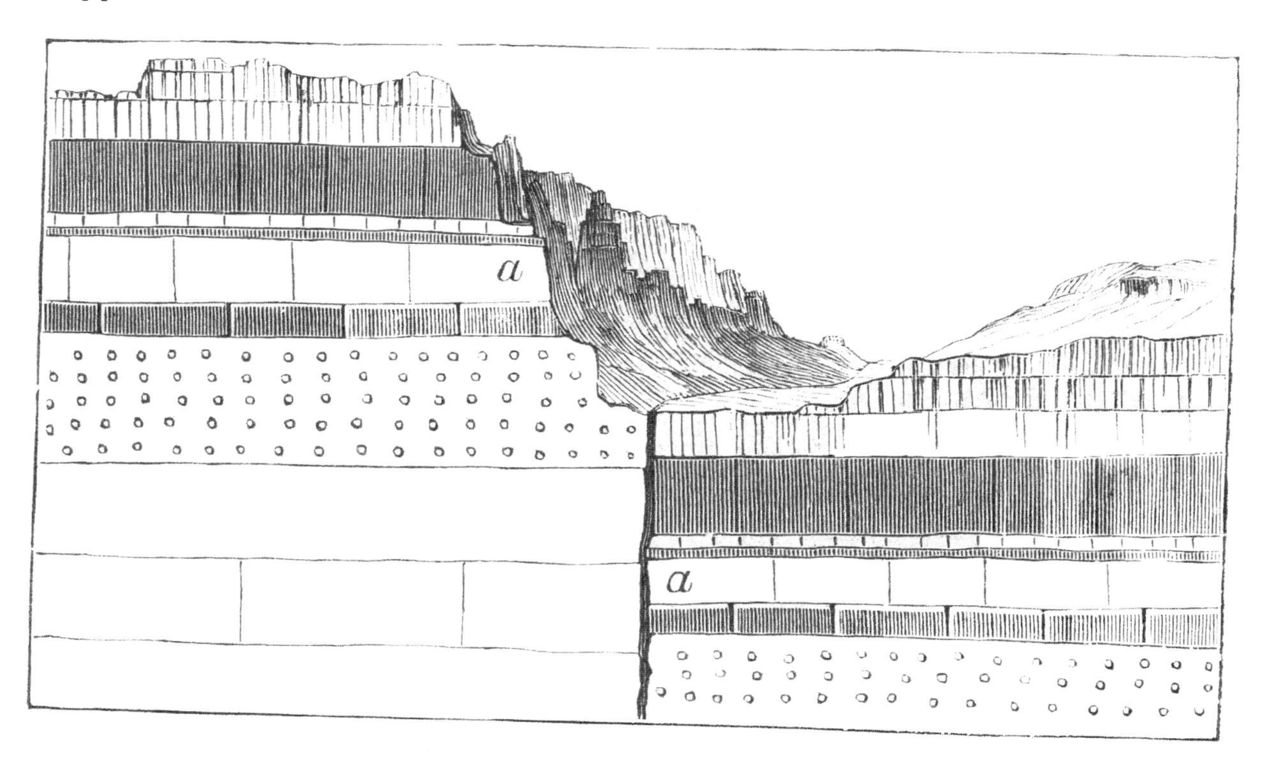

B

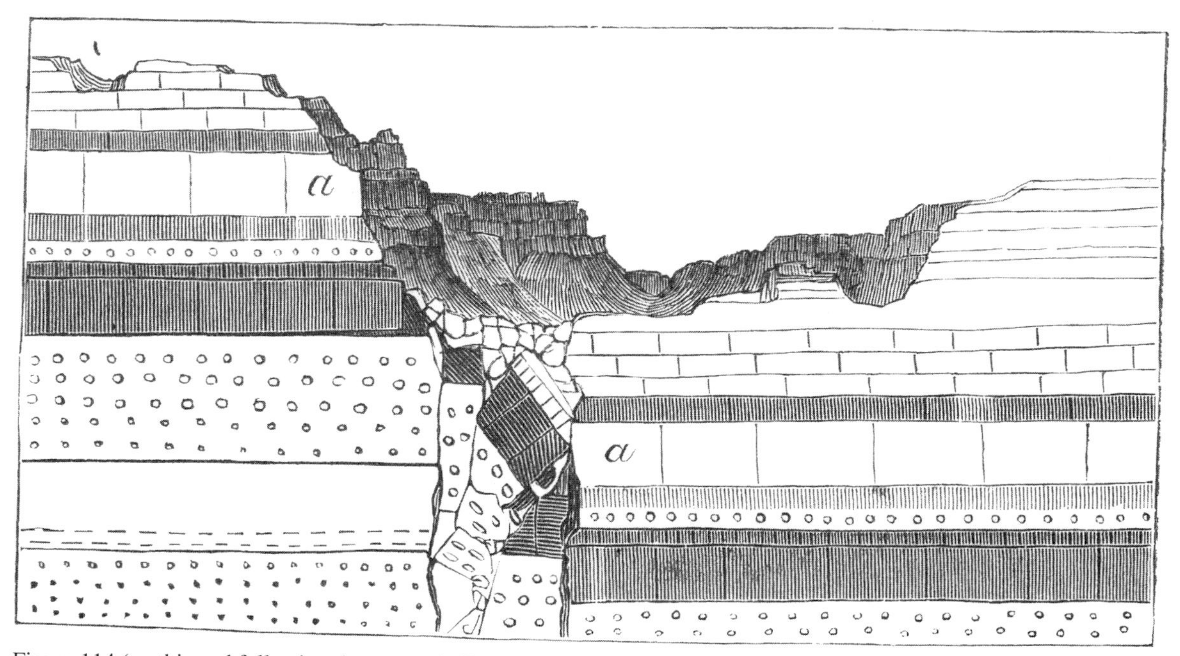

Figure 114 (on this and following three pages). Fault and monocline types recognized by Powell in the plateau country. A: Cross section of a simple fault with associated scenery (from Powell, 1875, fig. 64). B: Cross section across a fault with walls widely separated, the intervening space filled with broken rocks (structural rocks: cf. Şengör and Sakınç, 2001, p. 172) (from Powell, 1875, fig. 65).

C

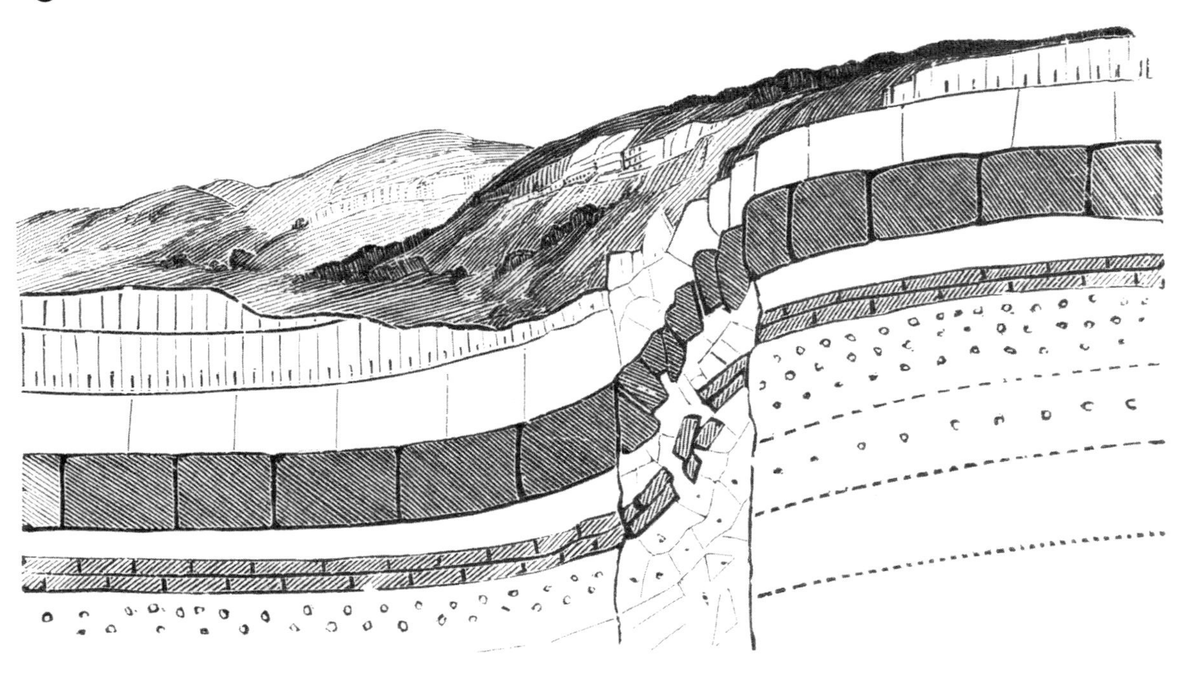

D

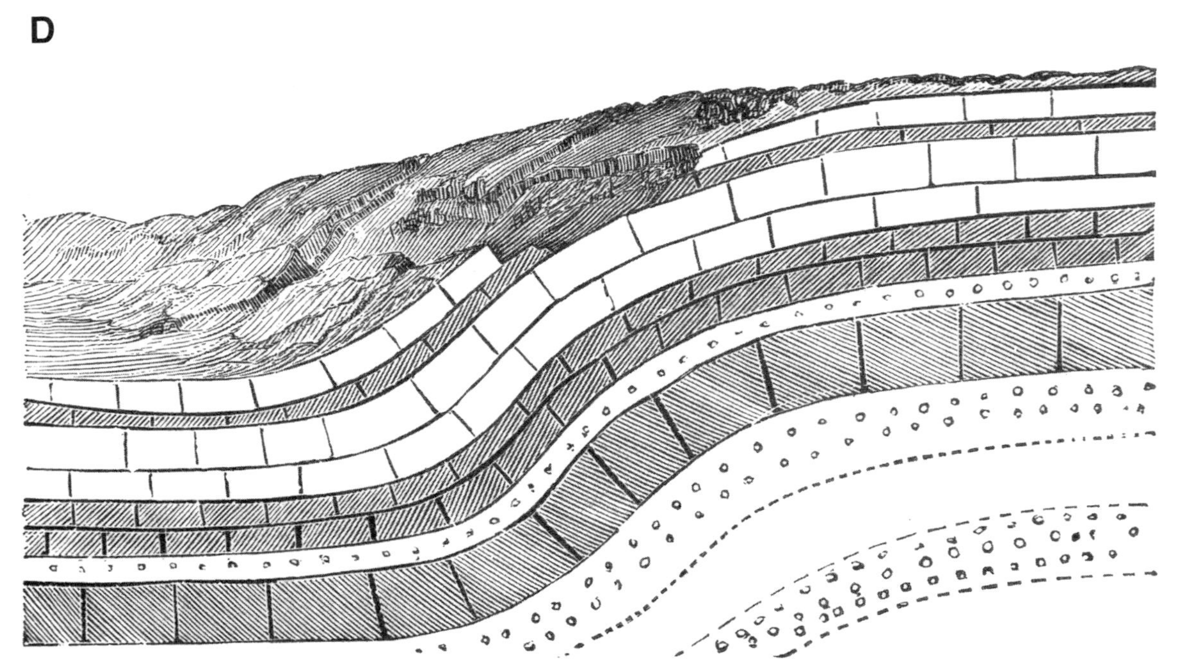

Figure 114 *(continued)*. C: Cross section across a fault with walls widely separated, the intervening space filled with broken rock, still exhibiting the original stratification (from Powell, 1875, fig. 66). D: Cross section of a monoclinal fold (from Powell, 1875, fig. 67).

E

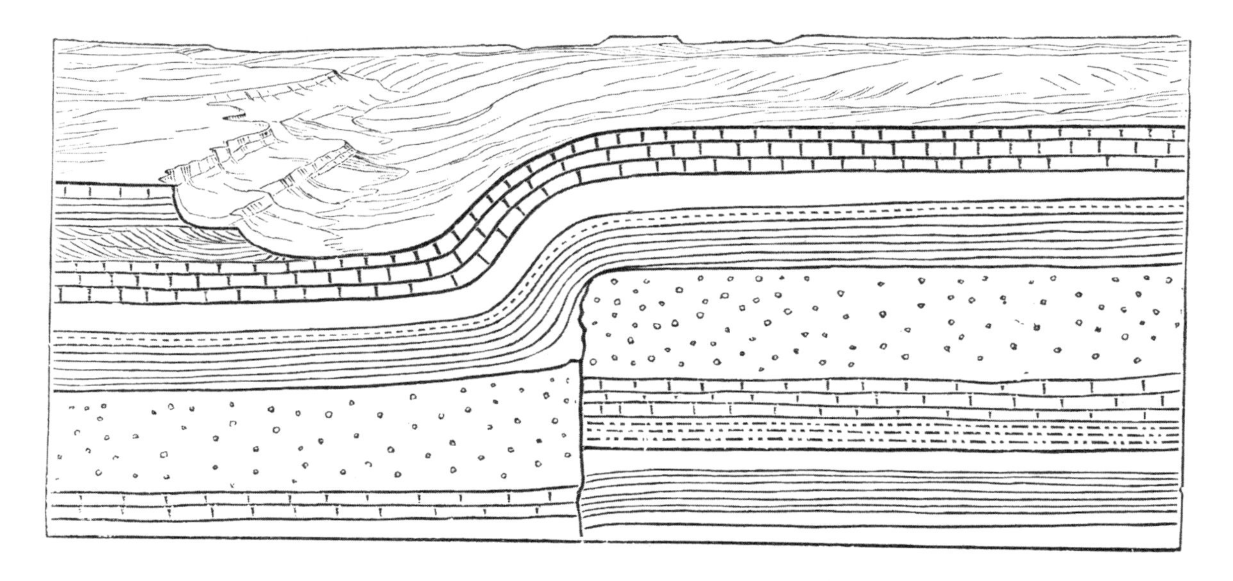

F

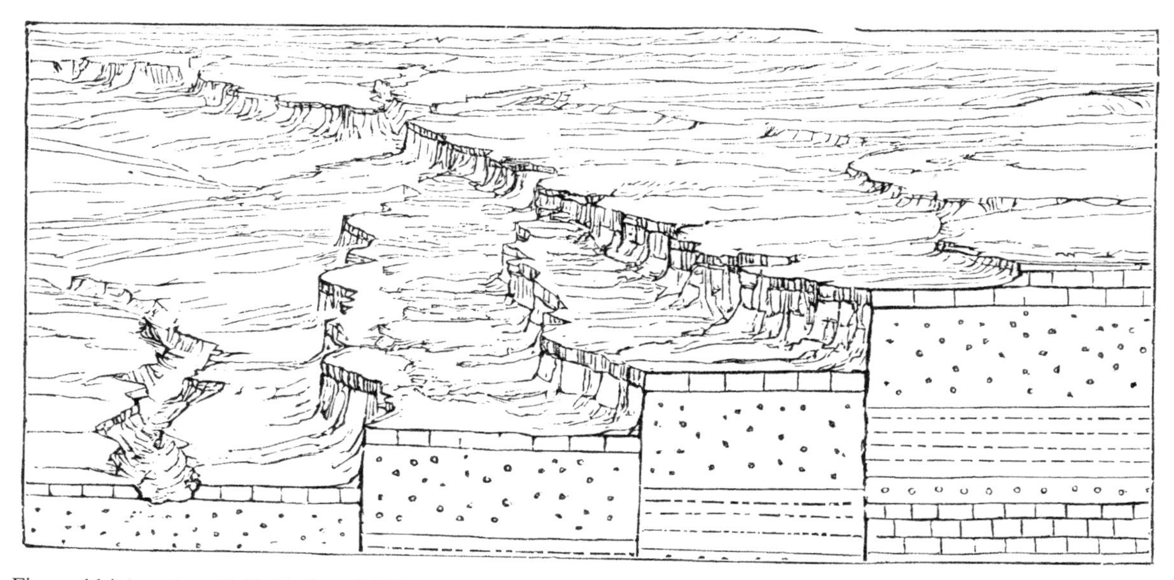

Figure 114 *(continued)*. E: Fault and fold in the same cross section (from Powell, 1875, fig. 68). F: Cross section across a branching fault (from Powell, 1875, fig. 69).

G

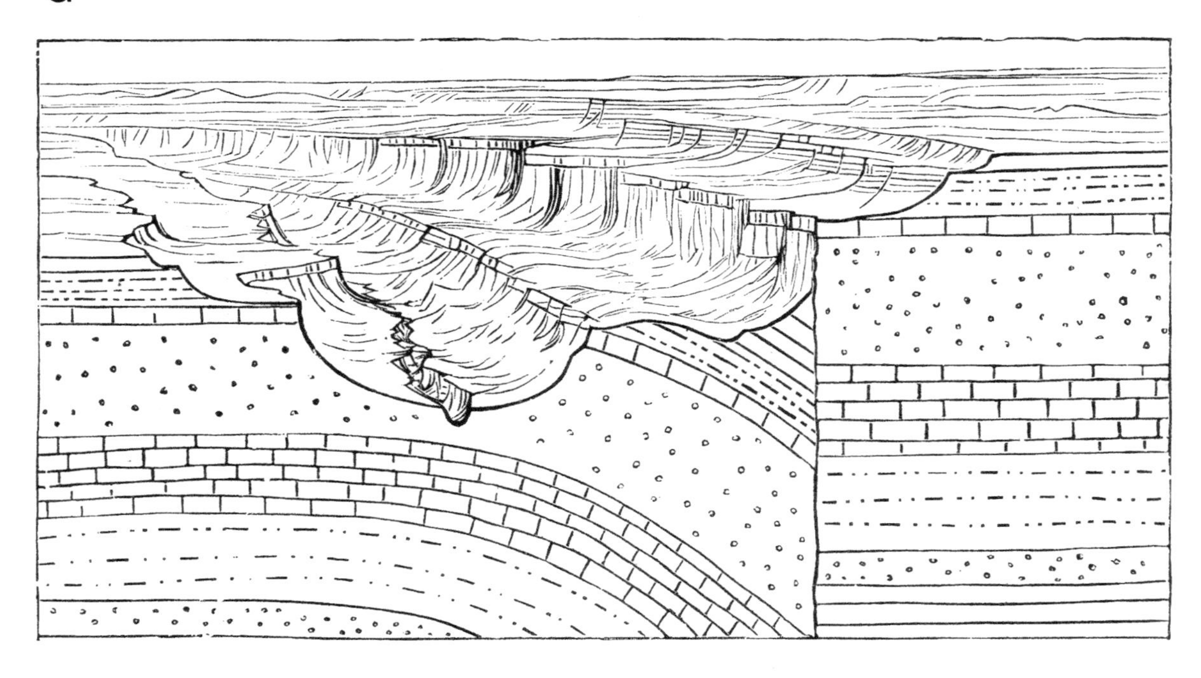

H

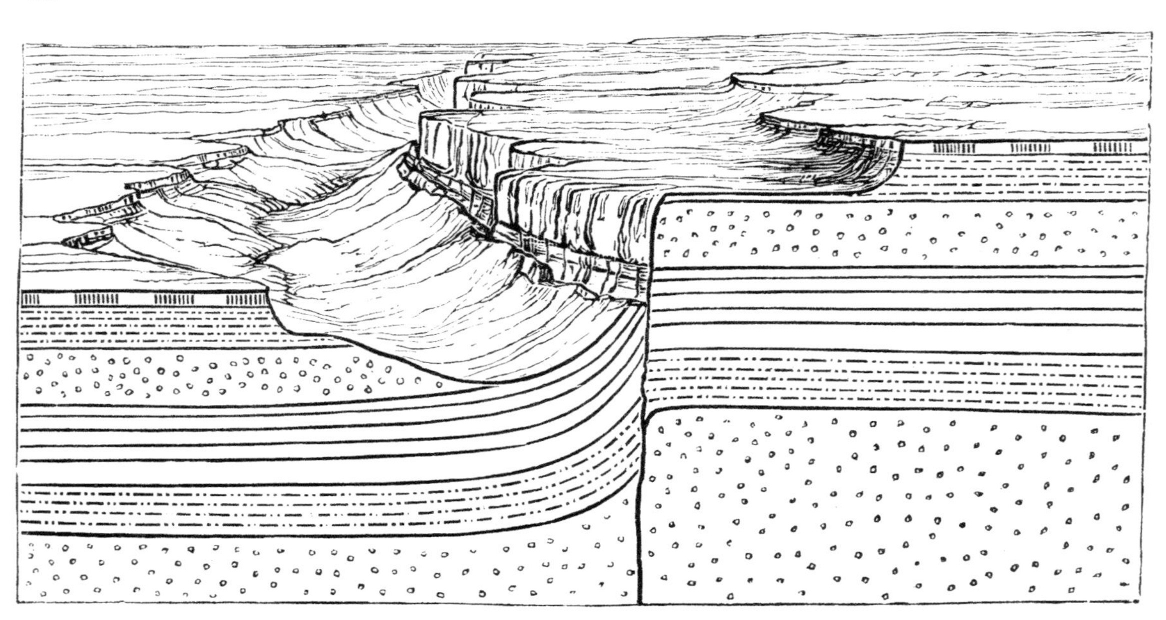

Figure 114 *(continued)*. G: Cross section of fault with thrown beds flexed downward (from Powell, 1875, fig. 70). H: Cross section of fault with thrown beds flexed upward (from Powell, 1875, fig. 71).

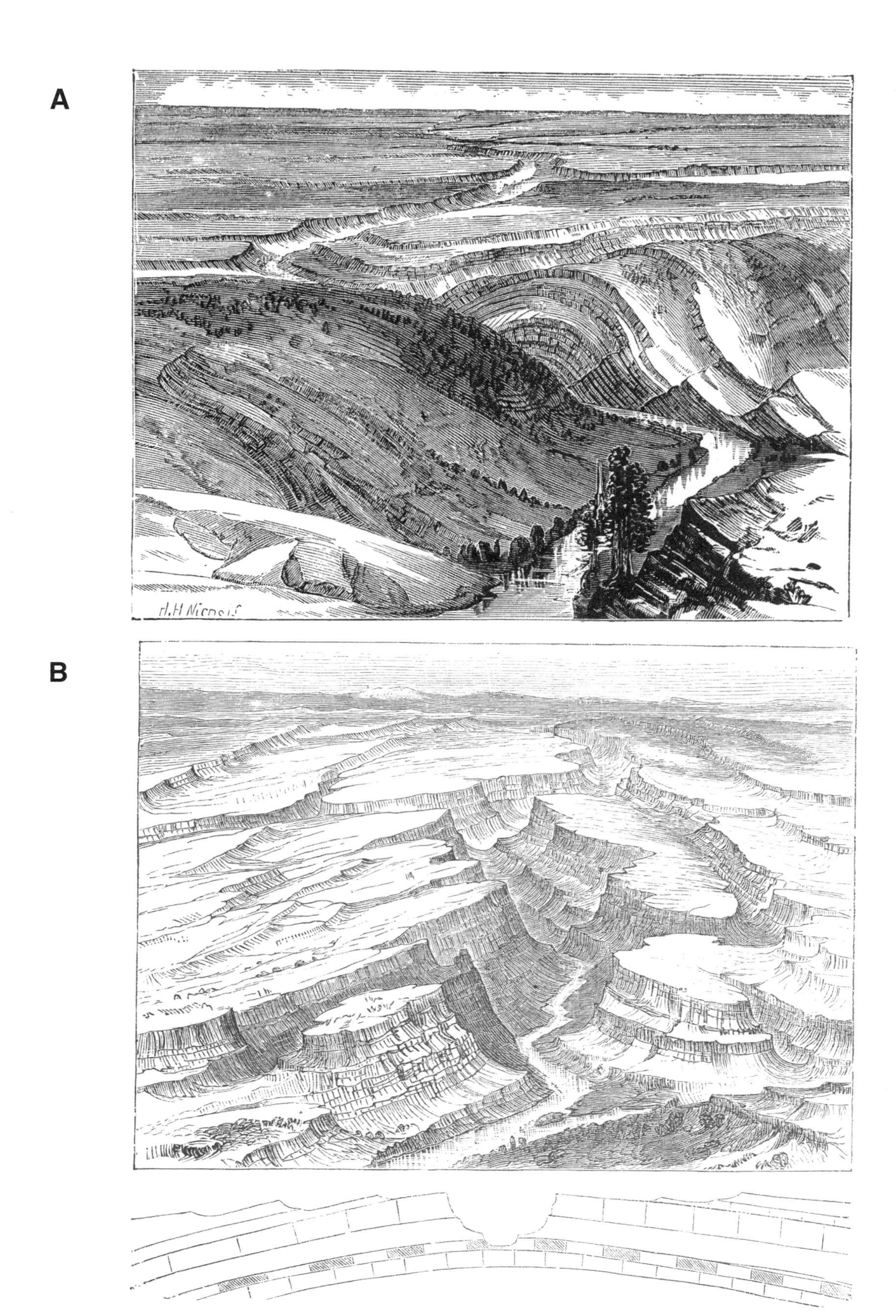

Figure 115 (on this and following two pages). Various types of valleys classified according to how they relate to the geological structure on which they are perched, according to Powell. A: The Green River Plains (Powell, 1895, p.130) with diaclinal valley, which pass through a fold (from Powell, 1875, fig. 53). B: An anticlinal valley that runs along the crest of an anticline (from Powell, 1875, fig. 55).

C

D

Figure 115 *(continued)*. C: A valley west of Green River (Powell, 1895, p. 125): a synclinal valley running along the trough of the syncline (from Powell, 1875, fig. 56). D: A cataclinal valley that runs in the direction of the dip (Powell, 1875, fig. 54; also in Powell, 1895, p. 124).

E

F

Figure 115 *(continued)*. E: An anaclinal valley, running against the dip of the beds (from Powell, 1875, fig. 57). F: Ridges on Bitter Creek (Powell, 1895, p. 123), a monoclinal valley that runs in the direction of the strike between the axes of the fold. One side of the valley is formed of the summits of the beds; the other is composed of the cut edges of the formation (from Powell, 1875, fig. 58).

Figure 116. Bird's-eye view of the Grand Canyon, looking east from the Grand Wash. One bird—Echo Cliffs; two birds—Kaibab Plateau; three birds—To-ro'-weap Cliffs; four birds—Hurricane Ledge; five birds—Shi-vwits Plateau (from Powell, 1875, fig. 72).

the rock beds in upheaval and subsidence here and there; and, second, by the gradual dessication due to the filling up of the basins by sedimentation and the erosion of their barriers; and the total result of this was to steadily diminish the lacustrine area. But the movements in the displacement extended over the Basin Province, for that region was then a comparatively low plain, constituting a general base level of erosion to which that region had been denuded in mesozoic and tertiary [*sic*] time when it was an area of dry land; for I think that from the known facts we may reasonably infer that the basin Ranges, though composed of Paleozoic and Eozoic rocks, are, as mountains, of very late upheaval. (Powell, 1876, p. 32–33)

How did Powell judge the age of the Basin Range uplift? He reasoned from relative erosion in the following way: he pointed out that it was almost certain that the Uintas began their upheaval at the end of the Mesozoic and continued intermittently almost to the present. This erosion accomplished some 30,000 ft. (~9000 m) of denudation. By contrast, in the Basin Ranges the faults were at the feet of the ranges and the erosion had not yet carried what Powell had called "the retreat of the cliffs" far enough to leave the faults out in the basins.

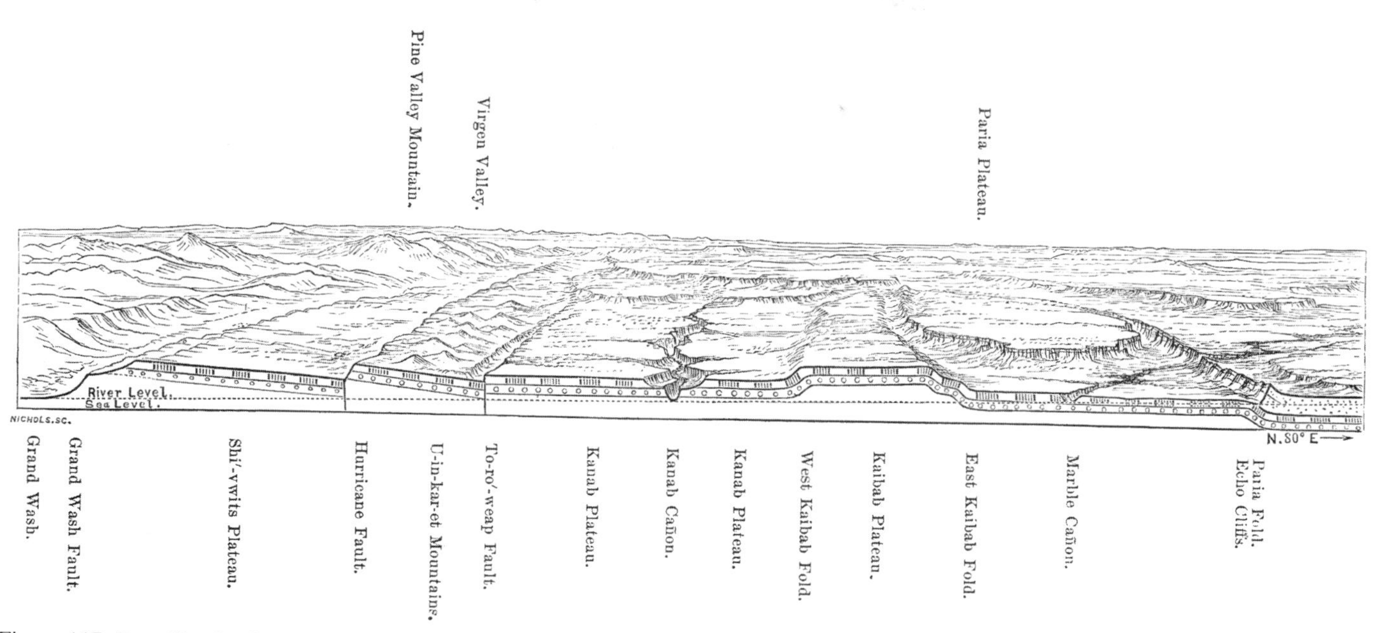

Figure 117. Powell's classical, oft-reproduced structural block diagram of the Colorado Plateau showing a section from west to east across the plateau north of the Grand Canyon, with a bird's-eye view of the terraces and plateaus above (from Powell, 1875, fig. 73; also in Powell, 1876, fig. 3; 1895, p. 90).

Basin Ranges had indeed suffered much erosion, and its products were now filling the intermediate fault troughs. As Powell explained:

But when we compare the erosion which these inclined blocks have suffered with that of many of the great blocks in the Plateau Province of the Kaibab structure, or with that of the Uinta uplift, or with the great uplifts in the Park Province, the erosion of the Basin Range sinks into insignificance. And, when we consider, further, that the erosion in the Plateau and Park Provinces which we are able to study has been performed during Cenozoic time, and that the conditions of maximum erosion were but intermittent during that time, we are forced to the conclusion that the conditions of great erosion now found in the Basin Ranges have existed but for a short period, i.e. the blocks were certainly not upheaved antecedent to Cenozoic time; and it would seem probable that it must have been in late Tertiary. (Powell, 1876, p. 33–34[373])

Powell indicated that the result of all these movements of displacement in the three provinces (most of the western United States) was general upheaval. But he underlined that the amount of upheaval in the three provinces was not equal: it was "great" in the Basin Province, "greater" in the Plateau Province, and "greatest" in the Park Province.

Powell thus painted the picture of a grand Cenozoic uplift in the western United States with its apex along the Rockies and flanks to their west and east, extending to the Sierra Nevada in the west and to the Mississippi River in the east. This picture was in accord with what generations of geologists had been saying since the beginning of the nineteenth century. But through the work of the Great Surveys, we now knew the age better, and especially through the efforts of Gilbert and Powell in studying its smaller and diverse structures, we also knew that the grand structure could not be a result of lateral shortening as Dana had imagined, but was the product of direct, vertical uplift. Powell once more summed up his conclusions as follows:

Throughout this great area, from the eastern slope of the Park Mountains on the east to the eastern slope of the Sierra Nevada on the west, and from the sources of the Green and Shoshoni Rivers on the north to the San Francisco Mountains on the south, the whole region is broken, flexed, and contorted along innumerable lines. But the great structure lines have a north and south trend; the ranges of the Plateau Province run from north to south; the great faults of the Plateau Province also run north and south, and the Park Ranges have a north and south trend. But these general outlines are broken by oblique and transverse displacements, usually of a minor magnitude, though in some cases, as in the Uinta Mountains, these transverse displacements assume as great proportions as the north and south flexures and faults. While the whole region is exceedingly complex by displacement, it is also exceedingly complex by reason of the unconformity of its sedimentary beds. And all this complexity is greatly increased by reason of the floods of lava which have been poured out here and there over the entire area, and now and then through Cenozoic up to the present time. And all these floods of lava, all these thousands of eruptive mountains, thousands of mesa sheets, thousands of volcanic cones, testify to a period of great volcanic activity while the region was in fact a great continental area, thus contradicting the generalization which has obtained in some quarters that volcanic activity is adjacent to the sea. And further, very much of this volcanic activity has been exhibited since the dessication of the lakes. (Powell, 1876, p. 36)

In the year in which the lines quoted above were published by Powell, his friend (and, by now a working associate) Captain Clarence Dutton followed up his 1874 paper on the refutation of the contraction theory by another in which he reiterated his arguments showing why the thermal contraction of the earth

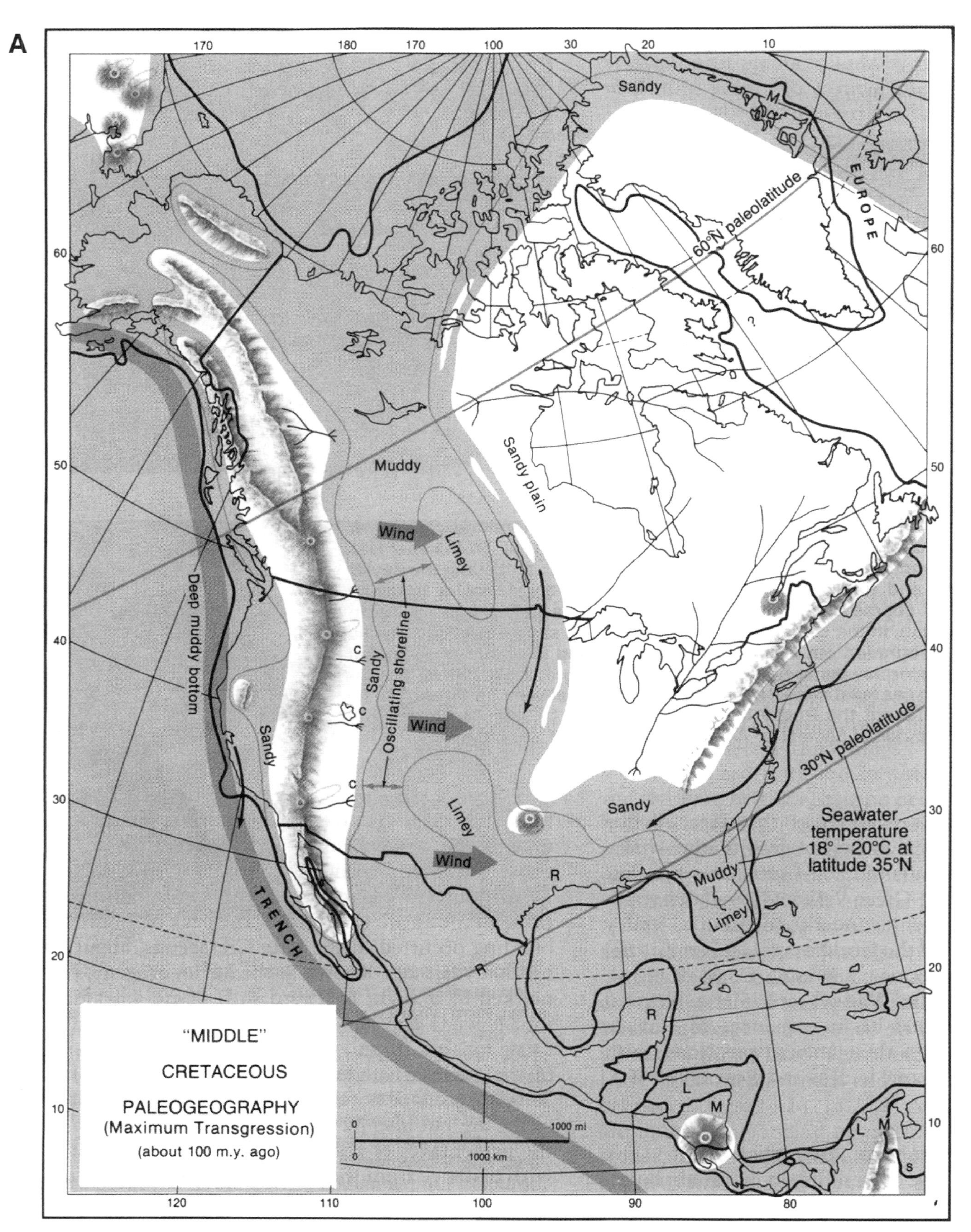

Figure 118 (on this and following two pages). Palaeogeographic evolution of North America from medial Cretaceous to medial Cenozoic shown on non-palinspastic maps (all frames from Prothero and Dott, 2002, figs. 14.27, 14.35, and 15. 2). A: Medial Cretaceous palaeogeography.

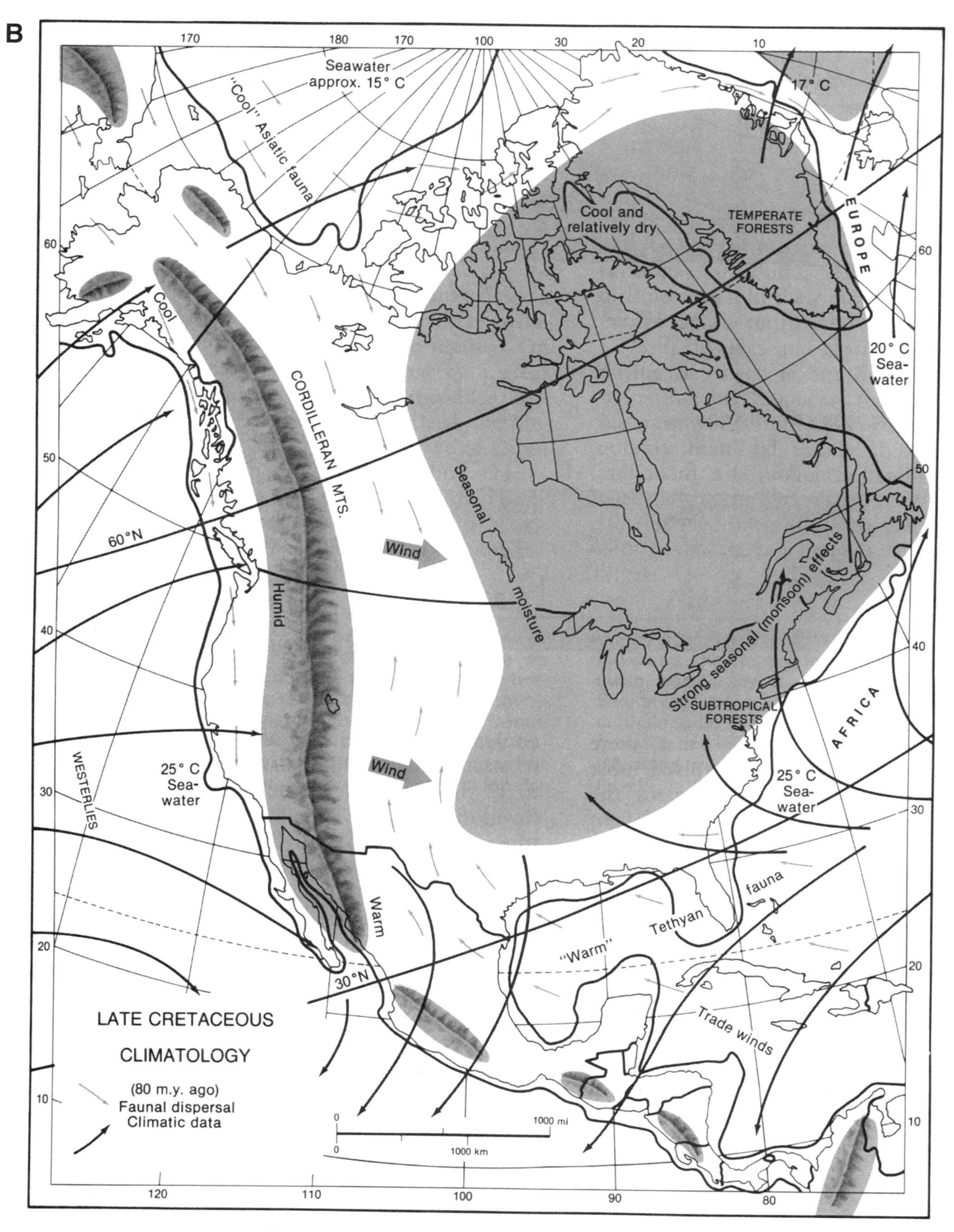

Figure 118 *(continued)*. B: Late Cretaceous palaeogeography.

C

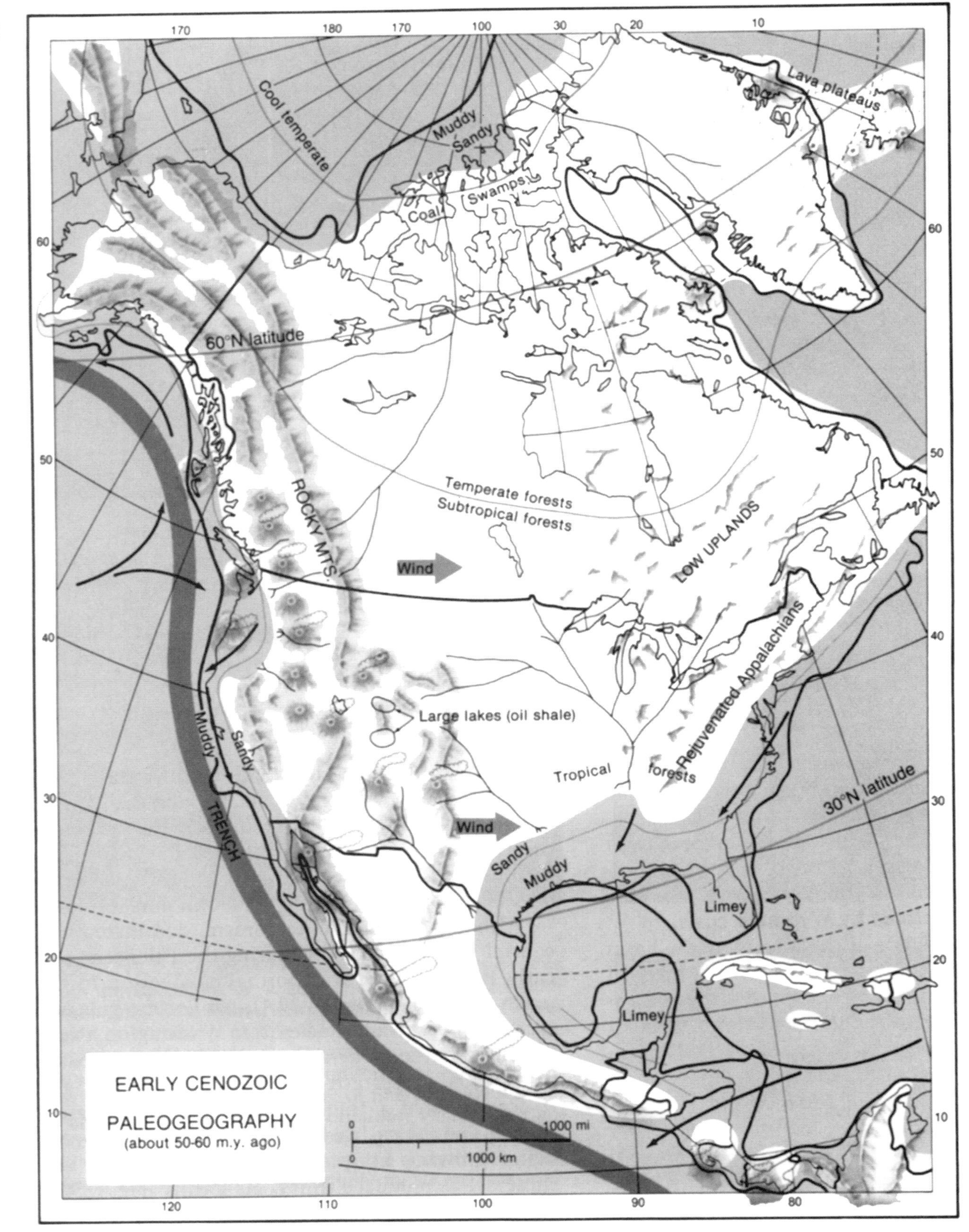

Figure 118 *(continued)*. C: Early Cenozoic palaeogeography.

was not even approximately adequate to create the structural forms one sees on the surface of our planet. In this new paper, Dutton argued that changing the loads on, or the density of, any piece of crust would either sink it or raise it. He thought that, although this was not a complete theory of tectonism, it went a far greater distance in explaining the structures the geologist sees in mountain belts, continental margins, and uplifted plateaus than did the contraction theory.

Dutton's paper was the first detailed theoretical treatment of terrestrial tectonism that came from the pen of a western worker. His dislike of the contraction theory, however, owed nothing to his western experience. It had been developed on the basis of purely physical considerations before he followed Powell's invitation to go West. By the time he was writing his 1876 paper, data from Powell's Plateau Province and the Park Ranges had started coming in, and Dutton was able to use some of them to constrain his hypothesis to replace the contraction hypothesis, which he believed had been thoroughly falsified.

Dutton started by stating the contraction theory as it had been advanced by Robert Mallet[374]. The reason Dutton chose Mallet's version was because it was "the only hypothesis which has ever been advanced upon this great question [*i.e., the question of the cause of terrestrial tectonism*] to a sufficient degree of explicitness and comprehensiveness to merit criticism; and its author is entitled to credit for having brought to bear upon the argument a series of experimental researches and a laborious mathematical analysis and has been throughout anxious to give weight to every objection which might be offered" (Dutton, 1876, p. 364–365). Mallet's hypothesis amounted to stating that the interior of the earth was hotter than its exterior and therefore must contract more, leaving the outer, colder shell unsupported. This unsupported shell would collapse onto the contracting interior. This process would create the surface inequalities and generate enough energy to cause rocks to melt giving rise to magmatism. In Dutton's eyes, Mallet's superiority to the earlier advocates of similar views was that he had actually performed experiments by crushing rocks to measure the heat thus generated.

But Dutton was not to be swayed. He showed, using Sir William Thomson's (1876, later Lord Kelvin) arguments that given reasonable initial conditions and thermal conductivities, the observed shortening in mountain ranges could not be obtained by thermal contraction. This was nothing more than restating what he had already said two years earlier. This time he also criticized Dana's idea, which implied that continents represented regions with a smaller thermal conductivity in contrast to that of the oceans. Dutton pointed out that areas that were now ocean had been a continent in the past (he alluded to the Appalachian case which showed that the Appalachian geosynclinal sedimentation had been fed from a continent to the southeast), and the present land areas had been oceans. He thus at once dismissed Dana's advocacy of the permanence of the continents and the oceans. His criticism was not entirely fair, for Dana had carefully pointed out that the interior of North America had been free of volcanism since the beginning of the Paleozoic. Yes, it had been flooded, but only by shallow, epeiric seas. So, former seas, for Dana at least, did not automatically mean former oceans. But Dutton, without further ado, equated any former sea with former ocean and argued that no cause was known that would alternately increase and decrease the thermal conductivity of a part of the earth's crust through geological time.

Another feature of the terrestrial globe that Dutton thought militated against the contraction theory was the fact that folded regions were generally confined to narrow and elongated belts. He rightly argued that contraction, had it been the cause of folding, would have acted from all directions, not just from one. He noted that the regions of folding were also the regions of what he called "maximum sedimentation" and somewhat unjustly complained that geologists had paid little attention to this fact (Dutton, 1876, p. 375). Perhaps he meant to say that geologists had not carefully thought about what it implied. He further observed that folding commonly commenced after the maximum sedimentary thicknesses had been accumulated. He reminded his readers that all of this was habitually accounted for by assuming that the loci of maximum sedimentation and folding were areas of weakness in the crust. Dutton had difficulty understanding what was meant by this assertion. He believed that if such areas had failed suddenly under the influence of the contraction-derived stresses, the contents of the sedimentary troughs would not have been nicely folded, but crushed and reduced to rubble. From this he deduced that the force that had acted to fold the sedimentary piles must have been the minimum necessary to accomplish the task.

Dutton concluded the first section of his paper by repeating the untenability of the contraction hypothesis. He thought it so untenable that further discussion of it would be superfluous. But:

> The task of opposing a theory which has no competitor, is not an agreeable one, and it is especially burdensome in the present instance. For, if the opposition be well founded, it leaves geologists without any explanation of the innumerable facts which they have accumulated at the expense of so much study and labor. The writer has no theory of his own to propose: believing that the true solution must be the work of a master mind, able to cope with the subject, both from the geological and physical side. Yet, with much diffidence and a consciousness of the great magnitude of the problem, an attempt will be made to set forth a few considerations of a simple character, which may possibly prove of some small service in suggesting certain limitations which must govern future inquiry. An attempt will also be made to indicate a few conditions, which any theory must conform to before it can claim even a conditional acceptance. (Dutton, 1876, p. 378)

In the second part of the paper, Dutton began by claiming that from the invalidation of the contraction hypothesis:

> … it would necessarily follow that those alternations of emergence and submergence which have occurred in some places, and those elevations and depressions which occasion the irregular profiles of the earth's surface, have not been *relative* movements, due to a variable amount of convergence towards the earth's centre, but have been *absolute*—now upwards, and now downwards. Plateaus and continents are true uplifts

and ocean basins are true subsidences. They have been regarded so for many years, and are still so regarded in the present discussion. (Dutton, 1876, p. 417, italics Dutton's)

He thus dismissed all the train of thought of the contractionists from Prévost to Dana. He then proceeded to show that, under these conditions, any uplift within a conical area with apex at the center of the earth and base at its surface, meant an increase of volume, decrease of density, or both. One obvious way to decrease the density was thermal expansion. Dutton quickly showed, however, that this path was inappropriate for the task at hand simply because there was no credible source to sufficiently increase the temperature in any one place in the earth to account for most of the observed uplifts. He then turned to some of the experimental petrological studies then being undertaken by such researchers as Auguste Daubrée[375] and underlined the importance of water in metamorphic reactions. He pointed out that, if rocks become hydrated during metamorphic crystalization, they expand and vice versa:

As between the amorphous and crystalline states of the minerals constituting the greater portion of the rocks, the variation in density would average at least from one tenth to one eighth of the original volumes, and this would be very materially increased by the addition or loss of water. The contraction and expansion thus produced would exceed many times that which could result from the greatest variations of temperature which could be reasonably granted. (Dutton, 1876, p. 422)

Dutton also noted that another effect of hydrothermal action on rocks would be to induce plasticity. "Wherever it has prevailed, the masses affected by it furnish indubitable evidence that they have been plastic … All crystalline rocks which have been disturbed, show a convolution of their layers or some equivalent 'implication' which could not possibly have occurred had they possessed always the hardness and inelastic rigidity which now characterizes them" (Dutton, 1876, p. 422). He then cited Daubrée who had said that the plasticity was attained at temperatures as low as 600 °F (~315.5 °C). He concluded that rocks which showed evidence of flow could not have been dry because he believed that causing dry rocks to flow would have required temperatures and pressures impossibly high.

Dutton then summarized his conclusions concerning metamorphism and deformation as follows:

(1) The hydrothermal theory of metamorphism is taken for granted. (2) The nature of this process, though not fully understood as yet, is presumed to be an intensified solvent power over silica, alumina and several other common minerals, by which it is enabled to break up the combinations in which those materials occur in many sedimentary rocks, and in any amorphous condition. In this action great pressure (tension) and a temperature approaching redness are essential conditions. (3) This state of silica, alumina, etc. is presumed to be the same essentially as that observed in the laboratory when those oxides are obtained in the soluble hydrous condition, where they are of considerably less specific gravity than the crystalline anhydrous forms.[376] This should be true if from no other cause than from the simple principle of alligation. (4) Hence it is inferred that the condition of hydrothermal solution is attended with a large diminution of specific gravity. (5) Inversely, the removal of any one or more of essential conditions, whether by a fall of temperature or decrease of pressure, is followed by the crystallization of the materials and an increase in density. (Dutton, 1876, p. 422–423)

Dutton reasoned that, if buried rocks are similar in composition to those exposed at the surface, then higher temperature and availability of circulating water would render the buried rocks less dense than the overlying rocks. The overlying rocks would therefore have a state of "unstable equilibrium" (Dutton, 1876, p. 423). Dutton did not think that the upper crust had much coherence. Any inequalities in loading would give rise to differential sinking or rising. At this point, he returned to the fundamental observation that folded regions had thicker piles of sediment than elsewhere and that lines of maximum sediment accumulation followed the trend of the folded belts. As he explained:

The directions of the lines of fracture would, in the absence of any other assignable cause, be determined by the inequalities in the distribution of deposit. The problem now becomes a hydrostatic one. The axes of maximum deposit become the axes of future synclinals, and the axes of minimum deposit mark the positions of future anticlinals. The heaviest portions sink into the lighter colloid mass below, protruding it laterally beneath the lighter portions, where by its lighter density it tends to accumulate. These movements are the plainest sequences [*sic*] of well-known hydrostatic laws, which we cannot hesitate to accept if we accept the premises. The resulting movements would be determined first by the amount of difference in the densities of the upper and lower masses, and second by inequalities in the thickness of the strata. The forces now become adequate to the building of mountains and the plication of strata, and their modes of operation agree with the classes of facts already set forth as the concomitants of those features. (Dutton, 1876, p. 424)

Dutton underlined that the nature of the deformation was critically dependent on the density contrast of the upper and lower masses. The one extreme, he pointed out, was represented by magmatism as reflected in intrusive and extrusive phenomena. At the other extreme, only gentle undulations of the upper crustal rocks would be produced. The Appalachians and the Jura Mountains, he wrote, would be the next member from the undulation end of the spectrum. More extreme cases would be generated as the top-heavy masses are steepened and even overturned "as is frequently seen in the mountains" (Dutton, 1876, p. 425). Dutton thought that the Appalachian structure as a whole exemplified this very nicely. He asked his reader to suppose that the foundations of the thick beds of the Appalachians to have been softened and expanded by the combined action of water, heat, and pressure, until their density became less (by a few percent) than the overlying beds. He then indicated that in the southeastern flank of the Appalachians, the hydrothermal activity must have been more advanced as shown by the more intense metamorphism and the beds of more irregular thickness. "Then the folding would have become a process of rotation in the branches of folds, continuing until the strata stood on their edges, and rested on harder unyielding rocks below the softened layer, or were caught and held by any intermediate position" (Dutton, 1876, p. 425). This description of steepening of beds and fold overturning must have been addressed directly to Dana (without specifying him as the addressee), in an attempt to show that Dana's elaborate description of formation, progressive shortening, and final overturning of folds (as depicted in Dana, 1847b, fig. 7), as an argument in favor of the contraction theory, could be better explained by the mechanism Dutton was introducing.

Dutton's theory was one of orogeny by means of vertical copeogenic tectonism. All horizontal motions were secondary to the primary vertical motions. As such it was a precursor of the theories of Erich Haarmann, Reinout van Bemmelen, and Vladimir V. Beloussov in the twentieth century (see below). But what about the long-wavelength falcogenic structures? In his 1876 paper, Dutton did not consider them, although he did point out that:

> The most abrupt and rugged range in the world—the Sierra Nevada—is a series of sharp pinnacles and ridges planted upon a broad expanse of high lands; and the same is true of the Andes, the Himalayas, the Alps, and of all ranges of granitoid mountains. And such should be their relation, from the argument here offered. The uplifting of the regional belts on which they occur by columnar expansion of the underlying magma involves at once the conclusion that the latter must ultimately reach a degree of density much below that of the overlying rocks, which break up into prisms or folds and sink or recoil away from the axis of the rising colloid mass. (Dutton, 1876, p. 426)

He thought that the whole mountain belt was one general area of upwelling of less dense material and its lateral spreading. The general upwelling produced a pedestal on which the second-order structures rested. Dutton did not, as yet, follow the implications of his theory for the highlands of the western United States where immense plateaus underlain by flat-lying sedimentary beds stood entire at stupendous elevations higher than the whole of the Appalachians.

But let us listen to Dutton's general conclusions, which he placed at the end of his 1876 paper. These comments clearly indicate the direction in which the fertile and learned mind of the great ordnance captain would travel in the next 15 years, when he came to study the great plateau country of the West:

> The considerations here offered as those which a comprehensive theory must take cognizance of and bring into correlation, are the following: 1. The regions of great disturbance are regions of great sediments, and those of least disturbance are regions of small sediments; regard being had to the rapidity with which any stratographic [*sic*] series has been accumulated. This order of facts appears to be general, so far as present knowledge extends.[377] 2. The epochs of disturbance have been those during and immediately following the deposition of thick strata. 3. The axes of displacements and vertical movement are parallel to, and probably coincident with, those of maximum and minimum deposit; where a series of the latter axes are parallel and have a definite direction, the plications and mountain forms have similar relations; and where there is no definite method in the variatons of thickness, the movements have no systematic trend or parallelism. 4. In the process of metamorphism, it is probable that great changes occur in the specific gravity of the materials metamorphosed, an absorption of water rendering them lighter and the elimination of water heavier. 5. All metamorphic rocks exhibit unquestionable evidence of having passed through a plastic or colloid condition; and if this condition prevails in any portion of the crust of the earth, the equilibrium of the parts so affected must be subject to hydrostatic laws. 6. The transfer of great bodies of sediment from one portion of the earth's surface to others, is tantamount to a disturbance of the earth's equilibrium of figure, which the force of terrestrial gravitation constantly tends to restore, and which it inevitably will restore wholly or in part if the materials of which it is composed are sufficiently plastic. (Dutton, 1876, p. 430–431)

Dutton's 1876 paper was a giant leap forward in understanding the relationships between deformation of the crust and the distribution of density, temperature, pressure, and water in the upper layers of the earth. However, few understood what he was trying to say. When he started doing theoretical tectonics, a vanishingly small number of people in his surroundings had sufficient knowledge and experience to help him. As we shall see in the next chapter, Dutton's novel views were regarded with hostiliy by the greatest European geologists of the time (especially by the continental ones, Eduard Suess being the most notable among them) and thus made little headway across the Atlantic Ocean. Neither were Dutton's concepts terribly popular within the Europe-orientated U.S. East Coast establishment. But his western friends read Dutton's ideas with interest, pondered them, and certainly took them seriously as they rode on their horses, hammer in one hand and a rifle in the other, out into the vast wilderness where both the variety of geological structures and the amount of actual rock exposure were much richer than anything with which either the Europeans or the East Coasters had so far become familiar.

Before Dutton could publish more on theoretical tectonics, now based on his own first-hand knowledge of the high plateaus of the West, the final reports of Clarence King's Fortieth Parallel Survey began coming out. Unlike Dutton, King was pessimistic about the amount of understanding that could be gained of the physical processes governing terrestrial tectonism with the meager knowledge that then existed of what he called "terrestrial thermodynamics" (King, 1878, p. 727). Like Dutton, however, he believed that "the suggestions of Herschel and Babbage as to the reactions upon the hot interior from superficial transportation will yet prove to be a key for unlocking some of the closed doors of geological dynamics..." (King, 1878, p. 727–728). King was firmly convinced that:

> the phenomena of geological section are expressive of two laws—the statics of the revolving sphere, and the dissipation of energy from its original and existing inner temperature, and granting the rigidity required by the tidal argument[378], I find, until the hypothesis of a critical shell within an immediately superficial region of the globe and the effect upon that shell by the processes of degradation and transportation are disposed of, no physical suggestions whose probable, not to say possible, application could account for the known operations of crust. Mere deformation of a solid globe under tangential strain is totally inadequate to account for a vertical fault of 40,000 feet [~*12 km*], nor does it explain the remarkable historic sequence by which loaded regions gradually subside foot for foot, while regions lately unloaded subside paroxysmally. No theory of the expansive force of imprisoned

elastic gases can account for the variability of upheaval and subsidence. And, lastly, no strictly chemical theory yet advanced, when brought into contact with stubborn facts, has the slightest shadow of applicability. I can plainly see that, were the critical shell established, its reactions might thread the tangled maze of phenomena successfully, but I prefer to build no farther till the underlying physics are worked out. (King, 1878, p. 728–729)

The idea of a critical shell had suggested itself to King from considerations about the effects of temperature and pressure distribution within the earth on its rigidity. He observed that if the empirical formula governing the augmentation of temperature with depth held, melting of rocks must commence at a depth of some 50 mi. (~80.5 km). He further observed, however, that rising pressure would prohibit such a reaction. At this point, King appealed to the arguments of Babbage, Herschel, and Dutton and pointed out that if this pressure could somehow be released, decompression melting would start at the base of the outer solid "shell." Some such mechanism, King believed, had to be operative because volcanism was localized, adjacent centers of volcanism showed no necessary sympathy with one another in terms of the composition of extravasated rocks or in timing, and great chemical diversity in successive and contemporaneous volcanic products was observed within one center. All of these observations precluded a common molten interior (as had been the fashionable belief until recently) or remnant magma pockets (such as William Hopkins had imagined). Only remelting of solid rock would cause volcanism. For that remelting, King thought, decompression was necessary (King, 1878, p. 700–705).

There was no lack of denudation. Vast tracts of land were depressed for tens of kilometers into the interior of the earth, and others were complementarily uplifted for similar amounts. King was impressed with the immense thicknesses of sediment he observed along his survey route. He believed that the Archaean (by which he meant Precambrian) successions totaled a *compacted* sedimentary thickness of some 60,000 ft. (~18 km). King believed that their original uncompacted thicknesses must have been some 36 km! These immense thicknesses were deformed once during the Precambrian and again before the deposition of the Cambrian sediments. In the area of the Fortieth Parallel Survey, the succeeding Paleozoic (until Carboniferous time), was "an age of subsidence, of sedimentation, and of rest from orographical disturbance" (King, 1878, p. 731). A tract of land existed west of western Nevada, to the east of which 40,000 ft. (~12 km) of sedimentary beds accumulated till the Carboniferous. King emphasized that none of the unconformities known in the Appalachian pre-Carboniferous successions were discovered in these western successions.

Then something extraordinary had happened: suddenly the western regions that had been "unloading" throughout the Paleozoic subsided, and the area where the thick sections were deposited had been uplifted. Here King felt the need to distinguish two sorts of subsidences: one was the slow, load-related subsidence (King credited James Hall for having recognized this sort some time ago; he was clearly unaware of Élie de Beaumont's publications on the subject.) The other kind was "catastrophic" (King, 1878, p. 732) and seemed unrelated to crustal loads. It was the unloading region that had suddenly gone down to accumulate 25,000 ft. (~ 7.5 km) of Mesozoic sedimentary rocks (King, 1878, p. 731). King believed this catastrophic subsidence to be:

analogous to that of the modern faults which are seen to form in earthquake regions. The sudden sinking of an area which has been relieved of a considerable portion of its load bears, of course, no relation to the equilibrium of the figure of the earth, but its origin must be sought in the obscurity of geological thermodynamics. With the subsidence and accompanying oceanic submergence of what had been the Palæozoic land, came the emergence of the thickest portion of the Palæozoic ocean beds, which was rapidly lifted above the water and became the first considerable land area of a new western continent. (King, 1878, p. 732)

The subsidence of the post-Carboniferous times, King likened in character to that of the Great Basin in Cenozoic time (King, 1878, p. 746).

On the available evidence, King could not decide what the nature of the post-Carboniferous disturbance was: did it produce folded ridges, like an ordinary mountain belt, or was the post-Carboniferous elevation "simply a plateau-like uplift," and all the folding between the Wasatch and the Sierra Nevada, including the latter, was a post-Jurassic phenomenon (King, 1878, p. 734). But King was sure that the land west of the Wasatch was "lifted gently from the mediterranean ocean, where, as we know, the Carboniferous beds lay undisturbed" (King, 1878, p. 747).

The uplifts that occurred after the Carboniferous had created two land areas: one east of the Mississippi River in the Appalachian region and the other west of the Wasatch. In between was an American mediterranean, covering the future prairies. After the Cretaceous, this central area was also upheaved, as King describes:

The effect of the post-Cretaceous action in the immediate Fortieth Parallel region was, first, the development of a broad level region, now occupied by the system of the Great Plains; secondly, the outlining of the basin of the Vermillion Creek Eocene lake; thirdly, the formation of distinct folds, of which the Wahsatch [*sic*] and Uinta are the most powerful examples; fourthly, the relative upheaval of the old Archæan ranges, whose highest points had through all geological time since Archæan ages existed as island-points lifted above the marine plain. (King, 1878, p. 748)

King thought that before the post-Cretaceous erosion had begun, the general topography of the presently mountainous western United States was dominated by "enormous arches, locally broken and dislocated into irregular blocks, and these folds were separated from each other by wide areas of gentle undulation or entire horizontality" (King, 1878, p. 749). He further noted that:

> One of the most interesting features in the whole orographical phenomena of America is the development of broad inclined planes south of the Fortieth Parallel work, in what is known as the Colorado Plateau. Here are areas which have been and are being ably described by Messrs. Powell and Gilbert, in which the sea-bed becomes an undisturbed plateau 5,000 and 6,000 feet [*~1500 and 1800 m*] above the level of the subsequent ocean. When we come to examine the relations of the post-Cretaceous folds with these adjacent undisturbed plateaus, it is evident that there were large regions in which no superficial contraction or diminution of area took place, whereas there were others in which occurred the most enormous and complex plications. Any theory, therefore, which attempts to account for the superficial results of geological dynamics will have to account for the existence of wide regions which, relatively to the sea, are suddenly upheaved without the slightest contraction, plication, fold, or fault, and of other regions within the same stratigraphical province which suffered the most extreme local compression, and all the complexities which can ensue from fold and fault. (King, 1878, p. 749)

King then made the interesting correlation that wherever there had been Archaean (i.e., Precambrian) folds, the newer deformations took the shape of folds; wherever there had been flat-lying sedimentary rocks, the area was subsequently uplifted en bloc (King, 1878, p. 749–750).

By contrast, areas previously elevated, now and then paroxysmally broke down along faults to generate large, faulted basins. King gave the example of areas west of the Wasatch Range: "Here the lofty country west of the Wahsatch [*sic*], which had formed the main source of supply for the Vermilion series, suddenly sank and permitted the waters of the lake to extend themselves over 200 miles westward into Nevada. This is another instance of that remarkable law of paroxsysmal subsidence taking place in the highest lands immediately after they have suffered extraordinarily rapid erosion" (King, 1878, p. 755). King was here seeing the immense Laramide intramontane basins of the Eocene (see Prothero and Dott, 2002, p. 452–454, for a quick review of the current interpretation), which are copeogenic structures and unrelated to the falcogenic history of the western United States.

Finally, King noted the faultless subsidence that gave rise to late Cenozoic lakes on the Great Plains. He recognized that they had formed by very gentle undulations of the surface of the Great Plains, which itself had become tilted shortly thereafter without any visible faulting or undulation (King, 1878, p. 756–757).

Summing up his observations in the 40th Parallel region, King underlined that there seemed to be two kinds of uplifts and two kinds of subsidences. Again, I shall let him speak for himself, for in the following two paragraphs, the great geologist gives a most succinct description of copeogenic and falcogenic events, a recognition that had grown out of his knowledge of the geological history of the western United States and that was to culminate in a dozen years in Gilbert's distinction of orogeny from epeirogeny:

> There are, therefore, two entirely different types of subsidence, one the gradual sinking of a region by loading, due to sedimentation, in which the most heavily loaded locality goes down deepest. This subsidence, from the nature of the sedimentary sections, is seen to be of the slowest and most gradual type. The other is a sudden paroxysmal subsidence on a plane of fault, in which the region lightened by erosion and removal is the one that goes down.[379]
>
> In the upheaval of wide areas there are two noticeable types of operation—one the lifting relative to sea-level of broad regions which, after upheaval, may be left horizontal or in gently inclined planes, their surface showing neither fault nor fold; the other, the well known operation of plication, by which actual deformation of the crust takes place, resulting in folds and faults and the tangential crushing of rocks. (King, 1878, p. 760)

King pointed out that the faultless type of deformation could simultaneously involve uplift and subsidence of different areas, and one area affected by this kind of deformation may pass laterally into another, in which deformation occurs by intense faulting and folding. "It is also a general law that those regions which experience elevation without local disturbance are the regions of relatively thin sediment superposed on a comparatively unaccidented Archæan foundation, whereas those which suffer the extreme plication are covered by the thickest deposits overlying and adjacent to the greatest Archæan mountain ranges" (King, 1878, p. 761). The historian may shun comparing these statements with our current understanding of the tectonics of the continents. The geologist, however, cannot help but feel delight in recognizing a conscious distinction, now so routinely made, of cratonic versus continental margin tectonism already in the interpretations of a great master of the geological science from the middle of the nineteenth century.

What was the cause of this demarcation that King so clearly and sharply drew between two modes of deformation? He himself was silent on this issue, but Dutton, after having studied the Plateau Country himself, made a go at understanding it. He said, "But we want something more than facts; we want their order, their relations, and their meaning" (Dutton, 1880, p. xvii). His friend and leader, Major Powell, called the result "an important contribution to geologic philosophy" (Powell, *in* Dutton, 1880, p. xii).

We already know with which bias Dutton rode out West. At the outset, he says that the first thing to do in treating geological phenomena is to look at the geological structure, "those attitudes of strata and the topographical forms which have been caused by the *vertical movements* of the rocks" (Dutton, 1880, p. xvi, italics mine). He was thus convinced that geological structures were caused by vertical movements. This was no doubt a result of his rejection of the then prevailing theory of contraction that emphasized, at least in the mountain belts, structures that formed as a consequence of horizontal motions.

Dutton thought that the Plateau Country of Utah, Arizona, Colorado, and New Mexico was "somewhat peculiar, especially when brought into comparison with displacements found in other regions." However, he believed that facts gathered from it had to find their place "in that branch of geological philosophy which treats of the evolution of the earth's physical features, the building of mountains, and the elevation of continents and plateaus" (Dutton, 1880, p. xvi). He did not think, though, that they appeared "to group themselves into the relation of effects to causes" (p. xvi). He pointed out that calling plateaus "mountains" was inappropriate. For that reason, he called the highland forming the continental water divide south of the Wasatch Range the "District of the High Plateaus of Utah" (Fig. 119). He admitted that "These uplifts displayed certain analogies to mountain ranges, but in most cases are distinguished by their well-marked tabular character" (Dutton, 1880, p. 2).

Dutton was mesmerised by the beauty and extent of the high plateaus. He wrote:

To the eastward of the High Plateaus is spread out a wonderful region. Standing upon the eastern verge of any one of these lofty tables where altitudes usually exceed 11,000 feet [~*3300 m*], the eye ranges over a vast expanse of nearly level terraces, bounded by cliffs of strange aspect, which are truly marvelous, whether we consider their magnitude, their seemingly interminable length, their great number, or their singular sculpture. They wind about in all directions, here throwing out a great promontory, there receding in a deep bay, but continuing on and on until they sink below the horizon, or swing behind some loftier mass, or fade out in the distant haze. Each cliff marks the boundary of a geographical terrace sloping gently backward from its crestline to the foot of the next terrace behind it, and each marks a higher and higher horizon in the geological scale as we approach its face. Very wonderful at times is the sculpture of these majestic walls. Panels, pilasters, niches, alcoves, and buttresses, needing not the slightest assistance from the imagination to point the resemblance; grotesque forms, neatly carved out of solid rock, which pique the imagination to find analogies; endless repetitions of meaningless shapes fretting the entablatures are presented to us on every side, and fill us with wonder as we pass. But of all the characters of this unparalleled scenery, that which appeals most strongly to the eye is the color. The gentle tints of an eastern landscape, the rich blue of distant mountains, the green of vernal and summer vegetation, the subdued colors of hillside and meadow, all are wanting here, and in their place we behold belts of fierce staring red, yellow, and toned white, which are intensified rather than alleviated by alternating belts of dark iron gray. The plateau country is also the land of cañons. Gorges, ravines, cañadas are found in every high country, but cañons belong to the region of the Plateaus. Like every other river, the Colorado has many tributaries, and in former times had many more than now, and every branch and every twig of a stream runs in cañons. The land is thoroughly dissected by them, and in many large tracts so intricate is the labyrinth and so inaccessible are their walls, that to cross such regions except in specified ways is a feat reserved exclusively to creatures endowed with wings. The region at levels below 7,000 feet [~*2100 m*] is a desert. A few miserable streams meander through its profound abysses. The surface springs will not average one in a thousand square miles [~*2600 km*^{2}], for the cañons in their lowest depths absorb the subterranean water-courses. But in the High Plateaus above we find a moist climate with an exuberant vegetation and many sparkling streams. (Dutton, 1880, p. 8–9).

This magical land was entirely under water during the Cretaceous. The waves of the sea extended from the Wasatch Mountains to Kansas and from the Gulf of Mexico to the Arctic. Dutton pointed out that the vast bodies of Cretaceous strata encountered in the High Plateaus corresponded in a general way with those of the Great Plains of Nebraska, Dakota, Montana, Wyoming, and Colorado (as described by Hayden and Meek), and with those in New Mexico and Arizona (as described by Newberry). The end of the Cretaceous marked a change, and the open sea gave way to brakish waters. Dutton remarked that although the details of the latest Cretaceous geography were hazy, it was clear that the area of the Great Plains and the high plateaus was covered with "Baltics or Euxines" (Dutton, 1880, p. 10). Finally, the brakish environments gave way to water bodies in which entirely freshwater molluscs flourished. Dutton cited Meek and Marsh who drew the Cretaceous-Tertiary boundary at the unconformity separating upturned Laramie Beds from the overlying series. He reminded that, away from the mountains (i.e., the U.S. Rockies), the boundary lay within an entirely conformable succession (here he was following Hayden, 1871). In such cases, the boundary was usually drawn either between different rock types, where this was feasible, or purely arbitrarily.

Dutton expressed great surprise while recording the presence of some 1800–4500 m thick shallow marine to lacustrine deposits of the Cretaceous and the Eocene in the Plateau Country. He underlined the shallow water to terrestrial nature of the deposits and stressed the absurdity of the suggestion that the Cretaceous sea could have been more than 2 km deep. The only conclusion that seemed reasonable was that the sea bottom had subsided as fast as the strata were laid down on it. The thicknesses of the shallow-water sedimentary rocks were reminiscent of those observed in the Appalachians, but the similarity between the two regions ceased there. Whereas the Appalachians were highly folded and faulted, the Plateau Country was a tableland formed from uplifted flat-lying sediments only here and there disturbed by rare faults, which were found near mountains and former shorelines.

If subsidences of such stupendous amounts did really occur, they presented "some ulterior questions" to Dutton's inquisitive mind: "By virtue of what condition of the underlying magmas was such a subsidence possible? If they sank, they must have displaced matter beneath them, and what became of the displaced matter?" (Dutton, 1880, p. 14). A problem of an "inverse order" seemed a possible key to an answer to his questions: "The Uintas, the Wasatch, the Great Basin have suffered an amount of degradation by erosion, which is perhaps one of the most impressive facts which the physical geologist has yet been brought to contemplate" (Dutton, 1880, p. 14). Some 9 km of strata had been removed from the Uintas and much more from the Wasatch. The Great Basin also had lost thousands of meters. These regions had risen concurrently with the removal of material from their top. But if they had risen, fresh material must have been pushed under them. "Whence came the replac-

Figure 119. Dutton's structural map of the "District of the High Plateaus of Utah" (from Dutton, 1880, Atlas) forming the western boundary region of the Colorado Plateau (cf. Baars, 2003, fig. 4). Heavy winding lines are faults and flexures. Compare this figure with Gilbert's east-west cross section across the Colorado Plateau (Fig. 113 herein) and with Powell's block diagram (Fig. 117 herein).

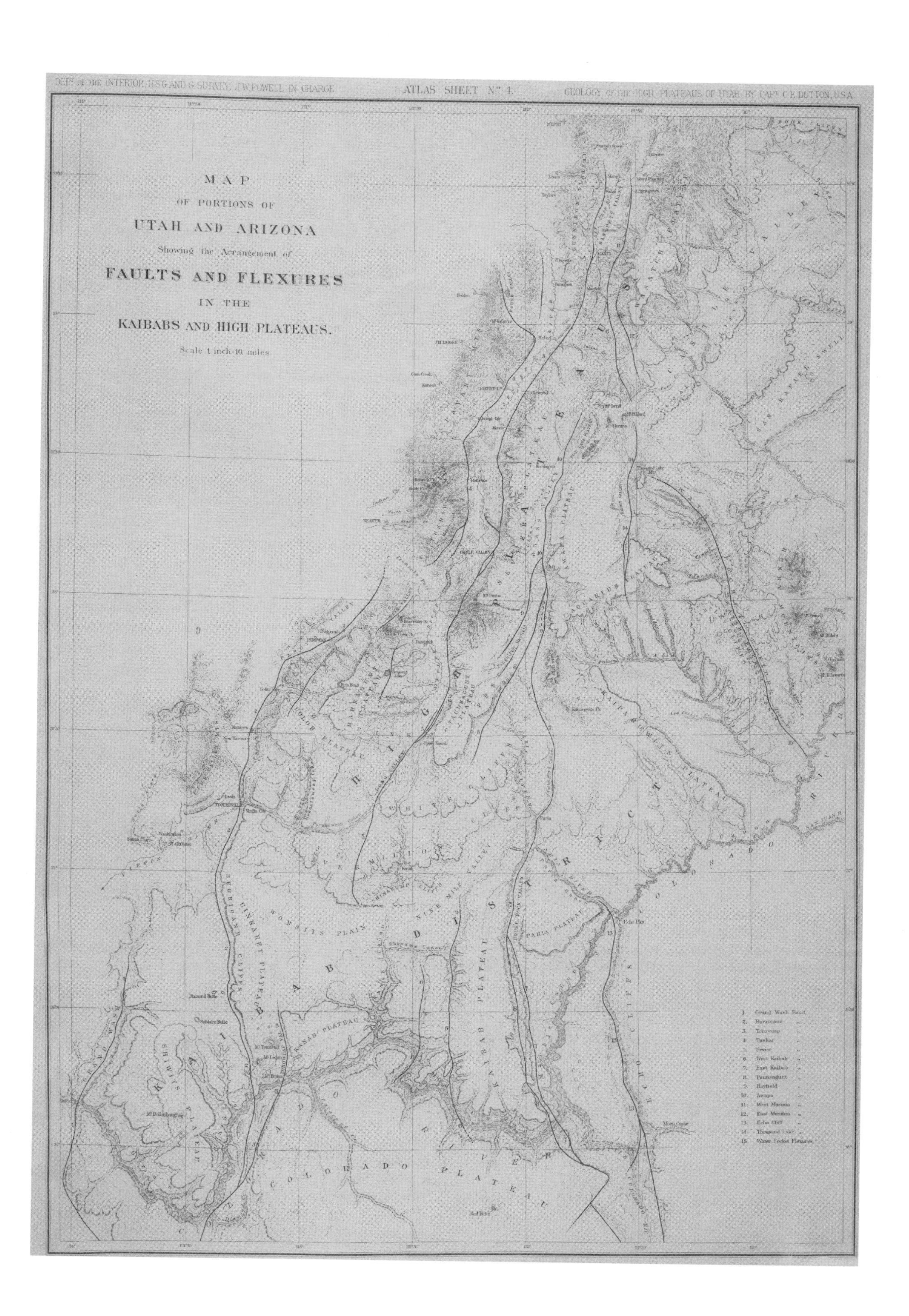
DEPT OF THE INTERIOR U.S.G. AND G. SURVEY. J.W. POWELL IN CHARGE
ATLAS SHEET No. 4.
GEOLOGY OF THE HIGH PLATEAUS OF UTAH, BY CAPT. C.E. DUTTON, U.S.A.
MAP
OF PORTIONS OF
UTAH AND ARIZONA
Showing the Arrangement of
FAULTS AND FLEXURES
IN THE
KAIBABS AND HIGH PLATEAUS.
Scale 1 inch-10 miles.
1. Grand Wash Fault
2. Hurricane "
3. Toroweap "
4. Tushar "
5. Sevier "
6. West Kaibab "
7. East Kaibab "
8. Paunsagunt "
9. Hayfield "
10. Awapa "
11. West Musinia "
12. East Musinia "
13. Echo Cliff "
14. Thousand Lake "
15. Water Pocket Flexures
COLORADO PLATEAU
WONSITS PLAIN
PARIA PLATEAU
COLOB PLATEAU
KAIBAB PLATEAU
SHIVWITS PLATEAU
KANAB PLATEAU
UINKARET PLATEAU
CHOCOLATE CLIFFS
NINE MILE VALLEY
HOUSE ROCK VALLEY
ECHO CLIFFS
SAN RAFAEL SWELL
COLORADO RIVER
Red Butte
Echo Peak
Diamond Butte

ing matter? It may be premature as yet to say that the elevation of the mountains and subsidence of the strata are correlated in the way which these inquiries suggest, but the juxtaposition of the facts must be regarded as significant" (Dutton, 1880, p. 14).

Finally, the Eocene lakes were also desiccated. Dutton explained, "Then, too, appears to have begun in earnest the gradual elevation of the entire region which has proceeded from that epoch until the present time, and which even yet may not have culminated. The two processes of uplifting and erosion are here inseparably connected, so much so, that we cannot comprehend the one without keeping constantly in view the other" (1868, p. 15). The upheaval continued slowly in the Miocene and led to continuing erosion (p. 17–18). Dutton estimated that since Eocene time the plateaus had risen some 3–3.5 km, while the adjoining basin areas some 1.5–1.8 km. He pointed out that the reason for the dominance of west-facing cliffs was because the plateau region had risen 1.5–2 km higher than the region to its west and was separated from it by faults dipping, and flexures facing west (Dutton, 1868, p. 24).

Dutton continued northward the structural geological work that had been started by Gilbert (1875) and Powell (1875) and documented the northern extension into Utah of the faults and flexures mapped by them in the Grand Canyon area (Fig. 119). He confessed that dating the faults was a problem because they cut both the Eocene and the Pleistocene (which Dutton called the Glacial period). He believed that the main phase of the activity had been during the Pliocene (Dutton, 1868, p. 35–36). The faulting and flexing were associated by a general upheaval not only of the High Plateaus, but also of the country to their south and east. Dutton thought it remarkable that no folding of the kind commonly known from the major mountain belts of the globe was associated with all this deformation and uplift. He presented this observation in the context of a brief historical review of ideas concerning mountain uplift to emphasize the peculiarity of the structure of the Plateau Country:

It is interesting to compare the structural forms produced by the displacements of the High Plateaus and Kaibabs with those observed in other countries and in other parts of the Rocky Mountain Region[380]. The earliest ideas acquired by geologists concerning mountain structure were derived from the study of the Alps and Jura. The conspicuous fact there presented is *plicaton*—waves of strata like the billows of the ocean rolling into shallow waters, and often more extreme flexing until the folds become closely appressed. With the extension of observation among other mountain belts of Europe, and wherever the traces of great disturbance among the strata were found, the same phenomenon of repetitive flexing was discerned, seldom amounting to "close plication," but undulating in greater or less degree. At a later period, when geology was colonized in America, its systematic researches were first prosecuted in the Appalachians, where the same order of facts was presented in a degree of perfection and upon a scale of magnitude far surpassing the original types of Switzerland. At a still later period the geologists who inaugurated in the Sierra Nevada and Coast Ranges the study of the Rocky system disclosed another grand example of the same relations. Thus the increase of observation has been for many years strengthening the original induction that plication and mountain-building are correlative terms.

But the rapid and energetic surveys of the remaining portions of the Rocky Mountain region have within a few years brought to light facts of a different order. From the eastern base of the Sierra Nevada to the Great Plains are very many mountain ranges, a large proportion of which have come under the scrutiny of geologists; and of those which have been hitherto studied sufficiently to justify any conclusions concerning their structure not one has been found to be plicated. Not one of them presents any recognizable analogy to the structure which is so remarkably typified in the Appalachians. It is certainly true that the study of these mountains has not been so minutely detailed nor so long continued as that of mountains situated in populous countries; that a considerable portion of them have not been examined geologically at all. But, on the one hand, the number of which we already possess a preliminary knowledge is considerable, and on the other hand the remarkable distinctness with which structural facts are there displayed, and the comparative ease with which they may be read, justify more confidence in our conclusions than might otherwise have been admissible. No one familiar with the progress of knowledge in this special direction can fail to recognize the conspicuous absence of plication in the mountain structures which are found east of the Sierra Nevada. (Dutton, 1880, p. 46–47, italics his)

A natural conclusion of this observation was that "As bearing upon the general hypothesis that the great features are produced by the action of tangential forces generated by the secular contraction of the earth's interior, it may be remarked that the displacements of the Plateau Province do not furnish any evidence of the operation of such forces. A careful study of the Kaibabs and High Plateaus has established the conviction that in those districts no such force has operated" (Dutton, 1880, p. 54). To the contrary, Dutton believed that especially the large monoclines traversing the Plateau Province mainly from north to south showed evidence of extension!

When Dutton looked around in the Plateau Province for compensating depressions to make up for the immense uplift, he found none. Neither in the Kaibabs nor in the High Plateaus of Utah "can we look for the same causation of faults and flexures as we might at first feel inclined to employ to explain those of Colorado and the Uintas" (Dutton, 1880, p. 53). So, although the block uplifts and their flanking basins making up Powell's Park Ranges looked remarkably different from the classical folded mountain ranges such as the Alps, the Jura, and the Appalachians, they were still not similar to the high plateaus. Both the Appalachians and the Park Ranges Dutton could explain with compensatory mass movements below the surface, brought about by sedimentation and erosion and metamorphism, as he had already outlined in his 1876 paper. But the high plateaus were different:

In the first chapter I have alluded to the possible effects attending the removal of great loads of strata from one locality of considerable area and the deposition of the same materials in adjoining areas; and while we may rationally suppose this transfer of loads to have important consequences in respect to vertical movements, we seem compelled to postulate additional forces, which for want of any definite conception as to their real nature we call Plutonic forces. The necessity for such a postulate seems perfectly obvious in the plateaus, and a little consideration will, I think, make its necessity apparent in the mountains of Colorado and the Uintas. (Dutton, 1880, p. 53)

Dutton thus was driven to consider the existence of "uplifting forces, almost pure and simple" (p. 53), of "plutonic" origin, by the enigmatic geology of the gorgeous high plateaus of the western United States. His considerations on the vulcanicity of the same area led him, through another path, to the same conclusion.

He started by testing Baron von Richthofen's (1867) suggestion—based mainly on the Cenozoic volcanism in Germany, Hungary, and the western United States, which he himself had observed, and on the volcanism from the Turkish-Iranian high plateau, Mexico, and South America reported by others—that the sequence of eruptions was as follows (from the earliest to the latest):

1. Propylite[381]
2. Andesite
3. Trachyte
4. Rhyolite
5. Basalt

Although Dutton had been initially inclined to disbelieve von Richthofen's ordering, as his work progressed on the High Plateaus, Dutton gradually came to agree with it. Once he convinced himself that the order was correct, he asked himself whether it correlated in any regular and progressive manner with the physical properties or with the composition of the rocks. At first, no correlation was found, as Dutton made the following tabular comparison

Arrangement by chemical constitution	Arrangement by order of eruption
1. Rhyolite	1. Propylite
2. Trachyte	2. Andesite
3. Propylite	3. Trachyte
4. Andesite	4. Rhyolite
5. Basalt	5. Basalt

(Dutton, 1880, p. 67)

He then tried to use sub-groups that he had earlier distinguished within the great groups to search for a correlation. The order of eruptions of the sub-groups in the High Plateaus turned out to be the following:

1. Hornblendic propylite
2. Hornblendic andesite
3. Hornblendic and augitic trachytes (less acid trachytes)
4. Augitic andesite (Richthofen)
5. Sanidin [*sic*] trachyte
6. Liparite
7. Dolerite
8. Rhyolite (proper)
9. Basalt (proper)

Then he made a new ordering as follows:

Place at the head of the series hornblendic propylite. Select from the list in the order given those rocks which are more acid than propylite. Take next those which are more basic than propylite, and write them also in the order in which they occur. We shall then obtain the following grouping:

	1. Hornblendic propylite	
3. Hornblendic trachyte		2. Hornblendic andesite
5. Sanidin [*sic*] trachyte		4. Augitic andesite
6. Liparite		7. Dolerite
8. Rhyolite		9. Basalt

(Dutton, 1880, p. 68)

Now Dutton asked himself how this grouping could explain the sequence of the eruptions. He pointed out that his division of the sub-groups into two series created one of advancing basic and another of advancing acidic composition. He reminded his readers that the basic rocks had high densities but low degrees of melting, whereas the acidic rocks exhibited low densities but high degrees of melting. So, acidic rocks may be light enough to be erupted in the beginning of the melting process but not yet melted completely, whereas the basic rocks may be melted but have not yet expanded sufficiently to be erupted. Hence, Dutton thought, the eruptions must commence by pouring out some intermediate composition rock. With the increase of temperature, the density of the basic rocks will diminish (thus their melts will be eruptable), and more of the acid rocks would be melted (also increasing their eruptability). Thus, the eruptions will produce the members of the two sequences (as defined by Dutton) ending in rhyolite and basalt.

The sequence of eruption of different types of rock and the explanation that Dutton devised to account for it appeared to Dutton to be a strong argument against the presence of a primordial magma reservoir within the earth: "Taking a generalized view of the subject, the objections against primordial liquids are insuperable" (Dutton, 1880, p. 140). He also repeated Clarence King's arguments against a liquid interior of the earth and against Hopkins' primordial maculae. It was clear to Dutton that magmatism required melting with limited temporal and spatial dimensions in the earth. This naturally raised the problem of finding a way to initiate melting.

To discuss the problem of melting, Dutton first considered the introduction of water into subterranean regions. He considered it "of importance," but of the precise nature of water's effects he was unsure (Dutton, 1880, p. 128). Dutton next contemplated King's mechanism of pressure relief resulting from erosion of superjacent rock piles, but found it inadequate:

This relief is effected through the removal of superincumbent strata by the process of denudation. Such removals have taken place upon a vast scale, and though geologists have possibly been suspected by other scientists of helping themselves very liberally to a supply of cause and effect of this kind, yet the surveys of our western domain have proven that they have been very modest and abstemious. But that such a process could have played a very important, much less a fundamental, part in causing volcanic eruptions seems to be negatived by facts. We do not find that eruptions always occur in localities which have suffered great denudation. We do not find even that they occur in such localities predominantly. Most of the existing volcanoes and most of those which have recently become extinct are situated in regions which have suffered very little denudation in recent geological periods, and many of them in regions of recent deposition. (Dutton, 1880, p. 128)

With water and release of pressure having been considered inadequate, only one other possibility was left for Dutton to employ in explaining the origin of melting at localized sites in the earth: raising the temperature locally:

But whatever the effects of the relief of pressure, and however essential the presence of water may be to the total process of eruptivity, something more is obviously needed, and this additional want is apparently well satisfied by a local rise of temperature in the rocks to be erupted. For it cannot be insisted upon too strenuously that from a dynamical standpoint the problem to be explained is the passage of lava-forming materials from a dormant to an energetic condition. And when we resolve this general statement into a more special and definite one, we find that it means the passage of solid materials into the liquid condition and … a decrease of density. Whatever may be the ulterior cause of volcanicity, a rise of temperature in the erupting masses seems to be an indispensable condition, and in assuming it we are apparently doing nothing more than taking the most obvious facts and giving them the plainest and simplest interpretation. (Dutton, 1880, p. 128–129).

Thus, the stratigraphic development, structural picture, and volcanic history of the Plateau Country of the western United States collectively led Clarence Dutton to the view that this area was formed by vertical motions that had been ultimately caused by what he called "plutonic forces" (Dutton, 1880, p. 53; see the quotation above on p. 222). One might think that Dutton, therefore, had not gone much beyond von Humboldt or Newberry in his understanding of the tectonics of the high plateaus. Nothing can be farther from the truth. Von Humboldt's and Newberry's interpretations were based on an assumed homology between misinterpreted volcanic cone-building and mountain-building, and they represented the culmination of a very long pedigree beginning with Eratosthenes in the second century B.C. (cf. Şengör, 1998). By the time Dutton was writing his book, von Humboldt's theory had long been falsified and given up by geologists. What Dutton did was not to resurrect von Humbold's theory but, by reconstructing the geological history of the High Plateaus based on new observations, combined with the older ones, to form a new problem situation. He then formulated a newer hypothesis consistent not only with that novel group of observations from the Plateau Province of the western United States, but also with the older observations from the folded mountain belts that had been studied in some detail—or so at least, Dutton believed. Although his western friends took his ideas seriously—especially his friend and leader Powell (1882) and Grove Karl Gilbert (1880, 1890) built upon them further—geologists coming from an orogenic background proved much more difficult to convince, as we shall see in the next chapter in the case of Eduard Suess. Lest anybody believe that Dutton had fooled himself into believing that he had solved forever the mystery of the great falcogenic uplifts, let us listen to what he said in his last paper on theoretical tectonics:

Geologic history discloses the fact that some great areas of the earth's surface which were in former ages below sea-level are now thousands of feet above it. It also gives us reason to believe that other areas now submerged were in other ages *terra firma*. Our western mountain region at the beginning of Cenozoic time was at sea level. It is now, on an average, 6,000 feet [~*1800 m*] above it. The great Himalayan plateau [*here he is probably referring to the Tibetan Plateau*] contains early Cenozoic beds full of marine fossils which now lie at altitudes of 14,000 feet [~*4250 m*] or more. The whole North American Continent has, since the close of the Paleozoic, gained in altitude. Now, it is sufficiently obvious that the theory of isostasy offers no explanation of these permanent changes of level. On the contrary, the very idea of isostasy means the conservation of profiles against lowering by denudation on the land and by deposition on the sea bottom, provided no other cause intervenes to change these levels. If, then, that theory be true we must look for some independent principle of causation which can gradually and permanently change the profiles of the land and sea bottom. And I hold this cause to be an independent one. It has been much the habit of geologists to attempt to explain the progressive elevation of plateaus and mountain platforms, and also the foldings of the strata by one and the same process. I hold the two processes to be distinct and having no necessary relation to each other. There are plicated regions which are little or not at all elevated, and there are elevated regions which are not plicated. Plication may go on with little or no elevation in one geologic age and the same region may be elevated without much additional plication in a subsequent age. This is in a large measure true of the Sierra Nevada platform, which was intensely plicated during the Paleozoic and early Mesozoic, but which received its present altitude in the late Cenozoic.

Whatever may have been the cause of these great regional uplifts, it in no manner affects the law of isostasy. What the real nature of the uplifting force may be is, to my mind, an entire mystery. (Dutton, 1892, p. 63–64.)

Having said this, he reiterated that the cause must somehow be related to a density decrease in the "subterranean magmas" underlying the uplifted region:

… but I think we may discern at least one of its [*i.e., the uplifting force's*] attributes, and that is a gradual expansion, or a diminution of the density, of the subtrerranean magmas. If the isostatic force is operative at all, this expansion is a rigorous consequence; … Hence I infer that the cause which elevates the land involves an expansion of the underlying magmas, and the cause which depresses it is a shrinkage of the magmas. The nature of the process is, at present, a complete mystery. (Dutton, 1892, p. 64)

Dutton came amazingly close to our present understanding of the causes of falcogenic movements. Yet, it was to take nearly three quarters of a century to bring the general opinion of the geological community into parallelism with his line of thinking[382]. The cause of this resistance was largely the great allergy created by the Buchian/Humboldian[383] vertical uplift theory of orogenic belts in those geologists who spent their professional lives in those mountain belts, such as the Alps, where the evidence of immense shortening could not be sensibly denied. The father of the modern global tectonics, the great Austrian geologist Eduard Suess, was one of those geologists.

CHAPTER XIII

EDUARD SUESS AND HIS CONTEMPORARIES: THE UPLIFT CONTROVERSY

Suess and the "No-Uplift" Model of the Evolution of the Lithosphere

Studer's enthusiastic application of the Buchian/Humboldtian vertical uplift theory to the Alps (see especially Studer, 1851, 1853, *passim.*) eventually became the undoing of this model. In 1872, but especially in 1875, Suess (1831–1914; Fig. 120)[384] attacked this application and argued, somewhat under the influence of Dana[385], that large mountain belts were more a result of horizontal shortening than vertical uplift: "It is seen ever clearer, already in these first considerations, that uniform movements of large masses in a horizontal sense have had a much more essential influence on the present structure of the Alpine System than the far too often emphasized vertical motions of individual parts, i.e. the direct uplifts through a force orientated radially outward from the inside of the planet and affecting its surface" (Suess, 1875, p. 25). In 1880 (the very year in which Dutton was driven to conclude that the only possible explanation for the post-Cretaceous tectonics of the Plateau Country was "plutonically" propelled vertical uplift!), Suess pointed out that, from the very beginning, he had thought primary uplift of any sort to be incompatible with the theory of thermal contraction and contrary to the available observations on sychronous and matching sea-level changes in large regions of the world[386]. He later even objected to Gilbert's (1884) interpretation that the Wasatch Range had been uplifted along a recent fault: "In addition to the inclination of the terraces Gilbert also describes an evidently recent "fault-scarp" close to the west foot of the Wahsatch [*sic*]; it is 30 to 40 feet [~9–12 m] high, and is regarded by Gilbert as proof that the Wahsatch is rising; this assumption is hard to reconcile with the inclination of the terraces. The fault scarp may also have arisen through the subsidence of the west wing" (Suess, 1885, p. 761–762, note 40).

Figure 120. Eduard Suess (1831–1914), whose immense knowledge of world-wide stratigraphy (which enabled him to recognize the presence of eustatic movements) and of world-wide structural geology (which impressed him about the dominance of horizontal motions) and his adherence to the contraction theory (which ascribed all tectonics to the constant diminution of the radii of the terrestrial globe) finally conspired against his recognition of the vertical uplifting forces within the earth. I was given Wilhelm Unger's print, shown here, by Professors Fritz Steininger and Alexander Tollmann, then both of the University of Vienna, to commemorate a Symposium on "Eduard Suess and the Development of Modern Geology and Austrian-Turkish Relationships" convened at the Istanbul Technical University in 1990; another copy of this print hangs in the Geological Institute in Vienna.

With such pronouncements, Suess fell out completely with the geological orthodoxy of his time: a perusal of the then commonly used textbooks indicates a widespread conviction concerning the existence of secular, slow, and broad-wavelength uplifts (e.g., Credner, 1872, p. 128 ff.; Dana, 1875, p. 582 ff., 739 ff.; von Hochstetter, 1875, p. 55–57; Geikie, 1882, p. 274 ff.; de Lapparent, 1883, p. 523–527; Le Conte, 1883, p. 127 ff.). Only a very few used a more cautious language allowing the implication that the sea level might also have been moving (e.g., von Hauer, 1875, p. 82–83). With the aid of the studies of his colleagues in Vienna (e.g., Neumayr, 1885, p., 129; also see under Alexander Bittner, Karl [*sic*] Diener, Theodor Fuchs, Rudolf Hoernes, Edmund Mojsisovics Edler von Mojsvar, Melchior Neumayr, Victor Uhlig, and Wilhelm Heinrich Waagen, *in* Anonymous, 1915, to get a rough idea of the variety and geographic spread of *only those* palaeogeographic studies of the major figures of the Viennese school which pertain to the Mesozoic and the Cenozoic and to the Tethyan realm), Suess questioned boldly all the evidence on which that conviction rested. That is why Suess, in 1888 (p. 680), invented the term *eustatic movements* (see Dott, 1992, for a brief review) and categorically denied *any* primary uplift (i.e., any increase of distance from the center of the earth) on the face of the earth.

On one point Suess won a decisive victory: he showed that the continents and oceans were not the expressions of giant, flat anticlines and synclines, as Élie de Beaumont (1852) claimed and as many geologists at the time believed. Even if the anticline-syncline interpretation of continents and oceans was then waning, still almost everyone believed that the ocean-continent transition was a gentle, normally unfaulted declivity (see especially Le Conte, 1883, p. 168, fig. 134 and 135). Suess showed that all ocean margins were faulted and that oceans clearly formed by subsidence along the bounding faults. This made impossible the origin of continents as gentle anticlinal crests and of oceans as gentle anticlinal valleys in a slowly but secularly compressed crust. Suess also saw that Hall's (1857a, 1857b; 1858; 1859; 1883) and Dana's ideas (Dana, 1873a, 1873b, 1873c, 1873d, 1873e, 1875) of the geosynclinal parentage of mountain belts, a derivative from Élie de Beaumont's model, could not be upheld (see especially Suess, 1888, p. 263–264; 1909, p. 722 and note 52). In *Die Entstehung der Alpen* (Suess, 1875), he had originally followed Élie de Beaumont's (and von Humboldt's?) idea of open folding to explain the secular rise of the Swedish coasts; Suess wrote: "the true motion of Scandinavia has the shape of a very long stretched fold with the shorter concave anti-fold in the south, or that a line drawn from the German coast to the Nordcap would in time turn into a ∽. But then it is not necessary to postulate a force different from that of the mountain-building, because in a homogeneous part of the earth's crust a north-directed contraction, which characterizes Europe, may generate also such structures" (Suess, 1875, p. 151). By 1888, however, Suess had come to deny completely the secular upheaval of Scandinavia—or any other part of the earth's lithosphere[387].

First and foremost, Suess had theoretical reasons: as I pointed out above, he thought that thermal contraction of the earth was incompatible with primary uplifts propelled by a radially centrifugal force from within the earth. In this he was under the strong influence of Constant Prévost[388]. Already in the *Entstehung der Alpen,* he had written, "Whereas Élie de Beaumont replaced the word 'élevation' with 'ridement,'[389] his ingenious opponent Const. Prévost denied expressly and definitely the existence of any centripetal[390], uplifting force. According to Prévost the prominences were only a secondary result of neighbouring subsidences, as it had been claimed before him by Deluc[391]" (Suess, 1875, p. 3).

The cause of the displacement of the strand Prévost had seen in the movements of the hydrosphere itself. Suess noted this with approval: "... he too did not base his position on the theory of elevation. On the contrary, he doubted whether the presence of intercalated fresh-water formations [*in the Paris Basin*] could be regarded as an indication of the complete withdrawal of the sea, and attempted to explain the whole stratified succession around Paris simply by a repeated subsidence of the waters, thus returning to the fundamental idea of Celcius, now based on other arguments and dressed in a new form" (Suess, 1888, p. 17).

Suess was sympathetic to the idea that the movement of the strand is a consequence of the movements of the hydrosphere for one other reason also; namely, because of the problem of the so-called "universal formations."[392] In the opening chapter of the *Antlitz,* this was one of the problems that Suess identified as being among the most fundamental in geology. He asked how it was possible that a stratigraphic terminology generated in Europe was found to be universally applicable everywhere. His *magnum opus* opens by supposing an observer were to approach the earth from outer space (possibly inspired by Élie de Beaumont's 1829 Oisans paper; see above p. 95). He invites this observer to regard the gross morphological features of the continents by pushing the clouds aside, and then the observer is ushered below the waves to behold the scenery of the ocean floor. Suess now leads him into a classroom:

> The leading features of this noble branch of science, the history of the earth, shall be expounded to him. He hears of the wonderful extension of human knowledge attained by examining the spectra of the celestial bodies, then of the various phases of cooling in which these bodies exist at the present day, of the conclusions which may thence be drawn concerning the formation of our solar system, and concerning that long and earliest period in the existence of our planet, during which the conditions necessary for organic life were not yet present; then he hears how later water, air, and life successively made their appearance, and how the period succeeding to these events is divided into geological formations[393], into eras, periods, and stages.
>
> Supposing the listener to have now reached this point, so that he stands in front of the porte of stratigraphical geology, and at the same time of the history of life: he will find himself surrounded by an overwhelming mass of details concerning the distribution, stratification, lithological character, technical utility, and organic remains of each subdivision of the stratified series. He stops to ask the question: what is a geological formation? What conditions determine its beginning and its end? How is it to be explained that the very earliest of them all, the Silurian formation[394], recurs in parts of the earth so widely removed from one another—from Lake Ladoga to the Argentine Andes, and from Arctic America to Australia—always attended by such characteristic features, and how does it happen that particular horizons of various ages may be compared with or distinguished from other horizons over such large areas, that in fact these stratigraphical subdivisions extend over the whole globe? (Suess, 1883, p. 10).

The first volume of the *Antlitz* is devoted essentially to a description of the deformational features of the globe. In the first part, Suess gives a review of the principal kinds of dislocations (including products of magmatism), and in the second part he takes his reader on a tour of the world to illustrate them. The emphasis of the tour is geographically on Eurasia (mainly the Alpine-Himalayan System including the Mediterranean) and topically on the contractional collapse of the earth producing, on the one hand, the large and small cauldron subsidences[395] and, on the other, the asymmetrically built mountain ranges verging away from the subsident areas onto forelands. This review sets the scene for the discussion of the causes of the displacement of the strand in the second volume[396].

The second volume of the *Antlitz,* which came out in 1888, only three years after the completion of the first, contains the third part of the work, entitled "The Seas of the Earth" (in the English translation simply entitled "The Sea"). The volume opens with a masterly review of all the ideas since antiquity concerning the displacement of the strand (see endnote 26)

Suess proposes near the end of this opening chapter a neutral terminology for describing the changes of level: *positive* for relative rise (i.e., with respect to land) of the strand, and *negative* for relative lowering of the strand. He ends the chapter with a description of a "real" uplift that occurred during the 23 January 1855 Cook Straits earthquake, whereby one side of a long fault remained stationary with respect to sea level while the other sank in the south and rose in the north. Suess points out that whereas in this particular case it was clear that land had moved with respect to sea level (*and, according to Suess, presumbly, its distance from the center of the earth increased!*), this clarity was not to be seen in the cases he was about to describe. He was setting the scene.

Chapters 2 and 3 of the second volume of the *Antlitz* are devoted to a meticulous description of the margins of the Atlantic and the Pacific Oceans, and chapter 4 provides for their comparison. Suess shows that the Atlantic is a younger ocean than the Pacific and was formed by subsidance along steep faults along the continental margins cutting right across the older fabric of the continent. The mode of the formation of the Atlantic Ocean is compared with that of the Mediterranean (which had been described in the first volume) and is shown to be the same: both had formed by coalescence of cauldron subsidences of various sizes and ages. The Pacific Ocean is also fault-bounded, but there the bounding faults parallel the fabric of mountain ranges that rim it. Whereas the circum-Atlantic faults are steep and have common dips oceanward, the Pacific-bounding faults dip with much gentler angles *away* from the ocean under the surrounding continents. Suess then undertakes a description of the Paleozoic, Mesozoic and Cenozoic seas in chapters 5 through 7, which is nothing less than a complete review of global stratigraphy. Potentially a topic of extreme monotony, Suess takes his reader through a breathtaking tour, chasing large transgressions and regressions through time and space. Meticulous documentation gradually unfolds a history of synchronous waxing and waning of the seas throughout the globe. He then puts the question: is it possible to explain this global synchroneity by local up and down motion of individual continents or of individual parts of a single continent?

Before answering this question, Suess takes his reader, in chapter 8, on a tour of the evidence in the Mecca of the elevation theory (Suess, 1888, p. 415 ff.; see also p. 487), namely the Scandinavian peninsula. In August 1885, with his friend Dr. L. Burgerstein, Suess himself went to the field in northern Norway (because of von Humboldt's observation? See above) to examine the evidence first-hand between the fjords Lungen and Bals, and from Tromsö in the north to almost Narvik in the south, following an itinerary recommended to him by the local expert (and customs official) Karl Pettersen (1826–1890)[397]. In his *Erinnerungen*, in which the touristic aspects of the journey are related (Suess, 1916, p. 365–372), Suess tells us candidly that he had already decided, through his studies of the literature, that the higher terraces had nothing to do with the movement of the strand (exactly as von Humboldt had said 40 years previously! See above). Much more likely, the terraces were probably the erosional terraces of temporary lakes formed between the receding ice-cap and the mountain slopes emerging from the ice.

Further in chapter 8 of the second volume of the *Antlitz*, Suess describes in great detail what he saw, supported with his immense knowledge of the relevant literature: his original guess is vindicated. The *seter* (or *setär*)[398] are like numerous benches cut into bedrock. Neither their number nor their elevations correlate from one fjord to another. Suess notes that they become numerous and higher towards the heads of the fjords. Seters turn around and in places join terminal moraines. As far as Suess could see from the existing maps and from the aneroid that he had wisely brought with him, the seters are also horizontal. Nowhere did Suess see any marine fossils on the seters, and neither had anybody else before him (Suess, 1888, p. 430). Seters thus cannot have been cut by the waves of the ice-age ocean. They are not marine terraces but erosional terraces of once existing lakes between the bedrock and the moraines containing them seaward (for the cross-sectional geometry of such lakes, see Flint, 1971, fig. 13-11).

Lower down are genuine marine terraces, but Suess ascribes them to sea level, which he believes stood higher during the ice ages: "...the terraces met with in very open bays, as for instance in Christiania fjord, which is essentially different from the narrow fjords of the north, may indeed be genuine vestiges of a sea-coast, like the terraces of western Patagonia, which also occupy an open situation; but the seter are not so, nor are many of the terraces of the west coast of Norway, especially those of considerable altitude" (Suess, 1888, p. 458–459). That the sea level was higher during the ice ages in the northern latitudes than it is now was a widely held view during Suess' time and was ascribed to the gravitational attraction of the ice masses that had gathered around the north pole. The rise of sea level along the border of an ice cap of 38° angular radius and a central thickness of 3300 m had been estimated at from 40 to 175 m (Geikie, 1903, p. 378 and the references given in footnote 4 therein). Thus, Suess' assumption was not only not unreasonable, but was perfectly along the lines of the best of the orthodox thinking of his day.

The richness of Suess' observations in northern Norway, the ingenuity displayed by him in interpreting those observations, with the aid of a vast array of both comparative and theoretical arguments about the terraces of northern, western, and southern Scandinavia, are truely awesome. His arguments range from fluvial and glacial geomorphology, through climatology to hydraulic engineering. When the reader reaches the end of the chapter 8, he or she feels that all Suess now would need to do is to turn the corner of Scandia and to take his reader into the Baltic Sea and into the Gulf of Bothnia to complete the argument against any Quaternary rise of Scandinavia. Instead, the reader is taken abruptly, in chapter 9, from the icy north to the sunny Mediterranean, to consider the evidence offered by the Temple of Jupiter Serapis in the Gulf of Pozzuoli, west of Naples[399].

Here the celebrated Temple of Jupiter Serapis had begun to be excavated in 1750, and its three standing columns were discovered to be marred by *Lithodomus* borings lower down. The inference was obvious: since the temple could not have been constructed underwater, sea level must have risen to bring the *Lithodomi* into contact with the columns to bore their holes. Then, again, the sea level must have dropped anew to expose the borings to view. How this actually happened had been the subject of a lively and fruitful debate, and Suess gives a superb review concluding with a presentation of his own interpretation (Suess, 1888, p. 486–494). The essence of Suess' interpretation, formulated after two visits to the area in the April 1872 and in the August 1878 (Suess, 1888, p. 485; 1916, p. 234–237) is to show that the Temple sits within a caldera (not dissimilar to the one described by Dana {1887, especially plate I} from Hawaii) and the level fluctuations are confined to it. Those fluctuations were probably caused by swelling and shrinking of subterranean magma chambers, although Suess is careful to point out an absence of correlation between the volume of the extruded material and the volume necessary to account for the changes of level (Suess, 1888, p. 493; see Issel's criticism resulting from a misunderstanding of Suess' position in Issel, 1896, p. 318). He had opened chapter 9 with a review of the youngest deposits and terraces along the Tyrrhenain coast of Italy, from north to south, to demonstrate that none of these deposits show any significant change of level (but Issel shows in his "*Carta bradisismica d'Italia*" *in* Issel, 1883, facing p. 177, mainly uplift in prehistoric times and subsidence in historical times. In two places only, he shows both up and down movements following each other in short intervals: in the lagoons and tombolos of Mt. Argentaria and in the Bay of Naples. Suess, 1888, p. 495, note 9, refers to Issel's detailed observations on the Cave of Capre, but not to his general map). Thus, when he finally comes to describe the case of Pozzuoli, the reader at once sees that it is a local phenomenon, not a general one. Thus Suess sweeps away the one piece of what he thinks to be pseudo-evidence that had been popularly used for many years in favor of continental oscillations, even by such careful and eminent observers as Lyell (1830, p. 449–459).

After the Pozzuoli obstacle is overcome, Suess returned to the north, this time to examine the historical record of the ongoing regression in the Gulf of Bothnia and the Baltic Sea. An elaborate argument is presented to show that the extremely low salinity and local meteorological conditions brought about a higher sea level in the Gulf of Bothnia in the seventeenth century that still was in the process of being equalized with the world-wide level. Suess believed that this emptying was rapid during the eighteenth century, when the most enthusiastic reports of the ongoing regression were published, and was now diminishing in rate (Suess, 1888, p. 524–525). He could see no such change of level around the margins of the Baltic and the North Seas. Suess ascribed subsidences along the British, Dutch and German coasts to local settling either of bogs or unconsolidated sediment.

Suess' elaborate argument to refute the rise of Scandinavia is the least convincing of any of his writings (it is amazingly similar to Urban Hiärne's naïve interpretation of nearly two centuries earlier, as we have seen above). Interestingly, it is the only part of his *magnum opus*, where he does not discuss the evidence step by step but deals with independent lines of evidence that do not naturally converge. Suess makes them converge by a careful arrangement of their sequence in his narrative and frequent—and uncharacteristic for his usual style—injections into the narrative declaring the impossibility of the uplift hypothesis. He never asks the question why so many lines of evidence—nearly all of which he thinks have been misinterpreted—seem to have conspired to fool so many excellent geologists in the same direction, namely the rise of Scandinavia. True, the picture he paints was fully compatible with his view of the behavior of the planet, but too many times he had to resort to local explanations to uphold his theory, giving the uncomfortable impression that this was special pleading. True also is the fact that his arguments were extremely well constructed, betraying an endless resource of knowledge concerning world regional geology in the largest sense (including geomorphology and climatology) and theoretical tectonics. Of course, Suess did not cite all of the relevant theoretical papers; for instance, he had never liked the idea of continents floating on a heavier substratum free to move up and down under loads. That is why, Jamieson's (1865) prophetic paper on glacial isostasy in Scotland never appears among Suess' references. Suess' argument might still have remained seductive to many had it not been that glacio-eustasy rapidly began to gain ground in the first decades of the twentieth century to show that not only was the sea level in the northern latitudes not high during the ice ages, as required by Suess' reasoning, but it was actually much lower[400]. Thus, the already disjointed argument of Suess lost the one critical battle, and his interpretation never gained wide acceptance.

But one conclusion Suess reached in the second volume of the *Antlitz* was independent of the Scandinavian observations:

> Provided with these provisional results, let us now return to the movements of the outer crust of the earth, and to the various kinds of dislocation that are produced by telluric stresses.
>
> Our observations on dislocations, taken as a whole, cannot be reconciled with any attempt to explain this class of phenomena by means of the elevation theory, i.e. by an undulation of the lithosphere. Neither the numerous small oscillations can be thus interpreted nor the great ones which embrace the whole globe. Movements of the lithosphere do not explain why the stratified series presents the same lacunae in the United States and in central Russia, nor do they explain the formation of long horizontal strand-lines in complete independence of the structure of the land. (Suess, 1888, p. 698)

Suess' formulation of the above conclusion is deceptive and does not do justice to his thinking. He did think that the lithosphere movement caused at least the negative movement of the strand. One reason he spent so much time elaborating the structure of the oceans was to show that they were inverted cone-shaped parts of the lithosphere that contracted more than the surrounding continents. This was an interpretation borrowed from Dana (1846, 1863, 1873), but much elaborated and placed in a different geometric/kinematic contraction model from Dana's. Owing to contraction, the bases of the cones subsided along circular or elliptical areas creating the ocean basins. Continents were simply high-standing wedges caught between subsident parts of the lithosphere. This also explained an old observation of Sir Francis Bacon, which Suess had mentioned in the opening chapter of the first volume of the *Antlitz*, namely the triangular southerly pointing ends of the continents[401] (Suess, 1883, p. 1). Now he tells us that these southerly pointing triangles are nothing more than wedges left standing between two intersecting elliptical subsidences. Every time a new subsidence takes place, the capacity of the world ocean basin is increased, and the waters recede from the continents into the enlarged basin. With time, sedimentation partially fills the basin and diminishes the basin's capacity causing transgressions. That is why, said Suess, regressions are much more sudden than transgressions. So sudden that they may cause large-scale reorganization of the terrestrial ecosystems and affect the biosphere as well. In a letter to Charles Schuchert, Suess speculated that the disappearance of the dinosaurs may have been related to the end-Cretaceous regression (Suess, 1911b; Şengör, 2000b).

Suess' model of the tectonic evolution of the globe was better than anything that was available at the time in terms of being compatible with the totality of the existing observations. The only thorn in its side was its author's hostility to primary uplifts; the Scandinavian problem was a direct consequence of that hostility. Suess' entire career as a tectonic geologist was in fact built on his belief in the fallacy of the primary uplift theory—whether of Lyell's continental uplifts, or von Buch's and Studer's mountain uplifts, or indeed Gilbert's normal-fault-block uplifts. His comprehensive synthesis was undertaken to combat this theory world-wide and to establish the victory of the contraction theory. The way this motif keeps cropping up amidst the rich and powerful stream of descriptions and discussions throughout the *Antlitz* is well familiar to Suess' readers. That is why he so intensely disliked the idea of isostasy (Suess, 1909, p. 700–716) and that is why he refused to accept the rise of Scandinavia. By the time Suess came to see that the contraction theory was not sufficient to explain the swelling amount of data in the diverse branches of geology (and see it he did: Suess, 1909, p. 721), he was already writing the last lines of his *magnum opus* in his 79th year (Suess, 1916, p. 323). He had not sufficient time left to rethink the whole model. Both the rise of Scandinavia and the idea of isostasy remained closed gates for him (see his letter to Schuchert: Suess, 1911b, especially p. 101). It is not an exaggeration to say that the only difference between Émile Argand's (especially in 1924 and 1936) and Suess' tectonic pictures is the former's acceptance of isostasy (cf. Şengör, 1998, p. 86).

Unlike Argand, many others who saw the weakness of Suess' argument also chose to disbelieve his masterly documentation of the evidence on the fault margins of the continents and the non-existence of geosynclines—now or ever. I think this may have been one reason a large number of geologists ignored Suess' model of the earth and stayed with Élie de Beaumont's model, well into the fifties of the twentieth century. But Suess himself confessed (in his above-mentioned letter to Charles Schuchert) that concerning the displacements of the strand, "after twenty-seven years I cannot offer you more than a loose heap of doubts regarding the explanation. I have learnt more and know less about it" (Suess, 1911b). Earlier, in the final chapter of the second volume of the *Antlitz*, had he not also cautioned his readers that our knowledge of the terraces in the high latitudes was still very incomplete (Suess, 1888, p. 699)?

However, there were also other kinds of observations at that time which seemed to some to militate against Suess' rejection of primary uplifts of the lithosphere.

Gilbert's Laccolites in the Henry Mountains: Half a Revival of von Buch's Craters of Elevation

Shortly before Suess stuck his neck out and denied any primary uplift in the lithosphere, the indefatigable Grove Karl Gilbert (see endnote 12), working in southern Utah[402], had announced the discovery of hypabyssal concordant intrusions of mostly diorite porphyries (Hunt, 1953; Johnson, 1970, p. 32; Jackson, 1998 and the references therein. For historical overviews, see Pyne, 1979, 1980, p. 83–95 and Hunt, 1980b, 1988) that domed the strata above them, much as Leopold von Buch earlier had imagined craters of elevation to have been generated. Gilbert's discovery showed how primary vertical uplift could create little dome structures above what he called laccolitic intrusions (from λάκκος = reservoir and λίθος = rock), recognizable at outcrop mainly by dip and strike changes (see especially Gilbert, 1880, fig. 2, 3, 11a, 11, 25, 26, and 34), yet causing little actual fabric change within the rocks (except some fracture) thus still comfortably within the province of copeogenic structures. Gilbert calculated how much the doming above a laccolite would stretch the superjacent strata. He took as an example the Lesser Mount Holmes (Fig. 121A, B) dome that has a structural uplift of ~500 m (Gilbert, 1880, p. 27) and a diameter of ~4.5 km. He calculated that the "Vermilion Cliff sandstone" (170 m thick: Gilbert, 1880, p. 75)[403], spanning the arch, had been stretched by only 100 m (Gilbert, 1880, p. 75), which is a stretching by 2.2% (Fig. 121 C, D). This calculation, like Élie de Beaumont's formulae, is helpful when wishing to evaluate the amount of extension generated by doming in the lithosphere, as shown in Şengör (2001a).

Suess had no problem with Gilbert's laccolites. They were of too small a scale to pose any threat to his "no-uplift" model of lithospheric evolution. But the nature of laccoliths encouraged others to formulate newer uplift hypotheses in the future. For example, "The coupola-structure of the American laccoliths gives the impression to any unprejudiced observer that here a vertically-working force was active" (Supan, 1911, p. 370).

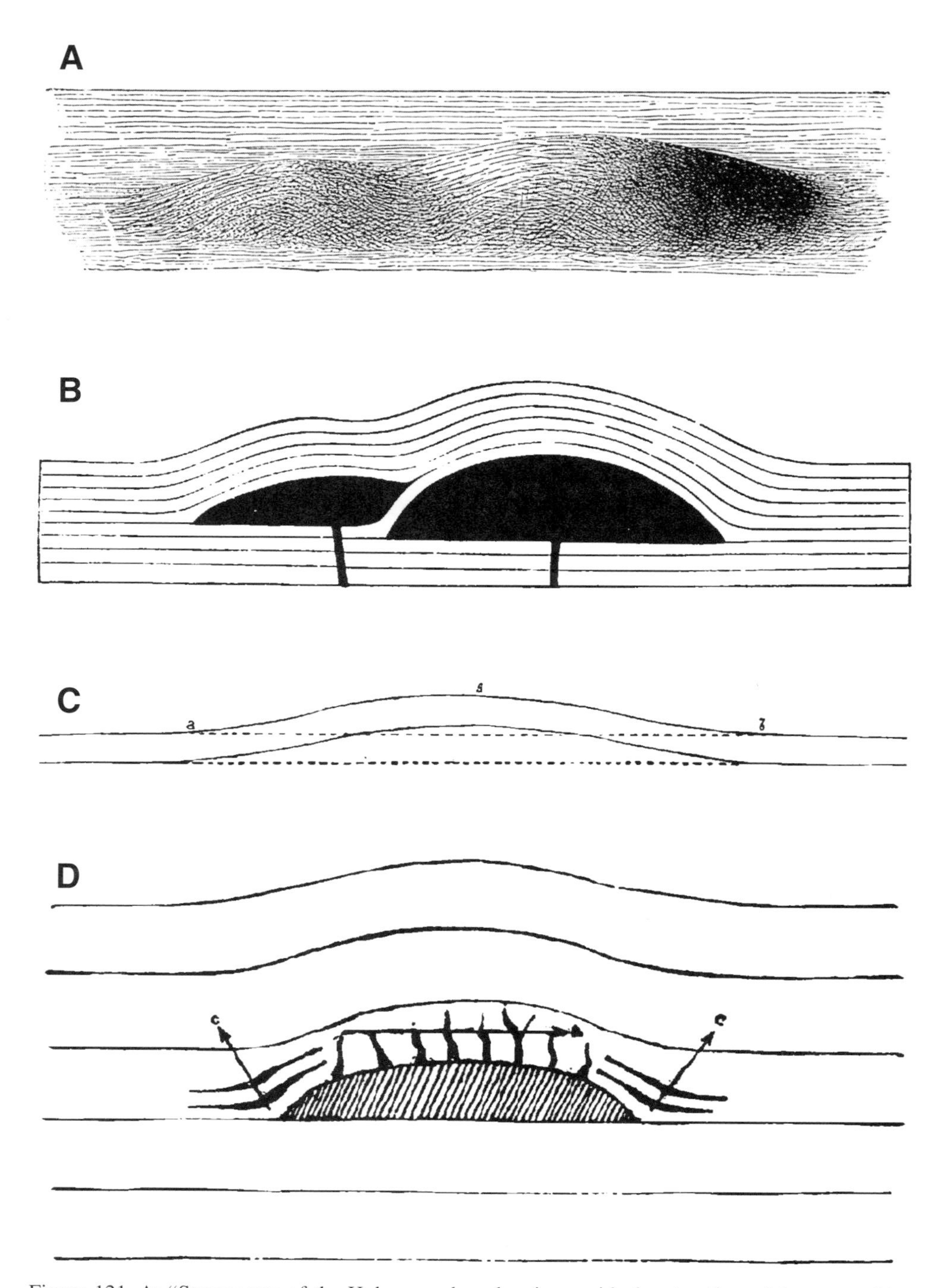

Figure 121. A: "Stereogram of the Holmes arches showing an ideal restoration of the overarching strata" (Gilbert, 1880, fig. 17). B: Ideal, restored cross section of the Mount Holmes laccoliths according to Gilbert (Gilbert, 1880, fig. 18). C: "Cross-sections of an uplifted dome. The dotted lines show the original position of a bed; the curved lines, the imposed" (Gilbert, 1880, fig. 52). D: "Diagram to illustrate the relation of Dikes and Sheets to the strains which are developed in the uplifting of laccolitic arches" (Gilbert, 1880, fig. 53). Note the stretching indicated by the dykes emanating from the domed surface.

Uplift Models by Suess' Contemporaries

Suess had an immense influence on tectonic thinking in the last quarter of the nineteenth century and the first quarter of the twentieth century. However, not even his influence could hide the obvious uplift of Scandinavia (which he tried to explain away as a result of climatic effects and of local tidal and marine currents: see Suess, 1888, p. 508–541, 700–701, and the discussion presented above) and parts of northern Canada and Patagonia. Consequently, large-scale, gentle continental deformations involving significant uplift remained on the agenda, very much in the form that Contejean (1824–1907)[404] had given them in his 1874 book.

Contejean had recognized three categories of movements of the crust (a synthesis of the mid-century views, especially of von Humboldt, Naumann, and Élie de Beaumont): (1) orogenic movements, (2) secular movements, and (3) earthquakes. Orogenic movements "have given birth to mountain chains, to faults and grand ruptures. They are distinguished by their energy, and because they effect a terrain much longer than it is wide" (Contejean, 1874, p. 272–273). By contrast, Contejean defined secular movements as:

> those slow oscillations which raise or depress the crust in some regions through hardly perceptible motions, the results of which become perceptible only after a long series of years. ... Movements described under the names of intumescences, undulations, and oscillations do not seem to me to be different from one another and all belong to the category of secular movements. They are distinguished from earthquakes by their slowness and from orogenic movements by affecting large areas in all directions and by their smaller amplitude. They are caused by local subsidences of the crust ordinarily counterbalanced by uplifts… (Contejean, 1874, p. 273)

Contejean was particularly careful to point out that the secular movements "produced vast undulations of the terrain, left the strata little inclined and that carried grand surfaces above the level of the sea. Most frequently, the dip of the beds are so little that one could recognise it in following the same bed for a long distance. At first it appears horizontal [*and*] only begins to climb near the edges of the basin of which it forms a part. The aspect of the sedimentary terrains is such that they look totally undeformed" (Contejean, 1874, p. 497–498). It is difficult not to think that, behind these words, there is a strong influence of Élie de Beaumont's ideas.

With Contejean, we finally obtain a clear textbook definition of two distinct tectonic processes: one creating long, curvilinear, and narrow mountain chains with much folding and faulting; another affecting very large, roughly equant areas with no folding and no faulting but only gentle, barely perceptible warping creating large undulations of the crust. As we saw above, King (1878) also clearly separated the processes, and it was Gilbert (1890) who had given them distinct names that persisted into our times: the process that created the faulted Wasatch front and with it the mountains, he called *orogenic* (applying a designation long used in Europe: e.g., Boué, 1874, p. 262; Contejean, 1874, p. 272); for the gentle bending of the terraces of the ancient Lake Bonneville, he created a new term, *epeirogenic*, burdened *in statu nascendi*. This term carried all the baggage of the Lyellian *and* the Beaumontian theories of continental elevation; it was to lead to much misunderstanding in the future:

> The displacements of the earth's crust which produce mountain ridges are called *orogenic*. For the broader displacements causing continents and plateaus, ocean beds and continental basins, our language affords no term of equal convenience. Having occasion to contrast the phenomena of the narrower geographic waves with those of the broader swells, I shall take the liberty to apply to the broader movements the adjective *epeirogenic*, founding the term on the Greek term ἤπειρος, a continent. The process of mountain formation is orogeny, the process of continent formation is epeirogeny, and the two collectively are diastrophism. It may be that orogenic and epeirogenic forces and processes are one, but so long at least as both are unknown it is convenient to consider them separately. (Gilbert, 1890, p. 340, emphasis his)

Gilbert must have been inspired to this distinction by his countryman and colleague, Dutton, who in his classical *Report on the Geology of the High Plateaus of Utah* (Dutton, 1880) emphasized the important structural distinction between what he called the Plateau Province (i.e., the Colorado Plateau and the Uintas) and the Basin and Range Province:

> In comparing the plateaus with the Basin Ranges we have to deal with the fact that the displacements of the latter are in the main older than those of the former, though younger than those of the eastern Rocky Ranges. Erosion has operated powerfully upon all the Basin Ranges, and *the aggregate displacements are greater than in the plateaus. The strata ordinarily incline at larger angles and exhibit a greater amount of subordinate fracturing and dislocation*. There is, however, some similarity between the plateau and basin uplifts. Both present a succession of inclined platforms, sloping in the same direction, with greater dislocations upon the uplifted sides. *In the Basin Ranges, the uplifting being greater, the inclination is correspondingly greater, so much so, that we pass from the notion of a plateau or platform to that of a mountain slope. The inclination of the plateau summits is rarely so great as 3°; the inclination of the structure slopes of the Basin Ranges is rarely so little as 8° or 10°.* (Dutton, 1880, p. 53–54; italics mine)

In the above paragraph, Dutton makes a clear *structural* distinction betweeen mountains, in which rock fabrics have been more clearly disrupted by "fracturing" and "dislocation" and plateaus, in which the fabrics of the rocks look almost undeformed. Mountains have short wavelengths, whereas plateaus have long wavelengths. The sharp mind of the ingenious ordnance captain also perceived a difference in the mode of formation: "…in the plateaus we have the result of uplifting forces, almost pure and simple, while elsewhere it is complicated, and generally reinforced by the effects of the transfer of great loads from the mountain platforms to the plains and the valleys around their bases, *followed by a readjustment of the plastic earth to a static equilibrium of profiles*" (Dutton, 1880, p. 53, italics mine).

In this remarkable passage, which so amazingly anticipates our modern views concerning flow below the brittle carapace of the crust in the Basin and Range province (e.g., see Wernicke and Axen, 1988; Block and Royden, 1990; Wernicke, 1990; McKenzie et al., 2000), Dutton also distinguished phenomenologically the plateaus from the Basin Ranges.

In 1883, the same year as the appearance of the first volume of Suess' *Antlitz*, the Italian geologist Arturo Issel (1842–1922) called the slow oscillations of the rocky rind of our planet *bradisismi* (Issel, 1883, p. 12, 1896, p. 304). Issel defined *bradisismi* as the slow movements of the earth's crust. The term *bradisismi* is made up of βρᾰδύς (= slow) and σεισμός (= shock, agitation, commotion; but Issel translates this inappropriately as movement; see his p. 12). Unfortunately, the second of the terms chosen by Issel is not terribly appropriate for the processes of very slow motions of the earth's rocky rind (this is probably at least a part of the reason his term never became popular). In his text-book, Issel explained the notion of *bradisismi* as follows:

> **Preliminary notions.**—From many observations one must conclude that the emergent lands are not only subjected to earthquakes, but also to slow oscillations, from which follow the progressive changes in the horizontal and vertical configurations of the continents and some islands. These osciallitons are called *bradisismi*.
>
> Owing to their slowness, these movements escape, most of the time, the attention of men. These, besides, can be ascertained in only some special cases, because their manifestations are distinguished with difficulty from those of other phenomena. Always, many indices allow the inference that *bradisismi* exercise their action in a great portion of the surface of the earth.
>
> ...
>
> The *bradisismi* are susceptible to manifestations in every sense, but because they principally obey gravity, which is exercised vertically, from the surface towards the interior, and the force acting in the inverse direction, they mostly consist in movements from the up to down and from the down to up. It would be helpful to occupy ourselves, with particular attention, with these two species, with which all others can easily be connected. (Issel, 1896, p. 304)

Issel's *bradisismi* and Gilbert's epeirogenic movements are thus broadly equivalent, although Issel did not emphasise the broad wave aspect of the *bradisismi* and Gilbert did not emphasize the slowness of the epeirogenic movements, but both the slowness and the wave-like nature of the movements were implicit in their respective writings. However, Gilbert was clearly unaware of Issel's work. (In fact, it was only in the subsequent Italian literature and in Suess' writings that I have encountered references to Issel's work: see, for example, Parona, 1903, p. 295).

In 1889, Joseph Le Conte, one of the elder statesmen of American geology[405], published a paper in which he noted the great difference between mountains formed through shortening, creating folding and thrust faulting, and those in which normal faults appeared predominant:

> The explanation of the *reverse* faults seems obvious enough. They occur, as we have already said, mostly in strongly folded regions. Such folds can only be produced by lateral compression. The pressure when extreme often produces overfolds. If such overfolds break, the dip of the fissure will be toward the direction from which the pressure came and the hanging wall be pushed forward and upward over the footwall by the sheer force of the lateral thrust. ... But the explanation of normal faults which are by far the most common is not so obvious. (Le Conte, 1889, p. 259, italics his)

The explanation that occurred to Le Conte was not original, but he was probably unaware of this. Dufrénoy and Élie de Beaumont (1841, p. 436–437, fig. and 28) half a century earlier, and Fraas (1867, p. 33) nearly two decades after the Frenchmen, had assumed that normal faults formed by doming up a part of the crust and thus distending it. That Le Conte could also find precursors of his idea in the English-language literature was so reminded a few months after the publication of his paper by the British engineer, T. Mellard Reade (1890, p. 51). In explanation of his model of normal-fault-making. Le Conte wrote:

> Suppose then the earth-crust in any place to be not crowded together by lateral pressure, as in the formation of mountains of the Appalachian type, but uplifted into an arch by intumescence of the subcrust liquid [*Le Conte believed that the crust floated on a liquid intermediary layer separating it from a solid core*]. Such local intumescence of the subcrust liquid may be the result (*a*) of elastic force of steam incorporated in the magma in more than usual quantity from above, or (*b*) of hydrostatic pressure transferred from a subsiding area in some other perhaps distant place. Such an arch being put upon a stretch would be broken by long fissures more or less parallel to one another and to the axis of uplift into oblong prismatic crust-blocks many miles in extent. After the outpouring of liquid lava or the escape of elastic vapors had relieved the tension, these crustal blocks would again be adjusted by gravity. If the blocks are *rectangular* prisms, some may float bodily higher and some sink bodily lower, giving rise to level tables separated by fault cliffs as in the Plateau region .... But if the fissures are more or less inclined, as is more commonly the case, then it is evident that the crust-blocks will be either rhomboidal [*Fig. 122(A) herein, a, b, f, g*] or wedge shaped [*Fig. 122(A) herein, c, d, e*]. These in the arching of the crust would be separated from one another, fig. 9B [*Fig. 122(B) herein*]. But after the relief of tension by outpouring of lava or by the escape of steam, they would of course readjust themselves by gravity in new positions. Now by the laws of floatation how would such blocks adjust themselves? It is quite evident that every rhomboid block would tip over on the *overhanging side* and heave up on the obtuse angle side producing in every case *normal* faults, and every wedge-shaped block would sink bodily lower or float bodily higher according as the base of the wedge were upward or downward, producing again in every case *normal* faults (fig. 9C) [*Fig. 122(C) herein*]. (Le Conte, 1889, p. 259–260)

Le Conte believed this model reflected how what he called the Basin system of Gilbert had formed. He assumed an arch-shaped uplift had affected the entire area of the Basin-and-Range region between the Sierra Nevada and the Wasatch front "at the end of the Tertiary" (Le Conte, 1889, p. 262) by intumescent lava (Fig. 123). The arch broke down and the broken parts readjusted themselves into the familiar basin-and-range topography.

Although nothing in Le Conte's paper was original in principle, it was new in its application of an old idea to the Basin-and-Range region and helped to accentuate the belief among the geologists working in the western United States that the

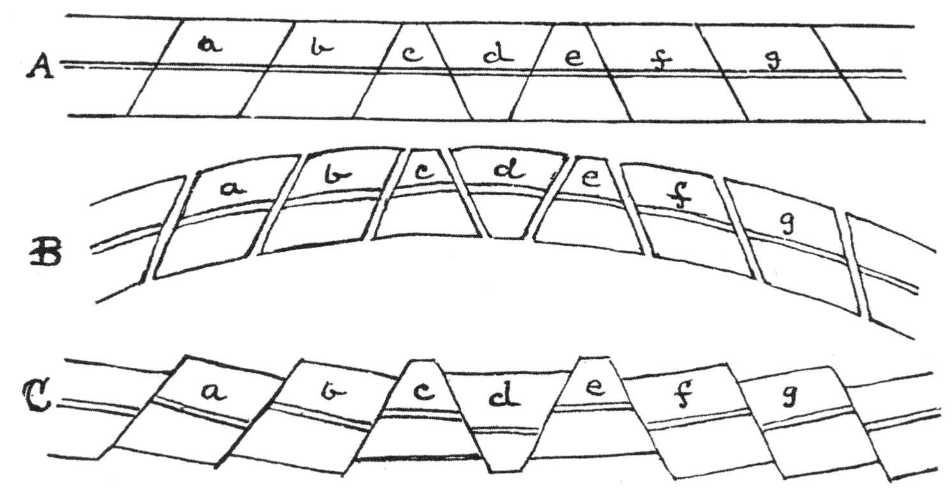

Figure 122. Le Conte's figure showing the origin of normal faults by doming-related stretching. A: Crust broken into blocks (presumably he meant crust having pre-existing planes of weakness). B: Crust arched and blocks separated. C: Crust re-adjusted by gravity (from Le Conte, 1889, fig 9).

Cenozoic tectonics in the western half of their country was indeed governed by primary vertical motions, much as Dutton had been urging.

After Contejean's clear definitions and Gilbert's terminological innovation, following in part Dutton's observations and a number of other suggestions such as Le Conte's, it was by the end of the nineteenth century clear that geologists interested in the tectonics of the earth's outer rocky rind had two distinct processes to deal with (although some, like Gilbert, still entertained the hope that ultimately they might turn out to be the manifestations of a single process). Most thought this ultimate single process to be the contraction of the globe, whereas others, such as Argand[406] (Fig. 124), thought it was continental drift. In fact, Argand thought that the distinction was greatly exaggerated in that both orogeny and epeirogeny were manifestations of crustal shortening. Reacting to Haug's[407] (Fig. 125) distinction between orogeny and epeirogeny and his ascription of the tectonic segmentation of the western Alps by *aires de surélevations* (axial culminations) and *aires d'ennoyages* (axial depressions) to epeirogeny (Haug, 1900), Argand wrote in protest: "By its enormous proportions, the active segmentation of the Penninic nappes between the Grisons and the Mediterranean is one of the grandest examples of this category of phenomena, one of the most generalized and least imperfectly known. As we have said, it does not furnish evidence in favour of an absolute dualism [*between orogeny and epeirogeny*] mentioned above" (Argand, 1912, p. 354). All of this dualism was an error of the thinking mind, thought Argand, a product of the psychology of the geologist who had difficulty in visualizing three dimensional objects being continuously deformed:

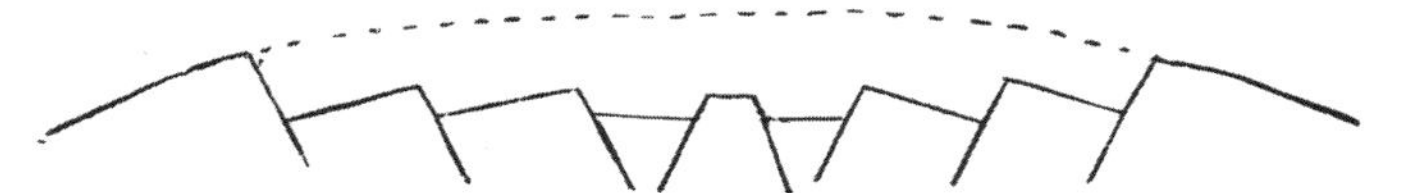

Figure 123. Le Conte's "ideal section showing mode of formation of Basin system" (from Le Conte, 1889, fig. 12).

> The [*entire*] geometry of the nappes, to be clarified, must be embraced. It is and will remain always difficult, unless one dwells on it for a long time, to imagine in a shapeless space the form of such complicated objects. It is easier, more common and less fruitful to consider things first in plan then in sections. The segmentation, which is seen more readily in the latter, is perceived in sections as a distinct and posterior phenomenon [*to orogeny*]. Unconsciously, the procession of images tends to impose itself as if it were a procession of facts. That, despite its disruptive and analytical character this double mental operation has become, for some, a double operation of Nature, this is what we should realize.
>
> The point is to conceive the objects in space, in a single image, and not in successive touches. It requires some attention, but beyond that one realizes very quickly that the segmentation seen in plan, with its protrusions and recesses, is completely the same thing as the segmentation seen in section with its axial depressions and elevations. These are the two aspects of one body deformed at the same time and by the same agents. (Argand, 1912, p. 355)

Falcogeny and Its Connection with Motions within the Interior of the Earth: Osmond Fisher and the Convecting Mantle

Only a year before the publiction of Gilbert's great memoir on Lake Bonneville, a suggestion was made that could have thoroughly revolutionized geological thinking had those interested in tectonics paid any attention to it. It might have satisfied those like Suess or Gilbert who sought intellectual satisfaction in a unifying, comprehensive, causative mechanism for all terrestrial tectonics. Or perhaps the suggestion might have channelized into more productive avenues the fer-

Figure 124. Emile Argand (1879–1940), whose view of primary vertical motions of the lithosphere were identical to that of Suess, but in a framework to Wegener's theory of continental drift. (Photograph by E. Sauser, Neuchâtel.)

Figure 125. Emile Haug (1861–1927), who was mainly responsible for re-erecting a Beaumontian world-view of tectonics after Eduard Suess. Reproduced from Lutaud (1958).

Figure 126. Reverend Osmond Fisher (1817–1914), whose remarkable theoretical deductions linking falcogeny with convection beneath the earth's crust remained unused by his contemporaries.

tile ideas of such researchers as Wettstein (1880, see especially his plate VI). The suggestion, by the Dorset mathematician and geologist, Reverend Osmond Fisher (1817–1914; Fig. 126)[408], consisted in assuming convection currents beneath the earth's crust, with ascending limbs under the oceans and descending limbs beneath the continents (Fisher, 1889, p. 77). This suggestion was not the result of idle speculation but of a careful consideration of the age of the earth as inferred from geological observations (Fisher assumed 100 m.y.[409]), the latent heat of rocks (another assumption he made on very scanty data), the evaluation of the plumb-line observations, and the style of the deformation visible at the outcrop. The latter two convinced Fisher that the crust was not as thick as many thought, especially in the English-speaking world (he assumed about 25 mi., which is exactly the value accepted today). The temperature history made it imperative that some of the crust be remelted at the same time as bits were being added to it by secular refrigeration: "Now the only way, in which this remelting can be accounted for, is by a quantity of heat coming up from below" (Fisher, 1889, p. 77). I do not think we would be completely amiss if we considered this point another step, most likely independent of Studer's statement and younger by 42 years towards Wilson's (1963) theory of mantle plumes. Fisher continued:

> Such convection implies upward and downward currents, and resulting local alterations of temperature *and level*, according to the play of the currents at different times; because convection is an action depending upon slight disturbing causes, so that the word "play," though not strictly scientific, well describes the changes in motion of the liquid. *Here then we find a key to the every varying changes of level in the earth's crust; elevation over extensive areas affecting sometimes one part of the surface, and sometimes another.* (Fisher, 1889, p. 77, italics mine).

After this statement, Fisher credits Alexander von Humboldt as having been perhaps the first to have thought of the role of currents in the fluid interior of the earth in affecting changes of level at its surface (von Humboldt, 1858b, p. 19–20; Fisher cites Sabine's English translation, v. IV, p. 19)[410].

Then Fisher continues to deduce geological consequences from his theory:

> Let us now consider what would happen beneath the oceans, where the ascending currents impinge. The liquid tending to spread laterally will produce a tensile stress in the central parts, which will become converted into a compressive stress as the continental areas are approached. In the central parts we may therefore expect that the crust will be fissured and that volcanic eruptions will be the consequence. This may explain why so many volcanic islands are found in mid ocean, and why so many eruptions take place in the bed of the sea, even where no permanent volcanic islands are formed. It is evident that whatever amount of compression is caused by this kind of action in the continental areas, must have its correlative extension in the width of the fissures beneath the oceans, which will become dykes of igneous rock in the suboceanic crust. (Fisher, 1889, p. 322).

I do not think this text needs commentary in the third year of the twenty-first century.

Fisher recognized the implications of his theory for explaining what are here called copeogenic and falcogenic structures: "The instability of convection currents, which shift their positions from slight disturbing causes, will go far to explain the instability of the earth's crust; for there must be slight changes of level produced by the ascending and descending currents,—slight that is as compared with the dimensions of the larger inequalities of the surface such as the height of mountains and the depth of oceans" (Fisher, 1889, p. 322). As Fisher demonstrated (1889, ch. XIII and p. 318), mountain ranges must arise above descending limbs of convection cells. The ascending currents are then left to generate the broad uplifts of small amplitude.

So far, Fisher's deductions are truly remarkable. It is a great pity that some geologist, let us say, of Suess' wide-sweeping knowledge, did not have an opportunity to collaborate with him. Fisher further concluded (erroneously!) that, in areas where the oceans are shallower, the ascending currents must be weaker, since he ascribed the shallowness to thicker crust beneath, not shaved away by an energetic ascending limb. With all this mobility of the substratum that he deduced, it is strange that continental drift did not suggest itself to him (one wonders whether he was aware of Wettstein's 1880 book that postulated the flow of continents?).

No geologist with whom I am familiar has made use of Fisher's conclusions. So far as I know, only Supan, in his textbook (1911, p. 373), gave a fair summary of Fisher's model, but with no commentary whatever. The mathematics may have put off the geologists and the geographers, but that excuse would have been invalid for Alfred Wegener, the geophysicist, who might have combined his own ideas with Fisher's to great benefit—but did not. Possibly owing to the lack of oceanic data[411] to check them, Osmond Fisher's great insights into the dynamics of the earth have remained unrecognized and unused.

CHAPTER XIV

FALCOGENIC AND COPEOGENIC EVENTS IN THE TWENTIETH CENTURY

Falcogenic Uplift Ideas Before the Death of Suess in the Twentieth Century

When the twentieth century opened, Eduard Suess was still alive and active—the last two volumes of the *Face of the Earth* had not yet come out—and his authority was generally undisputed. Yet, almost no one took seriously his denial of the possibility of independent vertical uplift of the lithosphere. The numerous textbooks, at various levels, published at the time in different countries, including his own, bear witness to this (e.g., von Toula, 1900 {p. 93–95}, 1906 {p. 55–58}; F.G.-M., 1902 {p. 51–53}; Witlarzil, 1902 {p. 40}; Geikie, 1903 {p. 379}; Parona, 1903, p. 294–303; Chamberlin and Salisbury, 1904 {p. 513 ff.}; Credner, 1906 {p. 55 ff.}; de Lapparent, 1906 {p. 575–591}; Löwl, 1906 {p. 136–145}; Haug, 1907 {p. 531}; Coupin and Boudret, 1909 {p. 265–270}; Meunier, 1909 {p. 46–50}; Eisenmenger, 1911 {p. 268–270}; Supan, 1911 {p. 438–463, especially p. 463}; M. Sadi, 1327 H. [1911–1912 A.D.] {p. 125–127}; Kayser, 1912 {p. 785–799}; Ficker and Trauth, 1913 {p. 105–106}), the very few exceptions being from Suess' circle of Alpine friends (e.g., Heim, 1908). Not that there was any agreement as to the causes of the uplifts, but all agreed that the lithosphere was undergoing slow, broad oscillations, very different from the movements one could infer from the internal structure of the mountain belts. Haug (1900), for example, thought that epeirogeny was confined to continents and orogeny to geosynclines. Only if broad movements crossed the geosynclines at considerable angles would he consider them epeirogenic. Empirically almost true, Haug's distinction greatly confused the fundamental issue and invited Stille's (1919) just criticism that Haug had grossly deviated from the original meanings of the terms as used by Gilbert (1890). When Kober (1928, p. 22) dubbed the continents as "epeirogens" and confined epeirogenic movements to any movement that took place on consolidated kratogens (even if they were nothing but extensive faulting![412]), these proposals reflected both the etymology of Gilbert's inappropriate term and the confusion caused by Haug.

Otto Ampferer and His Theory of Undercurrents

A loner in his own right was Suess' great countryman, Otto Ampferer (1875–1947: Fig. 127)[413], who did not belong to the inner circle of the Viennese giants in geology and who disagreed with almost everybody concerning the causes of terrestrial tectonism. The first of his papers to leave a permanent mark came out in 1906 and announced nearly everything Reverend Osmond Fisher had inferred less than two decades earlier and evidently in complete ignorance of them.

Ampferer began by showing why the contraction theory was inadequate. His main reasons were that the earth's crust was manifestly too weak to bear the implied stresses and that the plan-views of the orogenic belts were mostly impossible to account for by all-sided compression. He then considered local volume changes of rocks as bringing about orogeny (à la Reade, 1886)[414] but found this mechanism also inadequate. Finally he investigated the hypothesis of his countryman, Eduard Reyer (1849–1914)[415], of generating uplifts and gravity-driven slide-sheets off such uplifts (Reyer, 1888, especially p. 787; 1892). Though geometrically promising, this hypothesis did little to account for the well-documented deep deformation in mountain belts. The glide-horizon had to be carried deeper into the earth: "When we bring the glide-path from the area of upper, rigid rocks into those depths, where, owing to plasticity, sideways movements can be initiated with great ease, we have the theory of undercurrents before us, which includes gliding as a partial feature" (Ampferer, 1906, p. 601). From this, Ampferer concluded that the surficial movements we see reflected in the structural geology of the crust are nothing but passive motions, coupled to the active motion of the substratum. This active "flow" of the substratum can have a variety of causes. Ampferer first thought of compensating movements, material moving

Figure 127. Otto Ampferer (1875–1947) reached many of Revered Osmond Fisher's conclusions regarding subcrustal convection and falcogeny, but in apparent ignorance of Fisher's work.

away from areas of crustal thickening or from vast areas of subsidence, much along the lines of Dutton's thinking. He believed the existence of subcrustal density differences could generate such motions (influence of von Humboldt on this well-informed theoretician?). Finally he mentioned "thermal mass currents" (Ampferer, 1906, p. 603) as a possibility.

Ampferer took a simple laccolith as the simplest form of mountain-building and also as an illustration of what subcrustal currents might do. He pointed out that, at the next larger scale, every area of shortening must correspond with an area of extension, citing the close association of volcanism with orogeny to support his statement[416]. However, Ampferer says, volume changes in the subcrust can generate large areas of uplift or subsidence, as well as setting subcrustal currents into motion. These events would not only bring about mountain ranges such as the Alps or rift valleys such as those of East Africa, but also would cause gentle and slow up-and-down motions of entire continents and oceans, or parts thereof (Fig. 128). It is these slow and recurrent motions, Ampferer thought, that generate whole sequences in the stratigraphic record, which resemble each other so much. If it were only for mountain- and rift-building and the atmospheric and hydrospheric agencies, Ampferer believed that it would have been impossible to form stratigraphic sequences correlatable at inter-continental distances. Consequently, he thought that he answered Suess' emphatic question of 1883 (p. 10 ff.) as to the origin of the "universal formations."[417]

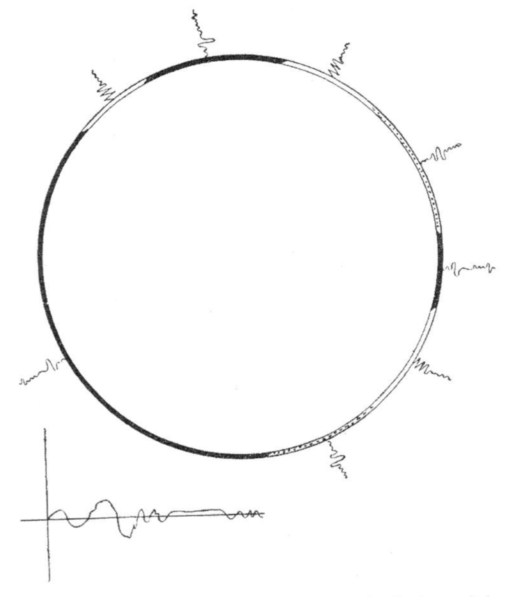

Figure 128. Ampferer's (1906, fig. 41) depiction of the oscillations of individual parts of the earth's crust in response to the movements of the subcrust. Each curve drawn above a crustal panel represents its history of oscillations about a mean sea level (shown in the small coordinate system drawn in the lower left-hand corner). Notice that, although each oscillation is different, there are some correspondences. It is these correspondences that create, according to Ampferer, both the universal formations and the universal unconformities. Ampferer thus thought that sea-level changes are not due to eustatic movements, as Suess had suggested, but to fortuitous correspondances of vertical movements of continents created by undercurrents.

Even now we have not yet been able to understand the relationship between the sublithospheric convection and terrestrial topography (see especially McKenzie, 1994). Mantle plume-generated uplifts are only one, and possibly a small, part of the whole thing. After Ampferer, no one looked at the same problem from the same broad perspective until the rise of plate tectonics. Although Ampferer continued his prophetic writings well into World War II, his theories made little impact on the international geological community simply because many of his predictions were hard to test, both technologically and theoretically. Ampferer needed to be rediscovered after plate tectonics (White et al., 1970; Davis et al., 1974; Şengör, 1977; Thenius, 1980, 1988).

Hans Stille: The Reincarnation of Élie de Beaumont

Hans Stille (1876–1966: Fig. 129)[418], perhaps the most influential tectonicist of the first half of the twentieth century, agreed with his contemporary Argand that both orogeny and epeirogeny were ultimately due to crustal shortening. However, he followed Élie de Beaumont, Contejean, and his French idol Emile Haug (1907, p. 531) in believing the distinction to be important. In a classical paper in 1919, Stille listed the differences as follows:

Orogeny	Epeirogeny
1. Orogeny has small wavelength but large amplitude.	1. Epeirogeny has large wavelength and small amplitude.
2. Orogeny alters the fabric of rocks.	2. Epeirogeny leaves the fabric of rocks intact.
3. Orogeny is episodic.	3. Epeirogeny is secular.

Stille certainly knew that he was standing in the footsteps of Élie de Beaumont, although I do not think that he was aware that his criteria for separating orogeny from epeirogeny were *exactly* the same as those which Élie de Beaumont had employed in separating his *bosselements* from his *ridements*! Stille, coming from a central-north German tradition (Carlé, 1988) remained completely attached to the Beaumontian view of the world and by-passed Suess (for a more detailed discussion of Stille's intellectual pedigree, see Şengör, 1991c, 1996, 1998). Nevertheless he did learn from Suess the value of worldwide comparative geology. One of his students, Hans-Joachim Martini, once said that Stille worked the international geological literature by the ton!

Stille was aware that his (and therefore Gilbert's) orogeny-epeirogeny distinction was not water-tight. He pointed out, for

Figure 129. Hans Wilhelm Stille (1876–1966), the "*Geologenpapst*" (the Pope of geologists) established and extended Gilbert's distinction of epeirogeny from orogeny in the twentieth century.

Figure 130. Leopold Kober (1883–1970) developed the orogen concept in the way that Stille later so effectively used and that helped to bring epeirogeny as a distinct process into much sharper relief.

example, that large orogenic folds would hardly alter the fabric of the rocks affected (Stille, 1919, p. 197–198). Yet he pleaded for a *total employment* of his criteria (see especially Stille, 1924, p. 16–17). This was not understood by Gilluly (1949; 1950, especially p. 103–104), who consequently misled generations of North American geologists about the significance of Stille's distinction. When one uses Stille's criteria as he recommended, it is surprising how successfully it distinguishes what we today consider orogenic[419] from epeirogenic[420] events.

From Kober[421] (1921, 1928; for a comprehensive contemporary review of 1921 in English, see Longwell, 1923; Fig. 130 herein) in Europe to Bucher[422] (1933; Fig. 131 herein) in North America, variants of Stille's ideas on long-wavelength structures as weaker expressions of horizontal shortening caused by contraction of the earth held sway in the 1920s and 1930s. Argand (1924) brushed them aside and considered all vertical motion to be a by-product of crustal thickening and thinning through horizontal motions *caused by continental drift*. The difference between Stille's and Argand's ideas was more than just a matter of degree. Argand envisioned extensive, semi-penetrative deformation within his giant basement folds, which not uncommonly appeared as a stack of basement nappes. The Austroalpine nappes and the Cenozoic Tien Shan were his prime examples; Stille would have considered them orogenic, not epeirogenic. That is why, as my quotation above from his 1912 paper shows, Argand did not think the orogeny-epeirogeny dualism fundamental. Though he was much admired and often quoted, he was seldom understood. Epeirogeny, in the minds of

Figure 131. Walter Herman Bucher (1888–1965), who dominated, together with Marshall Kay, the thinking on "classical tectonics" in North America in the mid-twentieth century.

most geologists, remained Gilbertian/Stillean until the rise of plate tectonics in the 1960s.

Africa: Birthplace of Magmarsis

While geologists active in Eurasia were holding fast to horizontal shortening (e.g., Stille) or to shortening and extension (e.g., Argand) to explain everything, those working in Africa were impressed with the *primary vertical motions* that had created a swell-and-basin topography during the Cenozoic (Krenkel, 1922, p. 163). Chavanne[423] (1879, p. 625) had already perceived that the Sahara, for example, was not a simple trough ("*Mulde,*" in reaction to such writings as Zimmermann's? See endnote 244—and Ansted's?—see endnote 424), but was divided into sub-basins by systems of elevation ("*Erhebungssysteme*")[424]. Chavanne's terminology betrays Élie de Beaumont's influence (see especially his section entitled *Geologie der Sahara, Ursprung der Wüste. Entstehung und Bildung der Dünen*: 1879, p. 625 ff.), but underlines the importance of uplift. This, Chavanne thought, was a result of the peculiar structure of the continent. He later wrote: "No other continent has such a massive structure. Neither the Himalayan System with the Tibetan Highland and the Pamir Plateau in Asia, nor the Cordilleras and the Rocky Mountains with the table-lands of South America and the high surfaces of North America and the Anahuac Plateau in Central America have a similar effect on the structure of the continent as the highland masses of South and East African highlands" (Chavanne, 1881, p. 38). In fact, when he drafted a hypsometric map of Africa and computed an average elevation of 661.8 ± 21 m, he thought Africa was the highest continent (Chavanne, 1881, p. 37). A decade earlier still, Fraas (1867, p. 33) had compared—no doubt with the famous figures 27 and 28 of Dufrénoy and Élie de Beaumont (1841, p. 436–437) in mind—the two sides of the Red Sea with the Vosges and Black Forest bordering the Upper Rhine Rift. In a summary map of the hypsometry of Africa taken from Chavanne's work, Baron von Schweiger-Lerchenfeld (1886, Plate II) was able to present to a wide public the whole of the basin-and-swell structure of Africa as outlined by Chavanne (1881, fold-out map) and almost as it is known today (see Burke[425], 1996, for an outstanding review). This inspired many to see an almost mosaic structure in the tectonics of Africa characterized by broad basins and narrower uplifts separating them (e.g. Arldt, 1919, fig. 69 on p. 557; Krenkel, 1925, fig. 4; Cloos, 1937; Holmes, 1944, fig. 223; Brock, 1955).

Against all attempts to explain these elongated uplifts as real anticlines and the depressions as synclines (see the superb review of all opinions on the origin of the east African Rift system and the uplift that accompanies it *in* Krenkel, 1922, p. 154–161), Krenkel[426] pointed out that these uplifts and depressions were, at best, results of "epeirogenic crustal movements" (p. 162), following a fairly common view in Germany then concerning relations between gentle but large-scale uplifts and fracturing of the lithosphere (e.g., Machatschek, 1918). The formation of these swells had not proceeded uniformly. One of the reasons of their formation "could be the upwards drive of magma: its Atlantic type rocks were involved in an upwards movement, carrying their ceiling upwards" (p. 177). This swell-building may have caused a summit (or "extrados") extension on top of the swells to form the east African rifts, *but Krenkel followed Suess (1891) and Machatschek (1918, p. 23) in doubting whether the uplifts could generate enough extension to create the rifts (Krenkel, 1922, p. 178).* He later concluded, contradicting another view of Suess, that the *uplift of the entire continent* and its division into basins and swells—a process which he called *magmarsis* (Krenkel, 1934, p. 1007; 1957, p. 427, 451)—must have been the reason for rifting, for which he coined the term *taphrogeny* (Krenkel, 1922, p. 181 and footnote 1)—and vulcanicity.

Around the time of Krenkel's publications, ideas that uplifts were caused by "magma" (almost à la von Buch) were becoming popular again (e.g., Salomon, 1925, and the rich literature references therein), and these ideas unfolded more productively in the 1930s.

The Rise of Primary Vertical Tectonism in the Twentieth Century: Erich Haarmann and Hans Cloos

Stille's fellow student and later "apprentice," Erich Haarmann[427] (Fig. 132), connects most directly the work of his master—and, in an indirect way, Krenkel's and other "magmatists" work—with the plume-generated uplift theme. Haarmann agreed with the distinctions in terms of structure families that Stille erected, (compare Stille, 1918, with Haarmann, 1926a; also see Haarmann, 1926b, p. 135, table; 1930, fig. 1 and p. 191–192), but disagreed entirely with the mechanism Stille proposed for their origin, namely universal crustal shortening. Haarmann pointed out that there was too much normal faulting all over the place, too much ambiguity in crustal movements, and too many large oval negative and positive structures, such as the ocean basins and continents, to be accounted for by all-sided shortening (although he did not discuss Diener's {1886, p. 398} peculiar model of dome-building by all-sided compression in platform regions, invented no doubt to be able to account for uplifts in Suess' no-primary-uplift-world[428]). Haarmann instead proposed, much like Dutton earlier, that all original relief was due to primary vertical movements, the growth of what he called *geotumors* and *geodepressions* (Fig. 133). Orogeny and taphrogeny were secondary affairs, caused by gravity-sliding off the rising geotumors (Fig. 134). Geotumors took the shape of elongated uplifts when they occurred along continental margins and were associated with sialic magmatism; this kind of geotumors created orogens. Oval or round tumors were encountered in forelands or hinterlands of orogens. Geotumors were mostly associated with sialic magmas, whereas geodepressions were associated with mafic magmas (Haarmann, 1930, especially p. 74 ff.). Haarmann called the phenomena associated with the vertical movements *primary tectogenesis* and those associated with derivative horizontal slidings *secondary tecto-*

Figure 132. Erich Otto Haarmann (1882–1945) inspired Hans Cloos to his famous study of uplift-fracturing-volcanism. (Photograph from the *Geologischen Rundschau*, v. 33, facing p. 6.)

genesis. He considered the main cause of tumor- and depression-building as being the swinging of the poles, which created a disturbance in the isostatic balance of the crust. Tumors and depressions formed while the earth tried to even out the differences caused by pole shifts.

Haarmann's thesis found little support and, following a thorough and damning criticism by nearly all the leading German geologists[429], would have been committed to total oblivion—notwithstanding some later Dutch and Russian followers such as M.M. Tetyayev (1882–1956)[430], his student V.V. Beloussov (1907–1990)[431], and R.W. van Bemmelen (1904–1983)[432]—had not one great geologist, Hans Cloos (1885–1951)[433] (Fig. 135), found much to recommend in Haarmann's analysis, dedicating to him a work that was to become an influential classic.

Cloos (1939) studied the result of the updoming of large areas (~1000 km in diameter, similar in dimensions to Haarmann's geotumors) in the field and through clay cake experiments. He believed that the generation of the dome would impart significant enough stresses onto the upper parts of the dome to cause rifts to originate[434]. He was aware of Élie de Beaumont's hypothesis concerning the origin of the Upper Rhine Rift and evidently liked it (Cloos, 1939, p. 428). Extending Élie de Beaumont's hypothesis, Cloos concluded, apparently independently (he does not cite Élie de Beaumont's idea of *étoilement* {~star-making}), that star-shaped rifts (*Grabenstern* {graben star}of

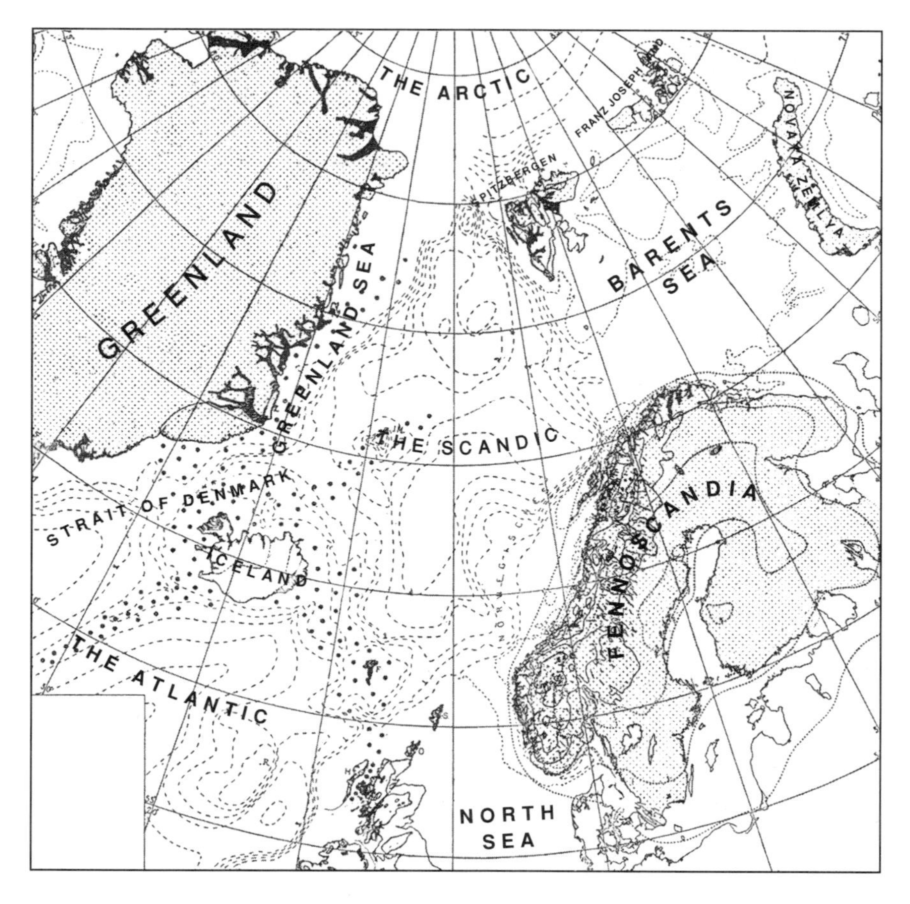

Figure 133. The Scandian zone of subsidence with marginal centers of uplift (after Haarmann, 1930, fig. 15; English lettering by me). The map shows, according to Haarmann, geotumors and geodepressions. Bordering continents, as well as Iceland, are zones of tumor-building. Haarmann, following his fellow student and teacher Stille, always maintained that the present *glacioisostatic* rise of Scandinavia had *also* a tectonic cause (see his fig. 7 *in* Haarmann, 1930). The oceanic deeps are considered geodepressions. This map thus gives a good idea of Haarmann's concept of geotumor and geodepression and of their dimensions, although some of his geotumors and geodepressions were more elongate (see his figs. 13 and 14 {*in* Haarmann, 1930}, depicting the deeps in the northern Caribbean, which he interpreted as geodepressions).

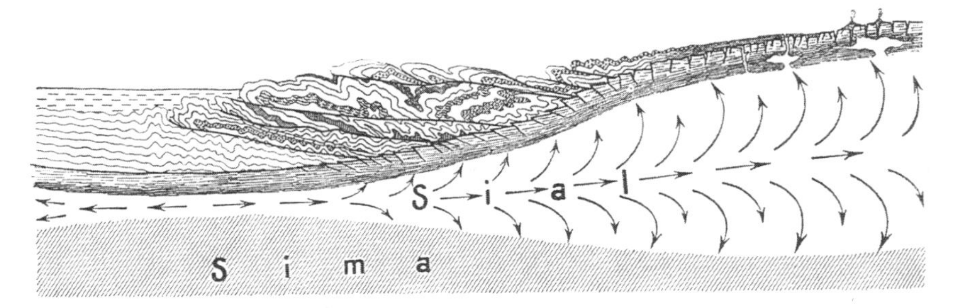

Figure 134. Anatomy of a geotumor causing tectonic denudation, crestal extension of the crust, and volcanicity (from Haarmann, 1930, fig. 34). Around the geotumor, tectonic denudation gives rise to superficial shortening, which Haarmann believed was the cause of the compressional structures seen in orogenic belts. This figure of a rising geotumor may have inspired Hans Cloos (see below).

Cloos, 1939, p. 513; also see the caption to Fig. 139 herein) are likely to form near the center of rising domes:

> Result: The calculation of the magnitudes in sufficiently well-known graben regions prove, first, that through crustal arching grabens (and fissures) can originate and under these conditions must originate. Further, they inversely show that under certain conditions, compatible with our present fundamental concepts, the amount of crustal extension expressed in the building of grabens is compensated by the doming of the area and that outside it an autonomous crustal extension may not be inferred. (Cloos, 1939, p. 435; see Fig. 139).

Although this fundamental statement is cautiously worded, it is still surprising, *for the only experiment Cloos conducted to true scale did not lead to graben formation at all* (Fig. 136)! The pictures, showing beautiful grabens dissecting the apical region of a dome in the clay cakes, were taken from experimental domes that were unrealistically steep (Fig. 137)[435]. Here it is important to realize that Cloos took the present elevation of the rift shoulders (see Fig. 138) as being very close to the elevation to which they had been uplifted prior to rifting. That this probably has never been so was discussed already in a superb paper by Stephen Taber in 1927, which followed the insight of Le Conte (1889), cited above. Cloos cites Taber with approval (Cloos, 1939, p. 426)[436], and he gives persuasive examples of how he dealt with this problem in East Africa (Cloos, 1939, p. 439 ff. and fig. 24). However, he still thought he found considerable updoming after this "antithetic rotation" (Cloos, 1928, 1932, 1939, p. 425) had been taken out.

Figure 135. Hans Cloos (1888–1951) was one of the three great German-speaking masters of tectonics in the mid-century (the others were Hans Stille in Berlin and Bruno Sander in Innsbruck). Cloos' work of falcogeny-rifting relationships has deeply influenced thought into the present.

Cloos calculated, with the help of a mathematical colleague, that the maximum amount of extension of an 117-mm-thick mud cake, uplifted by 1 mm in the center of a dome of radius 600 mm, was between 0.5 mm and 1 mm (Fig. 138). He showed that the same amount of doming would lead to greater amounts of extension as the thickness of the domed cake increased (Cloos, 1939, p. 432). So, if his mud cake were only 30 mm thick (instead of 117 mm), the amount of extension would have been only 0.1 mm! I cannot see how he hoped to generate the grand rift complexes he was studying by this method, or why he chose to ignore Suess' (1891), Machatschek's (1918), and Krenkel's (1922) earlier and well-justified misgivings about the ability of crustal doming to generate extensive taphrogens[437]. These rifts, he thought, cut through the crust and led to magmatism, the largest amount of magma being supplied to the surface near the center of the dome, thus coming very close to the crustal expression of Krenkel's magmarsis.

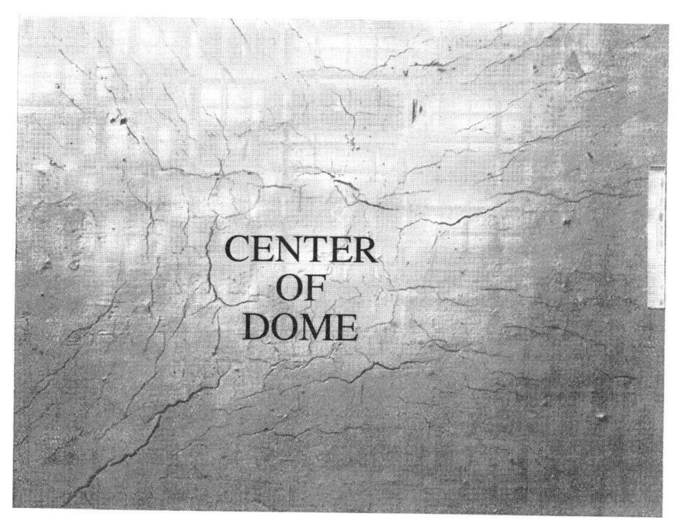

Figure 136. "Cracking by very little doming" (from fig. 17 and 18 *in* Cloos, 1939). The upper picture represents an uplift of 2 mm of a 117-mm-thick clay-cake forming a dome having a radius of 600 mm. In the lower picture, the dome rose another millimeter. The total maximum extension achieved was 1/600! The lithospheric domes pretty much represent these proportions, except the thickness. But extension in doming decreases as the thickness of the extended layer decreases. Thus, this experiment shows that doming cannot be the cause of rifting.

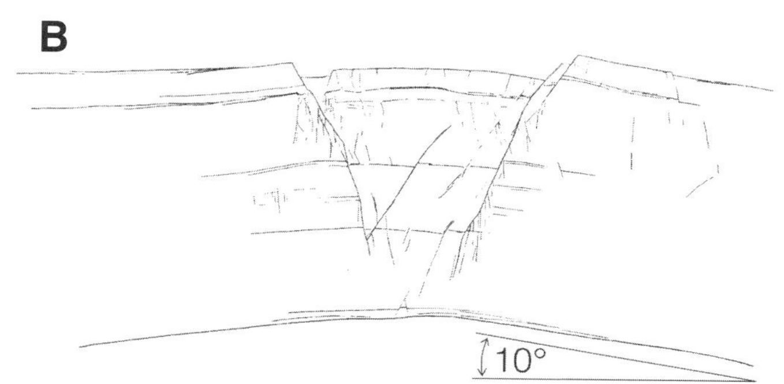

Figure 137. A: Photograph of a graben formed by updoming in a clay-cake experiment (Cloos, 1939). See Figure 137B herein. B: Sketch of a similar experiment (from fig. 16 *in* Cloos, 1939). The flank uplifts of this dome are 10 times steeper than any of the domes in Africa, with or without crowning rifts. Although Cloos does not give the flank dips of the experiment illustrated in Figure 137A here, it is clear that it too has flank dips of ~10°. The conditions of these experiments do not, therefore, reproduce the conditions in nature. It is surprising how seductive such pictures as Cloos reproduced were for the geological community. That is why their similarity to natural examples has not been challenged seriously (although both Dan McKenzie and Xavier Le Pichon told me that they had never taken seriously the idea of doming-related stretching as being the cause of rifting).

Cloos, very much like Krenkel before him, was emphatic in his insistence of the complete independence of the vertical forces that create the domes and the associated extensional structures from the dominantly horizontal tectonics seen in the mountain ranges (see especially Cloos, 1939, p. 496 ff.; Fig. 139 herein). First, he pointed out that regions that were former sites of orogeny were now parts of rising domes as in Africa (site of mostly late Precambrian orogeny) or in the Rhenic shield (site of late Paleozoic orogeny), so there was a clean *separation in time* between orogeny and epeirogenic uplift. The extensional structures were locally controlled by older structures, but there was absolutely no genetic association. Rifts were not influenced by orogens at high angles to them: both the angles and the distances of the alleged cases (e.g., Weber, 1921, 1923, 1927) were too varied to justify a causal link. Some geologists tried to assimilate Haarmann's tumors into the basement folds of Argand (1924), or the grand folds (*Großfalten*) of Abendanon (1914). That did not work either owing to the diverse geometric relations of the tumors to their allegedly associated mountain belts. There was therefore a *separation in space* between orogeny and epeirogenic uplift also, though not so clean as the temporal separation. At best, Cloos thought (as his great predecessor Gilbert had done earlier) that both mountain belts and the geotumors could be the horizontal and vertical manifestations of a common mechanism. He hoped the future would show what that mechanism might be.

Figure 138. "The southwest corner of Germany, as seen by the geologist" (Cloos, 1947, fig. 48). Cloos' famous panoramic view of the southern end of the Upper Rhine Rift as seen from the west-northwest. It was mainly this structure that he was trying to reproduce in his experiments in Cloos (1939; see Figure 137). For a block diagram that places this view in a broader context, see Cloos (1940, foldout *Texttafel*).

Cloos' work has been immensely influential, not only because it carried an air of precision and accuracy hitherto generally lacking in similar studies, but also because its author had an enviable reputation in the world geological community as a superb field geologist and a careful experimenter. In addition, Cloos, like Argand, was a consummate artist, as demonstrated in Figures 138 and 139 herein. As one reads through his text—written in clear and captivating German—and views his figures, it is not always easy to free oneself from his spell. Many of the textbook writers after World War II (from which he emerged as an anti-Nazi hero) fell under that spell. In consequence, his interpretation of grabens and rifts has become almost standard (e.g. Umbgrove, 1947, p. 308–311; Beloussov, 1948, p. 420–422; de Sitter, 1956, p. 144–146; Metz, 1957, p. 75–76; Hills, 1963, p. 190–191) and even when gentle criticism was raised, it hardly touched the essential point (e.g. Ashgirei, 1963, p. 260–261).

With the work of Hans Cloos, it was reemphasized that the terrestrial tectonics manifested itself in two distinct families of structures (as his fig. 60, reproduced here as Fig. 139, so dramatically illustrates). One family was related to the world's large orogenic belts and was clearly the result of great horizontal motions. The other, the large domes—in places crowned by rifts, in others by volcanoes, in some by both—looked as if it was independent of the horizontal commotion and was the expression of slow, but long-lived vertical motions.

The Barren Interlude Between Cloos and Wilson in the Study of Falcogenic Structures

The 1950s were a low ebb for the study of vertical motions, except in the Soviet Union, where Beloussov (see the dedication herein) used the mammoth Soviet work on the thickness and facies of sedimentary rocks in platform regions (the main results of which reached the world geological community in a number of high-quality paleogeographic atlases) to undertake a detailed analysis of the vertical oscillations of these otherwise serenely stable regions (Beloussov, 1948, 1954, translated into English, with slight modification in 1962, 1975, 1976, 1981; also see 1989). Beloussov's main field-work was in the Greater Caucasus where, particularly in the central and eastern parts of the chain, he had convinced himself of the absence of primary horizontal motions. His thinking followed a path not dissimilar to van Bemmelen's and their conclusions were coincident in all essential points.

Outside the Soviet Union, this mode of work made little impact, not only because the only other available large platform area from which there was a wealth of data was the North American craton, but also because of the mobile-belt-dominated tradition of tectonic research in the west. Indeed, at the Cloos memorial meeting in Bonn, Germany, in 1975, Rudolf Trümpy called Hans Cloos an "anomaly" because he had not been interested in the Alps! In fact, when at last serious interest on platforms arose in the west, it came from a North American stratigrapher (Sloss, 1963, 1964, 1966, 1972; also see 1988b), following the footsteps of two others (Schuchert, 1910, 1923, 1955; Pirsson and Schuchert, 1924; and Kay, 1951). I think King's (1955) brief but excellent review accurately outlined the common wisdom on orogeny and epeirogeny in the western world: he summed up the essence of orogeny and epeirogeny in his definition of the latter as "those broad upwarps and downwarps of regional dimensions which fail to produce folded structure" (King, 1955, p. 734). When we remember

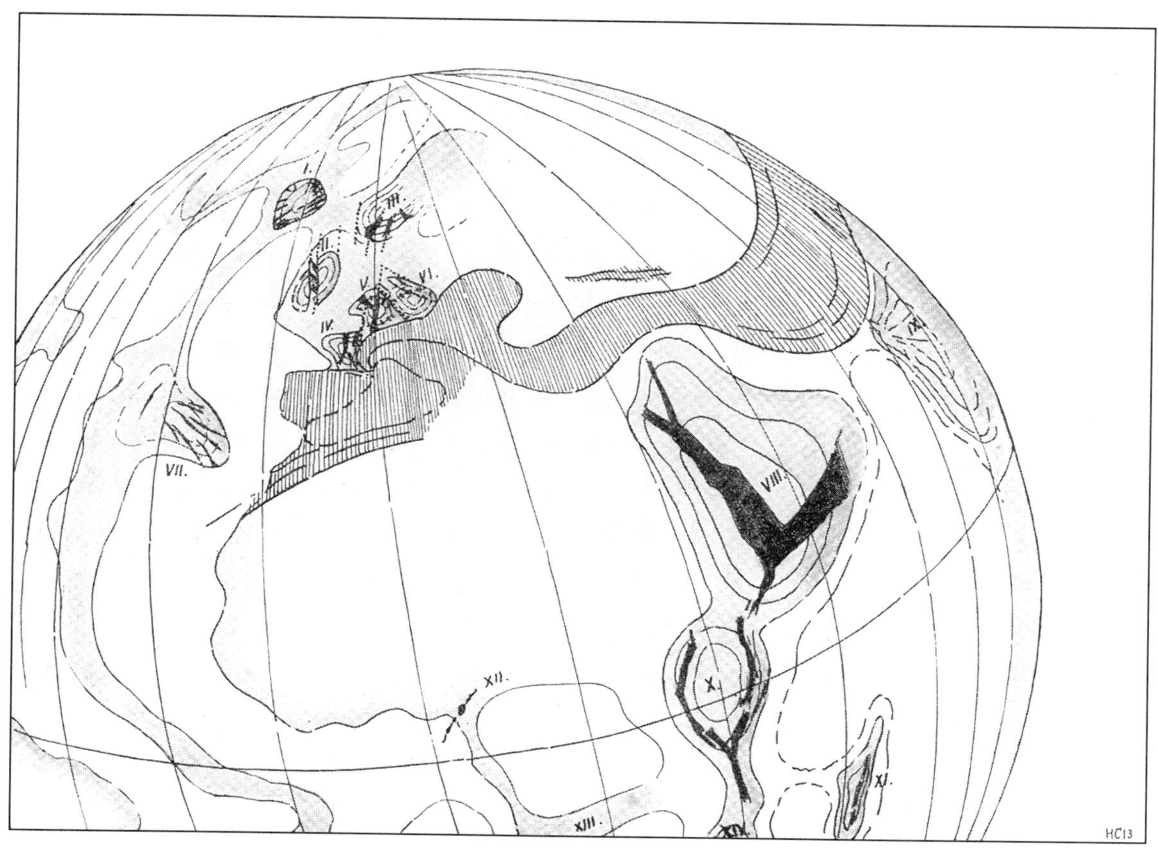

Figure 139. "Sketch showing the positions of a number of uplift-units (Tumors) of the Old World, in their relation to the young foldbelts and to other traces of horizontal motion. Highly simplified and schematised. Heavy lines and ruling: Late Mesozoic-Tertiary folded chains; weaker lines and shading: Positive fields (Domes, Shields, Tumors); black: Grabens and volcanic fractures in them; dotted lines (in Europe): Old sutures in and between fields, for example, with strike-slip faults. I. Iceland with volcanic fractures; II. British Shield with eruptive regions in the north; III. Scandinavian Shield with the grabens and fractures of Oslo, Vettern and others; IV. Shield of the French Central Plateau with grabens and volcanic fractures; V. The Rhenic Shield with grabens and volcanic fractures; VI. The Bohemian-Saxonian unit with NW-fractures and the Bohemian-Silesian volcanic arc; VII. The Azores-Shield with volcanic fractures branching off the Mid-Atlantic Ridge; VIII. The Arabian-Nubian Shield, with the large graben-star, strongly volcanic; IX. The Indian Shield with the feeder fissures of the Deccan eruptives; X. The East African Shield with peripheral, partly richly volcanic grabens; XI. The high region of Madagascar, with fractures and volcanicity; XII. The volcanic Cameroon-Line; XIII. The Kuanza Vertex, between the Congo- and the Kubango-Field; XIV. The Rhodesian Massif; XV. The Damara-Vertex between Kubango- and Oranjefeld (not visible in the figure), with the fractures of the Waterberg-Line and the volcano-plutons of Brandberg, Spitzkoppe, Erongo, and the others." (Cloos, 1939, fig. 60).

that for King's generation, "folded structure" meant folded, cleaved, thrust, and metamorphosed (cf. Şengör, 1990, p. 3, note 1; also see the quotation from Dutton above p. 196), we see that in the 1950s, one generally adhered to Gilbert's definitions of the two processes.

In Europe, it was also largely the opinion, owing to Stille's influence. King (1955) agreed with Gilluly (1949) in thinking that orogeny was only locally episodic, but King's emphasis was more on episodicity than on continuity, which he correlated with epeirogenic events elsewhere on the same continent. He cited Bucher (1933) in support of the view that both orogeny and epeirogeny were ultimately the result of crustal compression, emphasizing that the understanding of epeirogeny by 1955[438] had not come forward one centimeter since Argand and Stille. Cloos had not made much headway in that regard.

At this point, the historical survey of ideas on what I choose to call falcogenic movements ought to come to an end because the next significant step in understanding their nature was taken within the framework of plate tectonics. A perusal of the post-World War II tectonics texts (e.g., Umbgrove, 1947; Jacobs et al., 1959; Goguel, 1962) supports this conclusion. However, since I have undertaken the above review as a prelude to investigating mantle/plume-related falcogenic structures, I think I must bring this review to a close with the discovery of mantle plumes, which was essentially simultaneous with the discovery of plate tectonics.

CHAPTER XV

J. TUZO WILSON AND THE MANTLE PLUMES

Mantle plumes were originally postulated by J. Tuzo Wilson (1963)[439] (Fig. 140) because no other explanation then available accounted for the geometry and the age progression along the Hawaiian ridge, which by that time had become well-known (e.g., Chubb, 1957). Wilson's suggestion was to postulate a magma source, possibly within the stagnant center of a jet-stream-type convection cell within the mantle, that moved more slowly than did the jet stream of the main convection driving the continental drift (Fig. 141). Wilson wrote that ocean islands, the surface expression of the conduit rising from the mantle, were "in fact arranged like plumes of smoke which are thus being carried down-wind from their sources" (Wilson, 1965b, p. 158)[440]. In his early papers on plumes, Wilson was more concerned with using their traces as guides to plate reconstructions than with investigating all of their geological corollaries. This remained the main concern of all those interested in plumes until 1973 (e.g., Morgan, 1971; 1972a, 1972b).

Jason Morgan (1971, 1972a, 1972b), who focused mainly on the oceans, and Kevin Burke (Burke and Wilson, 1972; Burke and Whiteman, 1973; Burke and Dewey, 1973, Burke et al., 1973) who focused mainly on the continents, linked with Wilson's theory of mantle plumes (Wilson, 1963, 1965b, 1973) the large, roughly elliptical elevations characteristic of Hawaii or the Neogene development of Africa. Morgan was led to the postulate of oval uplifts over plumes of 1000 km in diameter, based on the "Hess Gravity Theorem" (i.e., that one does not need to have a gravimeter to measure gravity; one needs only to look at the topography: Morgan, 1972a). But Morgan was suspicious of continental topography. "Whether the unusually high Tibetan Plateau or southern Africa should be considered symptomatic of a subcontinental hot spot is open to question; the more uniform oceans are more amenable to this type of analysis" (Morgan, 1972a, p. 19). This statement expresses a fundamental truth concerning the differences between the oceans and the continents, yet it betrays also a fairly naïve view of continental topography. Kevin Burke later remarked (written communication, 2000) that "There is a more serious problem with submarine elevations. Erosion does not reduce elevation below sea level so it is hard to distinguish dynamically maintained from ancient elevations, especially where carbonate banks overlie volcanic material."

Burke developed his view about the plume-uplift association during the work he did with his co-workers on the evolution of the Benue Trough and the Gulf of Guinea (in the framework of the Benue Valley Project of the Geology Department of the University of Ibadan that he had initiated in 1966). That uplifts are associated with plumes and form one stage in a sequence of events eventually leading to continental disruption and ocean opening was first mentioned by Burke in a discussion remark made in December 1970 during the Conference on African Geology in Ibadan: "Rift valleys are not something unique and unrelated to ocean opening. Some rifts like the Cameroon and the East African are at too early a stage to show any evidence of opening and the only structures they reveal are

Figure 140. John Tuzo Wilson (1907–1993), inventor of the theories of plate tectonics and of mantle plumes.

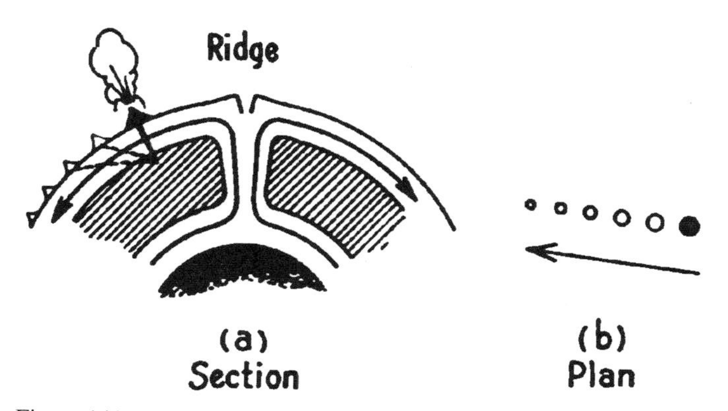

Figure 141. "Diagram to illustrate that if lava is generated in the stable core of a convection cell, and the surface is carried by the jet stream then one source can give rise to a chain of extinct volcanoes even if the source is not over a rising current. This is proposed as a possible origin of the Hawaiian chain of islands" (from Wilson, 1963, fig. 5).

related to uplift. Others ceased to be active before their development had gone very far, for example, the Mesozoic rift of southern East Africa. The Cretaceous Benue rift was one of the latter. It developed farther than its East African contemporary but still closed at an early stage" (Burke et al, 1972, p. 202).

In a paper written with Tuzo Wilson and William S.F. Kidd in 1973, we read that the authors "use the term 'hot spot' to describe succinctly a class of localized volcanism and associated uplift characteristically found within plates (Hawaii, Tibesti, for example), but also found on divergent plate boundaries (Iceland, for example)" (Burke et al., 1973, p. 133). Here, uplift is expressly associated with intra-plate volcanicity, but "The term [*hot spot*] is used to describe the surface feature, with no intended implications about processes below the surface" (Burke et al., 1973, p. 133).

In another paper in the same year, Burke, in co-authorship with another veteran of African geology, Arthur Whiteman, expressly associated mantle plumes with overlying oval uplifts (Burke and Whiteman, 1973). In that paper, Burke and Whiteman used Hans Cloos' model for uplift-rifting-volcanism to generate three or more armed rifts above lithospheric uplifts of several hundred kilometer width and almost universally 1 km basement uplift. If one assumes a 40-km-thick layer that would rift, the maximum amount of extension one could obtain for any of Burke and Whiteman's rifts using Cloos' mechanism is ~1.6 km, if uplift diameter is taken to be 50 km (which is the shortest dimension available in their Table 1 listing the "Physical characteristics of some African uplifts" {*in* Burke and Whitman, 1973}). It is allegedly the width of the Ahaggar uplift. In case of three-armed rifts, the dome-generated extension should be distributed among the three arms. In all of Burke and Whiteman's other uplifts, the amount of extension that can be generated by Cloos' mechanism is much less. Yet, Burke and Dewey (1973) employed the same mechanism to generate a wide variety of rifts atop present and past uplifts, as a prelude to continental fragmentation.

Work on falcogenic domes associated with igneous activity has shown in the latter half of the twentieth century (cf. also Şengör, 2001a) that (1) Cloos' mechanism does not work; (2) nevertheless, possibly owing to the relationship of the plume activity to the circulation driving the plates and the gravitational potential of plume-related domes, the sequence of events of uplift-volcanism-rifting-drifting (suggested by Burke and his numerous collaborators) seems to be valid; and (3) in the continents, large, round falcogenic domes are the best indicators of the presence of plumes.

CHAPTER XVI

CONCLUSIONS

Sea-level change has been inferred very early in the history of thought—so early that the origin of the idea is lost among mythic speculations. Vertical motions of the rocky surface, with respect to a reference fixed to the earth, have been much more difficult to come to grips with, owing to the difficulty of establishing a practicable point of reference and the selection of indicators showing distance to the selected reference in the past. The earliest models of the movement of land with respect to sea level were based on the observation that, in some areas land gained on the sea (as in western Asia Minor) and that in the past, some of the present land areas also appeared to have been covered by the seas (as indicated by fossils in western Asia Minor, Greece, southern Italy, and northern Africa). Early models of vertical land movement involved mechanisms inspired by few and disconnected observations, but these models made a clear distinction between structures of small wavelength and structures of large wavelength. This distinction remained disputable so long as means of observation of large-wavelength structures remained inadequate.

Only the initiation of detailed biostratigraphy and geomorphological methods of slope investigation eventually placed the presence of large-wavelength structures beyond doubt and illuminated their evolution. This has happened in a satisfactory way only in the latter half of the twentieth century. Mantle plumes, first recognized by Tuzo Wilson in 1963, are responsible for large domal uplifts in otherwise stable parts of continents, and mantle plumes seem to generate similar domes on ocean floors. The stratigraphic and structural record of such uplifts may be the least ambiguous indicators of past plume activity.

Sengör (2001a) picks up the consideration of the plume-related topography where this book has left off, mainly by using stratigraphy and geomorphology with a view to establishing guidelines for the identification of plume fossils. However, the study of falcogenic structures of the lithosphere has become a much wider field of study than it had ever been before, owing both to increased interest in falcogenic structures (mainly because of oil company interest in large cratonic, rift, and continental margin basins and because of the great ease with which oceanic basement can be studied with respect to its age and subsidence history) and to the possibility of studying them in greater detail than ever before because of advanced technology. Our understanding of the behavior of the continental crust also greatly expanded, including flow processes of scales much larger than ever before imagined (except by Wettstein, Suess, Taylor and Argand). Now, the rise and fall of considerable areas of the continental crust (on the order of million square kilometers) in some parts of the world are believed to be caused by intracrustal flow (Block and Royden, 1990; Wernicke, 1990; Kaufman and Royden, 1994; Royden, 1996; Royden et al., 1997; Clark and Royden, 2000; McKenzie et al., 2000; McQuarrie and Chase, 2000). Such intracustal flow greatly complicates the definition of falcogeny as leaving the fabric of the affected rocks intact. It seems as if one can have falcogeny in the upper, brittle crust, while orogeny is occurring at depth and is actually causing the falcogenic movements of the brittle carapace. This realization comes closer to Argand's view of the nature of "epeirogeny" than to that of anybody else in the past.

Nature indeed never functions in the neatly compartmentalized divisions we try to impose on her. The divisions we create are only our mental aids to comprehend hers. It is of great importance not to confuse the two. This is perhaps the most important lesson that has emerged from the desultory tour we have just completed among the ruins of older conceptual edifices erected to comprehend the nature of the long-wavelength deformations of the lithosphere.

REFERENCES CITED

Abdel-Monem, A., Watkings, N.D., and Gast, P.W., 1972, Potassium-argon ages, volcanic stratigraphy, and geomagnetic polarity history of the Canary Islands: Tenerife, La Palma, and Hierro: American Journal of Science, v. 272, p. 805–825.

Abel-Rémusat, [*J.P.*], 1820, Histoire de la Ville de Khotan, tirée des annales de la Chine et traduite du Chinois; Suivie de Recherches sur la substance minérale appelée par les Chinois PIERRE DU IU et sur le JASPE des anciens: Paris, Imprimerie de Doublet, XVI + 240 p.

Abel-Rémusat, [*J.P.*], 1836, Foë Kouë Ki ou Relation des Royaumes Bouddhiques: Voyage dans la Tartarie, dans l'Afhanistan et dans l'Inde, exécuté a la fin du IVe siècle, par Chÿ Fä Hian. Traduit du Chinois et commenté par M. Abel Rémusat. Ouvrage Posthume revu, complété, et augmenté d'éclaircissements nouveaux par MM. Klaproth et Landresse: Paris, A l'Imprimerie Royale, LXVI + [2] + 424 + 5 foldouts.

Abendanon, E.C., 1914, Die Grossfalten der Erdrinde: Leiden, E.J. Brill, X + 183 + 1 page of errata.

Adams, E.B., 1963, Fray Silvestre and the obstinate Hopi: New Mexico Historical Review, v. 38, p. 97–138.

Adams, E.B., 1976, Fray Francisco Atanasio Dominguez and Fray Silvestre Velez de Escalante: Utah Historical Quarterly, v. 44, p. 40–58.

Adams, F.D., 1938, The Birth and Development of the Geological Sciences: Baltimore, Maryland, The Williams & Wilkins Company, v + 506 p.

Adams, K.D., Wesnousky, S.G., and Bills, B.G., 1999, Isostatic rebound, active faulting, and potential geomorphic effects in the Lake Lahontan basin, Nevada and California: Geological Society of America bulletin, v. 111, p. 1739–1756.

[Adıvar], Abdülhak Adnan, 1964, Fârâbî, *in* İslâm Ansiklopedisi, v. 4: Millî Eğitim Basımevi, İstanbul, p. 451–469.

Adıvar, A.A., 1982, Osmanlı Türklerinde İlim (4th edition, much enlarged by rich footnotes added by A. Kazancıgil and S. Tekeli): Remzi Kitabevi, İstanbul, 243 p.

Adler, B.F., 1910, Karti Pervobitnikh Narodov: Trudi Geograficheskago Otdleniya, Izvestiya Imperatorskayo Lubitelei Estestvoznaniya, Antropologiya i Etnografiya, v. 119, p. VIII + 679 p.

Afnan, R., 1969, Zoroaster's Influence on Anaxagoras, the Greek Tragedians, and Socrates: New York, Philosophical Library, 161 p.

Agricola, G., 1544[1956], De Ortu et Causis Subterraneorum, *in* Fraustadt, G., and Prescher, H., eds., Georgius Agricola Schriften zur Geologie und Mineralogie I, Georgius Agricola—Ausgewählte Werke, Gedenkausgabe des Staatlichen Museums für Mineralogie und Geologie zu Dresden, Herausgeber Dr. rer. nat. Hans Prescher, VEB Verlag der Wissenschaften, Berlin, p. 69–211.

Agrippa, C., 1584, Sopra la Generatione de Venti, Baleni, Tuoni, Fulgori, Fiumi, Laghi, Valli, & Montagne: Roma, Apresso Bartholomeo Bonfadino, & Tito Diani, 47 p.

Ahmed, F., 1976, Geosyncline concept in geotectonics: 63rd Session of the Indian Science Congress, Section of Geology and Geography, Part II, Presidential Address, p. 1–34.

D'Ailly, P., 1992, Ymago Mundi y Otros Opúsculos, edited by A.R de Verger: Madrid, Biblioteca de Colón II, Alianza Editorial, XX + 356 p.

Airy, G.B., 1855, On the computation of the effect of the attraction of mountain-masses, as disturbing the apparent astronomical latitude of stations in geodetic surveys: Philosophical Transactions of the Royal Society of London, v. 145, p. 101–104.

Akyol, İ.H., 1940, Tanzimat devrinde bizde coğrafya ve jeoloji, *in* Tanzimat I, Maarif Matbaasi, İstanbul, p. 511–571.

Alexandre, M., 1988, Le Commencement du Livre Genèse I-V—La Version Grecque de la Septante et sa Réception: Paris, Beauchesne, 408 p. + 23 photographs.

Allaby, A., and Allaby, M., 1990, The Concise Oxford Dictionary of Earth Sciences: Oxford, Oxford University Press, xxi + 410 p.

Allègre, C., 1988, The Behaviour of the Earth—Continental and Seafloor Mobility (translated by D.K van Dam): Cambridge, Harvard University Press, xii + 272 p.

Allen, D.C., 1963, The Legend of Noah—Renaissance Rationalism in Art, Science, and Letters; Urbana, University of Illinois Press, vi + [i] + 221 p.

Allen, P.A., and Homewood, P., editors, 1986, Foreland Basins: Special Publication no. 8, International Association of Sedimentologists, Oxford, Blackwell, [III] + 453 p.

Almagià, R., 1948, Monumenta Cartographia Vaticana, ... v. II (Carte Geographiche a Stampa di Particolare Pregio o Rarita dei Secoli XVI e XVII Essistenti nella Biblioteca Apostlica Vaticana): Città del Vaticano, Biblioteca Apostolica Vaticana, 130 + [I] p. + XL plates.

Alter, J.C., 1941, Father Escalante's map: Utah Historical Quarterly, v. 9, p. 64–72.

Ambrose, S.E., 1997, Undaunted Courage—Meriwether Lewis, Thomas Jefferson, and the Opening of the American West: Simon & Schuster, 521 p.

Ampferer, O., 1906, Über das Bewegungsbild von Faltengebirgen: Jahrbuch der k. k. Geologischen Reichsanstalt, v. 56, p. 539–622.

Anderson, A.J., and Cazenave, A., editors, 1986, Space Geodesy and Geodynamics: London, Academic Press Harcourt Brace Jovanovich Publishers, x + 490 p.

Anderson, A.R., 1923, Alexander's Gate, Gog and Magog, and the Inclosed Nations: Monographs of The Medieval Academy of America, Cambridge, no. 5, viii + 117 p.

Anderson, F.H., editor, 1960, Francis Bacon ... The New Organon and Related Writings: New York, Macmillan Publishing Company, xli + 292 p.

Anderson, G.A., 1988, The cosmic mountain—Eden and its early interpreters, *in* Robbins, G.A., ed., Genesis 1-3 in the History of Exegesis: Lewiston, The Edwin Mellen Press, p. 187–224.

Anderson, J.A., 1979, First through the Canyon—Powell's lucky voyage in 1869: The Journal of Arizona History, v. 20, p. 391–408.

Anderson, J.A., 1983 John Wesley Powell's *Explorations of the Colorado River* ... Fact Fiction or Fantasy?: The Journal of Arizona History, v. 24, p. 363–380.

Anderson, R.S., 1977, A Biography of Clarence Edward Dutton (1841–1912), Nineteenth Century Geologist and Geographer [M.S. thesis]: Stanford University, vii + 126 p.

Andrade, E.N. da C., 1950, Robert Hooke (Wilkins Lecture): Proceedings of the Royal Society (London), series B (Biological Sciences), v. 137, p. 153–187, plates 12–14.

Andreossy, (Comte), 1818, Voyage a l'Embouchure de la Mer Noire ou Essai sur le Bosphore et la partie du Delta de Thrace Comprenant le Système des Eaux qui Abreuvent Constantinople; précéde de Considérations générales sur la Géographie physique: Paris, Plancher, lxiv + 334 p. plus a separate oblong volume of VIII + [one foldout map] of plates.

Angel, M., presentation, traduction et commentaries, 1995, Saint Albert le Grand Le Monde Minéral—Les Pierres De Mineralibus (livres I et II), *in* de Lestrange, A., ed., "Sagesse chrétiennes": Paris, Les Éditions du CERF, 443 p.

Anonymous, undated, Gerardus Mercator Atlas 1595 Einführung, *in* Mercatoris Theatrum Orbis Terrarum, Coron Verlag, Lachen am Zürichsee, p. 5–31.

Anonymous, 1750, A Dissertation upon Earthquakes their Causes and Consequences ...: James Roberts, London, 44 p.

Anonymous, 1859, Topographical maps, profiles, and sketches, to illustrate the various reports of surveys for railroad routes from the Mississippi to the Pacific Ocean, *in* Reports of Explorations and Surveys, to Ascertain the Most Practicable and Economical Route for a Railroad from the Mississippi River to the Pacific Ocean. Made under the Direction of the Secretary of War, in 1853–1854, According to Acts of Congress of March 3, 1854, and August, 5, 1854., Senate, 33d Congress, 2d Session, Ex. Doc. No. 78, Washington, D.C., v. XI, iv + 46 unnumbered plates.

Anonymous, 1869, Karl Friedrich Naumann: Leipziger Illustrierte Zeitung, v. 53, no. 1368 (18th September), p. 221–222.

Anonymous, 1873, Dr. Carl Friedrich Naumann: Leopoldina, v. 9, p. 83–87.

Anonymous, 1889, Vorwort, *in* Die Europäische Türkei von Ami Boué (La Turquie de l'Europe par A. Boué, Paris, 1840): Deutsch Herausgegeben von der Boué-Stiftungs-Commission der Kaiserlichen Akademie der Wissenschaften in Wien. I. Band, Tempsky, Wien, p. III-X.

Anonymous, 1909, Thomas Mellard Reade: Geological Magazine, v. 46, p. 333–336.

Anonymous, 1910, William Phipps Blake: American Journal of Science, v. 30, p. 95–96.

Anonymous, 1915, Verzeichnis der von der Kaiserlichen Akademie der Wissenschaften in Wien Herausgegebenen oder Subventionierten Schriften: Wien, Alfred Hölder, VI + [II] + 567 p.

Anonymous [Cloos, H.], 1942, Erich Haarmann: Geologische Rundschau, v. 33, p. 85–87 + portrait.

Anonymous, 1952, Geographical Exploration and Topographic Mapping by the U.S. Government, Catalog: National Archives Publication No. 53-2, The National Archives, National Archives and Records Service, General Services Administration, Washington, 52 p.

Anonymous, 1989, Atlas of the People's Republic of China: Beijing, Foreign Languages Press, China Cartographic Publishing House, 113 p.

Anonymous, 1991a, Die Schöpfungsmythen, 2. Ausgabe: Benziger, Zürich, 267 p.

Anonymous, 1991b, Gerhard Mercator—Der Weltgelehrte der Renaissance in Duisburg: Archäologisches Museum, Rethymnon and Archäologisches Nationalmuseum, Athen, Athina, 169 p.

Anonymous, 1999, Institut für Geschichte der Arabisch-Islamischen Wissenschaften an der Johann Wolfgang Goethe-Universität Frankfurt am Main Publications 1984-Summer 1999, Institut für Geschichte der Arabisch-Islamischen Wissenschaften an der Johann Wolfgang Goethe-Universität Frankfurt, [Frankfurt am Main], IV + 624 p.

Anonymous, 2000, Institut für Geschichte der Arabisch-Islamischen Wissenschaften an der Johann Wolfgang Goethe-Universität Frankfurt am Main Comprehensive Catalogue of Publications on Islamic Philosophy 1984-Summer 2000 Including New Publicaktions 1999-Summer 2000, Institut für Geschichte der Arabisch-Islamischen Wissenschaften an der Johann Wolfgang Goethe-Universität Frankfurt, [Frankfurt am Main], p. [I], 115 + [4].

Anonymous, 2003, Natural Sciences in Islam—Publications Available in January 2003: Johann Wolfgang Goethe-Universität Frankfurt am Main Institute for the History of Arabic-Islamic Science, 38 p.

Anspach, S., Behling, H., Meyer, D., and Schlagetter-Pellatz, S., editors, 1988, Sebastian Münster—Katalog zur Ausstellung aus Anlaß des 500. Geburtstages am 20. Januar 1988 im Museum—Altes Rathaus Ingelheim am Rhein: Stadt Ingelheim am Rhein, XIV + 139 p.

Ansted, D. T., 1863, Geological Gossip: or, Stray Chapters on Earth and Ocean: London, Routledge, Warne, and Routledge, vii + [I] + 325 p.

Antisell, T., 1856, Routes in California to connect with the Routes near the Thirty-Fifth and Thirty-Second Parallels, and Route near the Thirty-Second Parallel, between the Rio Grande and Pimas villages, explored by Lieutenant John. G. Parke, Corps Topographical Engineers, in 1854 and 1855—Geological Report, *in* Reports of Explorations and Surveys, to Ascertain the Most Practicable and Economical Route for a Railroad from the Mississippi River to the Pacific Ocean. Made under the Direction of the Secretary of War, in 1853–1854, According to Acts of Congress of March 3, 1854, and August, 5, 1854., Senate, 33d Congress, 2d Session, Ex. Doc. No. 78, Washington, D.C., v. VII (1857), p. 204, 13 plates + 1 folded geological map.

Appleman, D.E., 1985, James Dwight Dana and Pacific geology, *in* Viola, H.J., and Margolis, C., eds., Magnificent Voyagers—The U.S Exploring Expedition, 1838–1842: Washington, D.C., Smithsonian Institution, p. 88–117.

Argand, E., 1912, Sur la segmentation tectonique des Alpes occidentales: Bulletin de la Société vaudouise des Sciences Naturelles, v. 48, p. 345–356.

Argand, E., 1924, La tectonique de l'Asie: Congrés Géologiques International, Comptes Rendus de la XIIme session, Premier Fascicule, H. Vaillant-Carmanne, Liége, p. 171–372.

Argand, E., 1936, La Zone Pennique: Geologischer Führer der Schweiz herausgegeben von der schweizerischen Geologischen Gesellschaft bei Anlaß ihrer 50sten Jahresfeier, Fascicule III Allgemeine Einführungen: Basel, B. Wepf & Cie, p. 149–189.

Argyll, Duke of, 1874, Address delivered at the anniversary meeting of the Geological Society of London on the 20th February, 1874: Proceedings of the Geological Society of London, p. xxxiv–lxix.

Ariew, R., 1985, Preface, *in* Ariew, R., editor and translator, Medieval Cosmology—Theories of Infinity, Place, Time, Void, and the Plurality of the Worlds—Pierre Duhem: Chicago, The University of Chicago Press, p. xix–xxxi.

Arldt, T., 1919, Handbuch der Palaeogeographie, Band I: Palaeoaktologie: Leipzig, Gebrüder Bornträger, 679 p.

Ashgirei, G.D., 1963, Strukturgeologie: VEB Deutscher Verlag der Wissenschaften, XVI + 572 p. + 22 photographic plates

Assmann, J., 1984, Schöpfung, *in* Helck, W., and Otto E., eds., Lexikon der Ägyptologie: Wiesbaden, Otto Harrassowitz, v. 5, p. 678–690.

Aubouin, J., 1959, A propos d'un céntenaire: les aventures de la notion de géosynclinale: Revue de Géographie Physique et Géologie Dynamique, v. 2, p. 135–188.

Aubouin, J., 1965, Geosynclines: Developments in Geotectonics: Amsterdam, Elsevier, XV + 335 p.

d'Auboussin de Voisins, J.F., 1819, Traité de Géognosie, ou exposé des connaisances actuelles sur la constitution physique et minérale du globe terrestre, v. I: Levrault, Strasbourg, lxj + [i] + 496 p.

Auerbach, H.S., 1941a, Father Escalante's route (As depicted by the map of Bernardo de Miera y Pacheco): Utah Historical Quarterly, v. 9, p. 73–80.

Auerbach, H.S., 1941b, Father Escalante's itinerary: Utah Historical Quarterly, v. 9, p. 109–128.

Auerbach, H.S., 1943, Father Escalante's journal with related documents and maps: Utah Historical Quarterly, v. 11, p. 1–142.

Aujac, G., 1966, Strabon et la Science de Son Temps: Société d'Édition «Les Belles Lettres», Paris, 326 p. + 9 foldout figures.

Aujac, G., 1975, La Géographie dans le Monde Antique: Que sais-je, no. 1958: Paris, Presses Universitaires de France.

Aujac, G., 2001, Eratosthène de Cyrène, le Pionnier de la Géographie—Sa Mesure de la Circonférence Terrestre: Paris, CTHS, 224 p.

Aurand, H., 2000, Geology Terms in English and Spanish/Terminología Geológica en Español e Inglés, in Birnbaum, B.B., and Súarez, F., eds., Sunbelt Pocket Guide (L. Lindsay, ed.): San Diego, California, Sunbelt Publications, xi + 117 p.

Authors' Collective, 1969, Geology and Natural History of the Fifth Field Conference. Powell Centennial River Expedition 1969: Four Corners Geological Society, Durango, 212 p. + 1 foldout in pocket.

Averdunk, H., and Müller-Reinhard, J., 1914, Gerhard Mercator und die Geographen unter seinen Nachkommen: Gotha, V.E.B. Haack, VIII + 188 p. (reprinted in 1969 by Theatrum Orbis Terrarum Ltd. in Amsterdam).

Baars, D.L., 1995, Navajo Country—A Geology and Natural History of the Four Corners Region: Albuquerque, University of New Mexico Press, XII + 255 p.

Baars, D.L., 2000, The Colorado Plateau—A Geologic History, revised and updated: Albuquerque, University of New Mexico Press, xiii + 254 p.

Baars, D.L., 2002, A Traveler's Guide to the Geology of the Colorado Plateau: Salt Lake City, The University of Utah Press, viii + 294 p.

Babbage, C., 1838, The Ninth Bridgewater Treatise. A Fragment, second edition: London, John Murray, vii + [i] + xxii + 270 p.

Babbage, C., 1994, Passages from the Life of a Philosopher. Edited with a new Introduction by Martin Campbell-Kelly: New Brunswick, New Jersey, Rutgers University Press, 36 + viii + 383 p.

Babcock, R.S., 1990, Precambrian crystalline core, *in* Beus, S.S., and Morales, M., eds., Grand Canyon Geology: New York, Oxford University Press, and Museum of Northen Arizona Press, p. 11–28.

Baccou, R., 1951, Histoire de la Science Grecque de Thalès a Socrate: Paris, Aubier, 256 + [I].

Bacon, R., 1928, The Opus Majus of Roger Bacon—A Translation by Robert Belle Burke, v. I: Philadelphia, University of Pennsylvania Press, xiii + 418 p.

Bagchi, P.C., 1944, India and China—A Thousand Years of Sino-Indian Cultural Contact: Calcutta, China Press, Ltd., x + 240 p. + 1 foldout map.
Bagrow, L., 1945, The origin of Ptolemy's Geographia: Geografiska Annaler, v. XXVII, p. 318–387.
Bagrow, L., 1947, Supplementary notices to "The origin of Ptolemy's geography": Imago Mundi, no. IV, p. 71–72.
Bagrow, L., 1964, History of Cartography, revised and enlarged by R.A Skelton: Cambridge, Harvard University Press, 312 p.
Bahm, A.J., 1988, Comparing civilisations as systems: Systems Research, v. 5, p. 35–47.
Bailey, C., 1928, The Greek Atomists and Epicurus: Oxford, Clarendon, viii + [i] + 619 p.
Bailey, E., 1962, Charles Lyell: British Men of Science series: London, Thomas Nelson and Sons, x + 214 p.
Bain, D.H., 1999, Empire Express—Building the First Transcontinental Railroad: New York, Viking, xiii + [iii] + 797 p. + 32 unnumbered photographic plates.
Baker, J.N.L., 1963, The History of Geography: Basil Blackwell, Oxford, xxviii + 266 p.
Bakewell, R., 1823a. Travels Comprising Observations Made During A Residence in the Tarentaise and Various Parts of the Grecian and Pennine Alps, and in Switzerland and Auvergne, in the Years 1820, 1821, and 1822, v. I: London, Longman, Hurst, Rees, Orme, and Brown, xvi + 381 p.
Bakewell, R., 1823b, Travels Comprising Observations Made During a Residence in the Tarentaise and Various Parts of the Grecian and Pennine Alps, and in Switzerland and Auvergne, in the Years 1820, 1821, and 1822, v. II: London, Longman, Hurst, Rees, Orme, and Brown, vii + 447 p.
Bakewell, R., 1839, An Introduction to Geology, intended to Convey A Practical Knowledge of the Science, and Comprising the Most Important Recent Discoveries; with Explanation of The Facts and Phenomena Which Serve to Confirm or Invalidate Various Geological Theories; third American from the fifth London edition, edited, with an introduction by Prof. B. Silliman, Yale College: New Haven, B. & W. Noyes, xxxvi + 596 p. + 8 foldout plates.
Bala, M., 1967, Kirim, *in* İslâm Ansiklopedisi: İstanbul, Millî Eğitim Basimevi, v. 6, p. 741–746.
Baladié, R., 1980, Le Péloponnèse de Strabon—Étude de Géographie Historique: Paris, Société d'Édition «Les Belles Lettres», XXIII + 398 p. + XLIV photographic plates.
Baldwin, S.A., 1986, John Ray (1627–1705) Essex Naturalist: Witham, Baldwin's Books, [i] + 80 p.
Bally, A.W., Bender, P.L., McGetchin, T.R., and Walcott, R.I., editors, 1980, Dynamics of Plate Interiors: Geodynamics Series, v. 1: Washington, D.C., American Geophysical Union, Boulder, Colorado, Geological Society of America, [ii] + 162 p.
Bally, A.W., Catalano, R., and Oldow, J., 1985, Elementi di Tettonica Regionale: Bologna, Pitagora Editrice, X + 276 p.
Bally, A.W., and Snelson, S., 1980, Realms of subsidence, *in* Miall, A.D., ed., Facts and Principles of World Petroleum Occurrence: Canadian Society of Petroleum Geologists Memoir 6, p. 9–94.
Bandelier, A.F.A., 1886[1981], The Discovery of New Mexico by the Franciscan Monk, Friar Marcos de Niza in 1539, Madeleine Turrell Rodack, translator and editor: Tucson, Arizona, University of Arizona Press, 135 p.
Baragar, W.R.A., Ernst, R.E., Hulbert, L., and Peterson, T., 1996, Longitudinal petrochemical variation in the Mackenzie Dyke Swarm, Northwestern Canadian Shield: Journal of Petrology, v. 37, p. 317–359.
Barbier de Meynard, C., and de Courteille, P., 1861, Maçoudi—Les Prairies d'Or texte et traduction, tome premier: Paris, Imprimerie Impériale, XII + 408 p.
Barlow, N., editor, 1933, Charles Darwin's Diary of the Voyage of H.M.S. "Beagle": At the University Press, Cambridge, xxx + 451 p.
Barlow, N., editor, 1946, Charles Darwin and the Voyage of the *Beagle*: Philosophical Library, New York, [ii] + 279 p. + 1 fold-out map.
Barlow, N., editor, 1958, The Autobiography of Charles Darwin 1809–1882 with original ommissions restored, edited with Appendix and Notes by his granddaughter: London, Collins, 253 p.
Barnes, J., 1981, The Presocratic Philosophers, revised edition: Routledge & Kegan Paul, London, xxiii + 703 p.
Barnes, J., ed., 1984, The Complete Works of Aristotle—The Revised Oxford Translation, v I: Bollingen Series LXXI.2: Princeton, Princeton University Press, xi + [I] + 1250 p.
Barnes, J., 1987, Early Greek Philosophy: London, Penguin, 318 p.
Barnes, J., 2000, Aristotle—A Very Short Introduction: Oxford, Oxford University Press, [vii] + 160 p.
Barradas de Carvalho, J., 1974, La Traduction Espagnole du «De Situ Orbis» de Pomponius Mela par Maître Joan Faras et les Notes Marginales de Duarte Pacheco Pereira: Junta de Investigações Científicas do Ultramar Centra de Estudos Cartografia Antga, Lisboa, 248 p.
Barrell, J., 1918 [1973], A century of geology—the growth of knowledge of earth structure, *in* A Century of Science in America: New Haven, Yale University Press, p. 153–192 (reprinted by Scholarly Resources Inc., Wilmington, Delaware).
Barrett, P.H., Gautrey, P.J., Herbert, S., Kohn, D., and Smith, S., editors and transcribers, 1987, Charles Darwin's Notebooks, 1836–1844—Geology, Transmutation of Species, Metaphysical Enquiries: British Museum (Natural History): Ithaca, Cornell University Press, viii + 747 p.
Bartlett, K., 1940, How Don Pedro Tovar discovered the Hopi and Don Garcia Lopez de Cardenas saw the Grand Canyon, with notes upon their probable route: Plateau, v. 12, no. 3, p. 37–45.
Bartlett, R.A., 1962, Great Surveys of the American West: Norman, University of Oklahoma Press, xxiii + 408 p.
Bartolini, C., Buffler, R.T., and Cantú-Chapa, A., editors, 2001, The Western Gulf of Mexico Basin—Tectonics, Sedimentary Basins, and Petroleum Systems: American Association of Petroleum Geologists Memoir 75, xii + 480 p.
Bashford, H., and Wagner, H., 1927, A Man Unafraid—The Story of John Charles Frémont: San Francisco, Harr Wagner Publishing Co., [v] + 406 p.
Bashmakoff, A., 1948, La Synthèse des Périples Pontiques—Méthode de Précision en Paléo-Ethnologie: Études d'Ethnographie, de Sociologie et d'Ethnologie, v. III, Librairie Orientaliste Paul Geuthner, Paris, XI + 184 + [I] + 9 foldout maps.
Batyushkova, I.V., 1975, Istoriya Problemi Proiskhozhdeniya Materikov i Okeanov: Moskva, Akademiya Nauk SSSR Institut Istorii Estestvoznaniya i Tekhniki, Izdatelstvo "Nauka," 138 p.
Baysun, M.C., 1964, Evliya Çelebi, *in* İslâm Ansiklopedisi: İstanbul, Millî Eğitim Basimevi, v. 4, p. 400–412.
Bazanov, E.A., Pritula, Y.A., and Zabaluev, V.V., 1976, Development of main structures of the Siberian Platform: history and dynamics, *in* Bott, M.H.P., ed., Sedimentary Basins of Continental Margins and Cratons, Developments in Geotectonics, 12: Amsterdam, Elsevier, p. 289–300 (also Tectonophysics, v. 36).
Bean, G.E., 1989, Turkey's Southern Shore: London, John Murray, xxii + 154 p.
Beaumont, C., 1978, The evolution of sedimentary basins on a viscoelastic lithosphere: theory and examples: Geophysical Journal of the Royal Astronomical Society, v. 53, p. 471–497.
Beaumont, C., Keen, C.E., and Boutilier, R., 1982, A comparison of foreland and rift margin sedimentary basins: Philosophical Transactions of the Royal Society of London, v. A305, p. 295–317.
Beazley, C.R., 1897, The Dawn of Modern Geography, v. I: A History of Exploration and Geographical Science from the Conversion of the Roman Empire to A.D. 900, with an Account of the Achievements and Writings of the Christian, Arab, and Chinese Travellers and Students: London, Henry Frowde, xvi + 538 p.
Beazley, C.R., 1901, The Dawn of Modern Geography, part II: A History of Exploration and Geographical Science from the Close of the Ninth to the Middle of the Thirteenth century (c. A.D. 900–1260): London, John Murray, xix + 651 p.
Beazley, C.R., 1906, The Dawn of Modern Geography, v. III A History of Exploration and Geographical Science from the middle of the Thirteenth to the Early Years of the Fifteenth Century (c. A.D. 1260–1420): Oxford, Clarendon Press, xvi + 638 p.

de la Beche, H.T., 1837, Researches in Theoretical Geology: New York, F.J. Huntington & Co., 342 p.

Beck, H., 1959a, Alexander von Humboldt, v. I (Von der Bildungsreise zur Forschungsreise 1769-1804): Wiesbaden, Franz Steiner, XVI + 303 p.

Beck, H., 1959b, Graf Georg von Cancrin und Alexander von Humboldt, *in* von Humboldt, A., 14. 9.1769–6.5, 1859—Gedenkschrift zur 100, Wiederkehr seines Todestages, herausgegeben von der Alexander von Humboldt-Kommission der Deutschen Akademie der Wissenschaften zu Berlin: Berlin, Akademie-Verlag, p. 69–82.

Beck, H., 1961, Alexander von Humboldt, v. II (Vom Reisewerk zum »Kosmos« 1805–1859): Wiesbaden, Franz Steiner, XII + 439 p. + 28 plates and 6 maps.

Beck, H., 1966, Alejandro von Humboldt y Mexico—Aportaciones a una vision geográfica: Bad Godesberg, Inter Nationes, 54 p. + 5 plates.

Beck, H., 1982, Große Geographen—Pioniere-Außenseiter-Gelehrte: Berlin, Dietrich Reimer, 294 p.

Beck, H., 1985, Der unentdeckte Entdecker—Zum gegenwärtigen Stand der Alexander-von-Humboldt-Forschung: Frankfurter Allgemeine, v. 30, nr. 252 (30 October 1985), supplement "Geisteswissenschaften."

Beck, H., and Bonacker, W., editors and introduction, 1969, Alexander von Humboldt Atlas Géographique et Physique du Royaume de la Nouvelle-Espagne Vom Verfasser auch kurz benannt: Mexico-Atlas: Stuttgart, F.A. Brockhaus Komm.-Gesch. GmbH., Abt. Antiquarium, 34 + XCII p. + 28 maps.

Becker, G.F., 1912, Major C.E Dutton: American Journal of Science, 4th series, v. 33, p. 387–388.

Beckinsale, R.P., and Chorly, R.J., 1991, The History of the Study of Landforms or the Development of Geomorphology, v. 3: Historical and Regional Geomorphology 1890–1950: London, Routledge, xxiii + 496 p.

Beckwith, E.G., 1854a, Report of Explorations for a Route for the Pacific Railroad, by Capt. J.W. Gunnison, Topographical Engineers, near the 38th and 39th parallels of north latitude, from the mouth of the Kansas River, Mo., to the Sevier Lake in the Great Basin—Report by Lieut. E.G Beckwith, Third Artillery, *in* Reports of Explorations and Surveys, to Ascertain the Most Practicable and Economical Route for a Railroad from the Mississippi River to the Pacific Ocean. Made under the Direction of the Secretary of War, in 1853-4, According to Acts of Congress of March 3, 1854, and August, 5, 1854., Senate, 33d Congress, 2d Session, Ex. Doc. No. 78, Washington, D.C., v. II, part I, 128 p. (Maps and profiles associated with this report were published in Anonymous [1859]).

Beckwith, E.G., 1854b, Report of Explorations for a Route for the Pacific Railroad, on the line of the forty-first parallel of north latitude. Report by Lieut. E.G Beckwith, Third Artillery, *in* Reports of Explorations and Surveys, to Ascertain the Most Practicable and Economical Route for A Railroad from the Mississippi River to the Pacific Ocean. Made under the Direction of the Secretary of War, in 1853-4, According to Acts of Congress of March 3, 1854, and August, 5, 1854, Senate, 33d Congress, 2d Session, Ex. Doc. No. 78, Washington, D.C., v. II, part III, 131 p. + 10 botanical plates. (Maps and profiles associated with this report were published in Anonymous [1859]).

Bederke, E., 1967, Nachruf auf Hans Stille: Jahrbuch der Akademie der Wissenschaften in Göttingen für das Jahr 1967, Vandenhoeck & Ruprecht in Göttingen, p. 74–76.

de Beer, G., 1981, Darwin, Charles Robert, *in* Gillispie, C.C., editor-in-chief, Dictionary of Scientific Biography: New York, Charles Scribner's Sons, v. 3, p. 565–577.

Beghoul, N., Barazangi, M., and Isacks, B., 1993, Lithospheric structure of Tibet and western North America: mechanisms of uplift and a comparative study: Journal of Geophysical Research, v. 98, p. 1997–2016.

Behrmann, W., 1948, Die Entschleierung der Erde: Frankfurter Geographische Hefte, 16: Jahrgang, einziges Heft, 56 p + 12 plates.

Beloussov, V.V., 1948, Obshaya Geotektonika: Gosudarstvennoe Izdatelstvo Geologicheskoi Literaturi Ministerstva Geologii SSSR, Moskva, Leningrad, 598 p. + 1 folded map.

Beloussov, V.V., 1954, Osnovnie Voprosi Geotektoniki: Moskva, Gosgeoltekhizdat, 606 p. + 1 folded map.

Beloussov, V.V., 1961, Znachenie rabot M.M. Tetyayeva v razvitii geotektoniki, *in* Beloussov, V.V., and Sheinmann, Y.M., eds., Problemi Tektoniki, Gosudarstvennoe Nauchno-Tekhnicheskoe Izdatelstvo Literaturi po Geologii I Okhrane Nedr, Moskva, p. 7–17.

Beloussov, V.V., 1962, Basic Problems in Geotectonics: New York, McGraw-Hill Book Company, xvi + 809 p.

Beloussov, V.V., 1975, Osnovi Geotektoniki: "Nedra," Moskva, 260 + [II] p.

Beloussov, V.V., 1976, Geotektonika: Moskva, Izdatelstvo Moskovskogo Universiteta, 331 + [II] p.

Beloussov, V.V., 1981, Continental Endogenous Regimes, translated from the Russian by V. Agranat and Yu. Prizov: Moscow, Mir Publishers, 295 p.

Beloussov, V.V., 1989, Osnovi Geotektoniki: Moskva, "Nedra," 382 p. + 1 foldout map.

van Bemmelen, R.W., 1931a, Kritische beschouwingen over geotektonische hypothesen: Natuurkundig Tijdschrift voor Nederlandsch Indië, v. 91, p. 93–117.

van Bemmelen, R.W., 1931b, De bicausaliteit der bodembewegingen: Natuurkundig Tijdschrift voor Nederlandsch Indië, v. 91, p. 363–413.

van Bemmelen, R.W., 1932a, Ueber die möglichen Ursachen der Undationen der Erdkruste: Proceedings, Koninklijke Akademie van Wetenschappen te Amsterdam, v. 35, p. 392–399.

van Bemmelen, R.W., 1932b, De undatietheorie (hare afleiding toepassing op het telijk deel van den Soendaboog): Natuurkundig Tijdschrift voor Nederlandsch Indië, v. 92, p. 85–242.

van Bemmelen, R.W., 1933, Das Kräfteproblem in der Tektonik: Proceedings, Koninklijke Akademie van Wetenschappen te Amsterdam, v. 36, p. 197–202.

van Bemmelen, R.W., 1935, The undation theory of the development of the Earth's crust: Report of XVI International Geological Congress, Washington, D.C., 1935, v. 2, p. 965–982.

van Bemmelen, R.W., 1949, The Geology of Indonesia, v. IA (General Geology of Indonesia and Adjacent Archipelagos: XXIV + 732 p.), v. IB (Portfolio: 60 p. + 41 plates), v. II (Economic Geology: VIII + 265 p.): Government Printing Office, The Hague.

van Bemmelen, R.W., 1954, Mountain Building. A Study Primarily based on Indonesia region of the World's Most Active Crustal Deformations: Martinus Nijhoff, The Hague, XII + 177 p.

van Bemmelen, R.W., 1955, Tectogenèse par gravité: Bulletin de la Société belge de Géologie, Paléontologie et Hydrologie, v. 64, p. 95–123.

Benz, E., 1948, Emanuel Swedenborg—Naturforscher und Seher: Verlag Hermann Rinn, München, 585 + [III] p.

Berger, H., 1880 [1964], Die Geographischen Fragmente des Eratosthenes: Leipzig, Veit & Comp., 393 p. (reprinted in 1964 by Meridian Publishing Co., Amsterdam).

Berger, H., 1903 [1966], Geschichte der Wissenschaftlichen Erdkunde der Griechen: Leipzig, Veit & Comp., 662 p. (reprinted in 1966 by Walter de Gruyter & Co., Berlin).

Bernhardy, G., 1822, Eratosthenica: Ge. Reimer, Berolini, 272 p.

Bernoulli, D., Bertotti, G., and Zingg, A., 1989, Northward thrusting of the Gonfolite Lombarda ("South Alpine Molasse") onto the Mesozoic sequence of the Lombardian Alps: Implications for the deformation history of the Southern Alps: Eclogae Geologicae Helvetiae, v. 82, p. 841–856.

Bernoulli, D., Giger, M., Müller, D.W., and Ziegler, U.R.F., 1993, Sr-isotope stratigraphy of the Gonfolite Lombarda Group ("South-Alpine Molasse," northern Italy) and radiometric constraints for its age of deposition: Eclogae Geologicae Helvetiae, v. 86, p. 751–767.

Berry, W.B.N., 1987, Growth of a Prehistoric Time Scale Based on Organic Evolution, revised edition: Oxford, Blackwell, xiii + [ii] + 202 p.

Berthelot, A., 1930, L'Asie Ancienne Centrale et Sud-Orientale d'après Ptolémée: Paris, Payot, 427 p.

Bertrand, J., 1875, Élogue Historique de Élie de Beaumont: Paris, Institut de France, Typographie de Firmin-Didot et Cie, 28 p.

Bertrand, M., 1897, Préface, *in* La Face de la Terre (Das Antlitz der Erde) par Ed. Suess … traduit de l'Allemand, avec l'Autorisation de l'Auteur et Annoté sous la direction de Emm. de Margerie, v. I: Paris, Armand Colin et. Cie, p. V–XV.

von Beskow, B., 1857, Minne öfver Landshöfdingen, Vice Presidenten och Förste Arkiatern Urban Hjärne: Stockholm, P.A. Norstedt & Söner, 78 p.

Bettex, A., 1960, The Discovery of the World: New York, Simon and Schuster, 379 p.

Biardeau, M., 1991a, Vedic cosmogony, *in* Bonnefoy, Y., ed., Mythologies, v. II: Chicago, Chicago University Press, p. 806–807.

Biardeau, M., 1991b, Purânic cosmogony, *in* Bonnefoy, Y., ed., Mythologies, v. II: Chicago, Chicago University Press, p. 817–823.

Biardeau, M., 1991c, Matsya: The fish and the flood in the work of the mythic imagination, *in* Bonnefoy, Y., ed., Mythologies, v. II: Chicago, Chicago University Press, p. 853–854.

Bibby, G., 1969, Looking for Dilmun: London, Collins, ix + [v] + 383 + viii p.

Bierbaum, M., Faller, A., and Traeger, J., 1989, Niels Stensen—Anatom, Geologe und Bischof 1638-1686: Münster, Aschendorff, XII + 208 p. + 20 plates.

Bigelow, J., 1856, Memoir of the Life and Public services of John Charles Frémont, Including his Account of his Explorations, Discoveries and Adventures on Five Successive Expeditions across the North American Continent; Voluminous Selections from His Private and Public Correspondance; His Defence before the Court Martial, and full Reports of His Principal Speeches in the Senate of the United States: New York, Derby & Jackson, x + 480 p.

Bird, J.M., and Dewey, J.F., 1970, Lithosphere plate—Continental margin tectonics and the evolution of the Appalachian Orogen: Geological Society of America Bulletin, v. 81, p. 1031–1060.

Bird, J.M., and Isacks, B., editors, 1972, Plate Tectonics: Selected Papers from the Journal of Geophysical Research: Washington, D.C., American Geophysical Union, 563 p.

Bird, P., 1979, Continental delamination and the Colorado Plateau: Journal of Geophysical Research, v. 84, p. 7561–7571.

Birembaut, A., 1982, Élie de Beaumont, Jean-Baptiste-Armand-Louis-Léonce, *in* Gillispie, C.C., editor-in-chief, Dictionary of Scientific Biography: New York, Charles Scribner's Sons, v. 4, p. 347–350.

Birembaut, A., 1981b, Boué, Ami, *in* Gillispie, C.C., editor-in-chief, Dictionary of Scientific Biography: New York, Charles Scribner's Sons, v. 2, p. 341–342.

Black, A.H., 1989, Man and Nature in the Philosophical Thought of Wang Fu-chih: Seattle and London, University of Washington Press, xix + 375 p.

Blake, W.H., 1856, General report upon the geological collections, *in* Reports of Explorations and Surveys, to Ascertain the Most Practicable and Economical Route for a Railroad from the Mississippi River to the Pacific Ocean. Made under the Direction of the Secretary of War, in 1853–1854, According to Acts of Congress of March 3, 1854, and August, 5, 1854, Senate, 33d Congress, 2d Session, Ex. Doc. No. 78, Washington, D.C., v. III, part IV, No. 1, p. 1–119.

Blakey, R.C., 1990, Supai Group and Hermit Formation, *in* Beus, S.S., and Morales, M., eds., Grand Canyon Geology: New York, Oxford University Press, and Museum of Northen Arizona Press, p. 147–182.

Blei, W., 1991, Einige Bemerkungen zu Niels Stensens Geologie, zu seinen Vorgängern und zu seiner Nachwirkung, *in* Dewey, J.F., ed., Prof. Dr. *rer. nat.* İhsan Ketin and Tectonics: Bulletin of the Technical University of Istanbul, v. 44, p. 3–21.

Block, L., and Royden, L.H., 1990, Core complex geometries and regional scale flow in the lower crust: Tectonics, v. 9, p. 557–567.

Bluck, R.S., 1955, Plato's Phaedo, *in* Piest, O., general editor, The Library of Liberal Arts, no. 110: New York, The Liberal Arts Press, x + 208 p.

Blumenberg, H., 1987, Das Lachen der Thrakerin—Eine Urgeschichte der Theorie: Frankfurt am Main, Suhrkamp Verlag, 162 p.

Blundell, D., Freeman, R., and Mueller, S., editors, 1992, A Continent Revealed—The European Geotraverse: Cambridge, Cambridge University Press, xii + 275 p.

Blundell, D.J., and Scott, A.C., eds., 1998, Lyell: The Past is the Key to the Present: Geological Society (London) Special Publication 143, viii + 376 p.

Blunt, W., 1971, The Compleat Naturalist—A Life of Linnaeus: London, Collins, 256 p.

Bobb, B.E., 1962, The Viceregency of Antonio Maria Bucareli in New Spain, 1771–1779: Austin, University of Texas Press, [iii] + 313 p.

Boccaccio, G., 1978, Dizionario Geografico De montibus, silvis, fontibus, lacubus, fluminibus, stagnis seu paludibus, et de nominibus maris (traduzione di Nicolò Liburnio): Torino, Fògola Editore, XXIV + 248 + [I] p.

Bogdanov, A.A., and Khain, V.E., editors, 1964, G. Stille Izbrannie Trudi: Moskow, Mir, 887 p.

Bois, C., Bouche, P., and Pelet, R., 1982, Global geologic history and distribution of hydrocarbon reserves: Bulletin of the American Association of Petroleum Geologists, v. 66, p. 1248–1270.

Bollack, J., 1965, Empédocle, v. I, introduction a l'Ancienne Physique: Paris, Les Éditions de Minuit, 411 p.

Bollack, J., 1969a, Empédocle, v. II: Les Origines—Édition et Traduction des Fragments et des Témoignages: Paris, Les Éditions de Minuit, XXIV + 304 p.

Bollack, J., 1969b, Empédocle, v. III Les Origines—Commentaire 1: Paris, Les Éditions de Minuit, 305 p.

Bollack, J., 1969c, Empédocle, v. III Les Origines—Commentaire 2: Paris, Les Éditions de Minuit, p. 309–683.

Bolton, H.E., 1930a, Anza's California Expeditions, v. I, An Outpost of Empire: Berkeley, University of California Press, xxi + 529 p. + 55 illustrations + 10 maps.

Bolton, H.E., 1930b, Anza's California Expeditions, v. II, Opening a Land Route to California—Diaries of Anza, Díaz, Garcés, and Palóu: Berkeley, University of California Press, xv + 473 p. + 33 illustrations.

Bolton, H.E., 1930c, Anza's California Expeditions, v. III, The San Francisco Colony—Diaries of Anza, Font, and Eixarch, and Narratives by Palóu and Moraga: Berkeley, University of California Press, xxi + 436 p. + 22 illustrations.

Bolton, H.E., 1930d, Anza's California Expeditions, v. IV, Font's Complete Diary: Berkeley, University of California Press, xiii + 552 p. + 34 illustrations + 1 map.

Bolton, H.E., 1930e, Anza's California Expeditions, v. V, Correspondence: Berkeley, University of California Press, xviii + 426 p. + 12 illustrations.

Bolton, H.E., 1949, Coronado, Knight of Pueblos and Plains: New York, Whittlesey House, McGraw-Hill, xii + 491 p. + map as endpapers.

Bolton, H.E., 1950, Pageant in the Wilderness—The Story of the Escalante Expedition to the Interior basin, 1776, Including the Diary and Itinerary Father Escalante Translated and Annotated: Salt Lake City, Utah State Historical Society, [v] + 265 p.

Bonin, B., Dubois, R., and Gohau, G., 1997, Le Métamorphisme et la Formation des Granites—Évolutions des Idées et Concepts Actuels: Paris, Nathan, 320 p.

Bonnefoy, Y., 1991a, editor, Mythologies, a restructured translation of Dictionnaire des mythologies et des religions des sociétés traditionnelles et du monde antique, prepared under the direction of W. Doniger, v. 1: Chicago, University of Chicago Press, xxx + 646 p.

Bonnefoy, Y., 1991b, editor, Mythologies, a restructured translation of Dictionnaire des mythologies et des religions des sociétés traditionnelles et du monde antique, prepared under the direction of W. Doniger, v. 2: Chicago, University of Chicago Press, viii + [1] p. + p. 649–1267.

Bonney, T.G., 1895, Charles Lyell and Modern Geology: The Century Science series (Roscoe, H.E., Sir, ed.): London, Cassell and Company, 224 p.

Boorstin, D.J., 1983, The Discoverers—A History of Man's Search to Know His World and Himself: New York, Random House, xvi + 745 p.

Bork, K.B., 1990, Constant Prévost (1787–1856)—The life and contributions of a French uniformitarian: Journal of Geological Education, v. 38, p. 21–27.

Borodajkewycz, T., 1936, Konrad Millers Lebenswerk: Salzburg, Salzburger Hochschulwochen, 45 p. + 1 foldout in pocket.

Bottero, J., 1991, Intelligence and the technical function of power in the structure of the Mesopotamian pantheon: The example of Enki/Ea, *in* Bonnefoy, Y., ed., Mythologies, v. I: Chicago, Chicago University Press, p. 145–155.

Botting, D., 1973, Humboldt and the Cosmos: New York, Harper & Row, 295 p.
Boucot, A.J., 1968, Silurian and Devonian of the Northern Appalachians, *in* Zen, E-A., White, W.S., Hadley, J.B., and Thompson, J.B., Jr., eds., Studies of Appalachian Geology: Northern and Maritime, Interscience Publishers: New York, John Wiley & Sons, p. 83–94.
Bouillet-Roy, G., 1976, La Géologie Dynamique chez les Anciens Grecs et Latins d'aprés les Textes [Thèse de Docteur de l'Université de Paris (mention sciences)]: Paris, Université Pierre et Marie Curie—Paris 6, 438 p.
Boué, A., 1836, On the theory of the elevation of mountain chains, as advocated by M. Élie de Beaumont: The Edinburgh New Philosophical Journal, v. 21, p. 123–150.
Boué, A., 1874, Über den Begriff und die Bestandtheile einer Gebirgskette, besonders über die sogenannten Urketten, sowie die Gebirgssysteme-Vergleichung der Erde- und Mondes-Oberfläche: Sitzungsberichte der Akademie der Wissenschaften in Wien, mathematisch-naturwissenschaftliche Classe, v. 69, part I, p. 237–300.
Bowler, P.J., 1990, Charles Darwin—The Man and His Influence: Cambridge, Cambridge University Press, xii + 250 p.
Branagan, D., 1994, The Waistland—Three centuries of mineral exploration in the Pacific Region, *in* Branagan, D.F., and McNally, G.H., eds., Useful and Curious Geological Enquiries Beyond the World, Pacific-Asia Historical Themes: The 19th International INHIGEO Symposium, Sydney, Australia, 4–8 July, 1994, international Commission on the History of Geological Sciences (INHIGEO), Conference Publications, Springwood, NSW, p. 14–29.
Branagan, D.F., and McNally, G.H., editors, 1994, Useful and Curious Geological Enquiries Beyond the World, Pacific-Asia Historical Themes: The 19th International INHIGEO Symposium, Sydney, Australia, 4–8 July, 1994, international Commission on the History of Geological Sciences (INHIGEO), Conference Publications, Springwood, NSW, p. xxi + [i] + 358.
Branagan, D.F., and Townley, K.A., 1976, The geological sciences in Australia—A brief historical review: Earth-Science Reviews, v. 12, p. 323–346.
Brand, D.W., 1959, Humboldts Essai Politique sur le Royaume de la Nouvelle-Espagne, *in* Schulze, J.H., ed., Alexander von Humboldt Studien zu Seiner Universalen Geisteshaltung, Berlin, p. 123–141.
Brebner, J.B., 1964, The Explorers of North America, 1492–1806: Cleveland, World Publishing Co., 431 p.
Bréhier, L., 1950, La Civilisation Byzantine: Paris, Albin Michel, XXV + 627 p. + XXIV plates + 1 foldout map.
Briggs, W., 1976, Without Noise of Arms—The 1776 Domínguez-Escalante Search for a Route from Santa Fe to Monterey, with oil paintings by Wilson Hurley: Flagstaff, Arizona, Northland Press, ix + [i] + 212 p.
Brillante, C., 1990, History and the historical interpretation of myth, *in* Edmunds, L., ed., Approaches to Greek Myth: Baltimore, Maryland, Johns Hopkins University Press, p. 93–138.
Brinkmann, R., 1970, Memorial to Hans Stille: Proceedings of the Geological Society of America, Inc., for 1967, p. 263–267.
Brochant de Villiers, A.-J.-M., 1808, Observations géologiques sur les terrains de transitions qui se rencontrent dans la Tarantaise et autres parties de la Chaîne des Alpes: Journal des Mines, v. 23, p. 321–383.
Brock, B.B., 1955, Some observations on vertical tectonics in Africa: Transactions, American Geophysical Union, v. 36, p. 1044–1054.
Brock, B.B., 1972, A Global Approch to Geology—The Background of a Mineral Exploration Strategy based on Significant Form in the Patterning of the Earth's Crust: Cape Town, A.A Balkema, xix + 365 p.
Brock, S., (introduction and translation) 1990, Saint Ephrem Hymns on Paradise: Crestwod, St Vladimir's Seminary Press, 240 p.
Brockelmann, C., 1924, Altturkestanische Volkspoesie II: Asia Maior, v. I, p. 24–44.
Brodersen, K., 1994, Dionysios von Alexandria Das Lied von der Welt—Zweisprachige Ausgabe: Hildesheim, Georg Olms Verlag, 167 p.
Bromme, T., editor, 1851, Atlas zu Alex von Humboldt's Kosmos in zweiundvierzig Tafeln mit erläuterndem Texte: Stuttgart, Krais und Hoffmann, 136 p.
Brongniart, A., 1820, Rapport faite à l'Academie royale des Sciences, par M.A Brongniart, le 11 décembre 1820, sur un Mémoire de M. Constant Prévost, ayant pour titre: Essai sur la constitution physique et géognostique du bassin à l'ouverture duquel est située la ville de Vienne en Autriche: Extrait du Journal de Physique, Novembre 1820, 7 p.
Brooks, G.R., editor and commentator, 1977, The Southwest Expedition of Jedediah S. Smith—His Personal Account of the Journey to California 1826–1827: Glendale, California, The Arthur H. Clark Company, 259 p. + 1 foldout map.
Brown, L.D., and Reilinger, R.E., 1986, Epeirogenic and intraplate movements, *in* Active Tectonics, Studies in Geophysics, Geophysics Study Committee: Washington, D.C., National Academy Press, p. 30–44.
Brown, S.C., 1979, Benjamin Thompson, Count Rumford: Cambridge, The MIT Press, xii + 361 p.
Brown, S.C., 1981, Thompson, Benjamin (Count Rumford), *in* Gillispie, C.C., editor-in-chief, Dictionary of Scientific Biography: New York, Charles Scribner's Sons, v. 13, p. 350–352.
Browne, J., 1983, The Secular Ark—Studies in the History of Biogeography: New Haven and London, Yale University Press, x + 273 p.
Browne, J., 1995, Charles Darwin Voyaging: Princeton, Princeton University Press, xiii + [ii] + 605 p.
Bruhns, K., editor, 1872a, Alexander von Humboldt—Eine Wissenschaftliche Biographie, erster Band: Leipzig, F.A Brockhaus, XX + 480 + [1] p. + 1 portrait.
Bruhns, K., editor, 1872b, Alexander von Humboldt—Eine Wissenschaftliche Biographie, zweiter Band: Leipzig, F.A Brockhaus, VII + 552 p. + 1 portrait.
Bruhns, K., editor, 1872c, Alexander von Humboldt—Eine Wissenschaftliche Biographie, dritter Band: Leipzig, F.A Brockhaus, [I] + 314 p. + 1 portrait.
Brulé-Peronie, M., and Lécuyer, F., 1998, Volcanisme en Auvergne—Massif de Sancy & Monts Dore: Clermont-Ferrand, Miroir Nature, 34 p.
Bruun, O., and Kalland, A., editors, 1995, Asian Perceptions of Nature: Curzon Press, [II] + 276 + [1] p.
Buache, P., 1761, Essai de Géographie Physique: Seconde Suite des Mémoires de Mathematique et de Physique ... de l'Academie Royale des Sciences, de l'Anne'e M.D CCLII: Nouvelle Centurie, v. 3, p. 609–635.
von Bubnoff, S., 1931, Grundprobleme der Geologie: Berlin, Gebrüder Borntraeger, VIII + 237 p.
von Bubnoff, S., 1953, Requiem: Geologische Rundschau: v. 41, p. 1–10.
von Buch, L., 1809 [1867], Geognostische Beobachtungen auf Reisen durch Deutschland und Italien, zweiter Band—Mit einem Anhange von mineralogischen Briefen aus Auvergne an den Geh. Ober-Bergrath Karsten, Berlin (reprinted in Ewald, J., Roth, J., and Dames, W., eds., Leopold von Buch's Gesammelte Schriften, v. I: Berlin, G. Reimer, p. 341–523).
von Buch, L., 1810 [1870], Reise nach Norwegen und Lappland, zweiter Theil, *in* Ewald, J., Roth, J., and Dames, W., eds., Leopold von Buch's Gesammelte Schriften, v. 2: Berlin, G. Reimer, p. 357–563.
von Buch, L., 1820 [1877], Ueber die Zusammensetzung der basaltischen Inseln und über Erhebungs-Kratere: Abhandlungen der physikalischen Klasse der Akademie der Wissenschaften zu Berlin aus den Jahren 1818–1819, p. 51–86 (reprinted in Ewald, J., Roth, J., and Dames, W., eds., Leopold von Buchs Gesammelte Schriften, v. III: Berlin, G. Reimer, p. 3–19).
von Buch, L., 1824a [1877], Ueber die geognostischen Systeme von Deutschland—Ein Schreiben an den Geheimrath v. Leonhard: v. Leonhards Mineralogisches taschenbuch für das Jahr 1824, p. 501–506 (reprinted in Ewald, J., Roth, J., and Dames, W., eds., Leopold von Buch's Gesammelte Schriften, v. III: Berlin, G. Reimer, p. 218–221).
von Buch, L., 1824b [1877], Ueber geognostische Erscheinungen im Fassathal—Ein Schreiben an den Gehemrath von Leonhard: v. Leonhard's Mineralogisches taschenbuch für das Jahr 1824, p. 343–396 (reprinted in Ewald, J., Roth, J., and Dames, W., eds., Leopold von Buch's Gesammelte Schriften, v. III: Berlin, G. Reimer, p. 141–166).
von Buch, L., 1825 [1877], Physikalische Beschreibung der Kanarischen Inseln: Hofdruckerei der Königlichen Akademie, Berlin, v. I, p. XIV + 388, v. II,

p. 381 (reprinted in Ewald, J., Roth, J., and Dames, W., eds., Leopold von Buch's Gesammelte Schriften, v. III: Berlin, G. Reimer, p. 229–646; the editors augmented this reprint by some of the additions made by the author to the French translation of this book published in 1836).

von Buch, L., 1827 [1877], Ueber die Verbreitung grosser Alpengeschiebe: Poggendorff's Annalen der Physik und Chemie, v. 9, p. 575–588 (reprinted in Ewald, J., Roth, J., and Dames, W., eds., Leopold von Buch's Gesammelte Schriften, v. III: Berlin, G. Reimer,p. 659–668.)

Bucher, W.H., 1933, Deformation of the Earth's Crust—An Inductive Approach to the Problems of Diastrophism: Princeton, Princeton University Press, xii + 518 p.

Bucher, W.H., 1950, Megatectonics and geophysics: Transactions, American Geophysical Union, v. 31, p. 495–507.

Buchholz, E., 1871 [1970], Homerische Kosmographie und Geographie: Saendig Reprint Verlag Hans R. Wohlwend, Vaduz, XVI + 392 p.

Buck, W.R., 1988, Flexural rotation of normal faults: Tectonics, v. 7, p. 959–973.

de Buffon, [G.-L.L.], Comte, 1749, Histoire Naturelle Générale et Particulière, Avec la Description du Cabinet du Roi, v. 1: Paris, Imprimerie Royale, [iv] + 612 p.

de Buffon, [G.-L.L.], Comte, 1778, Histoire Naturelle Générale et Particulière, Supplement, Paris, v. 5, viii + 615 + xxviii p.

Bullard, F.M., 1976, Volcanoes of the Earth, revised edition: Austin and London, University of Texas Press, [xiv] + 579 p.

Bunbury, E.H., 1879a, A History of Ancient Geography, v. I: London, John Murray, xxx + 666 p.

Bunbury, E.H., 1879b, A History of Ancient Geography, v. II: London, John Murray, 743 p.

Burat, A., 1858, Géologie Appliquée—Traité du Gisement et de l'Exploitation des Minéraux Utiles, quatrième edition divisée en deux parties Géologie—Exploitation—première partie Géologie Pratique: Paris, L. Langlois, 550 p. + XXIX plates.

Burchfiel, B.C., Cowan, D.S., and Davis, G.A., 1992a, Tectonic overview of the Cordilleran orogen in the western United States, *in* Burchfiel, B.C., Lipman, P.W., and Zoback, M.L., eds., The Cordilleran Orogen: Conterminous U.S.: Boulder, Colorado, Geological Society of America, Geology of North America, v. G-3, p. 407–479.

Burchfiel, B.C., Lipman, P.W., and Zoback, M.L., 1992b, Introduction, *in* Burchfiel, B.C., Lipman, P.W., and Zoback, M.L., eds., The Cordilleran Orogen: Conterminous U.S: Boulder, Colorado, Geological Society of America, Geology of North America, v. G-3, p. 1–7.

Burde, A.I., Strelnikov, S.I., Mezhelovsky, N.V., and others, 2000, Tri Veka Geologicheskoi Kartografii Rossii: Ministerstvo Prirodnikh Resursov Rossiiskoi Federatsii, VSEGEI, Mezhregionalnii Tsentr po Geologiicheskoi Kartografii (Geokart), Moskva—Sankt-Peterburg, 438 p.

Burggraaf, P., 1997, The untold story of Grand Canyon's John Wesley Powell Memorial: The Journal of Arizona History, v. 38, p. 375–394.

Burke, J.G., 1981a, Dufrénoy, Ours-Pierre-Armand, *in* Gillispie, C.C., editor-in-chief, Dictionary of Scientific Biography: New York, Charles Scribner's Sons, v. 4, p. 217–218.

Burke, J.G., 1981b, Naumann, Karl Friedrich, *in* Gillispie, C.C., editor-in-chief, Dictionary of Scientific Biography: New York, Charles Scribner's Sons, v. 9, p. 620.

Burke, K., 1977, Aulacogens and continental breakup: Annual Review of Earth and Planetary Sciences, v. 5, p. 371–396.

Burke, K., 1996, The African plate: South African Journal of Geology v. 99, p. 339–409.

Burke, K.C., Dessauvagie, T.F.J., and Whiteman, A.J., 1972, Geological history of the Benue Valley and adjacent areas, *in* Dessauvagie, T.F.J., and Whiteman, A.J., eds., African Geology: Ibadan, Nigeria, Department of Geology, University of Ibadan, p. 187–205.

Burke, K., and Dewey, J., 1973, Plume-generated triple junctions: Key indicators in applying plate tectonics to old rocks: Journal of Geology, v. 81, p. 406–433.

Burke, K., Kidd, W.S.F., and Wilson, J.T., 1973, Relative and latitudinal motion of Atlantic hot spots: Nature, v. 245, p. 133–137.

Burke, K., and Whiteman, A.J., 1973, Uplift, rifting, and break-up of Africa, *in* Tarling, D.H., and Runcorn, S.K., eds., Implications of Continental Drift to the Earth Sciences: London, Academic Press, p. 735–755.

Burke, K., and Wilson, J.T., 1972, Is the African plate stationary?: Nature, v. 239, p. 387–390.

Burkert, W., 1992, The Orientalizing Revolution—Near Eastern Influence on Greek Culture in the Archaic Age (Translated by M.E Pinder and W. Burkert): Cambridge, Harvard University Press, ix + 225 p.

Burkhardt, F.H., and Smith, S., editors, 1985, The Correspondence of Charles Darwin, v. 1, 1821–1836: Cambridge, Cambridge University Press, xxix + 702 p.

Burkhardt, F.H., and Smith, S., editors, 1986, The Correspondence of Charles Darwin, v. 2, 1837–1843: Cambridge, Cambridge University Press, xxxiii + [iv] + 603 p.

Burmeister, K.H., 1963, Sebastian Münster—Versuch Eines Biographischen Gesamtbildes: Basler Beiträge zur Geschichtswissenschaft, v. 91: Basel, Verlag von Helbing & Lichtenhahn, XIX + 211 p.

Burnet, J., 1930, Early Greek Philosophy, fourth edition: London, Adam & Charles Black, vi + [ii] + 375 p.

Burnet, T., 1684, The Theory of the Earth: Containing an account of the Original of the earth, and of all the General Changes Which it hath already undergone, or is to undergo, Till the consummation of all things—The First Two Books Concerning The Deluge and Concerning Paradise: London, Walter Kettilby, [xiii] + 327 p.

Büttner, M., 1979a, Mercator und die auf einen Ausgleich zwischen Aristoteles und der Bibel zurückgehende „Klimamorphologie" vom Mittelalter bis ins frühe 17; Jahrhundert, *in* Büttner, M., ed., Wandlungen im geographischen Denken von Aristoteles bis Kant, Abhandlungen und Quellen zur Geschichte der Geographie und Kosmologie, Band 1, Ferdinand Schöningh, Paderborn, p. 139–150.

Büttner, M., 1979b, Die geographisch-cosmographischen Schriften des Aristoteles und ihre Bedeutung für die Entwicklung der Geographie in Deutschland, *in* Büttner, M., ed., Wandlungen im geographischen Denken von Aristoteles bis Kant, Abhandlungen und Quellen zur Geschichte der Geographie und Kosmologie, Band 1, Ferdinand Schöningh, Paderborn, p. 15–34.

Büttner, M., 1979c, On the changes of the geography from the 13th to the 16th century in Central Europe: A contribution to the history of geographical thought, *in* Büttner, M., ed., Wandlungen im geographischen Denken von Aristoteles bis Kant, Abhandlungen und Quellen zur Geschichte der Geographie und Kosmologie, Band 1, Ferdinand Schöningh, Paderborn, p. 51–58.

Büttner, M., 1979d, Philipp Melanchthon (1497–1560), *in* Büttner, M., ed., Wandlungen im geographischen Denken von Aristoteles bis Kant, Abhandlungen und Quellen zur Geschichte der Geographie und Kosmologie, Band 1, Ferdinand Schöningh, Paderborn, p. 93–110.

Büttner, M., 1979e, Bartholomäus Keckermann (1572–1609), *in* Büttner, M., ed., Wandlungen im geographischen Denken von Aristoteles bis Kant, Abhandlungen und Quellen zur Geschichte der Geographie und Kosmologie, Band 1, Ferdinand Schöningh, Paderborn, p. 153–172.

Büttner, M., 1989, The significance of the *Reformation* for the reorientation of geography in Lutheran Germany, *in* Büttner, M., ed., Religion/Umweltforschung im Aufbruch, Abhandlungen zur Geschichte der Geowissenschaften und Religion/Umwelt-Forschung, Band 2, Dr. N. Brockmeyer, Bochum, p. 380–412.

Büttner, M., 1992, Mercators Hauptwerk, der Atlas, aus theologischer und wissenschaftshistorischer Sicht—Zur Verhältnis Gott, Mench und Natur im Hauptwerk Mercators sowie zum Thema Gleichberechtigung der Geschlechter, *in* Büttner, M., ed., Neue Wege in der Mercator-Forschung Mercator als Universalwissenschaftler, Dr. N. Brockmeyer, Bochum, p. 8–97.

Büttner, M., and Burmeister, K.H., 1979, Sebastian Münster (1488–1552), *in* Büttner, M., ed., Wandlungen im geographischen Denken von Aristoteles bis Kant, Abhandlungen und Quellen zur Geschichte der Geographie und Kosmologie, Band 1, Ferdinand Schöningh, Paderborn, p. 111–128.

deBuys, W., editor, 2001, Seeing Things Whole—The Essential John Wesley Powell: Washington, Island Press/Shearwater Books, xiii + 388 p.

Cailleux, A., 1968, Histoire de la Géologie: Que Sais-Je?: Paris, Presses Universitaires de France, 128 p.

Carándell, J., 1928, Las ideas tectonicas de Argand: «Conferencias y Reseñas Cientificas» de la Real Sociedad Española de Historia Natural, v. II, no. 2 & 3, p. 1–15 (offprint pages repaginated).

Carlé, W.E.H., 1988, Werner, Beyrich, von Koenen, Stille—Ein Geistiger Stammbaum Wegweisender Geologen: Geologisches Jahrbuch, v. A108, p. 3–499.

Carozzi, A.V., 1970a, Robert Hooke, Rudolf Erich Raspe, and the concept of "earthquakes": Isis, v. 61, p. 85–91.

Carozzi, A.V., 1970b, New historical data on the origin of the theory of continental drift: Geological Society of America Bulletin, v. 81, p. 283–286.

Carozzi, A.V., 1970c, A propos de l'origine de la théorie des dérives continentales: Francis Bacon (1620), François Placet (1668), A. von Humboldt (1801) et A. Snider (1858): Compte Rendu de la Société de Phyisque et d'Histoire Naturelle de Genève, nouvelle série, v. 4, fascicule 3, p. 171–179.

Carozzi, A.V., 1977, Editor's Introduction, *in* Tectonics of Asia by Emile Argand translated and edited by A.V. Carozzi: New York, Hafner, p. xiii + xxvi.

Carozzi, A.V., and Carozzi, M., 1991, Reevaluation of Pallas' Theory of the Earth (1778): Archives des Sciences, v. 44, fascicule 1, [II] + 105 p.

Carter, H.L., 1968, Dear Old Kit—The Historical Christopher Carson with a new edition of the Carson Memoires: Norman, University of Oklahoma Press, xix + 250 p.

de las Casas, B., 1992, A Short Account of the Destruction of the Indies, edited and translated by Nigel Griffin with an introduction by Anthony Padgen: Penguin Classics, Penguin Books, xliii + 143 p.

Cassidy, J.G., 2000, Hayden—Entrepreneur of Science: Lincoln, University of Nebraska Press, xv + [ii] + 389 p.

Cassin, E., 1991, Divine sovereignty and the division of powers in Mesopotamian myths and poems, *in* Bonnefoy, Y., ed., Mythologies, v. I: Chicago, Chicago University Press, p. 172–181.

Cauvin, J., 1987, L'Apparition des premières divinités: La Recherche, v. 18, p. 1472–1480.

Cederborg, C.A., 1946, Urban Hjärne—Romantiserade Episoder ur en Märklig Mans Liv: Göteborg, Elanders Boktryckeri, 263 p.

Celsius, A., 1744, Oratio de Mutationibus Generalioribus quae in Superficie Corporum Coelestium Contingunt habita Upseliae Die XXII Junii, Anni MDCCXLIII in solemni promotione magisteriali, *in* Linneaus, C., Oratio de Telluris Habitabilis Incremento. Et Andreae Celsii ... Oratio de Mutationibus Generalioribus quae in superficie corporum coelestium contingunt: Cornelium Haak, Lugduni Batavorum, p. 85–104.

Cernajsek, T., Csendes, P., Mentschl, C., and Seidl, J., 1999, "... hat durch bedeutende Leistungen ... das Wohl der Gemeinde mächtig gefördert," Eduard Sueß und die Entwicklung Wiens zur modernen Großstadt: Österreichische Akademie der Wissenschaften Österreichisches Biographisches Lexikon—Schriftenrheie 5, Institut Österreichisches biographisches Lexikon und biographische Dokumentation, Wien, 28 p.

Chain, V.E., and Michajlov, A.E., 1989, Allgemeine Geotektonik: Leipzig, VEB Deutscher Verlag für Grundstoffindustrie, 303 p.

Challaye, F., 1956, Les Philosophes de l'Inde: Paris, Presses Universitaires de France, 330 + [II] p.

Challinor, J., 1967, A Dictionary of Geology, third edition: Cardiff, University of Wales Press, xv + 298 p.

Challinor, J., 1981, Playfair, John, *in* Gillispie, C.C., editor-in-chief, Dictionary of Scientific Biography: New York, Charles Scribner's Sons, v. 11, p. 34–36.

Chamberlin, R.T., 1939, Diastrophic behavior around the Bighorn Basin: Geological Society of America Bulletin, v. 50, p. 1903.

Chamberlin, T.C., and Salisbury, R.D., 1904, Geology, v. I—Geologic Processes and Their Results: New York, Henry Holt and Company, xix + 654 p.

Chamberlin, T.C., and Salisbury, R.D., 1909, Geology, v. I—Geologic Processes and Their Results, second edition, revised: New York, Henry Holt and Company, xix + 684 p.

[Chambers, R.], 1844, Vestiges of the Natural History of Creation: London, John Churchill, vi + 390 p.

Chambers, R., 1848, Ancient Sea Margins as Memorials of Changes in the Relative Level of Sea and Land: Edinburgh; W. & R. Chambers; London, W.S Orr & Co., vi + 337 p. + 1 p. of corrigenda.

de Chancourtois, M.B., 1874, Discours Prononcé le Vendredi 25 Septembre 1874 a Paris, aux Funérailles de M. Élie de Beaumont: Paris, Imprimerie Arnous de Rivière et Ce, 8 p.

de Chancourtois, M.B., 1876, Inauguration de la Statue de Élie de Beaumont: Paris, Imprimerie Arnous de Rivière et Ce, 11 p.

Chappell, J., 1974, Late Quaternary glacio- and hydro-isostasy on a layered earth: Quaternary Research, v. 4, p. 405–428.

Chase, C.G., Gregory-Wodzicki, K.M., Parrich, J.T., and DeCelles, P.G., 1998, Topographic history of the Western Cordillera of North America and controls on climate, *in* Crowley, J.J., and Burke, K.C., eds., Tectonic Boundary Conditions for Climate Reconstructions, Oxford monographs on geology and geophysics no. 39: Oxford, Oxford University Press, p. 73–99.

Chattopadhyaya, D., 1986, History of Science and Technology in India, v. I: The Beginnings: Calcutta, Firma KLM Private Ltd., xxiii + 556 p.

Chattopadhyaya, D., 1991, History of Science and Technology in India, v. II: The Formation of the Theoretical Fundamentals of Natural Science: Calcutta, Firma KLM Private Ltd., xxi + 593 p.

Chattopadhyaya, D., 1992, Lokayata—A Study in Ancient Indian Materialism: [New Delhi], People's Publishing House, xxvii + 696 p.

Chavanne, J., 1879, Die Sahara oder von Oase zu Oase. Bilder aus dem Natur- und Volksleben der grossen Afrikanischen Wüste: Wien, Pest und Leipzig, Hartleben, XVI + 639 p. + 7 colored plates.

Chavanne, J., 1881, Die mittlere Höhe Afrikas: Mittheilungen der kaiserlichen und königlichen geographischen Gesellschaft in Wien, v. 24, p. 340–377.

Chenoweth, W.L., 1997, John Strong Newberry—pioneer Colorado Plateau geologist: Canyon Legacy, no. 24, p. 2–4.

Chenoweth, W.L., 2000, Geology of Monument Valley Navajo Tribal Park, Utah-Arizona, *in* Sprinkel, D.A., Chidsey, T.C., Jr., and Anderson, P.B., eds., Geology of Utah's Parks and Monuments: Salt Lake City, Utah Geological Association Publication 28, p. 529–533.

Cherniss, H., and Helmbold, W.C., 1957, Plutarch Moralia volume XII, *in* Goold, G.P., ed., The Loeb Classical Library: Cambridge, Harvard University Press, xii + 590 p.

Chernov, V.G., 1989, Geologi Moskovskoyo Universiteta: [Moscow], Izdatelstvo Moskovskoyo Universiteta, 357 p.

Chesworth, W., 1975, Mantle plumes, plate tectonics, and the Cenozoic volcanism of the Massif Central: Journal of Geology, v. 83, p. 579–588.

Chevalier, F., 1997, Préface, *in* Alexandre de Humboldt Essai Politique sur le Royaume de la Nouvelle—Espagne du Mexique, v. 1: Paris, Utz, p. 9–40.

Chittenden, H.M., 1902a [1954], The American Fur Trade of the Far West—A History of the Pioneer Trading Posts and Early Fur Companies of the Missouri Valley and the Rocky Mountains and of the Overland commerce with Santa Fe, v. I: Stanford, Academic Reprints, xl + [i] + 482 p. + 1 foldout map.

Chittenden, H.M., 1902b [1954], The American Fur Trade of the Far West—A History of the Pioneer Trading Posts and Early Fur Companies of the Missouri Valley and the Rocky Mountains and of the Overland commerce with Santa Fe, v. II: Stanford, Academic Reprints, viii + [i] p. + p. 483–1029.

Chorley, R.J., Dunn, A.J., and Beckinsale, R.P., 1964, The History of the Study of Landforms or the Development of Geomorphology—volume one: Geomorphology before Davis: London, Methuen & Co., John Wiley & Sons, xvi + 678 p.

Chotard, H., 1860, Le Périple de la Mer Noire par Arrien—Traduction Étude Historique et Géographique Index et Carte: Paris, Auguste Durand, [i] + 240 p. + 1 colored foldout map.

Chronic, H., 1988, Pages of Stone: Geology of Western National Parks & Monuments—4: Grand Canyon and Plateau Country: Seattle, Washington, The Mountaineers, xiv + 158 p.

Chubb, L.J., 1957, The pattern of some Pacific island chains: Geological Magazine, v. 94, p. 221–228.

Clagett, M., 1994, Greek Science in Antiquity: Barnes & Noble, New York, xii + 217 p.

Clark, M.K., and Royden, L.H., 2000, Topographic ooze: building the eastern margin of Tibet by lower crustal flow: Geology, v. 28, p. 703–706.

Clarke, J., and Geikie, 1910, Physical Science in the Time of Nero Being a Translation of the *Quaestiones Naturales* of Seneca: London, Macmillan and Co., liv + 368 p.

Clarke, J.M., 1921, James Hall of Albany—Geologist and Palaeontologist: (publisher not indicated), Albany, 565 p.

Claval, P., 1995, Histoire de la Géographie: Que Sais-Je?: Paris, Presses Universitaires de France, 128 p.

Cloetingh, S., 1988, Intraplate stresses: A new element in basin analysis, *in* Kleinspehn, K.L., and Paola, C., eds., New Perspectives in Basin Analysis: Berlin, Springer-Verlag, p. 205–230.

Cloetingh, S., Kooi, H., and Groenewoud, W., 1989, Intraplate stresses and sedimentary basin evolution, *in* Price, R.A., ed., Origin and Evolution of Sedimentary Basins and Their Energy and Mineral Resources, Geophysical Monograph 48, IUGG Volume 3: Washington, D.C., American Geophysical Union, p. 1–16.

Cloetingh, S., McQueen, H., and Lambeck, K., 1985, On a tectonic mechanism for regional sealevel variations: Earth and Planetary Science Letters, v. 75, p. 157–166.

Cloos, H., 1916, Doggerammoniten aus den Molukken—Habilitationsschrift eingereicht bei der Hohen Philosophischen Fakultät der kgl. Universität Marburg zur Erlangung der Venia Docendi für Geologie und Palaeontologie I. Text [all published]: E. Schweizerbart'sche Verlagsbuchhandlung, Nägele und Dr. Sproesser, Stuttgart, 50 p.

Cloos, H., 1928, Über antithetische Bewegungen: Geologische Rundschau, v. 19, p. 246–251.

Cloos, H., 1932, Zur Mechanik großer Brüche und Gräben: Centralblatt für Mineralogie, Geologie und Paläontologie, Jahrgang 1932, Abteilung B, No. 6, p. 273–286.

Cloos, H., 1936, Einführung in die Geologie—Ein Lehrbuch der Inneren Dynamik: Berlin, Gebrüder Borntraeger, XII + 503 p.

Cloos, H., 1937, Zur Großtektonik Hochafrikas und seiner Umgebung—Eine Fragestellung: Geologische Rundschau, v. 28, p. 333–348 and plates VI and VII.

Cloos, H., 1939, Hebung-Spaltung-Vulkanismus—Elemente einer Geometrischen Analyse Irdischer Großformen: Geologische Rundschau, v. 30, Zwischenheft 4A, p. 405–527.

Cloos, H., 1940, Ein Blockbild von Deutschland—Erläuterung zu einer Tafel: Geologische Rundschau, v. 31, p. 148–153.

Cloos, H., 1947, Gespräch mit der Erde—Geologische Welt- und Lebensfahrt: München, R. Piper & Co., 410 p.

Cloos, H., 1948, Grundschollen und Erdnähte—Entwurf eines konservativen Erdbildes: Geologische Rundschau, v. 35, p. 133–154.

Cohen, M.R., and Drabkin, I.E., editors, 1958, A Source Book in Greek Science: Cambridge, Harvard University Press, xxi + 581 p.

Coleman, R.G., 1977, Ophiolites—Ancient Oceanic Lithosphere?: Minerals and Rocks series: Berlin, Springer-Verlag, IX + 229 p. + 1 map *hors-texte*.

Coleman, R.G., and Peterman, Z.E., 1975, Oceanic plagiogranite: Journal of Geophysical Research, v. 80, p. 1099–1108.

Colli, G., 1981, Die Geburt der Philosophie. Aus dem Italienischen von Reimar Klein. Mit einem Nachwort von Gianni Carchia und Reimar Klein: Frankfurt am Main, Europäische Verlagsanstalt, 128 p.

Collinson, C., Sargent, M.L., and Jennings, J.R., 1988, Illinois Basin region, *in* Sloss, L.L., ed., Sedimentary Cover—North American Craton; U.S.: Boulder, Colorado, Geological Society of America, Geology of North America, v. D-2, p. 383–426.

Colson, F.H., 1941, Philo, v. 9: The Loeb Classical Library: Cambridge, Harvard University Press, xii + 547 p.

Colton, G.W., 1970, The Appalachian Basin—its depositional sequences and their geologic relationships, *in* Fisher, G.W., Pettijohn, F.J. and Reed, J.C., Jr., eds., Studies of Appalachian Geology—Central and Southern: New York, Interscience Publishers, John Wiley & Sons, p. 5–47.

Columba, G.M., 1893, Gli Studi Geografici nel I Secolo dell'Impero Romano—Ricerche su Strabone, Mela e Plinio. Parte I. Le Dimensioni della Terra Abitata: Torina and Palermo, Carlo Clausen, VIII + 130 p.

Conche, M., 1991, Anaximandre—Fragments et Témoignages: Épiméthé: Paris, Presses Universitaires de France, 252 + [1] p.

Condie, K.C., 1989, Plate Tectonics & Crustal Evolution, third edition: Oxford, Pergamon Press, xii + 476 p.

Condie, K.C., and Sloan, R.E., 1998, Origin and Evolution of the Earth—Principles of Historical Geology: Upper Saddle River, Prentice Hall, viii + 498 p.

Coney, P.J., 1970, Geotectonic cycle and the new global tectonics: Geological Society of America Bulletin, v. 81, p. 739–747.

Contejean, C., 1874, Éléments de Géologie et de Paléontologie: Paris, J.-B. Bailliére, XX + 745 p.

Conybeare, W.D., and Phillips, W., 1822, Outlines of the Geology of England and Wales with an Introductory Compendium of the General Principles of that Science and comparative Views of the Structure of Foreign Countries: London, William Phillips, lxi + 470 p. + three foldout plates.

Cooley, M.E., Harshbarger, J.W., Akers, J.P., and Hardt, W.F., 1969, Regional Hydrogeology of the Navajo and Hopi Indian Reservations, Arizona, New Mexico, and Utah: [U.S.] Geological Survey Professional Paper 521-A: Washington, D.C., U.S. Government Printing Office, VI + 61 p.

Cordier, P., 1827, Essai sur la température de l'intérieur de la terre: Mémoires de l'Academie des Sciences pour l'année 1827, v. 7, p. 473–556.

Cornelius, H.-P., 1946–1948, Otto Ampferer: Mitteilungen der Geologischen Gesellschaft in Wien, v. 39–41, p. 195–213.

Cornford, F.M., 1912[1991], From Religion to Philosophy—A Study in the Origins of Western Speculation, with a foreword by Robert Ackerman: Princeton, Princeton University Press, xviii + 275 p.

Cornford, F.M., 1932, Before and After Socrates: Cambridge, Cambridge University Press, X + 113 p.

von Cotta, B., 1850, Geologische Briefe aus den Alpen: Leipzig, T.O., Weigel, VIII + 328 p. + V foldout plates.

Coues, E., 1895a, The Expeditions of Zebulon Montgomery Pike, To Headwaters of the Mississippi River, Through Louisiana Territory, and in New Spain, During the Years 1805–06–07. A New Edition, v. I: New York, Francis P. Harper, cxiii + 356 p.

Coues, E., 1895b, The Expeditions of Zebulon Montgomery Pike, To Headwaters of the Mississippi River, Through Louisiana Territory, and in New Spain, During the Years 1805–06–07. A New Edition, v. II: New York, Francis P. Harper, vi + p. 357–855.

Coues, E., 1895c, The Expeditions of Zebulon Montgomery Pike, To Headwaters of the Mississippi River, Through Louisiana Territory, and in New Spain, During the Years 1805–06–07. A New Edition, v. III, New York, Francis P. Harper, Index—Maps, p. [i] + 857–955, six folded maps + one page of map.

Coues, E., 1900a, On the Trail of a Spanish Pioneer—The Diary and Itinerary of Francisco Garcés (Missionary Priest) in his Travels through Sonora, Arizona and California 1775–1776 translated from an official contemporaneous copy of the original Spanish manuscript, and edited, with copious critical notes, v. I: New York, Francis P. Harper, xxx + 312 p. + 1 folded map.

Coues, E., 1900b, On the Trail of a Spanish Pioneer—The Diary and Itinerary of Francisco Garcés (Missionary Priest) in his Travels through Sonora, Arizona and California 1775–1776 translated from an official contemporaneous copy of the original Spanish manuscript, and edited, with copious critical notes, v. II: New York, Francis P. Harper, vi + [i] p. + p. 314–608.

Coulson, T., 1950, Joseph Henry—His Life and Work: Princeton, Princeton University Press, [ii] + 352 p.

Coupin, H., and Boudret, E., 1909, Géologie: Librairie Classique Fernand Nathan, [Paris], 352 p.
Cox, A., (editor), 1973, Plate Tectonics and Geomagnetic Reversals: San Francisco, W.H., Freemann and Company, vii + [ii] + 702.
Cox, A., and Hart, R.B., 1986, Plate Tectonics—How It Works: Oxford, Blackwell Scientific Publications, xxi + 392.
Crampton, C.G., 1958, Humboldt's Utah, 1811: Utah Historical Quarterly, v. 26, p. 269–281.
Crampton, C.G., and Griffen, G.G., 1956, The San Buenaventura, mythical river of the West: Pacific Historical Review, v. 25, p. 163–171.
Credner, H., 1872, Elemente der Geologie: Leipzig, Wilhelm Engelmann, XIV + 538 p.
Credner, H., 1906, Elemente der Geologie, zehnte unveränderte Auflage: Leipzig, Wilhelm Engelmann, XVIII + 802 p.
Crombie, A.C., 1952, Avicenna's influence on the medieval scientific tradition, *in* Wickens, G.M., ed., Avicenna: Scientist & Philosopher—A Millenary Symposium: London, Luzac & Company, p. 84–107.
Crombie, A.C., 1961, Augustine to Galileo, v. I Science in the Middle Ages: 290 p., v. II Science in the Later Middle Ages and early Modern Times 15th to 17th Centuries: Cambridge, Harvard University Press, 372 p.
Crombie, A.C., and North, J.D., 1981, Bacon, Roger, *in* Gillispie, C.C., editor-in-chief, Dictionary of Scientific Biography: New York, Charles Scribner's Sons, v. 1, p. 372–385.
Crombie, A.C., 1981, Descartes, René du Perron, *in* Gillispie, C.C., editor-in-chief, Dictionary of Scientific Biography: New York, Charles Scribner's Sons, v. 4, p. 51–55.
Crook, K.A., W., 1969, Contrasts between Atlantic and Pacific geosynclines: earth and Planetary Science Letters, v. 5, p. 429–438.
Crossette, G., 1946, Founders of the Cosmos Club 1878—A Collection of Biographical Sketches and Likenesses of the Sixty Founders: Cosmos Club, (no place of publication), 176 p.
Crossette, G., ed., 1970, Selected Prose of John Wesley Powell—A Collection of Essays & Articles Dogmas & Prophecies: Boston, David. R. Godine, [iii] + 122 p.
Crough, S.T., 1979, Hotspot epeirogeny: Tectonophysics, v. 61, p. 321–333.
Crough, S.T., 1983, Hotspot swells: Annual Review of Earth and Planetary Sciences, v. 11, p. 165–193.
Crumpler, L.S., 1982, Volcanism in the Mount Taylor region, *in* Grambling, J.A., and Wells, S.G., eds., New Mexico Geological Society Thirty-Third Annual Field Conference November 4–6, 1982, Albuquerque Country II, p. 291–298.
Cumming, W.P., Hillier, S.E., Quinn, D.B., and Williams, G., 1974, The Exploration of North America 1630–1776: G.P.: New York, Putnam's Sons, 272 p.
Currey, D.R., 1980, Coastal geomorphology of Great Salt Lake and Vicinity, *in* Gwynn, J.W., ed., 1980, Grat Salt Lake—A Scientific, Historical & Economic Overview: Utah Geological and Mineral Survey a division of the Utah Department of Natural Resources, Bulletin 116, p. 69–82 + 1 folded map in pocket.
Cuvier, G., 1796, Mémoire sur les espèces d'Eléphans tant vivantes que fossiles, lu à la séance publique de l'Institut national le 15 germinal, an IV: Magasin Encyclopédique, 2. année, no. 3, p. 440–445.
Cuvier, G., 1812, Recherches sur les Ossemens Fossiles de Quadrupeds où l'on rétablit les caractères de plusieurs espèces d'animaux que les révolutions du globe paroissent avoir détruites: Paris, Deterville, v. 1, not consecutively paginated.
Cuvier, G., 1825, Discours sur les Révolutions de la Surface du Globe et sur les Changements Qu'Elles Ont Produits Dans le Régne Animal: Paris, G. Dufour et Ed. D'Ocagne, ij + 400 p. + 6 plates.
Cuvier, G., 1827, Essay of the Theory of the Earth, with geological illustrations, by Professor Jameson—fifth edition: London, William Blackwood, Edinburg, and T. Cadell, xxiv + 550 p.
Cuvier, G., 1841, Histoire des Sciences Naturelles, depuis leur Origine jusqu'a nos Jours, chez tous les Peuples Connus ... complétée, rédigée, annotée et publiée par M.M., de Saint Agy, première partie, comprenant les siècles antérieurs au 16e de notre ère. Tome Premier: Paris, Fortin, Masson et Cie, III + 441 p.
Cuvier, G. and Brogniart, A., 1811, Essai sur la Géographie Minéralogique des Environs de Paris, avec une carte géognostique et des coupes de terrain: Paris, Baudouin, Imprimeur de l'Institut Impérial de France, viij + 278 p. + 2 plates and a colored foldout map.
Dale, H.C., 1941, The Ashley-Smith Explorations and the Discovery of a central Route to the Pacific 1822–1829—With the Original Journals Edited ..., revised edition: Glendale, The Arthur H. Clark Company, 360 p. + 1 colored foldout map of routes.
Dales, R.C., 1973, The Scientific Achievement of the Middle Ages: Philadelphia, University of Pennsylvania Press, ix + 182 p.
Daly, J.F., S.J., 1981, Sacrobosco, Johannes de (or John of Holywood), *in* Gillispie, C.C., editor-in-chief, Dictionary of Scientific Biography: New York, Charles Scribner's Sons, v. 12, p. 60–63.
Daly, R.A., 1940, Strength and Structure of the Earth: New York, Prentice Hall, ix + 434 p.
Dampier, W.C., (Sir), 1961, A History of Science and its Relations With Philosophy & Religion, fourth edition reprinted with a postscript by I.B., Cohen: Cambridge, Cambridge University Press, xxvii + 544 p.
Dana, J.D., 1843, On the areas of subsidence in the Pacific, as indicated by the distribution of coral islands: American Journal of Science and Arts, v. 45, p. 131–135.
Dana, J.D., 1846, On the volcanoes of the Moon: American Journal of Science and Arts, 2nd series, v. 2, p. 335–355.
Dana, J.D., 1847a, Geological results of the earth's contraction in consequence of cooling: American Journal of Science and Arts, 2nd series, v. 3, p. 176–188.
Dana, J.D., 1847b, Origin of the continents: American Journal of Science and Arts, 2nd series, v. 3, p. 94–100.
Dana, J.D., [1849], United States Exploring Expedition. During the Years 1838, 1839, 1840, 1841, 1842. Under the Command of Charles Wilkes, U.S.N.—Geology: New York, Geo. P. Putnam, xii + 756 p.
Dana, J.D., 1853, On Coral Reefs and Islands: New York, G.P. Putnam & Co., 143 p. + 2 foldout maps.
Dana, J.D., 1863, Manual of Geology: Treating of the Principles of the Science with special reference to American Geological History, for the use of colleges, academies and schools of science: Philadelphia, Theodore Bliss & Co., xvi + 798 p.
Dana, J.D., 1866, Observations on the origin of some of the Earth's features: American Journal of Science and Arts, second series, v. 42, p. 205–211.
Dana, J.D., 1873a, On the origin of mountains: American Journal of Science and Arts, third series, v. 5, p. 347–350.
Dana, J.D., 1873b, On some results of the earth's contraction from cooling, including a discussion of the origin of mountains and the nature of the earth's interior: American Journal of Science, ser. 3, vol. 5, p. 423–443.
Dana, J.D., 1873c, On some results of the earth's contraction from cooling, including a discussion of the origin of mountains and the nature of the earth's interior: American Journal of Science, ser. 3, vol. 6, p. 6–14.
Dana, J.D., 1873d, On some results of the earth's contraction from cooling, including a discussion of the origin of mountains and the nature of the earth's interior: American Journal of Science, ser. 3, vol. 6, p. 104–115.
Dana, J.D., 1873e, On some results of the earth's contraction from cooling, including a discussion of the origin of mountains and the nature of the earth's interior: American Journal of Science, ser. 3, vol. 6, p. 161–171.
Dana, J.D., 1875, Manual of Geology: Treating the Principles of the Science with special reference to American Geological History, second edition: New York, Ivison, Blakeman, Taylor, and Co., xvi + 828 p.
Dana, J.D., 1891, Characteristics of Volcanoes, with Contributions of Facts and Principles from the Hawaiian Islands, including a historical review of Hawaiian volcanic action for the past sixty-seven years, a discussion of the relations of volcanic islands to deep-sea topography, and a chapter on volcanic island denudation: New York, Dodd, Mead and Company, xvi + 399 p. + 1 foldout frontispiece.

Daniels, G.H., 1968[1994], American Science in the Age of Jackson: History of American Science and Technology Series: Tuscaloosa, The University of Alabama, XX + 282 p.

Dankoff, R., 1973, The Alexander romance in the Diwan Lughat at-Turk: Humaniora Islamica, v. I, p. 233–244.

Dankoff, R., 2002, Evliya Çelebi and the Seyahatname, *in* Güzel, H. C., ed., The Turks (Ottomans): Ankara, Yeni Türkiye, v. 3, p. 605–626.

Dankoff, R. and Kelly, J., 1982, Mahmûd al-Kasgarî Compendium of the Turkic Dialects. Edited and translated with Introduction and Indices, Part I, *in* Tekin, Ş. and Tekin G.A., eds., Sources of Oriental Laguages and Literatures, 7, Turkish Sources VII, printed at Harvard University, Office of the University Publisher[441], xi + 416 p.

Darmstaedter, E., 1926, Georg Agricola 1494–1555 Leben und Werk: Münchner Beiträge zur Geschichte und Literatur der Naturwissenschaften und Medizin, v. 1: München, Verlag der Münchner Drucke, 96 p.

Darmesteter, J., (Translator) 1887, The Zend-Avesta, Part I, The Vendîdâd, *in* Max Müller, ed., Sacred Books of the East, v. 4: Oxford, at the Clarendon Press, p. cii + 240.

Darmesteter, J., 1960, Le Zend-Avesta, Traduction Nouvelle avec Commantaire Historique et Philologique, deuxième volume, La Loi (Vendidad)—L'Épopée (Yashts) Le Livre de Prière (Khorda Avesta), reproduction photographique de l' édition princeps 1892–93: Librairie d'Amerique et d'Orient Adrien-Maisonneuve, Paris, XXXV + 747 p.

Darrah, W.C., 1951, Powell of the Colorado: Princeton, Princeton University Press, [vii] + 426 p.

Darwin, C.R., 1838[1977], On certain areas of elevation and subsidence in the Pacific and Indian Oceans, as deduced from the study of coral formations: Proceedings of the Geological Society of London, v. 2, p. 552–554 (reprinted in: Barrett, P.H., ed., The Collected Papers of Charles Darwin: Chicago, University of Chicago Press, p. 46–49).

Darwin, C.R., 1840[1977], On the connexion of certain volcanic phenomena in South America; and on the formation of mountain chains and volcanoes, as the effect of the same power by which continents are elevated: Transactions of the Geological Society of London, second series, part 3, v. 5, p. 601–631 (reprinted in: Barrett, P.H., ed., The Collected Papers of Charles Darwin, Chicago, University of Chicago Press, p. 53–86).

Darwin, C.R., 1842, The Structure and Distribution of Coral Reefs; Being the First Part of the Geology of the voyage of the *Beagle*, under the Command of Capt. Fitzroy, R.N., during the Years 1832 to 1836: London, Smith Elder, xii + 314 p. + 2 p. of errata + 3 folding maps.

Darwin, C., [1876], Geological Observations on the Volcanic Islands and Parts of South America: New York, D. Appleton and Co., xiii + 648 p. + 5 folded plates.

Darwin, C., 1889[1987], The Geology of the Voyage of H.M.S. *Beagle* Part I: Structure and Distribution of Coral Reefs, *in* Barrett, P.H., and Freeman, R.B., eds., The Works of Charels Darwin, v. 7: New York, New York University Press, 15 + xx + 344 p.

Darwin, C.R., 1962, Coral Islands by Charles Darwin with Introduction, map and remarks by D.R., Stoddart: Atoll Research Bulletin, v. 88, p. i–iv + 1–20.

Daubrée, A., 1860, Études et Éxperiences Synthétiques sur le Métamorphisme et sur la Formation des Roches Crystallines: Paris, Imprimerie Impériale, VII + 127 p.

Daubrée, A., 1867, Rapport sur les Progrès de la Géologie Expérimentale: Recueil de Rapports sur les Progrès des Lettres et des Sciences en France: Paris, Imprimerie Impériale, 142 p.

Daubrée, A., 1879, Études Synthétiques de Géologie Expérimentale: Paris, Dunod, III + 828 p.

Daubrée, A., 1880, Descartes, l'un des créateurs de la cosmologie et géologie: Extrait du Journal des Savants—Mars, Avril, 1880, 27 p.

Davidson, G., 1886, An examination of some of the early voyages of discovery and exploration on the northwest coast of America, from 1539 to 1603: United States Coast and Geodetic Survey, F.M., Thorn, Superintendent, Report for 1886, Appendix 7, p. 153–253 + 1 folded map.

Davies, F.F., and Pribac, F., 1993, Mesozoic seafloor subsidence and the Darwin Rise, past and present: American Geophysical Union Geophysical Monograph 77, p. 39–52.

Davies, G.L., [Herries], 1964, Robert Hooke and his conception of the earth history: Proceedings of the Geologists' Association, v. 75, p. 493–498.

Davies, G.L., [Herries], 1969, The Earth in Decay—A History of British Geomorphology: London, Macdonald Technical and Scientific, xvi + 390 p.

Davis, E.N., 1957, Die Jungvulkanischen Gesteine der Aegina, Methana und Poros und deren Stellung im Rahmen der Kykladenprovinz: Publikationen herasgegeben von der Stiftung «Vulkaninstitut Immanuel Friedländer», Mineralogisch-Petrographisches Institut der Eidgennösische Technische Hochschule Zürich: Zürich, Guggenbühl & Huber, Schweizer Spiegel Verlag, 74 p. + VI plates.

Davis, G.A., 1980, Problems of intraplate extensional tectonics, Western United States, *in* Continental Tectonics, Studies in Geophysics, Geophysics Study Committee: Washington, D.C., National Academy Press, p. 84–95.

Davis, G.A., Burchfiel, B.C., Case, J.E., and Viele, G.W., 1974, A defense of an "Old Global Tectonics," *in* Kahle, C.F., ed., Plate Tectonics—Assessments and Reassessments: American Association of Petroleum Geologists Memoir 23, p. 16–23.

Davis, G.H., and VandenDolder, E.M., editors, 1987, Geologic Diversity of Arizona and Its Margins: Excursions to Choice Areas: Field-Trip Guidebook, The Geological Society of America, 100th Annual Meeting, Phoenix, Arizona October 26–29, 1987: Arizona Bureau of Geology and Mineral Technology, Geological Survey Branch, Special Paper 5, vi + 422 p.

Davis, W.M., 1909, The lessons of the Grand Canyon: Bulletin of the American Geographical Society, v. 41, p. 345–354.

Davis, W.M., 1913, Dana's confirmation of Darwin's theory of coral reefs: American Journal of Science, fourth series, v. 35, p. 173–188.

Davis, W.M., 1915, Biographical Memoir, John Wesley Powell, 1834–1902: Washington, D.C., National Academy of Sciences, v. 8, p. 11–83.

Davis, W.M., 1924, The progress of geography in the United States: Annals of the Assocation of American Geographers, v. 14, p. 159–215.

Davis, W.M., 1927, Biographical Memoir, Grove Karl Gilbert, 1843–1918: Washington, D.C., National Academy of Sciences, v. 21, fifth memoir, v + 303 p.

Davis, W.M., 1928, The Coral Reef Problem: American Geographical Society Special Publication No. 9, [2] + 596 p.

Davison, C., 1927, The Founders of Seismology: Cambridge, Cambridge University Press, p. x + [ii] + 240 p.

Davison, J., and Kee, T.L., 1994, Mapping the Continent of Asia: Singapore, Antiques of the Orient, 88 p.

Dawdy, D.O., 1993, George Montague Wheeler—The Man and The Myth: Athens, Swallow Press/Ohio University Press, viii + 122 p.

Dawson, R., editor, 1964, The Legacy of China: Oxford, Clarendon, xix + 392 p.

Day, A.G., 1940, Coronado's Quest—The Discovery of the Southwestern States: Berkeley, University of California Press, xvi + 418 p. + 1 foldout map.

Deacon, M., 1971, Scientists and the Sea 1650–1900—A Study of Marine Science: London, Academic Press, xvi + 445 p.

Dean, D.R., 1975, James Hutton on religion and geology: the unpublished preface to his Theory of the Earth (1788): Annals of Science, v. 32, p. 187–193.

Dean, D.R., 1977, Essay Review R. Fox (ed.), *Lyell centenary issue: papers delivered at the Charles Lyell centenary Symposium, London 1975. (The British journal for the history of science*, 9 (1976), part 2.) London: British Society for the History of Science. 159 p. £ 4.00: Annals of Science, v. 34, p. 607–611.

Dean, D.R., 1980, Graham Island, Charles Lyell, and the craters of elevation controversy: Isis, v. 71, p. 571–588.

Dean, D.R., 1985, The rise and fall of the Deluge: Journal of Geological Education, v. 33, p. 84–93.

Dean, D.R., 1992, James Hutton and the History of Geology: Ithaca, Cornell University Press, xiii + [iii] + 303 p.

Dean, D.R., 1997, James Hutton in the Field and in the Study: Being an Augmented Reprinting of Vol. III of Hutton's Theory of the Earth as first published by Sir Archibald Geikie (1899)—A Bicentenary Tribute to the Father of Modern Geology: Scholars' Facsimiles & Reprints, Delmar, New York, irregularly paginated.

Debelmas, J., 1982, Alpes de Savoie: Guides Géologiques Régionaux: Paris, Masson, 182 p.

Debelmas, J., Lemoine, M., and Mattauer, M., 1966, Quelques remarques sur le concept de géosynclinal: Revue de Géographie Physique et de Géologie Dynamique, ser. 2, v. 8, p. 133–150.

Debelmas, J., Lemoine, M., and Mattauer, M., 1967, Essay Review, Geosynclines: American Journal of Science, v. 265, p. 292–300.

Debelmas, J., and Mascle, G., 1993, Les Grandes Structures Géologiques, 2e édition: Paris, Masson, VIII + 299 p.

Debenham, F., 1960, 6000 Jahre Mussten Vergehen ...—Entdeckung und Erforschung unserer Erde von den Anfängen bis Heute. Einführung von Edward Shackleton: Stuttgart, Chr. Belser Verlag, 270 p.

Decobecq, D., 1993, L'origine des cratères de la Lune selon G. Poulett Scrope (1864): Lave, no 44, p. 11–13.

DeFord, R.K., 1981, Gilbert, Grove Karl, *in* Gillispie, C.C., editor-in-chief, Dictionary of Scientific Biography: New York, Charles Scribner's Sons, v. 5, p. 395–396.

Deismann, A., 1933, Forschungen und Funde im Serai: Walter de Gruyter, Berlin, XI + 144 p.

Delage, E., 1930, La Géographie dans les Argonautiques d'Apollonios de Rhodes: Bibliothéque des Universités du Midi, Fascicule XIX, 310 + [1 p of errata].

Delatte, A., 1929/1930, Geographica: Byzantinische Zeitschrift, v. 30 (A. Heisenberg Festschrift), p. 511–518.

Delitzsch, F., 1881, Wo Lag Das Paradies? Eine Biblisch-Assyrologische Studie: J.C., Hinrich'sche Buchhandlung, Leipzig, X + [IV] + 346 p. + 1 foldout map.

Dellenbaugh, F.S., 1897, The true route of Coronado's march: Bulletin of the American Geographical Society, v. 29, p. 399–431.

Dellenbaugh, F.S., 1903, The Romance of the Colorado River—The Story of Its Discovery in 1540, with an Account of the later Explorations, and with Special Reference to the Voyages of Powell through the Line of the Grand Canyon: New York, G.P. Putnam's Sons, xxxv + 399 p.

Dellenbaugh, F.S., 1908[undated], A Canyon Voyage—The narrative of the Second Powell Expedition down the Green-Colorado River from Wyoming, and the Explorations on Land, in the Years 1871 and 1872: The University of Arizona Press, Tucson, xxxvi + 277 p.[442]

Dellenhbaugh, F.S., 1909, John Wesley Powell: A Brief review of his career, *in* The Romance of the Colorado River ..., 3rd edition: New York, G.P. Putnam's Sons, p. 371–386.

Demircioğlu, H., 1939, Der Gott auf dem Stier. Geschichte eines religiösen Bildtypus: Neue Deutsche Forschungen Abteilung Alte Geschichte (P. Strack, ed.), v. 6, Junker und Dünnhaupt Verlag, Berlin, XI + 151 p. + 1 map + 4 photographic plates.

De Mulder, M., 1981, Contribution à l'étude géomorphologique du volcan Karisimbi, Chaîne des Virunga (Republique du Rwanda): Musée Royal de l'Afrique Centrale, Tervuren (Belgique), Departement de Géologie et Minéralogie, Rapport Annuel pour l'Année 1980, p. 97–109.

Dennis, J.G., 1967, International Tectonic Dictionary—English Terminology: American Association of Petroleum Geologists Memoir 7, xi + 196 p.

Dennis, J.G., Murawski, H., and Weber, K., editors, 1979, International Tectonic Lexicon—A Prodrome, International Union of Geological Sciences, International Geological Correlation Program Project No. 100 E.M., Delany, Paris, Project Leader, E. Schweizerbart'sche Verlgsbuchhandlung (Nägele u. Obermiller), Stuttgart, V + 153 p.

Dennis, J.G., and Murawski, H., editors, 1988, International Tectonic Lexicon—A Prodrome, International Union of Geological Sciences, Commission for the Geological map of the World, E. Schweizerbart'sche Verlgsbuchhandlung (Nägele u. Obermiller), Stuttgart, VII + 119 p.

Dennis, J.G., editor, 1982, Orogeny: Benchmark Papers in Geology/62, Hutchinson Ross, Stroudsburg, xv + 379 p.

Descartes, R., 1644[1842], Les Principes de la Philosophie, *in* Aimé-Martin, L., ed., Oeuvres de Descartes Publiées d'après les Textes Originaux: Société du Panthoén Littéraire, Paris, p. 273–420.

DeVoto, B., 1952, The Course of Empire: Boston, Houghton Mifflin Company, xxii + [xii] + 647 p.

Desmond, A. and Moore, J., 1991, Darwin—The Life of a Tormented Evolutionist: New York, W.W. Norton & Company, xxi + 808 p.

Dewey, J.F., 1969a. Structure and sequence in paratectonic British Caledonides, *in* Kay, M., ed., North Atlantic—Geology and Continental Drift, American Association of Petroleum Geologists Memoir 12, p. 309–335.

Dewey, J.F., 1969b. Continental margins: a model for the transition from Atlantic type to Andean type: Earth and Planetary Science Letters, v. 6, p. 189–197.

Dewey, J.F., 1977, Suture zone complexities: a review: Tectonophysics, v. 40, p. 54–67.

Dewey, J.F., 1980, Episodicity, sequence and style at convergent plate boundaries, *in* Strangway, D.W., ed., The Continental Crust and Its Mineral Deposits (A Volume in Honour of J. Tuzo Wilson): Geological Association of Canada Special Paper Number 20, p. 553–573.

Dewey, J.F., 1982. Plate tectonics and the evolution of the British Isles. Journal of the Geological Society of London, v. 139, p. 371–412.

Dewey, J.F., 1987, Suture, *in* Seyfert, C.K., ed., Encyclopedia of Structural Geology and Plate Tectonics: New York, Van Nostrand Reinhold Co., p. 775–783.

Dewey, J.F., 2002, Transtension in arcs and orogens, international Geology Review (George A. Thompson Symposium volume), v. 44, p. 402–439

Dewey, J.F., and Bird, J.M., 1970, Plate tectonics and geosynclines: Tectonophysics, v. 10, p. 625–638.

Dewey, J.F., and Pitman, W.C., III., 1998, Sea-level changes: mechanisms, magnitudes and rates: Society of Economic Paleontologists and Mineralogists, Special Publication 58, p. 1–16.

Dewey, J.F., Pitman, W.C., III, Ryan, W.B.F. and Bonnin, J., 1973, Plate tectonics and the evolution of the Alpine System: Geological Soicety of America Bulletin, v. 84, p. 3137–3180.

Dewey, J.F., and Windley, B.F., 1988, Palaeocene-Oligocene tectonics of NW Europe: Geological Society of London Special Publication 39, p. 25–31.

Dickinson, W.R., 1971, Plate tectonic models of geosynclines: Earth and Planetary Science Letters, v. 10, p. 165–174.

Dickinson, W.R., 1974, Plate tectonics and sedimentation, *in* Dickinson, W.R., ed., Tectonics and Sedimentation, Society of Economic Paleontologists and Mineralogists, Special Publication 22, p. 1–27.

Dickinson, W.R., 2002, The Basin and Range Province as a composite extensional domain, international Geology Review, v. 44, p. 1–38.

Dicks, D.R., 1960, The Geographical Fragments of Hipparchus: London, The Athltone Press, xi + 215 p.

Dicks, D.R., 1970, Early Greek Astronomy to Aristotle: London, Thames and Hudson, 272 p.

Dicks, D.R., 1981, Eratosthenes, *in* Gillispie, C.C., editor-in-chief, Dictionary of Scientific Biography: New York, Charles Scribner's Sons, v. 4, p. 388–393.

Diehl, C., 1957, Byzantium—Greatness and Decline: New Brunswick, Rutgers University Press, xviii + 366 p.

Diels, H., 1951 [1996], Die Fragmente der Vorsokratiker, 6th edition, v. I (XI + [I] + 504 p.), v. II (428 p.), v. III (660 p.), edited by Walther Kranz: Zürich, Weidmann.

Diener, C., 1886, Libanon. Grundlinien der Physischen Geographie und Geologie von Mittel-Syrien: Wien, Alfred Hölder, X + 412 p.

Dieterich, J.H., 1988, Growth and persistence of Hawaiian volcanic rift zones: Journal of Geophysical Research, v. 93, p. 4258–4270.

Dieterici, F., 1861[1999], Die Naturanschauung und Naturphilosophie der Araber im zehnten Jahrhundert—Aus den Schriften der lautern Brüder: Verlag der Nicolai'schen Sort.-Buchhandlung (M. Jagielski), Berlin XVI + 216 p. (reprinted in: Sezgin, F., ed., 1966, Islamic Philosophy

v. 23 Friedrich Dieterici Die Naturanschauung und Naturphilosophie der Araber im zehnten Jahrhundert—Die Anthropologie der Araber im zehnten Jahrhundert n. Chr. Being a German Translation of the Raşâ'il Ikhwan aş-Şafa Chapters 14–30: Frankfurt am Main, Publications of the Institute for the History of Arabic-Islamic Science at the Johann Wolfgang Goethe University, not consecutively paginated).

Dietz, R.S., 1963, Collapsing continental rises: an actualistic concept of geosynclines and mountain building: Journal of Geology, v. 71, p. 314–333.

Dietz, R.S., 1972, Geosynclines, mountains and continent-building: Scientific American, v. 226, no. 3, p. 30–38.

Dietz, R.S., and Holden, J.C., 1966, Miogeoclines (miogeosynclines) in space and time: Journal of Geology, v. 74, p. 566–583.

Dilke, O.A.W., 1985, Greek and Roman Maps: London, Thames and Hudson, 224 p.

Diller, H., 1934, Wanderarzt und Aitiologe—Studien zur hippokratischen Schrift περι αερον γλατων τοπον: Philologus, Supplementband 26, no. 3, VIII + 121 p.

Diller, J.S., 1911, Major Clarence Edward Dutton: The Bulletin of the Seismological Society of America, v. 1, p. 137–142.

Diller, J.S., 1913, Memoir of Clarence Edward Dutton: Geological Society of America Bulletin, v. 24, p. 10–18.

Dillmann, F.-X., 1991, L'Edda—Récits de mythologie nordique par Snorri Sturluson: Paris, Gallimard, 233 p.

Dingle, H., 1958, Emanuel Swedenborgs Leistung für die Wissenschaft: Endeavour, v. 17, no. 67, p. 127–132.

Direktion der Deutschen Seewarte, 1882, Atlantischer Ozean. Ein Atlas von 36 Karten, die Physikalischen Verhältnisse und die Verkehrs-Strassen Darstellend, mit einer Erläuternden Einleitung und als Beilage zum Segelhandbuch für den Atlantischen Ozean: l, Hamburg, Friedrichsen & Co., [II] + 11 p. + 36 plates.

Direktion der Deutschen Seewarte, 1891, Indischer Ozean. Ein Atlas von 35 Karten, die Physikalischen Verhältnisse und die Verkehrs-Strassen Darstellend, mit einer Erläuternden Einleitung und als Beilage zum Segelhandbuch für den Indischen Ozean: l, Hamburg, Friedrichsen & Co., 16 p. + 35 plates.

Doblhofer, E., 1961, Voices in Stone—The Decipherment of Ancient Scripts and Writings (translated by Mervyn Savill): New York, The Viking Press, 327 p.

Dolnick, E., 2001, Down the Great Unknown—John Wesley Powell's 1869 Journey of Discovery and Tragedy Through the Grand Canyon: New York, HarperCollins Publishers, viii + [ii] + 367 p.

de Dolomieu, [D.], an VI[1797], Suite de Rapport fait à l'Institut National, par C.en Dolomieu, Ingénieur des mines, sur ses voyages de l'an V et de l'an VI: Journal des Mines, no. 42 (Ventôse), p. 406–432.

Dorobek, S.L., and Ross, G.M., eds., 1995, Stratigraphic Evolution of Foreland Basins: SEPM (Society for Sedimentary Geology) Special Publication 52, v + [i] + 310 p.

Dorobek, S.L., and Ross, G.M., 1995b, Stratigraphic evolution of foreland basins introduction, *in* Dorobek, S.L., and Ross, G.M., eds., Stratigraphic Evolution of Foreland Basins: SEPM (Society for Sedimentary Geology) Special Publication 52, p. iii–v.

Dott, R.H., Jr., 1969, James Hutton and the concept of a dynamic earth, *in* Schneer, C.J., ed., Toward A History of Geology: Cambridge, The M.I.T. Press, p. 122–141.

Dott, R.H., Jr., 1974, The geosynclinal concept, *in* Dott, R.H., Jr., and Shaver, R.H., eds., Modern and Ancient Geosynclinal Sedimentation: Society of Economic Paleontologists and Mineralogists Special Publication 19, p. 1–13.

Dott, R.H., Jr., 1979, The Geosyncline—first major geological concept "made in America," *in* Schneer, C.J., ed., Two Hundred Years of Geology in America: Hanover, University Press of New England, p. 239–264.

Dott, R.H., Jr., 1985, James Hall's discovery of the craton: Boulder, Colorado, Geological Society of America, Centennial Special Volume I, p. 157–167.

Dott, R.H., Jr., 1992, An introduction to the ups and downs of eustasy: Geological Society of America Memoir 180, p. 1–16.

Dott, R.H., Jr., 1997, James Dwight Dana's old tectonics—global contraction under divine direction: American Journal of Science, v. 297, p. 283–311.

Dournes, J., 1991, Indigenous Indo-Chinese Cosmogony, *in* Bonnefoy, Y., ed., Mythologies, v. II: Chicago, Chicago University Press, p. 981–985.

Doust, H. and Omatsola, E., 1989, Niger Delta, *in* Edwards, J.D., and Santogrossi, P.A., eds., Divergent/Passive Margin Basins: American Association of Petroleum Geologists Memoir 48, p. 201–238.

Dragoni, G., 1979, Eratostene e l'Apogeo della Scienza Greca: CLUEB-Bologna, 305 p.

Drake, E.T., 1996, Restless Genius—Robert Hooke and His Earthly Thoughts: Oxford, Oxford University Press, xiv + 386 p.

Driscoll, N.W., and Karner, G.D., 1994, Flexural deformation due to Amazon fan loading: A feedback mechanism affecting sediment delivery to margins: Geology, v. 22, p. 1015–1018.

Dubois, G., 1976, Emile Argand, *in* Naturalistes Neuchâtelois du XX Siècle, Cahiers de l'Institut Neuchâtelois, Ed. de la Baconnière, Neuchâtel, p. 99–100.

Duffield, W.A., 1997, Volcanoes of Northern Arizona—Sleeping Giants of the Grand Canyon Region, photographs by Michael Collier: Grand Canyon, Arizona, Grand Canyon Association, [ii] + 68 p.

Dufrénoy, [P.-A.], and Élie de Beaumont, [L.], 1834, Sur les groupes du Cantal, du Mont-Dore, et sur les soulèvemens auxquels ces montagnes doivent leur relief actuel, *in* Mémoires pour Servir a une Description Géologique de la France, rédigée … sous la direction de M. Brochant de Villiers, par MM. Dufrénoy et Élie de Beaumont, tome second: Paris, F.-G. Levrault, p. 223–337.

Dufrénoy, [P.-A.-O.], and Élie de Beaumont, [L.], 1840, Carte Géologique de la France exécuté sous la direction de Mr. Brochant de Villiers …[1/500,000], [Paris], 6 sheets.

Dufrénoy, [P.-A.], and Élie de Beaumont, [L.], 1841, Explication de la Carte Géologique de la France, rédigée sous la direction de M. Brochant de Villiers …tome premier: Paris, Imprimerie Royale, [XII] + [1] + 825 p.

Dufrénoy, [P.-A.], and Élie de Beaumont, [L.], 1848, Explication de la Carte Géologique de la France, rédigée sous la direction de M. Brochant de Villiers …tome deuxième: Paris, Imprimerie Nationale, XII + 813 p.

Duhem, P., 1906[1984], Études sur Léonard de Vinci—Ceux qu'Il a Lus et Ceux qui L'Ont Lu, première série: A. Hermann et Fils, Paris; reprinted by Éditions des Archives Contemporaines, (no place of publication), VII + 355 p.

Duhem, P., 1909[1984], Études sur Léonard de Vinci—Ceux qu'Il a Lus et Ceux qui L'Ont Lu, seconde série: Paris, A. Hermann et Fils; reprinted by Éditions des Archives Contemporaines, (no place of publication), 474 p.

Duhem, P., 1913, Le Système du Monde—Histoire des Doctrines Cosmologiques de Platon a Copernic, v. 1: Paris, Hermann, 512 p.

Duhem, P., 1958a, Le Système du Monde—Histoire des Doctrines Cosmologiques de Platon a Copernic, v. 9 (Cinquième Partie: La Physique Parisienne au XIVe Siècle (suit)): Paris, Hermann, 442 p.

Duhem, P., 1958b, Le Système du Monde—Histoire des Doctrines Cosmologiques de Platon a Copernic, v. 3 (Deuxième Partie: L'Astronomie Latine au Moyen Age (Suite)): Paris, Hermann, 549 p.

Duhem, P., 1973, Le Système du Monde—Histoire des Doctrines Cosmologiques de Platon a Copernic, v. 4 nouveau triage: Paris, Hermann, 597 p.

Dumas, [J.-B.-A.], 1874, Discours prononcés aux funérailles de M. Élie de Beaumont: Annales des Mines, 7e série, Mémoires, v. 6, p. 187–215.

Dureau-de-la Malle, A., 1807, Géographie Physique de la Mer Noire, de l'Interieur de l'Afrique et de la Méditerranée: Paris, Dentu, X + 401 p. + 2 foldout maps.

Durham, M.S., 1997, Desert Between the Mountains: Mormons, Miners, Padres, Mountain Men, and the Opening of the Great Basin 1772–1869: New York, Henry Holt Company, xiv + 336 p.

Durou, J.-M., 1993, L'Exploration du Sahara, *in* Nyssen, H. and Wespieser, S., eds., "Terres d'Aventure," Babel (no place of publication), 404 p.

Du Toit, A.L., 1937, Our Wandering Continents—An Hypothesis of Continental Displacement: Edinburgh, Oliver and Boyd, xiii + 366 p.

Dutton, C.E., 1874, A criticism of the contractional hypothesis: The American Journal of Science and Arts, third series, v. 8, p. 113–123.

Dutton, C.E., 1876, Critical observations on theories of the Earth's physical evolution: The Penn Monthly, v. 7, p. 364–378 and 417–431.

Dutton, C.E., 1880, Report of the Geology of the High Plateaus of Utah: Washington, D.C., Government Printing Office, xxxii + 307 p.

Dutton, C.E., 1882a, The physical geology of the Grand Cañon District, *in* 2nd Annual Report of the U.S. Geological Survey to the Secretary of the Interior 1880–1881 by John Wesley Powell, Director, p. 47–166.

Dutton, C.E., 1882b, *Physics of the Earth's Crust*, by the Rev. Osmond Fisher, M.A., F.G.S.: American Journal of Science, 3rd series, v. 23, p. 283–290.

Dutton, C.E., 1892, On some of the greater problems of physical geology: Bulletin of the Philosophical Society of Washington, v. 11, p. 51–64.

Duviols, J.-P. and Minguet, C., 1994, Humboldt—Savant-Citoyen du Monde: [Paris], Decouvertes, Gallimard, 144 p.

Eaton, A., 1820, An Index to the Geology of the Northern States, with Transverse Sections, Extending from Sesquehanna River to the Atlantic, Crossing Catskill Mountains. To which is Prefixed a Geological Grammar. Second edition, wholly written over anew: Troy, William S. Parker, XI + 286 p. + 2 plates.

Eaton, A., 1830, Geological Text-Book, prepared for Popular Lectures on North American Geology; with applications to Agriculture and Arts: Printed by Websters and Skinners, Albany, vii + 63 + [1] p.

Eaton, G.P., 1982, The Basin and Range Province: Origin and tectonic significance: Annual Review of Earth and Planetary Sciences, v. 10, p. 409–440.

Eaton, G.P., 1986, A tectonic redefinition of the Southern Rocky Mountains: Tectonophysics, v. 132, p. 163–193.

Eaton, G.P., 1987, Topography and the origin of the southern Rocky Mountains and Alvarado Ridge, *in* Coward, M.P., Dewey, J.F., and Hancock, P.L., eds., 1987, Continental Extensional Tectonics, Geological Society (London) Special Publication 28, p. 355–369.

Egan, F., 1977, Frémont—Explorer for a Restless Nation: Garden City, Doubleday & Company, xv + 582 p.

Ehrenberg, C.G., 1870, Gedächtnisrede auf Alexander von Humboldt im Auftrage der königlichen Akademie der Wissenschaften zu Berlin gehalten in der Leibniz-Sitzung am 7. Juli 1859: Berlin, Robert Oppenheim, 46 p.

Eichholz, D.E., 1965, Theophrastus De Lapidibus: Oxford, at the Clarendon Press, vii + 141 p.

Von Eicken, H., 1887, Geschichte und System der Mittelalterlichen Weltanschauung: Stuttgart, Verlag der J.G., Cotta'scher Buchhandlung, XVI + 822 p.

Einsele, G., 1992, Sedimentary Basins—Evolution, Facies, and Sediment Budget: Berlin, Springer-Verlag, X + 628 p.

Eisbacher, G.H., 1991, Einführung in die Tektonik: Stuttgart, Ferdinand Enke, VIII + 310 p.

Eisenmenger, G., 1911, La Géologie et ses Phénomènes 12 Conférences: Paris, Pierre Roger & Cie, 328 p.

Ekman, M., 1989, Impacts of geodynamic phenomena on systems for height and gravity: Bulletin Géodésique, v. 63, p. 281–296.

Élie de Beaumont, L., 1828a, Notice sur un gisement de Végétaux fossiles de belemnites, situé à Petit-Cœur près Moutiers, en Tarentaise: Annales des Sciences Naturelles, v. 14, p. 113–127.

Élie de Beaumont, L., 1828b, Sur un gisement de Végétaux fosilles et de Graphite, situé au col du Chardonet (département des Hautes Alpes): Annales des Sciences Naturelles, v. 15, p. 353–381.

Élie de Beaumont, L., 1829, Faits pour servir a l'histoire des montagnes de l'Oisans: Mémoires de la Société d'Histoire Naturelle de Paris, t. V, p. 1–32 (extrait).

Élie de Beaumont, L., 1829–1830, Recherches sur quelques-unes des Révolutions de la surface du globe, présentant différens exemples de coïncidence entre le redressement des couches de certains systèmes de montagnes, et les changemens soudains qui ont produit les lignes de démarcation qu'on observe entre certains étages consécutifs des terrains de sédiment: Annales des Sciences Naturelles, v. 18, p. 5–25, 284–417, v. 19, p. 5–99, 177–240.

Élie de Beaumont, L., 1830, Recherches sur quelques-unes des Révolutions de la surface du globe, présentant différens exemples de coïncidence entre le redressement des couches de certains systèmes de montagnes, et les changemens soudains qui ont produit les lignes de démarcation qu'on observe entre certains étages consécutifs des terrains de sédiment: Revue Française, no. 15, p. 1–58.

Élie de Beaumont, L., 1831, Researches on some of the Revolutions which have taken place on the Surface of the Globe; presenting various Examples of the coincidence between the Elevation of Beds in certain Systems of Mountains, and the sudden Changes which have produced the Lines of Demarcation observable in certain Stages of the Sedimentary Deposits: The Philosophical Magazine and Annals of Philosophy [New Series], v. 10, p. 241–264.

Élie de Beaumont, L., 1833, Recherches sur quelques-unes des Révolutions de la surface du globe, présentant différens exemples de coïncidence entre le redressement des couches de certains systèmes de montagnes, et les changemens soudains qui ont produit les lignes de démarcation qu'on observe entre certains étages consécutifs des terrains de sédiment, *in* Manuel Géologique par Henry T. De La Beche, seconde édition, traduction française revue et publiée par A.J., M. Brochant de Villiers: F.-G. Levrault, Paris, p. 616–665:

Élie de Beaumont, L., 1849, Systèmes de Montagnes, *in* d'Orbigny, C., ed., Dictionnaire Universelle d'Histoire Naturelle, tome douzième—première partie: Paris, Renard, Martinet, Langlois et Leclerc, Victor Masson, p. 167–311.

Élie de Beaumont, L., 1852, Notice sur les Systèmes de Montagnes, tome III: P. Paris, Bertrand, p. 1069–1543 + 4 foldout plates.

Ellenberger, F., 1958, Étude Géologique du Pays de Vanoise: Ministère de l'Industie et du Commerce, Mémoires pour Servir à l'Éxplication de la Carte Géologique Détaillée de la France: Paris, Imprimerie Nationale, 561 p. + 42 plates.

Ellenberger, F., 1970, Quelques remarques historiques sur le concept de géosynclinal: Comptes Rendus hébdomadaires de l'Academie des Sciences (Paris), v. 271, p. 469–472.

Ellenberger, F., 1984, L'histoire des idees sur les chaînes de montagnes de Hutton à Wegener: présentation d'un ouvrage récent, avec commentaire critique: Travaux du Comité Français d'Histoire de la Géologie (COFRHIGEO), deuxième série, v. II, n°6, p. 63–88.

Ellenberger, F., 1988, Histoire de la Géologie, t. 1: Paris, Technique et Documentation—Lavoisier, VIII + 352 p.

Ellenberger, F., 1989, Épeirogenèse: Paris, Encyclopædia Universalis, v. 8, p. 21–24.

Ellenberger, F., 1994, Histoire de la Géologie, t. 2: Paris, Technique et Documentation—Lavoisier, XIV + 383 p.

Elliger, W., 1961, Philipp Melanchthon—Forschungsbeiträge zur vierhundertsten Wiederkehr seines Todestges dargeboten in Wittenberg 1960: Göttingen, Vandenhoeck & Ruprecht, 204 p. + 14 plates.

Elston, D.P., Billingsley, G.H., and Young, R.A., 1989, Geology of Grand Canyon, Northern Arizona (with Colorado River Guides)—Lees Ferry to Pierce Ferry, Arizona: 28th International Geological Congress: Washington, D.C., American Geophysical Union, xv + 239 p.

Embree, A.T., editor, 1991, Sources of Indian Tradition, second edition, v. 1, From the beginning to 1800: New Delhi, Penguin, xxxii + 447 p. + 1 map.

Emory, W.H., 1848, Notes of a Military Reconnaisance, from Fort Leavenworth, in Missouri, to San Diego, in California, Including Parts of the Arkansas, Del Norte, and Gila Rivers: Senate 30th Congress, 1st session, Executive, no. 7: Washington, D.C., Wendell and Bethuysen, 416 p. with one separate folded map.

von Engelhardt, W. (Baron), 1980, Carl von Linné und das Reich der Steine: Veröffentlichungen der Joachim-Jungius-Gesellscaft der Wissenschaften zu Hamburg, v. 43, p. 81–96.

von Engelhardt, W. (Baron), in press, Goethe und Alexander von Humboldt—Bau und Geschichte der Erde, *in* Symposium der Leopoldina Akademie der Naturforscher über Goethe und Alexander von Humboldt, Halle, 29. 10. 1999.

England, P., and McKenzie, D., 1982, A thin viscous sheet model for continental deformation: Geophysical Journal of the Royal Astronomical Society v. 70, p. 295–321.

England, P., and McKenzie, D., 1983, Correction to: A thin viscous sheet model for continental deformation: Geophysical Journal of the Royal Astronomical Society, v. 73, p. 523.

Erarslan, K., 1968, Seydî Ali Reis'in Çağatayca gazelleri: İstanbul Üniversitesi, Edebiyat Fakültesi, Türk Dili ve Edebiyatı Dergisi, v. 16, p. 41–54.

Eren, M., 1960, Evliya Çelebi Seyahatnâmesi Birinci Cildinin Kaynaklari Üzerine Bir Araştırma: İstanbul, Nurgök Matbaası, VIII+133 p. + 1 p. of errata.

Erinç, S., 1969, Klimatoloji ve Metodları, genişletilmiş 2. baskı: İstanbul Üniversitesi Yayınları No 994, XII + 538 p.

Ernst, R.E., and Buchan, K.L., 1997, Giant radiating dyke swarms: Their use in identifying pre-Mesozoic large igneous provinces and mantle plumes, *in* Mahoney, J.J., and M.F., Coffin, eds., Large Igneous Provinces, Continental, Oceanic, and Planetary Fyood Volcanism: Washington, D.C., American Geophysical Union Geophysical Monograph 100, p. 297–333.

Ernst, R.E., Buchan, K., and Palmer, H.C., 1995a, Giant dyke swarms: Characteristics, distribution and geotectonic implications, *in* Baer, G., and Heimann, A., eds., Physics and Chemistry of Dykes: Rotterdam, Balkema, p. 3–21.

Ernst, R.E., Head, J.W., Parfitt, E., Grosfils, E., and Wilson, L., 1995b, Giant radiating dyke swarms on Earth and Venus: Earth-Science Reviews, v. 39, p. 1–58.

Ernst, R.E., Buchan, K.L., West, T.D., and Palmer, H.C., 1996, Diabase (Dolerite) Dyke Swarms of the World: First Edition: Geological Survey of Canada Open File 3241, 104 p.

Esin, E., 1979, Türk Kosmolojisi (İlk Devir Üzerine Araştırmalar) Early Turkish Cosmology: Supplement to the Handbook of Turkish Culture, series II, Edebiyat Fakültesi Matbaası, İstanbul, XV + 194 p.

'Espinasse, M., 1956, Robert Hooke: London, Heinemann, xii + 192 p.

Etheridge, M., McQueen, H., and Lambeck, K., 1991, The role of intraplate stress in Tertiary (and Mesozoic) deformation of the Australian continent and its margins: A key factor in petroleum trap formation: Exploration Geophysics, v. 22, p. 123–128.

Evans, D.S., 1981, Herschel, John Frederick William, *in* Gillispie, C.C., editor-in-chief, Dictionary of Scientific Biography: New York, Charles Scribner's Sons, v. 6, p. 323–328.

Evenhuis, N.L., 1990, Dating of the livraisons and volumes of d'Orbigny's *Dictionnaire Universel d'Histoire Naturelle*: Bishop Museum Occasional Papers, v. 30, p. 219–225.

Evliya Çelebi ibn Muhammed Zıllî Darviş, 1314H [1896AD], Evliya Çelebi Seyahatnamesi: Dersaadet'de İkdam Maatbaası, v. I, 674 + [10] p.

Ewald, J., 1867, Leopold von Buch's Leben und Wirken bis zum Jahre 1806, *in* Ewald, J., Roth, J., und Eck, H., eds., Leopold von Buch's Gesammelte Schriften, erster Band: Berlin, Georg Reimer, p. V–XLVIII.

Eyles, V.A., 1981, De la Beche, Henry Thomas, *in* Gillispie, C.C., editor-in-chief, Dictionary of Scientific Biography: New York, Charles Scribner's Sons, v. 4, p. 9–11.

Eyre, A., 1948, The Famous Fremonts and Their America: The Fine Arts Press (no place of publication indicated), [viii] + 374 p. + 6 unnumbered maps.

Fagan, B., 1999, Floods, Famines, and Emperors—El Niño and the Fate of Civilizations: Basic Books, [New York], xix + 284 p.

Fairclough, H.R., 1963, Love of Nature Among the Greeks and Romans: Our Debt to Greece and Rome series: New York, Cooper Square Publishers, ix + 270 p.

Fallot, M.P., 1939, Élie de Beaumont et l'Évolution des Sciences Géologiques au Collège de France—Leçon inaugurale donné le 7 décembre 1938: Paris, Dunod, (extait des Annales des Mines, livraison d'Avril 1939), 35 p.

Farquhar, F.P., 1965, History of the Sierra Nevada: Berkeley, University of California Press, xiv + [ii] + 262 p.

Farrington, B., 1944, Greek Science—Its Meaning For Us (Thales to Aristotle): Harmondsworth, Penguin Books, 143 p.

Farrington, B., 1949, Greek Science—Its Meaning For Us II Thephrastus to Galen: Harmondsworth, Penguin Books, 181 p.

Favre, A., 1867, Recherches Géologiques dans les Parties de la Savoie du Piémont et de la Suisse voisines du Mont-BLanc, v. III: Paris, Victor Masson, 587 + [4] p.

Feller, R., 1935, Die Universität Bern 1834–1934: Bern, Paul Haupt Verlag, 647 p.

Ferguson, S., 1961, Major John Wesley Powell's exploration of the Grand Canyon: The Journal of Arizona History, v. 2, p. 34–38.

Fernández-Armesto, F., 1991, editor, The Times Atlas of World Exploration: New York, HarperCollins Publishers, 286 p.

F.G.-M., 1902, Conférences de Géologie—classe de seconde: Paris, Librairie Générale, XI + 280 p.

Ficker, G., and Trauth, F., 1913, Grundlinien der Mineralogie und Geologie für die fünfte Klasse der österreichischen Gymnasien: Wien, Franz Deuticke, XX + 140 p. + 1 colored folded map.

Fiero, B., 1986, Geology of the Great Basin: Max Fleischmann Series in Great Basin Natural History: Reno, University of Nevada Press, xiv + 198 p.

Figueirôa, S., and Lopes, M., organisers, 1994, Geological Sciences in Latin America—Scientific Relations and Exchanges: Universidade Estadual de Campinas, Instituto de Geociências, Campinas, iv + 402 p.

Fischer, H., 1994, Georgius Agricola—Bilder aus dem Leben eines großen deutschen Humanisten: Privately published, Quedlinburg, 320 p. + 14 plates in an appendix.

Fischer, J., 1913, An important Ptolemy manuscript with maps, in The New York Public Library: United States Catholic Historical Society, Historical Records and Studies, v. 6(2), p. 213–234.

Fischer, J., 1932a, Clavdii Ptolemaei Geographiae Codex Vrbinas Graecvs 82 Phototypice Depictvs Consilio et Opera Cvratorvm Bibliothecae Vaticanae: Codices e Vaticanis Selecti, vol. XVIIII, E.J., Brill, Leiden and Otto Harrasowitz, Leipzig, Tomvs Prodromvs (de Cl.Ptolemaei Vita Operibvs Geographia Praesertim Eisvsque Fatis Pars Rior Commentatio: xvi + 605 p. + [1]), Pars Prior Textvs (facsimile of manuscript) cvm appendice critica Pii Franchi de Cavalieri (37 p. + [2] + 76 plates of facsimile), Part Altera (Tabvlae Geographicae LXXXIII Graecae-Arabicae-Latinae e Codicivs LIII Selectae).

Fischer, J., 1932b [1991], Introduction, *in* Geography of Claudius Ptolemy, New York Public Library, New York, p. 3–15 (reprinted in 1991 by Dover Publications).

Fischer, H., 1994, Georgius Agricola—Bilder aus dem Leben eines großen deutschen Humanisten: Privately published, Quedlinburg, 320 p. + 14 plates in an appendix.

Fisher, J.H., Barrat, M.W., Droste, J.B., and Shaver, R.H., 1988, Michigan Basin, *in* Sloss, L.L., ed., Sedimentary Cover—North American Craton; U.S.: Boulder, Colorado, Geological Society of America, Geology of North America, v. D-2, p. 361–382.

Fisher, O., 1879, On the inequalities of the earth's surface as produced by lateral pressure, upon the hypothesis of a liquid substratum: Transactions of the Cambridge Philosophical Society, v. 12, part 2, p. 434–454.

Fisher, O., 1889, Physics of the Earth's Crust, second edition, altered and enlarged: London, Macmillan and Co., xvi + 391 + 60 p.

Fisher, O., 1895, On the age of the world as depending on the condition of the interior: The Geological Magazine, new series, decade 4, v. 2, p. 244–246.

Fisk, H.N., and McFarlan, E., Jr., 1955, Late Quaternary deltaic deposits of the Mississippi River: Geological Society of America Special Paper 62, p. 279–302.

Fjeldskaar, W., and Cathles, L., 1991, Rheology of mantle and lithosphere inferred from post-glacial uplift in Fennoscandia, *in* Sabadini, R., Lambeck, K., and Boschi, E., eds., Glacial Isostasy, Sea Level and Mantle Rheology: NATO ASI Series, v. C334: Dordrecht, Kluwer, p. 1–19.

Flanagan, K.M., and Montagne, J., 1993, Neogene stratigraphy and tectonics of Wyoming, *in* Snoke, A.W., Steidtmann, J.R., and Roberts, S.M., eds., Geology of Wyoming, v. 2: Laramie, Wyoming, Geological Survey of Wyoming Memoir No. 5, p. 572–607.

Fleischer, H.L., 1851, Über das türkische Chatâi-name: Berichte der Königlichen Sächsischen Gesellschaft der Wissenschaften, Philologisch-historische Klasse, v. 3, p. 317–327 (reprinted in Fleischer's *Kleinere Schriften*, Leipzig, v. III, p. 214–225; also in *Texts and Studies on the Historical Geography and Topography of East Asia, in* Series Islamic Geography of the Institut für Geschichte der Arabisch-Islamischen Wisenschaften and der Johann Wolfgang von Goethe-Universität, Frankfurt am Main, Veröffentlichungen des Institutes für Geschichte der Arabisch-Islamischen Wisenschaften, v. 126, p. 96–107).

Flint, R.F., 1971, Glacial and Quaternary Geology: New York, John Wiley & Sons Inc., xii + 892 p.

Fontana, B., 1996, Biography of a Desert Church: The Story of Mission San Xavier del Bac (revised): The Smoke Signal, Spring 1996, no. 3, 68 p.

Foose, R.M., 1973, Rein W. van Bemmelen—An appreciation, *in* de Jong, K.A., and Scholten, R., eds., Gravity and Tectonics: New York, John Wiley & Sons, p. xxv–xxx.

Forbiger, A., 1844, Handbuch der Alten Geographie aus den Quellen bearbeitet, zweiter Band. Politische Geographie der Alten. Asia. Africa.: Leipzig, Verlag von Mayer und Wigand, X + [II] + 920 p. + 1 foldout map.

Forster, J.R., 1778, Observations Made During a Voyage Around the World, on Physical Geography, Natural History, and Ethic Philosophy. Especially on 1. The Earth and Its Strata, 2. Water and the Ocean, 3. the Atmosphere, 4. the Changes of the Globe, 5. Organic Bodies, and 6. the Human Species: London, G. Robinson, iv + 16 + 10–649 + [1] p.

Forsyth, D.W., 1979, Lithospheric flexure: Reviews of Geophysics and Space Physics, v. 17, p. 1109–1114.

Fortenbaugh, W.W., and Sharples, R.W., editors, 1988, Theophrastean Studies On Natural Science, Physics, and Metaphysics, Ethics, Religion and Rhetoric: Rutgers University Studies in Classical Humanities, v. III: New Brunswick, Transaction Books, ix + 348 p.

Fortenbaugh, W.W., Huby, P.M., and Long, A.A., editors, 1985, Theophrastus of Eresus—On His Life and Work: Rutgers University Studies in Classical Humanities, v. II: New Brunswick, Transaction Books, ix + 355 p.

Fortenbaugh, W.W., Huby, P.M., Sharples, R.W., and Gutas, D., editors, 1993a, Theophrastus of Eresus—Sources for His Life, Writings, Thought & Influence, Part One Life, Writings, Various Reports, Logic, Physics Metaphysics, Theology, Mathematics: Leiden, E.J. Brill, viii + 465 p.

Fortenbaugh, W.W., Huby, P.M., Sharples, R.W., and Gutas, D., editors, 1993b, Theophrastus of Eresus—Sources for His Life, Writings, Thought & Influence, Part Two Psychology, Human Physiology, Living Creatures, Botany, Ethics, Religion, Politics, Rhetoric and Poetics, Music, Miscellanea: Leiden, E.J. Brill, vii + 705 p.

Foster, M., 1994, Strange Genius—The Life of Ferdinand Vandeveer Hayden: Niwot, Colorado, Roberts Reinhart Publishers, xv + 443 p.

Fowler, D.D., Euler, R.C., and Fowler, C.S., 1969, John Wesley Powell and the Anthropology of the Canyon Country—A Description of John Wesley Powell's Anthropological Fieldwork, the Archeology of the Canyon Country, and Extracts from Powell's Notes on the Origins, Customs, Practices, and Beliefs of the Indians of that Area: [U.S.] Geological Survey Professional Paper 670, V + 30 p.

Fraas, O., 1867, Aus dem Orient—Geologische Beobachtungen am Nil, auf der Sinai Halbinsel und in Syrien: Stuttgart, Ebner & Seubert, VIII + 222 p.

Frängsmyr, T., 1990, Urban Hiärne and the problem of land uplift: Striae, v. 31, p. 7–9

Frängsmyr, T., editor, 1994a, Linneaus—The Man and His Work: Canton, Science History Publications/USA, xiv + 206 p.

Frängsmyr, T., 1994b, Linnaeus as a geologist, *in* Frängsmyr, T., ed., Linneaus—The Man and His Work: Canton, Science History Publications/USA, p. 110–155.

Frank, E., 1923, Plato und die Sogenannten Pythagoreer—Ein Kapitel aus der Geschichte des Griechischen Geistes: Halle, Max Niemeyer, X + 399 p.

Frankfort, H., Frankfort, H.A., Wilson, J.A., and Jacobsen, T., 1946[1949], Before Philosophy—The Intellectual Adventure of Ancient Man: Baltimore, Penguin Books, 275 p.

Franz, I., 2000, Die Tätigkeit an der Erstellung einer erneuerten Bibliographie Georg Agricolas (1494–1555), *in* Cultural Heritage in Geology, Mining and Metallurgy. Libraries-Archives-Museums. 3rd International "Erbe" Symposium, June 23–27-Saint-Petersburg, Russia, Berichte der Geologischen Bundesanstalt, v. 52, p. 27–32.

Frapolli, L., 1847, Réflexions sur la nature et sur l'application du caractère géologique: Bulletin de la Société Géologique de France, 2e série, v. 4, p. 604–646.

Fraser, P.M., 1971, Eratosthenes of Cyrene: Lecture on a Master Mind Series of the British Academy, Proceedings of the British Academy, v. 56: Oxford, Oxford University Press, 35 p. (p. 175–207).

Fraser, P.M., 1972a, Ptolemaic Alexandria, I. Text: Oxford, at the Clarendon Press, xv + [i] + 812 p. + 1 foldout map.

Fraser, P.M., 1972b, Ptolemaic Alexandria, II. Notes: Oxford, at the Clarendon Press, xiii + 1116 p.

Fraser, P.M., 1972c, Ptolemaic Alexandria, III. Indexes: Oxford, at the Clarendon Press, [i] + 157 p.

Fraustadt, G., and Prescher, H., 1956, Einführung in die geologischen Werke, *in* Fraustadt, G., and Prescher, H., eds., Georgius Agricola Schriften zur Geologie und Mineralogie I, Georgius Agricola—Ausewählte Werke, Gedenkausgabe des Staatlichen Museums für Mineralogie und Geologie zu Dresden, Herausgeber Dr. rer. nat. Hans Prescher: Berlin, VEB Verlag der Wissenschaften, p. 47–68.

Frazer, J.G., 1919, Adonis Attis Osiris—Studies in the History of Oriental Religion, v. I (The Golden Bough—A Study in Magic and Religion, Part IV), third edition: London, Macmillan and Co., xvii + 317 p.

Freccero, J., 1961, Satan's fall and the *Quaestio de aqua et terra*: Italica, v. 38, p. 99–115.

Freeman, K., 1949, The Pre-Socratic Philosophers—A Companion to Diels, *Fragmente der Vorsokratiker*: Oxford, Basil Blackwell, xiii + 486 p.

Freeman, K., 1950, Greek City-States: New York, W.W. Norton & Company, 273 p.

Freeman, K., 1962, Ancilla to the Pre-Socratic Philosophers, fourth impression: Oxford, Basil Blackwell, x + 162 p.

Freeman, R.B., 1977, The Works of Charles Darwin—An Annotated Bibliographic Handlist: Dawson: Kent, Archon Boks, 255 p. + 1 photographic plate.

Frémont, J.C., 1845, Report of the Exploring Expedition to the Rocky Mountains in the Year 1842, and to Oregon and North California in the Years 1843–1844: Printed by the Order of the Senate of the United States: Washington, D.C., Gales and Seaton, Printers, 693 p.

Frémont, J.C., 1887, Memoirs of My Life Including in the Narrative Five Journeys of Western Exploration During the Years 1842, 1843–4, 1845–6–7, 1848–9, 1853–4 together with a Sketch of the LIfe of Senator Benton, in Connection with Western Expansion: a Retrospect of Fifty Years Covering the Most Eventful Years of Modern American History, v. 1: Chicago, Belford, Clarke & Co., viii + iv + xix + 655 p.

French, R., 1994, Ancient Natural History: London, Routledge, xxii + 357 p.

French, R., and Greenaway, F., editors, 1986, Science in the early Roman Empire: Pliny the Elder, his Sources and Influence: [ii] + 287 p.

Frenzel, H.N., Bloomer, R.R., Cline, R.B., Cys, J.M., Galley, J.E., Gibson, W.R., Hills, J.M., King, W.E., Seager, W.R., Kottlowski, F.E., Thompson, S., III, Luff, G.C., Pearson, B.T., and Van Siclen, D.C., 1988, The Permian Basin region, *in* Sloss, L.L., ed., Sedimentary Cover—North American Craton: U.S.: Boulder, Colorado, The Geological Society of America, Geology of North America, v. D-2, p. 261–306.

Friedman, G.M., 1979, Geology at Rensselaer: a historical perspective: The Compass of Sigma Gamma Epsilon, v. 57, p. 1–15.

Friedman, G.M., 1998a, James D. Dana and Fay Edgerton; students and/or disciples of Amos Eaton: American Journal of Science, v. 298, p. 608–610.

Friedman, G.M., 1998b, Charles Lyell in New York State, *in* Blundell, D.J., and Scott, A.C., eds., Lyell: The Past is the Key to the Present: Geological Society [London] Special Publication 143, p. 71–81.

Friedman, G.M., 2000, Founders of American geology: Endeavour, new series, v. 24, p. 145–146.

Friedman, G.M., Sander, J.E., and Kopaska-Merkel, D.C., 1992, Principles of Sedimentary Deposits—Stratigraphy and Sedimentology: New York, Macmillan Publishing Company, 717 p. + [44] p. of Glossary.

Frisch, W., and Loeschke, J., 1990, Plattentektonik: Erträge der Forschung, v. 236: Darmstadt, Wissenschaftliche Buchgesellschaft, XI + [I] + 243 p.

Fritscher, B., 2001a, Meteorologie, *in* Landfester, M., ed., Der Neue Pauly Enzyklopädie der Antike—Rezeptions und Wissenschaftsgeschichte: Stuttgart and Weimar, J.B. Metzler, v. 15/1, columns 415–420.

Fritscher, B., 2001b, Geographie, *in* Landfester, M., ed., Der Neue Pauly Enzyklopädie der Antike—Rezeptions und Wissenschaftsgeschichte: Stuttgart and Weimar, J.B. Metzler, v. 14, columns 122–126.

Fritscher, B., 2001c, Geologie (und Mineralogie), *in* Landfester, M., ed., Der Neue Pauly Enzyklopädie der Antike—Rezeptions und Wissenschaftsgeschichte: Stuttgart and Weimar, J.B. Metzler, v. 14, columns 126–131.

Fritts, H.C., 1965, Tree-ring evidence for climatic changes in western North America: Monthly Weather Review, v. 93, p. 421–443.

Frost, C.D., and Frost, B.R., 1993, The Archean history of the Wyoming province, *in* Snoke, A.W., Steidtmann, J.R., and Roberts, S.M., eds., Geology of Wyoming, v. 1: Laramie, Wyoming, Geological Survey of Wyoming Memoir No. 5, p. 58–76.

Fung, Y.-L., 1976, A Short History of Chinese Philosophy: New York, The Free Press, xx + 368 p.

Furley, D., 1987, The Greek Cosmologists, v. 1 The Formation of the Atomic Theory and its Earliest Critics: Cambridge, Cambridge University Press, viii + 219 p.

Furley, D., 1989, Cosmic Problems—Essays on Greek and Roman Philosophy of Nature: Cambridge, Cambridge University Press, xiv + 258 p.

Furon, R., 1955, Histoire de la Géologie de la France d'Outre-Mer: Mémoires du Muséum National d'Histoire Naturelle, nouvelle série, série C, Sciences de la Terre, v. 5, 218 p.

Gadow, H., 1930, Jorullo—The History of the Volcano of Jorullo and the Reclamation of the Devastated District by Animals and Plants: Cambridge, at the University Press, xviii + 100 p. + 2 plates + 1foldout map.

Gaiser, K., 1985, Theophrast in Assos—Zur Entwicklung der Naturwissenschaft zwischen Akademie und Peripatos: Abhandlungen der Heidelberger Akademie der Wissenschaften, Philosophische-historische Klasse, Jahrgang 1985, 3. Abhandlung, 120 p. + VI Plates.

Galbraith, W.H., 1974, James Hutton: An Analytic and Historical Study [Ph.D. thesis]: University of Pittsburgh, Faculty of Arts and Sciences, xxii + 277 p.

Gallois, L., 1890, Les Géographes Allemands de la Renaissance: Bibliothèque de la Faculté des Lettres de Lyon, v. 13: Paris, Ernest Leroux, XX + 266 p. + 6 foldout plates.

Gallop, D., 1975, Plato *Phaedo* Translated with Notes: Oxford, Clarendon Press, vi + [i] + 245 p.

Gallop, D., 1993, Plato *Phaedo* Translated with an Introduction and Notes: Oxford, Oxford World's Classics, Oxford University Press, xxix + 105 p.

Galloway, W.B., 1900, The North Pole, The Great Ice Age and the Deluge with an Appendix on the Differing Magnetic Phenomena of the South: London, Sampson Low, Marston & Company, 48 p.

Galvin, J., translator and editor, 1965, A record of Travels in Arizona and California 1775–1776 Fr. Francisco Garces: San Francisco, John Howell-Boks, ix + 113 + [3] p. + 2 foldout plates.

Gansser, A., 1973, The Roraima problem (South America), *in* Contributions to the Geology and Paleobiology of the Caribbean Area and Adjacent Areas dedicated to Hans Kugler, Verhandlungen der Naturforschenden Gesellschaft zu Basel, v. 84, p. 80–100.

Gantz, T., 1993a, Early Greek Myth A Guide to Literary and Artistic Sources: Baltimore and London, The Johns Hopkins University Press, v. I, cxv + 466 p.

Gantz, T., 1993b, Early Greek Myth A Guide to Literary and Artistic Sources: Baltimore and London, The Johns Hopkins University Press, v. II, xiii + p. 467–873.

Gardet, L., 1979, Portraits of two savants and humanists—Bîrûnî and Albert the Great, *in* Hakim Muhammed Said, ed., Al-Bîrûnî Commemorative Volume. Proceedings of the International Congress Held in Pakistan on the Occasion of the Millenary of Abû Râihân Muhammad ibn Ahmad al-Bîrûnî, Hamdard Academy, Karachi, p. 195–203.

Gaudry, A., 1855, Résumé des travaux qui ont entrepris sur les terrains anthracifères des Alpes de la France et da la Savoie: Bulletin de la Société Géologique de France, 2ème série, v. 12, p. 580–678.

Gay, N.C., 1980, The state of stress in the plates, *in* Bally, A.W., Bender, P.L., McGetchin, T.R., and Walcott, R.I., eds., 1980, Dynamics of Plate Interiors, Geodynamics Series, v. 1: Washington, D.C., American Geophysical Union, and Boulder, Colorado, Geological Society of America, p. 145–162.

Gee, D.G., and Zeyen, H.J., editors, EUROPROBE 1996—Lithosphere Dynamics and Evolution of Continents: Published by the EUROPROBE Secretariate, Uppsala University, 138 p.

Geikie, A., 1875, Life of Sir Roderick I. Murchison, Based on His Journals and Letters, v. 2: London, John Murray, viii + 375 p.

Geikie, A., 1882, Text-Book of Geology: London, Macmillan and Co., xi + 971 p.

Geikie, A., 1903, Text-Book of Geology, fourth edition, revised and enlarged, v. I: London, Macmillan and Co., xxi + 702 p.

Geikie, A., 1905, The Founders of Geology, second edition: London, Macmillan and Co., xi + 486 p.

Genzmer, F. and Schier, K., 1997, Die Edda—Götterdichtung, Spruchweisheiten und Heldengesänge der Germanen: München, Diederichs Gelbe Reihe, Eugen Diederich, 472 p.

Gerhard, L.C., and Anderson, S.B., 1988, Geology of the Williston Basin (United States portion), *in* Sloss, L.L., ed., Sedimentary Cover—North American Craton; U.S.: Boulder, Colorado, Geological Society of America, The Geology of North America, v. D-2, p. 221–241.

Gerstner, P., 1994, Henry Darwin Rogers, 1808–1866—American Geologist: Tuscaloosa, The University of Alabama Press, xi + [ii] + 311 p.

Gèze, B., 1995, La géologie dans les romans de Jules Verne, *in* Essais sur l'Histoire de la Géologie en Hommage à Eugène Wegmann, Mémoires de la Société Géologique de France, nouvelle série, no. 168, p. 83–86.

Gibbs, A.K., and Barron, C.N., 1993, The Geology of the Guiana Shield: Oxford Monographs on Geology and Geophysics: New York, Oxford University Press, [ii] + 246 p. + one foldout map.

Gilbert, B., 1983, Westering Man—The Life of Joseph Walker: Norman, University of Oklahoma Press, viii + [i] + 339 p.

Gilbert, G.K., 1875, Report upon the geology of portions of Nevada, Utah, California, and Arizona, examined in the years 1871 and 1872, *in* Report upon Geographical and Geological Explorations and Surveys West of the One Hundredth Meridian, in Charge of First Lieut. Geo. M. Wheeler…, v. III—Geology: Washington, D.C., Government Printing Office, p. 17–187.

Gilbert, G.K., 1880, Report on the Geology of the Henry Mountains, second edition: Washington, D.C., Government Printing Office, xii + 170 p. + 5 foldout plates.

Gilbert, G.K., 1882, Contributions to the history of Lake Bonneville, *in* 2nd Annual Report of the U.S. Geological Survey to the secretary of the Interior 1880–'81 by John Wesley Powell, Director: Washington, D.C., Government Printing Office, p. 167–200 + plates XLII and XLIII.

Gilbert, G.K., 1884, A theory of the earthquakes of the Great Basin, with a practical application: American Journal of Science, 3rd series, v. 27, p. 49–53.

Gilbert, G.K., 1890, Lake Bonneville: Monographs of the U.S. Geological Survey, v. I: Washington, D.C., Government Printing Office, D.C., xx + 438 p.

Gilbert, G.K., 1893, Continental problems—Annual Address by the President, G.K. Gilbert: Geological Society of America Bulletin, v. 4, p. 179–190.

Gilbert, G.K., 1896, The origin of hypotheses. Illustrated by the discussion of a topographic problem, *in* Cross, W., and Hayes, C.W., eds., The Geological Society of Washington, Presidential Address … with Constitution and Standing Rules, Abstracts of Minutes and List of Officers and Members, 1895, p. 3–24.

Gilbert, G.K., 1902, John Wesley Powell: Science, N.S., v. 16, p. 561–567.
Gilbert, G.K., ed., 1903, John Wesley Powell—A Memorial to an American Explorer and Scholar: Chicago, The Open Court Publishing Co., [i] + 75 p.
Gilbert, O., 1907 [1967], Die Meteorologischen Theorien des Griechischen Altertums: Leipzig, B.G. Teubner, VIII + [I] + 746 p. (facsimile reprint in 1967 by Georg Olms Verlagsbuchhandlung, Hildesheim).
Giles, H.A., 1923 [1956], The Travels of Fa-hsien (399–414 A.D.,), or record of the Buddhistic Kingdoms Re-translated by H.A., Giles: London, Routledge & Kegan Paul, 96 p.
Gillispie, C.C., 1959, Genesis and Geology—The Impact of Scientific Discoveries upon Religious Beliefs in the Decades before Darwin: New York, Harper Torchbooks, Harper & Row Publishers, xiii + [i] + 306 p.
Gillispie, C.C., 1960, The Edge of Objectivity—An Essay in the History of Scientific Ideas: Princeton, Princeton University Press, ix + 562 p.
Gilluly, J., 1949, Distribution of mountain building in geologic time: Bulletin of the Geological Society of America, v. 60, p. 561–590.
Gilluly, J., 1950, Reply to discussion by Hans Stille: Geologische Rundschau, v. 38, p. 103–107.
Gilman, D.C., 1872, I. Annual Address. Subject: The last ten years of geographical work in this country: Journal of the American Geographical Society, v. 3, p. 111–133.
Gilman, D.C., 1873, II. Annual Address. Subject: Geographical work in the United States during 1871: Journal of the American Geographical Society, v. 4, p. 119–144.
Gilman, D.C., 1899, The Life of James Dwight Dana—Scientific Explorer, Mineralogist, Geologist, Zoologist, Professor in Yale University: New York, Harper & Brothers, xii + 409 p.
Ginzberg, L., 1909, The Legends of the Jews, v. I Bible Times and Characters From the Creation to Jacob: Philadelphia, The Jewish Publication Society of America, XVIII + 424 p.
Ginzberg, L., 1925, The Legends of the Jews, v. V Notes to volumes I and II from the Creation to Exodus: Philadelphia, The Jewish Publication Society of America, XI + 446 p.
Girke, D., editor, 1984, Eilhard Wiedemann Gesammelte Schriften zur arabisch-islamischen Wissenschaftsgeschichte v. I, Schriften 1876–1912: Veröffentlichungen des Institutes für Geschichte der Arabisch-Islamischen Wissenschaften, Reihe B: Nachdrucke, v. 1, 1: Frankfurt am Main, Institut für Geschichte der Arabisch-Islamischen Wissenschaften an der Johann Wolfgang Goethe-Universität, XX + 612 + 51 p. + portrait.
Glaessner, M.F., and Teichert, C., 1947, Geosynclines: a fundamental concept in geology: American Journal of Science, v. 245, p. 465–482 and 571–591.
Godwin, J., 1979, Athanasius Kircher—A Renaissance Man and the Quest for Lost Knowledge: Thames and Hudson, London, 96 p.
de Goër de Herve, A., 1972, La Planèze de Saint-Flour (Massif volcanique de Cantal—France), v. 1—Structure et Stratigraphie: Annales Scientifiques de l'Université de Clermont, nr. 47, [XIII] + 244 + [2] + VII p.
Goetzmann, W.H., 1991, Army Exploration in the American West, 1803–1863 (with a new introduction): Austin, Texas State HIstorical Association, The Fred and Ella Mae Moore Texas History Reprint Series, xxvii + 489 p.
Goetzmann, W.H., 1993, Exploration and Empire—The Explorer and the Scentist in the Winning of the American West: Monticello Editions: Austin, Texas State HIstorical Association, The Fred and Ella Mae Moore Texas History Reprint Series, xxii + [iii] + 656 + xviii p.
Goetzmann, W.H., 1995, New Lands, New Men—America and the Second Great Age of Discovery: Austin, Texas State Historical Association, The Fred and Ella Mae Moore Texas History Reprint Series, xxiii + 528 p.
Goetzmann, W.H., and Williams, G., 1992, The Atlas of North American Exploration—From the Norse Voyages to the Race to the Pole: Norman, University of Oklahoma Press, 224 p.
Goguel, J., 1962, Tectonics: San Francisco, W.H., Freeman and Company, viii + 380 p. (Translated by H. Thalmann).
Gohau, G., 1983a, Idées Anciens sur la Formation des Montagnes—Préhistoire de la Tectonique—Thèse pour le doctorat d'état: Université Jean-Moulin (Lyon III), Faculté de Philosophie, [Lyon], XIII + 776 + 1 errata page.
Gohau, G., 1983b, Idées anciens sur la formation des montagnes: Cahiers d'Histoire et de Philosophie des Sciences, nouvelle série, no.7, [I] + 86 p.
Gohau, G., 1987, Histoire de la Géologie: La Decouverte, Paris, 259 p.
Gohau, G., 1990, Les Sciences de la Terre aux XVIIe et XVIIIe Siècles—Naissance de la Géologie: Paris, Albin Michel, 420 p.
Gohau, G., 1995, Constant Prévost (1787–1856), géologue critique, *in* Essais sur l'Histoire de la Géologie en Hommage à Eugène Wegmann, Mémoires de la Société Géologique de France, nouvelle série, n°168, p. 77–82.
Gökmen, O. (translator), 1977, Türkiye Seyahatnamesi (1790 Yıllarında Türkiye ve İstanbul): Ankara, Ayyıldız Matbaası, XV + 183 p. + 2 maps.
Gökyay, O. Ş., ed., 1995, Evliya Çelebi Seyahatnamesi—Topkapı Sarayı 304 Yazmasının Transkripsiyonu-Dizini, 1. Kitap İstanbul: İstanbul, Yapı Kredi Yayınları, XXI + [V] + 498 p.
Gole, S., 1989, Indian Maps and Plans—From Earliest Times to the Advent of European Surveys: New Delhi, Manohar, 206 + [1] p.
Gomperz, T., 1901, Greek Thinkers, authorized edition, v. I: London, John Murray, xv + 610 p. (translated from the German by Laurie Magnus).
Gomperz, T., 1912, Greek Thinkers, authorized edition, v. IV: London, John Murray, xvii + 587 p. (translated from the German by G.G., Berry).
Goodchild, R., 1981, Cyrene and Apollonia—An Historical Guide: Tripoli, Department of Antiquities, Socialist People's Libyan Arab Jamahiriya, 120 p.
Goodman, N., 1967, Uniformity and simplicity: Geological Society of America Special Paper 89, p. 93–99.
Gortani, M., 1963, Italian pioneers in geology and mineralogy: Cahiers d'Histoire Mondiale, v. 7(2), p. 503–519.
Gosselet, J., 1896, Constant Prévost, Coup d'Œil Rétrospectif sur la Géologie en France Pendant de Première Moitié du XIXe Siécle: Annales de la Société Géologique du Nord, v. 25, p. 1–344.
Gottschalk, H.B., 1981, Strato of Lampsacus, *in* Gillispie, C.C., editor-in-chief, Dictionary of Scientific Biography, New York, Charles Scribner's Sons, v. 13, p. 91–95.
Götzinger, G., 1947, Otto Ampferer zur Erinnerung: Verhandlungen der Geologischen Bundesanstalt, Jahrgang 1947, p.1–3
Gould, S.J., 1965, Is uniformitarianism necessary? American Journal of Science, v. 263, p. 223–228.
Grabau, W.A., 1920, A Textbook of Geology—Part I General Geology: Boston, D.C., Heath & Co., xviii + 864 p.
Grant, E., 1971 [1977], Physical Science in the Middle Ages: The Cambridge History of Science Series: Cambridge, Cambridge University Press, xi + 128 p.
Grasso, M., 2001, The Apenninic-Maghrebian orogen in southern Italy, Sicily and adjacent areas, *in* Vai, G.B., and Martini, I.P., eds., Anatomy of an Orogen: the Apennines and Adjacent Mediterranean Basins: Dordrecht, Kluver Academic Publishers, p. 255–286.
Green, A.R., 1985, Integrated sedimentary basin analysis for petroleum exploration and production: 17th Annual Offshore Technology Conference, Houston, Texas, OTC 4842, p. 9–20.
Greene, M.T., 1982, Geology in the Nineteenth Century. Changing Views of a Changing World: Ithaca, Cornell University Press, 324 p.
Greene, M.T., 1992, Natural Knowledge in Preclassical Antiquity: Baltimore, The Johns Hopkins University Press, 182 p.
Gregory, H.E., 1917, Geology of the Navajo Country—A Reconnaisance of Parts of Arizona, New Mexico, and Utah: U.S. Geological Survey Professional Paper 93, 161 p.
Gregory, H.E., and Moore, R.C., 1931, The Kaiparowits Region—A Geographic and Geologic Reconnaisance of Parts of Utah and Arizona: U.S. Geological Survey Professional Paper 164, 161 p.
Gregory, K.M., and Chase, C.G., 1994, Tectonic and climatic significance of a late Eocene low-relief, high-level geomorphic surface, Colorado: Journal of Geophysical Research, 99, p. 20, 141–20, 160.
Gridgeman, N.T., 1981, Babbage, Charles, *in* Gillispie, C.C., editor-in-chief, Dictionary of Scientific Biography: New York, Charles Scribner's Sons, v. 1, p. 354–356.

Grønlie, A., 1981, The late and postglacial isostatic rebound, the eustatic rise of the sea level and the uncompensated depression in the area of the Blue Road Geotraverse: Earth Evolution Sciences, v. 1, p. 50–57.
Grout, A., 1995, Geology and India, 1770–1851. A Study in the Methods and Motivations of a Colonial Science [Ph.D. Dissertation]: London, School of Oriental and African Studies, London University, 357 p.
Grube, G.M.A., 1980, Plato's Thought: Indianapolis, Hackett Publishing Company, xxi + 346p.
Günther, S., 1897, Handbuch der Geophysik, zweite gänzlich umgearbeitete Auflage. Erster Band: Stuttgart, Ferdinand Enke, XII + 648 p.
Günther, S., 1900, A. v. Humboldt L. v. Buch: Berlin, Ernst Hofmann & Co., [III] + 271 p.
Günther, S., 1906, Varenius, *in* Brieger-Wasservogel, L., ed., Klassiker der Naturwissenschaften: Theod. Thomas, Leipzig, v. IV, [II] + 218 p.
Günther, S., 1904[1978]), Geschichte der Erdkunde, *in* Klar, M., ed., Die Erdkunde. Eine Darstellung ihrer Wissensgebiete, ihrer Hilfswissenschaften und der Methode ihres Unterrichtes, I. Teil, Franz Deuticke, Leipzig und Wien (reprinted by Saendig Reprint Verlag Hans R. Wohlwend, Vaduz), p. XI + 343 p.
Guthrie, W.K.C., 1962, A History of Greek Philosophy, v. I, The Earlier Presocratics and the Pythagoreans: Cambridge, Cambridge University Press, xv + 539 p.
Guthrie, W.K.C., 1965, A History of Greek Philosophy, v. II, The Presocratic Tradition from Parmenides to Democritus: Cambridge, Cambridge University Press, xix + 554 p.
Guthrie, W.K.C., 1975, A History of Greek Philosophy, v. IV, Plato The Man and His Dialogues: Earlier Period: Cambridge, Cambridge University Press, xviii + 603 p.
Guthrie, W.K.C., 1981, A History of Greek Philosophy, v. VI, Aristotle An Encounter: Cambridge, Cambridge University Press, xvi + 456 p.
Gwinner, M.P., 1978, Geologie der Alpen—Stratigraphie Paläogeographie Tektonik: Stuttgart, E. Schweizerbart'sche Verlagsbuchhandlung (Nägele u. Obermiller), VIII + 480 p.
Gwynn, J.W., ed., 1980, Great Salt Lake—A Scientific, Historical & Economic Overview: Utah Geological and Mineral Survey a division of the Utah Department of Natural Resources, Bulletin 116, viii + 400 p. + 16 photographic plates + 2 foldout maps in pocket.
Haarmann, E., 1926a, “Tektogenese“ oder “Gefügebildung“ statt “Orogenese“ oder “Gebirgsbildung“: Zeitschrift der Deutschen Geologischen Gesellschaft, v. 78, Monatsbericht Nr. 3–5, p. 105–107.
Haarmann, E., 1926b, Die Oszillationstheorie: Resumen de las Comunicaciones Anunciadas. XIV. Madrid, Internationaler Geologen-Kongress, p. 133–135.
Haarmann, E., 1930, Die Oszillationstheorie. Eine Erklärung der Krustenbewegunen von Erde und Mond: Stuttgart, Ferdinand Enke, XII + 260 p.
Hafen, L.R., and Hafen, A. W., eds., 1960, Fremont's Fourth Expedition, A Documentary Account of The Disaster of 1848–1849 With Diaries, Letters, and Reports by Participants in the Tragedy: Glendale, California, The Arthur Clark Company, 319 p.
Hagberg, K., 1940, Carl Linnæus—Ein Großes Leben aus dem Barock: Hamburg, H. Goverts Verlag, 287 + [I] p.
Haile, N.S., 1969, Geosynclinal theory and the organizational pattern of the North-west Borneo Geosyncline: Quarterly Journal of the Geological Society of London, v. 124, p. 171–194.
Hall, D.J., Cavanaugh, T.D., Watkins, J.S., and McMillen, K.J., 1982, The rotational origin of the Gulf of Mexico based on regional gravity data: American Association of Petroleum Geologists Memoir 34, p. 115–126.
Hall, J., 1842, Notes upon the geology of the western states: American Journal of Science and Arts, v. 42, p. 51–62.
Hall, J., 1843, Geology of New York—Part IV Comprising the Survey of the Fourth Geological District: Carroll and Cook, Printers to the Assembly, Albany, XII + [III] + 683 p. + [II] + a foldout colored geological map and a foldout o plate of sections + XIX plates (many of them foldout).
Hall, J., 1845, Appendix A. Geological Formations, *in* Frémont, J.C., Report of the Exploring Expedition to the Rocky Mountains in the Year 1842, and to Oregon and North California in the Years 1843–1844: Printed by the Order of the Senate of the United States: Washington, D.C., Gales and Seaton, Printers, p. 295–303.
Hall, J., 1857a, Direction of the currents of deposition and source of the materials of the older Palaeozoic rocks: The Canadian Naturalist and Geologist and Proceedings of the Natural History Society of Montreal, v. 2, p. 284–286.
Hall, J., 1857b, On the direction of ancient currents of deposition and the source of materials in the older palaeozoic rocks, with remarks on the origin of the Appalachian chain of Mountains: The Edinburgh New Philosophical Journal, v. 6 (New Series), 348–349.
Hall, J., 1857c, Palæontology and Geology of the Boundary, *in* Emory, W.H., Report of the United States and Mexican Boundary Survey, made under the direction of the Secretary of the Interior, House of Representatives, 34th Congress, 1st session, Ex. Doc. No. 135, v. I, p. 101–174, 21 plates + 1 colored foldout map.
Hall, J., 1858, General Geology, *in* Hall, J., and Whitney, J.D., Report of the Geological Survey of the State of Iowa, v. I, part I: Geology, published by the authority of the Legislature of Iowa, C. van Benthuysen, [Albany], p. 35–44.
Hall, J., 1859, Introduction, *in* Hall, J., Geological Survey of New York. Palæontology: volume III. …Part I: Text., p. 1–96.
Hall, J., 1865, Graptolites of the Quebec Group: Figures and Descriptions of Canadian Organic Remains. Decade II: Geological Survey of Canada, Dawson Brothers, Montreal; Ballière, London, New York, and Paris, iv + 151 p. + XXI plates.
Hall, J., 1883, Contributions to the geological history of the American continent: Proceedings of the American Association for the Advancement of Science, Thirty-first Meeting, Published by the Permanent Secretary, AAAS, Salem, p. 29–71.
Halleux, R. and Schamp, J. (editors and translators), 1985, Les Lapidaires Grecs—Lapidaire Orphique, Kérygmes Lapidaires d'Orphée, Socrate et Denys, Lapidaire Nautique, Damigéron-Évax (traduction latine): Paris, Collection des Universités de France, «Les Belles Lettres», XXXIV + 347 + [I] p.
Hamann, G., editor, 1983, Eduard Suess zum Gedenken: Österreichische Akademie der Wissenschaften, philologisch-historische Klasse, Sitzungsberichte, 422, Veröffentlichunden der Kommission für die Geschichte der Mathematik, Naturwissenschaften und Medizin, v. 41, 100 p.
Hamilton, E.L., 1956, Sunken Islands of the Mid-Pacific Mountains: The Geological Society of America Memoir 64, x + 97 p. + 2 maps in pocket.
Hammond, G.P., and Rey, A., 1940, Narratives of the Coronado Expedition 1540–1542, *in* Hammond, G.P., ed., Coronado Cuarto Centennial Publications, 1540–1940, v. II: Albuquerque, University of New Mexico Press, xii + 413 p.
Hansen, B., 1991, Bakewell, Robert, *in* Gillispie, C.C., editor-in-chief, Dictionary of Scientific Biography: New York, Charles Scribner's Sons, v. 1, p. 413.
Hansen, S. (Editor), 1991, Mythen vom Anfang der Welt: Augsburg, Pattloch Verlag, 447 p.
Hantzsch, V., 1898, Sebastian Münster—Leben, Werk, wissenschaftliche Bedeutung: Abhandlungen der philologisch-historischen Classe der Königlichen Sächsischen Gesellschaft der Wissenschaften, v. 18, no. III, p. 1–187.
Haq, B.U., Hardenbol, J. and Vail, P.R., 1988, Mesozoic and Cenozoic chronostratigraphy and eustatic cycles, *in* Wilgus, C.K., Hastings, B.S., Kendall, C.G., St. C., Posamentier, H.W., Ross, C.A., and Van Wagoner, J., eds., Sea-Level Changes: An Integrated approach, Society of Economic Paleontologists and Mineralogists, Special Publication No. 42, p. 71–108.
Harding, T.P., and Lowell, J.D., 1979, Structural styles, their plate-tectonic habitats, and hydrocarbon traps in petroleum provinces: Bulletin of the American Association of Petroleum Geologists, v. 63, p. 1016–1058.
Harland, W.B., 1967, Geosynclines: Geological Magazine, v. 102, p. 182–188.

Harland, W.B., 1996, A short history of time scales in geoscience, *in* Wang, H.Z., Zhai, Y.S., Shi, B.S., and Wang C.S., eds., Development of Geoscience Disciplines in China: Wuhan, China University of Geosciences Press, p. 34–41.

Harley, J.B., and Woodward, D., editors, 1987, The History of Cartography, volume one (Cartography in Prehistoric, Ancient, and Medieval Europe and the Mediterranean), Chicago, University of Chicago Press, xxi + 599 p.

Harley, J.B., and Woodward, D., editors, 1992, The History of Cartography, volume two, book one (Cartography in the Traditional Islamic and south Asian Societies), Chicago, University of Chicago Press, xxiv + 579 p.

Harris, A.G., Tuttle, E. and Tuttle, S.D., 1997, Geology of National Parks, fifth edition: Dubuque, Iowa, Kendall/Hunt Publishing Co., xvi + 758 p.

Harris, E.D., 1990, John Charles Frémont and the Great Western Reconnaisance, *in* Goetzmann, W.H., and Crouch, T.D., eds., World Explorers: New York, Chelsea House Publishers, 111 p.

Harris, W.R., 1909, The Catholic Church in Utah including An Exposition of Catholic Faith by Bishop Scanlan—A review of Spanish and Missionary Explorations. Tribal Divisions, names and regional habitats of the pre-European Tribes. The Journal of the Franciscan Explorers and discoverers of Utah Lake. The trailing of the Priests from Santa Fe, N.M., with Map of Route, Illustrations and delimitations of the Great Basin, intermountain Catholic Press, Salt Lake City, iii + iii + 350 p.

Hart, S.H., and Hulbert, A.B., 1932, Zebulon Pike's Arkansaw Journal, in search of the southern Louisiana Purchase Boundary Line (Interpreted by his Newly Discovered Maps): Overland to the Pacific, v. 1, The Stewart Commission of Colorado College and The Denver Public Library, no place of publication, xcvi + 200 p. + 7 plates

Hartmann, H., 1953, Georg Agricola, *in* Frickhinger, H.W., ed., Große Naturforscher, v. 13: Stuttgart, Wissenschaftliche Verlagsgesellschaft M.B.H., [II] + 134 p.

Harvey, P.D.A., 1991, Medieval Maps: London, The British Library, 96 p.

Hassert, K., 1941, Die Erforschung Afrikas: Leipzig, Wilhelm Goldmann Verlag, 248 + [II] p.

von Hauer, Ritter F., 1875, Die Geologie und ihre Anwendung auf die Kenntnis der Bodenbeschaffenheit der Österr.-Ungar. Monarchie: Wien, Alfred Hölder, VIII + 681 p.

Haug, E., 1900, Les Géosynclinaux et les aïres continentales. Contribution à l'étude des transgressions et regressions marines: Bulletin de la Société Géologique de France, 3e série, t. 28, p. 617–711.

Haug, E., 1907, Traité de Géologie, t. 1 (Les Phénomènes géologiques): Paris, Librairie Armand Colin, 538 p.

Haug, E., 1925, Contribution a une synthèse stratigraphique des Alpes occidentales: Bulletin de la Société Géologique de France, série 4, v. 25, p. 97–244.

Haussig, H.W., 1992, Die Geschichte Zentralasiens und der Seidenstrasse in Vorislamischer Zeit, 2. durchgesehene und um einen Nachtrag erweiterte Auflage: Darmstadt, Wissenschaftliche Buchgesellschaft, XII + 322 p. + 1 folded map.

Havemann, H., 1969, Die Entwicklung der Undationstheorie R.W., van Bemmelens: Geologie, v. 18, p. 775–793.

Hawgood, J.A., 1967, America's Western Frontiers: The Exploration and Settlement of the Trans-Mississippi West: New York, Alfred A. Knopf, xxiii + 440 + x p.

Hayden, F.V., 1857, Notes explanatory of a map and section illustrating the geologic structure of the country bordering on the Missouri River, from the mouth of the Platte River to Fort Benton: Proceedings of the Academy of Natural Sciences of Philadelphia, May 1857, p. 109–116 + 1 foldout colored geological map.

Hayden, F.V., 1858, Explanation of a second edition of a geological map of Nebraska and Kansas, based upon information obtained in an expedition to the Black Hills, under the command of Lieut. G.K., Warren: Proceedings of the Academy of Natural Sciences of Philadelphia, June, 1858, 139–158 + 1folded colored geological map.

Hayden, F.V., 1869, Geology of the Tertiary formations of Dakota and Nebraska accompanied with a map, *in* Leidy, J., The extinct Mammalian fauna of Dakota and Nebraska including an account of some allied forms from other localities together with a synopsis of the Mammalian remains of North America: Journal of the Academy of Natural Sciences of Philadelphia, v. 7, second series, p. 9–21 + folding colored geological map.

Hayden, F.V., 1871, Report of F.V., Hayden, part II, *in* Preliminary Report of the U.S. Geological Survey of Wyoming and Portions of Contiguous Territories (Being A Second Annual Report of Progress): Washington, D.C., Government Printing Office, p. 83–188.

Hayden, H.H., 1820, Geological Essays; or, An Enquiry into some of the Geological Phenomena to be Found in Various Parts of America, and Elsewhere: Baltimore, Printed by J. Robinson for the Author, viii + 412 p.

Hays, J.D., and Pitman, W.C., III, 1973, Lithospheric plate motion, sea-level changes and climatic and ecological consequences: Nature, v. 246, p. 18–22.

Hazen, R.M., 1974, The founding of geology in America: 1771–1818: Geological Society of America Bulletin, v. 85, p. 1827–1834.

Heather, D.C., 1979, Plate Tectonics: London, Edward Arnold, 80 p.

Hedberg, H.D., 1969a, Influence of Torbern Bergman (1735–1784) on stratigraphy: Acta Universitatis Stockholmiensis, Stockholm Contributions in Geology, v. 20, p. 19–47.

Hedberg, H.D., 1969b, The influence of Torbern Bergman (1735–1784) on stratigraphy: a résumé, *in* Schneer, C.J., ed., Toward A History of Geology: Cambridge, The M.I.T. Press, p. 186–191.

Heestand, R.L., and Crough, S.T., 1981, The effect of hot spots on the oceanic age-depth relation: Journal of Geophysical Research, v. 86, p. 6107–6114.

Heidel, A., 1949, The Gilgamesh Epic and Old Testament Parallels: Chicago, The University of Chicago Press, ix + 269 p.

Heidel, W.A., 1910, Περι Φυσεωs. A study of the conception of nature among the pre-Socratics: Proceedings of the American Academy of Arts and Sciences, v. 45, p. 79–132.

Heidel, W.A., 1921, Anaximander's book, the earliest known geographical treatise: Proceedings of the American Academy of Arts and Sciences, v. 56, p. 239–288.

Heidel, W.A., 1933, The Heroic Age of Science—The Conception, Ideals, and Methods of Science Among the Ancient Greeks: Baltimore, The Williams & Wilkins Company, vii + 203 p.

Heidel, W.A., 1937, The Frame of the Ancient Greek Maps With a Discussion of the Discovery of the Sphericity of the Earth: American Geographical Society Research Series no. 20, x + 141 p.

Heim, A., 1878a, Untersuchungen über den Mechanismus der Gebirgsbildung im Anschluss an die Geologische Monographie der Tödi-Windgällen-Gruppe, v. I: Basel, Benno Schwabe, XIV + 346 p.

Heim, A., 1878b, Untersuchungen über den Mechanismus der Gebirgsbildung im Anschluss an die Geologische Monographie der Tödi-Windgällen-Gruppe, v. II: Basel, Benno Schwabe, [I] + 246 p.

Heim, A., 1878c, Untersuchungen über den Mechanismus der Gebirgsbildung im Anschluss an die Geologische Monographie der Tödi-Windgällen-Gruppe, Atlas: Basel, Benno Schwabe, XVII plates.

Heim, A., 1905, Das Säntisgebirge: Beiträge der Geologischen Karte der Schweiz, Neue Folge XVI. Lieferng; des ganzen Werkes 46. Lieferung, Atlas, 42 Plates.

Heim, A., 1908, Autographie der Vorlesungen von Herrn Prof. Dr. Alb. Heim über Allgemeine Geologie. II. Band. Aufgenommen von Max Hottinger, stud. mech.: Zürich, Verband der Polytechniker, 114 p.

Hein, W.-H., editor, 1985, Alexander von Humboldt—Leben und Werk: C.H., Boehringer Sohn, Ingelheim am Rhein.

Helwig, J.A., 1985, Origin and classification of sedimentary basins: 17th Annual Offshore Technology Conference, Houston, Texas, OTC 4843, p. 21–32.

Henderson, J.B., 1994, Chinese cosmographical thought: The high intellectual tradition, *in* Harley, J.B., and Woodward, D., eds., The History of Cartography, volume two, book two (Cartography in the Traditional East and Southeast Asian Societies), Chicago, University of Chicago Press, p. 203–227.

Hendricks, J.D., and Stevenson, G.M., 1990, Grand Canyon Supergroup: Unkar Group, *in* Beus, S.S., and Morales, M., eds., Grand Canyon Geology:

New York, Oxford University Press, and Museum of Northen Arizona Press, p. 29–47.

Hentschel, C., 1969, Alexander von Humboldt's synthesis of literature and science, *in* Alexander von Humboldt, 1769–1969: Bonn/Bad Godesberg, Inter Nationes, p. 97–132.

Henze, D., 1978, Enzyklopädie der Entdecker und Erforscher der Erde, v. 1 A–C: Graz, Akademische Druck- und Verlagsanstalt, XIV + 767 p.

Henze, D., 1983, Enzyklopädie der Entdecker und Erforscher der Erde, v. 2 D–L: Graz, Akademische Druck- und Verlagsanstalt, XVI + 728 p.

Henze, D., 1993, Enzyklopädie der Entdecker und Erforscher der Erde, v. 3 K–Pallas: Graz, Akademische Druck- und Verlagsanstalt, XIV + 802 p.

Herbert, S., 1991, Charles Darwin as a prospective geological author: British Journal for the History of Science, v. 24, p. 159–192.

Hermes, J.J., 1968, The Papuan geosyncline and the concept of geosynclines: Geologie en Mijnbouw, v. 47, p. 81–97.

Herrmann, A., 1935, Die älteste türkische Weltkarte (1076 n. Chr.): Imago Mundi, no. I, p. 21–28.

Hermann, B.F., 1801, Mineralogische Reisen in Sibirien. Vom Jahr 1783 bis 1796. Dritter Theil: St. Petersburg, Bei der kaiserlichen Akademie der Wissenschaften, [4] + 312 p. + 6 copper-plates and foldout tables VIII through XIV.

Hershbell, J.P., 1971, Plutarch as a source for Empedocles: American Journal of Philology, v. 92, p. 156–184.

Hess, H.H., 1946, Drowned anicent islands of the Pacific basin: American Journal of Science, v. 244, p. 772–791.

Hiärne, U., 1702, Den Korta Anledningen Til Åthskillige Malm och Bergarters/Mineraliers och Jordeslags, &c. Beswarad och Förklarad Jämte Deras Natur Födelse och I Jorden Tilwerckande/ samt Vplösning och Anatomie, I Giörligaste Måtto Beskrifwen: Stockholm, J.H. Werner, [viii] + 132 p.

Hiärne, U., 1706, Den Beswarade och Förklarade Anledningens Andra Flock/ Om Norden och Landskap I Gemeen: Stockholm, Mich. Laurelio, p. [iv] + p. 133–416.

Hibbert, C., 1982, Africa Explored—Europeans in the Dark Continent 1769–1889: New York, W.W. Norton & Co., 336 p.

Hicks, R.D., translation, introduction and commentary, 1925, Diogenes Laertius Lives of Eminent Philosophers, v. I: Cambridge, The Loeb Classical Library, Harvard University Press, xxxvi + [i] + 549 p.

Hills, E.S., 1963, Elements of Structural Geology: New York, John Wiley & Sons, xi + 483 p.

Hilmi Ziya [Ülken], undated [1932], 1. Türk Kozmogonisi, 2. Türk Mitolojisi, 3. Türk Hikmeti, 4. Teknik Tefekkür: "Türk Tarihinin Anahatları" Eserinin Müsveddeleri, no. 41, Başvekâlet Müdevvenat Matbaası, [Ankara], 53 p.

Himmerich y Valencia, R., 1995, Foreword, *in* Vélez de Escalante (1995)

Hintze, L.F., 1988, Geologic History of Utah: Salt Lake City, Brigham Young University Geology Studies Special Publication 7, Department of Geology, Brigham Young University, [II] + 202 p.

Hinze, W., Braile, L.W., Keller, G.R., and Lidiak, E.G., 1980, Models for Midcontinent tectonism, *in* Continental Tectonics, Studies in Geophysics, Geophysics Study Committee: Washington, D.C., National Academy Press, p. 73–83.

Hitchcock, E., 1841, First Anniversary Address before the Association of American Geologists, at their second annual meeting in Philadelphia, April 5, 1841: The American Journal of Science and Arts, v. 41, p. 232–275.

Hitchcock, E., 1847, Elementary Geology, eighth edition: Cincinnati, William H. Moore & Co., 350 p. + 2 foldout plates.

Hobbs, W.H., 1909, The evolution and the outlook of seismic geology: Proceedings of the American Philosophical Society, v. 48, p. 1–44.

von Hochstetter, F., 1875, Die Erde nach ihrer Zusammensetzung, ihrem Bau und ihrer Bildung—Ein kurzer Leitfaden der Geologie: Prag, F. Tempsky, VIII + 195 p.

von Hoff, K.E.A., 1822, Geschichte der Durch Überlieferung Nachgewiesenen Natürlichen Veränderungen der Erdoberfläche. Ein Versuch. I. Theil: Gotha, Justus Perthes, XX + 489 p. + [1 p. of errata].

von Hoff, K.E.A., 1824, Geschichte der Durch Überlieferung Nachgewiesenen Natürlichen Veränderungen der Erdoberfläche. Ein Versuch. II. Theil—Geschichte der Vulcane und Erdbeben: Gotha, Justus Perthes, XXX + 560p. + [1 p. of errata].

von Hoff, K.E.A., 1840, Geschichte der Durch Überlieferung Nachgewiesenen Natürlichen Veränderungen der Erdoberfläche. Ein Versuch. IV. Theil—Chronik der Erdbeben und Vulcan-Ausbrüche mit Vorausgehender Abhandlung über die Natur Dieser Erscheinungen—Erster Theil. Vom Jahre 3460 vor, bis 1759 unserer Zeitrechnung: Gotha, Justus Perthes, IV + [II] + 470 p.

Hoffmann, F., 1837, Physikalische Geographie—Vorlesungen gehalten an der Universität zu Berlin in den Jahren 1834 und 1835, *in* Hinterlassene Werke von Friedrich Hoffmann, v. 1: Berlin, Nicolaische Buchhandlung, XI + 620 p.

Hoffmann, F., 1838, Geschichte der Geognosie, und Schilderung der Vulkanischen Erscheinungen—Vorlesungen gehalten an der Universität zu Berlin in den Jahren 1834 und 1835, *in* Hinterlassene Werke von Friedrich Hoffmann, v. 2: Berlin, Nicolaische Buchhandlung, VIII + [I] + 596 p.

Hoffmann, F., 1839, Geognostische Beobachtungen—Gesammelt auf einer Reise durch Italien und Sicilien in den Jahren 1830 bis 1832: Berlin, G. Reimer, 726 p. + 1 folded plate and 1 separate colored geological map in pocket (offprint from Archiv für Mineralogie, Geognosie, Bergbau und Hüttenkunde, v. 13).

Hofstede, G., 1996, Europe versus Asia: Truth versus Virtue: European Review, v. 4, p. 215–219.

Hoheisel, K., 1979a, Gregorius Reisch, *in* Büttner, M., ed., Wandlungen im geographischen Denken von Aristoteles bis Kant, Abhandlungen und Quellen zur Geschichte der Geographie und Kosmologie, Band 1: Paderborn, Ferdinand Schöningh, p. 59–67.

Hoheisel, K., 1979b, Johannes Stöffler (1452–1531) als Geograph, *in* Büttner, M., ed., Wandlungen im geographischen Denken von Aristoteles bis Kant, Abhandlungen und Quellen zur Geschichte der Geographie und Kosmologie, Band 1: Paderborn, Ferdinand Schöningh, p. 69–82.

Hohl, R., 1985, Die endogenen Vorgänge und Kräfte, *in* Hohl, R., ed., Die Entwicklungsgeschichte der Erde: Hanau, Werner Dausien, p. 169–280.

Holbrook, S.H., 1947, The Story of American Railroads: New York, Crown Publishers, x + 468 p.

Hölder, H., 1960, Geologie und Paläontologie in Texten und Ihrer Geschichte, *in* Wagner, F., and Brodführer, eds., Orbis Academicus—Problemgeschichten der Wissenschaft in Dokumenten und Darstellungen: Freiburg/München, Verlag Karl Alber, XVIII + 565 p.

Holm, R.F., 1987, San Francisco Mountain: a late Cenozoic composite volcano in northern Arizona, *in* Beus, S., ed., Centennial Field Guide, v. 2, Rocky Mountain Section of the Geological Society of America: Boulder, Colorado, Geological Society of America, p. 389–392.

Holm, R.F., and Ulrich, G.E., 1987, Late Cenozoic volcanism in the San Francisco and Mormon volcanic fields, Southern Colorado Plateau, Arizona, *in* Davis, G.H., and VandenDolder, E.M., eds., Geologic Diversity of Arizona and Its Margins: Excursions to Choice Areas: Field-Trip Guidebook, The Geological Society of America, 100th Annual Meeting, Phoenix, Arizona October 26–29, 1987, Arizona Bureau of Geology and Mineral Technology, Geological Survey Branch, Special Paper 5, p. 85–94.

Holmes, A., 1928, The Nomenclature of Petrology with reference to selected literature: London, Thomas Murby & Co., [I] + 284 p.

Holmes, A., 1944, Principles of Physical Geology: London, Thomas Nelson and Sons Ltd., xii + 532 p.

Holmes, A., 1965, Principles of Physical Geology, second edition completely revised: London, Nelson, xv + 1288 p.

Holmyard, E.J., and Mandeville, D.C., 1927, Avicennae de Congelatione et Conglutiatione Lapidum Being Sections of the Kitâb al-Shifâ' The Latin and Arabic Texts edited with an English Translation of the Latter with Critical Notes: Paris, Paul Geuthner, VIII + [I] + 86 p.

Hopkins, W., 1836, Researches in physical geology: Transactions of the Cambridge Philosophical Society, v. 6, part I, p. 1–84.

Hopkins, R.L., 1990, Kaibab Formation, *in* Beus, S.S., and Morales, M., eds., Grand Canyon Geology: New York, Oxford University Press, and Museum of Northen Arizona Press, p. 225–245.

Hoppe, G., 1994, Die Entwicklung der Ansichten Alexander von Humboldts über den Vulkanismus und die Meteorite, *in* Studia Fribergensia, Vorträge des Alexander-von-Humboldt-Kolloquiums in Freiberg vom 8. bis 10. November 1991 aus Anlaß des 200. Jahrestages von A. v. Humboldts Studienbeginn an der Bergakademie Freiberg, Beiträge zur Alexander-von-Humboldt-Forschung, v. 18, p. 93–106.

Hopper, R.J., 1976, The Early Greeks: New York, Barnes & Noble Books, A Division of Harper & Row Publishers, [ix] + 257 p.

Horst, U., 1955, Der Bierfwechsel Agricolas, ein Überblick, *in* Georgius Agricola 1494–1555 zu Seinem 400. Todestag 21. November 1955, published by the Zentrale Agricola-Kommission der Deutschen Demokratischen Republik im Agricola-Gedenkjahr 1955: Akademie-Verlag, Berlin, 257–265.

van Houten, F.B., 1969, Molasse facies: records of worldwide crustal stresses: Science, v. 166, p. 1506–1508.

Hsü, K.J., 1958, Isostasy and a theory for the origin of geosynclines: American Journal of Science, v. 256, p. 305–327.

Hsü, K.J., 1971, Franciscan mélanges as a model for eugeosynclinal sedimentation and underthrusting tectonics: Journal of Geophysical Research, v. 76, p. 1162–1170.

Hsü, K.J., 1972, The concept of the geosyncline, yesterday and today: Transactions of the Leicester Literary and Philosophical Society, v. 66, p. 26–48.

Hsü, K.J., 1973, The odyssey of geosyncline, *in* Ginsburg, R.N., ed., Evolving Concepts in Sedimentology: Baltimore, The Johns Hopkins University Press, p. 66–92.

Hsü, K.J., 1982, Geosynclines in plate tectonic settings: sediments in mountains, *in* Hsü, K.J., ed., Mountain Building Processes (Gansser and Trümpy Festschrift): London, Academic Press, p. 3–12.

Hsü, K.J., 1994, Why Isaac Newton was not a Chinese: Abschiedsvorlesung von Prof. Dr. Kenneth J. Hsü, gehalten am Freitag, 24. Juni 1994 im Auditorium maximum der ETH Zürich, 17 p.

Hsü, K.J., 1995, The Geology of Switzerland—An Introduction to Tectonic Facies: Princeton, Princeton University Press, xxv + 250 p.

Hsü, K.J., and Briegel, U., 1991, Geologie der Schweiz—Ein Lehrbuch für den Einstieg, und eine Auseinandersetzung mit den Experten: Basel, Birkhäuser, [II] + 219 p.

Hubbert, M.K., 1967, Critique of the principle of uniformity: Boulder, Colorado, Geological Society of America Special Paper, 89, p. 3–33.

Huet, P.D., 1691, La Situation du Paradis Terrestre: Paris, Jean Anisson, 240 + [XIX] P. + 1 foldout map.

Huggett, R., 1989, Cataclysms and Earth History: Oxford, Clarendon Press, xii + 220 p.

Hull, D.L., editor, 1973, Darwin and His Critics—The Reception of Darwin's Theory of Evolution by the Scientific Community: Cambridge, Harvard University Press, xii + 473 p.

Humbach, H. and Ziegler, S., 1998, Ptolemy Geography, Book 6 Middle East, Central and North Asia, China Part 1. Text and English/German Translations by Suzanne Ziegler: Wiesbaden, Dr. Ludwig Reichert Verlag, X + 260 p.

Humbach, H., and Ziegler, S., 2002, Ptolemy Geography, Book 6 Middle East, Central and North Asia, China Part 2: Wiesbaden, Dr. Ludwig Reichert Verlag, XIV + 137 p.

de Humboldt, A., 1805, Essai sur la Géographie des Plantes; accompagné d'un tableau physique des régions équinoxiales, fondé sur des mesures exécutées depuis le dixième degré de latitude boréale jusqu'au dixième degré latitude australe pendant les années 1799–1803: Paris, F. Schoell, an XIII (1805), 155 p.

de Humboldt, A., 1811a, Essai Politique sur le Royaume de la Nouvelle-Espagne … avec un Atlas Phsique et Géographique, Fondé sur des Observations Astronomiques, des Mesures Trigonométriques et des Nivellemens Barométriques: F. Schoell, tome premier (xcii + 356 p.), tome deuxième (351–905 p.).

de Humboldt, A., 1811b, Essai Politique sur le Royaume de la Nouvelle-Espagne: Paris, F. Schoell, v. I (456 p.), v. II (52 p. + one folded map), v III (420 p.), v. IV (568 p.), v. V (352 p.).

de Humboldt, A., 1811[1997]a, Essai Politique sur le Royaume de la Nouvelle-Espagne du Mexique, v. 1: Paris, Utz, 472 p.

de Humboldt, A., 1811[1997]b, Essai Politique sur le Royaume de la Nouvelle-Espagne du Mexique, v. 2: Paris, Utz, p. 473–906.

de Humboldt, A., 1812, Atlas Géographique et Physique du Royaume de la Nouvelle-Espagne, Fondé sur des Observations Astronomiques, des Mesures Trigonométriques et des Nivellemens Barométriques: Paris, G. Dufour, 20 sheets of maps, sections and graphs.

de Humboldt, A., 1816, Sur l'élévation des montagnes de l'Inde: Annales de Chimie et de Physique, Série 2, v. III, p. 297–317.

de Humboldt, A., 1820, Sur la limite inférieure des neiges perpétuelles dans les montagnes de l'Himâlaya et les régiones équatoriales: Annales de Chimie et de Physique, Série 2, v. XIV, p. 5–56.

de Humboldt, A., 1823, Essai Géognostique sur le Gisement des Roches dans les Deux Hémisphères: Paris, F.G. Levrault, viij + 379 p.

de Humboldt, A., 1831, Fragmens de Géologie et de Climatologie Asiatiques, tome premier: Gide, A. Pihan Delaforest, Delaunay, Paris, 309 p.

von Humboldt, A., 1832, A. v. Humboldt's Fragmente einer Geologie und Klimatologie Asiens aus dem Französischen mit Anmerkungen, einer Karte und einer tabelle vermehrt von Julius Loewenberg: Berlin, J.A. List, 272 p.

von Humboldt, A., 1835, Kritische Untersuchungen über die historische Entwicklung der geographischen Kenntnisse von der Neuen Welt und die Fortschritte der nautischen Astronomie in dem 15ten und 16ten Jahrhundert. Aus dem Französischen übersetzt von Jul. Ludw. Ideler, v. 1, Erste Lieferung: Berlin, Verlag der Nikolai'schen Buchhandlung, 560 p. + 2 p. of errata.

von Humboldt, A., 1837a, Ueber zwei Versuche, den Chimborazo zu besteigen: Schumacher's Astronomisches Jahrbuch für 1837, p. 176–206.

von Humboldt, 1837b, Ueber zwei Versuche, den Chimborazo zu besteigen: Berghaus' Annalen, dritte Reihe, v. 3, p. 199–216.

von Humboldt, A., 1843, Asie Centrale—Recherches sur les Chaînes des Montagnes et la Climatologie Comparée, tome deuxiéme: Paris, Gide, 558 p.

von Humboldt, A., 1845, Kosmos, Entwurf einer physischen Weltbeschreibung, Erster Band: Stuttgart and Tübingen, J.G. Cotta, XVI + 493 p.

von Humboldt, A., 1847, Kosmos, Entwurf einer physischen Weltbeschreibung, Zweiter Band: Stuttgart and Tübingen, J.G. Cotta, 544+[1] p.

von Humboldt, A., 1849, Ansichten der Natur, dritte verbesserte und vermehrte Ausgabe, v. I: Stuttgart and Tübingen, J.G. Cotta'scher Verlag, XVIII + 362 p.

von Humboldt, A., 1852, Kritische Untersuchungen über die historische Entwicklung der geographischen Kenntnisse von der Neuen Welt und die Fortschritte der nautischen Astronomie in dem 15ten und 16ten Jahrhundert. Aus dem Französischen übersetzt von Jul. Ludw. Ideler, v. 2: Berlin, Verlag der Nikolai'schen Buchhandlung, 528 p.

von Humboldt, A., 1853, Kleinere Schriften, erster Band. Geognostische und Physikalische Erinnerungen: Stuttgart und Tübingen, J.G. Cotta'scher Verlag, VIII + 474 p. + 5 foldout tables.

von Humboldt, A., 1858a, Vorwort, *in* H.B. Möllhausen, Tagebuch einer Reise von Mississippi nach den Küsten der Südsee: Leipzig, Hermann Mendelssohn, p. I–VIII.

von Humboldt, A., 1858b, Kosmos, Entwurf einer physischen Weltbeschreibung, Vierter Band: Stuttgart and Tübingen, J.G. Cotta, 649 + [1] p.

von Humboldt, A., 1869, Tablas Geográfico-Politicas del Reino de la Nueva España: Boletín de la Sociedad Mexicana de Geografía y Estadística, Epoca 2, v. 1, p. 635–657.

von Humboldt, A., and von Cancrin, G., 1869, Im Ural und Altai—Brifwechsel zwischen Alexander von Humboldt und Graf Georg von Cancrin: Leipzig, F.A. Brockhaus, XV + 170 p.

Hume, W.F., 1948, Terrestrial Theories—A Digest of Various Views as to the Origin and Development of the Earth and their Bearing on the Geology of Egypt: Cairo, Ministry of Commerce and Industry, Department of

Mines and Quarries, Geological Survey, Government Press, XLIX + 522 + 160 p.

Hunke, S., 1960, Allahs Sonne über dem Abendland—Unser Arabisches Erbe: Stuttgart, Deutsche Verlags-Anstalt, 376 p.

Hunt, T.S., 1873, On some points in dynamical geology: American Journal of Science and Arts, third series, v. 5, p. 264–270.

Hunt, C.B., 1953, Geology and Geography of the Henry Mountains Region Utah: [U.S.] Geological Survey Professional Paper 228, vii + 234 p. + 22 plates.

Hunt, C.B., 1969a, John Wesley Powell, his influence on geology: Geotimes, v. 14, p. 16–18.

Hunt, C.B., 1969b, Geologic History of the Colorado River—thirty million years of changes in the rivers and canyons that Powell was first to explore: [U.S.] Geological Survey Professional Paper 669-C, [IV] p. + p. 59–130.

Hunt, C.B., 1980a, G.K. Gilbert's Lake Bonneville studies, *in* Yochelson, E.L., ed., The Scientific Ideas of G.K. Gilbert—An Assesment on the Occasion of the Centennial of the U.S. Geological Survey: Boulder, Colorado, Geological Society of America Special Paper 183, p. 45–59.

Hunt, C.B., 1980b, G.K. Gilbert, on laccoliths and intrusive structures, *in* Yochelson, E.L., ed., The Scientific Ideas of G.K. Gilbert—An Assesment on the Occasion of the Centennial of the U.S. Geological Survey: Boulder, Colorado, Geological Society of America Special Paper 183, p. 25–34.

Hunt, C.B., 1982, Pleistocene Lake Bonneville, Ancestral Great Salt Lake, as described in the Notebooks of G.K. Gilbert, 1875–1880: Brigham Young University Geology Studies v. 29, part 1, viii + 225 p.

Hunt, C.B., 1988, Geology of the Henry Mountains, Utah, as Recorded in the Notebooks of G.K. Gilbert, 1875–1876: Boulder, Colorado, Geological Society of America Memoir 167, v + 229p.

Hutton, J., 1788, Theory of the Earth; or An Investigation of the Laws observable in the Composition, Dissolution, and Restoration of Land upon the Globe: Transactions of the Royal Society of Edinburgh, v. 1, p. 209–304.

Hutton, J., 1795a, Theory of the Earth with Proofs and Illustrations, v.1: London, Cadell, Junior and Davies, and Edinburgh, William Creech, viii + 620 p. + IV plates.

Hutton, J., 1795b, Theory of the Earth with Proofs and Illustrations, v. II: London, Cadell, Junior and Davies, and Edinburgh, William Creech, viii + 567 p. + 2 foldout plates.

Hutton, J., 1899[1997], Theory of the Earth with Proofs and Illustrations, v. III, edited by Sir Archibald Geikie: The Geological Societyof London (reprint of 1899 edition).

Hyman, R.A., 1987, Charles Babbage, 1791–1871—Philosoph, Mathematiker, Computerpionier, (aus dem Englischen übersetzt von Ulrich Enderwitz): Stuttgart, Klett-Cotta, 457 p.

İhsanoğlu, E., Şeşen, R., Bekar, M.S., Gündüz, G., and Furat, A.H., 2000a, Osmanlı Coğrafya Literatürü Tarihi, I. Cilt: İstanbul, İslâm Tarih, Sanat ve Kültür Araştırma merkezi (IRCICA), LXXXIX + 396 p. + 16 colored plates.

İhsanoğlu, E., Şeşen, R., Bekar, M.S., Gündüz, G., and Furat, A.H., 2000b, Osmanlı Coğrafya Literatürü Tarihi, II. Cilt: İstanbul, İslâm Tarih, Sanat ve Kültür Araştırma merkezi (IRCICA), [I] p. + p. 397–912 + 8 colored plates.

Institute of the History of Natural Sciences, 1983, Ancient China's Technology and Science: Beijing, China Knowledge Series, Foreign Languages Press, [III] + 632 p. + 8 photographic plates.

Interdepartmental Tectonic Committee and Editorial Committee of Geotektonika, 1987, Vladimir Vladimirovich Belousov (on his 80th Birthday): Geotektonika, v. 21, p. 483–485.

Ions, V., 1983, Indian Mythology: Library of the World's Myths and Legends: New York, Peter Bedrick Books, 144 p.

Issel, A., 1883, Le Oscillazioni Lente del Suolo—Bradisismi Saggio di Geologia Storica: Genova, Tipografia del R. Istituto de' Sordo-Muti, 422 p.+1 foldout map.

Issel, A., 1896, Compendio di Geologia (con concorso dell' Ingegnere S. Traverso) — prima parte: Torino, Unione Tipografico-Editrice, [I]+428 p. + 1 p. errata + 2 foldout tables.

Istituto Italiano di Cultura di Istanbul, 1994, XIV–XVIII Yüzyil Portolan ve Deniz Haritalari/Portolani e Carte Nautiche XIV–XVIII Secolo: İstanbul, Istituto Italino di Cultura, 170 p.

Ives, J.C., 1861, General Report, *in* Report upon the Colorado River of the West, Explored in 1857 and 1858 by Lieutenant Joseph C. Ives, Corps of Topographical Engineers, Under the Direction of the Office of Explorations and Surveys, A.A., Humphreys, Captain, Topographical Engineers, in Charge—By Order of Secretary of War—Senate, 36th Congress, 1st Session, Ex. Doc.: Washington, D.C., Government Printing Office, p. 11–131.

Jackson, B.D., 1923, Linneaus (afterwards Carl von Linné). The Story of His Life, Adapted from the Swedish of Theodor Magnus Fries, emeritus professor of Botany in the University of Uppsala, and brought down to the present time in the light of recent research: London, H.F. & G. Witherby, xv + 416 p.

Jackson, D., editor, 1962, Letters of the Lewis and Clark Expedition with Related Documents: Urbana, University of Illinois Press, xxi + 728 p.

Jackson, D., editor, 1966a, The Journals of Zebulon Montgomery Pike with Letters and Related Documents, v. 1: Norman, University of Oklahoma Press, xxviii + 464 p. + 60 photographic plates.

Jackson, D. editor, 1966b, The Journals of Zebulon Montgomery Pike with Letters and Related Documents, v. 2: Norman, University of Oklahoma Press, xiii + 449 p.

Jackson, J.A., editor, 1997, Glossary of Geology, fourth edition: Alexandria, Virginia, American Geological Institute, xii + 769 p.

Jackson, M.D., 1998, Processes of laccolithic emplacement in the southern Henry Mountains, southeastern Utah, *in* Friedman, J.D., and Huffman, C., Jr., coordinators, Laccolith Complexes of Southeastern Utah: Time of Emplacement and Tectonic Setting—Workshop Proceedings: U.S. Geological Survey Bulletin 2158, p. 51–59.

Jacob, C., 1990, La Description de la Terre Habitée de Denys d'Alexandrie ou la Leçon de Géographie: Paris, Albin Michel, 264 p.

Jacobs, J.A., Russell, R.D., and Wilson, J.T., 1959, Physics and Geology: New York, McGraw-Hill, xii + 424 p.

Jacobshagen, V., editor, 1986, Geologie von Griechenland: Berlin, Gebrüder Borntraeger, IX + 363 p.

Jaeger, W., 1948, Aristotle—Fundamentals of the History of His Development, translated with the author's corrections and additions by Richard Robinson: Oxford, Clarendon, [iv] + 475 p.

Jaeger, W., 1953, Die Theologie der Frühen Griechischen Denker: Stuttgart, Kohlhammer, 303 p.

Jaffe, M., 2000, The Gilded Dinosaur—The Fossil War Between E. D. Cope and O. C. Marsh and the Rise of American Science: New York, Three Rivers Press, [iv]+424 p.

James, E., 1823, Account of an Expedition from Pittsburgh to the Rocky Mountains Performed in the Years 1819, 1820, by Order of the Hon. J.C., Calhoun, Secretary of War, Under the Command of Maj. S.H., Long, of the U.S. Top. Engineers. Compiled from the Notes of Major Long, Mr. T. Say, and other Gentlemen of the Party: London, Longman, Hurst, Res, Orme, and Brown, v. I (vii + [i] + 344 p.), v. II (vii + [i] + 356 p.), v. III (vii + [i] + 347 + [i] p. + 8 plates + fold-out profile and map.

Jamieson, T.F., 1865, On the history of the last geological changes in Scotland: Quarterly Journal of the Geological Society of London, v. 21, p. 161–203.

Jenney, J.P. and Reynolds, S.J., 1989, Geologic Evolution of Arizona: Arizona Geological Society Digest 17, vi + 866 p. + 1 folded map in pocket.

Jettmar, K., 1964, Die Frühen Steppenvölker—Der Eurasiatische Tierstil Entstehung und Sozialer Hintergrund: Zürich, Kunst der Welt, Schweizer Druck- und Verlagshaus AG, 273 + [I] p.

de Jode, G., 1578[1965], Speculum Orbis Terrarum: Antwerpen (facsimiled by Theatrum Orbis Terrarum, Amsterdam, with an introduction by R.A., Skelton), 38 maps and unpaginated text.

John, E.A.H., 1988, The riddle of mapmaker Juan Pedro Walker, *in* Palmer, S.H., and Reinharz, D., eds., Essays on The History of North American Discovery and Exploration: College Station, Texas, Texas A&M University Press, p. 102–132.

Johnson, A.M., 1970, Physical Processes in Geology: San Francisco, Freeman, Cooper & Company, xiii + 577 p.

Johnson, D.D., and Beaumont, C., 1995, Preliminarry results from a planform kinematic model of orogen evolution: surface processes and the development of clastic foreland basin stratigraphy, *in* Dorobek, S.L. and Ross, G.M., eds., Stratigraphic Evolution of Foreland Basins: SEPM (Society for Sedimentary Geology) Special Publication No. 52, p. 3–24.

Johnson, H., and Smith, B.L., editors, 1970, The Megatectonics of Continents and Oceans: New Brunswick, Rutgers University Press, xii + 282 p.

Johnson, K.S., Amsden, T.W., Denison, R.E., Dutton, S.P., Goldstein, A.G., Rascoe, B., Jr., Sutherland, P.K., and Thompson, D.M., 1988, Southern Midcontinent region, *in* Sloss, L.L., ed., Sedimentary Cover—North American Craton: U.S.: Boulder, Colorado, Geological Society of America, Geology of North America, v. D-2, p. 307–359.

Johnson, V.E., [1890], Our Debt to the Past or Chaldean Science and an Essay on Mathematics and Fine Arts: London, Griffith Farran Okeden & Welsh, viii + 118 p.

Johnson, V.E., [1891], Egyptian Science from the Monuments and Ancient Books Treated as a General Introduction to the History of Science: London, Griffith Farran & Co., xvii + [ii] + 198 p.

Jones, H.L., 1917, The Geography of Strabo, v. I: Cambridge, The Loeb Classical Library, Harvard University Press, xlvi + 529 p.

Jones, O.T., 1938, On the evolution of a geosyncline: Proceedings of the Geological Society of London, v. 94, p. lx–cx, plates A–D.

Jones, W.H.S., 1923[1984], Hippocrates, v. I: Cambridge, The Loeb Classical Library, Harvard University Press, London, William Heinemann, lxx+360 p.

Jürss, F., editor, 1982, Geschichte des Wissenschaftlichen Denkens im Altertum: Berlin, Akademie-Verlag, 672 p. + 50 photographs on plates.

Kahn, C.H., 1960, Anaximander and the Origins of Greek Cosmology: Columbia University Press, xiii + [i] + 249 + [1] p.

Kahn, C.H., 1974[1993], Pythagorean philosophy before Plato, *in* Mourelatos, A.P.D., ed., The Pre-Socratics—A Collection of Critical Essays: Princeton, Princeton University Press, p. 161–185.

Kahraman, S. A., and Dağlı, Y., eds., 2003, Günümüz Türkçesiyle Evliya Çelebi Seyahatnamesi: İstanbul, 1. Cilt, 1. Kitap: İstanbul, Yapı Kredi Yayınları, XXV + 386 p.

Kainbacher, P., 2002, Die Erforschung Afrikas—Die Afrika-Literatur über Geographie und Reisen 1500–1945 Eine Bibliographie von A–Z: Privately Printed, Baden, 471 p.+1map and 1 map index in back pocket.

Kalb, J., Aneme, D., Kidane, K., Santur, M. and Kechrid, A., eds., 2000, Bibliography of the Earth Sciences for the Horn of Africa: Ethiopia, Eritrea, Somalia and Djibouti 1620–1993: Alexandria, Virginia, American Geological Institute, xxii+321+149 p.

Kaltenmark, M., 1991, The great flood in Chinese mythology, *in* Bonnefoy, Y., ed., Mythologies, v. II: Chicago, Chicago University Press, p. 1024–1026.

Karner, G.D., and Weissel, J.K., 1990, Compressional deformation of oceanic lithosphere in the central Indian Ocean: Why it is where it is? *in* Cochran, J.R., Stow, D.A.V. et al., Proceedings of the Ocean Drilling Program, Scientific Results: College Station, Texas, Ocean Drilling Program, v. 116, p. 279–289.

Karpytchev, M., 1997, Géoïde et Topographie Dynamique à Grandes Longeurs d'Onde, influence des Hétérogénéités Lithosphériques [Thèse présentée pour obtenir le grade de Docteur en Sciences]: Paris, l'Université Paris XI Orsay, 102 p.

Karrow, R.W., Jr., 1986, George M. Wheeler and the Geographical Surveys West of the 100th Meridian 869–1879, *in* Koepp, D.P., ed., Exploration and Mapping of the American West—Selected Essays, Map and Geography Round Table of the American Library Association, Occasional Paper No. 1: Chicago, Speculum Orbis Press, p. 120–157.

Karrow, R.W., Jr., 1993, Mapmakers of the Sixteenth Century and Their Maps Bio-bibliographies of the Cartographers of Abraham Ortelius, 1570 Based on Leo Bagrow's *A. Ortelii Catalogus Cartographorum*: Chicago, For the Newberry Library Speculum Orbis Press, xxx + 846 p.

Karte von Afrika als Ubersicht der Specialblätter zu C Ritters Erdkunde 2te Auflage Th. I, 1822, *in* Ritter, C., and O'Etzel, eds., F.A. Hand-Atlas von Afrika in vierzehn Blatt zur Allgemeinen Erdkunde, 1831: Berlin, G. Reimer, first unnumbered map.

Kasbeer, T., 1973, Bibliography of Continental Drift and Plate Tectonics: The Geological Society of America Special Paper 142, xi + 96 p.

Kasbeer, T., 1975, Bibliography of Continental Drift and Plate Tectonics volume II: The Geological Society of America Special Paper 164, v + 151 p.

Kâşgarlı Mahmud, 1990, Dîvânü Lûgati't-Türk, Tıpkıbasım/Facsimile: Kültür Bakanlığı Yayınları/1205, Klâsik Eserler Dizisi/11, [Ankara], [II] + 628 + [I] p.

Kaufman, P.S., and Royden, L.H., 1994, Lower crustal flow in an extensional setting: Constraints from the Halloran Hills region, eastern Mojave Desert, California: Journal of geophysical Research, v. 99, p. 15,723–15,739.

Kay, M., 1942, Development of the northern Allegheny synclinorium and adjoining regions: Geological Society of America Bulletin, v. 53, p. 1601–1658.

Kay, M., 1944, Geosynclines in continental development: Science, v. 99, p. 461–462.

Kay, M., 1947, Geosynclinal nomenclature and the craton: Bulletin of the American Association of Petroleum Geologists, v. 31, p. 1289–1293.

Kay, M., 1951, North American Geosynclines: Geological Society of America Memoir 48, ix + 143 p.

Kay, M., 1952, Paleozoic North American geosynclines and island arcs, international Geological Congress, Report of the Eighteenth Session, Great Britain, 1948, Part XIII, p. 150–153.

Kay, M., 1955, The origin of continents: Scientific American, September 1955, p. 2–6.

Kay, M., 1967, On geosynclinal nomenclature: Geological Magazine, v. 104, p. 311–316.

Kay, M., 1974, Reflections geosynclines, flysch and melanges, *in* Dott, R.H., Jr. and Shaver, R.H., editors, Modern and Anceint Geosynclinal Sedimentation, Society of Economic Paleontologists and Mineralogists Special Publication 19 (in Honor of Marshall Kay), p. 377–380.

Kay, M. and Colbert, E.H., 1965, Stratigraphy and Life History: New York, John Wiley & Sons, [v] + 736 p.

Kayser, E., 1912, Lehrbuch der Geologie, I. Teil: Allgemeine Geologie, vierte Auflage: Stuttgart, Ferdinand Enke, XII + 881 p.

Kearey, P., and Vine, F., 1990, Global Tectonics: Oxford, Blackwell, ix + 302 p.

Kearey, P., and Vine, F., 1996, Global Tectonics, second edition: Oxford, Blackwell, x + 333 p.

Keay, J., 2000, The Great Arc—The Dramatic Tale of How India Was Mapped and Everest Was Named: London, HarperCollins, xxi + [iii] + 182 p.

Keller, E.A., 1999, Gilbert's hydraulic experiments, *in* Moores, E.M., Sloan, D., and Stout, D.L., eds., Classic Cordilleran Concepts: A View from California: Boulder, Colorado, Geological Society of America Special Paper 338, p. 241–256.

Kelley, V.C., 1955, Regional tectonics of the Colorado Plateau and Relationship to the Origin and Distribution of Uranium: New Mexico University Publications in Geology, no. 5, 120 p.

Kelly, S.S., 1969, Theories of the earth in Renaissance cosmologies, *in* Schneer, C.J., ed., Toward A History of Geology, Proceedings of the New Hampshire Inter-Disciplinary Conference on the History of Geology, September 7–12, 1967: Cambridge, The M.I.T. Press, 214–225.

Kendall, M.B., 1981, Bucher, Walter Herman, *in* Gillispie, C.C., editor-in-chief, Dictionary of Scientific Biography: New York, Charles Scribner's Sons, v. 2, p. 558–559.

Kennedy, J.E., and Sarjeant, W.A.S., 1982, "Earthquakes in the Air": The seismological theory of John Flamsteed (1693): The Journal of the Royal Astronomical Society of Canada, v. 76, p. 213–223.

Kerényi, K., 1951, Die Mythologie der Griechen. Die Götter- und Menschheitsgeschichten: Zürich, Rhein-Verlag, 312 p.

Kerényi, K., 1958, Die Heroen der Griechen: Zürich, Rhein-Verlag, 476 p.

Kessel, J.L., 1976, Friars, Soldiers, and Reformers: Hispanic Arizona and the Sonora Mission Frontier 1767–1865: Tucson, The University of Arizona Press, xiv + [i] + 347 p.

Ketin, İ., 1952, Hans Cloos: Türkiye Jeoloji Kurumu Bülteni, v. 3, no. 2, p. 107–109.

Keyes, C.B., 1939, William Phipps Blake; pioneer of southwest: Pan-American Geologist, v. 72, p. 1–8 + portrait.

Keynes, G., 1960, A Bibliography of Dr. Robert Hooke: Oxford, Clarendon, xix + [ii] + 115 p.

Keynes, R.D., 1988, Charles Darwin's *Beagle* Diary: Cambridge, Cambridge University Press, xxix + 464 + 2 maps.

Khain, V.E., 1986, The doctrine of geosynclines and plate tectonics: Geotectonics, v. 20, p. 349–356.

Khain, V. E., 1997, Iz Vospominaniy Geologa: Rossiskaya Akademiya Nauk, Otdelenie Geologii, Geofiziki, Geokhimii i Gornikh Nauk, GEOS, Moskva, 188 p.

Khain, V.E., and Lomize, M.G., 1995, Geotektonika s Osnovami Geodinamiki: Moskva, Izdatelstvo Moskovskoyo Universiteta, 476 p.

Khanchuk, A. I., 1988, The relationships between the concepts of geosynclines and plate tectonics: Geotectonics, v. 22, p. 278–279.

Kimble, G.H.T., 1938, Geography in the Middle Ages: London, Methuen & Co., X + 272 p.

King, C., 1878, United States Geological Exploration of the Fortieth Parallel. Clarence King, Geologist-in-Charge, [v.I] Systematic Geology: Washington, D.C., Government Printing Office, p. xii + 803 p. + 28 plates.

King, L.C., 1959, Denudational and tectonic relief in south-eastern Australia: Transactions of the Geological Society of South Africa, v. 62, p. 113–138.

King, L.C., 1961, Cymatogeny: Transactions of the Geological Society of South Africa, v. 64, p. 1–20.

King, L.C., 1962, The Morphology of the Earth—A Study and Synthesis of World Scenery: Edinburgh and London, Oliver and Boyd, p. xii + 699 p.

King, L.C., 1967a, The Morphology of the Earth—A Study and Synthesis of World Scenery, second edition: Edinburgh and London, Oliver and Boyd, xiii + 726 p.

King, L.C., 1967b, South African Scenery—A Textbook of Geomorphology: Edinburgh, Oliver & Boyd, xxv + 308 p.

King, P.B., 1955, Orogeny and epeirogeny through time, *in* Poldervaart, A., ed., Crust of the Earth: Geological Society of America Special Paper 62, p. 723–739.

King, P.B., and Beikman, H.M., 1974, Geologic Map of the United States (Exclusive of Alaska and Hawaii), scale 1: 2,500,000: Reston, Virginia, U.S. Geological Survey, 2 sheets.

Kircher, A., S.I., 1657, Iter Extaticum II. Qui & Mundi Subterranei Prodromus Dicitur. Quo Geocosmi Opificum sive Terrestris Globi Structura, vnà cum abditis in ea constitutis arcanioris Naturæ Reconditorijs, per ficti raptus integumentum exponitur ad veritatem: Romæ Typis Mascardi, [xxii] + 237 + [13] p.

Kircher, A., S.I., 1665, Mundus Subterraneus in XIII Libros Digestus… v. I: Amsteldami, Joannes Janssonius and Elizeus Weyerstrat, [xxvi] + 220 p.

Kirfel, W., 1920, Die Kosmographie der Inder nach den Quellen Dargestellt: Bonn, Kurt Schroeder, 36* + 401 + [1] p. + 18 plates.

Kirk, G.S., Raven, J.E., and Schofield, M., 1983, The Presocratic Philosophers. A Critical History with a Selection of Texts, 2nd edition: Cambridge, Cambridge University Press, 501 p.

Kirthisinghe, B.P., 1993, Buddhism and Science: Delhi, Motilal Banarsidass Publishers, XII + 163 p.

Klaproth, J., 1826a, Mémoires Relatifs a l'Asie contenant des Recherches Historiques, Géographiques et Philologiques sur les Peuples de l'Orient, tome second: Paris, Librairie Oriental de Dondey-Dupré et Fils, 432 p.

Klaproth, J., 1826b, Tableaux Historiques de l'Asie Depuis la Monarchie de Cyrus Jusqu'a Nos Jours Accompagnés de Recherches Historiques et Ethnographieques sur cette Partie du Monde: Paris, Schubart; Londres, Treuttel et Wurz; Stuttgard, Cotta, 291 p. + Folio Atlas with 27 double-page colored maps.

Klaproth, J., 1928, Mémoires Relatifs a l'Asie contenant des Recherches Historiques, Géographiques et Philologiques sur les Peuples de l'Orient, tome troisiéme: Paris, Librairie Oriental de Dondey-Dupré et Fils, 520 p.

Klaproth, J., 1831, Asia Polyglotta, zweite Auflage: Paris, Heideloff & Campe, XV + [1] + 144 + 8 p. and Asia Polyglotta—Sprachatlas, LIX p. of tables + 1 colored map of Asia (dated 1923).

Klaproth, J., 1836, Carte de l'Asie Centrale Dressée d'après les cartes levées par ordre de l'Empereur Khian Loung par les Missionaires de Peking, et d'après un grand nombre de notions extraites de livres chinois par M. Jules Klaproth: L. Berthe, Paris, 4 sheets, scale 1/2,664,000.

von Klebelsberg, R., 1948, Otto Ampferer 1875–1947: Berge und Heimat, 3. Jahrgang, p. 85–87.

Klemme, H.D., 1975, Giant oil fields related to their geologic setting: a possible guide to exploration: Bulletin of Canadian Petroleum Geology, v. 23, p. 30–66.

Klemp, E., ed. and introduction, 1968, Africa auf Karten des 12. bis 18. Jahrhunderts—77 Lichtdrucke aus Europäischen Kartensammlungen: Edition Leipzig, 77 maps + 64 p. of explanatory text.

Klencke, H., 1876, Alexander von Humboldt's Leben und Wirken, Reisen und Wissen—Ein Biographisches Denkmal (fortgesetzt, vielfach erweitert und theilweise umgearbeitet von H. T. Kühne und E. Hintze): Leipzig, Otto Samper, VIII+494 p.

Klotz, A., 1931, Die geographischen commentarii des Agrippa und ihre Überreste: Klio, v. 24, p. 38–58 and 386–466.

Knight, E.A., 1930, Eratosthenes As a Representative of Hellenistic Culture [M.A. thesis]: University of London, v + 175 p.

Knopf, A., 1948, The geosynclinal theory: Geological Society of America Bulletin, v. 59, p. 649–670.

Knopf, A., 1960, Analysis of some recent geosynclinal theory: American Journal of Science (Bradley volume), v. 258-A, p. 126–136.

Knowles, D., 1988, The Evolution of Medieval Thought: London, Longmans, xxvi + 337 p.

Knox, A., 1905, Notes on the Geology of the Continent of Africa—with an Introduction and Bibliography: London, His Majesty's Stationary Office, 165 p. + 1 foldout map and 1 foldout of geological sections.

Kober, L. 1921, Der Bau der Erde: Berlin, Gebrüder Borntraeger, II + 324 p.

Kober, L., 1928, Der Bau der Erde, zweite neubearbeitete und vermehre Auflage: Berlin, Gebrüder Borntraeger, II + 500 p.

Koeberl, C., 2001, Craters on the Moon from Galileo to Wegener: A Short history of the impact hypothesis, and implications for the study of terrestrial impact craters: Earth, Moon and Planets, v. 85–86, p. 209–224.

Koepp, D.P., ed., 1986, Exploration and Mapping of the American West—Selected Essays, Map and Geography Round Table of the American Library Association, Occasional Paper No. 1: Chicago, Speculum Orbis Press, vi + 182 p.

Königsson, L.-K., 1990, Urban Hiärne, the history of Nordic Geology and its bearing for modern geology research: Striae, v. 31, p. 5–6

Konrad, A.N., 1972, Old Russia and Byzantium—The Byzantine and Oriental Origins of Russian Culture: Philologische Beiträge zur Südost- und Osteuropa-Forschung, v. 1: Wien, Wilhelm Braumüller, [I] + 390 p.

Kopal, Z., 1981, Michell, John, *in* Gillispie, C.C., editor-in-chief, Dictionary of Scientific Biography: New York, Charles Scribner's Sons, v. 9, p. 370–371.

Kramer, S.N., 1981, History Begins at Sumer—Thirty-Nine Firsts in man's recorded History: Philadelphia, The University of Pennsylvania Press, xxvii + 388 p.

Krätz, O., 2000, Alexander von Humboldt—Wissenschaftler, Weltbürger, Revolutionär, unter Mitwirkung von S. Kinder und H. Merlin, 2, korrigierte Auflage: München, Callwey, 214 p.

Krenkel, E., 1922, Die Bruchzonen Ostafrikas: Berlin, Gebrüder Borntraeger, VII + [I] + 184 p.

Krenkel, E., 1925, Geologie Afrikas, erster Teil: Berlin, Gebrüder Borntraeger, X + 461 p.
Krenkel, E., 1934, Geologie Afrikas, dritter Teil, erste Hälfte: Berlin, Gebrüder Borntraeger, VIII p. + p. 1003–1304.
Krenkel, E., 1939, Geologie der Deutschen Kolonien in Afrika: Berlin, Gebrüder Borntraeger, XXII + 272 p.
Krenkel, E., 1940, Der Geologische Bau der Deutschen Kolonien in Afrika und in der Südsee, *in* von Bubnoff, S., ed., Deutscher Boden, v. 11, Berlin, Gebrüder Borntraeger, VIII + 125 p.
Krenkel, E., 1957, Geologie und Bodenschätze Afrikas, 2. stark veränderte Auflage: Akademische Verlagsgesellschaft Geest & Portig K.-G., Leipzig, XV + 597 p.
Kretschmer, K., 1890, Die physische Erdkunde im christlichen Mittelalter: Geographische Abhandlungen, v. 4, p. 1–152.
Kretschmer, K., 1892, Die Entdeckung Amerikas in Ihrer Bedeutung für die Geschichte des Weltbildes: Berlin, W.H. Kühl, London, Sampson Low & Co., Paris, H. Welter, XXIII+471 p. + Atlas of [VI] + 40 p.
Kretschmer, K., 1912, Geschichte der Geographie: Sammlung Göschen, G.J. Göschen'sche Verlagsbuchhandlung, G.m.b.H., Berlin und Leipzig, 163 p.
Kukal, Z., 1990, The Rate of Geological Processes: Praha, Academia, 284 p.
Kumar, D., 1995, Science and the Raj—1857–1905: Delhi, Oxford University Press, xv + 273 p.
Lacroix, A., 1916, Le soi-disent granite de l'ile Bora-Bora: Compte Rendu sommaire des Sciences de la Société Géologique de France, 1916, p. 178.
Lajoie, K.R., 1968, Quaternary Stratigraphy and Geologic History of Mono Basin, Eastern California [Ph.D. thesis] Berkeley, University of California, xii + 271 p. + 19 plates in pocket.
Lambeck, K., 1988, Geophysical Geodesy—The Slow Deformations of the Earth: Oxford Science Publications: Oxford, Clarendon Press, xii + 718 p.
Lambeck, K., and Nakiboglu, S.M., 1981, Seamount loading and stress in the ocean lithosphere; 2, Viscoelastic and elastic-viscoelastic models: Journal of Geophysical Research, v. 86; p. 6961–6984.
Lambert, W.G., and Millard, A.R., 1970, Atra-hasîs the Babylonian Story of the Flood with the Sumerian Flood Story by M. Civil, corrected reprint: Oxford, Clarendon, xii + 198 p. + 11 plates.
Lang, G., 1905, Untersuchungen zur Geographie der Odyssee: Verlag der Hofbuchhandlung Friedrich Gutsch, Karlsruhe, 122 p.
Lang, H.S., 1998, The Order of Nature in Aristotle's Physics—Place and the Elements: Cambridge, Cambridge University Press, xii + 324 p.
Langley, S.P., Rathburn, R., Dall, W.H., Gilman, D.C., Walcott, C.D., Harris, W.T., Baker, M., and McGee, W.J., 1902, In Memory of John Wesley Powell: Science, N.S., v. 16, p. 782–790.
de Lapparent, A., 1883, Traité de Géologie: Paris, F. Savy, XVI + 1280 p.
de Lapparent, A., 1906, Traité de Géologie, cinquième édition, refondue et considérablement augmentée, v. I: Phénomènes Actuels: Paris, Masson et Cie, XVI + 591 p.
von Lasaulx, E., 1851, Die Geologie der Griechen und Römer—Ein Beitrag zur Philosophie der Geschichte: Abhandlungen der königlichen bayerischen Akademie der Wissenschaften, I. Classe, v. 6, part 3, p. 517–566.
Laskarev, V.D., 1924, Sur les équivalents du Sarmatien Supérieur en Servie: Belgrade, Recueil des Travaux Offert à M. Cvijic, p. 73–85.
de Launay, L., 1905, La Science Géologique. Ses méthodes, ses résultats—ses problèmes, son histoire: Paris, Armand Colin, 750 + [I] p. + 3 foldout maps.
Laurent, G., 1976, Actualisme et antitransformisme de Prévost: Histoire et Nature, N° 8, p. 33–51.
Laurent, G., 1981–1982, Lyell et Lamarck: Histoire et Nature, N° 19–20, p. 115–123.
Lawrence, P. 1978, Charles Lyell versus the theory of central heat: a reappraisal of Lyell's place in the history of geology: Journal of the History of Biology, v. 11, p. 101–128.
Lazzaro de Arregui, D., 1946, Description de la Nueva Galicia, edición y estudio por François Chevalier, prólogo de John van Horne: Consejo Superior de Investigaciones Cientificas Escuela de Estudios Hispano-Americanos, Publicaciones de la Escuela de Estudios Hispano-Americanos de la Universidad de Sevilla, XXIV, serie 3.ª no. 3, Sevilla, LXXI + 161 p. + 2 maps.
Le Conte, J., 1872a, A theory of the formation of the great features of the Earth's crust: American Journal of Science and Arts, third series, v. 4, p. 345–355.
Le Conte, J., 1872b, A theory of the formation of the great features of the Earth's crust: American Journal of Science and Arts, third series, v. 4, p. 460–472.
Le Conte, J., 1873, Formation of the Earth-surface: American Journal of Science and Arts, third series, v. 5, p. 448–453.
Le Conte, J., 1883, Elements of Geology: A Text-Book for Colleges and for the General Reader, revised and enlarged edition: New York, D. Appleton and Co., xiv + 633 p.
Le Conte, J., 1889, On the origin of normal faults and of the structure of the Basin region: American Journal of Science, 3rd series, v. 38, p. 257–263.
Lee, H.D.P., 1952, Aristotle in Twenty-Three Volumes VII Meteorologica: Cambridge, The Loeb Classical Library, Harvard University Press, xxxiv + 433 p.
Lees, G.M., 1952, Foreland folding: Quarterly Journal of the Geological Society [London], v. 108, p. 1–34, plates I–IV.
Le Goff, J., 1988, Medieval Civilization, translated by Julia Barrow: Oxford, Blackwell, viii + 393 p. + 6 maps.
Lehmann, K., 1971, Irrungen und Wirrungen in der Geologie: Verlagsdruckerei C. Th. Kartenberg (privately printed), Herne, 20 p.
Leibniz, G.G., 1749, Protogaea sive de prima facie tellvris et antiqvissimae historiae vestigiis in ipsis natvrae monvmentis dissertatio ex Schedis Manvscriptis in lvcem edita a Christiano Lvdovico Scheidio: Ioh. Gvil. Schmid, XXVIII + 86 + XII plates.
Leibniz, G.W., 1949, Protogaea, übersetzt von W. v. Engelhardt, *in* Leibniz Werke, Peuckert, W.E., ed., v. 1: Stuttgart, W. Kohlhammer, 182 p.
Leighton, M.W., 1996, Interior cratonic basins: A record of regional tectonic influences, *in* van der Pluijm, B.A., and Catacosinos, P.A., eds., Basement and Basins of Eastern North America: Boulder, Colorado, Geological Society of America Special Paper 308, p. 77–93.
Leighton, M.W., Kolata, D.R., Oltz, D.F., and Eidel, J.J., eds., 1990, Interior Cratonic Basins: American Association of Petroleum Geologists Memoir 51, xiv + 819 p.
Lelewel, J., 1850, Géographie du Moyen Age, Atlas: Bruxelles, V^e et J. Pilliet, xiv + [ii] + L plates + 30 p. (Reprinted in 1993 as Publications of the Institute for the History of Arabic–Islamic Science, series Islamic Geography, v. 133).
Lelewel, J., 1852a, Géographie du Moyen Age, v. I: Bruxelles, V^e et J. Pilliet, cxxxvj + 186 p. +5 map plates. (Reprinted in 1993 as Publications of the Institute for the History of Arabic–Islamic Science, series Islamic Geography, v. 129).
Lelewel, J., 1852b, Géographie du Moyen Age, v. II: Bruxelles, V^e et J. Pilliet, 243 p. + 1 map plate. (Reprinted in 1993 as Publications of the Institute for the History of Arabic–Islamic Science, series Islamic Geography, v. 130).
Lelewel, J., 1852c, Géographie du Moyen Age, v. III and IV: Bruxelles, V^e et J. Pilliet, 220 + 112 p. (Reprinted in 1993 as Publications of the Institute for the History of Arabic–Islamic Science, series Islamic Geography, v. 131).
Lelewel, J., 1857, Géographie du Moyen Age, Epilogue: Bruxelles, V^e et J. Pilliet, viij + 308 p. + 8 map plates. (Reprinted in 1993 as Publications of the Institute for the History of Arabic–Islamic Science, series Islamic Geography, v. 132).
Lempriere, J., 1984, Lempriere's Classical Dictionary: London, Bracken Books, xv + [i] + 736 p.
von Leonhard, C., 1832, Lehrbuch der Geognosie und Geologie: Stuttgart, E. Schweizerbart's Verlagsbuchhandlung, XVI + 869 p.
Le Pichon, X., Franchetau, J. and Bonnin, J., 1973, Plate Tectonics: Developments in Geotectonics, v. 6: Amsterdam, Elsevier, xix + [i] + 300 p.

Lestringant, F., 1994, Mapping the Renaissance World—The Geographical Imagination in the Age of Discovery, translated by David Fausett with a foreword by Stephen Greenblatt: Berkeley, University of California Press, xvii + 197 p.

Leuchs, K., 1927, Tiefseegräben und Geosynklinalen: Neues Jahrbuch für Mineralogie, Geologie und Paläontologie, supplement volume 58, Part B (Pompeckj volume), p. 273–294.

Levathes, L., 1994, When China Ruled the Seas—The Treasure Fleet of the Dragon Throne 1405–1433: Simon & Schuster, 252 p.

Levene, A., 1951, The Early Syrian Fathers on Genesis—From a Syriac MS. on the Pentateuch in the Mingana Collection; The First Eighteen Chapters of the MS. Edited with Introduction, Translation and Notes; and Including a Study in Comparative Exegesis: London, Taylor's Foreign Press, VIII + 352 + [2] p.

Li, J.L., and Xiao, W.J., 2001, The paradox of geosyncline hypothesis and orogenic analysis, *in* Briegel, U. and Xiao, W.J., eds., Paradoxes in Geology (Hsü Volume): Amsterdam, Elsevier, p. 7–13.

Lindberg, D.C., 1992, The Beginnings of Western Science—The European Scientific Tradition in Philosophical, Religious, and Institutional Context, 600 B.C. to A.D. 1450: Chicago, University of Chicago Press, xviii + 455 p.

Linneaus, C. (Carl von Linné), 1744, Oratio de Telluris Habitabilis Incremento. Et Andreae Celsii … Oratio de Mutationibus Generalioribus quae in superficie corporum coelestium contingunt: Cornelium Haak, Lugduni Batavorum, 17–84 p.

Lisitzin, E., 1974, Sea-Level Changes: Elsevier Oceanography Series 8: Amsterdam, Elsevier, VI + 286 p.

Littré, E., 1963, Dictionnaire de la Langue Française, édition integrale, v. 1: Gallimard/Hachette [Paris], 232 + 1541 p.

Livingstone, D.N., 1992, The Geographical Tradition—Episodes in the History of a Contested Enterprise: Oxford, Blackwell Publishing, viii + 434 p.

Lloyd, G.E.R., 1970, Early Greek Science Thales to Aristotle: New York, W.W. Norton & Co., [viii] + 156 p.

Lloyd, G.E.R., 1973, Greek Science after Aristotle: New York, W.W. Norton & Co., xv + 189 p.

Lloyd, G.E.R., 1991, Methods and Problems in Greek Science—Selected Papers: Cambridge, Cambridge University Press, xiv + 457 p.

Lloyd, G.E.R., 1996, Aristotelian Explorations: Cambridge, Cambridge University Press, ix + 242 p.

Lobitzer, H., 1981, Der Anteil Österreichs an der geologischen Erforschung Afrikas. 1. Teil: Bibliographie Vormärz bis zum Ende der Monarchie: Mitteilungen der Österreichischen Gesellschaft für Geschichte der Naturwissenschaften, v. 1, No. 3–4, p. 29–42

Lobitzer, H., 1982, Der Anteil Österreichs an der geologischen Erforschung Afrikas. 2. Teil: Bibliographie 1919–1982: Mitteilungen der Österreichischen Gesellschaft für Geschichte der Naturwissenschaften, v. 2, no. 2–3, p. 23–42.

Logan, W. (Sir), 1854–1855, Sur la formation silurienne des environs de Quebec: Bulletin de la Société Géologique de France, v. 12, p. 504–508.

Longwell, C.R., 1923, Kober's theory of orogeny: Bulletin of the Geological Society of America, v. 34, p. 231–242.

Longwell, C.R., 1928, Herschel's view of isostatic adjustment: American Journal of Science, v. 16, p. 451–453.

de Lorenzo, G., 1920, Leonardo da Vinci e la Geologia: Publicazioni dello Istituto di Studii Vinciani in Roma, v. III: Bologna, Nicola Zanichelli, 195+[I] p.

Lotze, F., 1956, Hans Stille geb. zu Hannover am 8. Oktober 1876 zur Vollendung seines 80. Lebensjahres, *in* Lotze, F., ed., Geotektonisches Symposium zu Ehren von Hans Stille: Stuttgart, Ferdinand Enke, p. III–VII.

Love, A.E.H., 1907, On the origin of continents and oceans: Nature, v. 76, p. 327–332.

Love, A.E.H., 1908, On the origin of continents and oceans (Presidential Address): British Association for the Advancement of Sciences 1907 (1908), Leicester Meeting, Section A (Mathematics and Physics), p. 427–438.

Löwl, F., 1906, Geologie, *in* Klar, M., ed., Die Erdkunde, Eine Darstellung ihrer Wissensgebiete, ihrer Hilfswissenschaften und der Methode ihres Unterrichtes, XI. Teil, Franz Deuticke, Leipzig und Wien, 332 p.

de Luc, J.A., 1798, Lettres sur l'Histoire Physique de la Terre adressées a M. le Professeur Blumenbach Renfermant de nouvelles Preuves géologiques et historiques de la Mission divine de Moyse: Paris, Nyon, cxxviij + 406 + 2 p. of unpaginated errata.

de Luc, J.A., 1809a, An Elementary Treatise on Geology: Determining Fundamental Points in that Science, and Containing An Examination of some Modern Geological Systems, and Particularly of the Huttonian Theory of the Earth. … translated from the French manuscript by the Rev. Henry de la Fite: F.C., and J. Rivington, London, xvii + [1 page of errata] + 415 p.

de Luc, J.A., 1809b, Traité Élémantaire de Géologie: Paris, Courcier, 395 p.

Lugeon, M., 1940, Emile Argand: Bulletin de la Société Neuchâteloise des Sciences Naturelles, v. 65, p. 25–53 + Portrait.

Lundberg, G., 1957, Linné—Botaniste Suédois, Nomenclateur et Poète de la Nature: Exposition au Muséum National d'Histoire Naturelle, Paris, [VIII] + 132 p.

Lutaud, L., 1958, Émile Haug: Bulletin de la Société Géologique de France, série 6, v. 8, p. 377–396.

Lyell, C., 1830, Principles of Geology, being an attempt to explain the former changes of the earth's surface, by reference to causes now in operation, v. 1: London, John Murray, xv + 511 p.

Lyell, C., 1832, Principles of Geology, being an attempt to explain the former changes of the earth's surface, by reference to causes now in operation, v. 2: London, John Murray, xii+330 p.

Lyell, C., 1833, Principles of Geology, being An Attempt To Explain the Former Changes of the Earth's Surface by Reference to Causes Now in Operation: London, John Murray, xxxi+[i of errata]+398+109 p. + V plates.

Lyell, C., 1835a, On the proofs of a gradual rising of the land in certain parts of Sweden: Philosophical Transactions of the Royal Society of London for the year MDCCCXXXV, Part I, p. 1–38.

Lyell, C., 1835b, Principles of Geology: Being An Enquiry How Far the Former Changes of the Earth's surface are referable to Causes Now in Operation, fourth edition, v. II: London, John Murray, 465 p.

Lyell, C., 1853, Principles of Geology: or, The modern Changes of the Earth and Its Inhabitants Considered As Illustrative of Geology, ninth and entirely revised edition: London, John Murray, xii + 835 p.

Lyell, [K.M.], 1881a, Life, Letters and Journals of Sir Charles Lyell, Bart.: London, John Murray, v. 1, xi+474 p.

Lyell, [K.M.], 1881b, Life, Letters and Journals of Sir Charles Lyell, Bart.: London, John Murray, v. 2, ix+482 p.

Lyon-Caen, H., Molnar, P., and Suárez, G., 1985, Gravity anomalies and flexure of the Brazilian Shield beneath the Bolivian Andes: Earth and Planetary Science Letters, v. 75, p. 81–92.

MacCurdy, E., translator and editor, 1954, The Notebooks of Leonardo da Vinci, v. 1: London, The Reprint Society, 610 p.

MacDonald, G.A., 1972, Volcanoes: Englewood Cliffs, Prentice-Hall, Inc., xii + 510 p.

Machatschek, F., 1918, Über epirogenetische Bewegungen, *in* Festband Albrecht Penck zur Vollendung des Sechzigsten Lebensjahrs gewidmet von seinen Schülern und der Verlagsbuchhandlung: Bibliothek Geographischer Handbücher, neue Folge: Stuttgart, Verlag von J. Engelhorns Nachf., p. 211–35.

Machette, N., editor, 1988, In the Footsteps of G.K. Gilbert——Lake Bonneville and Neotectonics of the Eastern Basin and Range Province—Guidebok for Field Trip 12, 100th Annual Meeting of the Geological Society of America, Denver, Colorado, October 31–November 3, 1988: Utah Geological and Mineral Survey Miscellaneous Publication 88–1, iv + 120 p.

Macomb, J.N., 1876, General Report, *in* Report of the Exploring Expedition from Santa Fé, New Mexico, to the Junction of the Grand and Green Rivers of the Great Colorado of the West, in 1859, under the Command of Capt. J.N. Macomb, Corps of Topographical Engineers (now Colonel of Enginers); with Geological report by Prof. J.S. Newberry, Geologist of the Expedition, Engineer Department, U.S. Army: Washington, D.C., Government Printing Office, p. 5–8

Macqueen, R.W., and Leckie, D.A., eds., 1992, Foreland Basins and Foldbelts: American Association of Petroleum Geologists Memoir 55, x + 460 p.

de Maillet, B., 1748[1968], Telliamed or Conversations Between an Indian Philosopher and a French Missionary on the Diminution of the Sea.-465: Urbana, University of Illinois Press (translated and edited by A.V. Carozzi).

Manchester, W., 1993, A World Lit Only By Fire—The Medieval Mind and the Renaissance. Portrait of an Age: Boston, Little Brown and Company, XVII + 322 p.

Manning, T.G., 1967, Government in Science—The U.S. Geological Survey 1867–1894: University of Kentucky Press, xiv + 257 p.

Mannsperger, D., 1969, Physis bei Platon: Berlin, Walter de Gruyter & Co., [iv] + 336 p.

Manquat, M., 1932, Aritote—Naturaliste: Cahiers de Philosophie de la Nature, v. 5: Paris, J. Vrin, 128 p.

Mansfeld, J., 1987, Die Vorsokratiker: Stuttgart, Philipp Reclam Jun., 682 p.

Maqbul Ahmad, S., 1995, A History of Arab-Islamic Geography (9th–16th Century A.D.,): Mafraq, AL al-Bayt University, xxxv + 454 p.

Marcou, J., 1853, A Geological Map of the United States and the British Provinces of North America with an Explanatory Text, Geological Sections, and Plates of the Fossils which Characterize the Formations: Boston, Gould and Lincoln, 92 p. + VIII plates + folded colored geological map [scale: approx. 1:5,367,000].

Marcou, J., [1854], Resumé of a geological reconnaisance extending from Napoleon, at the junction of the Arkansas with the Mississippi, to the Pueblo de los Angeles, in California, *in* Whipple, A.W., Report of Explorations for a Railway Route, near the Thirty-fifth Parallel of Latitude, from the Mississippi River to the Pacific Ocean, House Document 129, no publisher, no place of publication [Washington, D.C.?], p. 40–48. (reprinted in Marcou, 1856, p. 165–175).

Marcou, J., 1856, Resumé and field notes, *in* Reports of Explorations and Surveys, to Ascertain the Most Practicable and Economical Route for a Railroad from the Mississippi River to the Pacific Ocean. Made under the Direction of the Secretary of War, in 1853–4, According to Acts of Congress of March 3, 1854, and August, 5, 1854., Senate, 33d Congress, 2d Session, Ex. Doc. No. 78, Washington, D.C., v III, part IV, No. 2, p. 121–175, 2 foldout plates.

Marcou, J., 1858, Geology of North America; with two Reports on the Prairies of Arkansas and Texas, the Rocky Mountains of New Mexico, and the Sierra Nevada of California: Zurich, Zürcher and Furrer, vi + [ii] + 144 p. folding colored map and 9 plates.

Marcou, J., 1859, Dyas et Trias ou le Noveau Grès Rouge en Europe, dans l'Amerique du Nord et dans l'Inde: Archives des Sciences de la Bibliothèque Universelle, Ramboz et Schuchardt, Genéve, 63 p.

Marcou, J., 1888, American Geological Classification and Nomenclature: Printed for the Author by the Salem Press, Cambridge, Massachusetts, 75 p.+1 foldout table.

de Margerie, E., 1930, La Société Géologique de France de 1880 à 1929, *in* Centenaire de la Société Géologique de France—Livre Jubilaire: Paris, Société Géologique de France, p. 1–82.

de Margerie, E., 1946, Critique et Géologie—Contribution a l'Histoire des Sciences de la Terre (1882–1942), v. 3: Paris, Librairie Armand Colin, p. I–LVIII + 1157–1714.

de Margerie, E., 1952, Études Américaines—Géologie et Géographie, v. I: Paris, Librairie Armand Colin, XII + 294 p.

de Margerie: 1954, Études Américaines—Paysages, Régions, Explorateurs et Cartes, v. II: Paris, Librairie Armand Colin, p. XXI + [295]–812.

Marinelli, G., 1884, Die Erdkunde bei den Kirchenvätern (Deutsch von L. Neumann, mit einem Vorworte von S. Günther): Leipzig, B.G. Teubner, VIII + 87 p. + 2 foldout maps.

Markov, M.S., Mossakovskiy, A.A., Pushcharovskiy, Y.M., Khomizuri, G.P., and Shtreys, N.A., 1974, Main premises of the theory of geosynclines in the work of scientists at the USSR Academy of Sciences: Geotectonics, v. 8, p. 137–141.

Marshak, S., 2001, Earth—Portrait of a Planet (with contributions from Donald Prothero): New York, W.W. Norton & Company, xvi + 735 + [1] + A-12 + B-3 + G-23 + C-4 + I–48 p.

Marshak, S. and Paulsen, T., 1997, Structural style, regional distribution, and seismic implications of midcontinent fault-and-fold zones, United States: Seismological Research Letters, v. 68, p. 511–520.

Marsilli, L.F., Count, 1725, Histoire Physique de la Mer: Aux De'pens de la Compagnie, Amsterdam, XI + 173 p. + 12 + 40 plates.

Martin, H., 1968, Hans Cloos 1885–1951: 150 Jahre Rheinische Friedrich-Wilhelms-Universität zu Bonn 1818–1968 Mathematik und Naturwissenschaften, Bonn, p. 171–182.

Martini, H.-J., 1967, Abschied, Ansprache an die Trauergemeinde (bei der Beerdigung von Hans Stille Ende Dezember 1966): Geologisches Jahrbuch, v. 84, p. VIII–IX.

Marvine, A.R., 1874, Report of Arch. R Marvine, assistant geologist directing the Middle Park division, *in* Hayden, F.V., Annual Report of the United States Geological and Geographical Survey of the Territories, Embracinrg Colorado, Being A Report of Progress of the Exploration for the year 1873: Washington, D.C., Government Printing Office, p. 83–192.

Mason, S.F., 1962, A History of the Sciences, new revised edition: Macmillan USA, New York, 638 p.

Mather, K.F., and Mason, S.L., 1939, A Source Book in Geology: New York, McGraw-Hill Book Company, Inc., xxii + 702 p.

Matheney, R.K., Shafiqullah, M., Brookins, D.G., Damon, P.E., and Wallin, E.T., 1988, Geochronologic studies of the Florida Mountains, New Mexico, *in* Mack, G.H., Lawton, T.F., and Lucas, S.G., eds., Cretaceous and Laramide Tectonic Evolution of Southwestern New Mexico, Guidebook of the New Mexico Geological Society, Thirty-ninth Annual Field Conference, October 5–8, p. 99–107.

Mattauer, M., 1986, Les subductions intercontinentales des chaînes tertiaires d'Asie: leurs relations avec les déchrochements: Bulletin de la Société Géologique de France, sér. 8, v. 2, p. 143–157.

Mattes, M.J., 1988, Plate River Road Narratives—A Descriptive Bibliography of Travel Over the Great Central Overland Route to Oregon, California, Utah, Colorado, Montana and Other Western States and Territories, 1812–1866: Urbana, University of Illinois Press, xiv + 632 p.

May, H.G., and Metzger, B.M., eds., 1977, The New Oxford Annotated Bible with the Apocrypha, revised standard version: New York, Oxford University Press, xxviii + 1564 + xxiv + 340 p. + [i]p. + 12 maps+[ii] p.

Mayo, D.E., 1985, Mountain-building theory: The nineteenth century origins of isostasy and the geosyncline, *in* Drake, E.T., and Jordan, W.M., eds., Geologists and Ideas: A History of North American Geology, Centennial Special Volume 1: Boulder, Colorado, The Geological Society of America, p. 1–18.

Mayor, A., 2000, The First Fossil Hunters—Paleontology in Greek and Roman Times: Princeton, Princeton University Press, xx+361 p.

McBride, J.H., Sargent, M.L., and Potter, C.J., 1997, Investigating possible earthquake-related structure beneath the southern Illinois Basin from seismic reflection: Seismological Research Letters, v. 68, p. 641–649.

McBride, L.R., 1968, Pele—Volcano Goddess of Hawaii: Hilo, Hawaii, The Petroglyph Press, 48 p.

McCartney, P.J., 1977, Henry de la Beche—Observations on an Observer: Cardiff, Friends of the National Museum of Wales, viii + 77 p.

McGetchin, T.R., Burke, K., Thompson, G.A., and Young, R.A., 1980, Mode and mechanisms of plateau uplifts, *in* Bally, A.W., Bender, P.L., McGetchin, T.R., and Walcott, R.I., eds., 1980, Dynamics of Plate Interiors, Geodynamics Series, v. 1: Washington, D.C., American Geophysical Union, Boulder, Colorado, Geological Society of America, p. 99–110.

McIntyre, D.B., 1997, James Hutton's Edinburgh: The historical, social and political background: Earth Science History, v. 16, p. 100–157.

McIntyre, D.B., and McKirdy, A., 1997, James Hutton—The Founder of Modern Geology: Edinburgh, The Stationary Office, xi + 51 p.

McKee, E.D., 1938, The Environment and History of the Toroweap and Kaibab Formations of Northern Arizona and Southern Utah: Washington, D.C., Carnegie Institute Publication no. 492, 268 p.

McKee, E.D., 1969, Stratified Rocks of the Grand Canyon—A history of stratigraphic investigation in the Grand Canyon region: [U.S.] Geological Survey Professional Paper 669-B, [III] p. + p. 23–58.

McKenzie, D., 1984, A possible mechanism for epeirogenic uplift: Nature, v. 307, p. 616–618.

McKenzie, D., 1994, The relationship between topography and gravity on Earth and Venus: Icarus, v. 112, p. 55–88.

McKenzie, D., Nimmo, F., Jackson, J.A., Gans, P.B., and Miller, E.L., 2000, Characteristics and consequences of flow in the lower crust: Journal of Geophysical Research, v. 105, p. 11,029–11,046.

McQuarrie, N., and Chase, C.G., 2000, Raising the Colorado Plateau: Geology, v. 28, p. 91–94.

Medwenitsch, W., 1970, Leopold Kober: Mitteilungen der Geologischen Gesellschaft in Wien, v. 63, p. 207–216.

Meissner, R., Snyder, D., Balling, N. and Staroste, E., editors, 1992, The Babel Project—First Status Report: Commission of the European Communities Directorate-General XII Science, Research and Development, R&D Programme Non-Nuclear Energy Area: Deep Reservoir Gelogy, Brussels, vi + 155 p.

Melosh, H.J., and Ivanov, B., 1999, Impact crater collapse: Annual Review of Earth and Planetary Science, v. 27, p. 385–415.

Menard, H.W., 1973, Epeirogeny and plate tectonics: Eos (Transactions, American Geophysical Union), v. 54, p. 1244–1255.

Mercator, G., and Hondius, J., 1636 [1968], Atlas or A Geographicke Description of the Regions, Countries and Kingdomes of the World, through Europe, Asia, Africa, and America represented by new & exact Maps, The Second Volume: Henrici Hondij et John Janssonius, Amsterodami (Reprinted in 1968 by Theatrum Orbis Terrarum Ltd., Amsterdam with an Introduction by R.A., Skelton), 115 maps and p. 217–462.

Mercier, J., and Vergely, P., 1992, Tectonique: Paris, Dunod, VIII + 214 p.

Merrill, G.P., 1904, Contributions to the History of American Geology: Report of U.S. National Museum, Washington, D.C., p. 189–733 + Plate I.

Merrill, G.P., 1920, Contributions to a History of American State Geological and Natural History Surveys: Smithsonian Institution United States National Museum Bulletin 109, XVIII + 549 p. + 37 photographic plates.

Merrill, G.P., 1924, The First One Hundred Years of American Geology: New Haven, Yale University Press, xxi + 773 p. + 1 foldout plate.

Metz, K., 1957, Lehrbuch der Tektonischen Geologie: Stuttgart, Ferdinand Enke, VII + 294 p. + 1 folded map.

Metzler, J.S.I., 1941, Der apostolische Vikar Nikolaus Steno und die Jesuiten: Archivum Historicum Societatis Jesu, v. 10, p. 93–193.

Meunier, S., 1909, La Géologie Générale: Paris, Félix Alcan, XII + 344 p.

Meyer-Abich, A., 1969, Alexander von Humboldt, *in* Alexander von Humboldt, 1769–1969, Bonn/Bad Godesberg, Inter Nationes, p. 7–94.

Meyer-Abich, A., 1985, Alexander von Humboldt: Rororo Bildmonographien, Rowohlt, Reinbeck, 190 p.

Michell, J., 1761, Conjectures concerning the cause and observations upon the phenomena of earthquakes; particularly of that great earthquake of the first of November 1755, which proved so fatal to the City of Lisbon and whose effects were felt as far as Africa and more or less throughout all Europe: Philosophical Transactions of the Royal Society of London, v. 51, p. 566–634.

Michel-Lévy, A., 1905, La Chaire d'Histoire Naturelle des corps Inorganiques au Collège de France: Extrait de Revue Générale des Sciences, 30 Avril 1905: Paris, Armand Colin, 40 p.

Middlemost, A.K., 1972, Evolution of La Palma, Canary Archipelago: Contributions to Mineralogy and Petrology, v. 36, p. 33–48.

Middleton, L.T., Elliot, D.K., and Morales, M., 1990, Coconino Sandstone, *in* Beus, S.S., and Morales, M., eds., Grand Canyon Geology: New York, Oxford University Press, and Museum of Northen Arizona Press, p. 183–202.

Miethke, J., 1989, Zur sozialen Situation der Naturphilosophie im späteren Mittelalter, *in* Boockmann, H., Moeller, B., and Stackmann, K., eds., Lebenslehren und Weltentwürfe im Übergang vom Mittelalter zur Neuzeit, Politik-Bildung-Naturkunde-Theologie, Abhandlungen der Akademie der Wissenschaften in Göttingen, Vanderhoeck & Ruprecht in Göttingen, p. 249–266.

Miller, D.H., 1970, Balduin Möllhausen, A Prussian's Image of the American West [Ph.D. thesis]: Albuquerque, University of New Mexico, xii + 289 + [1] p.

Miller, D.H., 1972a, The Ives Expedition revisited—a Prussian's Impressions: The Journal of Arizona History, v. 13, p. 1–25.

Miller, D.H., 1972b, The Ives Expedition revisited: Overland into Grand Canyon: The Journal of Arizona History, v. 13, p. 177–196

Miller, H., 1992, Abriß der Plattentektonik: Stuttgart, Ferdinand Enke, VIII + 149 p.

Miller, K., 1916, Itineraria Romana—Römische Reisewege an der Hand der Tabula Peutingeriana…: Stuttgart, Strecker und Schröder, LXXV + 992 p.

Millhauser, M., 1959, Just Before Darwin—Robert Chambers and *Vestiges*: Middletown, Wesleyan University Press, ix + 246 p.

Milne, J., 1882, Earthquake vibrations: Nature, v. 25, p. 126.

Miquel, A., 1973, La Géographie Humaine du Monde Musulman jusqu'au Milieu du 11e Siècle—Géographie et géographie humaine dans la littérature arabe des origines à 1050, 2e éd.: Paris, Mouton, L + 426 p.

Miquel, A., 1975, La Géographie Humaine du Monde Musulman jusqu'au Milieu du 11e Siècle—La représentation de la terre et de l'étranger: Paris, Mouton, p. XXVII + 705 p.

Miquel, A., 1980, La Géographie Humaine du Monde Musulman jusqu'au Milieu du 11e Siècle—Le milieu naturel: Paris, Mouton, XX + 543 p.

Mirsky, J., editing and introduction, 1964, The Great Chinese Travelers: New York, Pantheon Books, A Division of Random House, vii + [ii] + 309 + [1] p.

Mitchell, A.H., and Reading, H.G., 1969, Continental margins, geosynclines and ocean floor spreading: Journal of Geology, v. 77, p. 629–646.

Modi, J.J., 1905–1908, Maçoudi on Volcanoes: Journal of the Bombay Branch of the Royal Asiatic Society, v. 22, p. 135–142.

Mohr, P.A., 1999, A Bibliography of the Discovery of the Geology of the East African Rift System (1830–1950), international Commission on the History of Geological Sciences, printed at The University of New South Wales, Australia, 24 p.

Möllhausen, B., 1858, Tagebuch einer Reise vom Mississippi nach den Küsten der Südsee … eingeführt von Alexander von Humboldt: Leipzig, Hermann Mendelssohn, XIV + 494 p. + 1 foldout map + 2 p. of explanatory text to map.

Möllhausen, B., 1861, Reisen in die Felsengebirge Nord-Amerikas bis zum Hoch Plateau von Neu-Mexico, unternommen als Mitgiled der im Auftrage der Regierung der Vereinigten Staaten ausgesandten Colorado-Expedition: Leipzig, Hermann Costenoble, v. I (XVI + 455 p. + 6 plates and one facsimile letter of Alexander von Humboldt), v. II (IX + [I] + 406 p. + 6 plates and one foldout map).

Moody, E.A., 1981a, Buridan, Jean, *in* Gillispie, C.C., editor-in-chief, Dictionary of Scientific Biography: New York, Charles Scribner's Sons, v. 1, p. 603–608.

Moody, E.A., 1981b, Albert of Saxony, *in* Gillispie, C.C., editor-in-chief, Dictionary of Scientific Biography: New York, Charles Scribner's Sons, v. 1, p. 93–95.

Moore, J.G., 2000, Exploring the Highest Sierra: Stanford, Stanford University Press, xv + 427 p.

Moorehead, A., 1960, The White Nile: New York, Harper & Brothers, [iv] + 385 p. with numerous illustrations and maps.

Moorehead, A., 1962, The Blue Nile: New York, Harper & Row, xii + 308 p. with numerous illustrations and 2 maps.

Moores, E.M. and Twiss, R.J., 1995, Tectonics: New York, W.H. Freeman and Company, xi + 415 p.

Morgan, D.L., 1953, Jedediah Smith and the Opening of the West: Lincoln, University of Nebraska Press, 458 p. + one foldout map.

Morgan, D.L., and Wheat, C.I., 1954, Jedediah Smith and His Maps of the American West: San Francisco, California Historical Society Special Publication No. 26, [I] + 86 p. + 7 maps.

Morgan, W.J., 1971, Convection plumes in the lower mantle: Nature, v. 230, p. 42–43.

Morgan, W.J., 1972a, Plate motions and deep mantle convection: Geological Society of America Memoir 132, p. 7–22.

Morgan, W.J., 1972b, Convection plumes and plate motions: American Association of Petroleum Geologists Bulletin, v. 56, p. 203–213.

Mörner, N.-A., 1979, The Fennoscandian uplift and late Cenozoic Geodynamics: Geological Evidence: GeoJournal, v. 3, p. 287–318.

Mörner, N.-A., ed., 1980, Earth Rheology, Isostasy and Eustasy: Chichester, John Wiley & Sons, xv + 599 p.

Moro, A.L., 1740, De Crostacei e degli Altri Marini Corpi Che Si Truovano su' Monti, Libri Due: Venezia, Stefano Monti, [xii] + 452 p.

Morris, J.M., 1997, El Llano Estacado—Exploration and Imagination on the High Plains of Texas and New Mexico: Austin, Texas State Historical Association, x + 414 p.

Morrison, R.B., 1965, Quaternary geology of the Great Basin, *in* Wright, H.E., and Frey, D.G., eds., The Quaternary of the United States—A Review Volume for the VII Congress of the International Association for Quaternary Research: Princeton, Princeton University Press, p. 265–285.

Moszkowski, A., 1921, Einstein—Einblicke in Seine Gedankenwelt: Hamburg, Hoffmann und Campe; Berlin, F. Fontane & Co., 240 p.

Moule, A.C., and Pelliot, P., 1935, Marco Polo The Description of the World, v. II: London, George Routledge & Sons Ltd., cxxxi p.

Moule, A.C. and Pelliot, P., 1938, Marco Polo The Description of the World, v. I: London, George Routledge & Sons Ltd., 595 p.

M. Sadi, 1327[1911–1912AD], İlm-i Arz, ikinci tabı: [İstanbul], Matbaa-i Hayriye ve Şürekası, 180 p. + 14 plates.

Muir-Wood, R., 1985, The Dark Side of the Earth: London, Allen & Unwin, ix + [ii] + 246 p.

Müllerus, C., 1883, Κλαυδιου Πτολεμαιου Γεωγραφικη Υφηγησις—Claudii Ptolemæi Geographia. e Codicibus Recognovit, Prolegomenis, Annotatione, Indicibus, Tabulis Instruxit, v. I/1: Paris, Firmin Didot, [II] + 570 p.

Müllerus, C., 1901, Κλαυδιου Πτολεμαιου Γεωγραφικη Υφηγησις—Claudii Ptolemæi Geographia. e Codicibus Recognovit, Prolegomenis, Annotatione, Indicibus, Tabulis Instruxit, v. I/2: Paris, Firmin Didot, p. II + 571–1023.

Mullins, J., 1977, The Goddess Pele: Honolulu, Hawaii, Aloha Graphics and Sales, 35 p.

Munro, D., editor, 1988, Chambers World Gazetteer—An A–Z of Geographical Information: Cambridge, Chambers/Cambridge, xviii + 733 p. + 112 maps.

[Münster, S., editor and commentator], 1583, C. Iulii Solini Polyhistor, Rerum Toto Orbe Memorabilium Thesaurus Locupletissimus Huic Ob Argumenti Similitudinem Pomponii Melae de Situ Orbis Libros Tres, Fide Diligentiaque Summa Recognitos, Adiunximus: Basileæ apud Michaelem Isingrinium et Henricum Petri, [xviii] + 230 p.

Münster, S. editor, 1540[1968], Geographia Vniversalis Vetus et Nova, Complectens Claudii Ptolemæi Alexandrini Enarrationis Libros VIII: Basileæ apud Henricum Petrum, (Dedication, Index and Book I unpaginated; the rest) 195 p.

Münster, S., 1544, Cosmographia—Beschreibûg aller Lender durch Sebastianum Munsterum in welcher begriffen Aller voelcker / Hercsthafften / Stetten / und nahmhaftiger flecken / herkomen: Sitten / gebreüch / ordnung /glauben / secten / und hantierung / durch die gantze welt / und fürnemlich Teücher nation. Was auch besonders in jedem landt gefunden und darin beschehen sey. Alles mit figuren / und für augen gestelt: Getruckt zu Basel durch Henrichum Petri, dclix p.

Münster, S., 1550[1968], Cosmographei oder beschreibung aller länder/ herschafften/fürnemsten stetten/geschichten/gebreüchē/hantierungen etc. ieß zum dritten mal trefflichsere durch Sebastianum Münsteru gemeret und gebessert/in weldtlichē und natürlichen historien. Jtē vff ein neuws mit hübschen figuren unnd landtaflen geizert/sunderlichen aber werden dar in contrafhetet sechs unnd vierßig stett/under welchē bey dreissig auß Teuscher nation nach jhrer gelegenheit dar zu kommē/ vnd von der stetten oberkeiten do hin sampt jrenn beschreibungen verordnet: [Heinrich Petri], Basel, Mccxxxiii p. (reprinted, with an introduction by R. Oehme, p. V–XXVIII, by Theatrum Orbis Terrarum Ltd., Amsterdam, MCMLXVIII).

Murchison, R.I., (Sir), 1849, On the geological structure of the Alps, Apennines and Carpathians, more especially to prove a transition from Secondary to Tertiary rocks, and the development of Tertiary deposits in Southern Europe: Quarterly Journal of the Geological Society of London, v. 5, p. 157–312 + plate VII.

Murchison, R.I., de Verneuil, E. and von Keyselring, A., 1845, The Geology of Russia in Europe and the Ural Mountains, v. I, Geology: London, John Murray, xxiv + 700 p.

Murray, G.E., 1970, Razvitie geologicheskikh znanii v pribrezhnikh raionakh vostoka Severnoi Ameriki (1750–1960), *in* Tikhomirov, V.V., E. Malkhasian, G., Mgrtchian, S.S., Ravikovich, A.I. and Sofiano, T.A., eds., Istoria Geologii, Akademia Nauk Armianskoi SSR, Mezhdunarodnii Komitet po Istorii Geologicheskikh Nauk, Izdatelstvo Akademii Nauk Armianskoi SSR, Erevan, p. 279–303 (English summary: History of development of geological knowledge in the coastal province of eastern North America (1750–1960), p. 304–306).

Murray, O., 1993, Early Greece, second edition: Cambridge, Harvard University Press, [ii] + 353 p.

Nakada, M., and Lambeck, K., 1986, Seamount loading of a compressible viscoelastic plate: an analytical solution: Journal of Geodynamics, v. 5, p. 103–110.

Nalivkin, V.D., 1976, Dynamics of the development of the Russian Platform structures, *in* Bott, M.H.P., editor, Sedimentary Basins of Continental Margins and Cratons, Developments in Geotectonics, 12: Amsterdam, Elsevier, (also Tectonophysics, v. 36), p. 247–262.

Nathorst, A.G., 1908, Carl von Linné as a geologist: Annual Report of the Smithsonian Institution for 1908, p. 711–743.

Natland, J.H., 1997, At Vulcan's shoulder: James Dwight Dana and the beginnings of planetary volcanology: American Journal of Science, v. 297, p. 312–342.

Naumann, C.F., 1850, Lehrbuch der Geognosie, erster Band: Leipzig, Wilhelm Engelmann, XI + [I] + 1000 p.

Naumann, F., ed., 1994, Georgius Agricola 500 Jahre—Wissenschaftliche Konferenz vom 25–27 März 1994 in Chemnitz, Freistaat Sachsen, 507 p.

Nazzaro, A., 1995, Vesuvius, Pompei, Campi Flegrei field excursions, *in* Volcanoes and History XX Symposium INHIGEO Naples-Aeolian Islands-Catania, September 19–September 25 1995—Guidebook: no place of publication, p. 7–26.

Nebenzahl, K., 1986, Mapping the Trans-Mississippi West: annotated selections, *in* Koepp, D.P., ed., Exploration and Mapping of the American West—Selected Essays, Map and Geography Round Table of the American Library Association, Occasional Paper No. 1: Chicago, Speculum Orbis Press, p. 1–23.

Needham, J., 1954, Science and Civilisation in China, v. 1, Introductory Orientations: Cambridge, Cambridge University Press, xxxviii + 318 p. + 2 foldout maps.

Needham, J., 1956, Science and Civilisation in China, v. 2 History of Scientific Thought: Cambridge, Cambridge University Press, XXII + [II] + 697 + 1 p.

Needham, J., 1959, Science and Civilisation in China, v. 3 Mathematics and the Sciences of the Heavens and the Earth: Cambridge, Cambridge University Press, xlvii + 877 p.

Nelson, C.M., 1996, Powell, John Wesley: Dasch, E.J., ed.-in-chief, Encyclopedia of Earth Sciences, v. 2: New York, Macmillan Reference USA, Simon & Schuster, Macmillan, p. 887–889.

Nelson, C.M., and Fryxell, F.M., 1997, Hayden, Ferdinand Vandiveer [*sic*], *in* Sterling, K.B., Harmond, R.P., Cevasco, G.A., and Hammond, L.F., eds., Biographical Dictionary of American and Canadian Naturalists and Environmentalists: Westport, Greenwood Press, p. 355–358.

Nelson, C.M., and Rabbitt, M.C., 1997, King, Clarence Rivers, *in* Sterling, K.B., Harmond, R.P., Cevasco, G.A., and Hammond, L.F., eds., Biographical Dictionary of American and Canadian Naturalists and Environmentalists: Westport, Greenwood Press, p. 431–434.

Nelson, C.M., Rabbitt, M.C., and Fryxell, F.M., 1981, Ferdinand Vandeveer Hayden: The U.S. Geological Survey years, 1879–1886: Proceedings of the American Philosophical Society, v. 125, p. 238–243.

Nelson, W.J., Denny, F.B., Deevra, J.A., Follmer, L.R., and Masters, J.M., 1997, Tertiary and Quaternary tectonic faulting in southernmost Illinois: Engineering Geology, v. 46, p. 235–258.

Neumayr, M., 1885, Die geographische Verbreitung der Juraformation: Denkschriften der kaiserlichen Akademie der Wissenschaften in Wien, mathematisch-naturwissenschaftliche Classe, v. 50, p. 57–86.

Neumann, C. and Partsch, J., 1885, Physikalische Geographie von Griechenland mit besonderer Rücksicht auf das Alterthum: Breslau, Wilhelm Koebner, XII + 475 p. + [1 p. of Errata].

Neumann, E., 1972, The Great Mother an Analysis of the Archetype, translated by Ralph Manheim: Bollingen Series XLVII: Princeton, Princeton University Press, xliii + 379 p. + 185 photographic plates.

Nevins, A., 1955, Frémont—Pathmarker of the West: New York, Longmans, Green and Co., xiv + 689 p.

Newberry, J.S., 1856, Geological report, *in* Abbot, H.L., Explorations for a Railroad Route, from the Sacrmento Valley to the Columbia River, made by Lieut. R.S., Williamson, Corps of Topographical Engineers, assisted by Lieut. Henry L. Abbot, Corps of Topographical Engineers, Reports of Explorations and Surveys, to Ascertain the Most Practicable and Economical Route for A Railroad from the Mississippi River to the Pacific Ocean. Made under the Direction of the Secretary of War, in 1853–4, According to Acts of Congress of March 3, 1854, and August, 5, 1854, Senate, 33d Congress, 2d Session, Ex. Doc. No. 78, Washington, D.C., v. VI, part II, 85 p.

Newberry, J.S., 1861, Geological Report, *in* Report upon the Colorado River of the West, Explored in 1857 and 1858 by Lieutenant Joseph C. Ives, Corps of Topographical Engineers, Under the Direction of the Office of Explorations and Surveys, A.A., Humphreys, Captain, Topographical Engineers, in Charge—By Order of Secretary of War—Senate, 36th Congress, 1st Session, Ex. Doc.: Washington, D.C., Government Printing Office, p. 1–154 + 3 plates.

Newberry, J.S., 1871, The ancient lakes of western America, *in* Preliminary Report of the U.S. Geological Survey of Wyoming and Portions of Contiguos Territories (Being A Second Annual Report of Progress): Washington, D.C., Government Printing Office, p. 329–339.

Newberry, J.S., 1876, Geological Report, *in* Report of the Exploring Expedition from Santa Fé, New Mexico, to the Junction of the Grand and Green Rivers of the Great Colorado of the West, in 1859, under the Command of Capt. J.N., Macomb, corps of topographical Engineers (now Colonel of Enginers); with Geological report by Prof. J.S., Newberry, Geologist of the Expedition, Engineer Department, U.S. Army: Washington, D.C., Government Printing Office, p. 9–118.

Newby, E., 1975, The World Atlas of Exploration: London, Artists House, 288 p.

Newcomb, S., 1990, Contributions of British experimentalists to the discipline of geology: Proceedings of the American Philosophical Society, v. 134, p. 161–225.

Newell, J.R., 1993, American Geologists and their Geology: The Formation of the American Geological Community, 1780–1865 [Ph.D. Dissertation]: Madison, University of Wisconsin, xvii + 371 p.

Newell, J.R., 1997, James Dwight Dana and the emergence of professional geology in the United States: American Journal of Science, v. 297, p. 273–282.

Niebuhr, B.G., 1811, Römische Geschichte, erster Theil: Berlin, Realschulbuchhandlung, XVI + 455 + [9] p. + 1 foldout map.

Nieuwenkamp, W., 1981, Buch, (Christian) Leopold von, *in* Gillispie, C.C., editor-in-chief, Dictionary of Scientific Biography: New York, Charles Scribner's Sons, v. 2, p. 552–557.

Nizan, P., 1938, Les Matérialistes de l'Antiquité—Démocrite–Épicure–Lucrèce, nouvelle édition: "Socialisme et Culture" Éditions Sociales Internationales, Paris, 178 p.

Nobbe, C.F.A., ed., 1843–1845[1990], Claudii Ptolemaei Geographia I–III: Hildesheim, Georg Olms Verlag, v. I (XXXVI + 284 p.), v. II (269 p.), v. III ([IV] + 207 p. + 1 foldout map).

Nordenskiöld, A.E., 1889, Facsimile-Atlas to the Early History of Cartography with Reproductions of the Most Important Maps Printed in the XV and XVI Centuries (translated from the Swedish original by J.A., Ekelöf and C.R., Markham): Stockholm, Privately Printed, 141 p. + LI plates.

Nordlind, A., 1918, Das Problem des gegenseitigen Verhältnisses von Land und Wasser und seine Behandlung im Mittelalter: Meddelanden från Lunds Universitets Geografiska Institution, serie B, no. 1, 56 + [1] p.

Nordlind, A., 1927, Okeanos: Meddelanden från Lunds Universitets Geografiska Institution, serie C, no. 20, p. 9–21.

North, F.J., 1965, Sir Charles Lyell—Interpreter of the Principles of Geology: Creators of the Modern World series: London, Arthur Barker, 128 p.

Norwich, J.J., 1997, Byzantium—The Decline and Fall: New York, Alfred A. Knopf, xxxvii + 488 p.

Oberhuber, K., 1990, Sumerisches Lexicon zu "Georg Reisner, Sumerisch-babylonische Hymnen nach Thontafeln griechischer Zeit (Berlin 1896)" (SBH) und verwandten Texten …, innsbrucker Sumerisches lexicon (ISL) des Instituts für Sprachen und kulturen des Alten Orients an der Universität Innsbruck Abteilung I: Sumerisches Lexicon zu den zweissprachigen literarischen Texten: Innsbruck, Verlag des Instituts für Sprachwissenschaft der Universität Innsbruck, [III] + 583 p.

Obruchev, V., and Zotina, M., 1937, Eduard Süss: Jizn Zamechatelnikh Lügei, no. 1, Jurnalno-Gazetnoe Obedinenie, Moskva, 231 p.+16 unnumbered plates.

Ocak, A.y., 1985, İslâm-Türk İnançlarında Hızır Yahut Hızır-İlyas Kültü: Türk Kültürünü Araştırma Enstitüsü Yayınları: Ankara, 54, Seri: IV, Sayı A. 16, 229 p.

Odens, P.R., 1980, Father Garces—The Maverick Priest: Privately printed by Sun Graphics, Yuma, Arizona, 45 p.

O'Hara, C.C., 1920[1976], The White River Badlands: Rapid City, South Dakota, South Dakota School of Mines, Bulletin No. 13, 181 p. + 96 plates.

Oldroyd, D.R., 1972, Robert Hooke's methodology of science as exemplified in his "Discourse of earthquakes": The British Journal for the History of Science, v. 6, p. 109–130.

Oldroyd, D.R., 1996, Thinking about the Earth: A History of Ideas in Geology: Cambridge, Harvard University Press, xxx + 410 p.

Oldroyd, D.R., and Hamilton, B., 1997, Geikie and Judd, and controversies about the igneous rocks of the Scottish Hebrides: Theory, practice, and power in the geological community: Annals of Science, v. 54, p. 221–268.

Oldroyd, D.R., and Howes, J.B., 1978, The first published version of Leibniz's Protogaea: Journal of the Society of the Bibliography of Natural History, v. 9, p. 56–60.

Olshausen, E., 1991, Einführung in die Historische Geographie der Alten Welt: Darmstadt, Wissenschaftliche Buchgesellschaft, X+232 p.+8 maps.

d'Orbigny, A., 1849, Cours Élémentaire de Paléontologie et de Géologie Stratigraphique, premier volume: Paris, Victor Masson, 299 p.

Oreskes, N., 1999, The Rejection of Continental Drift—Theory and method in American Earth Science: New York, Oxford University Press, ix + [i] + 420 p.

Orlov, V.P., responsible editor, 1999, Repressirovannie Geologi: Ministerstvo Prirodnikh Resursov Rossiskoi Federatsii (MFR RF), Federalnoe Gosudarstvennoe Unitarnoe Predpiyatie "Vserossiskii Nauchno-Issledovatelskii Geologicheskii Institut imeni A.P., Karpinskoyo" (VSEGEI), Rossiskoe Geologicheskoe Obshestvo (RosGeo), Moskva-Sankt Peterburg, 451 p.

Ortega-Gutiérrez, F., and Guerrero-García, J.C., 1982, The geologic regions of Mexico, *in* Palmer, A.R., ed., Perspectives in Regional Geological Synthesis: Planning for the Geology of North America: Boulder, Colorado, Geological Society of America, Geology of North America, DNAG Special Publication 1, iv + p. 99–104.

Ortelius, A., 1570[1964], Theatrum Orbis Terrarum: Antwerp (facsimiled by Theatrum Orbis Terrarum Ltd., Amsterdam; with an introduction by R.A., Skelton), 53 maps and unpaginated text.

Osberg, P.H., Tull, J.F., Robinson, P., Hon, R. and Butler, J.R., 1989, The Acadian Orogen, *in* Hatcher, R.D., Jr., Thomas, W.A., and Viele, G.W., eds., The Appalachian-Ouachita Orogen in the United States: Boulder, Colorado, Geological Society of America, The Geology of North America, v. F-2, p. 179–232.

Ospovat, A.M., 1960, Abraham Gottlob Werner and His Influence on Mineralogy and Geology [Ph.D. thesis]: Norman, Oklahoma, University of Oklahoma, vii + 259 p.

Ospovat, A., translator, introducer and commentator, 1971, Abraham Gottlob Werner, Short Classification and description of the Various Rocks: New York, Hafner Publishing Company, x + [ii] + 194 p.

Özgüç, N. and Tümertekin, E., 2000, Coğrafya—Geçmiş. Kavramlar. Coğrafyacılar: İstanbul, Çantay Kitabevi, [VIII]+434 p.

Pagani, L., 1990, [Einführung], *in* Ptolemäus Cosmographia—Das Weltbild der Antike, Parkland, p. III–XIV.

Page, L.E., 1981, Scrope, George Julius Poulett, *in* Gillispie, C.C., editor-in-chief, Dictionary of Scientific Biography: New York, Charles Scribner's Sons, v. 12, p. 261–264.

Pallas, P.S., An VII [1796], Tableau physique et topographique de la Tauride Suivi d'Observations sur la Formation des montagnes, et les Changemens Arrivés á Notre Globe par le Professeur Pallas, pour faire suite à son Voyage en Russie, *in* Voyage de Pallas, tome IX: Paris, Chez Guillaume and Guide, 172 p.

Pallas, P.S., 1803,[443] Travels Through the Southern Provinces of the Russian Empire, Performed in the years 1793 and 1794, vol. I: Printed for James Ridgway by B. McMillan, London, xxiv + 336 p.

Palmer, M., and Zhao, X.M., 1997, Essential Chinese Mythology: London, Thorsons, [I] + 195 p.

Parona, C.F., 1903, Trattato di Geologia con Speciale Riguardo alla Geologia d'Italia: Biblioteca delle Scienze Fisice e Naturali, Casa Editrice Dottor Francesco Vallardi, Milano, XIV+[I]+730+1 p.

Park, R.G., 1988, Geological Structures and Moving Plates: Glasgow, Blackie, vi + 337 p.

Parsons, T., Thompson, G.A., and Sleep, N.H., 1994, Mantle plume influence on the Neogene uplift and extension of the U.S. Western Cordillera?: Geology, v. 22, p. 83–86.

Pavlov, A.P., 1903, Ob izmeneniakh geografii Rossii v yurskoe I melovoe vremya: Nauchnoe Slovo, v. 1, p. 143–145.

Paxson, F.L., 1924, History of the American Frontier 1763–1893: Boston, Houghton Mifflin Co., vii + 598 p.

Payne, K., 1992, Sources for a study of the earth sciences in classical Greece: The Compass, v. 69, 313–319.

Peach, B.N., Horne, J., Gunn, W., Clough, C.T., and Hinxman, L.W., 1907, The Geological Structure of the North-West Highlands of Scotland (edited by A. Geikie): Memoires of the Geological Survey of Great Britain: Glasgow, His Majesty's Stationary Office, xviii + 668p. + 52 plates.

Pédech, P., 1976, La Géographie des Grecs: Paris, Presses Universitaires de France, 202 p.

Peltier, W.R., 1980, Models of glacial isostasy and relative sea level, *in* Bally, A.W., Bender, P.L., McGetchin, T.R., and Walcott, R.I., eds., Dynamics of Plate Interiors, Geodynamics Series: Washington, D.C., American Geophysical Union, Boulder, Colorado, Geological Society of America, v. 1, p. 111–128.

Penck, W., 1920, Der Südrand der Puna de Atacama (NW-Argentinien)—Ein Beitrag zur Kenntnis der Andinen Gebirgstypus und zu der Frage der Gebirgsbildung: Abhandlungen der Mathematisch-Physischen Klasse der Sächsischen Akademie der Wissenschaften, v. 37, No. 1, VI + 420 p. + 9 plates, + 1 foldout map.

Pennetier, G., 1911, Discours sur l'Évolution des Connaisances en Histoire Naturelle, première partie—L'Antiqutie et Moyen-Age: Actes du Muséum d'Histoire Naturelle de Rouen, no. 14: Rouen, J. Girieud, 56 p.

Pennetier, G., 1915, Discours sur l'Évolution des Connaisances en Histoire Naturelle, quatrième partie—XVIIIe–XIXe siècles: Actes du Muséum d'Histoire Naturelle de Rouen, no. 18–19–20, Rouen, J. Girieud, 319 p.

Perler, D., (translation, introduction and commentary) 1994, Dante Alighieri, Abhandlung über das Wasser und die Erde: Hamburg, Felix Meiner Verlag, LXXVII + 148 + [2] p.

Peschel, O., 1877, Geschichte der Erdkunde bis auf A. v. Humboldt und Carl Ritter (zweite vermehrte und verbesserte Auflage von S. Ruge): München, R. Oldenbourg.

Peterlongo, J.M., 1972, Massif Central—Limousin, Auvergne, Velay, *in* Pomerol, C., ed., Guides Géologiques Régionaux: Paris, Masson & Cie, 199 p.

Peters, F.E., 1968, Aristoteles Arabus—The Oriental Translations and Commentaries on the Aristotelian *Corpus*: New York University Department of Classics Monographs on Mediterranean Antiquity, Brill, Leiden, VIII + 75 p.

Petit, F., 1849, Sur la densité moyenne de la chaîne des Pyrénées et sur la latitude de l'Obsérvatoire de Toulouse: Paris, Comptes Rendus hébdomadaires de l'Académie des Sciences, v. 29, 2nd semestre, p. 729–734.

Pfeiffer, D., 1963, Die Geschichtliche Entwicklung der Anschauungen über das Karstgrundwasser: Hannover, Beihefte zum Geologischen Jahrbuch, v. 57, 111 p.

Phillimore, R.H., 1945, Historical Records of the Survey of India, volume I 18th Centruy: Dehra Dun, Office of the Geodetic Branch, Survey of India, xx + 400 p. + unpaginated index + 21 plates.

Phillimore, R.H., 1950, Historical Records of the Survey of India, volume II 1800 to 1815: Dehra Dun, Office of the Geodetic Branch, Survey of India, xxviii + 478 p. + 24 plates.

Phillimore, R.H., 1954, Historical Records of the Survey of India, volume III 1815 to 1830: Dehra Dun, Office of the Geodetic Branch, Survey of India, xxii + 534 p. + 24 plates.

Phillimore, R.H., 1958, Historical Records of the Survey of India, volume IV 1830 to 1843 George Everest: Dehra Dun, Office of the Geodetic Branch, Survey of India, xxiii + 493 p. + 23 plates.

Phillips, P.C., 1961a, The Fur Trade, with concluding chapters by J.W. Smurr, v. I: Norman, University of Oklahoma Press, xxvi + 686 p.

Phillips, P.C., 1961b, The Fur Trade, with concluding chapters by J.W. Smurr, v. II: Norman, University of Oklahoma Press, viii + 696 p.

Piccolomini, A., 1558, Del Trattato della Grandezza della Terra, e dell'Acqva: Venetia, Apresso Giordano Ziletti, [VI] + 43 + [1]p.

Pierson, H.D., and Wei, D.Y.L., 1992, Travelers from Ancient Cathay—An Account of China's Great Explorers: xi + 99 p. + 1 map.

Pilger, A., 1967, Ansprache zur Feier des 90. Geburtstages von Professor Dr. Hans Stille am 8. 10. 1966 in Hannover-Buchholz: Geologisches Jahrbuch, v. 84, p. I–VII.

Pilger, A., 1977, Laudatio auf Hans Stille zur Wiederkehr seines 100. Geburtstages: Zeitschrift der Deutschen Geologischen Gesellschaft, v. 128, p. 1–9

Pinneker, E.V., 1989, Eduard Suess als Hydrogeologe: Steirische Beiträge zu Hydrogelogie, v. 40, p. 165–174.

Pirsson, L.V., and Schuchert, C., 1924, Introductory Geology: New York, John Wiley & Sons, x + 693 p.

Pitcher, D.E., 1972, An Historical Geography of the Ottoman Empire from earliest times to the end of the sixteenth century: Leiden, E.J. Brill, x + 171 p. + 36 maps.

Pitman, W.C., III, 1978, Relationship between eustasy and stratigraphic sequences of passive margins: Geological Society of America Bulletin, v. 89, p. 1389–1403.

Pitman, W.C., III, and Golovchenko, X., 1991, Modelling sedimentary sequences, *in* D.W., Müller, McKenzie, J.A., and Weissert, H. eds., Modern Controversies in Geology (Proceedings of the Hsü Symposium): London, Academic Press, p. 279–309.

Playfair, J., 1802, Illustrations of the Huttonian Theory of the Earth: London, Cadell and Davies, and Edinburgh, William Creech, xx + 528 p.

Playfair, J., 1805, Biographical account of the late Dr. James Hutton, F.R.S. Edin.: Transactions of the Royal Society of Edinburgh, v. 5, part 3, p. 39–99.

Playfair, J., 1812, Essai sur la géographie minéralogique des environs de Paris by G. Cuvier and Alex. Brongniart. Paris 1811: Edinburgh Review, v. 20, p. 369–386.

Playfair, J.G., 1822, Biographical account of the late Professor Playfair, *in* Playfair, J.G., ed., The Works of John Playfair, Esq.: Edinburgh, Archibald Constable, v. 1, p. xi–cv.

van der Pluijm, B.A., and Marshak, S., 1997, Earth Structure—An Introduction to Structural Geology and Tectonics: WCB/McGraw-Hill, viii + 495 p.

Polaschek, E., 1959, Ptolemy's Geography in a new light: Imago Mundi, no. 14, p. 17–37.

Pompeckj, J.F., 1925, Die Auffassung von Vulkanismus seit Leopold von Buch: Sitzungsberichte der preussischen Akademie der Wissenschaften—22. Januar. Öffentliche Sitzung zur Feier des Jahrestags König Friedrichs II., Sonderabdruck, Berlin, 21 p.

Pomerol, C., Debelmas, J., Mirouse, R., Rat, P., and Rousset, C., 1980, Geology of France with twelve itineraries: Guides Géologiques Regionaux: Paris, Masson, 255 p.

[Popowitsch, J.S.V.], 1750, Untersuchungen vom Meere, die auf Veranlassung einer Schrift, de Colvmnis Hercvlis, welche der hochberühmte Professor in Altorf, Herr Christ. Gootl. Schwarz herausgegeben nebst andern zu derselben gehörigen Anmerkungen, von einem Liebhaber der Naturlehre und Philologie vorgetragen werden: Frankfurt und Leipzig, (no publisher), [XIV] + 38 + LXXVI + 49–432 + [44] p.

Popper, K.R., 1966, The Open Society and Its Enemies, v. 1, The Spell of Plato, fifth, revised edition: Princeton, Princeton University Press, xi + 361 p.

Popper, K.R., 1974, Replies to my critics, *in* Schilpp, P.A., ed., The Philosophy of Karl Popper: Library of Living Philosophers, v. 14, book II, p. 961–1197.

Popper, K.R., 1989, Back to the Presocratics, *in* Popper, K.R., Conjectures and Refutations, 5th revised edition: London, Routledge, p. 136–165.

Popper, K.R., 1998, The World of Parmenides—Essays on the Presocratic Enlightenment, edited by Arne F. Petersen, with the assistance of Jørgen Mejer: London, Routledge, x + 328 p.

Potier, M., 1875, Exposé des Travaux de M. Élie de Beaumont: Annales des Mines, 7e série, Mémoires, v. 8, p. 259–317 (includes a fairly complete bibliography of Élie de Beaumont's works prepared by A. Guyerdet).

Potochnik, A.R., and Reynolds, S.J., 1990, Side canyons of the Colorado River, Grand Canyon, *in* Beus, S.S., and Morales, M., eds., Grand Canyon Geology: New York, Oxford University Press, and Museum of Northen Arizona Press, p. 461–481.

Potter, C.J., Drahovzal, J.A., Sargent, M.L., and McBride, J.H., 1997, Proterozoic structure, Cambrian rifting, and younger faulting as revealed by a regional seismic reflection network in the southern Illinois Basin: Seismological Research Letters, v. 68, p. 537–552.

Powell, J.W., 1875, Exploration of the Colorado River of the West and its Tributaries Explored in 1869, 1870, 1871, and 1872 under the Direction of the Secretary of the Smithsonian Institution: Washington, D.C., Government Printing Office, xi + 291 p. + one map in pocket.

Powell, J.W., 1876, Report on the Geology of the Eastern Portion of the Uinta Mountains and A Region of Country Adjacent Thereto. With Atlas: Department of the Interior. U.S. Geological and Geographical survey of the territories. Second Division—J.W., Powell, Geologist in Charge: Washington, D.C., Government Printing Office, VII + 218 p.

Powell, J.W., 1878, Geological and geographical surveys—Letter from the Secretary of the Interior transmitting a report from Professor Powell in regard to surveys, in response to a resolution of the House of Representatives: House of Representatives, 45th Congress, 2d Session, Ex. Doc. No. 80, 19 p. + 1 foldout map.

Powell, J.W., 1895, Canyons of the Colorado: Flood &Vincent, xiv + 400 p. (reprinted in 1961 under the title *The Exploration of the Colorado River and its Canyons* by Dover Publications, Inc., New York, with one additional map showing the course of the Colorado River; facsimiled in a hard-cover edition in 1964 by Argosy-Antiquarian Ltd., New York; reprinted also in the Penguin Nature Classics series in 1987 under the same Dover title and with an introduction by Wallace Stegner).

Pratt, J.H., 1855, On the attraction of the Himalaya Mountains, and of the elevated regions beyond them, upon the plumb-line in India: Philosophical Transactions of the Royal Society of London, v. 145, p. 53–104.

Prendergast, M.L., 1978, James Dwight Dana: The Life and Thought of an American Scientist [Ph.D. dissertation] Los Angeles, University of California, xiii + 625 p.

Prescher, H. and Wagenbreth, O., 1994, Georgius Agricola—Seine Zeit und ihre Spuren: Leipzig, Deutscher Verlag für Grundstoffindustrie, 234 p.

Prévost, C., 1820, Essai sur la constitution physique et géognostique du bassin à l'ouverture duquel est située la Ville de Vienne en Autriche: Journal de Physique, v. 91, p. 347–364 and 460–473.

Prévost, C., 1821, Sur un nouvel exemple de la réunion des coquilles marines et fluviatiles: Journal de Physique, v. 92, p. 418–428.

Prévost, C., 1822, Observations sur les coquilliers de Beau-Champ et sur les mélanges coquilles marines et fluviatiles dans les couches inférieures de la formation du gypse des environs de Paris: Journal de Physique, v. 94, p. 1–18.

Prévost, C., 1831, Lettre du 3 Octobre 1831, lu par Desnoyers: Bulletin de la Société Géologique de France, (1e série) v. 2, p. 32–38.

Prévost, C., 1835, Notes sur l'Ile Julia, pour servir a l'histoire de la formation des montagnes volcaniques: Mémoires de la Société Géologique de France, v. 2, mém. 5, p. 91–124 + [4] p. of explications des planches.

Prévost, C., 1840, Sur la théorie des soulevements: Bulletin de la Société Géologique de France, (1e série) v. 11, p. 183–203.

Price, L.G., 1999, An Introduction to Grand Canyon Gelogy: Grand Canyon, Arizona, Grand Canyon Association, 63 p.

Priestley, H.I., 1946, Franciscan Explorations in California: Glendale, The Arthur H. Clark Company, 189 p.

Pritchard, J.B., (Editor), 1969, Ancient Near Eastern Texts Relating to the Old Testament, Third Edition with Supplement: Princeton, Princeton University Press, 710, 1969.

Privitera, F., 1998, Recent findings in the prehistory of Mt. Etna, *in* Morello, N., Volcanoes and History, Proceedings of the 20th INHIGEO Symposium, Napoli-Eolie-Catania (Italy) 19–25 September 1995: Genova, Brigati, p. 543–553.

Procelli, E., 1998, Prehistoric communities of the Etna area, *in* Morello, N., Volcanoes and History, Proceedings of the 20th INHIGEO Symposium, Napoli-Eolie-Catania (Italy) 19–25 September 1995: Genova, Brigati, p. 555–561.

Prontera, F., ed., 1983, Geografia e Geografi nel Mondo Antico: Guida Storica e Critica: Roma, Laterza, xxxii + 272 p.

Prothero, D.R., and Dott, R.H., Jr., 2002, Evolution of the Earth, sixth edition: Boston, McGraw Hill, 569 p. + A-12 + G-11 + I-10 + 2 full-scale end papers.

Ptolemy, C., undated, Géographie de Ptolémée traduction latine de Jacopo d'Angelo de Florence—Reproduction réduite des Cartes et Plans du Manuscript latin 4802 de la Bibliothéque Nationale: Paris, Catala Frères, 3 p. + LXXV plates.

Pumpelly, R., 1866, Geological Researches in China, Mongolia and Japan: Washington City, Smithsonian Contributions to Knowledge, 202, 143 p. + 9 Plates.

Pupo-Walker, E. and López-Morillas, F.M., 1993, Castaways—The Narrative of Alvar Nuñez Cabeza de Vaca: Berkeley, University of California Press, xxx + 158 p.

Puschcharovskiy, Y.M., 1987, The future of the geosynclinal theory in connection with the development of mobilism: Geotectonics, v. 21, p. 91–97.

Pyne, S.J., 1979, Certain allied problems in mechanics: Grove Karl Gilbert at the Henry Mountains, *in* Schneer, C.J., ed., Two Hundred years of Geology in America: Hanover, University Press of New England, p. 225–238.

Pyne, S.J., 1980, Grove Karl Gilbert—A Great Engine of Research: Austin, University of Texas Press, xiv + 306 p.

Pyne, S.J., 1998, How the Canyon Became Grand—A Short History: New York, Viking, xviii + 199 p.

Rabbitt, M.C., 1969, John Wesley Powell: Pioneer statesman of federal science: [U.S.] Geological Survey Professional Paper 669-A, [ii] + 21.p.

Rabbitt, M.C., 1979, Minerals, Lands, and Geology for the Common Defence and General Welfare, v. 1, Before 1879: U.S. Geological Survey, United States: Washington, D.C., Government Printing Office, x + 331 p.

Rabbitt, M.C., 1980, John Wesley Powell, Soldier, Explorer, Scientist: U.S. Department of the Interior/Geological Survey: [Washington, D.C.], U.S. Government Printing Office, 23 p.

Rabikovich, A.I., 1976, Charlz Laiel: Moskva, Izdatelstvo "Nauka," 199 p.

Rahman, A., 1982, Science and Technology in Medieval India—A Bibliography of Source Materials in Sanskrit, Arabic and Persian: New Delhi, Indian National Science Academy, xxxi + 719 p.

Rahman, A., ed., 1984, Science and Technology in Indian Culture—A Historical Perspective: New Delhi, National Institute of Science, Technology & Development Studies (NISTADS), [II] + 251 p.

Ranalli, G., 1982, Robert Hooke and the Huttonian Theory: Journal of Geology, v. 90, p. 319–325.

Rankin, D.W., Drake, A.A., Jr., Glover, L., III, Goldsmith, R., Hall, L.M., Murray, D.P., Ratcliffe, N.M., Read, J.F., Secor, D.T., Jr. and Stanley, R.S., 1989, Pre-orogenic terranes, *in* Hatcher, R.D., Jr., Thomas, W.A., and Viele, G.W., eds., The Appalachian-Ouachita Orogen in the United States: Boulder, Colorado, Geological Society of America, Geology of North America, v. F-2, p. 7–100.

Rashed, R., ed., 1996a, Encyclopedia of the History of Arabic Science, v. 1, Astronomy—Theoretical and Applied: London, Routledge, xiv + 330 p.

Rashed, R., ed., 1996b, Encyclopedia of the History of Arabic Science, v. 2, Mathematics and the Physical Sciences: London, Routledge, vii p. + p. 331–750.

Rashed, R., ed., 1996c, Encyclopedia of the History of Arabic Science, v. 3, Technology, Alchemy and Life Sciences: London, Routledge, vii p. + p. 751–1105.

vom Rath, G., 1871, Ein Ausflug nach Calabrien: Bonn, Adolph Marcus, VII + 157 p. + 1 folded plate.

Ray, J., 1692, Miscellaneous Discourses Concerning the Dissolution and Changes of the World, Wherein the Primitive *Chaos* and Creation, the General Deluge, Fountains, Formed Stones, Sea-Shells found in the earth, Subterraneous Trees, Mountains, Earthquakes, *Volcanoes*, the Universal conflagration and Future State, are largely Discussed and Examined: London, Samuel Smith, [xxi] + 259 p.

Reade, T.M., 1886, The Origin of Mountain Ranges Considered Experimentally, Structurally, Dynamically and in relation to their Geological History: London, Taylor and Francis, xviii + 359 p.

Reade, T.M., 1890, Origin of normal faults: American Journal of Science, 3rd series, v. 39, p. 51–52.

Reboul, H., 1835, Essai de Géologie Descriptive et Historique—Prolégomènes et Période Primaire: Paris, F.G. Levrault, 276 p.

Rebrik, B.M., 1987, Geologie und der Bergbau in der Antike: Leipzig, VEB Deutscher Verlag für Grundstoffindustrie, 185 p.

Reck, H., 1910, Über Erhebungskratere: Monatshefte der Deutschen Geologischen Gesellschaft, v. 62, p. 293–318.

Redfern, R., 1980, Corridors of Time—1,700,000,000 Years of Earth at Grand Canyon: New York, Times Books, 198 p.

Reichetzer, F., 1812, Anleitung zur Geognosie, Insbesondere zur Gebirgskunde—Nach Werner für die k. k. Berg-Akademie Bearbeitet: Wien, Camesinasche Buchhandlung, XII + 292 p.

Reinaud, J.-T., 1848[1985], Géographie d'Aboulféda traduite de l'Arabe en Français et accompagnée de notes et d'éclairsissements, Tome I, Introduction Générale a la Géographie des Orientaux: Paris, Imprimerie Nationale, CDLXIV p. (reprinted in 1985, by the Institut für Geschichte der Arabisch-Islamischen Wisenschaften and der Johann Wolfgang von Goethe-Universität, Frankfurt am Main, Veröffentlichungen des Instituts für Geschichte der Arabisch-Islamischen Wisenschaften, Reihe B. Nachdrucke Abteilung Geographie, v. I, edited by Fuat Sezgin).

Renou, L., 1925, La Géographie de Ptolémée l'Inde (VII, 1–4)—Texte Établi: Thèse Présentée a la Faculté des Lettres de l'Université de Paris, Librairie Ancienne Édouard Champion, Paris, XVI + 73 p. + 3 maps.

Rey, A., 1933, La Jeunesse de la Science Grecque: Paris, La Renaissance de Livre, XVII + 537 p.

Rey, A., 1939, La Maturité de la Pensée Scientifique en Grèce: Paris, Albin Michel, XXI + 574 p.

Rey, A., 1942, La Science Orientale Avant les Grecs, nouvelle édition avec des notes additionelles: Paris, Albin Michel, XVII + 519 p.

Rey, A., 1946, L'Apogée de la Science et Technique Grecque—Les Sciences de la Nature et de l'Homme, Les Sciences Mathématiques d'Hippocrate a Platon: Paris, Albin Michel, XVIII + 313 p.

Reyer, E., 1888, Theoretische Geologie: E. Schweizerbart'sche Verlagsbuchhandlung (E. Koch), Stuttgart, XIII + 867 p. + 3 maps.

Reyer, E., 1892, Ursachen der Deformationen und der Gebirgsbildung: Leipzig, Wilhelm Engelmann, 40 p. + 7 foldout plates.

Rice, M., 1994, The Archaeology of the Arabian Gulf c. 5000–323 BC: London, Routledge, xvii + 369 p.

Rice, W.N., 1915, The geology of James Dwight Dana, *in* Problems of American Geology—A Series of Lectures Dealing with Some of the Problems of the Canadian Shield and the Cordilleras, Delivered at Yale University on the Silliman Foundation in December, 1913: New Haven, Yale University Press, p. 1–42.

Richardson-Bunbury, J.M., 1992, The Basalts of Kula and Their Relation to Extension in Western Turkey [Ph.D. dissertation]: Cambridge, Cambridge University, [XIV] + 175 p.

Richer, J., 1994, Sacred Geography of the Greeks—Astrological Symbolism in Art, Architecture and Landscape (translated by Christine Rhone from the French): Albany, State University of New York Press, xli+319 p.

Richey, J.E., MacGregor, A.G. and Anderson, F.W., 1975, Scotland: The Tertiary Volcanic Districts, third edition: British Regional Geology, Department of Scientific and Industrial Research, Geological Survey and Museum: Edinburgh, Her Majesty's Stationary Office, viii+120 p.+9 plates.

von Richthofen, F. (Baron), 1867, The natural system of volcanic rocks: Memoirs of the California Academy of Sciences, v. 1, pt. 2, p. 1–94.

von Richthofen, F., 1902, Die morphologische Stellung von Formosa und den Riukiu-Inseln: Sitzungsberichte der königlich preussischen Akademie der Wissenschaften zu Berlin, Jahrgang 1902, Zweiter Halbband, Stück XL, p. 944–975 + plate III.

van Rijn, N., 1993, Celebrated geologist, J. Tuzo Wilson, 84: The Toronto Star, Saturday, April 17, 1993, p. A19 (Obituaries).

Ritter, C., 1822, Die Erdkunde im Verhältniß zur Natur und zur Geschichte des Menschen, oder Allgemeine Vergleichende Geographie as Sichere Grundlage des Studiums und Unterrichts in Physikalischen und Historischen Wissenschaften ... —Erster Theil, Erstes Buch. Afrika—Zweite stark vermehre und verbesserte Ausgabe: Berlin, G. Reimer, XXVII + [I p. of Errata] + 1084 p.

Robertson, J.W., 1927, Francis Drake & Other Early Explorers Along the Pacific Coast: San Francisco, Grabhorn, [iii] + 290 p.

Robin, L. (editor and translator), 1941, Phédon, *in* Platon Oeuvres Complètes, v. IV, part I, deuxième édition, revue et corrigée, Collection des Universités de France, «Les Belles Lettres», Paris, LXXXVI + [I] + 118 p.

Robinson, H.W., and Adams, W., eds., 1935, The Diary of Robert Hooke, M.A., M.D., F.R.S. 1672–1680—Transcribed from the Original in the Possession of the Corporation of the City of London (Guildhall Library) with a Foreword by Sir Frederick Gowland Hopkins, O. M., President of the Royal Society: London, Taylor & Francis, xxviii + 527 p.

Rodgers, J., 1970, The Tectonics of the Appalachians: New York, Wiley Interscience, xii + 271 p.

Rodier, G., 1890, La Physique de Straton de Lampsaque: Thèse pour le doctorat présentée a la Faculté des Lettres de Paris: Paris, Felix Alcan, [I] + 133 + 1 page of errata + [I].

Rodolico, F., 1981, Marsili (or Marsigli), Luigi Ferdinando, *in* Gillispie, C. C., editor-in-chief, Dictionary of Scientific Biography: New York, Charles Scribner's Sons, v. 9, p. 134–136.

Roeder, D., 1992, Thrusting and wedge growth, southern Alps of Lombardia (Italy): Tectonophysics, v. 207, p. 199–243.

Rolle, A., 1991, John Charles Frémont—Character as Destiny: Norman, University of Oklahoma Press, xv + [i] + 351 p.

Ronca, I., 1971, Ptolemaios Geographie 6, 9–21, Ostiran und Zentralasien Teil I Greichischer Text neu herausgegeben und ins Deutsche Übertragen: Istituto Italiano per il Medio ed Estremo Oriente, Centro Studi e Scavi Archeologici in Asia, Consiglio Nazionale delle Ricerche, Rome, XIX + 118p. + 3 foldout maps.

Ross, E.D., 1931, Introduction, *in* Benedetto, L.F., Travels of Marco Polo translated into English … by Professor Aldo Ricci: The Broadway Travellers: London, George Routledge & Sons, Ltd., p. vii + xvii.

Ross, W.D., 1923, Aristotle: London, Methuen & Co., vi + [i] + 300 p.

Rousseau-Lissens, A., 1961, Géographie de l'Odyssée—La Phéacie: Bruxelles, Brepols, 102 p. + 32 photographs.

Rousseau-Lissens, A., 1962, Géographie de l'Odyssée—Les Récits I: Bruxelles, Brepols, 133 p. + 32 photographs.

Rousseau-Lissens, A., 1963, Géographie de l'Odyssée—Les Récits II Mycenes, Sparte, Pylos: Bruxelles, Brepols, 245 p. + 1 reproduction + 44 photographs.

Rousseau-Lissens, A., 1964, Géographie de l'Odyssée—Les Récits III L'Atlantide, Troie, Ithaque: Bruxelles, Brepols, 263 p. + 32 photographs.

Roux, G., 1980, Ancient Iraq, second edition: London, Penguin, 496 p.

Rowe, C.J., editor and commentator, 1993, Plato Phaedo, *in* Kenney, E.J., and Easterling, P.E., eds., Cambridge Greek and Latin Classics: Cambridge, Cambridge University Press, xi + 301 p.

Royden, L.H., 1996, Coupling and decoupling of crust and mantle in convergent orogens: implications for strain partitioning in the crust: Journal of Geophysical Research, v. 101, p. 17,679–17,705.

Royden, L.H., Burchfiel, B.C., King, R.W., Wang, E., Chen, Z.L., Shen, F., and Liu Y.P., 1997, Surface deformaton and lower crustal flow in eastern Tibet, Science, v. 276, p. 788–790.

Rudwick, M.J.S., 1971, Uniformity and progression: reflections on the structure of geological theory in the age of Lyell, *in* Roller, D.H.D., ed., Perspectives in the History of Science and Technology: Norman, University of Oklahoma Press, p. 209–237.

Rudwick, M.J.S., 1974, Poulett Scrope on the volcanoes of Auvergne: Lyellian time and political economy: British Journal for the History of Science, v. 7, p. 205–242.

Rudwick, M.J.S., 1976, The meaning of fossils—Episodes in the history of paleontology, 2nd edition: Chicago, University of Chicago Press, [X] + 287 p.

Rudwick, M.J.S., 1981, Prévost, Louis-Constant, *in* Gillispie, C.C., editor-in-chief, Dictionary of Scientific Biography: New York, Charles Scribner's Sons, v. 11, p. 133–134.

Rudwick, M.J.S., 1990, Introduction, *in* Principles of Geology—First Edition, volume I Charles Lyell With a new Introduction by Martin J.S. Rudwick: Chicago, University of Chicago Press, p. vii–lviii.

Rudwick, M.J.S., 1997a, Georges Cuvier, Fossil Bones, and Geological Catastrophes—New Translations and Interpretations of the Primary Texts: Chicago, University of Chicago Press, xvi + 301 p.

Rudwick, M.J.S., 1997b, Smith, Cuvier et Brogniart, et la reconstitution de la géohistoire, *in* De La Géologie À Son Histoire—Ouvrage édité en hommage à François Ellenberger sous la direction de Gabriel Gohau … Coordinateur: Jean Gaudant, Comité des Travaux Historiques et Scientifiques Section des Sciences, Paris, 119–128.

Runes, D.D., 1959, Pictorial History of Philosophy: New York, Philosophical Library, Inc., x + 406 p.

Russell, B. [A.W.], 1945 [1972], A History of Western Philosophy: New York, A Touchstone Book, Simon & Schuster, xxiii + 895 p.

Russell, B., 1998, Autobiography: London, Routledge, xv + 750 p.

Rutten, L., 1923, Development of geological knowledge in the Dutch East Indies, *in* Geology—The History and Present State of Scientific Research in the Dutch East Indies, Koninklijke Akademie van Wetenschappen, (Amsterdam). Internationale Circumpacifische Onderzoek Commissie. (I.C.O.-Commissie). Amsterdam, p. 2–21.

Ryabukhin, A.G., 1999, Geologicheskie idei L. Eli de Bomona: Istoria i sovremennost (k 200-letiyu so dnaya rojlennia): Byullten Moskovskoyo Obshestva Ispitatelei Prirodi. Otdel Geologi, v. 74, no. 4, p. 63–68.

Ryan, W.B.F., and Pitman, W.C., III, 1998, Noah's Flood—The New Scientific Discoveries About the Event That Changed History: New York, Simon & Schuster, 319 p.

Ryan, W.B.F., Pitman, W.C., III, Major, C.O., Shimkus, K., Moskalenko, V., Jones, G.A., Dikitrov, P., Görür, N., Sakınç, M., and Yüce, H., 1997, An abrupt drowning of the Black Sea shelf: Marine Geology, v. 119–129.

Sabadini, R., Lambeck, K., and Boschi, E., editors, 1991, Glacial Isostasy, Sea Level and Mantle Rheology: NATO ASI Series: Dordrecht, Kluwer, v. C334, vii + 708 p.

Sachau, E.C., 1910, Alberuni's India—An Account of the Religion, Philosophy, Literature, Geography, Chronology, Astronomy, Customs, Laws and Astrology of India About A.D. 1030, An English Edition with Notes and Indices, v. I: London, Kegan Paul, Trench, Trübner & Co., l + [i] + 408 p.

St. John, B., Bally, A.W., and Klemme, H.D., 1984, Sedimentary Provinces of the World—Hydrocarbon Productive and Nonproductive: Tulsa, Oklahoma, American Association of Petroleum Geologists, 35 p.

Sainte-Claire Deville, C., 1878, Coup-d'Œil Historique sur la Géologie et sur les Travaux d'Élie de Beaumont: Paris, G. Masson, VII + 597 + [1] p.

Sallmann, K.G., 1971, Die Geographie des Älteren Plinius in Ihrem Verhältnis zu Varro—Versuch einer Quellenanalyse: Berlin, Walter de Gruyter, XII + 296 p. + 4 plates.

Salomon, W., 1925, Magmatische Hebungen (Mit besonderer Berücksichtigung von Calabrien): Sitzungsberichte der Heidelberger Akademie der Wissenschaften, Mathematisch-naturwissenschaftliche Klasse, Jahrgang 1925. 11. Abhandlung, p. 1–28.

Salomon, W., 1926, Gibt es Gesteine, die für bestimmte Erdperioden charakteristisch sind? Sitzungsberichte der Heidelberger Akademie der Wissenschaften, Mathematisch-naturwissenschaftliche Klasse, Jahrgang 1926. 9. Abhandlung, p. 1–7

Salomon, W., 1928, Geologische Beobachtungen des Leonardo da Vinci: Sitzungsberichte der Heidelberger Akademie der Wissenschaften, Mathematisch-naturwissenschaftliche Klasse, Jahrgang 1928. 8. Abhandlung, p. 1–13.

Salvador, A., editor, 1994, International Stratigraphic Guide—A Guide to Stratigraphic Classification, Terminology, and Procedure, second edition: Boulder, Colorado, The International Union of Geological Sciences and The Geological Society of America, xix + 214 p.

Sambursky, S., 1956 [1987], The Physical World of the Greeks: London, Routledge & Kegan Paul, xv + 255 p.

Sambursky, S., 1959 [1987], Physics of the Stoics: London, Routledge & Kegan Paul, xii + 153 p.

Sambursky, S. 1962 [1987], The Physical World of Late Antiquity: London, Routledge & Kegan Paul, xii + 189 p.

San, J.H., 1984, Ancient China's Inventions: Hong Kong, The Commercial Press, 99 s.

de Santarem, [M.F., de Barros y Sousa], le Vicomte, 1842, Recherches sur la Priorité de la Découverte des Pays Situés sur la Cote Occidentale d'Afrique au–dela du Cap Bojador, et sur les Progrés de la Science Géographique, après la Navigations des Portugais, au XVe siècle: A la Librairie Orientale de Ve Dondey-Dupré, Paris, cxiv + 335 p. + 1 p. of errata + Atlas of 30 plates.

de Santarem, M.F., de Barros y Sousa, Vicomte, 1849a[1985], Atlas de Santarem, facsimile of the final edition: Amsterdam, Rudolf Muller, [II] + 78 sheets + 80 p. of explanatory text by H. Wallis and A.H. Sijmons.

de Santarem, M.F., de Barros y Sousa, Vicomte, 1849b, Essai sur l'Histoire de la Cosmographie et de la Cartographie Pendant le Moyen-Age et sur les Progrès de la Géographie après les Grandes Découvertes du XVe Siècle, pour servir d'Introduction et d'Explication a l'Atlas Composé de Mappemondes et de Portulans, et d'Autres Monuments Géographiques, depuis le VIe Siècle de Notre Ère jusqu'au XVIIe, tome premier: Paris, Maulde et Renou, LXXXVII + 515 p. + [2 p. errata].

de Santarem, M.F., de Barros y Sousa, Vicomte, 1850, Essai sur l'Histoire de la Cosmographie et de la Cartographie Pendant le Moyen-Age et sur les Progrès de la Géographie après les Grandes Découvertes du XVe Siècle, pour servir d'Introduction et d'Explication a l'Atlas Composé de Mappemondes et de Portulans, et d'Autres Monuments Géographiques, depuis le VIe Siècle de Notre Ère jusqu'au XVIIe, tome deuxième: Paris, Maulde et Renou, XLV + [1 p. errata] + 592 p.

de Santarem, M.F., de Barros y Sousa, Vicomte, 1852, Essai sur l'Histoire de la Cosmographie et de la Cartographie Pendant le Moyen-Age et sur les Progrès de la Géographie après les Grandes Découvertes du XVe Siècle, pour servir d'Introduction et d'Explication a l'Atlas Composé de Mappemondes et de Portulans, et d'Autres Monuments Géographiques, depuis le VIe Siècle de Notre Ère jusqu'au XVIIe, tome troisième: Paris, Maulde et Renou, LXXVI + 646 p. + [1 p of errata].

de Santarem, M.F., de Barros y Sousa, Vicomte, 1855,[444] Note sur la Publication de l'Atlas Composé de Mappemondes et de Portulans et D'Autres Monuments Géographiques depuis le VIe Siècle de Notre Ère jusqu'au XVIIe: Nouvelles Annales des Voyages, v. 2, [I] + 20 p. (offprint pages repaginated).

de Santillana, G., 1961, The Origin of Scientific Thought: New York, A Mentor Book, The New American Library, 320 p.

Sapper, K., 1917, Katalog der Geschichtlichen Vulkanausbrüche: Schriften der Wissenschaftlichen Gesellschaft in Straßburg 27. Heft, Karl J. Trübner, Straßburg, IX + [I] + 358 p.

Sarton, G., 1927, Introduction to the History of Science, v. I, From Homer to Omar Khayyam: for the Carnegie Institution of Washington by William Wilkes and Company, Baltimore, xi + 839.

Sarton, G., 1931a, Introduction to the History of Science, v. II, part I, From Rabbi ben Ezra to Ibn Rushd: for the Carnegie Institution of Washington by William Wilkes and Company, Baltimore, xxxv + 480 p.

Sarton, G., 1931b, Introduction to the History of Science, v. II, part II, From Robert Grosseteste to Roger Bacon: for the Carnegie Institution of Washington by William Wilkes and Company, Baltimore, xvi + p. 485–1251.

Sarton, G., 1947, Introduction to the History of Science, v. III, Science and Learning in the Fourteenth Century, part I First Half of the Fourteenth Century: for the Carnegie Institution of Washington by William Wilkes and Company, Baltimore, xxxv + 1018 p.

Sarton, G., 1948, Introduction to the History of Science, v. III, Science and Learning in the Fourteenth Century, part II Second Half of the Fourteenth Century: for the Carnegie Institution of Washington by William Wilkes and Company, Baltimore, x + [i] + p. 1019–2155.

Sarton, G., 1952, A History of Science—Ancient Science Through the Golden Age of Greece: Cambridge, Harvard University Press, xxvi + 646 p.

Sarton, G., 1955, Appreciation of Ancient and Medieval Science during the Renissance (1450–1600): Philadelphia, University of Pennsylvania Press, xvii + 233 p.

Sarton, G., 1959, A History of Science—Hellenistic Science and Culture in the Last Three centuries B.C.: Cambridge, Harvard University Press, xvii + 554 p.

de Saussure, H.B., 1779, Voyages dans les Alpes, Précédés d'un Essai sur l'Histoire Naturelle des Environs de Genève, v. 1: Neuchâtel, Samuel Fauche, 541 p.

Schachermeyr, F., 1983, Die Griechische Rückerinnerung im Lichte Neuer Forschungen: Österreichische Akademie der Wissenschaften, Philosophisch-Historische Klasse, Sitzungsberichte, v. 404, 415 p. + 28 plates.

Schaer, J.-P., 1991, Emile Argand 1879–1940. Life and portrait of an inspired geologist: Eclogae Geologicae Helvetiae, v. 84, p. 511–534.

Schaffer, F.X., 1916, Grundzüge der Allgemeinen Geologie: Leipzig, Franz Deuticke, VIII + 492 p.

Schatski, N.S., 1961, Vergleichende Tektonik Alter Tafeln, *in* Teschke, H.-J., ed., Fortschritte der Sowjetischen Geologie, Heft 4, 220 p.

Schertz, G., editor, 1969, Steno—Geological Papers (Translation by Pollock, A.J.): Acta Historica Sceintiarum Naturalium et Medicinalium: Odense University Press, v. 20, 370 p.

Schertz, G., 1986a, Niels Stensen—Eine Biographie, Band I 1638–1677: Leipzig, St. Benno-Verlag, 376 p. + 26 figures on plates.

Schertz, G., 1986b, Niels Stensen—Eine Biographie, Band II 1677–1686: Leipzig, St. Benno-Verlag, 318 p.

Scheuchzer, J.J., 1731, Kupfer-Bibel In welcher Die Physica Sacra Oder Geheiligte Natur-Wissenschaft Derer In Heil. Schrifft vorkommenden Natürlichen Sachen Deutlich ertlärt und bewährt ... Erste Abtheilung Von Ta. I–CLXXIV: Christian Ulrich Wagner, Augspurg und Ulm, [XXV] + 672 p.

Scheuer, J., 1991, Ruda/Siva and the destruction of the sacrifice, *in* Bonnefoy, Y., ed., Mythologies, v. II: Chicago, Chicago University Press, p. 813–817.

Schleucher, K., undated, Alexander von Humboldt—der Mensch, der Forscher, der Schriftsteller: Darmstadt, Eduard Roether, 735 p.

Schmaler, M., 1904, Die Entwicklung der Ansichten über den Gebirgsbau Zentralasiens von der Wiedergeburt der Erdkunde bis zum Beginne der wissenschaftlichen Exploration dieses Gebietes—Inaugural-Dissertation zur Erlangung der Doktorwürde der Philosophischen Fakultät der Universität Leipzig: Druck von Selmar von Ende, Königsee i. Thür., 125 p. + One foldout map plate + Vita.

Schmeckebier, L.F., 1904, Catalogue and Index of the Publications of the Hayden, King, Powell, and Wheeler Surveys namely Geological and Geographical Survey of the Territories, Geological Exploration of the Fortieth Parallel, Geographical and Geological Surveys of the Rocky Mountain Region, Geographical Surveys West of the One Hundredth Meridian: Washington, D.C., Government Printing Office, 208 p.

Schmidt, E.G., 1981, Himmel—Meer—Erde im frühgriechischen Epos und im alten Orient: Philologus, v. 125, p. 1–24.

Schmidt, M.G., 1999, Die Nebenüberlieferung des 6. Buchs der Geographie des Ptolemaios—Griechische, Lateinische, Syrische, Armenische und Arabische Texte: Wiesbaden, Dr. Ludwig Reichert Verlag, VIII + 291 p. + 1 map.

Schmidt, K., and Hoppe, P., 1971, Syneklise: Deutsches Handwörterbuch der Tektonik, 3. Lieferung, Hannover (no publisher, no page numbers).

Schmidt-Thomé, P., 1960, Zur Titeltafel: Geologische Rundschau, v. 50, p. II.

Schmidt-Thomé, P., 1972, Tektonik, *in* Brinkmann, R., ed., Lehrbuch der Allgemeinen Geologie, v. II: Stuttgart, Ferdinand Enke Verlag, XIX + 579 p.

Schmitt, C.B. and Knox, D., 1985, Pseudo-Aristoteles Latinus—A Guide to Latin Works Falsely Attributed to Aristotle Before 1500: London, The Warburg Institute, University of London, viii + 103 p.

Schmitz, H., 1988, Anaximander und die Anfänge der Griechischen Philosophie: Bonn, Bouvier, V + 79 p.

Schmöckel, H., 1962, Das Land Sumer—Dei Wiederentdeckung der ersten Hochkultur der Menschheit, 3. Auflage: W. Kohlhammer, Stuttgart, 195 p. + 48 photographic figures on plates.

Schneer, C.J., 1970, Ebenezer Emmons i istoriya formirovanniya osnov Amerikanskoi geologii, *in* Tikhomirov, V.V.E., Malkhasian, G., Mgrtchian, S.S., Ravikovich, A.I., and Sofiano, T.A., eds., Istoria Geologii, Akademia Nauk Armianskoi SSR, Mezhdunarodnii Komitet po Istorii Geologicheskikh Nauk, Izdatelstvo Akademii Nauk Armianskoi SSR, Erevan, p. 307–319 (English summary: Ebenezer Emmons and the foundations of American geology, p. 319–323).

Schönenberg, R., 1975, Die Entstehung der Kontinente und Ozeane in Heutiger Sicht: Darmstadt, Wissenschaftliche Buchgesellschaft, VIII + 351 p.

Schott, G., 1912, Geographie des Atlantischen Ozeans: Hamburg, C. Boysen, XII + 330 p. + 28 plates.

Schubert, F.N., 1980, Vanguard of Expansion—Army Engineers in the Trans-Mississippi West 1819–1879: Historical Division, Office of Administrative Services, Office of the Chief of Engineers, Superintendent of Documents, U.S.: Washington, D.C., Government Printing Office, D.C., xii + 160 p.

Schuchert, C., 1910, Paleogeography of North America: Bulletin of the Geological Society of America, v. 20, p. 427–606, plates XLVI–CI.

Schuchert, C., 1915, Preface, *in* Problems of American Geology—A Series of Lectures Dealing with Some of the Problems of the Canadian Shield and the Cordilleras, Delivered at Yale University on the Silliman Foundation in December, 1913: New Haven, Yale University Press, p. vii–x.

Schuchert, C., 1918[1973], A century of geology—the progress of historical geology in North America, *in* A Century of Science in America: New Haven, Yale University Press, p. 60–121 (reprinted by Scholarly Resources Inc., Wilmington, Delaware).
Schuchert, C., 1923, Sites and nature of the North American geosynclines: Geological Society of America Bulletin, v. 34, p. 151–230.
Schuchert, C., 1926, Stille's analysis and synthesis of the mountain structures of the Earth: American Journal of Science, fifth series, v. 12, p. 277–292.
Schuchert, C., 1943, Stratigraphy of the Eastern and Central United States: New York, John Wiley & Sons, xv + [i] + 1013 p.
Schuchert, C., 1955, Atlas of Paleogeographic Maps of North America: New York, John Wiley & Sons, ix + 177 p.
Schulten, A., 1914, Begriff und Wort «Hochebene»: Dr. A. Petermann's Mitteilungen aus Justus Perthes' Geographischer Anstalt, 60. Jahrgang, I. Halbband, p. 18–20.
Schütte, H., 1939, Sinkendes Land an der Nordsee? Zur Küstengeschichte Nordwestdeutschlands: Schriften des Deutschen Naturkundevereins/ Neue Folge, Hohenlohesche Buchhandlung Ferd. Rau, Öhringen, 144 p.
Schvarcz, J., 1862, On the Failure of the Geological Attempts in Greece Prior to the Epoch of Alexander, Part I: London, Taylor and Francis, xiv + 75 p
Schvarcz, J., 1868, The Failure of Geological attempts Made by the Greeks from the Earliest Times Down to the Epoch of Alexander, revised and enlarged edition: London, Trübner & Co., xx+153 p.
Schwab, F.L., ed., 1982, Geosynclines—Concept and Place within Plate Tectonics: Benchmark Papers in Geology: Stroudsburg, Hutchinson Ross, v. 64, xv + 411 p.
Schwan, W., 1986, Erinnerungen an späte Berliner Jahre mit Hans STILLE—1945–1964: Nachrichten der Deutschen Geologischen Gesellschaft, v. 35, p. 128–138.
Schwarzberg, J.E., 1992a, Introduction to South Asian cartography, *in* Harley, J.B., and Woodward, D., eds., 1992, The History of Cartography, volume two, book one (Cartography in the Traditional Islamic and south Asian Societies): Chicago, University of Chicago Press, p. 295–331.
Schwartzberg, J.E., 1992b, Geographical mapping, *in* Harley, J.B., and Woodward, D., eds., 1992, The History of Cartography, volume two, book one (Cartography in the Traditional Islamic and south Asian Societies): Chicago, University of Chicago Press, 388–493.
Schwartzberg, J.E., 1992c, Nautical maps, *in* Harley, J.B., and Woodward, D., eds., 1992, The History of Cartography, volume two, book one (Cartography in the Traditional Islamic and south Asian Societies): Chicago, University of Chicago Press, 494–503.
Schwartzberg, J.E., 1992d, Conclusion, *in* Harley, J.B., and Woodward, D., eds., 1992, The History of Cartography, volume two, book one (Cartography in the Traditional Islamic and south Asian Societies): Chicago, The University of Chicago Press, p. 504–509.
von Schweiger-Lerchenfeld, A., 1886, Afrika. Der Dunkle Erdtheil im Lichte unserer Zeit, v. 1: Wien, A. Hartleben's Verlag, VI + 464 p.
Scobel, A., 1910, Andrees Allgemeiner Handatlas, Jubiläumsausgabe, vierter revidierter Abdruck: Bielefeld und Leipzig, Verlag von Velhagen & Klasing, [IV] + 207 + 188 p.
Scott, W.B., 1908, An Introduction to Geology, second edition revised throughout: New York, The Macmillan Company, xxvii + 816 p.
Scrope, G.P., 1825, Considerations on Volcanos, the Probable Causes of their Phenomena, the Laws which Determine their March, the Disposition of their Products, and their Connexion with the Present State and Past History of the Globe; Leading to the Establishment of A New Theory of the Earth: London, W. Phillips, xxxi+270 p.+2 foldout plates.
Sears, J.W., 1990, Geologic structure of the Grand Canyon Supergroup, *in* Beus, S.S., and Morales, M., eds., Grand Canyon Geology: New York, Oxford University Press, and Museum of Northen Arizona Press, p. 71–82.
Secord, J.A., 1986, Controversy in Victorian Geology—The Cambrian-Silurian Dispute: Princeton, New Jersey, Princeton University Press, xvii + 363 p.
Secord, J.A., 1994, Introduction, *in* Secord, J.A., ed., Vestiges of the Natural History of Creation and Other Evolutionary Writings Robert Chambers: Chicago, University of Chicago Press, p. ix–xlviii.
Sedgwick, [M.K.],[445] 1926, Acoma, the Sky City—A Study in Pueblo-Indian History and Civilisation: Cambridge, Harvard University Press, xiv + 318 p.
See, T.J.J., 1907, On the temperature, secular cooling and contraction of the earth, and on the theory of earthquakes held by the ancients: Proceedings of the American Philosophical Society, v. 46, p. 191–299.
von Seebach, K., 1869, Die Eruption bei Methana im 3. Jahrh. v. Chr.: Zeitschrift der Deutschen Geologischen gesellschaft, v. 21, p. 275–280.
Seidl, J., 2000, Einige Inedita zur Frühgeschichte der Geowissenschaften an der Universität Wien. Die Bewerbung von Eduard Sueß um die *Venia legendi* für Paläontologie (1857): Geschichte der Erdwissenschaften in Österreich, 2. Symposium, Abstracts, Berichte des Intitutes für Geologie und Paläontologie der karl-Franzens-Universität Graz, v. 1, p. 55.
Selley, R.C., editor, 1997, African Basins: Amsterdam, Elsevier, XVII + 394 p.
Semper, M., 1914, Die Geologischen Studien Goethes: Leipzig, Verlag von Veit & Comp., 389 p.
Şengör, A.M.C., 1977, New historical data on crustal subduction: Journal of Geology, v. 85, p. 631–634.
Şengör, A.M.C., 1979, Classical Theories of Orogenesis, *in* Miyashiro, A., and Aki, K., eds., Mobile Earth, III, Orogeny (v. 12 of the Iwanami Shoten Earth Science Series): Tokyo, Iwanami Shoten Publishers, p. 1–34 (in Japanese).
Şengör, A.M.C., 1982a, The classical theories of orogenesis, *in* Miyashiro, A., Aki, K. and Şengör, A.M.C., eds., Orogeny: Chichester, John Wiley & Sons, p. 1–48.
Şengör, A.M.C., 1982b, Eduard Suess' relations to the pre-1950 schools of thought in global tectonics: Geologische Rundschau, v. 71, p. 381–420.
Şengör, A.M.C., 1983, Gondwana and "Gondwanaland": A discussion: Geologische Rundschau, v. 72, p. 397–400.
Şengör, A.M.C., 1987, Tectonic subdivisions and evolution of Asia: İstanbul, Bulletin of the Technical University, v. 40 (Ratip Berker Volume), p. 355–435.
Şengör, A.M.C., 1990, Plate tectonics and orogenic research after 25 years: A Tethyan perspective: Earth-Science Reviews, v. 27, p. 1–201.
Şengör, A.M.C., 1991a, The Evolution of the Universe: Diogenes, no. 155, p. 17–24.
Şengör, A.M.C., 1991b, Our home the Planet Earth: Diogenes, no. 155, p. 25–51.
Şengör, A.M.C., 1991c, Difference between Gondwana and Gondwana-Land: Geology, v. 19, p. 287–288.
Şengör, A.M.C., 1991d, Timing of orogenic events: A persistent geological controversy, *in* Müller, D.W., McKenzie, J.A., and Weissert, H., eds., Controversies in Modern Geology (Hsü-Festschrift): London, Academic Press, p. 405–473.
Şengör, A.M.C., 1992, Batlamyüs'ü Batıya tanıtan adam: İstanbul'lu Emanuel Chrysoloras: Bilim Tarihi, Temmuz 1992 (no. 9), p. 11–12.
Şengör, A.M.C., 1993, Some current problems on the tectonic evolution of the Mediterranean during the Cainozoic, *in* Boschi, E., Mantovani, E., and Morelli, A., eds., Recent Evolution and Seismicity of the Mediterranean Region, NATO ASI Series, Series C: Mathematical and Physical Sciences, v. 402: Dordrecht, Holland, Kluwer Academic Publishers, p. 1–51.
Şengör, A.M.C., 1994, Eduard Suess, *in* Eblen, R.A., and Eblen, W.R., eds.,The Encyclopaedia of the Environment: Boston, Houghton Mifflin Co., p. 676–677.
Şengör, A.M.C., 1995, Sedimentation and tectonics of fossil rifts, *in* Busby, C.J., and Ingersoll, R.V., eds., Tectonics of Sedimentary Basins: Oxford, Blackwell, p. 53–117.
Şengör, A.M.C., 1996, Eine Ergänzung der Carlé'schen Liste der Veröffentlichungen von Hans Stille und einige Schlüsse: Ein Beitrag zur Geschichte und Philosophie der tektonischen Forschung: Zentralblatt für Geologie und Paläontologie, no. 9/10 (1994), p. 1051–1106.
Şengör, A.M.C., 1997, The mountain and the bull: The origin of the word "Taurus" as part of the earliest tectonic hypothesis, *in* Başgelen, N., Çelgin, G., and Çelgin, V., eds., Festschrift für Taşlıklıoğlu, Arkeoloji ve Sanat Yayınları, İstanbul, p. 1–48 (first published in 1992 as an offprint).

Şengör, A.M.C., 1998, Die Tethys: vor hundert Jahren und heute: Mitteilungen der Österreichischen Geologischen Gesellschaft, v. 89, p. 5–176.

Şengör, A.M.C., 1999a, Continental interiors and cratons: any relation? Tectonophysics, v. 305, p. 1–42.

Şengör, A.M.C., 1999b, Some salient European contributions to Cordilleran tectonics, *in* Moores, E.M., Sloan, D., and Stout, D.L., eds., Classic Cordilleran Concepts: A View from California: Boulder, Colorado, The Geological Society of America, Special Paper 338, p. 31–36.

Şengör, A.M.C., 2000a, Die Bedeutung von Eduard Suess (1831–1914) für die Geschichte der Tektonik: Berichte der Geologischen Bundesanstalt, v. 51, p. 57–72.

Şengör, A.M.C., 2000b, Die Ansicht von Eduard Sueß über das Aussterben der Dinosaurier, *in* Geschichte der Erdwissenschaften in Österreich 2. Symposium Abstracts, *Berichte des Intitutes für Geologie und Paläontologe der Karl-Franzens-Universität Graz*, v. 1, p. 56.

Şengör, A.M.C., 2001a, Elevation as indicator of mantle plume activity, *in* Ernst, R., and Buchan, K., eds., Mantle plumes: Their identification through time: Boulder, Colorado, Geological Society of America Special Paper 352, p. 183–225.

Şengör, A.M.C., 2001b, Is the Present the Key to the Past or the Past the Key to the Present? James Hutton and Adam Smith versus Abraham Gottlob Werner and Karl Marx in Interpreting History: Boulder, Colorado, Geological Society of America Special Paper 355, x + 51 p.

Şengör, A.M.C., 2002a, Is the "Symplegades" myth the record of a tsunami that entered the Bosporus? Simple empirical roots of complex mythological concepts, *in* Aslan, R., Blum, S., Kastl, G., Schweizer, F. and Thumm, D., eds., Mauerschau, Festschrift für Manfred Korfmann: Remshalden-Grunbach, Verlag Bernhard Albert Greiner, v. 3, p. 1005–1028.

Şengör, A.M.C., 2002b, On Sir Charles Lyell's alleged distortion of Abraham Gottlob Werner in *Principles of Geology* and its implications for the nature of the scientific enterprise: Journal of Geology, v. 110, p. 355–368.

Şengör, A.M.C., 2003, Naomi Oreskes, The Rejection of Continental Drift: Theory and Method in American Earth Science (book review): Isis, v. 94, p. 186–188.

Şengör, A.M.C., and Burke, K., 1978, Relative timing of rifting and volcanism on Earth and its tectonic implications: Geophysical Research Letters, v. 5, p. 419–421.

Şengör, A.M.C., and Kidd, W.S.F., 1979, The post-collisional tectonics of the Turkish-Iranian Plateau and a comparison with Tibet: Tectonophysics, v. 55, p. 361–376.

Şengör, A.M.C., and Natal'in, B.A., 1996, Palaeotectonics of Asia: Fragments of a Synthesis, *in* Yin, A., and Harrison, M., eds., The Tectonic Evolution of Asia, Rubey Colloquium: Cambridge, Cambridge University Press, p. 486–640.

Şengör, A.M.C., and Natal'in, B.A., 2001, Rifts of the world, *in* Ernst, R., and Buchan, K., eds., Mantle plumes: Their identification through time: Boulder, Colorado, Geological Society of America Special Paper 352, p. 389–482.

Şengör, A.M.C., and Sakınç, M., 2001, Structural Rocks: Stratigraphic Implications, *in* Briegel, U., and Xiao, W.J., eds., Paradoxes in Geology (Hsü Volume): Amsterdam, Elsevier, p. 131–227.

Serbin, A., 1893, Bemerkungen Strabos über den Vulkanismus und Beschreibung der den Griechen Bekannten Vulkanischen Gebiete—Ein Beitrag zur Physischen Geographie der Griechen, inaugural-Dissertation zur Erlangung der Doktorwürde der Hohen Philosophischen Fakultät der Friedrich-Alexander-Universität Erlangen, Buchdruckerei Alb. Sayffaerth, Berlin, 63 p.

Şeşen, R., 1998, Müslümanlarda Tarih-Coğrafya Yazıcılığı: İstanbul, İslâm Tarih, Sanat ve Kültürünü Araştırma Vakfı (ISAR), XIII + 451 p. + 3 maps.

Seydî [Ali] Reis, 1313H[1897–98AD], Mirat al Memâlik: Dersaadet, İkdam Matbaası, 99 p.

Seydî Ali Reis, undated, Mir'at-ül Memalik, (Necdet Akyildiz, ed.): Tercüman 1001 Temel Eser, Kervan Kitapçılık, İstanbul, 175 p.

Seyfert, C.K., editor, 1987, The Encyclopedia of Structural Geology and Plate Tectonics: New York, Van Nostrand Reinhald Company, xii + 876 p.

Sezgin, F., 1971, Geschichte des Arabischen Schrifttums, Band IV, Alchimie–Chemie, Botanik–Agrikultur bis ca. 430 H: Leiden, E.J. Brill, X+[I]+398 p.+1 p. of errata.

Sezgin, F., 1974, Geschichte des Arabischen Schrifttums, Band V, Mathematik bis ca. 430 H: Leiden, E.J. Brill, XIV + [I] + 514 + [1] p.

Sezgin, F., 1978, Geschichte des Arabischen Schrifttums, Band VI, Astronomie bis ca. 430 H: Leiden, E.J. Brill, XV + 521 + [1] p.

Sezgin, F., 1979, Geschichte des Arabischen Schrifttums, Band VII, Astrologie – Meteorologie und Verwandtes bis ca. 430 H: Leiden, E.J. Brill, XIII+486 p.

Sezgin, F., 1987a, The Contribution of the Arabic-Islamic Geographers to the Formation of the World Map: Veröffentlichungen des Institutes für Geschichte der Arabisch-Islamischen Wissenschaften, Rheide D Kartographie, Bd. 2, 50 p. + 48 maps.

Sezgin, F., 1987b, Editor's Introduction, *in* Sezgin, F., ed., Klaudios Ptolemaios Geography Arabic Translation (1465 A.D.) Reprint of the Facsimile Edition of the MS Ayasofya 2610, in collaboration with M. Amawi, C. Ehrig-Egert, A. Jokhosha, E. Neubauer, I. Schuboltz, Veröffentlichungen des Institutes für Geschichte der Arabisch-Islamischen Wissenschaften, Rheide D Kartographie, Bd. 1, p. 1–16.

Sezgin, F., 1995, Geschichte des Arabischen Schrifttums, Gesamtindices zu Band I–IX, Institut für Geschichte der Arabisch-Islamischen Wissenschaften an der Johann Wolfgang Goethe-Universität, Frankfurt am Main, [I] + 604 p.

Sezgin, F., 2000a, Geschichte des Arabischen Schrifttums, Band X, Mathematische Geographie und Kartographie im Islam und Ihr Fortleben im Abendland—Historische Darstellung, Teil I, Institut für Geschichte der Arabisch-Islamischen Wissenschaften an der Johann Wolfgang Goethe-Universität, Frankfurt am Main, XXX + 634 p.

Sezgin, F., 2000b, Geschichte des Arabischen Schrifttums, Band XI, Mathematische Geographie und Kartographie im Islam und Ihr Fortleben im Abendland—Historische Darstellung, Teil 2, Institut für Geschichte der Arabisch-Islamischen Wissenschaften an der Johann Wolfgang Goethe-Universität, Frankfurt am Main, 716 p.

Sezgin, F., 2000c, Geschichte des Arabischen Schrifttums, Band XII, Mathematische Geographie und Kartographie im Islam und Ihr Fortleben im Abendland—Kartenband, Institut für Geschichte der Arabisch-Islamischen Wissenschaften an der Johann Wolfgang Goethe-Universität, Frankfurt am Main, 362 p.

Sezgin, F., 2000d, Der Kalif al-Ma'mûn und sein Beitrag zur Weltkarte—Arabischer Ursprung europäischer Karten: Forschung Frankfurt, Jahrgang 18, Nr. 4, p. 22–31.

Shatskiy, N.S., 1940[1964], O sineklizakh A.P. Pavlova, *in* Shtreis, N.A. (editor for 2nd volume), Sherbakov, D.I. (responsible editor) and nine others (editors), Akademik N.S. Shatskiy Izbrannie Trudi: Moscow, Akademiya Nauk SSSR, Izdatelstvo "Nauka," v. 2, p. 271–277.

Shatskiy, N.S., 1941, R.I. Murchison 1792–1871: Moskva, Moskovskoe Obshestvo Ispitateley Prirodi 1805–1940, 67+[I] p. + 1 plate.

Shatskiy, N.S., 1945[1964], Ocherki tektoniki Volga-Uralskoi neftenosnoi oblasti i smejnoi chasti zapadnoyo sklona Yujnoyo Urala, *in* Shtreis, N.A. (editor for 2nd volume), Sherbakov, D.I. (responsible editor) and nine others (editors), Akademik N.S. Shatskiy Izbrannie Trudi: Moscow, Akademiya Nauk SSSR, Izdatelstvo "Nauka," v. 2, p. 288–368.

Shatskiy, N.S., 1967, Syneclises, anticlises, placanticlines, and associated structures, *in* Mather, K.F., ed., Source Book in Geology 1900–1950: Cambridge, Harvard University Press, xv + 435 p.

Shaw, S.J., 1976, History of the Ottoman Empire and Modern Turkey, v. I, Empire of the Gazis/The Rise and Decline of the Ottoman Empire 1280–1808: Cambridge University Press, xiii + 1map + 351 p.

Shendge, M.J., 1991, Some Vedic myths in a new light: The Quarterly Journal of the Mythic Society, v. 82, p. 70–84.

Shirley, R.W., 1993, The Mapping of the World. Early Printed World Maps 1472–1700: London, New Holland, XLVI + 669 p.

Sholpo, V.N., responsible editor, 1999, Vladimir Vladimirovich Belousov: Rossiskaya Akademia Nauk Obedinenniy Institut Fiziki Zemli im. O.Y., Schmidta, Moskva, 399 p. + 20 photographic plates.

Sholpo, V.N., 2000, Destiny self-fulfilled: Science in Russia, no. 3 (117), p. 92–97.

Shukhardin, C.V., 1955, Georgiy Agrikola: Moskva, Izdatelstvo Akademii Nauk SSSR, 206 + [I] p.

Sichtermann, H., 1984, Mythologie und Landschaft: Gymnasium, v. 91, p. 289–305, plates I–VIII.

Sidi Ali Reïs, 1899, The Travels and Adventures of the Turkish Admiral Sidi Ali Reïs in India, Afghanistan, Central Asia, and Persia during the Years 1553–1556, translated from the Turkish, with notes by A. Vambéry: London, Luzac & Co., XVIII + 123 p. + 1 p. of errata.

Siegfried, R., and Dott, R.H., Jr., (editing and introduction), 1980, Humphry Davy on Geology—The 1805 Lectures for the General Audience: Madison, University of Wisconsin Press, xliv + 169 p.

Sieh, K. Ward, S.N., Natawidjaja, D., and Suwargadi, B.W., 1999, Crustal deformation at the Sumatran subduction zone revealed by coral rings: Geophysical Research Letters, v. 26, p. 3141–3144.

Sigstedt, C.O., 1952, The Swedenborg Epic—The Life and Works of Emanuel Swedenborg: New York, Bookman Associates, xvii + 517 p.

Sigurdsson, H., 1999, Melting the Earth: The History of Ideas on Volcanic Eruptions: New York, Oxford University Press, x + 260 p.

Sijmons, A.H., 1985, The arrangement of plates in Santarem's Atlas, *in* Wallis, H., and Sijmons, A.H., Atlas de Santarem—Facsimile of the Final Edition 1849 Explanatory Notes: Amsterdam, Rudolf Muller, p. 30–40.

Silberman, A., 1988, Pomponius Mela Chorographie, Texte établi, traduit et annoté: Collection des Universités de France, Société d'Édition "Les Belles Lettres," Paris, 347 p.

Simkin, T., Siebert, L., McClelland, L., Bridge, D., Newhall, C., and Latter, J.H., 1981, Volcanoes of the World: Stroudsburg, Hutchison Ross Publishing Company, viii + 232 + [1] p.

Singer, C., 1959[1996], A History of Scientific Ideas: New York, Barnes & Noble Books, xviii + 525 p.

Singleton, C., 1977, Dante Alighieri, The Divine Comedy ... Inferno, v. 2: Commentary: Bollingen Series LXXX: Princeton, Princeton University Press, [v] + 683 p. + 9 photographic plates + 4 maps.

de Sitter, L.U., 1956, Structural Geology: London, McGraw-Hill Publishing Co., [iv] + 552 p.

Skelton, R.A., 1968, Bibliographical note, *in* Mercator, G. and Hondius, J., 1636[1968], Atlas or A Geographicke Description of the Regions, Countries and Kingdomes of the World, through Europe, Asia, Africa, and America represented by new & exact Maps, The Second Volume: Amsterodami, Henrici Hondij et John Janssonius (reprinted in 1968 by Theatrum Orbis Terrarum Ltd., Amsterdam, with an introduction by R.A. Skelton), p. V–XXVII.

Sloss, L.L., 1963, Sequences in the cratonic interior of North America: Geological Society of America Bulletin, v. 169, p. 449–460.

Sloss, L.L., 1964, Tectonic cycles of the North American Craton, *in* Symposium on Cyclic sedimentation: Kansas Geological Survey Bulletin, v. 169, p. 449–460.

Sloss, L.L., 1966, Orogeny and epeirogeny: the view from the craton: Transactions of the New York Academy of Sciences, series II, v. 28, p. 579–587.

Sloss, L.L., 1972, Synchrony of Phanerozoic sedimentary-tectonic events of the North American craton and the Russian platform: 24th International Geological Congress, Montréal, Section 6 (Stratigraphy and Sedimentology), p. 24–32.

Sloss, L.L., 1988a, Tectonic evolution of the craton in Phanerozoic time, *in* Sloss, L.L., ed., Sedimentary Cover—North American Craton; U.S.: Boulder, Colorado, Geological Society of America, The Geology of North America, v. D-2, p. 25–51.

Sloss, L.L., 1988b, Forty years of sequence stratigraphy: Geological Society of America Bulletin, v. 100, p. 1661–1665.

Smiley, T.L., Nations, J.D., and Péwé, T.L., and Schafer, J.P., 1984, Landscapes of Arizona—the Geological Story: Lanham, University Press of America, xxvi + 505 p.

Smith, C., 1989, William Hopkins and the shaping of dynamical geology: 1830–1860: British Journal of the History of Science, v. 22, p. 27–52.

Smith, G., 1875, Assyrian Discoveries; An Account of Explorations and Discoveries on the Site of Nineveh, During 1873 and 1874: London, Sampson Low, Marston, Low and Searle, xviii + 461 p. + 1 foldout map.

Smith, G., 1876, The Chaldean Account of Genesis: London, Sampson Low, Marston, Searle, and Rivington, xvi + 319 p.

Smith, G.O., 1918[1973], A century of government geological surveys, *in* A Century of Science in America: New Haven, Yale University Press, p. 193–216 (reprinted by Scholarly Resources Inc., Wilmington, Delaware).

Smith, J.C., 1993, Georges Cuvier. An Annotated Bibliography of His Published Works: Washington and London, Smithsonian Institution Press, xx + 251 p.

Smith, R.J., 1996, Chinese Maps—Images of "All Under Heaven": Hong Kong, Oxford University Press, viii + 88 p.

Snead, R.E., 1980, World Atlas of Geomorphic Features: Huntington, Robert E. Krieger, IX + 301 p.

Snell, B., 1946, Die Entdeckung des Geistes—Studien zur Entstehung des Europäischen Denkens bei den Griechen: Claassen & Goverts Verlag, 264 p.

Snell, D.C., 1997, Life in the Ancient Near East 3100–332 B.C., E.: New Haven, Yale University Press, xvii + 270 p.

Snoke, A.W., 1993, Geologic history of Wyoming within the tectonic framework of the North American Cordillera, *in* Snoke, A.W., Steidtmann, J.R., and Roberts, S.M., eds., Geology of Wyoming: Laramie, Wyoming, Geological Survey of Wyoming Memoir No. 5, v. 1, p. 2–56.

Snow, J.K., and Wernicke, B., 2000, Cenozoic tectonism in the Central Basin and Range: magnitude, rate and distribution of upper crustal strain: American Journal of Science, v. 300, p. 659–719.

Von Soden, W., 1985, The Ancient Orient—An Introduction to the Study of the Ancient Near East (translated from the German by Donald G. Schley): Grand Rapids, Michigan, William B. Erdmans Publishing Co., xviii + 262 p. + 1 map.

Solomon, S.C., 1987, Secular cooling of the Earth as a source of intraplate stress: Earth and Planetary Science Letters, v. 83, p. 153–158.

Sorauf, J.E., 1962, Structural Geology and Stratigraphy of the Whitmore Area, Mohave County, Arizona [Ph.D. dissertation]: Lawrence, Kansas, University of Kansas, 361 p.

Sørensen, V., 1984, Seneca—Ein Humanist an Neros Hof: München, C.H. Beck, 320 p.

Spencer, J.E., 1996, Uplift of the Colorado Plateau due to lithosphere attenuation during Laramide low-angle subduction: Journal of Geophysical Research, 101, p. 13,595–13,609.

Stafford, R.A., 1989, Scientist of Empire. Sir Roderick Murchison, scientific exploration & Victorian imperialism: Cambridge, Cambridge University Press, 293 p.

Stahl, W.H., (translation, introduction and notes), 1952, Macrobius Commentary on the Dream of Scipio: Columbia University Press, xi + [i] + 278 p.

Stahl, W.H., 1953, Ptolemy's Geography A Select Bibliography: The New York Public Library, 86 p.

Stahl, W.H., 1959, Dominant traditions in early Medieval Latin science: Isis, v. 50, p., 95–124.

Stahl, W.H., 1962[1978], Roman Science: Westport, Greenwood Press, X + 308 p.

Stahl, W.H., Johnson, R. and Burge, E.L., 1977, Martianus Capella and the Seven Liberal Arts, v. II The Marriage of Philology and Mercury: Columbia University Press, [iii] + 389 p.

Stamp, L.D., 1962, A Glossary of Geographical Terms: London, Longmans, xxix + [i] + 539 p.

Stansbury, H., 1852, Exploration and Survey of the Valley of the Great Salt Lake of Utah including A Reconnaisance of A New Route through the Rocky Mountains: Senate, Special session, March 1851, Executive. No. 3: Philadelphia, Lippincott, Grambo & Co., 487 p.

Stanton, W., 1975, The Great United States Exploring Expedition of 1838–1842: Berkeley, University of California Press, x + 433 p.

Stapleton, D.H., 1985, Accounts of European Science, Technology, and Medicine by American travelers Abroad, 1735–1860, in the Collections of the American Philosophical Society: American Philosophical Society Library, Philadelphia, 48 p.

Stcherbatsky, F.T., 1930[1962]a, Buddhist Logic, v I: New York, Dover Publications, XII + 558 + [1] p.

Stcherbatsky, F.T., 1930[1962]b, Buddhist Logic, v II: New York, Dover Publications, VI + 468 p.

Steel, R., Gjelberg, J., Helland-Hansen, W., Kleinspehn, K., Nøttvedt, A., and Rye-Larsen, M., 1985, The Tertiary strike-slip basins and orogenic belt of Spitsbergen, *in* Biddle, K.T., and Christie-Blick, N., eds., Strike-slip Deformation, Basin Formation, and Sedimentation: Society of Economic Paleontology and Minerology Special Publication 37 (in honor of J.C. Crowell), p. 339–359.

Stearns, R.P., 1970, Science in the British Colonies of America: Urbana, University of Illinois Press, xx + 760 + [1] p.

Stegner, W., 1935, Clarence Edward Dutton; Geologist and Man of Letters [Ph.D. thesis]: University of Iowa, [iii] + 104 p. + frontispiece.

Stegner, W., undated [1936], Clarence Edward Dutton: An Appraisal: Salt Lake City, University of Utah Press, 23 p.

Stegner, W., 1937, C.E., Dutton—explorer, geologist, nature writer: Scientific Monthly, v. 45, p. 82–85.

Stegner, W., 1953, Beyond the Hundredth Meridian—The Exploration of the Grand Canyon and the Second Opening of the West: Boston, Sentry Edition, Houghton Mifflin Company, xxiii + 438 p. + 7 maps + 12 plates + 1 foldout.

Stegner, W., 1981, Dutton, Clarence Edward, *in* Gillispie, C.C., editor-in-chief, Dictionary of Scientific Biography: New York, Charles Scribner's Sons, v. 4, p. 265–266.

Steinschneider, M., 1960, Die Arabischen Übersetzungen aus dem Griechischen: Graz, Akademische Druck- und Verlagsanstalt, 381 p.

Steneck, N.H., 1976, Science and Creation in the Middle Ages—Henry of Langenstein (d. 1397) on Genesis: Notre Dame, University of Notre Dame Press, xiv + 213 p.

Stenonis, N., 1667[1969], Elementorum Myologiae Specimen/A Specimen of the Elements of Myology, *in* Scherz, G., ed., Steno—Geological Papers (Translation by Pollock, A.J.): Acta Historica Scientiarum Naturalium et Medicinalium: Odense University Press, v. 20, p. 65–131

Stenonis, N., 1669[1969], De Solido intra Solidum Naturaliter Contento Dissertationis Prodromus/The Prodromus to a Dissertation on Solids Naturally Contained Within Solids, *in* Scherz, ed., Steno—Geological Papers (Translation by Pollock, A.J.): Acta Historica Scientiarum Naturalium et Medicinalium: Odense University Press, v. 20, p. 133–234.

Stephens, L.D., 1982, Joseph Leconte—Gentle Prophet of Evolution: Louisiana State University Press, 340 p.

Stevens, H.N., 1908, Ptolemy's Geography. A Brief Account of all the Printed Editions Down to 1730: Henry Stevens, Son and Stiles, 62 p. (undated reprint by Theatrvm Orbis Terrarvm, Ltd., Amsterdam).

Stevens, I.I., 1855, Report of Explorations for a Route for the Pacific Railroad near the Forty-Seventh and Forty-Ninth Parallels of North Latitude, from St. Paulto Puget Sound, *in* Reports of Explorations and Surveys, to Ascertain the Most Practicable and Economical Route for a Railroad from the Mississippi River to the Pacific Ocean. Made under the Direction of the Secretary of War, in 1853–4, According to Acts of Congress of March 3, 1854, and August, 5, 1854, Senate, 33d Congress, 2d Session, Ex. Doc. No. 78, Washington, D.C., v I, vii + 651 p.

Stewart, J.A., 1990, Drifting Continents and Colliding Paradigms—Perspectives on the Geoscience Revolution: Bloomington, Indiana University Press, xii + [i] + 285 p.

Stewart, J.H., and Poole, F.G., 1974, Lower Paleozoic and uppermost Precambrian Cordilleran miogeocline, Great Basin, western United States, *in* Dickinson, W.R., ed., Tectonics and Sedimentation: Society of Economic Mineralogists and Palontologist Special Publication 22, p. 28–57.

Stille, H., 1918, Über Hauptformen der Orogenese und ihre Verknüpfung: Nachrichten der königlichen Gesellschaft der Wissenschaften in Göttingen, mathematisch—physikalische Klasse, Jg. 1918, p. 362–393.

Stille, H., 1919, Die Begriffe Orogenese und Epirogenese: Zeitschrift der Deutschen Geologischen Gesellschaft: Monatsberichte, v. 71, p. 164–208.

Stille, H., 1920, Über Alter und Art der Phasen variscischer Gebirgsbidung: Nachrichten der Königlichen Gesellschaft der Wissenschaften zu Göttingen, Mathematisch-physikalische Klasse, 1920, p. 218–224.

Stille, H., 1922, Die Schrumpfung der Erde—Festrede gehalten zur Jahresfeier der Georg August-Universität zu Göttingen am 5. Juli 1922: Berlin, Gebrüder Borntraeger, 37 p.

Stille, H., 1924, Grundfragen der Vergleichenden Tektonik, Gebrüder Borntraeger, Berlin, VIII + 443 p.

Stille, H., 1936a, Wege und Ergebnisse der geologisch-tektonischen Forschung, *in* Hartmann, M., ed., 25 Jahre Kaiser-Wilhelm-Gesellschaft zur Förderung der Wissenschaften, zweiter Band Die Naturwissenschaften: Berlin, Julius Springer, p. 77–97.

Stille, H., 1936b, Tektonische Beziehungen zwischen Nordamerika und Europa: Comptes Rendus International Geological Congress, Report of the XVI Session, United States of America 1933, v. 2, p. 829–838.

Stille, H., 1939a, Kordillerisch-atlantische Wechselbeziehungen: Geologische Rundschau, v. 30, p. 315–342.

Stille, H., 1939b, Geotektonische Probleme im atlantischen Raume: Bericht über den 250 jährige Jubiläumsfeier der Kaiserlich Leopoldinisch-Carolingischen Deutschen Akademie der Naturforscher, Halle (Saale), p. 129–139.

Stille, H., 1940, Einführung in den Bau Amerikas: Berlin, Gebrüder Borntraeger, XX + 717 p.

Stille, H., 1948, Ur- und Neuozeane: Abhandlungen der Deutschen Akademie der Wissenschaften zu Berlin, Mathematisch-naturwissenschaftliche Klasse, Jahrgang 1945/46, Nr. 6, 68 p. + 2 foldout maps.

Stille, H., 1949, Die Jungalkonkische Regeneration im Raume Amerikas: Abhandlungen der Deutschen Akademie der Wissenschaften zu Berlin, Mathematisch-naturwissenschaftliche Klasse, Jahrgang 1948 Nr. 3, Abhandlungen zur Geotektonik Nr. 1, 39 p.

Stoćes, B., and White, C.H., 1935, Structural Geology with special reference to economic deposits: London, Macmillan and Co., xiv + 460 p.

Stoddart, D.R., 1976, Darwin, Lyell, and the geological significance of coral reefs: British Journal for the History of Science, v. 9, p. 199–218.

Stoddart, D.R., 1985, Darwin, Jukes and the theory of reef development in Australia in the nineteenth century: Geological Society of America Abstracts with Programs, v. 17, p. 728.

Stoddart, D.R., 1994, "This Coral Episode"—Darwin, Dana, and the coral reefs of the Pacific, *in* MacLeod, R., and Rehbock, P.F., eds., Darwin's Laboratory—Evolutionary Theory and natural History in the Pacific: Honolulu, University of Hawai'i Press, p. 21–48.

Stoever, D.H., 1794, The Life of Sir Charles Linnæus…translated from the original German by J. Trapp: London, E. Hodson, Bell-Yard, xxxviii + 435 p.

Strassberg, R.E., translation, annotation and introduction, 1994, Inscribed Landscapes—Travel Writing from Imperial China: Berkeley, University of California Press, xxv + [ii] + 580 p.

Streeter, T.W., 1960, Bibliography of Texas 1795–1845, Part III, United States and European Imprints Relating to Texas, 1795–1837: Cambridge, Harvard University Press, v. I, xlii + 278 p.

Ströhle, A., 1921, Die den Alten Bekannten Vulkanischen Gebiete, inaugural-Dissertation zur Erlangung der Doktorwürde einer Hohen Philosophischen Fakultät der Universität Tübingen: Stuttgart, Jung & Sohn, [I] + 53 p.

Strohm, H., 1935, Untersuchungen zur Entwicklungsgeschichte der Aristotelischen Meteorologie: Philologus, Supplementband 28, Heft 1, [II] + 85 p.

Strohm, H., 1970, Aristoteles Meteorologie Über Die Welt übersetzt von Hans Strohm, *in* Grumach, E., and Flashar, H., eds., Aristoteles Werke in

Deutscher Übersetzung, Bd. 12, Teil I Meteorologie, teil II Über die Welt: Darmstadt, Wissenschaftliche Buchgesellschaft, 352 p.

Strohmeier, G., (translator, commentator, and editor) 1988, Al-Bîrûnî In den Gärten der Wissenschaft: Leipzig, Reclam, 318 p.

Struik, D.J., 1948, Yankee Science in the Making: Boston, Little Brown and Company, xiii + 430 p.

Stückelberger, A., 1988, Einführung in die Antiken Naturwissenschaften: Darmstadt, Wissenschaftliche Buchgesellschaft, X + 214 + [3] p. + VII plates.

Studer, B., 1847, Lehrbuch der Physikalischen Geographie und Geologie, zweites Kapitel, enthaltend: Die Erde im Verhältnis zur Wärme: Bern, J.F.J. Dalp, VII + 526 p.

Studer, B., 1851, Geologie der Schweiz, erster Band. Mittelzone und südliche Nebenzone der Alpen: Bern, Stämpflische Verlagshandlung, IV + [II] + 485 p. + 1 foldout map.

Studer, B., 1853, Geologie der Schweiz, zweiter Band. Nördliche Nebenzone der Alpen. Jura und Hügelland: Bern, Stämpflische Verlagshandlung, VII + 497 p.

Studtmann, J., 1934, Nikolaus Steno—Der größte Naturforscher seiner Zeit ein Apostel der norddeutschen Diaspora: Hildesheim, Franz Borgmeyer Verlag, 30 p.

Sturges, P., 1984, A Bibliography of George Poulett Scrope Geologist, Economist and Local Historian: Baker Library, Harvard Business School, Boston, 83 + [1] p.

Suess, E., 1858, Brachiopoden der Stramberger-Schichten, *in* von Hauer, F., ed., Beiträge zur Palæontographie von Österreich, v. I, p. 15–58.

Suess, E., 1859, Über die Wohnsitze der Brachiopoden: Sitzungsberichte der kaiserlichen Akademie der Wissenschaften, mathematisch-naturwissenschaftliche Klasse, v. 37, no. 18, p. 185–248.

Suess, E., 1862, Der Boden der Stadt Wien nach seiner Bildungsweise, Beschaffenheit und seinen Beziehungen zum Bürgerlichen Leben: Wien, W. Braumüller, [VI] + 326 p.

Suess, E., 1872, Über den Bau der italienischen Halbinsel: Sitzungsberichte der kaiserlichen Akademie der Wissenschaften, mathematisch-naturwissenschaftliche Classe, v. 65, part I, no. 3, p. 217–221.

Suess, E., 1873, Die Erdbeben Nieder-Österreichs: Denkschriften der mathematisch-naturwissenschaftlichen Classe der Kaiserlichen Akademie der Wissenschaften, v. 33, p. 1–38 + 2 foldout maps.

Suess, E., 1875, Die Entstehung der Alpen: Wien, W. Braumüller, IV + 168 p.

Suess, E., 1880, Ueber die vermeintlichen säcularen Schwankungen einzelner Theile der Erdoberfläche: Verhandlungen der kaiserlichen und königlichen Reichsanstalt, Nr. 11, p. 171–180.

Suess, E., 1883, Das Antlitz der Erde, v. Ia (Erste Abtheilung): F. Tempsky, Prag and G. Freytag, Leipzig, 310 p.

Suess, E., 1885, Das Antlitz der Erde, v. Ib: F. Tempsky, Prag and G. Freytag, Leipzig, IV + 311–778 + [1] p.

Suess, E., 1888, Das Antlitz der Erde, v. II: F. Tempsky, Prag and Wien, and G. Freytag, Leipzig, IV + 704 p.

Suess, E., 1891, Beiträge zur geologischen Kenntniss des östlichen Afrika, Theil IV, Die Brüche des östlichen Afrika: Denkschriften der mathematisch-naturwissenschaftlichen Classe der kaiserlichen Akademie der Wissenschaften in Wien, v. 58, p. 111–140, 1 foldout profile plate + 3 plates of photomicrographs.

Suess, E., 1904, The Face of the Earth, translated by Hertha B.C. Sollas under the direction of W.J. Sollas, v. I: Oxford, Clarendon, p. xii + 604 p.

Suess, E., 1906, The Face of the Earth, translated by Hertha B.C. Sollas under the direction of W.J. Sollas, v. II: Oxford, Clarendon, vi + 556 p.

Suess, E., 1909, Das Antlitz der Erde, v. III.2: F. Tempsky, Wien and G. Freytag, Leipzig, IV + 789 p. + 3 foldout maps.

Suess, E., 1911a, La Face de la Terre (Das Antlitz der Erde), traduit de l'Allemand, avec l'autorisation de l'auteur et annoté sous la direction de Emm. De Margerie, v. III (2e partie): Paris, Armand Colin, XI + [I] + p. 531–956.

Suess, E., 1911b, Synthesis of the paleogeography of North America: American Journal of Science, 4th series, v. 31, p. 101–108.

Suess, E., 1916, Erinnerungen: Leipzig, S. Hirzel, IX + 451 p.

Suess, F.E., 1914, Eduard Reyer: Mitteilungen der Geologischen Gesellschaft in Wien, v. 7, p. 327–329.

Suess, F.E., 1937, Bausteine zu einem System der Tektogenese—I. Periplutonische und enorogene Regionalmetamorphose in ihrer tektogenetischen Bedeutung, *in* Soergel, W., ed., Fortschritte der Geologie und Paläontologie: Berlin, Gebrüder Borntraeger, v. XIII, no. 42, VIII + 238 p.

Suhr, E.G., 1959, The Ancient Mind and its Heritage, v. I Exploring the Primitive, Egyptian and Mesopotamian Cultures: New York, Exposition Press, 175 p.

Suhr, E.G., 1960, The Ancient Mind and its Heritage, v. II Exploring the Hebrew, Hindu, Greek and Chinese Cultures: New York, Exposition Press, 307 p.

Sullivan, M.S., 1934[1992], The Travels of Jedediah Smith—A Documentary Outline, Including His Journal: Lincoln, University of Nebraska Press, [ix] + 195 p.

Supan, A., 1911, Grundzüge der Physischen Erdkunde, fünfte, umgearbaitete und verbesserte Auflage: Leipzig, Veit & Comp., IX + [I] + 969 + [1] p. + XX colored maps.

Süssmilch, C.A., 1909, Notes on the physiography of the southern tableland of New South Wales: Journal and Procedings of the Royal Society of New South Wales, v. 43, p. 331–354, plates IX–XIII.

Swedenborg, E., 1847, Miscellaneous Observations Connected with the Physical Sciences, translated from the Latin by Charles Edward Strutt: London, William Newberry, xvi + 168 p.

Taber, S., 1927, Fault troughs: Journal of Geology, v. 35, p. 577–606.

Taeschner, F., 1923, Die Geographische Literatur der Osmanen: Zeitschrift der Deutschen Morgenländischen Gesellschaft, Neue Folge, v. 2, no. 1, p. 31–80, 144.

Takahashi, H., 2003, Observations on Bar cEbroyo's marine geography: Hugoye: Journal of Syriac Studies [http://syrcom.cua.edu/syrcom/Hugoye] v. 6, no. 3, pars. 1–60.

Takahashi, T.J., and Griggs, J.D., 1987, Hawaiian volcanic features: A photoglossary, *in* Decker, R.W., Wright, T.L., and Stauffer, P.H., eds., Volcanism in Hawaii, U.S. Geological Survey Professional Paper 1350: Washington, D.C., Government Printing Office, v. 2,. p. 845–902.

Takmer, B., 2002, Lykia Orografyası, *in* Şahin, S. and Adak, M., eds., Likya İncelemeleri I, (Epigrafi ve Tarihsel Coğrafya Dizisi, v. I, Şahin, S. and Başgelen, N., eds.): İstanbul, Arkeoloji ve Sanat Yayınları, p. 33–51.

Talbert, R.J.A., editor, 2000, Barrington Atlas of the Greek and Roman World: Princeton, Princeton University Press, xxviii + [I] + 102 maps + unpaginated gazetteer.

Taton, R., editor, 1963, Ancient and Medieval Science—from the beginning to 1450 (translated from the French by A.J. Pomerans): London, Thames and Hudson, xx + 551 p.

Taylor, A.E., 1928[1972], A Commentary of Plato's Timaeus: Oxford, Clarendon, xvi + 700 p.

Taylor, A.E., 1936, Plato—The Man and His Work: New York, The Dial Press, xi + 522 p.

Taylor, E.G.R., 1934, Late Tudor and Early Stuart Geography 1583–1650—A Sequel to Tudor Geography: London, Methuen & Co. Ltd., ix + [ii] + 322 p. + 8 plates.

Tekeli, S., 1975, Modern Bilimin Doğuşunda Bizans'ın Etkisi: Privately Published, Ankara, XI+142 p.

Tekin, Ş., Tekin, G.A., and İz, F., 1989, The Seyahatname of Evliya Çelebi, Book One: İstanbul—Facsimile of *Topkapı Sarayı Bagdad 304*. Part 1: folios 1a–106a: Sources of Oriental Languages & Literatures, 11, printed at Harvard University, Office of the University Publisher, 12 p. + 106 folios.

Temple, R., 1999, The Genius of China—3,000 Years of Science, Discovery, and Innovation: Prion, 248 + [8] p.

de Terra, H., 1955, Humboldt—The Life and Times of Alexander von Humboldt: New York, Alfred A. Knopf, xiii + 386 + ix p.

Terrell, J.U., 1969, The Man Who Rediscovered America—A Biography of John Wesley Powell: New York, Weybright and Talley, 281 p.

Thalmann, H., 1943, Memorial to Émile Argand: Proceedings of the Geological Society of America, Annual Report for 1942, p. 153–165 + portrait.
Thenius, E., 1980, Der Beitrag österreichischer Geowissenschaftler zum "sea-floor spreading" und "plate tectonics"—Konzept: Verhandlungen der Geologischen Bundesanstalt, Jahrgang 1979, Heft 3, p. 407–414.
Thenius, E., 1988, Otto Ampferer, Begründer der Theorie der Ozeanboden-spreizung: Die Geowissenschaften, 6. Jahrgang, Nr. 4, p. 103–105.
Thierry, S., 1991, Southeast Asian origin myths and founding myths, *in* Bonnefoy, Y., ed., Mythologies, v. II: Chicago, Chicago University Press, p. 916–918.
Thomas, A.B., 1941, Teodoro de Croix and the Northern Frontier of New Spain 1776–1783—From the Original Document in the Archives of the Indies, Seville, Translated and Edited …: Norman, University of Oklahoma Press, xiii + 273 p.
Thomasian, R., 1981, Moro, Antonio-Lazzaro, *in* Gillispie, C.C., editor-in-chief, Dictionary of Scientific Biography: New York, Charles Scribner's Sons, v. 9, p. 531–534.
Thompson, C., 1936, A Dictionary of Assyrian Chemistry and Geology: Oxford, Clarendon Press, xlviii + 266 p.
Thompson, G.A., and Zoback, M.L., 1979, Regional geophysics of the Colorado Plateau: Tectonophsics, v. 61, p. 149–181.
Thompson, S.J., 1988, A Chronology of Geological Thinking from Antiquity to 1899: Metuchen, The Scarecrow Press, vii + 320 p.
Thomson, G., and Missner, M., 2000, On Aristotle: Belmont, Wadsworth Philosophers Series, Wadsworth Thomson Learning, [i] + 100 p.
Thomson, J.O., 1948[1965], History of Ancient Geography: Cambridge University Press (reprinted in 1965 by Biblo and Tannen, New York), 427 p.
Thomson, W. (Sir), 1876, Opening address in Section A: Mathematical and Physical: Nature, v. 14, p. 426–431.
Thorndike, L., 1923a, A History of Magic and Experimental Science During the First Thirteen Centuries of Our Era, v I: New York, Columbia University Press, xl + 835 p.
Thorndike, L., 1923b, A History of Magic and Experimental Science during the First Thirteen Centuries of Our Era, v II: New York, Columbia University Press, vi + 1036 p.
Tibbetts, G.R., 1992, The beginnings of a cartographic tradition, *in* Harley, J.B., and Woodward, D., eds., The History of Cartography, volume two, book one (Cartography in the Traditional Islamic and south Asian Societies): Chicago, University of Chicago Press, p. 90–107.
Tietze, E., 1917, Einige Seiten über Eduard Suess—Ein Beitrag zur Geschichte der Geologie: Jahrbuch der kaiserlich und königlichen geologischen Reichsanstalt, v. 66, p. 333–556.
Tikalsky, F.D., 1982, Historical controversy, science and John Wesley Powell: The Journal of Arizona History, v. 23, p. 407–422.
Tobien, H., 1981, Studer, Bernhard, *in* Gillispie, C.C., editor-in-chief, Dictionary of Scientific Biography: New York, Charles Scribner's Sons, v. 13, p. 1232–124.
Tollmann, A., 1983, Leopold Kober zum 100, Geburtstag: Mitteilungen der Österreichischen Geologischen Gesellschaft, v. 76, p. 19–25.
Tollmann, A., 1984, Reinout Willem van Bemmelen: Mitteilungen der Österreichischen Geologischen Gesellschaft, v. 77, p. 369–372.
Tollmann, A., 1990, Eduard-Sueß-Feier der Österreichischen Geologischen Gesellschaft zu seinem 75. Todestag: Mitteilungen der Österreichischen Geologischen Gesellschaft, v. 82, p. 1–17.
Tomkeieff, S.I., 1943, Megatectonics and Microtectonics: Nature, v. 152, p. 347.
Tomkeieff, S.I., 1983, Dictionary of Petrology, edited by E.K. Walton, B.A.O. Randall, M.H. Battey, and O. Tomkeieff: Chichester, John Wiley & Sons, 680 p.
von Toula, F., 1900, Lehrbuch der Geologie—ein Leitfaden für Studierende: Wien, Alfred Hölder, XI + [I] + 412 p. + XXX plates + 2 colored fold-out maps.
von Toula, F., 1906, Lehrbuch der Geologie—ein Leitfaden für Studierende, zweite Auflage: Wien, Alfred Hölder, XI + 492 p. + XXX plates
von Toula, F., 1918, Lehrbuch der Geologie—ein Leitfaden für Studierende, dritte Auflage: Wien, Alfred Hölder, XI + 556 p.
Toulmin, S., 1990, Cosmopolis—The Hidden Agenda of Modernity: Chicago, University of Chicago Press, xii + 228 p.
de Tournefort, J.P., 1717, Relations d'un voyage du Levant: Paris, Imprimerie Royale, v. I (544 p.), v. II (526 p. + unpaginated *Table des matieres principales contenües dans les deux tomes*).
Toynbee, A.J., 1947, A Study of History, abridgement of volumes I–VI by D.C. Somervell: New York, Oxford University Press, xiii + 617 p.
Tozer, H.F., 1882[1974], Lectures on the Geography of Greece: Chicago, Ares Publishers, xvi + 405 p.
Tozer, H.F., 1893, Selections from Strabo: Oxford, Clarendon, x + [1] + 376 p.
Tozer, H.F., 1935, A History of Ancient Geography, second edition with additional notes by M. Carey: Cambridge, Cambridge University Press, xxi + [i] + 387 p.
Tricot, J., 1955, Aristote—Les Météorologiques—Nouvelle Traduction et Notes, deuxième édition: Librairie Philosophique J. Vrin, Paris, XVII + 299 + [I] p.
Trümpy, R., 1955, Die Wechselbeziehungen zwischen Paläogeographie und Deckenbau: Vierteljahresschrift der Naturforschenden Gesellschaft zu Zürich, v. 100, p. 217–231.
Trümpy, R., 1984, Des géosynclinaux aux océans perdus: Bulletin de la Société Géologique de France, sér. 7, v. 26, p. 201–206.
Tuchman, B.W., 1978, A Distant Mirror—The Calamitous 14th Century: New York, Ballantine Books, XX + 667 p. + 62 black-and-white and 9 colored figures on plates + one map.
Turcotte, D.L., and Angevine, C.L., 1982, Thermal mechanisms of basin formation: Philosophical Transactions of the Royal Society of London, v. A305, p. 283–294.
Turcotte, D.L., and Burke, K., 1978, Global sea-level changes and the thermal structure of the earth: Earth and Planetary Science Letters, v. 41, p. 341–346.
Turcotte, D.L., and Schubert, G., 2002, Geodynamics, second edition: Cambridge, Cambridge University Press, xiii + 456 p.
Türkay, C., 1959, Osmanlı Türklerinde Coğrafya: İstanbul, Maarif Basımevi, [I] + 50[2] p.
Turner, A.K., 1993, The History of Hell: New York, Harcourt Brace & Company, 275 p.
Turner, C.E., 1990, Toroweap Formation, *in* Beus, S.S., and Morales, M., eds., Grand Canyon Geology: New York, Oxford University Press, and Museum of Northen Arizona Press, p. 203–223.
Turner, F.J., and Weiss, L.E., 1963, Structural Analysis of Metamorphic Tectonites: New York, McGraw-Hill Book Company, ix + 545 p.
Tweto, O., 1979, Geologic Map of Colorado: Reston, Virginia, Department of the Interior, U.S. Geological Survey, 1 sheet plus a sheet of legend, scale 1:500,000.
Umbgrove, J.H.F., 1947, The Pulse of the Earth, second edition: The Hague, Martinus Nijhoff, XXII + 358 p. + 8 foldout plates + 2 foldout tables.
Ünver, İ., 1975, Türk Edebiyatında Manzum İskender-Nâmeler [Ph.D. Thesis]: Ankara Üniversitesi, Dil ve Tarih Coğrafya Fakültesi, Eski Türk Edebiyatı Kürsüsü, no. 205, VIII + 364 p.
Vail, P.R., Mitchum, R.M., Jr., Todd, R.G., Widmier, J.M., Thompson, S., III, Sangree, J.B., Bubb, J.N., and Hatlelid, W.G., 1977, Seismic stratigraphy and global changes of sea-level, *in* Payton, C.E., ed., Seismic Stratigraphy—Applications to Hydrocarbon Exploration: American Association of Petroleum Geologists Memoir 26, p. 49–212.
Vance, J.E., Jr., 1995, The North American Railroad—Its Origin, Evolution, and Geography, *in* Creating the North American Landscape Series: Baltimore, The Johns Hopkins University Press, XVI + 348 p.
Van Wagoner, J.C., and Bertram, G.T., editors, 1995, Sequence Stratigraphy of Foreland Basin Deposits—Outcrop and Subsurface Examples from the Cretaceous of North America: American Association of Petroleum Geologists Memoir 64, xxi + 487 p.
Varenius, B., 1664, Geographia Generalis In qua affectiones generales Telluris explicantur: Ex Oficina Elzeviriana, Amstelodami, [XXXVIII] + 748 p.
Varenne, J., 1991, Pre-Islamic Iran, *in* Bonnefoy, Y., ed., Mythologies, v. II: Chicago, University of Chicago Press, p. 877–890.

Vater, F., 1845a, Der Argonautenzug—Aus den Quellen Dargestellt und Erklaert, erstes Heft: Kasan, Universitaets-Druckerei, VIII + 168 p.

Vater, F., 1845b, Der Argonautenzug—Aus den Quellen Dargestellt und Erklaert, zweites Heft: Kasan, Universitaets-Druckerei, VII + [I] + 166 p.

Veatch, A.C., 1935, Evolution of the Congo basin: Geological Society of America Memoir 3, vii + 183 p. + 10 plates.

Veevers, J.J., editor, 1984, Phanerozoic Earth History of Australia: Oxford, Clarendon Press, xv + 418 p.

Vélez de Escalante, F.S., 1995, The Domínguez-Escalante Journal—Their Expedition Through Colorado, Utah, Arizona, and New Mexico in 1776; translated by Fray Angelico Chavez, edited by Ted J. Warner: Salt Lake City, University of Utah Press, xxii + 153 p.

Verbeke, G., 1981, Simplicius, *in* Gillispie, C.C., editor-in-chief, Dictionary of Scientific Biography: New York, Charles Scribner's Sons, v. 12, p. 440–443.

Vernant, J.-P., 1982, The Origins of Greek Thought: Ithaca, New York, Cornell University Press, 144 p.

Verne, J., undated[1864], Voyage au Centre de la Terre: Paris, J. Hetzel et C[ie], 219 p.

Verne, J.-J., 1976, Jules Verne—A Biography: New York, An Arnold Lent Book, Taplinger Publishing Company, xii + 245 p.

Veron, J.E.N., 1986, Corals of Australia and the Indo-Pacific: Honolulu, University of Hawai'i Press, xi + 644 p.

Ver Wiebe, W.A., 1936, Geosynclinal boundary faults: Bulletin of the American Association of Petroleum Geologists, v. 20, p. 910–938.

Villemur, J.-R., 1967, Reconnaisance Géologique et Structurale du Nord du Bassin de Taoudenni: Mémoires du Bureau de Recherches Géologiques et Minières, No. 51, [viii]+151 p.+ 19 photographic plates.

Vincent, P.M., 1963, Les Volcans Tertiaires et Quaternaires du Tibesti Occidental et Central (Sahara du Tchad): Mémoires du Bureau de Recherches Géologiques et Minières, No. 23, 307 p.

Viola, H.J., 1987, Exploring the West: Smithsonia Institution, [Washington, D.C.], 256 p.

Viola, H.J., and Margolis, C., eds., 1985, Magnificent Voyagers—The U.S. Exploring Expedition, 1838–1842: [Washington, D.C.], Smithsonian Institution, 303 p.

Vivien de Saint Martin, [L.], 1873, Histoire de la Géographie et des Découvertes Géographiques Depuis les Temps les Plus Réculé jusqu'a nos Jours: Librairie Hachette et Cie, Paris, XVI + 615 p.

Vivien de Saint Martin, [L.], 1874, Atlas dressé pour l'Histoire de la Géographie et des Découvertes Géographiques Depuis les Temps les Plus Réculé jusqu'a nos Jours: Librairie Hachette et Cie, Paris, 4 p. + 12 map plates.

Vlastos, G., 1952, Theology and philosophy in early Greek thought: Philosophical Quarterly, v. 2, p. 97–123.

Vodden, C., 1992, No Stone Unturned—The First 150 Years of the Geological Survey of Canada: [The Geological Survey of Canada, Ottawa], ii + 52 p.

Vogel, K., 1967, Byzantine science, *in* The Cambridge Medieval History, v. IV, part II, The Byzantine Empire: Cambridge, Cambridge University Press, p. 267–305.

Vogt, C., 1846. Lehrbuch der Geologie und Petrefaktenkunde Zum Gebrauche bei Vorlesungen und zum Selbstunterrichte Theilweise nach L. Élie de Beaumont's vorlesungen an der Ecole des Mines, Erster Band.: Braunschweig, Friedrich Vieweg und Sohn, XIX + 436 p.

Vose, G.L., 1866, Orographic Geology; or, the Origin and Structure of Mountains. A Review: Boston, Lee and Shepard, 134 + [1] p.

Wagenbreth, O., 1979a, Leopold von Buch (1774–1853) und die Entwicklung der Geologie im 19. Jahrhundert: Abhandlungen des Staatlichen Museums für Mineralogie und Geologie zu Dresden, v. 29, p. 41–57.

Wagenbreth, O., 1979b, Der sächsische Mineraloge und Geologe Carl Friedrich Naumann (1797–1873): Abhandlungen des Staatlichen Museums für Mineralogie und Geologie zu Dresden, v. 29, p. 313–396.

Wagner, H., 1929, Spanish Voyages to the Northwest Coast of North America: San Francisco, California Historical Society, VIII + 572 p. + 5 foldout maps.

Wagner, H.R., and Camp, C.L., 1982, The Plains & the Rockies—A Critical Bibliography of Exploration, Adventure and Travel in the American West 1800–1865, fourth edition, revised, enlarged and edited by Robert H. Becker: San Francisco, John Howell-Books, xx + 745 p.

Walker, A.M., 1946, Francisco Garcés—Pioneer Padre of Kern, illustrations by Joan Cullimore: Kern County Historical Society and The County of Kern through its Chamber of Commerce, Kernville, Merchants Printing & Lithographing Company, Bakersfield, California, viii + 97 p.

Wallace, E., 1955, The Great Reconnaisance—Soldiers, Artists and Scientists on the Frontier 1848–1861: Boston, Little, Brown and Company, xii + [iv] + 288 p.

Wallace, W.A., O.P., 1981, Albertus Magnus, Saint, *in* Gillispie, C.C., editor-in-chief, Dictionary of Scientific Biography: New York, Charles Scribner's Sons, v. 1, p. 99–103.

Waller, R. [editor], 1705, The Posthumous Works of Robert Hooke, ... containing his Cutlerian Lectures, and other Discourses ...: London, Sam. Smith and Benj. Walford, xxviii + 572 p. + 9 plates + 11 p.

Walzer, R. and Frede, M., translators, 1985, Galen—Three Treatises on the Nature of Science ... with an Introduction by M. Frede: Indianapolis, Hackett Publishing Company, xxxvi + 112 p.

Wang, C.S., 1972, Geosynclines in the new global tectonics: Geological Society of America Bulletin, v. 83, p. 2105–2110.

Wang, C.S., 1979, Concept of geosyncline: Geologische Rundschau, v. 68, p. 696–706.

Warmington, E.H., 1934, Greek Geography: London, J.M. Dent & Sons, xlviii + 269 p.

Warmington, E.H., 1981, Strabo, *in* Gillispie, C.C., editor-in-chief, Dictionary of Scientific Biography: New York, Charles Scribner's Sons, v. 13, p. 83–86.

Warren, K., 1855, Memoir to Accompany the Map of the United States from the Mississippi River to the Pacific Ocean, Giving a Brief Account of the Exploring Expeditions since A.D. 1800, with a Detailed Description of the Method Adopted in Compiling the General Map, *in* Reports of Explorations and Surveys, to Ascertain the Most Practicable and Economical Route for a Railroad from the Mississippi River to the Pacific Ocean: [United States] Senate 33d Congress, 2d session, Ex. Doc. no. 78, Beverley Tucker, Washington, D.C., 120 p.

Warsi, W.E.K., and Molnar, P., 1977, Gravity anomalies and plate tectonics in the Himalaya, *in* Jest, C., ed., Colloques internationaux du C.N.R.S. No. 268, Écologie et Géologie de l'Himalaya, Himalaya, Sciences de la Terre, Editions du CNRS, Paris, v. I, p. 463–478.

Waschkies, H.-J., 1989, Die Protogaea von Leibniz—Ein Beitrag zur rationalen Ausdeutung des Schöpfungsmythos und der Ausarbeitung des Cartesischen Programms zu einer rationalen Kosmogonie, *in* Büttner, M., ed., Religion/Umwelt-Forschung im Aufbruch, Abhandlungen zur Geschichte der Geowissenschaften und Religion/Umwelt-Forschung, Band 2, Studienverlag Dr. N. Brockmeyer, Bochum, p. 60–100.

Watkins, R.T., 1986, Volcano-tectonic control on sedimentation in the Koobi Fora sedimentary basin, Lake Turkana, *in* Frostick, L.E., Renaut, R.W., Reid, I., and Tiercelin, J.J., eds., Sedimentation in the African rifts: Geological Society of London Special Publication no. 25, p. 85–95.

Watkins, T.H., 1969, The Grand Colorado—The Story of a River and its Canyons: American West Publishing Co. (no place of publication), 310 p.

Watson, E.S., compiler and editor, 1954, The Professor Goes West—Illinois Wesleyan University—Reparts of Major John Wesley Powell's Explorations: 1867–1874: Bloomington, Illinois, Illinois Wesleyan University Press, vii + 138 p.

Watts, A.B., ten Brink, U.S., Buhl, P., and Brocher, T.M., 1985, A multichannel seismic study of lithospheric flexure across the Hawaiian-Emperor seamount chain: Nature, v. 315, p. 105–111.

Watts, A.B., Karner, G.D., and Steckler, M.S., 1982, Lithospheric flexure and the evolution of sedimentary basins: Philosophical Transactions of the Royal Society of London, v. A305, p. 249–281.

Weber, M., 1921, Zum Problem der Grabenbildung: Zeitschrift der Deutschen Geologischen Gesellschaft, v. 73, p. 238–291.

Weber, M., 1923, Bemerkungen zur Bruchtektonik: Zeitschrift der Deutschen Geologischen Gesellschaft, v. 75, p. 184–192.

Weber, M., 1927, Faltengebirge und Vorlandbrüche: Zentralblatt für Mineralogie, Geologie und Paläontologie, v. 5, p. 235–245.

Webster, C., 1981, Ray, John, *in* Gillispie, C.C., editor-in-chief, Dictionary of Scientific Biography: New York, Charles Scribner's Sons, v. 12, p. 313–318.

Wegener, A., 1915, Die Entstehung der Kontinente und Ozeane: Braunschweig, Friedrich Vieweg & Sohn, V + 94 p.

Wegmann, E., 1981, Suess, Eduard, *in* Gillispie, C.C, ed., Dictionary of Scientific Biography, New York, Charles Scribner's Sons, v. 13, p. 143–149.

Wehrli, F., 1969, Straton von Lampasakos, *in* Die Schule des Aristoteles—Texte und Kommentar, v. 5 (zweite, ergänzte und verbesserte Auflage): Basel/ Stuttgart, Schwabe & Co., 85 p.

Welsch, J., 1908, Notice nécrologique sur Charles Contejean: Bulletin de la Société Géologique de France, série 4, v. 8, p. 204–208.

Wensinck, A.J., 1918, The Ocean in the Literature of the Western Semites: Verhandelingen der Koninklijke Akademie van Wetenschappen te Amsterdam, Afdeeling Letterkunde, nieuwe reeks, v. 19, no. 2, XI + 66 p.

Wernicke, B., 1985, Uniform-sense normal simple shear of the continental lithosphere: Canadian Journal of Earth Sciences, v. 22, p. 108–125.

Wernicke, B., 1990, The fluid crustal layer and its implications for continental dynamics, *in* Salisbury, M.H., and Fountain, D.M., eds., Exposed Cross-Sections of the Continental Crust, NATO ASI: Dordrecht, Kluwer, p. 509–544.

Wernicke, B., and Axen, G.J., 1988, On the role of isostasy in the evolution of normal fault systems: Geology, v. 16, p. 848–851.

Wescher, C., editor, 1874, Dionysii Byzantii De Bospori Navigatione quae Supersunt Una Cum Supplementis in Geographos Græcos Minores Aliisque Ejusdem Argumenti Fragmentis: Paris, Firmin Didot, XXXIII + [I] + 152 + [II] p.

West, E.W., (translator), 1880, Pahlavi Texts, Part I The Bundahis-Bahman Yast, and Shâyast Lâ-Shâyast, *in* Max Müller, editor, Sacred Books of the East, v. 5: Oxford, at the Clarendon Press, lxxiv + 438 p.

Westervelt, W.D., 1916, Hawaiian Legends of Volcanoes: Ellis Press, Boston, Constable & Co., London, xv + 205 p. + [9] p.

Westfall, R.S., 1981, Hooke, Robert, *in* Gillispie, C.C., editor-in-chief, Dictionary of Scientific Biography: New York, Charles Scribner's Sons, v. 6, p. 481–488.

Wettstein, H., 1880, Die Strömungen des Festen Flüssigen und Gasförmigen und Ihre Bedeutung für Geologie, Astronomie, Klimatologie und Meteorologie: Zürich, J. Wurster & Cie, [ii] + 406 p.

Weyl, R., 1958, Leonardo da Vinci und das geologische Erdbild der Renaissance: Nachrichten der Gießener Hochschulgesellschaft, v. 27, p. 110–121.

Wheat, C.I., 1957, 1540–1861 Mapping the Transmississippi West, v. I, The Spanish *Entrada* to the Louisiana Purchase 1540–1804: San Francisco, The Institute of HIstorical Cartography, xiv + 264 p. + 50 map reproductions.

Wheat, C.I., 1958, 1540–1861 Mapping the Transmississippi West, v. 2, From Lewis and Clark to Fremont 1804–1845: San Francisco, The Institute of Historical Cartography, xiii + 281 p. + 58 map reproductions.

Wheatley, P., 1961[1966], The Golden Khersonese. Studies in the historical geography of the Malay Peninsula before A.D. 1500: Kuala Lumpur, University of Malaya Press, xxxiii + 388 p. (reprinted photographically in 1966 by Craftsman Press Ltd., Singapore).

Wheeler, G.M., undated, Geological Atlas Projected to Illustrate Geographical Explorations and Surveys West of the 100th Meridian of Longitude Prosecuted in accordance with acts of Congress under the authority of The Honorable The secretary of War, under the direction of Brig. Gen'l. A.A., Humphreys, Chief of Engineers, U.S. Army. Embracing the results of the different expeditions under the command of 1st Lieutenant Geo. M. Wheeler, Corps of Engineers.: no publisher, no place of publication, 30 maps.

Wheeler, G.M., 1889, Report upon U.S. Geographical Surveys West of the One Hundredth Meridian, in charge of Capt. Geo. M. Wheeler ..., v. I.—Geographical Report: Engineer Department, U.S. Army: Washington, D.C., Government Printing Office, 780 p.

Whewell, W., 1832, Principles of Geology, v. II, by Charles Lyell: Quarterly Review, v. 47, p. 103–132.

Whipple, A.W., Ewbank, T., and Turner, W.W., 1855, Report upon the Indian tribes, *in* Reports of Explorations and Surveys, to Ascertain the Most Practicable and Economical Route for A Railroad from the Mississippi River to the Pacific Ocean. Made under the Direction of the Secretary of War, in 1853–4, According to Acts of Congress of March 3, 1854, and August, 5, 1854., Senate, 33d Congress, 2d Session, Ex. Doc. No. 78: Washington, D.C., v. III, part III, 127 p.

White, D., Roeder, D.H., Nelson, T.H., and Crowell, J.C., 1970, Subduction: Geological Society of America Bulletin, v. 81, p. 3431–3432.

White, D.A., 2001, Plains & Rockies 1800–1865 One hundred twenty additions to the Wagner-Camp and Becker bibliography of travel and adventure in the American West With 33 selected reprints—A supplemental volume to the series News of the Plains and Rockies: Spokane, The Arthur H. Clark Company, 544 p.

White, G.W., 1970, Rannie Amerikanskie nauchnie trudi po istorii geologii 1803–1835 gg, *in* Tikhomirov, V.V.E. Malkhasian, G., Mgrtchian, S.S., Ravikovich, A.I. and Sofiano, T.A., eds., Istoria Geologii, Akademia Nauk Armianskoi SSR, Mezhdunarodnii Komitet po Istorii Geologicheskikh Nauk, Izdatelstvo Akademii Nauk Armianskoi SSR, Erevan, p. 114–126 (English summary: Early American publications on the history of geology, p. 126–129).

White, G.W., 1973, The history of geology and mineralogy as seen by American writers, 1803–1835: A bibliographic essay: Isis, v. 64, p. 197–214.

White, R.S., and McKenzie, D., 1989, Magmatism at rift zones: the generation of volcanic continental margins and flood basalts. Journal of Geophysical Research, v. 94, p. 7685–7729.

Whitrow, G.J., 1988, Time in History—The evolution of our general awareness of time and temporal perspective: Oxford, Oxford University Press, x + [i] + 217 p. + 1 plate.

Wiedemann, E., 1890, Inhalt eines Gefässes in verschiedenen Abständen vom Erdmittelpunkte nach Al Khâzinî und Roger Baco: Annalen der Physik, v. 39, p. 319.

Wiedemann, E., 1912a, Beiträge zur Geschichte der Naturwissenschaften XXVII: Sitzungsbrichte der Physikalisch-medizinischen Sozietät in Erlangen, v. 44, p. 1–40.

Wiedemann, E., 1912b, Beiträge zur Geschichte der Naturwissenschaften XXVIII: Sitzungsbrichte der Physikalisch-medizinischen Sozietät in Erlangen, v. 44, p. 113–118.

Wiedemann, E., 1912c, Beiträge zur Geschichte der Naturwissenschaften XXIX: Sitzungsbrichte der Physikalisch-medizinischen Sozietät in Erlangen, v. 44, p. 119–125.

Wiedemann, E., 1912d, Beiträge zur Geschichte der Naturwissenschaften XXX: Sitzungsbrichte der Physikalisch-medizinischen Sozietät in Erlangen, v. 44, p. 205–256.

Wiens, D.A., 1985/86. Historical seismicity near Chagos: a complex formation zone in the equatorial Indian Ocean: Earth and Planetary Science Letters, v. 76, p. 350–360.

Wiens, D.A., DeMets, C., Gordon, R.G., Stein, S., Argus, D., Engeln, J.F., Lundgren, P., Quible, D., Stein, C., Weinstein, S., and Woods, D.F., 1985, A diffuse plate boundary model for Indian Ocean tectonics: Geophysical Research letters, v. 12, p. 429–432.

Wightman, W.P.D., 1950, The Growth of Scientific Ideas: Edinburgh, Oliver and Boyd, x + [ii]+495 p.

von Wilamowitz-Moellendorf, U., 1919a, Platon—erster Band Leben und Werke: Berlin, Weidmannsche Buchhandlung, V + [I] + 756 p.

von Wilamowitz-Moellendorf, U., 1919b, Platon—zweiter Band Beilagen und Textkritik: Berlin, Weidmannsche Buchhandlung, [II] + 452 p.

von Wilamowitz-Moellendorf, U., 1955a, Der Glaube der Hellenen, Band I: Darmstadt, Wissenschaftliche Buchgesellschaft, VII + 405 p.

von Wilamowitz-Moellendorf, U., 1955b, Der Glaube der Hellenen, Band II: Darmstadt, Wissenschaftliche Buchgesellschaft, XI + 617 p.

Wilding, R., 1996, Scrope vs. Mallet—a battle of heavyweights: Geology Today, v. 12, p. 110–114.

Wilkins, T., 1988, Clarence King—A Biography, revised and enlarged edition: Albuquerque, University of New Mexico Press, xiii + 524 p.

Williams, W.C., 1981, Chambers, Robert, *in* Gillispie, C.C., editor-in-chief, Dictionary of Scientific Biography: New York, Charles Scribner's Sons, v. 3, p. 191–193.

Willis, B., 1912, Index to the Stratigraphy of North America: U.S. Geological Survey Professional Paper 71: Washington, D.C., Government Printing Office, 894 p. + colored foldout geological map.

Wilsdorf, H., 1956, Georg Agricola und Seine Zeit, *in* Georgius Agricola—Ausgewählte Werke, Gedenkausgabe des Staatlichen Museums für Mineralogie und Geologie zu Dresden, Herausgeber Dr. rer. nat. Hans Prescher: Berlin, VEB Verlag der Wissenschaften, XVI + 335 p. + 70 plates.

Wilsdorf, H., and Friedrich, J., 1954, Die Bergbaukunde und ihre Nachbargebiete in der Cosmography des Sebastian Münster, *in* Wilsdorf, H., Präludien zu Agricola, Freiberger Forschungshefte, Kultur und technik, D 5, p. 65–222.

Wilson, L.G., 1962, The development of the concept of uniformitarianism in the mind of Charles Lyell: Ithaca-26 VIII–2 IX: Paris, Hermann, p. 993–996.

Wilson, L.G., 1967, The origins of Charles Lyell's uniformitarianism: Geological Society of America Special Paper 89, p. 35–62.

Wilson, L.G., 1970, Sir Charles Lyell's Scientific Journals on the Species Question: New Haven, Yale University Press, lxi + 572 p.

Wilson, L.G., 1972, Charles Lyell—The Years to 1841: The Revolution in Geology: New Haven and London, Yale University Press, xiii + 553 p.

Wilson, L.G., 1998, Lyell in America—Transatlantic Geology, 1841–1853: Baltimore, The Johns Hopkins University Press, xii + 429 p.

Wilson, J.T., 1963, A possible origin of the Hawaiian Islands: Canadian Journal of Physics, v. 41, p. 863–870.

Wilson, J.T., 1965a, A new class of faults and their bearing on continental drift: Nature, v. 207, p. 907–910.

Wilson, J.T., 1965b, Convection currents and continental drift. XIII. Evidence from ocean islands suggesting earth movement, *in* Blackett, P.M.S., Bullard, E., and Runcorn, S.K., eds., A Symposium on Continental Drift (Philosophical Transactions, v. 258A): London, The Royal Society, p. 145–167.

Wilson, J.T., 1973, Mantle plumes and plate motions: Tectonophysics, v. 19, p. 149–164.

Wilson, J.T., editor, 1976, Continents Adrift and Continents Aground: Readings from Scientific American: San Francisco, W.H. Freeman and Company, vii + [ii] + 230 p.

Wilson, J.T., 1982, Early days in university geophysics: Annual Review of Earth and Planetary Sciences, v. 10, p. 1–14.

Windley, B.F., 1995, The Evolving Continents, 3rd edition: Chichester, John Wiley & Sons, xvi + [i] + 526 p.

Winship, G.P., 1896, The Coronado expedition, 1540–1542, *in* Fourteenth Annual Report of the Bureau of Ethnology to the Secretary of the Smithsonian Institution 1892–1893 by J.W. Powell, Director: Washington, D.C., Government Printing Office, p. 329–637.

Winter, J.G., translator and commentator, 1916, The Prodromus of Nicolaus Steno's Dissertation Concerning a Solid Body Enclosed By Process of Nature Within A Solid: University of Michigan Studies, Humanistic Series, v. XI—Contributions to the History of Science—Part II: New York, The Macmillan Company, p. 166–283.

Witlarzil, E., 1902, Geschichte der Erde zunächst für Mädchenlyzeen: Wien, Alfred Hölder, IV + 72 p.

Wolf, A., and Wolf, H.-H., 1983, Die Wirkliche Reise des Odysseus—zur Rekonstriktion des Homerischen Weltbildes: München, Langen Müller, 304 p. + 39 photos on plates.

Wolfer, E.P., 1954, Eratosthenes von Kyrene als Mathematiker und Philosoph: Groningen, Nordhoff, 67 p.

Wolff, H., ed., 1995, Vierhundert Jahre Mercator Vierhundert Jahre Atlas—"Die ganze Welt zwischen zwei Buchdeckeln" Eine Geschichte der Atlanten: Weißenhorn, Anton H. Konrad Verlag, 384 p.

Wolkstein, D. and Kramer, S.N., 1983, Inanna Queen of Heaven and Earth—Her Stories and Hymns from Sumer: New York, Harper & Row Publishers, xix + 227 p.

Wood, F.J., 1976, J.G., Kohl and the "lost maps" of the American coast: The American Cartographer, v. 3, p. 107–115.

Wood, M., 2000, Conquistadors: Berkeley, University of California Press, 288 p.

Wood, R.G., 1966, Stephen Harriman Long 1784–1864—Army Engineer Explorer Inventor: Glendale, California, The Arthur H. Clark Company, 292 p.

Woodcock, N.H., 1986, The role of strike-slip fault systems at plate boundaries: Philosophical Transactions of the Royal Society of London, v. A317, p. 13–29.

Woodroffe, C.D., and Falkland, A.C., 1997, Geology and Hydrogeology of the Cocos (Keeling) islands, *in* Vacher, H.L., and Quinn, T., eds., Geology and Hydrogeology of Carbonate Islands, Developments in Sedimentology 54: Amsterdam, Elsevier Science, p. 885–908.

Woodroffe, C.D, McLean, R.F., 1994, Reef Islands of the Cocos (Keeling) Islands: Atoll Research Bulletin, v. 403, p. 1–36

Woodroffe, C., McLean, R., Polach, H., and Wallensky, E., 1990a, Sea-level and coral atolls: Late Holocene emergence in the Indian Ocean: Geology, v. 18, p. 62–66.

Woodroffe, C., McLean, R., and Wallensky, E., 1990b, Darwin's coral atoll: geomorphology and recent development of the cocos (Keeling) Islands, Indian Ocean: National Geographic Research, v. 6, p. 262–275.

Woodward, D, 1987, Medieval *mappaemundi*, *in* Harley, J.B., and Woodward, D. (eds.), The History of Cartography, v. 1 (Cartography in Prehistoric, Ancient, and medieval Europe and the Mediterranean): Chicago, University of Chicago Press, p. 286–370.

Woodward, J., 1695, An Essay Toward Natural History of the Earth and Terrestrial Bodies, Especially Minerals: As Also of the Sea, Rivers, and Springs, with an Account of the Universal Deluge and the Effects that it Had Upon the earth: London, Richard Wilkins, [xii] + 277 p.

Worster, D., 2001, A River Running West—The Life of John Wesley Powell: Oxford, Oxford University Press, XIII + 673 p.

Wright, J.K., 1925, Geographical Lore of the Time of the Crusades: American Geographical Society Research Series no. 15, xxi + 563 p.

Wyllie, P.J., 1971, The Dynamic Earth: Texbook in Geosciences: New York, John Wiley & Sons, xiv + 416 p.

Wyllie, P.J., 1976, The Way the Earth Works: An Introduction To The New Global Geology and Its Revolutionary Development: New York, John Wiley & Sons, viii + 296 p.

Yang, Z.Y., and Li, F.l., 1996, On the study of invertebrate paleontology in China—a retrospect, *in* Wang, H.Z., Zhai, Y.S., Shi, B.S., and Wang C.S., eds., Development of Geoscience Disciplines in China: Wuhan, China University of Geosciences Press, p. 71–86.

Yee, C.D.K., 1994a, Reinterpreting traditional Chinese geographical maps, *in* Harley, J.B., and Woodward, D., eds., The History of Cartography, volume two, book two (Cartography in the Traditional East and Southeast Asian Societies): Chicago, The University of Chicago Press, p. 35–70.

Yee, C.D.K., 1994b, Taking the world's measure: Chinese maps between observation and text, *in* Harley, J.B., and Woodward, D., eds., The History of Cartography, volume two, book two (Cartography in the Traditional East and Southeast Asian Societies): Chicago, University of Chicago Press, p. 96–127.

Yee, C.D.K., 1994c, Chinese cartography among the arts: Objectivity, subjectivity, representation, *in* Harley, J.B., and Woodward, D., eds., The History of Cartography, volume two, book two (Cartography in the Traditional East and Southeast Asian Societies): Chicago, University of Chicago Press, p. 128–168.

Yee, C.D.K., 1994d, Traditional Chinese cartography and the myth of westernization, *in* Harley, J.B., and Woodward, D., eds., The History of Cartography, volume two, book two (Cartography in the Traditional East and Southeast Asian Societies): Chicago, University of Chicago Press, p. 170–202.

Yee, C.D.K., 1994e, Concluding remarks: Foundations for a future history of Chinese mapping, *in* Harley, J.B., and Woodward, D., eds., The History of Cartography, volume two, book two (Cartography in the Traditional East and Southeast Asian Societies): Chicago, University of Chicago Press, p. 228–231.

Yerasimos, S., 1994, Olivier, Guillaume-Antoine: Dünden Bugüne İstanbul Ansiklopedisi: İstanbul, T.C., Kültür Bakanlığı ve Türkiye Ekonomik ve Toplumsal Tarih Vakfı, v. 6, p. 128–129.

Yılmaz, Ö., 1987, Seismic Data Processing: Tulsa, Oklahoma, Society of Exploration Geophysicists, Investigations in Geophysics, v. 2, xii + 526 p.

Yochelson, E.L., 1980, The Scientific Ideas of G.K. Gilbert—An Assesment on the Occasion of the Centennial of the U.S. Geological Survey: Boulder, Colorado, Geological Society of America Special Paper 183, viii + 183 p.

Yonge, C.M., 1958, Darwin and coral reefs, *in* Barnett, S.A., ed., A Century of Darwin: London, Heinemann, p. 245–266.

Young, C.R., editor, 1969, The Twelfth-Century Renissance: European Problem Studies: New York, Holt, Rinehart and Winston, [ii] + 116 p.

Yule, (Sir) H., 1903, The Book of Ser Marco Polo, the Venetian, Concerning the Kingdoms and Marvels of the East, 3rd edition, revised by Henri Cordier, v. 2: London, John Murray, xxii + 662 p. (reprinted by Dover Publications, Inc., in 1993, together with Henri Cordier's *Ser Marco Polo Notes and Addenda to Sir Henry Yule's edition, containing the results of recent research and discovery*, x + 161).

Yule, (Sir) H., 1914[1966], Cathay and the Way Thither Being a Collection of Medieval Notices of China. New edition, revised throughout in the light of recent discoveries by Henri Cordier. Vol III (Missionary Friars—Rashiduddin—Pegelotti—Marignolli): Hakluyt Society Publications, 2nd series, No. 37: Cambridge, Cambridge University Press, xvi + 269 p. (reprinted in 1966 by the Cheng-Wen Publishing Company, Taipei).

Yule, (Sir) H., 1916[1966], Cathay and the Way Thither Being a Collection of Medieval Notices of China. New edition, revised throughout in the light of recent discoveries by Henri Cordier. Vol IV (Ibh Batuta — Benedict Goës — Index): Hakluyt Society Publications, 2nd series, No. 41: Cambridge, Cambridge University Press, xii + 359 p. (reprinted in 1966 by the Cheng-Wen Publishing Company, Taipei).

Zachariasen, J., Sieh, K., Taylor, F.W., and Hantoro, W.S., 2000, Modern vertical deformation above the Sumatran subduction zone: paleogeodetic insights from coral microatolls: Bulletin of the Seismological Society of America, v. 90, p. 897–913.

Zachariasen, J., Sieh, K., Taylor, F.W., Edwards, R.L., and Hantoro, W.S., 1999, Submergence and uplift associated with the giant 1833 Sumatran subduction earthquake: Evidence from coral microatolls: Journal of Geophysical Research, v. 104, p. 895–919.

Zafiropulo, J., 1953, Empédocle d'Agrigente: «Les Belles Lettres», Paris, 305 + [1] p.

Zaitsev, Y.A., 1990, Areageosynclines and their role in geotectogenesis: Acta Universitatis Carolinae—Geologica, no. 1, p. 55–73.

Zeller, E., 1919, Die Philosophie der Griechen in ihrer Geschichtlichen Entwicklung. Erster Teil: Allgemeine Einleitung, Vorsokratische Philosophie. Erste Hälfte—Sechste Auflage. Mit Unterstützung von Dr. Franz Lortzing herausgegeben von Dr. Wilhelm Nestle: Leipzig, O.R. Reisland, XVI + 782 p.

Zhong, S.Z., 1984, Ancient China's Scientists: Hong Kong, The Commercial Press, 111 p.

Zimmermann, W.F.A., 1861, Die Wunder der Urwelt—Eine Populäre Darstellung der Geschichte der Schöpfung und des Urzustandes unseres Weltkörpers so wie der verschiedenen Entwicklungsperioden seiner Oberfläche, seiner Vegetation und seiner Bewohner bis auf die Jetztzeit. Nach den Resultaten der Forschung und Wissenschaft: Fünfzehnte Auflage: Berlin, Gustav Hempel, VIII + 548 p.

von Zittel, K.A., 1899, Geschichte der Geologie und Paläontologie bis Ende des 19. Jahrhunderts: München, R. Oldenbourg, XI + 868 p.

Zoback, M.L., and Zoback, M.D., 1989, Tectonic stress field of the continental United States: Boulder, Colorado, Geological Society of America Memoir 172, p. 523–539.

Zoback, M.L., and Zoback, M., 1997, Crustal stress and intraplate deformation: Geowissenschaften, v. 15, p. 116–123.

Zoback, M.L., Zoback, M.D., Adams, J., Assumpcao, M., Bell, S., Bergman, E.A., Bluemling, P., Brereton, N.R., Denham, D., Ding, J., Fuchs, K., Gay, N., Gregersen, S., Gupta, H., K., Gvishiani, A., Jacob, K., Klein, R., Knoll, P., Magee, M., Mercier, J.L., Mueller, B.C., Paquin, C., Rajendran, K., Stephansson, O., Suarez, G., Suter, M., Udias, A., Xu, Z.H., and Zhizin, M., 1989, Global patterns of tectonic stress: Nature, v. 341, p. 291–298.

ENDNOTES

[1]I quote below two paragraphs from Popper (1974, p. 977, italics Popper's), in whose spirit this book was written:

> The great scientists, such as Galileo, Kepler, Newton, Einstein, and Bohr (to confine myself to a few of the dead) represent to me a simple but impressive idea of science. Obviously, no such list, however much extended, would *define* scientist or science *in extenso*. But it suggests for me an oversimplification, one from which we can, I think, learn a lot. It is the working of great scientists which I have in mind as my paradigm for science. Not that I lack respect for the lesser ones; there are hundreds of great men and great scientists who come into the almost heroic category.
>
> But with all respect for the lesser scientists, I wish to convey here a heroic and romantic idea of science and its workers: men who humbly devoted themselves to the search for truth, to the growth of our knowledge; men whose life consisted in an adventure of bold ideas. I am prepared to consider with them many of their less brilliant helpers who were equally devoted to the search for truth—for great truth. But I do not count among them those for whom science is no more than a profession, a technique: those who are not deeply moved by great problems and by the oversimplifications of bold solutions.

It is science in this heroic sense that I wish to study. In that regard, the book reflects my bias, but I agree with Lord Russell that "a man without bias cannot write interesting history—if, indeed, such a man exists. I regard it as mere humbug to pretend to lack of bias. ... Since I do not admit that a person without bias exists, I think the best that can be done with a large-scale history is to admit one's bias and for dissatisfied readers to look for other writers to express an opposite bias. Which bias is nearer to the truth must be left to posterity." (Russell, 1998, p. 465–466).

[2]Stewart (1990, p. 119–121) analyzed this debate in a context of the sociology of science. I regret that Professor Beloussov died before he, Kevin Burke, and I had a chance to meet and analyze that analysis of our exchange.

[3]From the Greek κῦμα, meaning anything swollen. King (1961, p. 1) translates κῦμα as a wave or undulation. *Cymatogen* and the associated term *cymatogeny* (King, 1961) are uncommon terms. King (1961 p. 1) defines cymatogeny as a phenomenon "wherein differential movement of the surface takes place in the vertical sense with a production of smooth arching amounting to thousands of feet though there is little or no deformation of rock strata by folding or faulting. The earth's surface is thrown into gigantic undulations or waves, sometimes measuring hundreds of miles across, and hence the name 'cymatogeny' ... is preferred for this type of deformation. Cymatogeny is the 'undulating ogeny.'" I have not seen cymatogen or cymatogeny used in the continental European or in the Russian literature, nor do they appear in their geological dictionaries that I have been able to consult. However, the *Glossary of Geology* (of the American Geological Institute) has cymatogen in all of its four editions (1972, 1980, 1987, 1997), but its predecessor, the *Glossary of Geology and Related Sciences—with Supplement*, published in 1960 (second edition) does not. Neither the Challinor (1967) dictionary nor *The Concise Oxford Dictionary of Earth Sciences* (Allaby and Allaby, 1990) include the terms. They are similarly missing from the *Spisok Tektonicheskikh Terminov (List of Tectonic Terms)* prepared as a working document by A.A. Bogdanoff in 1961 following the 21st Session of the International Geological Congress in Copenhagen to produce a hexalingual tectonic dictionary. The second iteration of the same document (anonymous), prepared for the 22nd Session of the International Geological Congress in 1964 in India does not have them either. Neither the *International Tectonic Dictionary—English Terminology* (Dennis, 1967), nor any of the published volumes of the *International Tectonic Lexicon* (Dennis et al., 1979; Dennis and Murawski, 1988) has cymatogen or any associated term as an entry. Although the 1984 *English-Chinese Dictionary of Geology* has both cymatogenic and cymatogeny (p. 232), neither its smaller successor, the *Yin Han Zhong He Di Zhi Xue Chi Hui* (English-Chinese Comprehensive Dictionary, Science Press, Beijing, 1985), nor its German-Chinese counterpart (*Deutsch-Chinesisches Wörterbuch der Geologie*, Science Press, Beijing, 1987) has those terms or any of their derivatives. In none of the geographical dictionaries and lexicons have I come across either term. The second edition of the *Oxford English Dictionary* (1989) does not have an entry under cymatogen or cymatogeny.

[4]The following are publications that I have found useful to introduce the uninitiated into the history of science in antiquity and in the Middle Ages in general. The subjects of the history of science, and the philosophy of science (which is inseparable from the history), in antiquity and in the Middle Ages are so vast that I can only scratch the surface for the beginner here. Because I do not discuss extra-European ideas for those times (except for the Muslim world in the Middle Ages), I do not cite books and papers that are relevant to them. For history of science in general, refer to Wightman (1950), Singer (1959[1996]), Gillispie (1960), Dampier (1961), and Mason (1962). An excellent survey of man's intellectual development in the light of science and exploration is Boorstin's (1983) wonderfully readable book. For science in antiquity and the Middle Ages, refer to Cuvier (1841), Sarton (1955), Taton (1963), and Lindberg (1992); for science in antiquity: Jürss (1982), Furley (1989) and Clagett (1994); and for Greek science: Sarton (1952, 1959), Heidel (1933), Rey (1933, 1939, 1946), Farrington (1944, 1949), Sambursky (1956[1987], 1959[1987], 1962[1987]), Cohen and Drabkin (1958), de Santillana (1961), Lloyd (1970, 1973, 1991), Vernant (1982), Walzer and Frede (1985), and Furley (1987). Peter Fraser's monumental *Ptolemaic Alexandria* (1972a, 1972b, 1972c) contains a comprehensive review of the scientific literature generated in Alexandria when it was the foremost center of scientific research in the world. As Fraser reviews not only science but all aspects of learning and social life, his book is indispensable for those who wish to understand the development of antique and medieval science after the third century B.C. For the pre-Socratic science in general, refer to Baccou (1951), Freeman (1949, 1962), Barnes (1981, 1987), Kirk et al. (1983) and Mansfeld (1987). For Roman science, see Thorndike (1923a), Stahl (1962), and French and Greenaway (1986). For medieval science, refer to Stahl (1959), Crombie (1961), Grant (1971[1977]), Dales (1973), and Steneck (1976). For the history of atomism and materialism in antiquity, which is relevant for the history of the earth sciences especially owing to Democritus' influence on Plato's and Lucretius' pronouncements on geological phenomena, see Bailey (1928) and Nizan (1938). Byzantine science is generally given scanty treatment by most authors writing about medieval science. For an account of Byzantine civilization that includes a summary of Byzantine science, see Bréhier (1950). Vogel (1967) presents a more extensive summary of Byzantine science. Volumes II and III of Sarton's annotated catalogue (Sarton 1931a, 1931b; 1947; 1948) include perhaps the most extensive information anywhere about Byzantine science, but that work, intended strictly for the specialist, is not always reliable for the earth sciences (e.g., he does not mention Buridan's pseudo-isostatic theory, and Sarton's entry on Maximus Planudes includes no reference to his work on the Ptolemaic maps, for which Planudes is perhaps best known). Sarton's works contain references to more specialized literature. Tekeli (1975) discusses the role of Byzantine science in the birth of modern science. She does so mainly on the basis of Sarton's data, but in an uncritical way and thus preserves many of Sarton's errors and omissions. For the history of Muslim science, the most important phase of which is confined to the Middle Ages, see Rashed (1996a, 1996b, 1996c).

For the history of geology during the same period, our sources are meager. An almost universally neglected source of information by historians of geology is the history of geography (e.g., Vivien de Saint Martin, 1873, 1874; Peschel, 1877; Günther, 1904[1978]; Kretschmer, 1912; Livingstone, 1992; Claval, 1995; Özgüç and Tümertekin, 2000), which is richly documented and is very relevant to the topic treated in this book. In some of the more recent histories of geography, however, geography itself is regrettably sacrificed to a social history with a view to providing a historical context for the development of geography. Livingstone's book is a prominent example of this genre (and contains an excellent bibliography including other representatives of the genre), which I find less useful than the traditional histories of the subject in teaching us how actually the thinking on geographical problems developed. An excellent recent overview of ancient geography that provides abundant literature references is Olshausen (1991); also, see Stückelberger (1988). For the mythical geography of the Greeks, which is important as the source of many later scientific ideas, see Kerényi (1951, 1958) for a superb summary of the mythology; also see Wilamowitz-Moellendorff (1955a, 1955b) and Gantz (1993a, 1993b). For the associated geography, refer to Vater (1845a, 1845b), Buchholz (1871[1970]), Lang (1905), Delage (1930), Rousseau-Liessens (1961, 1962, 1963, 1964), Wolf and Wolf (1983), and Richer (1994). Rousseau-Lissens' interpretations are unorthodox, but not his documentation. For mythology-landscape relationships, see Schmidt (1981), Sichtermann (1984), and Frazer (1919, p. v–vi). Dicks' (1970) scholarly book on early Greek astronomy up to the time of Aristotle has much that is relevant both to geography and to the understanding of the pre-Socratic science. I have found the following useful to trace the origin and evolution of ideas in the scientific geography in antiquity in general: Bunbury (1879a, 1879b; reprinted in 1959 by Dover with a new introduction by W.H. Stahl), Tozer (1935), Thomson (1948[1965]), and Aujac (1975). Kretschmer

(1892, ch. 1) presents a shorter, but well-documented summary. Fritscher's two good reviews (2001a, 2001b) cover the history, influence, and historiography of antique geography to the end of the nineteenth century and provide excellent bibliographies so typical of the Pauly tradition. For Grecian geography, in addition to the general histories of antique geography just cited, see Berger (1903[1966]), Warmington (1934), Heidel (1937), Dicks (1960), and Pédech (1976). For Roman geography, refer to Columba (1893), K. Miller (1916), and Klotz (1931). For medieval geography, the best sources are still de Santarem (1842, 1849a[1985], 1849b, 1850, 1852, 1855), Marinelli (1884), Lelewel (1850, 1852a, 1852b, 1852c, 1857), Kretschmer (1890, 1892, ch. 2, 3, and 4), Beazley (1897, 1901, 1906), Wright (1925), and Kimble (1938). For Byzantine geography, in addition to the general sources just cited, see Delatte (1929/30).

The history of cartography is commonly an invaluable (and also underused) source of information concerning the history of geographical and geological ideas. For the history of cartography in ancient and medieval Europe and the Mediterranean, the best one-volume source that I can recommend is Harley and Woodward (1987), which also contains an excellent bibliography. For Greek and Roman maps, Dilke (1985) is very instructive. For concise presentations of medieval maps, see Lelewel (1850), Kretschmer (1892), Kimble (1938, ch. VIII) and Harvey (1991). Sezgin (1987a and 2000a, 2000b, and 2000c) discusses and illustrates many medieval maps of European make. Konrad Miller's great publications on medieval maps address the professional history of cartography (for an annotated list of his publications and a summary of Miller's career, see the German-English bilingual publication by Borodajkewycz, 1936). Neither the late medieval geography nor its cartography in Europe is completely intelligible without a knowledge of the contributions from the Muslim world. There is regrettably no satisfactory treatise summarizing the Muslim physical earth sciences during the Middle Ages. Anyone wishing a quick survey must make do with what is contained in Duhem (1958a) and Miquel (1973, 1975, 1980). Miquel deals with physical geography and to some extent geology only as frameworks for the human milieu, which is his main interest; he does such an outstanding job, though—because, as Hentschel, 1969, p. 102, once commented, "total involvement" is an essential part of *géographie humaine*—that his book ably supplants the meager historiography of the Arabic-Muslim physical earth sciences. His summary chapter in Rashed (1996c, p. 796–812) is, however, regrettably far too short with no references or notes. Maqbul Ahmad's (1995) and Şeşen's (1998) recent volumes are unsatisfactory for physical geography and cartography. Sezgin's (1987a) slim volume on cartography is, as he himself says, more a prodromus than a mature work, yet it is very instructive, especially for the beginner. Harley and Woodward (1992), dealing with the history of the cartography in traditional Islamic and South Asian societies, lacks reliable authors on Islamic cartography. Sezgin's recent volumes (Sezgin, 2000a, 2000b, 2000c) represent a major leap forward in the history of Arabic-Muslim mathematical geography and cartography, thoroughly eclipsing anything that had come before and fulfilling the promise of his 1987 volume, yet these volumes contain little on physical geography. One should keep Sezgin (1971 {Alchemy, Chemistry, Botany, Agriculture}, 1974 {Mathematics}, 1978 {Astronomy}, 1979 {Astrology, Meteorology, and related subjects} and 1995 {general indices of volumes I–IX}) at hand while reading his volumes on geography. Especially relevant is the introduction to Sezgin (1978), in which he elaborates his views on the origin of Arabic natural sciences. For an abstract of his 2000 book from his own pen, see Sezgin (2000d). For the meager contribution of the Ottoman Turks to geography, see also Taeschner (1923), Akyol (1940), Türkay (1959), Adivar (1982) and the annotated bibliography by İhsanoğlu et al. (2000a, 2000b). In the absence of an authoritative summary on Arabic-Muslim physical geography and geology, all I can do is to refer to the most recent publication catalogues of the Institute of the History of the Arabic-Muslim Sciences of the Johann Wolfgang von Goethe University in Frankfurt am Main. The reader will find there lists of facsimilies and reprints (many that have useful introductions) both of the original works and of the secondary material concerning the originals and their authors, dating from the eighteenth century to the fifties of the twentieth century (Anonymous, 1999, 2000, 2003). Regrettably, the volumes of facsimiles and reprints dealing with Arabic-Muslim geology have just started coming out and the ones so far published are all devoted to mineralogy with few papers on lithology (see Anonymous, 2003).

Antique natural science is treated in Pennetier (1911), Stückelberger (1988), Greene (1992), and French (1994), although French denies that there was any science in antiquity, a view which is difficult to share. Rey (1942) calls even the pre-Greek mathematics and nature observation "science," but with such qualifications (see especially p. 439–442) as to oblige me to refuse that appellation to them. The same applies to Schmöckel's (1962, p. 118–121) "Sumerian science," Johnson's (1891, p. 138) "Egyptian science" and "Chaldaean science" (Johnson, [1890], e.g., p. 82), and Thompson's (1936, p. xiii ff) "Babylonian science." Von Soden (1985, ch. XI) also wrote about "Sumerian and Babylonian science," despite that he appreciated what he considered under that heading "we would scarcely be prepared to regard as sciences. It could thus be considered sensible to set the word 'science' in the following treatment in quotation marks; I would not want to do this, however, since this too frequently signifies a degree of denigration which would be inappropriate here." (p. 153). For science/non-science boundary in the past, see also Sarton, 1927, p. 8–10. For the medieval times, see especially Pennetier (1911) and Crombie (1961).

The only book known to me that is devoted exclusively to ancient geology is that by Rebrik (1987) but it deals more with mining than geology. An older monograph by von Lasaulx (1851) and the unpublished doctoral dissertation by Bouillet-Roy (1976) deal with the geology of the Greeks and the Romans only. Fritscher (2001c) is a more recent review of antique geology and its influence and historiographic tradition to the end of the nineteenth century. Schvarcz (1862, 1868) treats only the Greek geology. For a discussion and examples of Greek lapidaries, see Halleux and Schamp (1985). The work of Payne (1992) is devoted to the sources of Greek geology but is inadequate, omitting many important anthologies and secondary sources. Gilbert (1907[1967]) has much that pertains to antique Greek geology. For aspects of Roman geology, see Clarke and Geikie (1910). Adams (1938) and Duhem (1958a, p. 79–323) are still the best treatments of ancient and medieval geology, despite the fact that both are seriously inadequate. De Lorenzo's book (1920) on Leonardo da Vinci's geology presents a fine review of the geology that preceded Leonardo, reaching down into the mythologies of Mesopotamia, the Greeks, Scandinavians, and Indians, although, regrettably, without detailed references to the literature (the few, but very useful, references he cites are scattered in the body of his text). Adrienne Mayor's recent book (2000) on "antique palaeontology" is excellent and contains valuable documentation of mythology–legend–observation relationships and of the mythological roots of science. Most of the modern histories of geology are also deficient in the historiography of ancient and medieval geology, to which they characteristically devote about 10% or less of their total space. The best recent treatment is in Ellenberger (1988, p. 11–69 for antiquity; p. 71–110 for the Middle Ages). Another fair treatment is in a small but excellent book by Cailleux (1968, p. 7–43). The Byzantine manuscript edited and published by Delatte (1929/1930) has sections on the origin of thermal waters and earthquakes. For more, see the relevant entries in Sarton's volumes under "Eastern Christianity" (1927; 1931a, 1931b; 1947; 1948).

[5]For concepts of *geodynamics* and *lithosphere,* as here understood, see: *elementary* (also for the layman): Heather (1979), Allègre (1988), Frisch and Loeschke (1990), Şengör (1991a, 1991b), and Miller (1992); *elementary undergraduate texts*: Wyllie (1976), Condie (1989), Kearey and Vine (1996), van der Pluijm and Marshak (1997), and Marshak (2001); *advanced textbooks*: Wyllie (1971), Le Pichon et al. (1973), Cox and Hart (1986), Khain and Lomize (1995), and Turcotte and Schubert (2002). The following anthologies include the most important papers (or their popular versions) that originally introduced the modern incarnations of these concepts: Bird and Isacks (1972), Cox (1973), Schoenenberg (1975), and Wilson (1976). For two introductory bibliographies, see Kasbeer (1973, 1975). Also see, as supplement to the anthologies cited, *Bulletin of the American Association of Petroleum Geologists*, v. 56, no. 2 (February, 1972).

For the development in time of the structures described in this book and their tectonic environment, see Windley (1995; with 2200 references, it is an excellent handbook of historical tectonics) and Condie and Sloan (1998).

Historians who have no geological background will do well to skim the two following books to familiarize themselves with the *time and length scales* of the processes and structures dealt with in this book: Snead (1980) and Kukal (1990).

[6]Even if I included the deformations associated with extra-terrestrial sources (i.e., "exodynamic" ones, such as those resulting from bolide impact, as discussed in Melosh and Ivanov {1999}, the classification offered here would not have been affected. All deformations that cut the lithosphere (e.g., within-crater and near-crater field) would be categorized as short wavelength; deformations that only flex the lithosphere (away from the crater and later isostatic adjustments including the entire crater plus environs) would be categorized as large wavelength category. (See the model of the Sudbury impact crater in Melosh and Ivanov, 1999, fig. 2).

[7]Thus, structures of small wavelength cover the following scales of geologic bodies, as defined by Turner and Weiss (1963, p. 15–16): *submicroscopic, microscopic, and mesoscopic*. Some small wavelength structures are also included in

the macroscopic scale of Turner and Weiss (e.g., some large nappes in mountain belts). That is why it is useful to employ yet a larger category, that of *megascopic structures*, which Turner and Weiss (1963) do not employ. The term *mega* (from the Greek μέγας, meaning big, great) has been used to imply a scale covering parts of or entire continents and ocean basins (e.g., Bucher, 1950; Johnson and Smith, 1970; van der Pluijm and Marshak, 1997, p. 9), although others have used it as a synonym for macro (e.g., Holmes, 1928, p. 152 {although Holmes uses mega- and macroscopic simply as the opposite of microscopic}; Brock, 1972, p. 11; Tomkeieff, 1943, 1983, p. 565; Jackson, 1997, p. 383 and 397). I suggest to reserve the megascopic category for structures whose wavelength spans thousands of kilometers. At such a scale, the structures of large wavelength are mostly megascopic structures. Only some of their smallest members may fall into the larger end of the macroscopic structure category.

[8]Terminology for such saucer- or bowl-shaped basins include: Umbgrove (1947, p. 44): "*discordant basins*"; Klemme (1975; also Klemme's *type-I basins*; see p. 32–33) and Bally and Snelson (1980): "*cratonic basins*"; Harding and Lowell (1979): "*basement downwarps*" and "*sags*"; Bois et al. (1982): "*intracratonic basins*"; Green (1985): "*interior basins*"; Einsele (1992, p. 4–5, 459–460): "*oceanic and continental or interior sag basins*"; and Debelmas and Mascle (1993, p. 81 ff.): "*basins proper*," which is equivalent to Pavlov's (1903) "*syneclises*" (also see Shatskiy, 1940[1964], Dennis, 1967, p. 147–148, and Schmidt and Hoppe, 1971), Haarmann's (1930, p. 13–14, 41 ff) "*geodepressions*," and Kay's (1944; 1947; 1951, p. 20 ff., 107) "*autogeosynclines*."

Syneclise is a little-known term outside Russian geological circles and former socialist block countries that were strongly influenced by the Russian geological tradition (for instance, the term is not in the second edition of the *Oxford English Dictionary*, 1989). Following its introduction by Pavlov in 1903, Schatski (1940 [1964] and 1945[1964], p. 290–292; for an English translation of the cited passages in the 1945 paper, see Shatskiy, 1967, p. 257–267) discussed the concept at length, pointing out equivalent concepts and analogous structures. So far as I am aware, the first somewhat detailed discussion of syneclises in a western European language was given in the German translation of a collection of his articles on the comparative tectonics of old platforms (Schatski, 1961, p. 125–130; also see Ashgirei, 1963, p. 364–365). Schatski (1961, p. 125; for a parallel passage, see 1967, p. 259) wrote that he designated "as syneclises very shallow bendings with hardly noticeable dip of beds along the flanks (few decimeters to at most 3 or 4 meters per 1 km). These bendings always cover large platform regions and commonly exhibit oval, round or angular, in places very irregular forms." (Schmidt and Hoppe, 1971, do not properly cite this quotation. They ascribe it to Schatski and Bogdanow, 1958. Such a publication does not exist.) But, both Shatskiy and other Soviet geologists have come to apply the term to structures of diverse tectonic settings and origins (most of which have been recognized to be so long after Shatskiy's death!) such as intracratonic basins, trapped oceans within continents (such as the Pre-Caspian depression: Burke, 1977; Şengör and Natal'in, 1996). Consequently, the term has lost its usefulness. It is now rapidly fading away from the geological literature.

Claiming equivalences between various technical terms across different theories has inherent dangers and should be made with great caution.The historian of geology who is not a geologist is particularly prone to err in such claims or in their critical evaluation. For instance, equating Haarmann's geodepressions with Kay's autogeosynclines provides a fine illustration of this danger. For Haarmann, all major depressions were a result of gentle oscillations of the crust. Thus, for him, there was no fundamental tectonic difference between the Michigan Basin and the Atlantic Ocean. That was not so for Kay. He would agree that the Michigan Basin, an autogeosyncline according to his scheme (Kay, 1944, 1947, 1951), formed through a mechanism not so different from the way Harmann claimed his geodepressions formed. Here the two terms would be perfectly equivalent. But Kay would not attribute the same mechanism as forming the Atlantic Ocean, which, for him, was a subsident craton, at least before the late 1960s (see Kay, 1947, p. 1289, where he points out, following Stille {1936} that orthogeosynclines lie between "cratons, whether higher continental or lower oceanic"; Kay and Colbert {1965, p. 713} repeat essentially the same after two decades). Although Stille's and Kay's views seem coincident on nearly every point concerning the tectonics of basins (after all, the latter was inspired by the former: Kay, 1942, p. 1641 ff.; 1967, p. 312–313; 1974, p. 377 ff.), I do not think Kay would have agreed with Stille's insistence that all structures, including the normal fault-bounded rifts, are ultimately compressional. Therefore, the "equivalences" that I list above must be taken to indicate broad analogies of structures and mechanisms across different tectonic theories and/or schools of thought.

[9]Both problems of tectonic inheritence from fault-dominated older structures and the history of Phanerozoic faulting in these structures are more complex than implied in these books and papers. For some of the complexities, which cannot be discussed in this book, see Marshak and Paulsen (1997), McBride et al. (1997), Nelson et al. (1997), and Potter et al. (1997). Despite these fault-related complications, the *dominantly faultless* Phanerozoic subsidence of the major U.S. interior basins (such as Michigan) is clear.

[10]Distinction of foredeep from backdeep naturally rests on the distinction of foreland from backland or hinterland. Only with plate tectonics has a clear, generally accepted distinction become possible, namely that a subducting plate (regardless of its composition) always carries the foreland, and the overlying plate (regardless of its composition) always represents the hinterland. Before plate tectonics, the distinction was based on diverse, sometimes mutually contradictory criteria such as structural vergence, migration direction of orogenic deformation, polarity of eugeosycline/miogeosyncline couples, character and volume of magmatism, "dominant direction of movement" in an orogenic belt (e.g., Lees, 1952, p. 4), and the like. Even the man who introduced the foreland/hinterland distinction himself became confused about how to use them where a mountain belt displayed a structural symmetry at the crustal level. See Suess, 1885, p. 775, where the Andean foreland and the foreland of the "Asiatic structure" was identified to be the Pacific and in agreement with the concept that the Rocky Mountains—considered a part of the "Asiatic structure" and transitional to the Andes—were an area of backfolding (Suess, 1909, p. 717); then in Suess (1909, p. 535), there is talk of the Brasilian foreland of the Andes! Although Suess considered the Andean structure as separate from the Asiatic structure, to which he attached the Rockies, the parallelisms he drew between the structure of the Cordillera of the North and South Americas make clear that he considered both to have similar structures. Only in the last volume of the *Antlitz* (Suess, 1909) did he emphasize that the west-vergent part of the Andes was not exposed. That is why Kober (1921, p. 164–165, especially fig. 29) later assumed a sunken west wing of the Andean orogen.

In the Americas, misuse of the term foreland has been universal and at least in large part owing to James D. Dana's and Hans Stille's cumulative influence. See, for example, the inappropriate, incorrect, and obviously second-hand historical references in Dorobek and Ross (1995). Johnson and Beaumont (1995), clearly aware of the importance of the fore- and backdeep distinction, went so far as to invent the internally inconsistent terms *pro-foreland basin* vs. *retro-foreland basin*. They seem to think that foreland basin is a term for any asymmetric flexural basin adjacent to an orogen! In pre-plate tectonic days in the twentieth century, only Stille developed an internally consistent set of criteria for separating forelands from hinterlands, which when viewed in retrospect with plate tectonic spectacles, appears generally sound (see especially Stille, 1940, p. 614–616; 1948, p. 26–31). He insisted that structural vergence is an unreliable indicator of the location of the foreland. Instead, Stille pointed out that migration of folding within one *folding era* (such as Caledonian, Variscan, or Alpine according to his theory of episodic and simultaneous world-wide orogeny {Stille, 1940, p. 653}) and within one geosynclinal system almost always occurred in the direction of foreland. Moreover, miogeosynclines always lie on the foreland side of a major orthogeosynclinal system.

Where Stille did go wrong was his advocacy of the two-sidedness of all orogenic belts. That is why he commonly too readily identified a legitimate area of backfolding as one of forefolding against a different foreland (e.g., the southerly backfolding of the South Alpine molasse between Como and Varese {Stille, 1924, p. 270 ff}. Suess {1875, p. 86–95} had interpreted the southern Alps already as an exceptional thrust in the opposite direction—to the generally north-vergent structure of the Alps; 10 years later, he interpreted it as backfolding and backthrusting Suess {1885, p. 852}. Modern tectonic interpretations of the region between Como and Varese follow Suess {e.g., Roeder, 1992, with minor and insignificant complications; Bernoulli et al., 1989, 1993; but again see Hsü and Briegel, 1991, p. 148 ff; Hsü, 1995, p. 119 ff.})

[11]For a remarkable exception, but in a sedimentology textbook, see Friedman et al. (1992, especially p. 643–644 and the three items under *epeirogeny* in the "Glossary"). Both geographers and sedimentologists have traditionally been more interested in epeirogenic structures and movements than structural geologists have been (e.g., Stille, 1919, p.165).

[12]Grove Karl Gilbert (1843–1918) is one of the grandest figures in the history of geology and one with great relevance to the subject of the present book. For an overview of Gilbert's life, Pyne (1980) is still the most comprehensive account,

which may be consulted also for references to previous accounts (the best being William Morris Davis's {1927} great U.S. National Academy of Sciences biographical memoir, which is very valuable owing to material that Davis utilized and that since has perished). The papers in Yochelson (1980) pertain to the multifarious facets of Gilbert's amazing research activity. An excellent historical study of Gilbert's great Henry Mountains work is provided by Hunt (1988), a modern master of the Henry Mountains geology (with an edition of Gilbert's relevant field notebooks). Hunt also published a most valuable edition of Gilbert's Lake Bonneville notebooks (Hunt, 1982). For Gilbert's geomorphological studies in general, see Chorley et al. (1964, ch. 28). For Gilbert's epistemological views, see G.K. Gilbert (1896). DeFord's (1981) Gilbert entry in the *Dictionary of Scientific Biography* is disappointingly brief and provides only limited sources. Also see Crossette (1946, p. 67–69), de Margerie (1952, p. 263–268), and Keller (1999).

[13]From ἤπειρος and γένεσις (meaning an origin, source, a productive cause) from the root γεν-. G.K. Gilbert (1890, p. 340) translates ἤπειρος simply as continent, but that word originated before Homeros' time when there was hardly the concept of a continent as it was understood at Gilbert's time. The Liddell and Scott *Greek-English Lexicon* renders ἤπειρος as "*terra-firma, the land*, as opposed to the *sea*." The award-winning Greek scholar (Pyne, 1980, p. 9) would not have chosen ἤπειρος out of ignorance. I suggest that, in Gilbert's mind, ἤπειρος meant exactly what *Land* meant in Suess' mind when he coined Gondwana-*Land*. For many years I have protested against claims that *Land* in German meant *country* in Suess' mind, and thus, Gondwana-Land was a redundant construction because *wana* in Dravidian allegedly meant kingdom, country (Şengör, 1983, 1991c; *wana*, in Dravidian, actually means *forest, woodland*). Both *Land* in Suess' mind and ἤπειρος in Gilbert's meant land *and therefore* continent (see especially Gilbert's presidential address to the Geological Society of America where he explains a change of view owing to recent data from ocean floors: "It is at once evident that … we must substitute for the continents, as limited by coasts, the continental plateau, as limited by the margins of the continental shoals" {Gilbert, 1893, p. 181}). Suess viewed the oceans basins as tops of crustal cones that subsided along faults to create the basins (see below), whereas Gilbert originally thought of them as gentle synclinal troughs between gentle anticlinal platforms forming the continents (following his older compatriots James Dwight Dana and Joseph Le Conte; see below). In both cases, the structure of the land was the cause of its being land and not sea. It is clear that Gilbert had in mind only the continent-forming part of the process when he coined a term for it. For a history of ideas on continents and oceans from a geographic-tectonic viewpoint, see Batyushkova (1975).

[14]From ὄρος (meaning a mountain, hill; also a boundary, landmark if accented differently) and γένεσις (meaning an origin, source, a productive cause), from the root γεν-. Orogeny was a term that had long been in use in Europe when employed by Gilbert (e.g., Boué, 1874, p. 262; also see Şengör, 1990, p. 8–11).

[15]From διαστροφή (meaning distortion).

[16]I write "apparently" because diastrophism occurs in none of Powell's writings that predate Gilbert's Bonneville monograph, despite numerous instances of later authors misreferencing a number of Powell's papers as the source of this term. I have come to think that Powell must have suggested this term in conversation or in an unpublished document and that Gilbert must have adopted it from there. The only earlier usage with which I am familiar is by John Milne who, in commenting on the kinds of Japanese earthquakes to *Nature*, wrote, "Others again are compounded of direct and transverse motions, and might be therefore called diastrophic." (Milne, 1882).

[17]Ellenberger (1989). For example, witness Stille's (1919, p. 176–185) justified criticism of Haug's abuse of the two terms or Bucher's "progressive confusion" of Stille's clear statements in Bucher (1933, p. 403 ff.). For one piece of evidence of modern confusion, see Hohl (1985, p. 169–173).

[18]From which derives the Latin *falx* (a sickle, reaping hook, a pruning hook, scythe) and *falco* (a falcon, so-called owing to the bent-looking beak and the claws). Also the Latin verb *flecto*, which means to bend, bow, curve, turn, and turn round. This is the root of *flex* in English. Falcogenic structures indeed flex the lithosphere.

[19]From this we have in medieval Greek, κοπτερός (meaning sharp, like a knife).

[20]Clarence Edward Dutton (1841–1912), U.S. Army officer, whose diverse interests of "leisure" ranged from geomorphology, structural geology, volcanology, and theoretical geophysics to seismology, is the man to whom we owe, among other contributions, the term and the concept of isostasy. For information on Dutton's life, see J.S. Diller (1911, 1913), Becker (1912), Stegner (1935, undated[1936], 1937, 1981, 1953, especially p. 158–174), Crossette (1946, p. 44–46), and Anderson (1977). For an extensive analysis and summary of Dutton's work on the region of the Colorado River in Utah and Arizona, see de Margerie (1954, p. 627–685). For his contributions to geomorphology, see Chorley, et al. (1964, ch. 29).

[21]I consider dynamic uplift by plumes, as hypothesised by White and McKenzie (1989), to be one of the thermal isostatic processes.

[22]I here follow my habit of using Stille's (1920) useful classification of structures generated by orogenies: *germanotype* for the domainal, mainly brittle, non-penetrative structures in extra-orogenic areas; *alpinotype* for those formed within orogenic zones proper by penetrative to semi-penetrative deformation. In English-language literature, these terms are commonly rendered as *paratectonic* for germanotype and *orthotectonic* for alpinotype (see, for example, Dennis, 1967, p. 154; Jackson, 1997, p. 18, 267), but Dewey's improper usage (especially in Dewey, 1969a, 1969b) of orthotectonic for subduction-related orogens (cf. Jackson, 1997, p. 454) and paratectonic for collision-related orogens, has completely blurred their original meanings. Although Dewey's usage never gained popularity, at least not outside Britain, I avoid the terms ortho- and paratectonic structures and return to Stille's original terms not to give rise to any possible confusion.

[23]Here I mean the history of the myth-to-science transition among only the Greeks because it was only in the Greek culture that such a transition occurred directly (in addition to the publications listed in endnote 4, see Cornford, 1912[1991]; Frankfort et al., 1946[1949]; Snell, 1946; Fairclough, 1963; Colli, 1981; Blumenberg, 1987). In all others, including the great riverine cultures of Asia, it occurred under the direct or indirect influence of the Greeks—or it did not occur at all, as, for instance, in the pre-Columbian cultures of the Americas. American, Chinese, and Indian cultures do indeed display exceptions in terms of individuals or small groups inventing a way of scientific thought. Von Humboldt mentioned the flicker of individual intellectual efforts in societies as yet untouched by civilization on the basis of his own experience along the Orinoco: "But also in an uncivilized state, one recognizes with surprise here and there individual traces of the awakening of the self-motivated intellectual power" (von Humboldt, 1858a, p. II), but they lack the tradition of continuity in science that the Greek culture has enjoyed in Europe from the beginning to the present, even through the darkest times of the Middle Ages (e.g., Ariew, 1985, p. xix: "[*Pierre Duhem*] is said to have single-handedly destroyed the myth of the 'scientific night' of the Middle Ages"). For readers who wish to see for themselves, I recommend the following sources for the Chinese and the Indian cultures, with which I am somewhat familiar:

General: Suhr (1959, 1960), Bruun and Kalland (1995). For mythology, see Bonnefoy (1991a). For science, see Rey (1942). For comparison of the oriental cultures with the European culture with respect to points important for the development of natural science, see Bahm (1988), Hsü (1994), and Hofstede (1996).

China: For the mythology, Palmer and Zhao (1997) offer a convenient summary. Joseph Needham's monumental multi-volume treatise *Science and Civilisation in China* (1954, …; Cambridge Univeristy Press) is the fundamental work for understanding the development of science and scientific thinking in China, but it addresses the specialist. Of this source, the student of the history of geology should perhaps consult volumes 1–3 (Needham, 1954, 1956, 1959). Needham has summarized the history of science in China in a very readable chapter in Dawson (1964). For other summary accounts of the history of science in China, see Institute of the History of Natural Sciences (1983), San (1984), Zhong (1984), and Temple (1999). For cosmography and cartography, see Yee (1994a, 1994b, 1994c, 1994d, 1994e) and Henderson (1994). For a popular summary, see Smith (1996). For the history of Chinese travel accounts and travelers, see Mirsky (1964), Pierson and Wei (1992), Levathes (1994), and Strassberg (1994). Levathes' account of the great voyages of the Muslim admiral Zhang He is particularly revealing from the viewpoint of the lack of a scientific research tradition in China and its reasons (cf. also Hsü, 1994). The best summary account of the history of the philosophy in China is still Fung (1976). Black's (1989) study of man and nature in the philosophical thought of the great seventeenth century Chinese thinker Wang Fu-chih (1619–1692) illustrates the effects both of the absence of a scientific tradition on philosophical thinking in China and of the reception by Chinese thinkers of European science introduced by the Jesuits. Boorstin (1983, *passim*) has many useful things to say about the development of science, technology, and geography in China in comparison with that in the west.

India: General: see Embree (1991). For the mythology, Ions (1983) gives a fine summary. For the history of science, see Kirfel (1920), Rahman (1982, 1984), Chattopadhyaya (1986, 1991), and Kirthisinghe (1993). For Indian philosophy, Das Gupta's multi-volume treatise is strictly for the specialist; instead, see Challaye (1956). For the history of materialism (*lokayata*: literally "that which is found among the people") in India, see Chattopadhyaya (1992). Stcherbatsky's (1930[1962]a, b) immortal *Buddhist Logic* is indispensible for any student of the history of science in India. For Indian cartography and cartographic traditions (with much mythologic information), see Gole (1989) and Schwartzberg (1992a, 1992b, 1992c, 1992d). Deepak Kumar's (1995) *Science and the Raj* has an excellent introductory chapter entitled "Science in a Colony: Concept and Contours" that compares and contrasts European science and attitude to science with their counterparts in the Indian cultures before the arrival of the Europeans.

I do not include the Arabic-Muslim science here because it properly belongs in the Greco-European tradition as explained in endnote 4. The pastoralists and the hunters living north of the great water cultures of Asia are too little known in terms of those components of their thinking and cultural legacy that might be termed scientific to be considered here, despite such early comprehensive studies as Adler (1910) and other works on their cosmology (e.g., Hilmi Ziya [Ülken], undated [1932]; Esin, 1979). We do possess some remarkable products from them, however, such as the eleventh century world map of Mahmud of Kashgar (for a facsimile, see Kaşgarlı Mahmud, 1990, folios 22–23; for the most authoritative recent translation, see Dankoff and Kelly, 1982, foldout map between p. 82 and 83; for general description and assessment, see Herrmann, 1935; on nature description by Mahmud of Kashgar, see Brockelmann, 1924). Mahmud of Kashgar's world map, however, was so heavily influenced by the Arabic-Islamic (and therefore by the Greco-European) culture (e.g., Dankoff, 1973) that it seems difficult to discover the size of the pre-Islamic core in it.

For a brief history of Sino-Indian cultural contact, see Bagchi (1944).

[24]For the lives and the social and economic environments of the people who generated these myths, see Daniel C. Snell's (1997) outstanding scholarly book, which he managed to address also to the educated layperson.

[25]See, for example, Assmann (1984), Huggett (1989, ch. 2), Anonymous (1991), Hansen (1991). For a comprehensive account of all mythologies involving flood legends, see Bonnefoy (1991a, 1991b). For English translations of relevant Middle Eastern text fragments and commentary, see Pritchard (1969). For a novel and very different view of at least some of the flood myths, but one that still involves the emergence of land, see Ryan and Pitman (1998).

[26]The first chapter of the second volume of Suess' *Das Antlitz der Erde*, entitled "Conflict of Opinion Regarding the Displacement of the Strand: Terminology and General Observations" (Suess, 1888, p. 2–41; in the English translation, Suess,1906: *The Face of the Earth*, p. 1–29) is a delightful and very informative summary of the important and influential ideas put forward to explain the movement of the strand. It ought to be compulsory reading for all those interested in falcogenic or eustatic movements. In Suess' masterly summary one sees the constant swinging of opinion on moving the sea to moving the land and back, from the Middle Ages to the end of the nineteenth century. Little could Suess have known that the same conflict was to survive him (see the following late- to post-Suess geology textbooks: Geikie, 1903 {p. 377–397, with numerous references}; de Lapparent, 1906, {p. 757–591}; Haug, 1907 {p. 491–510}; Scott, 1908 {p. 29–36}; Chamberlin and Salisbury, 1909 {p. 537–541, 544–545}; Supan, 1911 {p. 438–466, with numerous references}; Kayser, 1912 {p. 777–799}; Schaffer, 1916 {p. 103–109}; von Toula, 1918 {p. 64–68}; Grabau, 1920 {p. 691–696}), only to mutate successively into the questions of "which moves faster?" and then "which changes its rate of motion faster?" (see Pitman, 1978; Pitman and Golovchenko, 1991; Dewey and Pitman, 1998). For a brief history of the ideas on sea level changes from the viewpoint of a contemporary of Suess, who held the motions of the lithosphere responsible for them, see Issel (1883, p. 15–31). Lisitzin (1974, Appendix) gave a brief review of the older ideas on sea-level changes from a geophysical viewpoint.

[27]See Cuvier (1827, p. 145, and the long footnote * there), where he gives a detailed discussion of how in the Greek mythology Deucalion's deluge story had been introduced and stepwise enriched in detail to conform to the Mesopotamian versions. For the original French, see Cuvier (1825, p. 145 ff., footnote 1). This long footnote is not in the original 1812 publication, which appeared only as the *Discours Préliminaire* of Cuvier's epoch-making *Recherches sur les Ossemens Fossiles de Quadrupèds* (Cuvier, 1812). The 1825 publication, commonly stated to be the first separate printing of the *Discours*, is actually the third edition (Smith, 1993, p. 153)

[28]Although both his geology and mythological information are much dated, Frazer (1919), chapter VIII entitled "Volcanic Religion," still provides instructive reading in connection with this topic. For the bull motif, also see Demircioğlu (1939), not cited in Şengör (1997).

[29]See also Popowitsch (1750, p. 131–132) for a description of the Mediterranean storms created by the south wind.

[30] "Flood" according to Lambert and Millard (1970, p. 176). Oberhuber (1990, p. 547) is more specific: *Sturmflut*, i.e., flood with storm or storm-flood. See also p. 27 herein, note 16.

[31]After a century of scholarship, these lines are now read as:

"99 While Shullat and Hanish go in front,
100 Moving as heralds over hill and plain." (see Table 1).

[32]In the modern interpretation the same line reads: "131 The sea grew quiet, the tempest was still, the flood ceased." (Table 1).

[33]Haupt became the director of the Oriental Seminary of Johns Hopkins University in Baltimore, Maryland, in 1883 (see Suess, 1916, p. 323).

[34]Κυάνεαι πέτραι or νῆσοι (i.e., dark, literally "dark blue") rocks or islands; also called the *Kylai* (κύλεαι meaning "hollow" [Arrian's *Periplus*: Chotard, 1860, p. 36; cf. Bashmakoff, 1948, p. 156–159], probably on account of the wave-eroded hollows that characterize these volcaniclastic rocks). The Symplegades (Ξυμπληγάδες πέτραι, i.e., the "justling rocks" that were supposed to close in on all who sailed between them: Strabo, III. 2. 12), were claimed to have existed on both sides of the entrance to the Black Sea from the *Bosporus Thracicus* (i.e., the Bosphorus as this word is now understood; see Apollonius Rhodius, *Argonautica*, I. 3, II. p. 317–340, but especially p. 549–610; Dureau-de-la Malle, 1807; Chotard, 1860, p. 212 ff; and Wescher, 1874, especially p. 28, for the description of the physical geography of the northern mouth of the Bospohorus as known in antiquity. For a synthesis of the antique peripli of the Black Sea, see Chotard's {1860} commentary, with emphasis on the discrepancies in the antique reports, and the posthumous work of Bashmakoff {1948}). For an account of these rocks and a history of the opinions expressed about them, together with a new interpretation of their mythology, see Şengör (2002a).

[35]Hellespontus (the Sea of Helle), the daughter of Athamas and Nephele, who fled from her father's house with her brother to avoid the unbearable treatment of her mother-in-law. She is described variously riding a golden ram, or a cloud, or just a ship, from which she fell into the sea and drowned (see Lempriere, 1984, p. 298–299 under *Helle* for the classical authorities). Thus, the sea was called *Hellespontus* (i.e., the Dardanelles, a name given to the Hellespontus after the seventeenth century).

[36]Strato of Lampsacus (present-day Lâpseki in northwestern Turkey along the Asiatic shore of the Dardanelles) was the successor of Theophrastus as the head of the Lyceum in Athens. He was probably born around 340–330 B.C. and died sometime around 270 B.C (Rodier, 1890, p. 42, note 2 continued from p. 41; Wehrli, 1969, p. 47; for an English summary, see Gottschalk, 1981; also see the notes on Strato in Clagett, 1994, p. 68–72). Rodier gives a fine summary, with testimony, of Strato's views on natural sciences, including geology. Wehrli (1969) is the best collection (with commentary) of Strato's Fragments.

[37]The second (?) director of the Museion in Alexandria, Eratosthenes of Cyrene (284 {or 274} to 202 {or 194} B.C.), the geographer, was one of the greatest scientists of antiquity and probably the one who gave geography its character as a science as we know it today. He is famous for his amazingly accurate calculation of the earth's circumference and the tilt of the earth's axis to the ecliptic. For his life and contributions, see Knight (1930), Wolfer (1954), the books on

the history of geography given in endnote 4 (including Fraser's book), Dicks (1960, 1981), Fraser (1971), Dragoni (1979), and Aujac (2001). One of his works was the *Geography* in three books (for the various names of Eratosthenes' book as recorded by later authors, see Berger, 1903[1966], p. 387, note 2), which is regrettably lost. Strabo preserved large sections of it, and Bernhardy (1822) and Berger (1880[1964]) collected most of its fragments. There is also an unexamined thesis manuscript in the University of London Library, entitled *The Fragments of Eratosthenes of Cyrene* by R. M. Bentham, written in 1948. It consists of 468 pages of typescript. I have not seen it. For a reference to the oldest of the Eratosthenian geographical fragments by G. Seidel in 1789, see Aujac, 2001, p. 220. Aujac's book contains the most up-to-date references concerning Eratosthenes and his geography. For a succinct and illustrated account of Eratosthenes' birthplace, Cyrene, see Goodchild (1981).

[38]Strabo of Amasya (64/63 B.C.–ca. 25 A.D.) was Eratosthenes' ablest successor as geographer in antiquity, whose great geographical treatise in 17 books reached us almost intact. Although his interest was more in human geography than in physical geography, he, like his great predecessor Eratosthenes, regarded geography as a whole consisting of both social and physical elements. For his life, milieu, and contributions, see the books on the history of geography given in endnote 4, plus Aujac (1966) and Warmington (1981).

[39]Cuvier is here criticizing Pallas (without naming him) who, in his book on the southern provinces of Russia, following Tournefort (1717, v. II, p. 212 ff.), adhered to the ancient hypothesis that the Bosphorus had been formed by a violent earthquake and the consequent bursting of the Black Sea into the Aegean:

> Tournefort, by arguments very cogent indeed, has endeavoured to ascertain, that the mountains of the Thracian Bosphorus were formerly connected, and formed the natural limits which separated the Black Sea from the Mediterranean, so that the waters of the former, flowing from the great rivers Danube, Dniestr, Don, and Kuban, became a prodigious lake, much higher than the Mediterranean Sea, and even higher than the ocean; that after the destruction of this strong boundary, either by an earthquake, or the weight and pressure of accumulated water, the Black Sea disembogued itself impetuosly into the Mediterranean, till it acquired the due equilibrium; and that on the first impetus of this deluge, a part of Greece, and the islands of the Archipelago, were overwhelmed and desolated. Indeed this inundation appears to have taken place, according to the most authentic historical evidence. (Pallas, 1803, p. 97)

Count Andreossy (1818, p. 59) pointed out that Tournefort had given a description of the Bosphorus without studying it first-hand himself and that Pallas had become his victim, although knowing (1) that the Bosphorus was originally indeed a fluvial valley indicating flow from north to south, and (2) the geological theories of the time of Pallas. I find it difficult to imagine how Pallas could have avoided reaching his conclusions even if he himself had studied the Bosphorus first-hand. What Pallas describes here and in the paragraphs that follow the one cited above, is the first description in the scientific literature (to my knowledge) of what we now call Paratethys or *Lac Mer* (Laskarev, 1924). The only difference is the time scale of the inferred events. So Pallas basically reached conclusions that were found to be reasonable by not only many of his contemporaries (e.g., Dureau-de-Lamalle, 1807, ch. XXVI and XXVII and the references cited there, plus the foldout map drawn by J.N. Buache and entitled *Carte pour servir à l'ouvrage de Mr. Dureau-de-Lamalle sur la Géographie des mers intérieurs*), but also by us to this day.

I must mention, however, that a still earlier description of a similar story is given in the first volume of Evliya Çelebi ibn Muhammed Zıllî Darviş's (1611-?1682; see Baysun, 1964, and İz's introduction in Tekin et al., 1989) *Seyahatname* (*editio princeps*: 1314H{1896AD}). In the second chapter of this travel book, Evliya talks about the "opening of the Black Sea" (p. 37–40 of the *editio princeps*: "Bahr-i Siyah Fethi Beyanındadır"; p. 13–18 of the new Yapı Kredi Yayınları transliteration by Gökyay, 1995, and p. 9–20 of the modern Turkish version by Kahraman and Dağlı, 2003):

> According to the true words of the historians familiar with astronomy, the Black Sea is a remnant of Noah's Flood. Its depth is 80 fathoms. It is a deep black sea. Before the Flood it did not mix with the Mediterranean and ended near the Black Sea straits near İslambol [*İstanbul*]. At that century, the fields of Salanta [*a field near Budapest*], Dobraçin [*Debrecen: 47°30′N 21°37′E*], Keçkement [*Kecskemét: 46°56′N 19°43′E*], Kinkos [*Gyöngös: 47°46′N 20°00′E*] and Pest [*Pest of Budapest: 47°30′N, 19°03′E*] and the valleys of Sirem Semendire [*Semendria, Smederevo: 44°40′N, 20°56′E*] in Hungary were entirely [*parts of*] the Black Sea. In the province of Dodushka [*province of Carinthia in southeastern Austria; north-northeast of Venice*] near Venice, the places where the Black Sea used to mix with the waters of the Gulf of Venice are still visible. In fact, near Silistre [*Silistra: 46°06′N, 27°17′E*] the Fortress of Pravadi is a high burg reaching the skies. In that century, this fortress was at the sea shore. There are still iron rings to tie ships. Places on the rocks abraded by the rails of the bulwarks and the sterns of ships are still obvious. Another sign of the Black Sea is the Fortress of Menkub near [*i.e., SSW of: see Pitcher, 1972, maps XIII-C2, XVI-C2, XXXB2*] Bahçesaray [*Bakhchisaray: 44°44′N, 33°53′E*] in the Crimea, which reaches the blue clouds. There too are ports to put the ships and colums to tie them. The Crimean island [*Evliya uses the Arabic word* jezireh *for the Crimea that may mean both an island and a peninsula; but for the designation "island" for the Crimea, see Bala, 1967, p. 741, col. 2*], the field of Heyhât [*literally region of suffering, desert: region roughly between 47° and 48°N and 32° and 36°E; about equivalent to the present-day Pricher-nomorskaya Nizmennost' (Pre-Black Sea Low Plain): see Sezgin, 2000c, map 155a*], the Kipchak steppe [*Dasht-i Kipchak: the steppe region between the rivers Dnyestr and Donetz*] and the entire land of Sakalibe [*i.e, Slavs*] were [*parts of*] the Black Sea. In fact a part of it reached the Caspian Sea, i.e., the Sea of Gilan [*from the north Iranian province of Gilan located between the Alborz and the Talesh Mountains*] and Demirkapi [*Iron Gate: this is the present-day pass of Derbent in the Eastern Greater Caucasus; see Anderson, 1932, p. vii*]. In fact, this humble man [*i.e, the author*] found signs of marine creatures when, during the Moscow campaign in the era of İslâm Giray Khan, ... he was digging trenches in the field of Heyhât and in the places called Kerneli and Biym and Ashm for watering the ... horses. For instance, he dug out the shells of such insects [*sic*] as crabs, crawfish, mussels and oysters. From this it is understood that the valley of Heyhât was also [*a part of*] the Black sea. (Evliya Çelebi ibn Muhammed Zıllî Darviş, 1314H {1896 AD}, p. 37–38)

Evliya then proceeds to relate a legend from an Islamic version of the Alexander Romance in order to describe how the Black Sea and the Mediterranean were put in communication. According to this version, Alexander had conquered the whole world except the country of "Macedonia and Izmirne" ruled by a woman named Kaydafe.

(Her name is spelled either Kaydafe or Kaydefa, depending on whether the letter *alif* is interpolated between the letters *dal* and *fe*; in the *editio princeps* of the *Seyahatname*, which is based on a manuscript in the Library of Pertev Pasha {Ms. 458–462: see İz's introduction in Tekin et al., 1989, p. 4}, there is an *alif* and hence the reading is Kaydafe {with emphasis on the second a}. This is the reading adopted by Eren, who collated a number of manuscripts. In the facsimile of the mansucript, known as Bağdat 304 in the Topkapi Palace Library {Tekin et al., 1989}, taken as the basis of the new Yapı Kredi Yayınları transliteration {Gökyay, 1995}, there is no *alif* between *dal* and *fe*, hence the reading would be Kaydefa {with emphasis on the last a}. I here adopt Eren's, 1960, choice, because doubts have been expressed whether the Bağdat 304 is indeed an author's copy as had been previously claimed {see İz's introduction in Tekin et al., 1989, p. 1}.)

To learn more about this lady, Alexander went to her court incognito, but was recognized and captured. She imprisoned him, but later released him upon his promise that he would not wage war against her or raise a sword against her. Alexander then went back to his capital at the foot of the Alborz Mountain, called Iraq-ı Dadyân, and consulted with the wise men. They all advised him to take the army and storm Kaydafe's land. Alexander refused, not wishing to break his promise and asked them to find other means for him to avenge himself. At this point, the prophet Hizr (for an extensive study, with a good bibliography, of this character, prominent in many Turco-Islamic legends, see Ocak, 1985) raised his head and said, "O Alexander! If you wish to avenge yourself without fighting a war and murderous combat, immediately cut the Black Sea [*sic*] near Macedonia and make it flow to the Mediterranean. The entire land of Kaydafe will be submerged under the waters. Thus you will have avenged yourself and also will have kept your promise." Alexander liked this plan. His engineers found that the Black Sea was higher than the Mediterranean and 700,000 workers were immediately set to work to cut the Black Sea [*sic*]. The prophet Hizr was the overseer of the work. Evliya says that the prophet Hizr was the cause of the mixing of the Black Sea and the Mediterranean. Apparently, the "cutting" lasted three years and finally the Black sea flooded the low lying regions in Kaydafe's land, submerging 1700 cities. Since then, Evliya says, Hungary, the fields of Siram and Smederovo, Polish, Czech, and Kipchak steppes, and the Prichernomorskaya Nizemnnost' became habitable regions. (For further references to fossil finds reported by Evliya in the former larger Black Sea, see Dankoff, 2002, p. 614).

It is clear that a paleo-Black Sea idea is a very ancient one. Eren (1960, p. 38–45), in her study of the sources of the first book of Evliya's *Seyahatname*, showed that Evliya probably learned of its Alexander romance version from Abu'l Qasım Firdawsi's (?932-?941—?1020-?1026) *Shah-Nâmeh* (completed 1010), Abu Ğaffar Muhammad ibn Ğerir al-Tabari's (839–923) *Tarih-i al Ümem ve al-Mulûk* and from some source on the Muslim version of the Alexander Romance, i.e. some *İskendername*. Eren (1960, p. 42), wrote that Evliya mentions an İskendername by Figani (late fifteenth century) in the *Seyahatname*, although she was unable to locate a copy of this work for comparison. Ünver (1975, p. 321–322), in his doctoral thesis on the *Iskendername* poems in the Turkish literture, confirmed that the *İskendername* by Figani, which, Gibb, in his *A History of Ottoman Poetry,* said (for reference, see Ünver, p. 322, footnote 1) had been a failure and quickly had been forgotten, can no longer be located.

The Black sea story in the Alexander Romance probably originated in connection with the flood legends (in the way claimed by Ryan and Pitman, 1998?) and was supported by the chance finds of fossils similar to the creatures now inhabiting the Black sea in the lowlands surrounding it. The geomorphology of the Bosphorus most probably further encouraged its support. Eventually, by weeding out the legendary component and by the increase of observations, it became transformed into a scientific hypothesis.

[40]The Strait of Gibraltar.

[41]The Black Sea.

[42]The Aegean Sea.

[43]For the French original, see Cuvier (1825, p. 177–178, continuation of footnote 1 on p. 175). M. Olivier (mentioned in the text) is the French physician Guillaume-Antoine Olivier (1756–1814), who was sent to the Ottoman Empire at the head of a delegation by the Convention government in France. Olivier toured much of the empire plus Iran between the years 1793–1798 and wrote an important book entitled *Voyage dans l'Empire Ottoman, Egypte et la Perse, fait par Ordre du Gouvernement pendant les Six Premières Annés de la République* that included an atlas and a map of the Bosphorus (published in Paris, 1801). Cuvier is making reference to this book. For Olivier and his voyage, see Gökmen (1977, p. V–VII) and Yerasimos (1994).

[44]Data in the *Septuagint* (*Genesis*, 5:26–28; 7:6) suggest that Methuselah (Ginzberg, 1909, p. 141–142) survived the Flood by 14 years, which is impossible since he had not boarded the ark. By contrast, the *Vulgate* makes him die in the year of the Flood (cf. Alexandre, 1988; for St. Augustine's discussion of this "very celebrated problem," see his *City of God* {*De Civitate Dei Contra Paganos*; XV, 11}). Indeed, Nicholas of Lyra imagined that Noah must have spent the seven days before the Flood in mourning Methuselah's death (Allen, 1963, p. 75). Does the Greek version in fact reflect the influence of the Greek flood legend, or at least did this seeming impossibility not bother the 70 inspired translators of the *Septuagint*, because it was compatible with at least one flood legend with which they were familiar? (Actually, the *Septuagint* translation is a collection of translations spanning a considerable time period: Allen, 1963, p. 58.) Ginzberg (1925, p. 165), on the other hand, quotes sources for an extremely interesting solution to the problem that will gain additional relevance to the topic of this book when we encounter the neptunistic earth histories of the eighteenth century. Since Methuselah could not have survived the flood on earth and yet lived beyond it, he must have been taken by God into the Paradise, at least for the duration of the flood. But, it has long been thought, and Anderson (1988) has summarized the recent thinking on it, that Paradise is located atop the highest mountain on earth! What would be more natural than to run to the highest land when there is a flood? As Cerambus was carried on the wings of supernatural beings, so it is believed that Methuselah was taken into the Garden of Eden (by winged angels?). What I read into these myths is that people fled to high areas when the waters were driven in either by the wind or by something that the south wind accompanied. Was Ziusudra equivalent to Utnapishtim and, thus, was the earliest Methusela as well as Noah?

[45]Also the following remarks: "It is said by the natives, especially by their monks who stay at the foot of the mountain, men of very holy life though without the faith, that the deluge never mounted to that point, and thus the house [of Adam] never been disturbed." (Yule, 1914[1966], p. 233–234). Marignolli writes that this mountain is opposite to the Paradise, which is supposedly located on our earth:

> … I proceeded by sea to SEYLLAN, a glorious mountain opposite to Paradise. And from Seyllan to Paradise, according to what the natives say after the tradition of their fathers is a distance of forty Italian miles; so that, 'tis said, the sound of waters falling from the fountain of paradise is heard there.
>
> Now Paradise is a place that (really) exists upon the earth surrounded by the Ocean Sea, in the regions of the Orient on the other side of Columbine India and over against the mountain of Seyllan. 'Tis the loftiest spot on the face of the earth, reaching as Johannes Scotus hath proven, to the sphere of the moon; a place remote from all strife, delectable in balminess and brightness of atmosphere, and in the midst whereof a fountain springeth from the ground, pouring forth its waters to water, according to the season, the Paradise and all the trees therein. And there grow all the trees that produce the best of fruits; wondrous fair are they to look upon, fragrant and delicious for the food of man. Now that fountain cometh down from the mouth and falleth into a lake, which is called by the philosophers EUPHIRATTES. Here it passes under another water which is turbid, and issues forth, on the other side, where it divides into four rivers which pass through Seyllan. (Marignolli *in* Yule, 1914[1966], p. 220–221)

These rivers Marignolli identified as Gyon or Gihon (Jaxartes or Syr Darya or Saihûn or Sihon), Phison (Oxus or Amu Darya or Jaihûn or Jihon), Tygris, and Euphrates. But the way he described their courses was incrediblely confused and included most rivers of south Asia and northeastern Africa! Thus, Gihon "circleth the land of Ethiopia where there are now negroes, and which is called the Land of Prester John" (Yule, 1914[1966], p. 222). Yule points out that *Septuagint* has Geon (Ghon) for the Nile in *Jeremiah*, ii, 18 and in *Ecclesiasticus*, xxiv, 37 (Yule, 1914[1966], p. 222, note 1). Phison, is supposed to go to Cathay and turn into Caramuran, which is the Turkish name for the Huang He (i.e., the Yellow River). It is supposedly lost in the sands (here the reference may be to the Oxus) and then it is supossed to reappear as Thana (i.e., the Don!). The Tigris and Euphrates are tolerably correctly described.

It is perhaps worthwhile to remind the reader here that many medieval Muslim geographers (such as al-Makdisi, al-Batini, and al-Biruni) followed the Indian designation of Odjein or Ozein for a town in India, mostly described to be in Ceylon, as the "Dome of the Earth," which was believed to be equidistant from the eastern and the western ends of the inhabited world. Through the peculiarities of the Arabic alphabet and orthographic errors, this Odjein or Ozein soon began to be written as Azin or Azyn or Arin or Aryn (cf. Reinaud, 1848[1985], p. CCXL ff.; see especially note 1 on p. CCXLI). It was in the form of Arin that this concept was taken by the European geographers of the Middle Ages (e.g., in Roger Bacon's *Opus Majus*, 1928, p. 319–329; see also Sezgin, 2000a, p. 163, 219, 241, 246, 264) and the Renaissance and through Pierre d'Ailly's *Ymago Mundi* (ca. 1483) influenced Christopher Columbus' concept of a dome-shaped tumescence in the regions he had explored, with the alleged culmination point in the island of Trinidad owing to its milder temperatures despite the low latitude (see d'Ailly, 1992, especially p. 60 with Columbus' marginalia; also von Humboldt, 1852, p. 44; Thorndike, 1923b, p. 645–646; Sezgin, 2000a, p. 219). Columbus thought that this tumescence was the Paradise!

While describing Marignolli's account of the areas that escaped the flood in Ceylon, Yule also draws attention to a passage in Masoudi's *Prairies of Gold*, where the great Muslim scholar pointed out that a race of Indians living in the country of Komar (present-day Assam) trace their lineage to Cain (i.e., they imply that their line also escaped the flood {Barbier de Meynard and de Courteille, 1861, p. 72}).

[46]Galloway (1900) implies that the animals listed in *Leviticus* (11) and *Deuteronomy* (14:4–20) were the only ones taken into the ark and therefore such animals as lions, tigers, leopards, hyenas, gorillas, rhinoceroses, naked elephants, kangaroos, emus, boas, cobras, rattlesnakes, vipers, scorpions, "and all the rest, an innumerable multitude not included in the Noahic list of clean and unclean animals" (p. 43) were left out and yet escaped the Deluge in equatorial regions. Galloway uses this inference to support his theory that a Quaternary axial shift of the earth may have been responsible both for the ice age and the Biblical deluge. He claims that at two points where the old equator and the present equator coincide, some prominences would have remained above the waters and thus provided asylums to the animals in the surrounding areas. He quotes a contemporary of his, Edward Harold Browne (1811–1891), the Bishop of Ely and later of Winchester, concerning the animals mentioned in *Genesis* that survived the Flood and yet which had not been taken onboard by Noah. I have not come across other references to that effect and therefore do not think that the presence of extra-ark animals may have played a role in generating views about pieces of land surviving as such during the Flood. *Genesis* 7 seems to me to imply that *all* kinds of animals were taken on board.

[47]For the discovery and history of decipherment of this monumental inscription, see Doblhofer (1961, ch. III, entitled "Ahura Mazda came to my aid"). The fourth (unnumbered) plate of that book is a magnificent photograph of the inscription.

[48]"Wise Lord" in the Avestan language. Also transliterated as *Ormizd* or *Ormazd*. From the descriptions of his attributes, he seems to be the Iranian version of the volcano-god.

[49]The motive of this disaster is somewhat reminiscent of the Mesopotamian flood myths.

[50]This is amazingly similar to von Linné's (1744) theory of biological dispersion. See below.

[51]The oldest Vedic texts, which constitute *all* that has remained of the original Aryan Vedic culture in India, can be grouped into two classes: those of revelation (*sruti*) and those of tradition (*smrti*, i.e., "remembrence"). The texts of revelation are called Veda (knowledge) and are divided into three plus one categories: (1) The Veda of Verses (*Rgveda*), (2) The Veda of Melodies (*Sâmaveda*), (3) The Veda of Ritual Formulae (*Yajurveda*), plus the later and less prestigious (4) Veda of Incantations (*Atharveda*). All these Vedas contain a hierarchy of texts in themselves that are (from the primary source to later appendices and explanations): (I) *Samhitas*, forming the earliest texts probably dating from the fifteenth century B.C., if not earlier; (II) *Brahmanas*, injunction (*vidhis*) and explication (*arthavadas*) texts; (III) *Aranyakas*, containing appendices to *Brahmanas*, and (IV) *Upanishads*, extensions of *Brahmanas* and *Aranyakas*. These groups I–IV are comprehended within the revelation (*sruti*) texts. In addition, there are *Vedangas*, literally "limbs" forming technical treatises of how to read the above-listed revealed texts that form a part of the corpus of the remembered tradition (*smrti*). The *smrti* tradition includes another group of texts called the *Purânas*, dealing mainly with the things of the past. There is a flood story in each of the two traditions of the *sruti* and *smrti*.

[52]Cretan seer, one of the founders of the Orphic sect. Dates for his existence are uncertain. Some think he may have lived in the late seventh century B.C., but Freeman (1962, p. 9; 1949, p. 26) dates him to the time of late sixth to the early fifth centuries, although even his very existence has been doubted by some. He is famous for the paradoxical saying, reported by St. Paul in his letter to Titus (I, 12) according to the testimony by Clement of Alexandria, that all Cretans were liars (Freeman, 1949, p. 31). On the strength of the same testimony, we learn that in some lists he appears as one of the Seven Sages in place of Periander. St. Jerome says that the saying about the Cretans comes from the *Oracles of Epimenides* (Freeman, 1948, p. 9).

[53]DK3B18. This shorthand indicates that the fragment cited is in the *Fragmente der Vorsokratiker* of Hermann *D*iels (as revised by Walther *K*ranz, 1951, sixth edition). The first number is the number assigned to the pre-Socratic author in that book. The letter B implies that the fragment is a direct quote and not a testimony (A), and the last number is the fragment number. In this book, I have used mostly Freeman's (1962) translation of the fragments into English. Those who wish to see the original Greek must go to the standard work of Diels (1951[1996]).

[54]Some doubt has been expressed about the authorship of this statement: Hicks (1925, p. 116, note *a*) points out that the long poem *Theogony* by Epimenides, from which the Rhodes statement is generally believed to have been quoted, may have in fact been written by Lobon of Argos (called a "disreputable stichometrist of the second century B.C." by Kirk et al., 1983, p. 87), or Lobon may simply have just affirmed the existence of Epimenides' poem *Theogony* in his book *On Poets*.

[55]5; also DK4B8; *Astronomia* or *Astrologia* (as Plutarch called it) is one of the Hesiodic poems preserved only in fragments and may have been originally attached to the end of another fragment called *Divination by Birds*.

[56]Pre-eighth century B.C. See Freeman (1962, p. 7–9; 1949, p. 19–25).

[57]518–438 B.C. Greek lyric poet from Boeotia, who was educated and lived in Athens.

[58]ca. 370 to ca. 288 B.C. Greek philosopher from Eresus, Lesbos; friend and pupil of Aristotle and his successor as the head of the Lyceum in Athens. As we shall see below, Theophrastus is of great importance for our understanding of the tectonic thought in the Peripatetic school. For details about his life, work and thoughts, see Fortenbaugh et al. (1985), Fortenbaugh and Sharples (1988), and Fortenbaugh et al. (1993a, 1993b).

[59]See also the texts and translations in Colson (1941, p. 268–269) and Fortenbaugh et al. (1993a, p. 342–343).

[60]For the social background of the Ionian Enlightenment, one must understand the evolution of the Greek people and its institutions as a whole. The best book that I know for an introduction to the evolution of the early Greeks is Murray (1993), in which the birth of the Ionian Enlightenment is also discussed (see especially ch. XIV). Another excellent account is the shorter description given in Gomperz (1901, p. 3–42). Also, see Freeman (1950), Hopper (1976), and Burkert (1992).

[61]See Kahn (1960, p. 102–109), Guthrie (1962, p. 101–102), Schmitz (1988, p. 45 ff.), and Conche (1991, p. 204–207 and note 26); for the best narrative of his theory, see Gomperz (1901, p. 52–53). Suess (1888, p. 16) calls this the "*Dessication theory*."

[62]If the earlier accounts are to be read to imply motion of land. See above for my reservations.

[63]For the evolution of the concept of "ocean," see the books on the history of geography listed in endnote 4 above and Nordlind (1918, 1927) and Wensinck (1918).

[64]Φῦσικόι (*phusikoi)*—as Aristotle, I think justifiably, calls them (Barnes, 1987, p. 13). See especially Heidel (1910) for an excellent treatment of the meaning of the term Φύσις (*phusis*) for the pre-Socratics. Mannsperger (1969) is an excellent philological study of the word *phusis*.

[65]Not all historians of the thought of Ionia agree on how religious or irreligious the *phusikoi* were. For two extreme viewpoints, see Gomperz (1901), Burnet (1930), and Popper (1989) for irreligiosity; and Jaeger (1953; originally published in English in 1947 on the basis of the Gifford Lectures given in 1936) for religiosity. Vlastos' (1952) excellent paper takes an intermediate position, though I quite frankly find the evidence more to be in favor of the irreligious group. Also see von Wilamowitz-Moellendorff (1955b, especially p. 106–114, 202–220). Characteristic is the sort of distinction that Wilamowitz-Moellendorff makes between religion and science during the fourth century B.C. concerning the thought of Plato: "Plato imagined to have established scientifically what he considered religious, although he continued to ponder. Science, in the meantime, advanced right past him" (Wilamowitz-Moellendorff, 1955b, p 256). The rationalist Ionians took the irrational mythology and rationalized it—made it into science. It is therefore quite likely that the retreat of the sea was an idea Anaximander took from the mythology and simply rationalized it.

[66]Note what Playfair wrote nearly two-and-a-half millenia later:

> Whether this great change of relative place [*i.e., strata deposited beneath the ocean now being in the mountains*] can be best accounted for by the depression of the sea, or the elevation of the strata themselves, remains to be considered. Of these two suppositions, the former, at first sight, seems undoubtedly the most probable, and we feel less reluctance to suppose, that a fluid, so unstable as the ocean, has undergone the great revolution here referred to, than that the solid foundations of the land have moved a single fathom from their place. This, however, is a mere illusion. (Playfair, 1802, p. 40–41)

And ten years later: "The successive changes of level that must have taken place, are very hard to be understood; and whether they are to be ascribed to the alternate rising and falling of the land, or to the alternate falling and rising of the sea, are discussions on which we have not leisure to enter, and about which we are not prepared to decide" (Playfair, 1812, p. 381). See also Büttner (1979a) for man's natural inclination to exogenic geodynamic models.

[67]All of the references given about the pre-Socratics in endnote 4 deal with this remarkable man. In addition, see especially Popper (1998, Essay 2 entitled "The Unknown Xenophanes—An Attempt to Establish His Greatness," p. 33–67).

[68]This is precisely the same view that was entertained as late as Celsius' famous oration delivered in Uppsala on 22 June 1743 (Celsius, 1744). Celsius was the last scientific representative of this theory. Suess (1888, p. 38, note 22) cites many sources to show that it lived on in the popular opinion until the first half of the nineteenth century!

[69]See Popper (1998, p. 37) for the conjecture that Xenophanes may have had discussions with his teacher Anaximander and his fellow "pupil" Anaximenes to elaborate the theories of the former.

[70]For Zoroaster's influence on Greek thinking, see Afnan (1969 and the references cited therein).

[71]Macrobius essentially provides an exodynamic theory of historical geology, which was probably inherited from the ancient Greeks. Compare his climatic view of the coming and going of civilizations with its modern version in Fagan (1999), and consider Suess' (1909, p. 777) closing sentence of the *Antlitz* in their light: "In the face of these open questions let us rejoice in the sunshine, the starry firmament and all the manifold diversity of the Face of the Earth, which has been produced by these processes, recognizing, at the same time, to how great a degree life is controlled by the nature of the planet and its fortunes." Neither Xenophanes nor Macrobius would have objected to any part of this poetic statement.

[72]See the references to the pre-Socratics in endnote 4.

[73]In addition to the publications concerning the pre-Socratics in endnote 4, Bollack's four-volume (1965; 1969a, 1969b, 1969c) *Empedocles* is the most comprehensive study devoted to the thought of the Agrigentian. Also see Zafiropulo (1953) and Guthrie (1965, p. 122–265).

[74]An Orphic idea! See especially DK1B13 (Athenagoras only!), and DK1B16.

[75]See Bollack (1969b, p. 166) for the justification as to why ἀνόπαιον (upward) implies fire.

[76]Tutor, confidant and, eventually, a victim of Nero, Lucius Annaeus Seneca (ca. 4 B.C. to 65 A.D.) was a shady character. A philosopher, scientist, and humanist of considerable merit, he may have been involved in crimes in Nero's time that cannot be excused as unavoidable misfortunes forced on him by the insane emperor. For an excellent (and sympathetic) account of his life, see Sørensen (1984). Seneca's *Quaestiones Naturales*, which he composed during the most unfortunate years of his life after 62 A.D., constitute his chief contribution to geology. The English translation by Clarke and Geikie (1910) contains a detailed geological commentary.

[77]We know, on the testimony of Aetius, a doxographer of the second(?) century A.D. (according to Kirk et al., 1983, p. 5; Freeman, 1949, p. 427 assigns him only "to later than the 4th century B.C."), that this theory actually belongs to a fifth century Pythagorean, Philolaus of Croton (Kirk et al., 1983, p. 342 ff.). It is unclear how much of it he may have inherited from the Pythagorean school.

[78]Neo-Platonic philosopher from Cilicia (born ca. 500 A.D.). Taught at Athens in the Academy and took refuge at the court of the Persian king Chosroes (531–579 A.D.) from the persecution of Justinian in 529 A.D. Simplicius returned to Athens in 533 and became an author. For a brief biography, see Verbeke (1981).

[79]Pythagoras and the Pythagoreans constitute an extremely problematic group from the viewpoint of the history of science owing to the dearth of firm data and the abundance of conflicting opinions. In addition to the sources concerning the pre-Socratics in endnote 4, see Guthrie (1962, p. 146–340) and Kahn (1974[1993]) for Pythagoras and the Pythagoreans.

[80]Today Mavros (Scobel, 1910, map 23) or Lakkiotikos (Neumann and Partsch, 1885, p. 158). Tozer (1882[1974], p. 120–122) gives a good discussion of its actual geography and mythological significance, including the idea that it was the "model inferno."

[81]A river in Epirus in the vicinity of the Acheron (Lempriere, 1984, p. 178).

[82]One of the branches in the upper course of the Crathis emanating from the Araonia Mountains, which are located south of the Gulf of Corinth in Arcadia (Tozer, 1882[1974], p. 32–33, 92, 117–120, 310–311; Neumann and Partsch, 1885, p. 180; Kerényi, 1951, p. 39; Scobel, 1910, map 23). It was identified with the Homeric Styx by William Martin Leake. Its 150-m-high waterfall, disappearing in winter into the snow, inspired the image of a river whose origin and mouth were unknown and represented the primordial darkness, although Kerényi (1951, p. 39) proposed that the geographical river acquired its name from the mythological one, rather than the other way around.

[83]Dante's Hell in the *Divine Comedy* has the same rivers (e.g., *Inferno*, Canto XIV, lines 116, 119) adopted from Virgil's *Aeneid* (VI, journey into Hades begins at line 264), but those rivers were common knowledge in classical antiquity (Singleton, 1977, p. 244, lines 115–19), the concept having been inherited from still earlier hell images in the Middle Eastern mythologies (cf. Kramer, 1981, ch. 11). Konrad (1972, ch. V) has shown that Dante was also greatly influenced by the eastern and Byzantine apocryphic Christian literature and particularly by the stories *The Journey of the Mother of God Through Hell* and *The Journey of St. Paul.* These stories also have very long pedigrees going back to the visits the Sumerian sky goddess, Inanna, and the servant of the king-hero Gilgamesh, Enkidu, separately pay to the nether-world (cf. Pritchard, 1969, p. 52–57; Kramer, 1981, p. 196–198; Wolkstein and Kramer, 1983). In addition, the stories carry a strong Zoroastrian influence and form threads connecting the late medieval geocentric Hell image with the Zoroastrian image from antiquity, showing a recurrent east-west interaction in shaping the image of the fiery interior of the earth. The infernal rivers also occur in Milton's Hell in *Paradise Lost.* For the geometry of Dante's Hell and the positioning of these rivers, see Singleton (1977, fig. 4); for Milton's, see Turner (1993, map on p. 185). Compare Dante's view of the interior of the earth with Figs. 11A and 11B of the present book. Suess (1888, p. 37–38, note 10) pointed out that, to many medieval authors, a central fire appeared as a paradox impossible to resolve: that the lightest of all elements, namely fire, should be in the center! But Lucifer was cast headlong down to earth (*Isaias* 14, 12–15). In Dante's version, his point of fall is at the antipode of Jerusalem, in the uninhabited hemisphere (which Dante called "*mondo senza gente*" i.e., world without people: {*Inferno*, Canto XXVI, line 117}). The land shirks from the accursed projectile and "makes a veil of the sea" and regathers itself on the opposite hemisphere to form the inhabited world with Jerusalem in the center (Canto, XXXIV, line 123; the "veil" is presumably made either by subsiding beneath the sea or by removing itself entirely from under the sea by horizontal motion on the surface of the globe. Given Dante's image of the earth then, probably the former {see Suess, 1888, fig. 2; Perler, 1994, fig. on p. 12} certainly not by "volcanic activity" as interpreted by Sigurdsson, 1999, p. 76). As Satan continues his fall to the center of the earth, "perhaps in order to escape from him that which appears on this side [i.e., *the side opposite to the inhabited world: the water hemisphere*] left here the empty space [*which formed the cavity of Hell*] and rushed upwards" (Canto XXXIV, lines 124–126). Suess says that here we see the principle of evil being identified with the force of gravity bringing fire to the center and providing a Scriptural—and a poetic—solution to the dilemma. This is certainly in keeping with the world picture developed by Aristotle, which remained popular among geographers and "proto-geologists" until Newton, in which God resides outside the sphere of the fixed stars (for Aristotle's naturalistic reasons for placing God, "the first mover," at the outer edge of the universe in terms of his physics, see Ross, 1923, p. 94). Thus, any movement upwards is "worthier" than any downwards (Büttner, 1979a, p. 17). Even some Renaissance scientific treatises placed Hell into the center of the earth (e.g., the *Margarita Philosophica* of Gregorius Reisch in 1503 {Hoheisel, 1979a, p. 60, points out that reports about earlier printings have proved wrong} and Giovanni Paolo Galuccio in his *Theatrum Mundi, et Temporis*, Venetiis, in 1558 {see Kelly, 1969, p. 219}) and even the illustrious Galileo Galilei is said to have written a student thesis on Hell based on Dante's geographic description of it (Sigurdsson, 1999, p. 76; Sigurdsson refers to Kelly {1969} as his source, but there is no mention of such a thesis in Kelly's paper. I thus remain ignorant as to where Sigurdsson obtained his information). For more on the relation between Lucifer's fall in the *Divine Comedy* and Dante's geology as expounded in his lecture on water and the earth, see von Humboldt (1852, p. 93–94), Günther (1897, p. 12 and the literature cited there), Kimble (1938, Appendix entitled "Dante's Geographical Knowledge"), Freccero (1961), and Konrad (1972, ch. IV, p. 72–73, and ch. V, p. 96). Sezgin (2000a, p. 223–225) cites further references on Dante's geography and its sources. For Columbus' alleged dependence on Dante for his estimate of the circumference of the earth, see Sezgin (2000a, p. 220).

[84]cf. Darmesteter (1887, p. 24, note 1); also see West (1880, p. 15, note 3; p. 34), where it is mentioned that Arezûr-bûm is in the Alborz; also see p. 223, note 7. In the book *Shâyast Lâ-Shâyast* (ch. XIII, section 19), Arezûr is expressly called "the gate of hell" (West, 1880, p. 361). In his final translation of the *Avesta*, Darmesteter (1960, p. 35, note 11) also supports his theory based on various manuscripts of the *Bundahish* (XII, 8) to identify the Demâvend as the Arezûra: For the influence of the Zorostrian image of a subterranean Hell on medieval European thinking (via Byzantine, Bogomil, and Russian literature), see Konrad (1972, ch. V).

[85]Compare this with the passage in the *Old Testament*: "Now when all the people perceived the thunderings and the lightenings and the sound of the trumpet and the mountain smoking, the people were afraid and trembled" (*Exodus*, 20:18). The Shofar (or Shophar), a ritual musical instrument of the Jews, made from the horn of a ram or some other animal and used on important public and religious occasions, produces a similar sound (at most, the fundamental octave

and twelfth) and was probably inspired by the sound of the mountain. In the Hebrew text of the Bible, the word translated as "trumpet" is indeed *shofar*. It was used in battle more to signal the presence of God to give the Israelites the victory than an actual military instrument. The Titanomachia in the Greek mythology has been interpreted as a description of volcanic action (e.g. Greene, 1992). It is thus interesting to note that there too Aigikeros invents the trumpet by using the conch shell to help Zeus in the battle (Gantz, 1993a, p. 45).

[86]See Kahn (1974[1993]) against the view that all science called Pythagorean may have started with Hippasus and Philolaus in the fifth century B.C. What Kahn says is compatible with my reading of the evidence.

[87]The female element was the prime guard and tender of fire. Neumann (1972, p. 284) says, "As in the house round about, the female domination is symbolized in its center, the fireplace, the seat of warmth and food preparation, the 'hearth,' which is also the original altar. In ancient Rome this basic matriarchal element was most conspicuously preserved in the cult of Vesta and its round temple. This is the old round house or tent with a fireplace in the middle." Of course, Vesta or Vestia is no other than the daughter of Kronos and Rhea, Hestia, the eldest sister of Zeus, who was the Greek goddess of the hearth (ἑστία means the hearth of a house, fireside: cf. Kerényi, 1951, p. 92). There is little doubt that the myth of Hestia is based on an even older Indo-European goddess, knowledge about whom is preserved for us as the head of the Scythian pantheon, who was also the protector of the hearth (Jettmar, 1964, p. 19). It is extremely interesting that the central fire of the earth may have been a model deduced from the female model of the earth and *not* from observations that we normally tend to associate with it. Such observations as volcanoes, hot springs, and increase of heat as one descends into the earth may have become associated with the central fire model only subsequently. Sir James Frazer's discussion of the religious significance of volcanoes and hot springs seems to lend some support to this view (Frazer, 1919, ch. VIII). This goes to support Popper's long-held view that *any* excuse to generate a testable hypothesis is legitimately scientific.

[88]Helmbold (*in* Cherniss and Helmbold, 1957, p. 275); see also DK31A69, Guthrie (1965, p. 189), and Bollack (1969a, p. 96–97). On Plutarch as a source for Empedocles, see Hershbell (1971). Schvarz (1862, p. 6) renders the same passage thus: "As touching those rocks, crags and cliffs which we see to appear out of the earth: Empedocles is of opinion that they were there set, driven up, sustained, and supported by the violence of certain boiling and swelling fire within the bowels of the earth." I prefer the translation by Helmbold (*in* Cherniss and Helmbold, 1957, p. 275) with two changes I have introduced myself. Of these, the way I render ἀνέχεσθαι is important. Ἀνέχω may be translated both as "to hold up" or "to lift up." I prefer the latter here because otherwise it becomes almost tautological with ἵστημι, "to make to stand up."

[89] "Everything connected with Empedocles' cosmology is now controversial" (Barnes, 1981, p. 308). Guthrie (1965, p. 189) thinks it significant, as does Bollack (1969b, p. 248). Bollack says, though, that Empedocles would have attached the same significance to rocks precipitated from hot waters because for him the involvement of heat was the main thing. Also, see the other references in Guthrie.

[90]These lines of Lucretius remind me of Leonardo da Vinci's notes:

"The summits of the mountains in course of time rise continually.
The opposite sides of mountains always approach one another.
The bases of mountains are always drawing close together.
During the same period of time the valleys sink much more than the mountains rise."

(MacCurdy, 1954, p. 309; for a geological assessment, see Şengör, 1991d, p. 418). I wonder whether Leonardo had on his lap an open copy of his great countryman's immortal poem while writing down these thoughts?

[91]We know very little about the life of this man, except that he lived in the latter half of the fifth century B.C. and was one of the more notable "rivals" of Socrates in instructing the youth in Athens. Already in antiquity, he was confused with Antiphon the rhetorician, and with Antiphon the tragedist. There is little doubt that Antiphon the Sophist and Antiphon the Seer are the same man. See the extended discussion on his identity in Freeman (1949, p. 391 ff.).

[92]Sigurdsson (1999, p. 36) writes that "Another early scholar who pondered volcanic eruptions was the Greek Anaxagoras of Clazomenae (ca. 500–428 B.C.), who said that a mysterious substance called ether sank into the hollow interior of the Earth, where it mixed with vapours causing lightening and fire, which were forced towards the surface." His source for this quote is Thompson (1988, p. 3), who in turn cites Geikie (1905, p. 13) and Adams (1938, p. 400). Only Adams says (without citing his source) what Thompson seems to be quoting, and I think it is a misunderstanding of what Hippolytus says in his *Refutatio Omnium Haeresium* (DK59A42): "The winds originate by thinning out of air under the influence of the Sun and by streaming of heated parts in the direction of the heavens whereby they are pushed away. Thunder and lightening form by the heat that enter into the clouds. Earthquakes originate when the upper air and the air under the earth collide; because when these move, the earth encountered by them shakes."

The following passage, cited by Aetius in his *Placita* (DK59A71), may have contributed to the confusion into which Adams fell: "Anaxagoras [says] that the surrounding aether is fundamentally a fiery substance. Through the force of its rotation it violently tears pieces of crags from the earth and raises them high. It ignites these and thus makes them into stars." Nowhere in these passages do I see a suggestion of volcanism. (However, the unmistakable resemblance of these words with Musaeus' statement that "Shooting stars are borne up from the ocean and generated in the Aether" {DK2B17} may possibly invite suspicion in terms of volcanism as suggested above {Ch. II}.) Neither do I see a coherent theory of tectonism save for an earthquake mechanism. Anaxagoras was mentioned to have been a student of Anaximenes (Gomperz, 1901, p. 209, finds this tradition contradicted by the evidence of dates), but he also incorporated into his teachings the views of the Italian Greeks, including Empedocles. Anaxagoras was a thinker and observer of remarkable talent and considerable originality. I should not be surprised if much of the Socratic/Platonic earth model, which was to have such an immense and long-lived influence later, had originally come out of his brain (possibly without the religious overtones placed on it by the Athenian couple). After all, it was Anaxagoras who brought the Ionian natural philosophy to Athens. It is regrettable that so little of his writings came down to us.

[93]See Freeman (1949, p. 396), where she says that these statements by Antiphon were derived from observations on volcanic eruptions, but Fraustadt and Prescher (1956, p. 51) point out that his view was based on similar, earlier views.

[94]Plato (428/7–348/7 B.C.), according to Karl Popper (and many others), the greatest philosopher of all times, hardly needs an introduction, for we continue to live with so many of his concepts. I here cite only two biographies that I find very useful: von Wilamowitz-Moellendorf (1919a, 1919b) and Taylor (1936). No study of Plato can be satisfactory without Popper's immortal *Open Society and Its Enemies*, v. I (Popper, 1966). Also see Guthrie (1975). Grube's (1980) little book, although it is outdated and contains nothing on Plato's geology, is still a very useful introduction to Plato's thinking for nonphilosophers and for those who cannot read Plato in his original Greek. The 1980 edition has a useful introduction by Donald J. Zeyl (p. xvii–xxi), who also added to it a new bibliography (p. 321–331) and a bibliographic essay (p. 307–319) to highlight some of the newer developments in Platonic scholarship. Mannsperger (1969) is a philological study of the concept of "nature" (φῦσις) in Plato's philosophy, which I have found useful in illuminating Plato's view of nature in general.

[95]*Phaedo* is thought to belong to those Platonic dialogues classed between the "Socratic" ones, written early in Plato's life when he was a writer and under the direct influence of Socrates, and those written towards the end of his life. Bluck (1955, p. 144) thinks that 392 B.C. would be the earliest date for the composition of *Phaedo*. Gallop (1975, p. 74) agrees and puts the time of its writing "more than a decade after the events it purports to describe" (thus, after 389 B.C.). These ascriptions should be considered, however, in the light of Rowe's (1993, p. 11–12) comments on some of the usual contextual arguments used to chronologically order Plato's dialogues. I ascribe everything in *Phaedo* that I quote to Plato without further ado because (1) it does not matter for my historical reconstruction even if Socrates did have a major contribution, and (2) the earth description (Frank, 1923) has rather too much detail for Socrates' taste and it is clearly Pythagorean (also see Gallop, 1975, p. 223; 1993, p. x). Moreover, we know that about a decade after Socrates' demise and probably at about the time while Plato was engaged in the composition of the *Phaedo*, later authorities such as Diogenes Laertius (III.18) recorded that Plato went to Sicily with the express desire of viewing the Mount Etna (cf. Guthrie, 1975, p. 18, footnote 1; p. 336, footnote 3). T.J. See (1907, p. 240) implicitly also ascribes all the geological discussions in *Phaedo* to Plato by saying that Plato "was an excellent geologist for his time."

[96]One of the characters in the dialogue. He was a passionate Theban youth who offered money for Socrates' release.

[97]ῥύαξ (meaning a rushing stream, a torrent) is the word that Plato here employs. The meaning of lava is clear from the context, as it is, for example, in Tuchydides' ὁ ῥύαξ τού πῦρός.

[98]Gallop (1975, p. 66) translates the same passage as follows: "… and continuous underground rivers of unimaginable size, with waters hot and cold, and abundant fire and great rivers of fire, and many of liquid mud, some purer and some more miry, like the rivers of mud in Sicily that flow ahead of the lava-stream, and the lava stream itself; with these each of the regions is filled, as the circling stream happens to reach each one on each occasion."

[99]As late as the early nineteenth century, Sir Humphrey Davy thought that air could freely circulate in the earth, going in and out of volcanic craters, and thus contribute to the oxidation of its crust. He developed this opinion mainly on the basis of observations he and others made of south Italian volcanoes (see Davy's 1805 lectures in the Royal Institution: Siegfried and Dott, 1980, especially p. 133) and thought that the caverns underneath Solfatara communicated with Vesuvius and that the two volcanoes were alternately active. He threw a piece of paper into Solfatara's crater. As the paper was sucked in, Davy thought this indicated an ingoing current of air fueling the oxidizing and heat-producing chemical reactions beneath Vesuvius (Sigurdsson, 1999, p. 163). Is Jules Verne's *Journey to the Centre of the Earth*, in which the hero, Professor Lidenbrock, an ardent follower of Davy's theory, not a product of a similar thought? And is that not why Michel Serres calls that wonderful novel "the perfect novel of the Empedocles complex!,"and in which Axel, the nephew of Professor Lidenbrock, exclaims: "Oh, what a journey! What a wonderful journey! We had gone in by one volcano and come out by another, and this other was more than three thousand miles from Sneffels, from that barren country of Iceland thrust into the middle of the world!" (Verne, undated[1864], p. 216)? Verne had been introduced by his publisher, Jules Hetzel (1814–1886), to the geologist Charles Sainte-Claire Deville, Élie de Beaumont's successor in Collège de France. Sainte-Claire Deville was probably his source for Davy's ideas (Verne, 1976, p. 69; Gèze, 1995, p. 83; also see Sigurdsson, 1999, p. 167). Gay-Lussac showed the free communication of air within volcanic channels to be not possible, because the pressure of the high-density magmatic liquid is directed outward (Sigurdsson, 1999, p. 164).

[100]I follow Robin (1941, p. LXXI) in thinking that Plato here must mean along the full *radius* and *not* the diameter, mainly because of the reference to Homer's statement that the greatest chasm connects the surface of the earth with the Tartarus. But Tartarus in a spherical earth can only be in the center as already acknowledged at least since Empedocles (see Fig. 11B). Rowe (1993, p. 281) introduces this possibility, but is incredulous about it: "'Right through the earth' might just mean 'right through to the center,' but this hardly seems the most obvious interpretation." Why not? Rowe does not say.

[101]Gallop (1975, p. 223) calls these speculations "geophysical theories" and points out with some measure of surprise that Socrates of the dialogue shows more knowledge of such speculatons than his professed ineptitude for natural science might have suggested. This is the main reason, as I said above, for my ascribing all of these natural scientific theories to Plato.

[102]See text below (Ch. VIII). Some may wish to put the later limit of the reign of Plato's theory into the middle of the seventeenth century, when Descartes published his *Principles of Philosophy* (see below). Although I believe Descartes' theory of the earth to have had a profound effect on the development of geology by influencing Steno, its direct contribution to the invention of theories on the internal geodynamics of the globe seems to have been limited.

[103]In this book, the "reversibility" of falcogenic structures must be understood within the restriction von Bubnoff gave it: "No earth-historical process is reversible *sensu stricto*" (von Bubnoff, 1931, p. 40).

[104]The *Phrygia Katakekaumene* of Strabo (XII. 8. 18–19). For an excellent modern study of the geology of this remarkable region, see Richardson-Bunbury (1992).

[105]For Aristotle in general, see Gomperz (1912), Ross (1923), and Guthrie (1981), although Guthrie's volume is somewhat disappointing in its coverage. Jonathan Barnes' *Aristotle* in Oxford's "A Very Short Introduction" series (Barnes, 2000) is outstanding. Also see Thomson and Missner (2000). For Aristotle's intellectual development, see Jaeger (1948). For Aristotle as a scientist, see especially Duhem (1913, ch. IV), Manquat (1932), and Lloyd (1996). For Aristotle's physics, which was the basis of his geology and for almost all of the tectonic theories of the Middle Ages, see Lang (1998). For Aristotle's geology, the best sources are still Adams (1938, p. 15–19) and Duhem (1958a, p. 240–245), and yet both are very inadequate. Also see Duhem (1909[1984], p. 284–285), Strohm (1935, 1970), Ellenberger (1988, p. 16–18), and Büttner (1979a, 1979b).

[106]See endnote 58 for references regarding Theophrastus' life and works in addition to those given here.

[107]For the various arguments for and against authenticity, see Lee (1952, p. xiii–xxiii). The current consensus is that it is not Aristotelian, although it continues to be included in the Aristotelian corpus (e.g., in the new *Revised Oxford Translation* of *The Complete Works of Aristotle*, v. I, p. 608–625: Barnes, 1984). Perhaps it should now be transferred to the Theophrastian corpus.

[108]Aristotle also mentions a freshwater source in the Pontus: "Here at about three hundred stades' distance from the shore fresh water comes up in a large area, an area not continuous but falling into three divisions" (*Meteorologica*, 351a). Since the Caspian is saline, it is unclear what the relevance of this observation is to the alleged subterranean connection between the Caspian and the Black Sea.

[109]This is precisely what Kircher thought some two millenia later, in 1665, mainly owing to Aristotle: see Kircher's (1665) *Tabula geographico hydrographica motus oceani, currentes, abyssos, montes igniuomos*, v. I, p. 124; the *Mappa maris mediterranei fluxus currentes et naturam motionum explicans*, p. 152; and Book II, ch. 19 and the whole of Book III. (*Note*: All page numbers of Kircher's books are given in the present work according to the copies I have in my private library and which I reference. Slight differences in pagination have been detected among individual copies even of the same edition {cf. Godwin, 1979, p. 2}. In such cases, my page number references can only serve as general guides rather than as precise locations.)

[110]Some may attempt to justify this by reference to Aristotle's passage in *Meteorologica* (367b9–12), where the Stagirite ascribes the internal heat of the earth to disintegration of air into small particles in subterranean channels and their subsequent collision, which causes ignition. Regardless of its origin, the resultant "fire" is internal to the earth and, according to Aristotle, generates phenomena "from within" which we today also think are related to the internal processes of the planet. That is why, See (1907) likens the earthquake theories of Plato and Aristotle to his own, based on the idea that ocean water leaks into the earth and produces channels of steam and lava. It is allegedly the violent motions of the latter two that produce earthquakes.

[111]*Ruah* in Hebrew and *Rûh* in Arabic, both meaning breath (Ar: *nefes*), exhalation (Ar: *tabahhûr*), and soul!

[112]Pfeiffer cites an appendix (*Kitab ad-Dalâil*) to an Arabic manuscript of the Apocalypse visions of Daniel (*Kitab as-Sudâ*) dating from 1446/47 A.D. The *Kitab ad-Dalâil* was originally written by the Syrian Nestorian scholar Al-Hassan ibn Baklul (Al-Hasan ibn {al-} Bahlûl: second half of the tenth century A.D.: Sezgin, 1979, p. 332). It was discovered, edited, and translated by Gotthelf Bergsträsser, who published it in 1918 (for the reference to Bergsträsser's paper; see Sezgin, 1979, p. 217, footnote 1). The 42nd chapter of this book includes the Arabic translation from the Syriac of the *Meteorology* by Theophrastus. Bergsträsser regarded the book as an important document for the history of meteorology, and Sezgin agrees (see Sezgin, 1979, p. 217, for a detailed discussion of Bergsträsser's discovery and the literature about it; for the contents of the *Kitab ad-Dalâil*, p. 282–283). In that book Theophrastus is supposed to say the following concerning earthquakes: "…when water is imprisoned in a cavity of the earth and when it moves owing to a narrow outlet it has found or some other reason, it thus moves the earth as a ship is moved by the waves" (Pfeiffer, 1963, p. 22). I have been unable to find a reference to Bergsträsser's paper or to this fragment of Theophrastus' *Meteorology* in Fortenbaugh et al. (1993a).

[113]More correctly, of "new breaks," for many of the subterranean galleries and passageways may be regarded as pre-existing breaks that are parts of the earth's fabric.

[114]Pliny does not cite Theophrastus among the sources for his book II. But we know that he knew of him (see his *Preface*, section 29). He could have also easily learned Theophrastus' opinions second-hand. We do know, however, that some of Pliny's knowledge of the Mediterranean islands does go back directly to Theophrastus (e.g., Sallmann, 1971, p. 216, footnote 52).

[115]I owe my knowledge of the existence of this "exception" to Serbin's excellent doctoral dissertation (1893, p. 19–22).

[116]For a geological map (~1/50,000 scale) and a brief description of the geology of the Methano Peninsula, see Davis (1957), in which bookplate VI illustrates the tectonic setting of the Methano volcanism. See Jacobshagen (1986) for a modern interpretation of the palaeogeographic setting (p. 223–229) and tectonics (p. 250–256).

[117]Neumann and Partsch (1885, p. 307), Sapper (1917, p. 45), but see von Seebach (1869). Tozer (1882[1974], p. 135) says "about the year B.C. 282" but without citing reasons. Baladié (1980, p. 163, footnote 119) simply cites "the 3rd century eruption." Reck (1910, p. 293) impossibly ascribes the first description of this event to Aristotle!

[118]See Jones (1917, p. 219, footnote 3) about Strabo's odd definition of the Hermionic Gulf.

[119]Strabo uses the Eratosthenian stadia, i.e., one stade being considered here equivalent to 157.50 m (1/250,000 of the length of the equator) following Berthelot (1930). Thus, for 7 stadia we obtain a height of 1102.5 m, which is excessive. But the modern equivalent of the Greek stade is a matter of much dispute today—as it was already for the Arab geographers of the ninth century A.D. (Sezgin, 2000a, p. 94)—and in the modern literature of the history of cartography it ranges from 185 m to 148 m. Even if one takes the minimum value, the height reported by Strabo still remains above 1000 m, as opposed to the 743 m which marks the highest spot on the Methano Peninsula. For literature on the Greek stade, see Harley and Woodward (1987, p. 148, note 3).

[120]For an excellent overview of the medieval European civilization, which is so often almost completely ignored in histories of geology, see von Eicken (1887) and Le Goff (1988). The great monastic historian, Dom David Knowles, in his *The Evolution of Medieval Thought*, provides an excellent introduction to the philosophic thinking in the Middle Ages as it evolved from the Greco-Christian synthesis. One must read this book, however, with the knowledge that Knowles was a conservative Catholic theologian. The second edition of this classic book (Knowles, 1988) has been updated by means of an introduction and new references by Christopher Brooke and David Luscombe. Russell (1945[1972]) remains an indispensable guide to all readers interested in medieval thought, as he is for the whole of the history of philosophy.

[121]It is difficult to agree with Ellenberger, who sees the description of a newly discovered geostrophic cycle presented in Chapter 18 of the *Rasâ'il Ikhwan as-Safâ'*, "which we would in vain search with the Greeks and the Romans" (Ellenberger, 1988, p. 80). There is not one idea, nor a single observation there that had not been expressed by the Greeks before, especially by Aristotle, Theophrastus, Eratosthenes, and Strabo, and commonly in much more sophisticated contexts.

[122]*Aeneid*, VI. 581: "Here, the ancient sons of the Earth, the Titan's brood, hurled down by the thunderbolt, writhe in the lowest abyss."

[123]For William of Auvergne, see Thorndike (1923b, p. 338–371) and Duhem (1958a, p. 109–110; 1958b, p. 249–260). William was the Bishop of Paris from 1228 to his death in 1249. He was a man who respected science and favored scientific investigation. This attitude, however, got him into trouble because it ultimately clashed with his conviction that God's will was above all. Duhem (1958b, p. 250 ff.) presents a fine discussion of the importance of William's *De Universo*.

[124]Like Sarton (1931b, p. 934), I prefer the earlier date for his birth, as does Thorndike (1923b, p. 517–592), despite alleged childhood memories of Albert. See also Wallace (1981) and Angel (1995, p. 18–47).

[125]We know very little about Bacon's life, and what little we do know mostly comes from his own writings. For good summaries and references, see Thorndike (1923b, p. 616–691), Burke (*in* Bacon, 1928, p. xi–xiii), Sarton (1931b, p. 952–967), and Crombie and North (1981).

[126]Former Paris lawyer and secretary to Louis IX, Guy de Foulques (or Fulcodi), entered the Church upon his wife's death and was made a cardinal. He was elected Pope on 5 February 1265 and died in 1268. Burke (in Bacon,1928, p. xi) says that while in Paris, de Foulques had every opportunity to hear of Bacon and his work. He was evidently so impressed that after his elevation to the Papacy as Clement IV he wrote a letter to Bacon in 1266 directing him to transmit all his writings without delay. Upon this invitation, Bacon rapidly composed the *Opus Maius*, *Opus Minus*, and *Opus Tertium* and sent them to the Pope together with a copy of his previously written *De Multiplicatione Specierum*. The *Opus Maius* was also accompanied by a map (unfortunately now lost), which Crombie and North (1981) believe was in a rectangular projection, whereas Sezgin (2000a, p. 216) thinks it had a globular projection. We do not know whether the pontiff ever received the books and the map, as he died a few months after Bacon had dispatched them. But he must have, for shortly after Bacon sent off his parcel, he was released from close watch and allowed to return to Oxford. That Bacon was able to compose in such a remarkably short time three major treatises dealing with diverse topics is a testimony to his great learning.

[127]The hostility of the church to profane learning began to wane in the eleventh century following the ecclesiastical reforms (cf. Russel, 1945[1972], p. 385; for a fine history of the twelfth century awakening, see especially ch. IX–XII). For a collection of widely ranging and more modern essays on the twelfth century renaissance, see Young (1969). I note that the twelfth century Renaissance was in large part caused by the teaching in the University of Constantinople, whose students came from nearly all quarters of the Old World and carried back to their homes the knowledge they acquired in the Byzantine capital (Diehl, 1957, p. 105–107), and by infiltration into Europe (via Spain, Sicily, and Constantinople) of ancient learning preserved in Syriac and Arabic translation plus the huge amount of new knowledge developed by the Muslims themselves since the ninth century A.D. For an excellent book on the European debt to the Muslims in civilization, see the standard volume by Hunke (1960). For geography, Miquel (1973, 1975, and 1980) and the recent volumes by Sezgin (2000a, 2000b, 2000c) are the most outstanding sources I know.

[128]Twice the *Rector* of the University of Paris (1328 and 1340), commonly known for "Buridan's ass," i.e., when placed between a stack of hay and a bucket of water, dies of thirst and hunger for not being able to make up its mind as to which one it should go to first (Sarton, 1947, p. 540–546; Moody, 1981a). Buridan was one of the most original thinkers of the European Middle Ages.

[129]See endnote 136.

[130]This is a book that was known in the Middle Ages under the various names of *De Elementis*, *De proprietatibus elementorum*, *De causis libellus proprietatum elementorum*, and *De naturis rerum*, and was translated by the famous Italian translator Gerard of Cremona (1114–1187) from the Arabic (*editio princeps* 1496; Steinschneider, 1960, p. 204). Sarton (1927, p. 135) considers it most definitely a Muslim work, probably part of a larger, encyclopaedic book, although it was credited to Aristotle probably by Gerard (Schmitt and Knox, 1985, p. 20), possibly because of mentions of "our" book "on the heavens and the earth" (see Steinschneider, 1960, p. 204). The real author must have been one of the better-known Muslim philosopher-astronomer-cosmologists, as he knew Bagdad (Steinschneider, 1960, p. 205). Steinschneider (1960, p. 204) points out that the great Turkish philosopher-scientist Abu Naşr Muhammed ibn Muhammed ibn Tarhan ibn Uzlug al-Farabi (870?–950) had been also considered among its possible authors, although Adıvar (1964) does not mention this. Schmitt and Knox (1985) pointed out that S.L. Vodraska, in an unpublished doctoral dissertation entitled *Pseudo-Aristotle, De causis proprietatum et elementorum* (University of London, 1969; I have not seen this thesis), presented evidence to argue that the original work had probably been written sometime between 830 and 875 somewhere in Iraq, possibly in Basra. This "geological treatise" (Peters, 1968, p. 57) is apparently only the first part of a larger work as previously supposed (Schmitt and Knox, 1985, p. 20). Duhem (1909[1984], p. 299–302) discusses its basic content. The unknown author denies that seas and lands have exchanged places in the past because such a change, he believes, must be a result of astral motions, and these are far too slow to bring about the alleged changes in historical times. Mountain-building in this book is ascribed to internal processes. The unknown author criticizes the ideas that considered the earth to be a perfect sphere before the deluge, and he ascribes the present irregular topography to diluvial and post-

diluvial processes; if the earth had really been a perfect sphere before the flood, it should have been covered by a uniform bed of water and any rain would not have been able to affect the subaqueous floor.

[131]*Meteorologica* (352 b):

For the land of the Egyptians, who are supposed to be the most ancient of the human race, appears to be all made ground, the work of a river. This is clear to anyone who looks at the country itself, and further proof is afforded by the facts about the Red Sea. One of the kings tried to dig a canal to it. (For it would be of no little advantage to them if this whole region was accessible to navigation: Sesostris is said to be the first of the ancient kings to have attempted the work.) *It was however found, that the sea was higher than the land*: and so Sesostris first and Darius after him gave up digging the canal for fear that the river should be ruined by an admixture of sea-water." (italics mine)

Tricot (1955, p. 79) says that Aristotle's error of thinking the Red Sea to be higher than Egypt was shared by all antiquity and the modern times.

[132]*Quaestiones Naturales* (III, 27 and 28), where Seneca describes the "rise of the outer sea" where these is a "heap" of water that will overrun the world during the coming flood that will wipe out the world.

[133]The Neoplatonic philosopher, who lived in the sixth century and is known through his commentaries on Plato and Aristotle. He is the last of the Alexandrinian philosophers whose writings on Plato and Aristotle have reached us.

[134]Daly (1981). *De Sphaera* was published ca. 1220, certainly before the publication of Robert Grosseteste's book entitled *Compendium Sphaerae*.

[135]Over-estimation of the heights of mountains was commonplace until close approximations by barometric observations became frequent in the eighteenth century (see, for instance, Zimmermann, 1861, p. 332, for a comment to this effect).

[136]Albertus de Saxonia or Albertus de Helmstede (Albert of Helmstaedt): see Duhem (1906[1984], p. 319–320); he seems to be the same person as Albertutius and Albert de Ricmestorp (Sarton, 1948, p. 1428: *contra* Duhem!) and whom the philosopher Augustin Nipho called Albertillus and Albertilla (Duhem 1906[1984], p. 331–334). For what we know of his life, see Duhem (1906[1984], p. 319–338), Sarton (1948, p. 1428–1432), and Moody (1981b).

[137]For a vivid picture of Europe during these times that provides a cultural backdrop to the intellectual developments discussed herein, see the following two superb books: Tuchman (1978) and Manchester (1993). See also Le Goff (1988).

[138]Also known as the Jacopo Angelo of Florence (Münster, 1540, in *Epistola Nuncupatoria*, third unnumbered page); also referred to as such by Ptolemy (undated, title page) and Gallois (1890, p. 197).

[139]I think the Reformation and the Counter-Reformation can be best understood by the explanation that Lord Russell gave to their essence: rebellion of less civilized nations against the intellectual—and financial, I might add—domination of Italy (Russell, 1945[1972], p. 522; for financial aspects, see Manchester, 1993, especially p. 131–145). But Russell's formulation expresses an old idea: see, for instance, Gallois (1890, p. XVIII), where he says that the Renaissance Germans did not like Italy because they hated being treated as barbarians.

[140]"On Mountains, Forests, Springs, Lakes, Rivers, Swamps or Marshes, and on the Names of the Sea."

[141]I cite here Nicolò Liburnio's Italian translation listed under Boccaccio 1473[1978].

[142]1508 is the date of the *editio princeps*, which is the only text we have. I cited here a critical edition and translation by Perler (1994, p. LXXV–LXXVII for the text here cited.), which also has an extensive introduction. Also see Adams (1938, p. 341–342). Adams cites an English translation by C.H. Bromby entitled *A Question of the Water and of the Land*, published in 1897 in London. I have not seen this translation, and I do not here discuss the question of authenticity of Dante's text, becuse it is immaterial for my purpose.

[143]This title is commonly translated as *Introduction to Geography* (e.g., Pagani, 1990, p. IV) or *Manual of Geography* (Dilke, 1985, p. 76), but Ὑφήγησις (*uphegesis*) I think is best rendered into English by "guide," which is what the word actually means and which was what Ptolemy intended his book to be (see especially Berthelot, 1930, p. 9, 113, 117; on p. 146, Berthelot expressly translates Γεωγραφιχὴ Ὑφήγησις as *Guide géographique*. Also see Sezgin, 1987b, p. 2; and Istituto Italiano di Cultura di Istanbul, 1994, p. 39: "*la guida geografica di Tolomeo*").

For the Greek text of Ptolemy's book, the only comprehensive authority is still the edition of Nobbe (1843–1845) originally published in three volumes but reprinted in a single volume and enlarged with a new introduction by Aubrey Diller in 1990 by the Georg Olms Verlag publishing house. To judge Ptolemy's impact on the Renaissance world, I have examined the two Latin Ptolemy facsimiles produced by the Theatrvm Orbis Terrarvm Ltd., Amsterdam. These are the Bologna (1477) and the Rome (1478) editions. The same company also reproduced in facsimile the enlarged Ptolemy atlases with *tabulae novellae*, which I also examined, *viz.*, Florence (1482, in Italian verse), Ulm (1482), Venice (1511), Strassburg (1513), and the Basle (1540). In addition, I also consulted the magnificent facsimile of the *Codex Valentianus Latinus* (University of Valencia Library, no.1895 in the catalogue of Gutiérrez del Cano; facsimile in 3000 copies by Vicent Garcia Editores S.A. in 1983, Valencia) translated by Jacopo Angelo da Scarperia, associated English translation and commentary by Victor Navarro Brotons and his associates (no date, no place of publication). For the reduced reproduction of the maps of another Latin codex with Jacopo Angelo's text, but with a different cartography in the Bibliothèque Nationale in Paris (*Manuscript Latin* 4802), see Ptolemy (undated); these maps resemble in the style of their decoration those of *Codex Urbinas Latinus 277* in the Vatican Library. To be able to form an opinion on Ptolemy's possible influence on the Ottoman world during the Renaissance, I also looked at the facsimile of the Arabic "adaptation" of Ptolemy's *Geography* (1465 A.D.; original in the Ayasofya collection in the Süleymaniye Library, Istanbul, under MS Ayasofya 2610), edited by Fuat Sezgin and published by the *Institut für Geschichte der Arabisch-Islamischen Wisenschaften and der Johann Wolfgang von Goethe-Universität, Frankfurt am Main, Veröffentlichungen des Institutes für Geschichte der Arabisch-Islamischen Wisenschaften, Rheie D (Kartographie) Band 1, 1987.*

Here I must stress that both Ptolemy's text and his maps are considered extremely problematic owing to divergences in some 40 extant manuscripts, the oldest of which goes no further back in time than the twelfth century (Dilke, 1985, Table 11.1). Opinions widely diverge as to the authenticity of the texts available, ranging from those who consider certain manuscripts fairly reliably Ptolemaic, while others deny almost the whole book to Ptolemy! Berthelot (1930) and Fischer (1932a) pointed out that the best texts are naturally those that agree with the excerpts quoted by authors closest to Ptolemy's time (such as Ammianus Marcellinus and Martianus of Heraklea). Of these, the best are said to be the *Codex Urbinas Graecus 82* published and commented upon by Fisher (1932a), the *Codex Florentinus Laurentianus Graecus Plut. XXVIII*, and the Paris Codex 1404, plus the now lost manuscript that formed the basis of the 1513 Strassburg edition published by Jacobus Eszler and Georgius Ubelin (Berthelot, 1930, p. 112; Fischer, 1932a, p. 208–415). The great geographer Martin Waldseemüller ("Hylocamilus") drew the maps of this edition. Fischer (1913; also see Fischer, 1932a, p. 340–342) also pointed out that the *Codex Ebnerianus* in the New York Public Library collections was an important manuscript (Fischer, 1932b[1991]). Bagrow (1945, 1947, 1964), by contrast, is willing to concede to Ptolemy only Books I, II (only ch. 1), VII (only ch. 5), and VIII (only ch. 1 and 2), i.e., only those giving the techniques of map-making! Bagrow thought that the geographical parts had been added later probably on the basis of some later atlas (also see Polaschek, 1959). Therefore, we are not entirely sure if it was the geographical knowledge of the second century A.D. that was being revived in the Renaissance by the Ptolemy translations and editions, but the people of the Renissance thought it was.

Since there still is no critical edition of the entire text of this milestone work (although we now have *almost* the entire text in critical editions (see, Müllerus {1883, 1901}, Renou {1925}, Ronca {1971}, Humbach and Ziegler {1998, 2002}, and Schmidt {1999}), such controversy is difficult to resolve. Until some expert in classical Greek, cartography, and history of geography takes it upon himself or herself to attempt a complete critical edition, all one can do is to sift through the massive and not always useful literature (cf. Stevens, 1908; Stahl, 1953) and distill the texts considered best and to work with them, comparing them with the oldest maps available. While writing this book, I think I managed to see a good part of the published material, including one complete (*Codex Valentianus Latinus*), one partial (maps only: *Codex Ebnerianus*) manuscript facsimile, and the originals of the great codices *Urbinas Graecus 82* in the Vatican library and of the *Seragliensis 57*, which is housed in the Topkapi Palace Library in Istanbul, my home city (cf. Deismann, 1933, p. 89–93).

[144]Russell (1945 [1972], p. 523) calls the century following the Reformation "barren" in philosophy; it was equally barren in the earth sciences until Descartes, despite Agricola. In geography, the second moitie of the century was the era of "*local* maps" (Gallois, 1890, p. 238–239, italics mine).

[145]Sigurdsson (1999, p. 47) attributes the idea that sulphur was the main agent of combustion in volcanoes to Seneca. But the association of fire with sulphur in volcanic action had already been hinted at by Strabo while describing the Campi Flegrei. That this is a very old idea is seen in the reference in the *Iliad* to "[ὄρνῡμι] Θείου καιομένοιο" ([to incite] the flame of sulphur) where Θεῖου stands for sulphur (VIII, 135).

[146]For Agricola, see Darmstaedter (1926, which has a good bibliography of writings on Agricola prior to 1926), Adams (1938, p. 183–195, 342–344), Hartmann (1953), Shukhardin (1955, with a helpful bibliography especially for the former "socialist block" literature), Wilsdorf (1956, p. 82 ff.), Fischer (1994), and Prescher and Wagenbreth (1994). Also see the contributions in Naumann (1994) and the news of ongoing work on a new Agricola bibliography in Franz (2000).

[147]It is not exactly clear where these mountains are. The reference Agricola uses is Pliny, who in the second book of his *Historia Naturalis* writes the following in paragraph CX: "Mount Chimaera in the country of Phaselis is on fire, and indeed burns with a flame that does not die by day or night … Also the Mountains of Hephaestus in Lycia flare up when touched with flaming torch, and so violently that even the stones of the rivers and the sands actually under water glow." From this one would think that the Chimaera is near Phaselis (near present Tekirova in Turkey, 36°31′ N, 30°32′ E) and that the Hephaistian Mountains are elsewhere, though also in Lycia. However, Strabo (XIV. 3.5) unhesitatingly puts the myth of the Chimaera into the vicinity of the Cragus Mountains (southern part of the present-day Baba Dağ, some 75 km due W of Phaselis as the crow flies; this is the identification adopted by the monumental *Barrington Atlas of the Greek and Roman World*: Talbert, 2000, map 65). Forbiger (1844, p. 252f.) cites a passage in the *Iliad* (VI. 177) implying that the Chimaera is indeed near the Cragus Mountains, although I find that passage not nearly as clear as Forbiger thinks, because of the mention of Bellerophon's war with the "glorious Solymi" (i.e. the people of Phaselis! See Takmer, 2002, for more geographical detail on the Solymi) right after he slays the Chimaera (we might, of course, also consider that Homer may not have been very clear concerning the geography he was singing about!). There is a mention of the Hephaistian land and/or town near Phaselis (see Forbiger, 1844, p. 253), so Forbiger prefers to follow Strabo in putting the Chimaera into the western part of Lycia and placing the Hephaistian Mountains near Phaselis. Although all modern popular historical geographical texts and tourist guides put Chimaera near Phaselis, the great epigrapher George E. Bean was of the opinion that this name is allochthonous and that it must have been transferred here from its original location in the west as repoted by Strabo, owing to the presence of fires fed by natural gas: "The transference of the name to Yanar in the extreme east seems to be due merely to the fact that the Chimaera breathed fire" (Bean, 1989, p. 138). So, this great authority on west Anatolian ancient geography also follows Forbiger. Indeed, in a recent study of the orography of Lycia, Takmer (2002) found sufficient reasons for identifying the Cragus Mountains not with Baba Dağ (36°32′ N, 29°10′ E), but with the higher Ak Dağ (36°32′ N, 29°33′ E). This brings the Chimaera farther eastward, but not nearly enough to make it coincident with the present Yanar or Yanartaş. He also cites the periplus of the Pseudo-Scylax to the effect that above the Siderous Promontory (present-day Adrasan Burnu: 36°21′ N, 30°32′ E) there is the holy ground of Hephaistos with eternal fire emanating from the ground (see Takmer, 2002, p. 46). In the orography of Lycia, however, Takmer does not list a separate Hephaistos Mountain. On the basis of this somewhat imprecise database, I am unable to form a firm independent opinion as to the whereabouts of the Hephaistian Mountains referred to by Agricola, but the majority opinion seems to favor the view that they probably are a part of the Tahtalı Mountains, i.e., the Lycian Olympus. It is likely that Agricola did not know either where the burning mountain in Lycia was located.

[148]The Aristotelian theory for the origin of earthquakes remained current at least until the eighteenth century as shown by the publications that followed the "100-fold increase" in earthquake frequency in 1750 in England and the famous 1755 earthquake in Lisbon! (e.g., see Kennedy and Sarjeant, 1982; Anonymous {*Dissertation upon Earthquakes*}, 1750; and Michell, 1761. See also Oldroyd, 1996, ch. 10). Association of winds with earthquakes was a current question even in the nineteenth century! I quote the following passages from Friedrich Hoffmann's posthumous *Vorlesungen* to give an idea of the kind of observations and ideas that were around in the first half of the nineteenth century:

> Many reports of unusual weather conditions that accompany earthquakes merit no further consideration. Only the oft-repeated remark that sudden winds and electrical meteors immediately preceding earthquakes or, according to some reports, accompanying them, which are compared with sudden winds that imediately precede showers, should be mentioned.
>
> These events were noticed especially in the case of the 1805 earthquake near Naples. The reporter Colares claims to have noticed that the wind suddenly started blowing during the Lisbon earthquakes of 1st, 20th and 24th November. During the 1795 earthquake in England, sudden winds that came from above were curiously noticed in the upper parts of the mines in Derbyshire. Terrible winds were reported accompanying a number of strong shocks in Upper Italy in December 1810. The atmosphere was most unusally disturbed, according to many reports, during the earthquake on the island of Zante on 20 December 1820. Strong storms raged for days, terrible, threatening, dark clouds accumulated and, finally, soon after the shocks, heavy rain followed that created great havoc.
>
> These events deserve indeed to be noticed, but we should not forget that we always find ourselves in a field, in which, for the present, it is impossible to sort the coincidental from the necessary and in which wrong deductions concerning the association of the phenomena are inevitable. In fact a number of cases that became known to us show how easy it is to err here. (Hoffmann, 1838, p. 364–365)

Do not let us forget that Davy's theory of volcanism and earthquakes (Siegfried and Dott, 1980, ch. 9 and 10) was not only compatible with an earthquake-wind association, but actually required it. Hobbs (1909), in his somewhat skewed history of seismology, considered the volcanic theory of earthquakes that remained dominant until the second half of the nineteenth century only a modernized version of Aristotle's theory, which, in a sense, is true.

[149]Many of his contemporaries called him the "Strabo of Germany" (Gallois, 1890, p. 221). Gallois says that Münster was the last of the students of the German school of geography in the Renaissance, but that he takes the first place in it (Gallois, 1890, p. 236). Münster traveled little and probably was never even in France. However, he did climb the Alps and was most impressed with what he saw (Gallois, 1890, p. 226). Münster was more than just a Hebraist and a geographer, though. In 1551, he published his *Rudimenta mathematica* in Basel, which established his reputation also as an accomplished astronomer and mathematician.

[150]For the details of Münster's life and work, see Gallois (1890, p. 190–236), Hantzsch (1898), Burmeister (1963), Büttner and Burmeister (1979), and Anspach et al. (1988).

[151]I here cite the 1550 edition of this book (except where indicated otherwise). It, together with the Latin and the French editions of 1552, is considered the definitive edition (Gallois, 1890, p. 217). Also see Oehme (*in* Münster 1550[1968]).

[152]Claudius Julius Solinus was a third-century Roman grammarian (acme *ca.* 230 A.D.). His book consists mainly of material culled from Pliny the Elder's *Natural History*, but worked in a geographical framework. Karrow (1993) ranks Solinus among the four giants of geography and cosmography in antiquity, the others being Strabo, Pomponius Mela, and Ptolemy, despite the fact that Solinus' *Polyhistor* is commonly considered to be a shallow and unoriginal treatise (Woodward, 1987, p. 299).

[153]All we know of Mela, which is very little, comes from his book. He was a Spaniard and began composing his book sometime in 43 or 44 A.D. For Mela and his work, see Silberman (1988). For an outstanding example of Mela's persistent influence during the Renaissance, see Barradas de Carvalho (1974).

[154]Münster's editorship of these two books published in one volume is only known from a reference to himself as "I, Münster" in a passage of the book (Davison and Kee, 1994, p. 8).

[155]Münster's Ptolemy edition and his commentary were not original. He took Pirckeymer's Latin text, as amended by Villanovus, and used the commentary of Werner. As Gallois (1890, p. 198) points out, his edition was not original, but was "the most accurate."

[156]For a review of Agricola's correspondence, see Horst (1955).

[157]Oehme (*in* Münster 1550[1968], p. VIII) wrote "One of his main sources was the Bible, and he considered all that was said in it to be above doubt." Also see

Lestringant (1994, p. 7). However, it seems that Münster was also aware of the Syriac and Arabic geographical literature. Some similarities between his relation of the geography of the seas and those of Bar ᶜEbroyo (which means Bar Hebraeus) seem to support this conjecture (for comparison with *Cosmographei*, see Takahashi's, 2003, analysis of Bar ᶜEbroyo's marine geography and its sources). It is hardly surprising that an accomplished Hebrist should be familiar with the literature of the Semitic peoples neighboring the Hebrews.

158That is, flatland.

159For the survival of the very same reasoning well into the nineteenth century (i.e., nearly four centuries after Münster) and in geology books that have since become classics of this science, to correlate the pre-diluvian and post-diluvian earth—see, for instance, Conybeare and Philipps (1822, p. lx; note*: "The notice with regard to the rivers flowing from Eden appear to indicate at least partial identity between antideluvian and postdiluvian continents …"

160Mainly after Pliny the Elder (II. 89). Münster shared this source with Georgius Agricola for the same topic (see Agricola, 1544[1956], p. 39[127]).

161"The origin of materials in the earth in five books." I quote here the historical-critical edition by Fraustadt published in 1956 in the series *Georgius Agricola—Ausgewählte Werke*, Memorial Edition published by the *Staatlichen Museum für Mineralogie und Geologie zu Dresden* and edited by Hans Prescher. The pagination I give is that of the first edition of 1544, also given in the edition of Fraustadt.

162In the *Cosmographei,* there are many references to Agricola by name and also quotations from his books without reference to his name. A comparison of the first edition of Münster's book (1544) and the main 1550 edition shows that Münster added much geological material from the books that Agricola had published in the meantime. For a detailed study of the relationships between Münster and Agricola, as reflected in the *Cosmography* of Münster (and its later editions), see Wilsdorf and Friedrich (1954).

163About the *Atlas*, see Anonymous (undated) and the "Bibliographical Note" by R.A. Skelton *in* Mercator and Hondius (1636[1968], p. V–XXVII) and the references cited therein. See also Wolff (1995).

164Büttner pointed out that the English translation of the *Atlas* text may not have been made from the published Latin text of 1595 (see Büttner, 1992, p. 7 for explanation of the Latin title-page).

165French philosopher, mathematician, natural scientist, and soldier, Descartes is one of those few people whose thoughts have never ceased to actively engage the posterity in a beneficial way. As Gillispie (1960, p. 94) rightly observed, Descartes stood at the divide between the antique and the modern (he certainly did so in the history of geology). The literature about him and his ideas is understandably huge. For a brief sketch of his life, see Crombie (1981). For the geologist Descartes, see especially Daubrée (1880) and Gohau (1983a, 1983b, *passim*; 1990, p. 71–74). I must note here that I entirely disagree with Toulmin's (1990) view that had the European world followed Montaigne's wishy-washy and "practical" inclinations rather than Descartes' logical and theoretical thoughts, we would have avoided the problems of "modernity." What Toulmin longs to see had been tried in the pre-Cartesian societies, and the results were seldom short of barbarism. On this extremely important topic of scientific versus the—incorrectly labeled—"humane" (in reality *romantic*) view of nature in general, see especially p. 52 and 53 of Gillispie (1960).

166I emphasize that Descartes' "contractionist" scheme is not a product of *thermal* contraction as Daubrée (1880) and many since mistakenly have asserted.

167Nicolaus Stenonis after his conversion to Catholicism. The literature on Steno is immense and many-faceted, reflecting the many-sided interests of this great man. The most comprehensive biography is that in two volumes by the late Father Gustav Scherz, which he had almost completed by the time of his tragic death in an automobile accident on 29 March 1971 (Scherz, 1986a, 1986b; see Blei, 1991, for a criticism of some of the aspects concerning the history of geology in this biography). A more up-to-date and concise biography is that by Bierbaum et al. (1989). The first volume of the *Stenoniana* (Nova Series 1, 1991, Lægeforeningens forlag, Copenhagen, 159 p.) reviews various aspects of Steno's life, science, and religion. For a short biography, see Studtmann (1934). Steno's complete works are gathered in six volumes published in the first half of the twentieth century. I chose not to refer to them in this book, as the editions that I cite herein are, I believe, more readily accessible, but any serious student of Steno would find having them at hand helpful: *Opera Philosophica* (edited by Vilhelm Maar in two volumes, both published in 1910 by Vilhelm Tryde in Copenhagen), which contains the scientific works; *Opera Theologica* (edited by Knud Larsen and Gustav Scherz in two volumes, published in 1944 and 1947 by Nyt Nordisk Forlag); and *Epistolae* (edited by Gustav Scherz, in two volumes, both published in 1952 by Nyt Nordisk Forlag and Herder in Freiburg). Metzler's (1941) study, showing the difficulties Steno had with the Jesuits in Hamburg, is an extremely interesting document illustrating the gullible part of the great scientist when his faith was involved.

168For the life and works of Hooke, see the contemporary assessment entitled *The Life of Dr. Robert Hooke* (*in* Waller, 1705, p. i–xxviii), which was also the source for Ward's description of Hooke's life in his *Lives of Gresham Professors* (1740). Andrade (1950) is an extremely well-written and very competent account of Hooke's work and his character. Andrade also mentions (p. 153, footnote *) an unpublished thesis completed in 1930 and entitled *The Contributions of Robert Hooke to the Physical Sciences*, which I have not seen. Keynes's (1960) bibliography of Hooke includes a fine list of writings on Hooke to 1960. Hooke's diary has been edited and published by Robinson and Adams (1935). Between its pages xiii and xxviii, the book contains a biography of Hooke entitled "Robert Hooke: A Brief sketch of His Life." For a modern account of Hooke's life, see Espinasse (1956). See Westfall (1981) for more recent material. The most detailed treatment of Hooke's geological work to date is the wonderful book by Drake (1996). For Hooke's geology, also see Davies (1964), Carozzi (1970a), and Ranalli (1982). Oldroyd (1972) discusses Hooke's methodology of science, which he concludes (as did Gillispie {1960, p. 136} before him), was essentially Baconian, but with a great deal of role conceded to hypotheses. From Oldroyd's analysis (see especially his fig. 1), I conclude that Hooke, although a dedicated experimentalist, was closer to a critical rationalist position than to a Baconian one. Oldroyd's paper also has, in its footnote 1, a useful list of sources to introduce the reader to the history of Hooke's life and thoughts. Hooke's studies concerning the sea were reviewed by Deacon (1971, ch. 8).

169For other references to fire inside the earth during the Renaissance, see Sigurdsson (1999, ch. 7).

170One of the last of the truly "Renaissance men," Athanasius Kircher (Fig. 24) was a polymath of great merit. See Godwin (1979) for a summary of his life and work. Of his numerous books, many have portions relevant to the earth sciences, but two deal with the subject directly: the *Iter Extaticum II* of 1657 and and the *Mundus Subterraneus* of 1665.

171Kircher's *Arte Magnetica* Steno had read as an undergraduate and excerpted (Schertz (1969, p. 227, note 69). There is little doubt that the Dane was familiar with the German's very popular writings about the earth.

172Here Kircher compares the earth with the human body, following Seneca (III, 15), and probably also Agricola, and points to the existence of many veins and organs such as the stomach and the heart. He further points out the existence of pores on the skin. All of this must have appealed to the anatomist Steno.

173For a concise view of the nature of submarine commotion and topography while Steno was working on his geology, see Kircher (1657, p. 153). Superficially, there does not seem to be much difference between Kircher's and Steno's world views. In reality, Kircher's description was not scientific in that his scheme was deduced from the Platonic/Aristotelian world view and called for no tests to see whether it corresponded to reality. By sharp contrast, Steno's is the description of a new method whereby statements such as those by Kircher can be tested. All other statements by Steno are based on discoveries he made or on tests he performed on the basis of his new method, which is *stratigraphy*. Although contemporaries, Kircher was the end of the antique geology; Steno, the beginning of the modern.

174The *Edda* has to be used with extreme caution for any local motifs, owing to very clear parallels with the mythologies of the other Indo-European peoples. However, long and severe winters clearly indicate a northern clime, and some passages may well be descriptions of ebb and flood, which do not occur in the Mediterranean to any remarkable extent (<2 m; where they do occur, they give rise to currents in confined places, such as the strong currents in the Messina Strait, which most probably have given rise to the myth of the monsters of

Scylla and Charybdis described by Homeros {*Odyssey*, XII}: see Lisitzin, 1974, p. 258). I am disturbed, however, that clearer references to very conspicuous *ongoing retreat of the sea* are not found in the mythology of a sea-going people, notwithstanding Darwin's consoling words: "In barbarous and semi-civilized nations how long might not a slow movement, even of elevation such as that now affecting Scandinavia, have escaped attention!" (Darwin, 1889[1987], p. 129), because Suess (1888, p. 453) reports that Greenlanders do retain a memory of the Tasersuak ice-dammed lake (near 62°30′ N and 50° W) being a fjord. For the *Edda*, see also Dillmann (1991).

[175]Swedish naturalist and mystic (1688–1772). For his life and contributions, see Benz (1948), Sigstedt (1952), and Dingle (1958). Regrettably, the literature on Swedenborg as the mystic is far larger than that on Swedenborg the scientist.

[176]For von Linné's life and work, see Stoever (1794), Jackson (1923), Hagberg (1940), Lundberg (1957), Blunt (1971), and Frängsmyr (1994a). Both Stoever and Lundberg have a list of von Linné's publications. Lundberg also provides an excellent bibliography of works on von Linné and an annotated list of his students.

[177]One of the best books documenting the rationalization of the flood legend in the Renaissance, which prepared the way to von Linné's cataclysmic earth history and from there influenced Torbern Bergman (who was a pupil of von Linné) and Abraham Gottlob Werner, see Allen (1963). For von Linné's and Bergman's influence on Werner, see Nathorst (1908) and Hedberg, (1969a, 1969b); also see von Engelhardt (1980) and Oldroyd (1996, p. 84 ff).

[178]For the life of this remarkable man, see the excellent introduction in Brock (1990).

[179]The actual word Count Marsilli uses for Moro's *essential* is "*veritable*."

[180]I wish to point out that this discussion is contained in the chapter 12 of Moro's book II. The chapter is entitled "On the Uniform Manner of the Birth of Mountains" (*Nascimento Uniforme di Tutti i Monti*).

[181]The Reverend John Michell (1761), polymath and Woodwardian professor at Cambridge, thought that volcanoes are made by this means only (cf. Sigurdsson, 1999, p. 159 and fig. 12.1), yet he envisaged other deformations, of a much larger wavelenth, to make mountains such as the Andean Cordillera (see Michell, 1761, table XIII, p. 585). This is not dissimilar to Moro's world picture. For Michell's (1724/1725(?)–1793) life and work, see Davison (1927, ch. II). Kopal's (1981) *Dictionary of Scientific Biography* entry discusses only Michell's astronomical work.

[182]The literature on Hutton is vast. No one book does him complete justice owing to his multifarious interests which one individual now finds difficult to embrace. From a geological viewpoint, the combination of writings by Dean (1992) and McIntyre and McKirdy (1997) provide the best introduction. McIntyre's (1997) outstanding description of Hutton's Edinburgh should also be read to understand the milieu in which Hutton worked and thought. Galbraith's (1974) thesis contains much material, but his over-emphasis on Hutton's alleged deism distracts, and in places, I think, misleads the reader. Apart from these, one should read Playfair's (1805) wonderful memorial to his great friend and Hutton's own geological writings, which, contrary to the conventional view, I find to be well-written and extremely clear. Of the newer revisionist literature attempting to shrink Hutton's place in the founding of modern geology, I can hardly recommend any except Davies' (1969) superb book—if it is to be considered revisionist.

[183]For accounts of the life and work of Playfair, see his nephew's biographical memoir (Playfair, 1822) and Challinor (1981).

[184]His friend Alexander von Humboldt called Leopold von Buch "the greatest geognost of our era" (1853, dedication) and Julius Ewald, in his unfinished biography of von Buch, agreed. It is surprising that a major biography of von Buch has never been published. In the first volume of his collected works (Ewald et al., 1867), Julius Ewald promised that each volume covering a certain period in von Buch's life would be accompanied by the corresponding segment of a comprehensive biography. Only the first volume had such an accompaniment, however. As the original promise failed, the newer promise of a separate publication was also never fulfilled, and as Ewald ordered the destruction of all his unfinished manuscripts when he died (von Zittel, 1899, p. V), no complete von Buch biography from his pen ever emerged. (Sir Roderick Murchison once referred to Ewald, whom he regarded "an excellent palæontologist" {Murchison, 1849, p. 223}, as a researcher who seemed "almost to shun publication" {Murchison, 1849, footnote *}). As long as Ewald lived, nobody else attempted to produce a von Buch biography out of deference to Ewald, and when he died, the fashion of "Victorian biographies" was past. Thus, one has to make do with the tremendously scattered von Buch-literature. The following are items I have frequently consulted: Ewald (1867), Günther (1900, p. 183–271), and Wagenbreth (1979a), in addition to von Buch's own works. For an English summary, see Nieuwenkamp (1981). Also see the special memorial colloquium volume in the *Zeitschrift der Geologischen Wissenschaften,* 1974, (Berlin, GDR), v. 2 , no. 12.

[185]Natland (1997, p. 322) calls Von Buch's view "an updated conception of Wernerian doctrine." In fairness to von Buch, many basalt and dolerite outcrops in the region of the Puys in the Massif Central do indeed look as if "laid down" rather than "flowed," to an eye unfamiliar with volcanic terrains (see, for example, de Goër de Herve, 1972, Plate IX, photo 4; and Plate XI, photo 1). Since von Buch was one of the earliest pioneers of modern volcanological research, he is easy to forgive for misinterpreting what he saw, especially when we remember that here he had in his mind the memory of Scheibenberg, which was for him (despite the leucite in the basalts of Rome {cf. Sigurdsson, 1999, p. 208}), still neptunian in origin (see Pompekj, 1925).

[186]Günther (1897, p. 369) corrects this as *barrancas*; see his section on *barrancas* concerning their modern interpretation. However, both orthographies seem to be used, and *barranco* is apparently preferred on La Palma (Middlemost, 1972, p. 35). Aurand (2000, p. 57) implies that there is also a shade of difference between their meanings: "**barranca** ravine, escarpment **barranco** gully, ravine, gorge, fissure, gulch." It thus seems that von Buch's usage is the more correct one.

[187]The structural interpretation of barrancos is also current today, although the dominant role of fluvial erosion is not denied (e.g., De Mulder, 1981).

[188]Although Scrope made a significant impact on the development of geology and wrote the first systematic book on volcanology, the long interruption in his active geological pursuits owing to his political involvement may have been the reason he never received a detailed biography. All existing necrological notices are unsatisfactory. Sturges' (1984) book containing a bibliography of Scrope also has the most detailed biography of him that I know (Sturges, (1984, p. 11–32). Page's (1981) account is another good biographical piece with which I am familiar. Rudwick (1974) contains an excellent analysis of Scrope's geological thinking and its relations to Lyell's thinking. Also, see Decobecq (1993) for Scrope's comparison of the volcanoes of the earth and the craters of the moon. Scrope's remarks may well have influenced Élie de Beaumont and, later, James Dwight Dana (see below). For Scrope's controversy with Robert Mallet (1810–1881), one of the founding fathers of seismology, regarding the nature of the earth's interior and of its causes, see Wilding (1996).

[189]1797–1875: The standard sources of information on Lyell are: Lyell (1881a, 1881b), Bonney (1895), Bailey (1962; Dean, 1992, p. 202, footnote 1, calls it superficial; I do not agree), North (1965), Wilson (1972, 1998) and Rabikovich (1976, with quite a good bibliography containing only a few inaccuracies). Dean's Lyell chapter in Dean (1992, ch. 10) is excellent and wets one's appetite for his book-length Lyell biography, which is in preparation. A major Lyell bibliography, in the style of Freeman's for Darwin's works, is currently in preparation by Stuart A. Baldwin. Volume IX, part 2, no. 32 of the *British Journal for the History of Science*, the Lyell Centenary Issue, contains a set of fine papers indicating new viewpoints regarding Lyell (but also such "revisions" as the late R. Porter's with which I strongly disagree: see Şengör, 2002b), including D.R. Stoddard's on Darwin, Lyell, and the tectonics of coral reefs (cf. Dean, 1977). Blundell and Scott (1998) is the bicentenary volume celebrating Lyell's birth and contains a number of good papers on Lyell and his legacy. Lyell is one of those giants of geology about whom there is a huge and rapidly growing literature that is quite impossible to introduce here in any greater detail.

[190]Friedrich Wilhelm Heinrich Alexander, Baron von Humboldt is one of the greatest natural scientists of all times, but is also most justly hailed by all as the creator of modern geography together with his friend Carl Ritter. Albert Einstein is the only one I know who did not regard him as a genius (cf. Moszkowski, 1921, p. 60–61), except for the great poet Friedrich von Schiller,

who entertained the extreme view that von Humboldt was downright stupid (he wrote to his friend Korner in 1797 that von Humboldt displayed a "poverty of intellect": Botting, 1973, p. 40). There is an overabundance of biographical material on Alexander von Humboldt. The ones I cite below are those I think would be the most useful for the uninitiated: Ehrenberg (1870), Bruhns (1872a, 1872b, 1872c), Klencke (1876), de Terra (1955), Beck (1959a, 1961, 1982), Botting (1973), Hein (1985), Meyer-Abich (1985), Duvoils and Minguet (1994), and Schleucher (undated). Beck's monumental two-volume biography is the most comprehensive. Its first volume deals with the life of the great geographer to the end of his America voyage, and the second contains a detailed treatment of his trip to Asia. Krätz's book is the best-illustrated von Humboldt biography I know. For a summary of the modern state of the research on Alexander von Humboldt, see Beck (1985). Livingstone (1992, p. 134, footnote 67) lists some further, shorter notices on von Humboldt in English.

[191]During the writing of this book, I only had access to the second edition of Klaproth's *Asia Polyglotta*. But the Asia map in the *Sprachatlas* bears the date of the first edition (1923) and appears unchanged for the second edition.

[192]Abel-Rémusat (1820) translated a number of passages from the Chinese Annals supporting the interpretation that at least the surroundings of Khotan (Hotan in the Pinyin transliteration: see Anonymous, 1989, map 26) could not possibly be located on a plateau of excessive height. I quote selected passages here to give the reader an idea of the kind of information that was available to the tectonicians of the early nineteenth century and specifically to von Humboldt regarding the hypsometry of central Asia.

Abel-Rémusat (1820) provides the following translated accounts: according to the Liang Dynasty Annals (sixth century A.D.): "The air is temperate and is convenient for the raising of grain, wheat, and wine" (Abel-Rémusat, 1820, p. 16); the Annals of the Northern Wei (fifth century A.D.): "The soil is fertile in all sorts of grains, in mulberry trees, in hemp" (p. 19); the Annals of the Sui Dynasty (seventh century A.D.): "The soil produces hemp, wheat, millet, all sorts of fruits. There are many gardens and forests" (p. 32); the Annals of the Tang Dynasty (seventh century A.D.): to the east of Khotan "In the surroundings of this swamp, it is hot and humid" (p. 64); the Annals of the Second Chin: "In this country wine is made by raisins. There is another wine of violet color and another of a blue color. I do not know from what they are produced, but the taste is excellent. The inhabitants eat rice accommodated with honey and millet cooked in cream. Their costumes are made of linen and silk. They have gardens where fruit trees are cultivated" (p. 80); and the Annals of the Sung Dynasty (tenth century A.D.): "The soil produces raisin and a great quantity is fermented to make wine, which is excellent" (p. 84).

As is seen, none of these descriptions rise above the level of ordinary notices kept by Chinese bureaucrats. People such as Klaproth, Stanislas Julien, and Abel-Rémusat culled this sort of information mainly from three types of Chinese sources: dynastic histories, encyclopedias, and travels and topographies. As Wheatley (1961[1966], p. 1) pointed out, although the Annals vary considerably in the nature of their information, the great majority of these texts recorded geographical lore from second-hand sources, which is especially true of the official histories.

Exceptions are the travel accounts. Von Humboldt had access to the translation by his linguist friends of the accounts of Chinese Buddhist pilgrims' voyages. For example, in the translation of the first known Chinese Buddhist pilgrim Fa-Hsien's travels through central Asia, which were begun by Abel-Rémusat and completed by Klaproth and Landresse (Abel-Rémusat, 1836), we read the following information, which von Humboldt no doubt employed in his phytogeographical considerations of the hypsometry of central Asia:

> After stopping here [*i.e., Hotan*] for fifteen days, the party went through south for four days, and entering upon the Bolor-Tagh range [*Tsoung ling Mountains; Abel-Rémusat, 1836, p. 22*], arrived at the country of Tâsh-Kurghân, where they went into retreat. When this retreat was finished, they journeyed on twenty-five days and reached the country of Kâshgar [*kingdom of Kie tcha; Abel-Rémusat, 1836, p. 22*]... This country is mountainous and cold; and with the exception of wheat, no grain will grow and ripen. ... From the hills eastwards [*should this be westwards? Because eastwards from Kashgar is nothing but flat desert save for a few low hills to the south of Maral Bashi; Abel-Rémusat points out {1836, p. 25, notes 5 and 7} that Chinese geographers provide no data as to the whereabouts of this kingdom, except that it is located amidst the mountains of Tsoung ling {Bolor; i.e., Pamir and westernmost Kuen-Lun}*]... is the middle of the Bolor-Tagh range; and from this onwards all plants, trees, and fruits are different from those of China, with the exception of the bamboo, pomegranate, and sugar cane. (Giles, 1923[1956], p. 7–8)

[193]In the original *Nouvelle-Hollande*.

[194]Largely coincident with the present Russian Trans-Baykal domains.

[195]Little or Lesser Bukharia, or Little Bokhara (refer to Fleischer, 1851, p. 317 ff.; Yule, 1916[1966], p. 187) is the southwestern and western extremity of the Tarim basin, where the towns of Khotan, Yerkiang (Yarkant, but now called Sache: see Anonymous, 1989, map 26), and Kashgar (or Kasigar; now Kashi: see Anonymous, 1989, map 29) are located (cf. Abel-Rémusat, 1820, p. III). Fleischer refers to it as "Tarfân oder [die] kleine Bucharei." He further indicates that it was the southwestern frontier province of China and had a capital of the same name (i.e., Turfan). I think that Yule (1916[1966], p. 187) is thoroughly justified in writing with some impatience that he does not know why this region, "perhaps best designated as Eastern Turkestan," was called Little Bokhara!

[196]This sentence appears in a hand-written paragraph that von Humboldt added to the letter. Owing to the illegibility of von Humboldt's handwriting, the main part of the letter had been written in another hand.

[197]For the von Humboldt-Cancrin correspondence, see von Humboldt and von Cancrin (1869). Bruhns, von Humboldt's biographer (see endnote 190 above), had at his disposal still unpublished von Humboldt–von Cancrin letters, which were property of Professor Gustav Rose, von Humboldt's travel companion in Russia. For the background on the correspondence, see Beck (1959b).

[198]Both the eighteenth century geographers and the geologists placed emphasis on the elongate aspects of mountain ranges and even depicted their assumed submarine extensions (e.g., Buache, 1761; Pallas, 1779 {quoted *in* Schmidt-Thomé, 1960}; see Hoffmann, 1837, p. 139–149 for a brief historical review). Before them, the same idea, in different contexts, had always been current, beginning at the latest by Eratosthenes in the second century B.C. (see Şengör, 1998, p. 1–20). What was new with von Buch was a self-contained theory that allegedly explained the internal structure, the timing of formation, and the geographic aspects of mountain belts at once. He *documented* that not all mountains had formed simultaneously.

[199]Biographical material on Élie de Beaumont, one of the leading geologists of the nineteenth century, and especially material elucidating his tectonic ideas, the influence of which is still closely felt, are surprisingly limited even in his native France. Birembaut's (1981) article in the *Dictionary of Scientific Biography* is both inadequate and misleading. For biographical material, see Chancourtois (1874, 1876), Dumas (1874), Bertrand (1875), and Fallot (1939). For an analysis of his work, the best source is still Sainte-Claire Deville (1878, p. 381–582); also see Potier (1875), Greene (1982, ch. 3 and 4; for a thorough critique of these chapters, see Ellenberger, 1984, especially p. 72–75), Ryabukhin (1999), and the very brief account in Michel-Lévy (1905, p. 6–9). A modern biography of Élie de Beaumont is both one of the most urgent desiderata in the history of geology and a most promising field of exploration for a young historian of geology well-versed both in geology *and* geometry.

[200]This interesting man, whose book *Introduction to Geology* (its second edition) fired the enthusiasm of young Lyell, was never admitted to the Geological Society of London, although he worked as a professional geologist for the second half of his life. For a brief sketch of Bakewell's life, see Hansen (1981) and, on the basis of that work, a much shorter summary by Newcomb (1990, p. 213).

[201]Biographical information about Dufrénoy is no more abundant than that about his friend, Élie de Beaumont. Almost all the existing published information on Dufrénoy is referred to in Burke's (1981a) article in the *Dictionary of Scientific Biography* and in the Dufrénoy entries in W.A.S. Sarjeant's *Geologists and the History of Geology* (v. II and the first supplement). To those descriptions I am able to add only the very meager information in Pennetier (1915, p. 67) and de Margerie (1946, p. 1499–1501).

[202]I learned the story of this "affair" from my former teacher, Professor Rudolf Trümpy.

[203]Bonin et al. (1997, p. 28) tell us that Élie de Beaumont distinguished a normal metamorphism from an abnormal one. They quote Frapolli (1846[*sic*], p. 615[*sic*] and 625) as their authority. [*Note:* this Frapolli reference should be 1847 instead of 1846, and p. 615 should be p. 616.] Moreover, Frapolli refers to Vogt's textbook (1846) as his source. Vogt, however, points out that his book is based on Élie de Beaumont's lectures in the École des Mines (Paris). Frapolli indicates that Élie de Beaumont had spoken of these ideas already in 1833 in his

lectures in the very year Lyell coined the term "metamorphic rocks" (Lyell, 1833, p. 374–375). On p. 247 of his 1846 textbook, Vogt says expressly that Élie de Beaumont spoke of an accidental metamorphism and this could be contrasted with a normal metamorphism. On p. 625 of Frapolli's (1847), the text reads as if the term "abnormal metamorphism" is *not* due to Élie de Beaumont. Şengör and Sakınç (2001) took accidental metamorphism as Élie de Beaumont's term and I follow them here.

[204]This idea was reintroduced by Babbage and Herschel (see p. 99) and, on the basis of their work, by Le Conte (1872a, p. 467–468). It is possible that Le Conte was also aware of Élie de Beaumont's idea, as he cites him by name without quoting a specific work (Le Conte, 1872b, p. 352).

[205]Von Buch's map of La Palma (Fig. 40A) shows that barrancos become fewer and narrower as one ascends the cone. This is exactly the opposite of what his theory of the extensional origin of the barrancos requires. It is interesting that von Buch did not notice this discrepancy in his evidence.

[206]Indeed it was a clay cake that von Buch's great countryman, Hans Cloos, used more than a century later to support the uplift origin of radial cracks! See my discussion of Cloos' (1939) ideas in chapter XIV.

[207]William Hopkins (1793–1860), British mathematician, physicist and geophysicist who proposed the term "physical geology" for what we today call geophysics. For Hopkins and his contributions to geology, see Smith (1989).

[208]The theory of elevation craters has received much bad press since the nineteenth century (see Dean, 1980), mainly owing to the dogmatic character of its two main proponents. However, at the time, there appeared much to be in the theory's favor, and geologists such as Dana found it valid even in the middle of the century (Dana, 1846, p. 345 and footnote; but see Natland, 1997). Darwin, for one, never entirely gave up the theory of elevation craters, as his *Autobiography*, written between May and August 1876, testifies (Barlow, 1958, p. 5). Even Lyell himself was struck by the evidence for elevation since the Miocene (in its Lyellian sense!) in the Madeiras and the Canaries (Wilson, 1970, p. 109), probably including La Palma, which was von Buch's showpiece of elevation craters (Wilson, 1970, p. 110–111). Natland (1997, p. 323) has summarized the recent evidence from the Canaries in favor, at least in part, of von Buch's theory. When Scrope sent Archibald Geikie down to Italy to look at some of the volcanoes on his behalf in 1870, Geikie thought he could see near Naples evidence for elevation associated with vulcanicity after all (Oldroyd and Hamilton, 1997, p. 238; see also the remarkable description of the swelling of the ground in Puzzuoli during the 1538 birth of Monte Nuovo by S. Portius in his *De conflagratione agri Puteolani* {Florentiae, 1551: the description is quoted in Sapper, 1917, p. 6, note 2}). A small victory for the craters of elevation came with Gilbert's theory of laccoliths in the Henry Mountains of the western United States (Gilbert, 1880; although the first edition was published in 1877, I cite the second edition, owing to some corrections that Gilbert made and an appendix he added on similar laccoliths elsewhere; also see von Zittel, 1899, p. 411, where he stresses that the similarity between the elevation craters and the "laccolites" of Gilbert is *partial*. This important nuance was regrettably omitted in the English translation of his book; also Ampferer, 1906, p. 603, and Johnson, 1970, p. 31–72). Today, von Buch's theory *partially* lives on under various other designations such as "volcano-tumescence" (e.g., Watkins, 1986, p. 92–93 and especially fig. 5). Reck's (1910) paper is a short historical review of the idea of elevation craters and a defense of von Buch's view that magma indeed has the power of uplifting superjacent strata and in places does give rise to structures not distinguishable from von Buch's elevation craters. The best evidence for an actual elevation crater that fulfills Lyell's requirement (that sedimentary rocks be uplifted and stretched atop a subvolcanic intrusion with a summital crater: {Lyell, 1830, p. 387}: "Had von Buch and Humboldt, for instance, ... discovered a single cone composed exclusively of marine or lacustrine strata, without a fragment of any igneous rock intermixed; and in the centre a great cavity, encircled by a precipitous escarpment; then we should have been at once to concede, that the cone and crater configuration, whatever be its mode of formation, may sometimes have no reference whatever to ordinary volcanic eruptions."), is probably the Paoha Island in Mono Lake in eastern California. The island rose as a dome above the waters of Mono Lake less that 4000 years ago. For a description of the geology of this island and its geological history, see Lajoie (1968, especially p. 112–130 and 141–147, and fig. 20 and 21). That von Buch's model was geometrically eminently plausible is shown by pingos, which are laccolithic forms generated by the freezing of groundwater beneath a frozen carapace. These peculiar features, also known as hydrolaccoliths or cryolaccoliths, have summital "elevation craters" and numerous genuinely structural "*barrancos*" radiating away from them. For a photograph of a superbly developed example, see Holmes (1965, fig. 305). Note here that the "extensional *barrancos*" indeed become wider towards the summit crater as I mentioned above and as Prévost (1835) illustrates in his theoretical sketches. However, the very seductive "rifts" seen on the Hawaiian volcanoes have nothing to do with volcano-blistering (see Dieterich, 1988, and the literature cited therein).

[209]For Anaximander's and all of the Ionian Pre-Socratics' actualism, see Vernant's (1982, p. 103 ff.) important statements enlarging upon Cornford (1932, p. 20). Cornford and Vernant essentially say that science was born in Ionia, once it was assumed that the past had been no different from today. However, Gould's (1965) claim that saying geology is *uniformitarian* is equivalent to saying that it is a *science*, thus the term uniformitarianism is not necessary (a claim later repeated in a very similar form by Goodman, 1967) results from too narrow a view of what the uniformitarian/catastrophist debate was about and how we study geological history today. "Science was born from a uniformitarian stand" is not a statement equivalent to "all science is uniformitarian," which not even Sir Charles Lyell claimed (notwithstanding some recent assertions to that effect). For the origins of Lyell's uniformitarianism, see Wilson (1962; 1967; and 1972, p. 118–124, 156–157, 161–162, 169–182) and North (1965, ch. 4 and 5). For Lyell's fascination with Baron de Ferussac's earth theory, which is very similar to Anaximander's, and its influence on Lyell's views, see Wilson (1972, p. 122–123). For the relations of Lyell's views to Lamarck's, see also Laurent (1981–1982).

Rudwick's (1971, 1990) analysis of the essence of Lyell's position in the history of geology—that it was a non-directionalist view against a former directionalist view—reflects only a small, and I think insignificant, part of the truth. I still think that Whewhell's terms *uniformitarian* versus *catastrophist* (Whewell, 1832, p. 126) reflect the essence of the difference between Lyell and his opponents much better when, for example, one considers (1) von Buch's concept of the overnight upheaval of the Alps that allegedly triggered a water avalanche and drove cannon-ball–like erratic blocks northward across the Molasse plain (von Buch, 1827[1877]; even the "catastrophist" Murchison found von Buch's ideas difficult to take seriously {see Geikie, 1875, p. 75; for another excellent account of von Buch's catastrophist ideas, see Semper, 1914, p. 184–188}); or (2) Pallas' uplift of the Himalaya and the Sunda and Philippine archipelagos to sweep the mammoth carcasses into Siberia (Pallas, An VII[1796], p. 158 ff.; *in* Carozzi and Carozzi, 1991, p. 32 ff.); or (3) indeed Cuvier's arguments of the freezing of the mammoths and their environment so rapidly as to prevent the putrefaction of flesh (Cuvier, 1812, p. 11 and footnote); or (4) Élie de Beaumont's rising of the Andes so as to cause the last great deluge, "of which the traditions of many peoples still retain the memory" (Élie de Beaumont, 1829–1830, p. 232; 1830, p. 55–56; 1831, p. 260–261); or (5) Alcide d'Orbigny's universal catastrophes that repeatedly destroyed the entire biosphere (d'Orbigny, 1849, p. 125–129). None of these processes could have been seen in Lyell's world (nor in ours!) within the reasonable limits of the operation of the known natural laws. Arguments and evidence presented in the defense of catastrophes looked *at the time* barely plausible for some of them (such as Élie de Beaumont's contraction-driven mountain building), while for others, Lyell showed that most of the interpretations were based more on faith than on observation (such as when Lyell showed that Élie de Beaumont's consideration that *mountain ranges that were uplifted in the same period were to be seen as simultaneous*, amounts to "an abuse of language" (Şengör, 1991d). Do not let us forget that when Hans Stille turned to a Beaumontian interpretation of tectonics in the twentieth century, he was conscious of *turning away from uniformitarianism to catastrophism* (Stille, 1922, p. 11).

Most of Lyell's arguments may have had the eloquence of those of a lawyer, but they had also the reasonableness and the backing of a scientist—more, at any rate, than those of most of his opponents. Had Sedgwick not seized upon Élie de Beaumont's interpretation of the Biblical deluge to defend it against the rising tide of uniformitarianism? (cf. Gillispie, 1959, p. 143). Those, such as Rudwick, who stress Lyell's allegedly non-directionalist, almost Aristotelian, stand, consistently ignore a proviso that Hubbert (1967) had introduced, namely that both Lyell and Hutton considered the Phanerozoic evidence and that the world of the Precambrian had remained closed to them. Within the Phanerozoic, life was the only thing that appeared to have a directional history. Lyell could combat all others effectively (and in almost all confrontations, he proved to be right in the long run), and he had a passable argument even at interpreting life as non-directional until Darwin tore him away from that assumption.

[210]I think that Rudwick (1971, p. 225; 1990, p. xxvii) misplaces the emphasis in saying that Lyell rejected von Buch's theory, because it contradicted "his own (more catastrophic!) theory. ..." Lyell was unable to accept that tectonic movements can so significantly uplift the coast as to render old beaches permanently dry (Lyell, 1830, p. 231). Accepting such a theory would have also given support to the theory of craters of elevation, which Lyell found unacceptable. His remark about the absence of earthquakes in Scandinavia was made to emphasize, according to him, the complete absence of any agencies to accomplish what von Buch was talking about. Thus, not the absence or presence of earthquakes (and therefore *not* the "impeccable actualism and extreme gradualism" of von Buch's view), but the *unidirectionality* of substantial tectonic movement (and, therefore, its *non-uniformitarian* character; if it were uniformitarian, it would have been similar in effect to Buridan's model, which would not have been acceptable to either von Buch or Lyell), which resembled what had been claimed to have uplifted the elevation craters, was what Lyell was finding unacceptable.

[211]The first director of the Geological Survey of Great Britain. See McCartney (1977) and Eyles (1981).

[212]De la Beche's book was originally published in London in 1834. I am here citing the American edition.

[213]Born in Hamburg, Germany of French parents, studied in part in Scotland, and married and settled in Vienna, this citizen of the world wrote the first regional geology books of Scotland and of the European provinces of the Ottoman Empire. He was one of the four founders of the Geological Society of France (one of the remaining three was Constant Prévost; see below) and was the author of numerous papers and the first geological maps of Europe and of the world. Despite all this, biographical material on Ami Boué is extremely flimsy. One reason is that he out-lived almost all his friends. Secondly, he never belonged to any organization for any length of time, except for professional societies, but he lived far away from the main seats of most of them. What little was written about him and his still carefully tended grave near Vienna testify to an amiable and generally liked character. He was not an original thinker, but was a compiler of immense capacity and a sharp critic. Birembaut's (1981b) article in the *Dictionary of Scientific Biography* is totally inadequate, and the reader would do well to consult the papers listed by Sarjeant in his *Geologists and the History of Geology*, v. II. To that source, I can only add Anonymous (1889).

[214]To recommend sources for biographic information about Charles Darwin has long gone beyond the competence of one science historian. I therefore confine myself here to the absolute essentials. His bibliography was put together by Freeman (1977) in a celebrated list. Paul H. Barrett and R.B. Freeman have edited a collection of his works under the collective title of *The Works of Charles Darwin* in 29 volumes published between 1987 and 1989 by the New York University Press, New York. The final volume contains Lady Nora Barlow's (Darwin's granddaughter's) restoration of the text of his autobiography. My only quarrel with this wonderful collection is that the editors did not always choose the first editions. Darwin's correspondence is still being published (1985–) in a sumptuous series entitled *The Correspondence of Charles Darwin* by the Cambridge University Press. This series has now reached its thirteenth volume, which brings the correspondence to the year 1864. My one problem with it is that the editors now and then improve the original sketches by introducing printed letters that replace the originals. This practice not infrequently mars other lines and destroys the originality of the figures. Barlow (1933) contains the original text of Darwin's diary that he kept on board the *Beagle* and later furnished the basis of the two editions of the *Journal of Researches* (see Barlow, 1933, p. xxvii–xxx, for the publication history of this book; Keynes, 1988, published an updated version). It is shorter than the *Journal*, which was embellished by much scientific material. That is why Barlow says in the Preface that "the present volume can lay small claim to scientific importance, but rather should be regarded as part of the history of Charles Darwin's apprenticeship to science" (Barlow, 1933, p. ix). Barlow (1946) contains letters and excerpts from notebooks that Darwin kept on board the *Beagle* between 1832 and 1839. Barrett et al. (1987) contains the edited contents of 15 Darwin manuscripts, including 11 notebooks, and four related handwritten documents written during the interval 1836 to 1844, when Darwin generally discontinued using notebooks to store information. Of the notebooks published therein, the "Red Notebook" contains geological material, whereas notebooks "A" and "Glen Roy" are devoted primarily to geology. Desmond and Moore (1991) is possibly the best one-volume Darwin biography. A multi-volume biography by Browne is currently being published. At this writing, only the first volume (Browne, 1995) is out. For a short biography, see Bowler (1990); for a short article on Darwin's life, see de Beer (1981). Finally, Hull (1973) is a useful book that is a collection of the writings by Darwin's various critics. Among them, William Hopkins' critique is by far the most interesting for the purpose of the present book, as it attacks Darwin's method of reasoning by comparing it unfavorably with that of Newton and many other physicists. The same mistrust between "physicists" and "naturalists" still haunts the natural scientist.

[215]For competent summaries of Darwin's studies of the coral islands, see Yonge (1958), Stoddart (1976, 1985, 1994; and *in* Darwin, 1962), Burkhardt and Smith (1985, p. 567–571). Veron (1986) is a superbly illustrated review of the coral reefs and corals in the Indo-Pacific realm. It is recommended to those wishing to see the primary material that Darwin had a chance to observe personally.

[216]See also his letter to Lyell containing a sharper criticism of Élie de Beaumont's view (*in* Burkhardt and Smith, 1986, p. 105).

[217]Lyell (1853, p. 436) mentioning the very same white rock on the basis of Darwin's description to emphasize the extreme youth of the origin of some of the volcanic islands in the Atlantic, omits the uplift interpretation despite his own earlier statement to Darwin (see above the citation from Wilson {1970, p. 109–111})!

[218]In a letter he sent to W.D. Fox from Lima in August 1835 (between the 9th and 12th), Darwin wrote, "I am become a zealous disciple of Mr. Lyells views, as known in his admirable book.—Geologizing in S. America, I am tempted to carry parts to a greater extent, even than he does" (Burkhardt and Smith, 1985, p. 460).

[219]Stoddart (1994, p. 40, note 6) wrote that "Sandra Herbert has shown from his manuscript geological notes that Darwin continued to speculate on alternative atoll origins while in the Galapagos during September and October 1835." This is not true. What Herbert (1991, p. 188–189) quoted and discussed was Darwin's attempt at weighing the existing theories in the light of the evidence that he was able to collect. In the Galápagos, he emphasized the asymmetry of the craters of "sandstone" being similar to atolls (what he called "lagoon islands"). He felt compelled to point this out, a circumstance Lyell had used to support the theory that atolls had nucleated on top of sunken craters, because "I am no believer in the theory of Lagoon Islds. being [illeg.] on the circular ridges of submarine craters" (Herbert, 1991, p. 188). Darwin mentioned the competing theories and observations that could be taken to be adverse to his theory to emphasize how superior his theory was despite them.

[220]As we now know, no granitic rocks occur within the ring of fire that surrounds the Pacific and the floor of the ocean, and all of its islands are entirely basaltic in composition, with the exception of the so-called plagiogranites (Coleman and Peterman, 1975, adapted from the Russian "plagioclase granite") that are nothing more than trondjhemites and leucocratic tonalites containing oligoclase or andesine, quartz and less that 10% biotite, pyroxene, or hornblende. Magnetite and ilmenite occur as accessory minerals (Coleman, 1977, p. 47–53). The rocks that Darwin refers to as granites were shown in the beginning of the twentieth century by Lacroix (1916) largely to be olivine gabbros, thus putting to rest the "myth" of the occurrence of granites within the Pacific.

[221]The emergence in the East and West Indian Islands (the Indonesian and Philippine Archipelagos and the islands of the Caribbean) is mostly a result of thrust stacking in accretionary complexes and the consequent rise of undertucked outer non-volcanic rises. It is thus copeogenic. Trenchward of these rises, the corals commonly submerge as a result of elastic bending because of the subduction drag due to locking of the understhrusting and overriding plates (for examples of modern studies, see Sieh et al., 1999; Zachariasen et al., 1999, 2000). Woodroffe et al. (1990a) concluded that the Cocos (Keeling) atoll, located far away from the subduction zone on the downgoing plate, has also been emerging in the last 3000 years resulting from late Holocene eustatic sea-level lowering, or a regional lithospheric response of the floor of the Indian Ocean to post-glacial melting, or a local hydro-isostatic response of the Cocos Rise to increased postglacial ocean volume, or indeed a combination of these factors. With the exception of the purely eustatic sea-level lowering, all other causes that Woodroff et al. (1990a) ascribe to the emergence of the Cocos (Keeling) islands are falcogenic.

[222]Darwin added a free and somewhat incomplete translation of von Humboldt's passage to his manuscript's fair copy (Darwin, 1962, p. 17). Here is a

complete translation: "As a consequence, the epoch of subsidence of western Asia coincides largely with the uplifting of the plateau of Iran, the plateau of Central Asia, of Himalaya, of Kuen Lun, of Tien Shan and all the ancient systems of mountains orientated east-west. Perhaps also with the uplift of the Caucasus and the knot of the mountains of Armenia and Erzurum."

[223]I must point out that, with the exception of Hawaii, Galapagos, and the Juan Fernandez, all active volcanoes Darwin depicted to illustrate his thesis—that volcanoes coincide with regions of uplift—are magmatic arc volcanoes (see his Plate 3, "Shewing the Distribution of the Different Kinds of Coral Reefs, Together with the Position of the Active Volcanos in the Map" *in* Darwin, 1889[1987]). He thus presaged the locations of the three active plumes (cf. the hot-spot catalogue in Şengör, 1995), but not that of the South Pacific Superswell (Davis and Pribac, 1993). But from the passage I quoted above (Darwin, 1962, p. 5, p. 103 this book), it is clear that Darwin sensed that the linear island chains within the Pacific basin were somehow different tectonically from the island festoons that surround it. Here again one sees his incredible ability to grasp very quickly the essential features of a natural object that was previously unfamiliar to him and to deduce testable inferences from them to understand its nature.

[224]This idea of alternating areas of rise and fall may have been inspired by the observation he jotted down in his notebook on 19 July 1835, in the Port of Callao, near Lima, Peru (that I quoted above): "On the Atlantic side [*of South America*] my proofs of recent rise become more abundant at the very point where on the other side they fail" (Barlow, 1946, p. 245).

[225]Gosselet (1896) gives the most comprehensive biography of this outstanding geologist, an ardent supporter of uniformitarianism. Von Zittel (1899, especially p. 288, footnote * and *passim*) presents a competent summary of his life and contributions in the context of the nineteenth century geology. Gohau's (1995) short article, however, is the best summary I know of Prévost's paleontological, stratigraphic, and tectonic ideas. Also see Laurent (1976) for Prévost's paleontological views. Bork (1990) gives the best English account of Prévost's life and work of which I am aware. Rudwick's (1981) short piece in the *Dictionary of Scientific Biography* is unfortunately very inadequate both as a summary of, and with respect to the secondary sources concerning, Prévost's life and work.

[226]Fast enough to drive their overlying marine waters northwards in cascades, which were responsible for driving and, in some cases, shooting across the molasse plain (!) the erratic boulders onto the Jura Mountains. Even Murchison, whose position in the catastrophist/uniformitarian distance was far closer to von Buch's than to Lyell's, found von Buch's ideas hard to swallow (Geikie, 1875, p. 75), as I mentioned above (see endnote 209).

[227]A better translation of *contre-coup* as employed here would be "reaction," meaning the reaction of the units around the subsiding unit to its subsidence.

[228]Although Dana is commonly considered the greatest figure in the history of American geology, there is still no satisfactory modern published biography of him. Prendergast's (1978) unpublished doctoral dissertation is currently the best source for Dana's life and work. So far as I know, Gilman (1899) remains the only published book-size biography on Dana. Rice's (1915) review is comprehensive and presents Dana's geology from the viewpoint of a generation who remembered him in person and heard his lectures. The James Dwight Dana special issue of the *American Journal of Science* (v. 297, no. 3, 1997) contains four good papers on different aspects of Dana's work. Also see Friedman (1998a) for early influences in Dana's life.

[229]For the history of the U.S. Exploring Expedition, see Stanton (1975) and Viola and Margolis (1985).

[230]For the relationship of Dana with Darwin with respect to the coral islands quesiton, see Stoddart (1994).

[231]In the first edition of his *Manual of Geology* (Dana, 1862, corrected printing 1863; I here cite the latter whose pagination is identical with the 1862 printing), Dana reiterated his views on terrestrial tectonism that he first put forth in the 1846 Moon paper and in the 1847 papers. He wrote the following concerning the origin and nature of what he called "geoclinals":

Formation of valleys.—The plication of the earth's crust produces alternating depressions and elevations, unless the folds are pressed together into a close mass. The depressions are *synclinal valleys*. The minor valleys of this kind are generally obliterated by subsequent denudation; and often even the summits of ridges, under this latter agency, may consist of the rocks of a synclinal axis. Besides synclinal valleys, there are often also *monoclinal* valleys (p. 720). In addition there are wider depressions lying between distant ranges of elevations which were produced through a gentle bending of the earth's crust (made up of plicated strata or not); and these great valleys or depressions (like the Mississippi or Connecticut valleys) may be called *geoclinal*, the inclination on which they depend being in the mass of the crust, and not in its *strata*. (Dana, 1863, p. 722, italics his)

Dana further mentioned that the sinking of such geoclinal valleys may lead to "trap eruptions." Having argued that because metamorphism tends to seal off fractures and to obliterte them, eruptions rarely occur simultaneously with metamorphism, he gave the Appalachian case as an illustration:

In periods of metamorphism, the lateral pressure causing the plications appears in general to have so closed up the fractures made, that igneous ejections were rare. It is not certain that any took place during the metamorphism of the Appalachian region; though subsequently, after the rocks had been stiffened by crystallisation, the sinking of the geoclinal valleys occupied by the Mesozoic sandstone formation gave origin to a great profusion of trap ejections (p. 430). (Dana, 1863, p. 726; see further his description of the Connecticut Valley Mesozoic geology on p. 415)

The simplified geological map of the United States Dana gave in his *Manual* (his fig. 135 on p. 133; Fig. 57 herein) illustrates the scale of the geoclinals he describes.

The word *geocline* (as geomiocline) was used later by Dietz and Holden (1966) for continental margins, for what Cloos had earlier (and more appositely) called *geomonoclines* ("*Geomonoklinal*": Cloos, 1936, p. 460), possibly inspired by his late friend Alfred Wegener's earlier criticism of Émile Haug's remark that mountain chains originate from geosynclines: "I hold 'shelves' more apposite than 'geosynclines,' as one cannot very well describe a marginal shelf, such as the one from which the Andes of South America had been built up, as a trough" (Wegener, 1915, p. 35, footnote 1). Later, Stewart and Poole (1974) introduced, inappositely, the term *eugeocline*.

[232]"One could also assimilate the entire Atlantic Ocean into an immense geosyncline" (Haug, 1907, p. 164).

[233]I wonder whether it was this statement that prompted Suess to go to the northern shores of Norway to check out the terraces, instead of following the footsteps of von Buch, Brongniart, and Lyell? As we shall see below in Chapter XIII, Suess initially (in 1875) may have also followed von Humboldt's view concerning the mechanism of the uplift of Scandinavia, interpreting it as the anticlinal crest of a very broad and flat fold.

[234]See Tobien's (1981) short article in the *Dictionary of Scientific Biography* for all the important sources on Studer of which I am aware. I am able to add to it only the half-page account in Feller (1935, p. 42–43).

[235]I am not ignoring Lazzaro Moro's, or Michell's, or indeed Hutton's models, or de Dolomieu's (an VI[1797]) descriptions and inferences. However, those pioneers were more concerned with emphasizing the role of internal heat in mountain-building and volcanism than with the precise geometry of the heat-generated intumescences.

[236]Robert Chambers (1802–1871). For his life, see Williams (1981). Regrettably, this short article, devoted almost entirely to an analysis and the influence of the *Vestiges*, does not deal with Chambers' geology. For a much more extended account, see Millhauser (1959), which includes a list of all of Chambers' writings, including his scientific papers (Chambers published four papers on sea-level change: two before and two after the *Ancient Sea Margins*). But Millhauser's excellent book too does not discuss Chambers' geology except only as it relates to the *Vestiges* and even that very briefly (see Millhauser, 1959, p. 88–90; Millhauser gives his reader a brief glimpse of Chambers' geological activity after the *Vestgies* on p. 167–168).

[237]The influence of Naumann's book outdid, in geological circles, even Lyell's *Principles*. Both remained in print in subsequent editions in the 1870s. Naumann's book was more systematic and more comprehensive than Lyell's and thus more suitable as a teaching aid. In terms of the basic ideas they propounded, neither Lyell nor Naumann claimed originality; it was the mass of evidence they presented and the fabric of the teaching material they weaved that made the impact. In many topics—such as the nature of mountain-building and

the methods of its study, progression of life through time, and use of index fossils in stratigraphy—posterity followed interpretations preferred by Naumann, not Lyell. This is evident, both in such epoch-making writings as Suess' *Die Entstehung der Alpen* and in authoritative near-contemporary histories of geology such as von Zittel's, from the fact that every time a reference textbook is mentioned or implied, it is generally Naumann's. Needless to say, the impact made by Lyell's first edition of the *Principles on Geology* cannot be matched by any book in the history of geology after Steno, with the sole exceptions of Hutton's 1788 paper and Cuvier's 1796 memoir on living and fossil elephants (for the latter's importance, see Rudwick, 1997a {especially p. 13–24, where an English translation is also presented} 1997b; Şengör, 1998, footnote 25; Şengör and Sakınç, 2001, p. 134–145). The comparison here made is between the *Principles* as a textbook (i.e., a teaching aid) and Naumann's *Lehrbuch*. Naumann's book seems to have been for geologists of the latter half of the nineteenth century what Arthur Holmes' *Principles of Physical Geology* has been to the geologists of the latter half of the twentieth century.

[238]The article on Naumann by Burke (1981b) in the *Dictionary of Scientific Biography* is most inadequate. In addition to the references cited there, one should consult: Anonymous (1869; on the occasion of the golden jubilee of Naumann's academic profession), Anonymous (1873), Duke of Argyll (1874), and especially Wagenbreth (1979b, with a complete list of publications).

[239]The publication history of this extraordinary book is as extraordinary as the book itself. It is a story worth telling as the two accounts I know to exist in English give misleading impressions about it (Suess, 1904, p. iv; and Greene, 1982, p. 116–117). In my rendering of it, I largely follow Élie de Beaumont's own account in the Preface to the *Notice* (Élie de Beaumont, 1852, p. I–VI) with cross-checks with the *Dictionnaire* itself.

In 1904, Eduard Suess wrote to Professor W.J. Sollas, under whose direction the *Antlitz* was being translated into English, pointing out that "In 1852 Élie de Beaumont's '*Notice sur les Systèmes de Montagnes*' appeared in a form not very easily accessible, the *Dictionnaire Universel d'Histoire Naturelle*" (Suess, 1904, p. iv). This is incorrect. What did appear in the *Dictionnaire* was a long but truncated article simply entitled "Systèmes de Montagnes" (Élie de Beaumont, 1849). The article actually appeared in the 137th installment (*livraison*) of the *Dictionnaire* on 1 September 1849 (Élie de Beaumont, 1852, p. V. See Evenhuis {1990} for the problems associated with the dating of the individual *livraisons* of the *Dictionnaire*; Evenhuis {1990} was in fact unable to find the date of publication of the 137th *livraison*). This installment is in the twelfth volume. The cover of this volume bears the date 1849, whereas the title page shows 1848!

Charles d'Orbigny, the editor of the *Dictionnaire Universel d'Histoire Naturelle* invited Élie de Beaumont in 1841 to contribute a number of articles on geology. At the time, Élie de Beaumont was engaged with his friend Dufrénoy in the publication of the geological map of France, so could only promise "a cooperation on the long run and with little activity" (Élie de Beaumont, 1852, p. II). However, he did commit himself to writing the article on *montagnes*. Yet when the time came in 1846 for him to deliver the article on *montagnes*, Élie de Beaumont found himself in the middle of writing the second volume of the explanatory notice of the geological map of France and totally unable to do anything else. D'Orbigny, deferred the article to a later letter and substituted *Montagnes—Voyez Soulèvements et Révoltions du globe* ("Mountains—See uplifts and Revolutions of the globe": v. 8, 1847[1846 on the title page], p. 340). As Élie de Beaumont's other commitments dragged on, poor d'Orbigny found himself in the necessity of putting *Revolutions du Globe—Voyez Systèmes de Montagnes* (v. 11, 1849[1848 on the title page], p. 84) and *Soulevements—Voyez Systèmes de Montagnes et Terrains* (v. 11, 1849; 1847[1846 on the title page], p. 696) as their time came in turn. In 1848, when the second volume of the explanatory notice of the geological map of France was already at the publisher, Élie de Beaumont thought he could finally get on with fulfilling his promise to d'Orbigny. He had gathered much material on the subject, some of which he had published in the interim. Alas, other occupations still prevented Élie de Beaumont from writing his article. Further deferments in the *Dictionnaire* not being possible, d'Orbigny asked the printer to leave enough space for Élie de Beaumont to fit his article into and continue printing the rest of the issue. When Élie de Beaumont finally finished his article, he found that it was too long for the space he originally had agreed to. Thus, what finally appeared in the *Dictionnaire* on 1 September 1849 was a version truncated at the "Système de l'Erymanthe et du Sancerrois" (i.e., at p. 311 of volume 12 of the *Dictionnaire*, which corresponds with p. 528 of the *Notice*)! But p. 528 brings the *Notice* only to the end of volume I. Needless to say, the most interesting parts of the *Notice* are in volume 3, where Élie de Beaumont gets into the theoretical questions of global tectonics, which the readers of d'Orbigny's *Dictionnaire* never got to see. It is clear that there was no question of Élie de Beaumont's material, which eventually did not make it into the *Dictionnaire*, not being "necessary or appropriate" for the *Dictionnaire*, as Greene (1982, p. 116) thought. Rather, Élie de Beaumont had simply become a victim of deadlines!

A note about the availability of the *Notice* today: In my 30 years of ardent book-collecting on four continents, I have never seen the *Notice* offered for sale in any antiquarian or second-hand book catalogue, nor have I come across it in a shop. In fact, I owe my personal copy to the kindness and generosity of my friend Dr. Nazario Pavoni of Zurich.

[240]Even Littré's *Dictionnaire de la Langue Française* illustrates this word with a quotation from Auguste Daubrée concerning Élie de Beamont's *bosselements* (Littré, 1963, p. 1128)!

[241]The best account is still that of Sainte-Claire Deville (1878, p. 381–582) as I pointed out above (I do not mean, however, that it is an adequate account!). All subsequent accounts, including the two chapters in Greene's monograph (1982, ch. 3 and 4; for a thorough critique of these chapters, see Ellenberger, 1984, especially p. 72–75), are even less adequate abstracts and clearly do not do justice to this great man or to his theory. I devoted one lecture to him in my course in the Collège de France in the May 1998 and hope to give a full account in my book based on my course, which is in preparation.

[242]Dufrénoy and Élie de Beaumont (1848, p. 610). Now we know that the maximum thickness of the Jurassic sedimentary rocks in the Paris Basin is 1500 m (Pomerol et al., 1980).

[243]Also, see the first section of Ellenberger's note (1970, p. 469–470).

[244]Zimmermann (1861, p. 336) credits Friedrich Hoffmann (1797–1836), regrettably without citing a published reference, with the suggestion that the salt deposits of the Sahara, Sicily, and Palestine once formed a coherent sheet *that also extended beneath the Mediterranean*, which was supposedly nothing more than a trough-shaped depression, with the Sahara forming one of its shallower flanks. As far as I know, this is the first suggestion that salt underlies the Mediterranean. Hoffmann, however, thought the salt was Cretaceous in age (Hoffmann, 1839, especially p. 380–383).

[245]Three quarters of a century later, Hans Stille claimed, in opposition to Haug (1900, p. 626; 1907, fig. 38), that geosynclines begin to fail along their margins creating mountains with internal vergence toward the continental margin of the geosyncline (Stille, 1924, figs. 4, 5, 6). See Stoćes and White (1935, fig. 558) for a more elegant graphic depiction than provided by Stille of this assumed process.

[246]Dana repeatedly cited Élie de Beaumont's mountain systems in the first edition of his *Manual of Geology* (Dana, 1863, p. 502, 533, 720). In the 1873 paper, there is no reference to Élie de Beaumont, but there is a detailed one to George Vose's (1866) *Orographic Geology*. Élie de Beaumont's theory is the first one discussed in that book, and Vose gives very detailed references to the *Notice sur les Systèmes de Montagnes* and to the discussions of Élie de Beaumont's views in English (Vose, 1866, p. 13, note 1). It is extremely improbable that Dana had not read it before writing the 1873 paper.

[247]It is clear that the great Frenchman had recognized that flexure calculations could be used as a guide to the (elastic) thickness of the earth's crust.

[248]With the exception of three (Ellenberger, 1970; Şengör, 1990, 1998), as far as I am aware, none of the writings that include an account of the history of the geosyncline idea takes the history farther back in time than Hall, or at most than Babbage and Herschel. Even French authors have repeatedly credited the geosyncline idea entirely to Hall or to Dana. Following Albert de Lapparent's crediting Dana with the geosyncline idea (de Lapparent, 1883, p. 1222, 1225; and in all subsequent editions before 1900), Haug, for example, also gave the credit to Americans in his influential 1900 paper on geosynclines and continental areas, but wrote that "the notion of the geosyncline goes back incontestably to James Hall" (Haug, 1900, p. 618). Even the scholarly de Launay held Hall responsible for "introducing into science the notion of 'geosyncline'" (de Launay, 1905, p. 83, footnote 1). Haug repeated the same view in the first volume of his *Traité*, the founding text of the Kober-Stille school of the twentieth century (Şengör, 1998). Kober (1921, p. 18) spoke of the "geosyncline theory of Hall and Dana." Stille (1924), in his *Grundfragen der Vergleichenden Tektonik* (Fundamental Questions of Compara-

tive Tectonics), perhaps the most influential book on tectonics in the twentieth century before the rise of plate tectonics, wrote that Dana was the first to use the term "geosyncline" but acknowledged that the notion went back to Babbage and Herschel and that Hall had pointed out that folding was a natural accompaniment of subsidence and sedimentation. Stille nevertheless said that the type example was the Appalachian geosyncline (Stille, 1924, p. 7). Such examples may be multiplied many-fold. Ellenberger (1970) was the first to acknowledge Élie de Beaumont's contribution but credited his 1829 Oisans paper for having introduced the "first European intuitions of the geosyncyline concept," such as geosynclinal metamorphism. This was a gross understatement, and those reading Élie de Beaumont's 1829 paper without having read his 1828 papers would have difficulty understanding what Ellenberger was getting at.

The following additional writings, together with the ones cited above, would give the reader a tolerable background in the history of the geosyncline concept beginning with 1857. Some of these references are about the history of the concept, others themselves have helped to create a part of the history of the geosyncline concept in the twentieth century. None (except Şengör 1998) acknowledges Élie de Beaumont's role: Haug (1900), Barrell (1918, especially p. 176–183), Schuchert (1923), Leuchs (1927), Longwell (1928), Ver Wiebe (1936), Jones (1938), Kay (1944, 1947, 1951, 1952, 1955, 1967), Glaessner and Teichert (1947), Beloussov (1948, p. 181–230; 1962, p. 311–391 and 500–512; 1975, p. 82–92; 1976, p. 138–147; 1981, p. 21–120), Knopf (1948, 1960), Trümpy (1955, 1984), Hsü (1958, 1971, 1972, 1973, 1982), Aubouin (1959, 1965; for excellent critiques of Aubouin's concept of geosyncline, see Debelmas et al., 1966 and 1967; for an application of Aubouin's version of the geosynclinal theory to Northwest Borneo, see Haile, 1969), Dietz (1963, 1972), Dietz and Holden (1966), Harland (1967), Hermess (1968), Crook (1969), Mitchell and Reading (1969), Coney (1970), Dewey and Bird (1970), Dickinson (1971), Wang (1972, 1979), Dott (1974, 1979, 1985), Markov et al. (1974), Stewart and Poole (1974), Ahmed (1976), Schwab (1982), Şengör (1982, 1998), Mayo (1985), Khain (1986), Pushcharovskiy (1987), Khanchule (1988), Zaitsev (1990), and Li and Xiao (2001).

[249]Hall was called the "Founder of American Geology" by Joseph Le Conte and the "Founder of American Stratigraphy" by the great Quaternary geologist and ethnographer W.J. McGee (Schuchert, 1943, p. 12–13). Friedman (1979, p. 6, caption to fig. 8) endorsed McGee's view. Clarke (1921, reprinted in 1923 and 1978) provides a comprehensive biography of Hall. For Hall's relationships with the founding figures of American geology, in addition to Clarke's book, see Friedman (1979, 1998b, 2000). For a brief review and assessment of Hall's stratigraphic work, see Schuchert (1918, p. 85–89). For the history of his tectonic ideas, see Dott (1979, 1985), Mayo (1985), and Şengör (1998).

[250]Horace Henry Hayden (1769–1844), a native of Connecticut, was the main architect of American dental education. A multi-faceted scientist, he did research in mineralogy, geology, and botany in addition to dentistry (including physiology and pathology). He was one of the founders of the Maryland Academy of Sciences and the *American Journal of Dental Science*. His geological sympathies were with the Neptunism of Werner. Referring to North America, he remarked that:

> In its various parts are exhibited all the different formations, that are mentioned by geologists in support of the Neptunian theory: such as primitive transition, secondary, or floetz, &c. At the same time few or no indications occur that can favor in the least possible degree, the Huttonian theory; or, in other words, that any known part within the present limits of the United States, can owe its origin to 'Intestinal motion' of Patrin, or volcanick agency; … as not an indication of the kind, I believe, has ever been found east of the Mississippi river. (Hayden, 1820, p. 2–3)

For a short biography (with portrait), see the Pierre Fauchard Academy International Hall of Fame in Dentistry at www.fauchard.org/awards/fame07.htm. Also Merrill (1904, p. 259–260; 1924, p. 85).

[251]For the presently used nomenclature of the sequences described by Hall (with the exception of the Carboniferous), see the useful summaries in Bird and Dewey (1970, especially fig. 5 and appendix), Colton (1970), and Rankin et al. (1989). Bird and Dewey (1970), Colton (1970), and Rankin et al. (1989) may be conveniently connected with Hall's nomenclature via Schuchert (1943).

[252]Hall is here referring to the 1850s work of the Canadian Survey on the lower Palaeozoic rocks of the Canadian Appalachians. He was in contact with Sir William Logan, the head of the Canadian Survey, and received fossils from him to describe. See, for example, Logan (1854–1855), and especially Hall (1865).

[253]Under pressure of criticism, Hall changed his interpretation in 1883, without openly saying so (see the quotation on p. 131). But, by that time, his peculiar interpretation had long fallen into obscurity, and few noticed his grudging retraction.

[254]This interpretation may well have been influenced by Hall's teacher, A. Eaton, whose sections in the influential *Index* (Eaton, 1820, plates 1, 2) and *Text-Book* (Eaton, 1830, fig. 3), and in addition by what Hall was able to glean from contemporary European models, such as those of Élie de Beaumont.

[255]For Henry's life, see Coulson (1950). Hall functioned as a favorable witness for Henry in the notorious telegraph controversy with Samuel F.B. Morse (cf. Coulson, 1950, p. 231).

[256]It is characteristic that Hall makes no reference to the fact that almost nobody among his hearers and readers had managed to interpret what he had implied!

[257]This description completely anticipates and may well have inspired Émile Haug's interpretation of the Mid-Atlantic swell as a geanticline: "One could also assimilate the entire Atlantic ocean into an immense geosyncline on its way to becoming doubled, the axial ridge … corresponding with a median geanticline" (Haug, 1907, p. 164). Also see Stočes and White (1935, fig. 557).

[258]This organization later became the American Association for the Advancement of Science.

[259]"Esta Sierra es el Espinazo de esta America Septentrional" (map-text, in upper center position on the famous *Miera Map of the A-type*: Wheat {1957, p. 100}); see Fig. 72 herein and the discussion of the various versions of the Miera map in the section on the Domínguez-Escalante expedition. In the B-type (tree and serpent-type), the same text reads "Esta Sierra es el Espinazo de esta America Septemtrional," whereas in the C-type (bearded Indian-type) it reads "Esta Sierra es el Espinazo de esta America Setemtrional." In the much less decorated copy kept in the Yale University's Bienecke Library (Goetzmann, 1995, p. 109), this remark does not occur at all.

[260]The one that has ingrained itself into my memory is that wonderful John Ford movie entitled "She Wore a Yellow Ribbon" starring John Wayne, in which the plateau country appears in its full beauty displayed by its diverse types of mesas and natural monuments in the Monument Valley of Arizona/Utah. The flank dips of the very gentle folds affecting this area generally range from 1 to 4 degrees, with a maximum of 14 degrees having been measured on the northwest flank of the Agathla anticline southeast of Wetherill Mesa (Chenoweth, 2000).

[261]See Thomas Jefferson's instructions to André Michaux, French botanist and secret agent, for the exploration of the westen part of what is now the conterminous United States (*in* Ambrose, 1997, p. 70–71). For the state of geographical knowledge and exploration in the North American Cordillera that falls within the frontiers of the conterminous United States at the beginning of the nineteenth century, see especially Warren (1855), Wheeler (1889), Brebner (1964, especially ch. 18–24), De Voto (1952), Cumming et al. (1974, ch. 5, 6, 7), and Goetzmann (1991, 1993). For graphic display on maps of the progress of exploration, see Wheeler (1889), Wheat (1957, 1958), and Goetzmann and Williams (1992). For an excellent and exceptionally complete bibliography of the geographical and geological explorations west of the 100th meridian in the present-day United States, see the classic work by Wagner and Camp as revised, enlarged, and edited by Robert H. Becker (1982) and the supplement published by White (2001). For travel accounts on the central-western emigrant routes (i.e., the ones that go over the most pronounced falcogenic bulge of the western United States), see also the rich bibliography with 2082 items by Mattes (1988).

The accounts of the explorations of Jedediah Smith (1799–1831: see Morgan, 1953, especially the foldout map showing Smith's travels in the years 1822–1831; also Morgan and Wheat, 1954)—the first civilized man to enter what Frémont was later to call the Great Basin (except the brief foray of Étienne Provot to the Great Salt Lake area; see below) and the territory west of the Colorado (Goetzmann, 1993, p. 135)—give a clear idea how thoroughly unknown were the present-day United States Cordillera and the great basins enclosed by it. When Alexander von Humboldt published his map entitled *Carte Générale du Royaume de la Nouvelle Espagne* in 1812 (see Fig. 75 herein), the most recent information he could find for what is the present-day Utah and Arizona was the Miera map of 1776 (Alter, 1941, especially p. 65; Bolton, 1950, colored facsimile of Miera's map in pocket; Wheat, 1957, ch. VI;

see Fig. 72 herein). On the basis of similar information, de Buffon (1749, p. 319) already knew that "the directions of great mountains are from north to south in America and from the occident to the orient in the Old World." In 1778, de Buffon thought he had to revise this opinion and claimed that "in general, the grandest eminences of the globe are disposed from north to south and those that trend in other directions must be regarded as nothing more than collateral branches of the primary mountains. It is in part by this disposition of the primitive mountains that the pointed contours of continents present themselves in the direction from north to south" (de Buffon, 1778, p. 305–306, 309–310). This idea was also in part inherited from Kircher's ideas on the "mountain skeleton of the globe" (Kircher, 1665, p. 69; see Şengör, 1998, fig. 6), which in turn was conceived on the basis of the reports of his Jesuit brothers especially in the New World. The echoes of de Buffon's idea were to be felt as late as 1949 (cf. Stille, 1949).

[262]Between 1500 and 1800, some 100 voyages of exploration were undertaken to the western parts of what is now the territory of the conterminous United States of America. Of these, 26 were Spanish, 23 English, 21 French, at least 15 by Jesuits and Franciscans who were mostly Spanish, 6 Russian, 3 German, 2 Italian, 2 Danish, 1 Portuguese, and 1 Hungarian (Wheeler, 1889, p. 483–484). The dominance was thus with the Spanish. This domination was much more so in the interior parts of the continent that are of more direct interest to the contents of this section. Regrettably, the Spanish results were not widely publicized because of the colonial policy of Spain (which was not terribly different from its imperial rivals, such as England, in keeping discoveries secret: see Boorstin, 1983, ch. 35). Nevertheless, much information became public and influenced both mapmakers and geographers well until the middle of the nineteenth century, as shown by von Humboldt's (1811a, 1811b[1997]; 1812) grand synthesis.

The literature on these explorations is vast and makes extremely interesting and rewarding reading from the viewpoint of the history of geology. But this is not the place to delve into it. Wheeler (1889, p. 488–496) gives a very useful tabulated summary, and both DeVoto (1952) and Brebner (1964) provide easily readable and accessible (but from the viewpoint of the history of science not terribly rewarding) résumés. Cumming et al. (1974, p. 10–13 and ch. 5) is another excellent and well-illustrated account covering a time period from the middle of the seventeenth century to the end of the third quarter of the eighteenth century. For the coastal strip, see Davidson (1886), Robertson (1927), Wagner (1929), and Priestley (1946). Davidson's account is particularly valuable as he presents a column-by-column comparison of the texts and the terminology of various Spanish authors among themselves, and their terminology with the terminology presently used.

For definitive narratives of Anza's California expeditions, enriched with maps and illustrations, Bolton's monumental volumes (1930a–e) remain unsurpassed. For collections of maps relating in some way or other to the Spanish explorations, see Nordenskiöld (1889), Wheeler (1889), Kretschmer (1892, Atlas), Winship (1896), Robertson (1927), Almagià (1948), Wheat (1957), and Nebenzahl (1986). For a graphic display of the various exploration routes, see Goetzmann and Williams (1992). Cumming et al. (1974) also has many reproductions of old maps.

[263]For Cabeza de Vaca's narrative, presented in his *La Relación que dió Aluar Núñez Cabeça de Vaca de lo Acaescido en las Indias en la Armada donde yua por Gouernador Pãphilo de Narvaez* (first published in 1542, in Zamora—an error-strewn edition not controlled by the author—and then a corrected edition published in Valladolid in 1555), I used the text edited by Enrique Pupo-Walker and translated into English by Frances M. López-Morillas (Pupo-Walker and López-Morillas, 1993). Bolton (1949, p. 472) lists, without citing a source, a map prepared by Cabeza de Vaca and Dorantes and a report by Cabeza de Vaca at the request of Antonio de Mendoza, the Viceroy of New Spain, among the lost documents belonging to the history of geographic exploration of southern North America. However, Wheat (1957, p. 17) says, referring to Cabeza de Vaca, that "So far as we know, the earliest western wanderer made no map, nor did any of the names used by him for his stopping points find their way onto contemporary European maps." However, given the Spaniards' penchant for asking for maps even from the natives, I find it highly unlikely that Mendoza did not request Cabeza de Vaca to make at least a sketch map of his route (cf. Day, 1940, p. 18). Bandelier (1886[1981], p. 46) pointed out that "Cabeza de Vaca's reports are sometimes precise, but more often they become confused, under the influence of an imagination overstimulated by long suffering. Unfortunately, the points concerning geography and ethnography are sometimes the ones that are treated the most vaguely. The route of the journey is, therefore, subject to interpretations that are mere conjectures, and from this have resulted historical errors that have been perpetuated for several centuries." DeVoto agrees (1952, p. 18): "Since Cabeza de Vaca was living a myth, his account is majestically unregardful of landmarks and geography." Careful detective work by Wood (2000) resulted in the most reliable itinerary so far (see the map on his p. 241; Fig. 68 herein).

[264]Dean William R. Harris wrote that "Coronado ... accomplished the most wonderful exploring expedition ever undertaken on the American Continent" (Harris, 1909, p. 33), a judgment with which I agree. Coues' remark that "Coronado's march from Culiacan to Kansas is a singular climax of fame and futility" (Coues, 1900b, p. 513) reveals a complete lack of appreciation of the history of geography. Before de Oñate retraced his steps half a century later (Vivien de Saint Martin, 1873, p. 341), Coronado's geography had already influenced European geographers. A simple comparison of pre- and post-Coronado maps in the sixteenth century (e.g., the 1520 German globe with the 1560 Bolognino Zaltieri map or with the 1570 Atlas of Abraham Ortelius or with the 1589 Ortelius North America map {reproduced *in* Wheeler, 1889}) would suffice to show how uninformed Coues' remark is (also see endnote 272). In the explanatory text of his 1570 map of America, Ortelius expressly cited Coronado (as Franciscus Vasquez) as one of his sources (Ortelius, 1570[1964], p. 2). Coronado's terminology also appears in another sixteenth century atlas, namely that of Gerard de Jode. North America in de Jode's *Speculum Orbis Terrarum* (1578[1965]) appears only as a part of the world map and contains Coronado's topography and toponymy. Day (1940, p. xiv) rightly emphasized that Coronado's expedition was to change the ignorance that had prevailed about the territory of the present-day southwestern United States, "to bring light from the darkness, to widen the girdle of the known."

For Coronado's life and excellent accounts of his expedition, see especially Day (1940) and Bolton (1949). Bolton points out (p. vii; also see Morris, 1997, p. 114) that to his contemporaries, Coronado was known as Francisco Vázquez (as is witnessed by Ortelius' reference mentioned above; I follow Bolton's example of referring to him as Coronado). Von Humboldt (1852, p. 463 and footnote ***) refers to him as Cornado following the orthography of Antonio de Herrera y Tordesillas, the official historian of the Spanish possessions in the New World, but does point out that in Antonio de Leon's *Biblioteca Oriental y Occidental* (1629, p. 76), his name is presented as Coronado. The name comes from the village of Cornado, near Coruña, in the province of Galicia in northern Spain, where the family had first settled after they had emigrated from France in the fourteenth century (Day, 1940, p. 21). Day notes that the family surname was *de Cornado* or *de Coronado*. See Henze (1978, p. 721–727) for a short account of Coronado and his accomplishments as an explorer. For English translations of the reports of the Coronado expedition, I have used Winship (1896), which also gives the Spanish originals of Castañeda's narrative (p. 413–469) and the *Relación Postrera de Sívola* (p. 566–568; this anonymous account was probably written by Fray Torobio de Benavente {called Father Motolinía: Morris, 1997, p. 44}) and Hammond and Rey (1940). The quotes below are usually from Hammond and Rey's translation, except where Winship's rendering is deemed superior. Neither Winship nor Hammond and Rey seem to have paid much attention to the physical geography and geology of the features described by Coronado and his men and consequently in many places their translations are faulty (for a few of such examples documented by Morris, 1997, see his p. 50, 119, 120). I have tried to correct these, mainly with the kind help of Professor Joann Stock of the California Institute of Technology, who knows both the geology of the terrain traversed and the language used to describe it, to decipher the conquistadores' meaning. Both Winship (1896) and Hammond and Rey (1940) have useful historical introductions in their books. See also chapter VI in Whipple et al. (1855), in which Fray Marco de Niza's and Coronado's expeditions are summarized on the basis of Hakluyt's relation. The April 1984 issue of the *Arizona Highways* magazine (v. 60, no. 4, 47 p.) is entirely devoted to Coronado. In that issue, Stewart Udall presents new evidence for corrections to Bolton's route map of Coronado, accompanied by excellent illustrations including artwork by Bill Ahrendt and photography by Jerry Jacka. Dellenbaugh (1897) gave Coronado's route a different course from the normally accepted one that was thought not simply erroneous, but actually preposterous by Coues (1900b, p. 514; Coues' pitiless exclamation, however, may in part have been fed by the old Hayden-Powell rivalry {i.e., a close Hayden ally pounding on the errors of a Powell man}; for the rivalry, see especially Foster, 1994, ch. 22). In my account, I follow the usual interpretation as represented by the route map in Goetzmann and Williams (1992, p. 37; with Morris' correction of the trail in Texas: see Morris, 1997, map on p. 101), mainly because of Castañeda's statement that, in going from Cíbola to Tiguex, they had the north to their left (Winship, 1896, p. 517 {Spanish text, p. 450};

Hammond and Rey, 1940, p. 252) and because of the subsequent archaeological discoveries. Dellenbaugh's (1897) reconstruction of the route (essentially from the Florida Mountains in the present Luna County in southern New Mexico towards Albuquerque; for location, map, and brief geology of the Florida Mountains, see Matheney et al., 1988) hardly would have necessitated such an emphasis. In addition to these sources, I have used Harris (1909, especially ch. II), Day (1940), Bolton (1949), DeVoto (1952, especially p. 34–55), Wheat (1957, ch. II), and Morris (1997) for the historical background.

265Along with these paintings, Coronado sent a cattle skin, some turqoises, two turqoise earrings, 15 Indian combs, two boards decorated with turquoises, two baskets made of wicker, two rolls worn by Indian women when they carried water, and samples of weapons used by the natives. He further reported on the mineral wealth and the fabrics produced. He regretted that he could not report on what the women wore, as they were kept away from the Spaniards (Winship, 1896, p. 562–563; Hammond and Rey, 1940, p. 176–177). He even had some of the natives measured (see Coronado's letter to the King: Winship, 1896, p. 582; Hammond and Rey, 1940, p. 188). When combined with the detailed information on the physical aspects of the country, commonly supported with maps, his observations and collections amount to a fine scientific sampling of the general geography of the explored terrain. Indeed, in the sixteenth century, very competent and informative books were written by Spanish authors on the lands, rocks, animals, and plants of the New World, which made it plain that the knowledge inherited from antiquity in such books as Aristotle's and Theophrastus' treatises on plants, animals, and minerals and Pliny's widely read *Natural History* did not contain a complete description of the natural riches of our planet and that clearly they had known nothing of the New World. New information on the natural history (in its old sense including all three of the kingdoms of Nature) of the New World made known in the books of such authors as Gonzalo Fernandez de Oviedo y Valdés (1478–1557), Nicolás Bautista Monardes (1493–1588), and José d'Acosta (1539?–1600) was based on observations made by the conquistadores, which still surprise us with their wide scope and great detail (Stearns, 1970, ch. 2). Both in his *Examen Critique* and in his immortal *Kosmos*, Alexander von Humboldt praised the sharpness and richness of the observations of the conquistadores—especially Columbus—and the enrichment their accounts brought about in the literature devoted to the verbal depiction of Nature, out of which, eventually, the modern scientific language gradually developed in part (von Humboldt, 1847, p. 53–65, 1852, p. 103 ff., 156–157 and159–160; also see commentary on von Humboldt's views by Hentschel, 1969, p. 116–117 and notes 182 and 183 on p. 163).

Now contrast this with the custom of the sixteenth century geographers in general: "Briefly it may be said that the scholar of the sixteenth century, with his exclusively classical education, looked backward in time almost as a matter of course. To him a region was merely of interest as the seat of its human population, and only one section of that population, namely the pedigreed nobility and landed gentry, was deserving his close attention" (Taylor, 1934, p. 9), and:

> As has already been suggested, the interests of the Elisabethans were centered on men rather than on places, and hence the reading public looked to travel literature for accounts of the laws, religions, manners and customs of foreign peoples. ... But of description, especially topographical description, and description of natural scenery, climate, vegetation and rural economy, there was very little in Hakluyt's masterpiece: nor was it to be expected. The unity of nature had yet to be recognized, and the collection of scientific material for its own sake, or for the elucidation of some principle of causation, had not yet begun. (Taylor, 1934, p. 21)

Most contemporary translations and epitomes of the Spanish books describing the New World in fact displayed more interest in its exploitation that in its natural history (e.g., Stearns, 1970, p. 23, note 7).

One of the most extraordinary overland journeys in the sixteenth century was that undertaken by the Ottoman Admiral and justly celebrated author of the famous nautical handbook *Kitab al Muhit*, Seydî (or Sîdî) Ali Reis, from Gujarat in India to Edirne in Turkey. His book describing the journey—the *Mirât al Memalik* ("Mirror of Countries": Seydî Ali Reis, 1313H [1895 A.D.] {for an English translation of this book, with introduction and commentary, see Sidi Ali Reïs, 1899}; and undated), completed December 1557 A.D.—contains not one passage describing a landscape or reporting on the fauna or the flora of the extraordinary regions through which he traveled. He talks exclusively about the rulers and their courts and some passages concerning the literature (dominantly poetry: e.g., Erarslan, 1968) of the cultures encountered (cf. Seydî Ali Reis, 1313H [1897-1898 A.D.] and undated). Against a backround such as this, the Spaniards' careful landscape and flora and fauna descriptions in the sixteenth century sound very modern.

266To his contemporaries he was simply García López, but I follow Bolton (1949, p. vii) and the present custom in referring to him as Cárdenas. For the life of this fine soldier, see Hammond and Rey (1940, p. 340 ff., footnote 1). For the documents related to his trial in Spain, see Hammond and Rey (1940, p. 337–368). Bolton (1949, ch. 33) rightly expresses disgust at how unjustly Cárdenas was persecuted to make an example out of him, in part under the influence of the zealous "Apostle to the Indies," Father Bartolomeo de las Casas, who in his famous book incriminted the Spanish conquistadores of inhuman crimes in the New World (for an easily accessible English translation of Father de las Casas' book, with notes by the translator, Nigel Griffin, and with a good introduction by Anthony Pagden, see de las Casas, 1992).

267For the geographical results of the Coronado expedition and their contemporary impact, see especially Bolton's (1949) excellent assessment in his chapter 34. The Coronado expedition got underway only two years after Mercator had dubbed the continent he was exploring *America pars septentrionalis* (i.e., North America {Boorstin, 1983, p. 254}). The late sixteenth century knowledge of the shape of the southern half of the continent—much closer to our present knowledge of it than anything that had been available earlier—and the main outlines of its morphology were due entirely to Coronado's expedition and to his men, who took the trouble of combining the information gathered by his expedition and that by De Soto's expedition in the southeastern part of the continent. See endnote 265 above and also below for some of the details.

268We know almost nothing about this chronicler. The meager available information has been summarized in Day (1940, p. 382–383) and in Hammond and Rey (1940, p. 191, footnote 1). We also do not know when he wrote his narrative, but it was probably around 1555 (DeVoto, 1952, p. 52).

269Florida in the sixteenth century comprised nearly all of the southern federal states of the United States, namely North Carolina, South Carolina, Georgia, Florida, Alabama, Tennessee, Mississippi, Kansas, and parts of Texas and Louisiana. It corresponded with the terrain explored by Hernando De Soto's expedition (see Goetzmann and Williams, 1992, p. 34–35).

270From 1492 to about 1532, the general belief was that North America was a peninsula of Asia plus an assortment of islands to the east of Asia (Wheeler, 1889, p. 484; Kretschmer, 1892, Atlas, plate XVIII). As pointed out in endnote 268 above, North America was named by Mercator in 1538. In the early 1500s, South America began to emerge from the mist more rapidly than North America. Paolo Forlani, in his 1560 map for example, still does not separate North America from Asia (Forlani (1) *in* Shirley, 1993, p. 121, entry 106), but by 1561(?), we see the "Straits of Anian" in Giacomo Gastaldi's insecurely dated world map (Shirley, 1993, plate 92, entry 107) and in the 1566 Zaltieri map (Wheeler, 1889, p. 504; Kretschmer, 1892, p. 440 and Atlas, plate XIX, no. 3). In his great *Examen Critique*, von Humboldt confessed ignorance as to where the name *Anian* stemmed from (von Humboldt, 1832, p. 477, footnote *, continued on p. 478). Day (1940, p. 325, note 3) quotes Wagner (1929, p. 358) that the name *Anian* derived from a typographical error in Marco Polo's book. This is not quite right. Kretschmer (1892, p. 440 ff.) pointed out that the German historian of geography, Sophus Ruge, first drew attention to a passage in Marco's relation (Book III, 5) stating (for reference, see Peschel, 1877, p. 273–274, note 2) that:

> When one leaves Zayton [*medieval Arabic designation for present-day Quanzhou at 25° 53′ N, 118° 36′ E*] and sails for 1,500 miles towards the sunset so one arrives at the Bay of Hainan [*present-day Gulf of Tonkin*]. It is so large that one needs two months to sail through it from its northern shore. From there one comes to the southern parts of the province of Mangi [*South China in general; in Marco's days it was more specifically the areas occupied by the Song dynasty soon to be eclipsed by the Mongols; see Yule, 1903, p. 144, note 3*] and from there till one arrives at the lands of Ania [*identified with present-day Annam: see Peschel, 1877, p. 273, note 2*], Toloman [*Yule (1903, p. 123), reads Toloman as Coloman and places it in eastern Guizhou, near the Wumeng Shan*] and many other already mentioned countries. (Kretschmer, 1892, p. 441)

This passage is one that occurs only in the manuscript called "Z," an eighteenth century copy of which was originally discovered by Professor L.F. Benedetto in the Ambrosiana Library in Milan, Italy, in the twenties of the twentieth century while he was engaged in creating his great edition of Marco Polo's travels for the Comitato Geographico Nazionale Italiana (Moule and Pelliot, 1938, p. 366). This copy originally belonged to Cardinal Franciscus Xaverius de Zelada. Later Sir Percival David managed to locate the parent manuscript dating from an interval from the thirteenth to the fifteenth century in the

Chapter Library of the Cathedral at Toledo, Spain, on 7 December 1932, which was then transcribed and published by Moule (in Moule and Pelliot, 1935). Ruge thought the passage was a later interpolation, in which he was followed by Kretschmer (1892, p. 441). The orthography of *Ania* is written as *Amu* by Moule (in Moule and Pelliot, 1935, p. lxii) and Moule and Pelliot (1938, p. 366). Kretschmer (1892, p. 440) gives other variants according to different editors. He thought that the passage quoted (and thus the name Anian) had been inserted on the basis of maps made in the fifteenth century (see his Atlas, plates XXVII and XXX). This, however, is not possible, because the original manuscript from which "Z" was copied in 1795 was written in the interval from the eighth to the fifteenth centuries (Ross, 1931, p. x). The source of the word Anian is thus still the text of the "Z" manuscript or some parent of it.

How this Ania, originally meaning Annam, came to be applied to regions at the extreme northeast of Asia, is explained by Kretschmer (1892, p. 442): He pointed out that in the light of the great ignorance of the European cartographers of the sixteenth century, Marco's provinces wandered north-northeast as southeastern Asia became better known with its own proper names. Initially, Asia and North America had been thought to be connected. Finally, it was felt that Asia had to be separated from North America. The "pinching" took place at what had been called "Bay of Asia," which then became "Strait of Anian."

[271]Winship (1896, p. 513, footnote 1) identifies this as the region of Newfoundland. Hammond and Rey (1940, p. 247) seem to think it is Cape Cod.

[272]Many of the sixteenth century cartographers depicted the mountain ranges of North America as trending east-west (e.g., see the Zaltieri map reproduced in Wheeler, 1889, map facing p. 504 and the text of p. 504, also the North America map of Battista Agnese {second half of the sixteenth century: Kretschmer, 1892, Atlas, plate XXV}). This North America map of Battista Agnese, Wytfielt's 1597 map of the Americas (Kretschmer, 1892, Atlas, plate XIX, no. 6), and the 1609 Hondius map (Wheeler, 1889, map facing p. 506) are among the earliest that honor the geographical observations of the Coronado expedition on orography. The continental outlines Castañeda here describes was most likely taken from the Ptolemy Atlas published in 1548 in Venice, because the "Carta Marina Nova Tabula" in that Atlas (for facsimiles, see Nordenskiöld, 1889, plate XLV, left-hand side; Kretschmer, 1892, Atlas, plate XVIII, no. 3) corresponds precisely with his description, which has led me to think that he must have had that map open before him.

[273]For the history and the variants of the name Mississippi and the different appellations of the great river, see Coues (1895a, p. 287–289, footnote beginning on p. 287).

[274]For an account of this province written in the early seventeenth century (manuscript completed on 21 December 1621), see Lazaro de Arregui (1946). This account contains a contemporary map displaying the mountains of New Galicia province in a style similar to the magnificent *Codex Valentianus Latinus* of Ptolemy's *Cosmographia* in the University of Valencia library (no.1895 in the catalogue of Gutiérrez del Cano).

[275]Winship and Hammond and Rey translate the Spanish word *montes*, which Melchior Díaz used in his report, differently. Winship (1896, p. 550), translates it as "mountains," whereas Hammond and Rey (1940, p. 159) as "forests," but they indicate in a footnote that the word usually means "wooded hills." Hammond and Rey may have chosen to ignore the hilliness of the country obviously implied by Díaz's informants (and by the Spanish word he chose to express it) because of Coronado's later eyewitness account that Cibola was a flat place not hemmed in by mountains (see below). Coues (1900b, p. 329, footnote) commented on Father Garcés' diary relating to his travels on the Colorado Plateau near the Colorado River approaching the South Rim that "*En montes* is not 'on mountains'; I have set 'over highlands,' which is true of the ground, but 'through woods' would be as correct a translation." It is clear that the translation of the Spanish word *montes* must be made in light of a knowledge of the geography *and* historical geography of the described terrain.

[276]Powell, in his historic report of the *Exploration of the Colorado River of the West and its Tributaries*, remarked that on the Kaibab Plateau, "clouds yield their snows even in July" (Powell, 1875, p. 189). A century earlier, Father Garcés (see below) had noted that even in late July, the San Francisco peaks near Flagstaff, Arizona (his Sierra Napac), were snowy (Coues, 1900b, p. 353). When Cárdenas reached the Grand Canyon, the region was nearing the end of a dry episode that had been going on since the beginning of the sixteenth century. It was not unlike the situation in the mid-nineteenth century when Powell visited the Canyon (Fritts, 1965).

[277]Dellenbaugh points out, on p. 35, that Cárdenas' itinerary is poorly known, although I find Dellenbaugh's identification of the point where Cárdenas must have reached the Grand Canyon (1903, p. 34; also see Dellenbaugh, 1897, p. 416–418), on the South Rim, right across from the mouth of the Andrus Canyon, convincing. It agrees much better with the Spaniards' descriptions (see especially the *Relación del Suceso* in Winship, 1896, p. 575 and also *in* Hammond and Rey, p. 288) than does the conventional location at the Canyon (or Grand) View or Desert View just south of the Cardenas Butte (e.g., Bartlett, 1940; Pyne, 1998, p. 6). Goetzmann and Williams (1992, p. 37) also seem to follow Dellenbaugh's opinion here. Day's suggestion (1940, p. 344–345) that the location of the Spaniards' first view was at the point across from the mouth of Kanab Creek is also possible, but the length of the march of the Spaniards and the direction of the river turning to south-southwest from where they reached it, agree better with Dellenbaugh's suggestion.

[278]I think Hammond and Rey (1940, p. 252) mistranslate Castañeda's sentence ("esta tierra es un ualle entre sierras a manera de peñones": Winship, 1896, p. 450) as "This land is a valley between sierras that rise like boulders." I think Winship's translation is closer to Castañeda's meaning: "This country is a valley between rocky mountains" (Winship, 1896, p. 518), as the word *peñones* denotes craggyness. I believe that what is meant here is the craggy sides of the mesas.

[279]Regarding the mesa of the Acoma, Alvarado recorded that its pueblo "was one of the strongest ever seen, because the city is built on a very high rock. The ascent is so difficult that we repented climbing to the top" (Winship, 1896, p. 594; Hammond and Rey, 1940, p. 182). Castañeda also recorded that the soldiers who went to the top found it a difficult climb (Winship, 1896, p. 494 {Spanish text: p. 433}; Hammond and Rey, 1940, p. 223).

[280]For descriptions of the great difficulties encountered by Cárdenas' three men (Captain Melgosa, Juan Galeras, and another one whose name is not recorded; for more about the first two, see Bolton, 1949, p. 140) in their attempt to descend to the river from the South Rim of the Grand Canyon, see Castañeda's account *in* Winship (1896, p. 489 {Spanish text, p. 429}), Hammond and Rey (1940, p. 215), and Bolton (1949, p. 139–140).

[281]See endnote 278 and the photographs facing p. 44, 88, 166, 172, 284 in Sedgwick (1926) to appreciate the accuracy of the statements of the Spaniards. Also see the beautiful oil painting by Wilson Hurley depicting a winter panorama of Acoma (*in* Briggs, 1976, p. 167).

[282]Sedgwick presents the variants of the name as used by various explorers and two other etymologies: *Hakukue* (in Zuñi): "Drinkers of the dew" and *A-ko-kai-obi* (in Hopi): "The place of the ladle." These apparently refer to the two great natural reservoirs on the top of the mesa and the way they are used by the inhabitants. For the variants of the name Acoma, also see Coues (1900b, p. 368).

[283]This accurate assessment by Captain Jaramillo is in stark contrast to those of Lieutenant Zebulon Montgomery Pike (1779–1813; Coues, 1895b, p. 525; also see Goetzmann, 1993, p. 51, 62) and of Major Stephen H. Long (1784–1864; see Wood, 1966, p. 116), who thought of the Great Plains as wholly unfit for cultivation and uninhabitable by a people who depend on agriculture. This view greatly hampered the settlement of the plains and was prevalent almost into the last quarter of the nineteenth century, when it was finally dispelled by the work of the Great Surveys.

[284]Morris (1997, ch. 8) presents a marvellous discussion of Castañeda's perception of the Great Plains obtained from his west Texas Llano Country viewpoint, based on a careful physical geographical exegesis of his text. Castañeda was not among the 30 who accompanied Coronado northwards to Quivira, so he never saw the real prairie land.

[285]For biographies of Newberry—in addition to those listed by Sarjeant in his *Geologists and the History of Geology* (v. III and Supplement, v. I)—see Chenoweth (1997).

[286]Both Alvarado and Jaramillo noted the presence of the water divide in the high plateau: see Winship (1896, p. 575) and Hammond and Ray (1940, p. 289) for Alvarado's report, and Winship (1896, p. 587) and Hammond and Rey, (1940,

p. 299) for that of Jaramillo. Bandelier (1886[1981], p. 97) thought Jaramillo's report of the continental divide very significant. Dellenbaugh's (1897, p. 412) dismissal of it is nothing but special pleading in support of his own alternative route for Coronado's march. Both Coues (1900b, p. 517) and Morris (1997, p. 47) note Jaramillo's reliability for terrain observation and location.

[287]For brief biographies of Father Francisco Tomás Hermenegildo Garcés (1738–1781), see Coues (1900a, p. 2–24), Walker (1946), Odens (1980), Henze (1983, p. 320–321), Kessel (1976, especially ch. 1–6; for a portrait of Garcés, see p. 144), and Fontana (1996; this booklet has a fine bibliography and a brief history of Garcés' mission of San Xavier del Bac). Garcés' memory is today kept alive in Yuma by the Garcés Celebration of the Arts, initiated in 1968. It has evolved into one of Arizona's most important art festivals, complementing the other three at Tucson, Flagstaff, and Phoenix. In this book, I quote Garcés from Coues' 1900 edition, which has been the standard edition for nearly a century. For another translation of his diary, based on a different manuscript, see Galvin (1965). The advantage of Galvin's book is that it contains a facsimile of Father Font's map and a map Galvin himself generated showing Garcés' travels in 1775–1776 to complement Font's map. Garcés said in his "Preliminary Remarks" that Father Font's map shows his "entire route." This is incorrect, as Galvin pointed out in a note he appended to the translation of the legend of Font's map (*in* Galvin, 1965). The Font map omits the following routes: (1) Garcés' travel with the Anza expedition from Tubac to the Yuma crossing; (2) his excursion to the mouth of the Colorado River; and (3) his return home from the Yuma crossing (Father Font did not accompany Father Garcés on the route "shown by dots"; he continued with Anza's expedition). Governor Crespo of Sonora suggested routes but did not make the journey. The line of presidios shown on the Font map is too rigidly straight and is incomplete.

[288]For brief biographies of Father Francisco Silvestre Vélez de Escalante, see Harris (1909, p. 94–100) and Henze (1983, p. 179–180). Also see Adams (1963, 1976) and Briggs (1976, *passim*).

[289]A variant of this map, with a different title, has been found in the British Museum by Lowery (classified as Additional Manuscript, 17,651-9; see Wheat, 1957, p. 225, note to item 169). I have not seen this variant. Galvin (1965) provides a facsimile of a version of Father Font's map facing p. 102 in his book.

[290]For the Domínguez-Escalante journey, the best accounts are those by Bolton and Briggs. Bolton (1950) contains a translation of Escalante's diary with identification of most of the places mentioned, plus a redrawn and in part corrected (cf. Wheat, 1957, p. 112, footnote 36) colored copy of a Miera map of the "bearded Indian" (or "C-type") that Bolton found in the *Archivo General* in Mexico City (cf. Wheat, 1957, p. 111–112) and a modern map showing the route. The diary is also translated in its entirety in Harris (1909, p. 125–242, also with a route map) and in Auerbach (1943, p. 27–113). However, the best translation is no doubt that by Father Angelico Chavez (Vélez de Escalante, 1995), which includes Bolton's route map (p. 144–145). For the route, also see Auerbach (1941a, 1941b, and 1943, which includes a foldout facsimile, facing p. 24, of a "bearded Indan type" Miera map showing the travel route). For the historical context, in addition to the sources just cited, see Auerbach (1943), Adams (1963, 1976), and DeVoto (1952, p. 290–297, with a route map on p. 295). DeVoto's statement that "Nothing came of this sunnily stupendous journey" (p. 297) is flatly contradicted by the great use made of the diary and the map that resulted from it by von Humboldt in drawing his influential map of the western United States (see below). When one considers that W.H. Emory's (1848) great map was the first which improved on von Humboldt (Goetzmann, 1993, p. 255), one realizes how great and lasting indeed the results of this "sunnily stupendous journey" were!

[291]Adams (1976) provides some biographical information about Father Domínguez and further references. Also see Vélez de Escalante (1995, p. 3, footnote 2).

[292]For a complete listing of the members of the Domínguez-Escalante expedition, see Himmerich y Valencia (1995, p. vii–viii).

[293]For a biographical outline of this remarkable Spaniard, see Wheat (1957, p. 97).

[294]This manuscript map is uncatalogued in the Bienecke Library. Its call number is Uncat.WA MS.Miera. *Miera y Pacheco, B. de*, is indicated as author and the title/description is "Ms map Illustrating Escalante's Exploration of the Colorado, Utah & North Arizona." As place of production of the map, San Felipe, Mexico, is indicated. The date of the map is 1776. The Meriden Gravure Company produced an excellent facsimile of the Yale copy of the Miera map in 1970, which is sold by the Bienecke Library. I am grateful to Professor William H. Goetzmann for alerting me to the existence of this copy and of its facsimile, and to the Bienecke Library's Public Services personnel for supplying me the meager information that exists on this manuscript.

[295]For Viceroy Bucareli's life and administration, see Bobb (1962); for his role in the explorations, also see Briggs (1976).

[296]For Teodoro de Croix and his administration of the internal provinces of New Spain, see Thomas (1941). This book is particularly useful as it contains a translation of Caballero de Croix's 1781 General Report (*in* Thomas, 1941, part II, p. 64–243), which reviews the state of each of the internal provinces of New Spain, namely Texas, Coahuila, New Mexico, New Vizcaya, Sonora, and California only five years after the historic journey of Escalante and his comrades.

[297]Wheat (1957, p. 227) gives the file number as *L. M. 8a-1a-a* and indicates it being in the *Archivo de Mapas* of the Ministry of War in Madrid, Spain.

[298]Johann Georg Kohl (1808–1878) was a German historian of cartography who emigrated to the United States in 1854, bringing with him a considerable collection of old maps relating to America that he had facsimiled by hand. In 1856, he was commissioned by the U.S. Congress to duplicate his maps, which were eventually housed in the Library of Congress. In 1886, Justin Winsor created a catalogue of this collection containing 474 Kohl facsimiles. It was reprinted in 1904 with an index and preface by Philipp Lee Phillipps (reprinted again in 2002 by Martino Publishing Company, Mansfield Center, Connecticut). Kohl also made a collection of coastal charts. See Wood (1976).

[299]Goetzmann (1995, p. 109) reproduces the Yale version of the Miera map and then comments "Note how the Spanish seem more interested in Indian tribes (reservoirs of heathen soul) than in landforms." Given his meticulous scholarship, this statement from Goetzmann is most surprising because the Yale version of the Miera map is the only version in which topographic detail is on a cursory level, characteristic of most maps of its age. In all its other versions, the Mirea map stands out by its author's unusually careful attention to topographic detail. As I discuss below, the diary of Father Escalante also shows how carefully the Spaniards observed the natural surroundings, animate and inanimate. How easily do we tend to overlook the fact that their observations of the land in Utah, Colorado, Arizona, and parts of New Mexico were the only basis to draw maps of these regions well into the 40s of the nineteenth century! (See below my discussion of von Humboldt's map of these regions; see also Goetzmann, 1993, p. 69, where he says, "…as late as 1824, Miera y Pacheca's [*sic*] map, …, was the basis for all maps of that region, including Humboldt's map of New Spain of 1811, Pike's maps of 1810, and the great Lewis and Clark maps of 1810–1820.") If Emory's 1848 map was the first improvement on von Humboldt's, as Goetzmann (1993, p. 255) himself points out, and since Pike largely plagirized von Humboldt (see below), it means that the Miera map was used as late as 1848! In any case, the Miera map shows more topographic detail than do the Lewis and Clark maps for the area it covers (cf. Wheat, 1957, map reproductions 176 and 185, with Wheat, 1958, map reproduction 284).

[300]Compare the sierras on the map of Miera with the major folds on the map of the Colorado Plateau presented in Cooley et al. (1969, fig. 7), forming topographic prominences and controling drainage on the plateau. The correlation is remarkable. For more on the geographical information and the errors contained in the Miera map, see Crampton and Griffen (1956).

[301]For this remarkable individual and his work, see Wheat (1958, p. 65, footnote 5) and John (1988). Also see Coues (1895b, p. 656–667) for supplementary information on Walker not cited in John.

[302]Wheat (1958, p. 24–25) writes that neither this copy nor the copy of it made surreptitiously by Aaron Burr have yet been found. This regrettably leaves unresolved the sources of some geographic blunders seen in the map of Zebulon Montgomery Pike (see below), as pointed out by Wheat (1958, p. 2526). The French title I cite for this map is, of course, the title of its published version. We do not know in which language the American copy was prepared. Jackson (1966b, p. 378, note 2) suspects that the "report" von Humboldt refers to in his

letter to Jefferson must be his "tableau statistique," a 19-page summary of statistics about Mexico that he had given to Jefferson during his visit. This study of von Humboldt's was published in von Humboldt (1869). Beck (1866) pointed out that the *Essai Politique* grew out of its enlargement.

[303]A recent inexpensive reprint of this great (but now very rare and very expensive) book was published by Éditions Utz in Paris in two volumes (de Humboldt, 1811a[1997a] and 1811b[1997b]), with an informative preface by François Chevalier reviewing von Humboldt's viewpoint and the subjects dealt with in the book. All my page references are to this easily obtainable reprint edition. Regrettably, the Atlas is not reprinted, and the few maps from it that are reprinted in the two volumes are barely legible. For a facsimile of the Atlas, see Beck and Bonacker (1969). The book and the atlas were also reprinted as part of the magnificent complete facsimile edition of von Humboldt's *Voyage aux Régions Equinoxiales du Nouveau Continent* (1805–1834) in 30 volumes published jointly in Amsterdam by Nico Israel and in New York by Da Capo Press (1971–1973). Unfortunately, this grand collection is sold for 11,345 Euros, which rather restricts its accessibilty. The 1811 edition was also published as an octavo book in five volumes (de Humboldt, 1811b), but without the Atlas. Brand (1959) is a study on this important book by von Humboldt. Also see Beck (1966) for an overall evaluation of von Humboldt's Mexico trip from the viewpoint of the history of geography.

[304]This atlas, also called the Mexico-Atlas by von Humboldt himself (Beck and Bonacker, 1969, title page), was published with various dates (1808, 1811, and 1812) and variously under the imprints of F. Schoell and G. Dufour, both in Paris. Von Humboldt himself states that it was published in 1811 (von Humboldt, 1857 *in* Möllhausen, 1861, v. I, p. VIII). Beck and Bonacker (1969, title page) also state that it was published in 1811 (although the copy they facsimilied, printed by Dufour, carries the date of 1812). Beck and Bonacker is an excellent black-and-white facsimile of this very rare and very expensive atlas with an informative introduction to von Humboldt's cartographic work plus eight more plates from other von Humboldt atlases pertaining to his work in America. For detailed bibliographic references to the Mexico-Atlas, see Wagner and Camp (1982, items 7a:1a, 7a:3, 7a:3a:1, 7a:3a:2). Streeter (1960, p. 17–21) presents a detailed discussion of the publication history of the atlas, its variants, and its features as they relate to the geography of Texas. Wheat (1957, p. 132–138) discusses the maps of the Kingdom of New Spain, von Humboldt's route map from Mexico City to Santa Fé, map of Mexico, the Pacific coast from Cape San Sebastian to Cape San Lucas, as well as the Mississippi Valley and the Atlantic seaboard. Beck and Bonacker (1969) also comment on the maps of the Atlas.

[305]In two manuscripts of the Hippocratic treatise *Airs, Waters, Places*, namely in the *Codex Vaticanus Graecus* 276 (twelfth century A.D.) and *Codex Barberinus* (fifteenth century A.D.), we read in paragraph (or chapter: see Diller, 1934, p. 35 ff.) XVIII: "ἡ δὲ Σκυθέων ἐρημίη καλευμένη πεδιάς ἐστι καὶ λειμακώδης καὶ ὑψηλή καὶ ἔνυδρος πετριώς" instead of the more common "ἡ δὲ Σκυθέων ἐρημίη καλευμένη πεδιάς ἐστι καὶ λειμακώδης καὶ ψηλή καὶ ἔνυδρος πετριώς" (see Jones, 1923[1984], p. 118). Jones translates "ὑψηλή" as "plateau" (from the root ὕψι, high, upward; ὑψηλή would thus mean high, highland (Jones, 1923[1984], p. 118, 119; ὑψηλός means high). Thus, the first sentence would read in English "What is called the Scythian desert is level grassland, a plateau [ὑψηγή], and fairly well-watered," whereas the second, Jones' preferred reading, "What is called the Scythian desert is level grassland, without trees (ψιγή, meaning bare), and fairly well-watered." If the ὑψηγή reading is correct, then the *concept* of a plateau as a highland would have to be dated some five centuries before Strabo, but even in that case we do not see a special technical term. As far as we know, the distinction of having introduced a *special term* to express the concept of a plateau in all cases belongs to Strabo (although Schulten {1914} seems to think that even the term may have predated Strabo. Note that L. Edelstein thought that the second part of *Airs, Waters, Places*, {ch. 12–24}, could have been written by a geographer rather than by a physician: Diller, 1934, p. 3). Diller's book is a particularly helpful guide for understanding the place *Airs, Waters, Places* has in the geographical literature of the Greeks.

[306]Nearly 500 km in width!

[307]Von Humboldt's footnote: "According to Bruce (vol. III, p. 642, 652 and 7121), the sources of the Nile, in the Gojam, are elevated 3,200 metres above the level of the Mediterranean." The Gojam region is ~61,000 km^2 in area and is located in western Ethiopia. In that region, individual peaks, such as Birhan (4154 m) and Amedamit (3619 m), rise above 3200 m (Munro, 1988, p. 233). Needless to say, the sources of the Nile are much farther south, but this was not known when von Humboldt was writing. However, it is true that the Gojam region is a part of the immense falcogenic east African plateau crowned by the rift valleys (cf. Şengör, 2001a).

[308]Owing to the great rarity of von Humboldt's Atlas, I give here the captions of these figures and their translations (also see Beck and Bonacker, 1969, which itself has become a bibliographic rarity):

12. Tableau physique de la pente orientale du plateau de la Nouvelle-Espagne (chemin de Mexico à la Vera-Cruz, par Puebla et Xalapa), dressé d'après des mesures barométriques et trigonométriques prises en 1804, par M. de Humboldt. [*Physical picture of the eastern slope of the plateau of New Spain (road from Mexico to Vera-Cruz, via Puebla and Xalapa), drawn according to the barometric and trigonometric measurements of Mr. de Humboldt made in 1804.*]

13. Tableau physique de la pente occidentale du plateau de la Nouvelle-Espagne (chemin de Mexico à Acapulco), dressé d'après des mesures barométriques et trigonométriques prises en 1803, par M. de Humboldt. [*Physical picture of the western slope of the plateau of New Spain (road from Mexico to Acapulco), drawn according to the barometric and trigonometric measurements of Mr. de Humboldt made in 1803.*]

14. Tableau du plateau central des montagnes du Mexique, entre les 19° et 21° de latitude boréale (chemin de Mexico à Guanaxuato), dressé d'après le nivellement barométrique de M. de Humboldt. [*Picture of the central plateau of the mountains of Mexico, between 19° and 21° northern latitude (road from Mexico to Guanaxuato), drawn according to the barometric leveling of Mr. de Humboldt.*]

[309]"Filons ouverts," literally "open dykes" in French (in German, it would have been "offene Gänge"); a very Wernerian terminology used by this illustrious pupil of Werner!

[310]This appellation has its inspiration in the text of Captain Bernardo de Miera y Pacheco. On his map, where, after having pointed out that the present-day Rockies form the backbone of the continent, he says that cranes breed in this chain: "… en ella se crian las grullas" (see Wheat, 1957, p. 109).

[311]Von Humboldt's description of the origin of Jorullo as an elevation crater was wrong because he did not have access to the eye-witness reports, dated 8 and 13 October 1759 by the administrator Sáyago, submitted to the Viceroy by the Governor of Michoacan on 13 October 1759. For English translations of these, see Gadow (1930, p. 77–83). Gadow also reproduces (some in translation, some in the original language) excerpts from other reports from 1759 to 1908 concerning the origin of the Jorullo.

[312]For a modern description of the geology of San Francisco Mountain, see Holm (1987) and Holm and Ulrich (1987). For those without previous preparation in geology, the popular book reviewing the geology of the volcanoes of northern Arizona, including San Francisco Mountain, Duffield (1997) is recommended. This award-winning book is delightfully and very instructively illustrated by the wonderful photographs of Michael Collier.

[313]This is Mt. Taylor in New Mexico, west of Albuquerque at 35°14′ N and 107°36′ W. It is about 3424 m high and consists of Pliocene and Pleistocene basalts and differentiated volcanic rocks capping high mesas. Von Humboldt's interest was kindled both by the colored foldout geological map in Jules Marcou's report in the Whipple survey (Marcou, 1856; Fig. 96A herein), that shows Mt. Taylor and San Francisco Mountain on a single east-west traverse that exposes other volcanic rocks as well, and by Möllhausen's descriptions (1858, p. 323–335 and notes 23, 24, 25 on p. 492–493; also the colored plate showing San Francisco Mountain {facing p. 324}). The position of Mt. Taylor and San Francisco Mountain defining an east-west axis would have reminded von Humboldt of the Mexican volcanic line trending east-west. For a modern review of the geology of Mount Taylor, see Crumpler (1982). Baars (2000, p. 204) points out that Mount Taylor is the Sacred Mountain of the south according to Navajo legends.

[314]The literature on the exploration of the American West during the first half of the nineteenth century is immense. The best introduction is still Goetzmann's two books (1991, 1993). Viola (1987) is a lavishly illustrated history of the exploration of the western United States. Both Warren (1855) and Wheeler (1889) give more detail but less context. Some of the papers in Koepp (1986) deal with this period. Schubert (1980) is a concise and well-illustrated history of the contributions the U.S. Army Engineers made to the westward expansion of the United States and hence to the increase of our knowledge of the geography

and geology of the West. Wheat (1958) has the best summary of the cartographic history of the period under consideration. Also see Anonymous (1952) and Morgan and Wheat (1954). Goetzmann (1995, ch. III–V) gives an overall summary within the context of nineteenth century exploration of America and also elsewhere by Americans. All these should be read with Goetzmann and Williams (1992) at hand and should be compared with the progress of geology in particular and science in general in America in the first half of the nineteenth century as told by Merrill (1904, 1920, 1924), Struik (1948, especially p. 264–317), Ospovat (1960, p. 198–215), Murray (1970), Schneer (1970), Hazen (1974), Prendergast (1978), Rabbitt (1979), Stapleton (1985), Newell (1993, 1997), and Daniels (1968[1994]). One indicator of the state of a science in a society is how the scientists of that society view the history of their own subject. For the case of the American science for the earlier part of the first half of the nineteenth century, see White (1970, 1973).

[315]The exploration of the western part of what is today the conterminous United States by U.S. citizens or those in her service is inevitably a part of the moving frontier of this young and enterprising nation westwards across the continent. For the history of this frontier in the time period we are interested in, see Paxson (1924) and Hawgood (1967).

[316]For Pike's life, see Coues (1895a, p. xix–cxiii) and Hart (*in* Hart and Hulbert, 1932). In heatedly defending Pike against accusations of spying, Hulbert (*in* Hart and Hulbert, 1932) also gives a fairly good sketch of Pike as a man and as an officer. The three volumes of Coues (1895a, 1895b, 1895c) constitute the edition of Pike's narrative of his journey that I used for his text, enriched by numerous scholarly footnotes on historical geography and history of individuals, nations, and places. However, the Coues edition had been undertaken before Mexico returned Pike's confiscated papers and maps to the United States in 1910 (except items numbered 19 and 20, which had perished). The returned documents remained unnoticed for 15 years and were saved from being lost when Stephen H. Hart wanted to consult them for research. After an adventurous search in the archives of the State and the War Departments in Washington, D.C., they were finally located in 1927. In 1932, S.H. Hart and A.B. Hulbert used them in re-editing Pike's journal of his western journey (with the reproductions of three map fragments and one sketch by Pike of Pike's Peak). Donald Jackson (1996a, 1996b) published the best edition of Pike's journals together with all the relevant documents including his original maps. Any serious study of Pike's expedition to the West must consider all these editions. Goetzmann (1991, p. 36–39; 1993, p. 43–53) provides brief summaries of Pike's expedition, and Goetzmann and Williams (1992, p. 145) give a route map. Both DeVoto (1952, p. 423–431) and Goetzmann (1993) and, following them, Viola (1987), imply that Pike may indeed have been involved in Wilkinson's scheming, although I find Pike's character unsuitable for intriguing against his fatherland, for which he ended up giving his life at the relatively young age of 34 (see also Jefferson's letter about Pike to von Humboldt, dated 6 December 1813, *in* Jackson, 1966b, {p. 387–388}; also Jackson's verdict in his Foreword: 1966a, p. vii). That Pike may have been used by Wilkinson without knowing the latter's real intentions is most likely. See Wheat (1958, ch. XII) for a discussion of Pike's published maps and reproductions of two of the maps.

[317]For the Hunter et al. expedition, which was sent out by President Jefferson on 16 October 1804, see Viola (1987, p. 26) and Goetzmann (1993, p. 41–42).

[318]Pike adapted von Humbodt's "Carte du Mexique et des Pays Limitrophes Situés au Nord et à l'Est" without acknowledgment, which annoyed the Baron, who complained in a letter to Thomas Jefferson (Jackson, 1966b, p. 377). But it is wrong to assume that Pike's map is a copy of von Humboldt's, as Goetzmann (1993, p. 37) unfortunately implies by reproducing Pike's map and citing it as von Humboldt's. Also unjust is Coues' (1895a, p. xlii–xliii) statement that "I have reluctantly satisfied myself that Pike's map of New Spain is no other than Humboldt's *Carte Générale du Royaume de la Nouvelle Espagne*, with Nau's errors and some little further modification." We know that while Pike was still in Mexico, he had shown Don Joaquin del Real Allencaster, the Governor of New Mexico, a sketch-map he had made *en route* containing "all the rivers and countries he had explored" (Allencaster's report to General Salcedo, quoted in Coues, 1895a, p. xlvi). We now possess Pike's manuscript maps (see Jackson, 1966a, plates 9 to 31 inclusive). A comparison of the manuscript maps published by Jackson (1966a) and the published Pike map of "New Spain" would readily convince anybody that the published map had much of Pike's own data in it (compare, for example, the published "New Spain" map in Jackson {1966b, map 5} with the manuscript shown in Jackson {1966a, plate 28}). Wheat (1958, ch. XII) discusses the differences of Pike's published map with that of von Humboldt's. Streeter (1960, p. 20) points out that for the Texas portion, the rivers on Pike's map are an improvement on von Humboldt's, whereas von Humboldt's coastline is more accurate. Jackson (1966b, p. 378, note 2) also comments on the difference, noting that Pike's map is superior with respect to the upper areas of Louisiana. All in all, it seems that Pike erred owing to his inexperience rather than to malice. I think that Jefferson's reply to von Humbodt's complaint is still the best that can be said on Pike's behalf: "I am sorry he ommited even to acknoledge [*sic*] the source of his information. It has been an oversight, and not at all in the spirit of his generous nature. Let me sollicit [*sic*] your forgiveness then of a declared hero, of an honest and zealous patriot, who lived and died for his country ..." (Jackson, 1966b, p. 388).

[319]For a brief statement of Tanner's *vita*, see Wheat (1958, p. 82, footnote 1).

[320]For a biographical sketch of Ashley, see Chittenden (1902[1954], p. 247–251).

[321]For a biographical sketch of Henry, who is said to have been a good violin player, see Chittenden (1902[1954], p. 251–252).

[322]Phillips (1961b, p. 396) says that the advertisement was in the *Missouri Republican* of 20 March 1822. I have not checked these details as they are immaterial for the purpose of the present book.

[323]For an account with documentation of the main geographical results of General Ashley's activities, see Dale (1941); Goetzmann (1993, ch. IV) provides an excellent summary with sources. For a general background on the fur trade and its geographical aspects, see Chittenden (1902[1954]a, b) and the monumental book by Phillips (1961a, 1961b), the last chapters of which were written by J.W. Smurr after Phillips' death.

[324]For a brief biographical sketch of Stuart, with no vital dates, see Chittenden (1902[1954]b, p. 908–909). There is much additional information in Phillips (1961b).

[325]Jedediah Strong Smith was perhaps the greatest explorer of the American West, a kind of Mozart of the exploration business (upon his death, the *Illinois Monthly Magazine*, June 1832, called him "the greatest American traveller": Brooks, 1977, p. 10)! He lived only to 32 years of age, when he was brutally murdered by savage Comanches while searching for water. At the time of his death, he had seen more of the American West than anybody alive. His untimely death was possibly the greatest disaster in the history of the exploration of the American West. At the time, he was engaged in the preparation of a geographical account and an atlas of the Rocky Mountain region (Chittenden, 1902[1954], p. 254), which consequently was never completed. In a wonderful example of outstanding scholarship, Morgan and Wheat (1954) scraped together all the extant cartographic information that Smith was able to put on record and published it in a superbly illustrated book. For a detailed depiction of Smith's life and times, see Morgan (1953). In 1934, Maurice S. Sullivan published fragments of Smith's journals and notes, which had been copied by Samuel Parkman, concerning his entry into the fur trade, his walk across the Utah desert, and his second journey to Califonia in 1827–1828 (Sullivan 1934[1992]). Brooks (1977) edited, with introduction and commentary, Smith's subsequently discovered southwest journal containing the story of his first journey to California and back across the Great Basin in 1826. Both Chittenden (1902[1954]a, b) and Phillips (1961b) contain much additional information on the trapping and geographical activities of this great man. Also see Goetzmann (1993, p. 112, and in later pages, to p. 144).

[326]Gilbert's (1983) book is the best biography in existence of Joseph Rutherford Walker's life. Walker was one of the greatest and undoubtedly the most civilized of the great mountain men who helped the opening up of the American West to civilization. For a history of the exploration and settlement of the Great Basin area, see Durham (1997).

[327]For a sketch of Benton's life with emphasis on his western interests, see Frémont (1887).

[328]For a summary of the history of the U.S. Army Bureau of Topographical Engineers, see Schubert (1980). Also see Viola (1987, especially the chapter entitled "The Great Reconnaisance," p. 87–119), and Goetzmann (1991, p. 7–21 and his Epilogue and the references cited). The importance of this Bureau and

of its heroic officers for the growth of our knowledge of the geography and tectonics of the western United States, especially in the mid-nineteenth century before the American Civil War, which led to the dissolution of the Bureau, cannot be overestimated.

[329]For a summary of Nicollet's explorations, see Warren (1855, p. 40–42). Goetzmann (1993, p. 242) points out that Nicollet was the first to introduce into western exploration the technique of stratigraphic correlation by means of fossils. As a well-educated Frenchman, Nicollet could hardly have not known Cuvier's powerful method! He was also the first in the American West to make extensive use of the barometer for altitude measurements (Wheeler, 1889, p. 550). For a picture depicting Nicollet at a trading post, see Schubert (1980, p. 11).

[330]Being a national hero both in his own lifetime and afterwards, Frémont's life story has been chronicled often. He himself wrote an autobiography, but only its first volume was ever published (Frémont, 1887). For modern biographies, see Nevins (1955) for by far the best biography of Frémont, Bashford and Wagner (1927), and Egan (1977). Alice Eyre's *The Famous Fremonts and Their America* (1948) is a well-illustrated and documented biography giving more details of Frémont's wife Jessie than his other biographies ("the documentary history of these two persons is but a single subject of study": Rolle, 1991, p. 283). Bigelow (1856) is a contemporary account, written by the co-owner of the *New York Evening Post* and the principal election campaign advisor to Frémont (Bigelow, 1817–1911, became a celebrated author and historian later), with the help of Frémont's talented and intelligent Jessie, as a propaganda piece for the 1856 presidential election campaign, for which Frémont stood as the anti-slavery Republican candidate. Harris (1990) is a very readable, well-illustrated, brief, and popular account. Rolle's (1991) excellent biography presents an interesting psychoanalysis of Frémont, the value of which I am unable to assess. Goetzmann (1991, *passim* and 1993, especially p. 240–252, but also elsewhere) contain brief accounts of his life and expeditions. Goetzmann and Williams (1992, p. 158–159) has the route map for the first four of his expeditions.

The fairly negative picture painted in Gilbert's chapter on Frémont (1983, p. 198–216) is not to be taken very seriously as it reflects Joseph Walker's personal dislike of a man whom he considered a coward. History disagrees with the great mountain man's impression. Bashford and Wagner (1927) felt justified to entitle their biography of Frémont "A Man Unafraid." Rolle (1991) repeatedly emphasized Frémont's courage. In fact, Harris (1990, p. 31) went so far as to claim that "One criticism that was never leveled at Frémont was that he was afraid to act." Walker did just that, but only because he never bothered to revise a judgment made rashly in California during the Mexican War. Another famous mountain man, Frémont's faithful scout and friend Kit Carson, had a diametrically opposed view of Frémont, to whom, he once said, he owed "more than any other man alive" (quoted *in* Harris, 1990, p. 33). In 1856, Carson put on record the following tribute to Frémont: "I was with Frémont from 1842 to 1847. The hardships through which we passed, I find impossible to describe, and the credit which he deserves I am incapable to do him justice in writing … I have heard that he is enormously rich. I wish to God that he may be worth ten times as much more. All that he has or may ever receive, he deserves. I can never forget his treatment of me while in his employ and how cheerfully he suffered with his men when undergoing the severest hardships" (quoted *in* Nevins, 1955, p. 616). Another close associate of Frémont, Alexis Godey, rendered a similar, but much more detailed, defense of Frémont against the charges published in the Los Angeles *Star* of 6 September 1856 and in *The Washington Post* of 31 July 1856 (Hafen and Hafen, 1960, Appendix E).

Alexander von Humboldt had the highest opinion of Frémont as a geographer (von Humboldt, 1849, p. 51–52; also see his letter to Frémont, dated 7 October 1850 {*in* Bigelow, 1856, p. 327–328}, transmitting the grand gold medal from the Prussian King for Frémont's labors in science and the news that Frémont had been elected, upon Carl Ritter's suggestion, as an honorary member of the Geographical Society in Berlin. The Bigelow biography is dedicated to von Humboldt as being "among the first to discover and acknowledge" Frémont's genius) and in this history agrees with the Baron (with the exception of those smitten by the deplorable late-twentieth century historiographical fad of belittling great men). Warren (1855) also presents summaries of Frémont's exploring expeditions emphasizing their principal scientific results (see especially p. 42–43, 46–52).

[331]For the correct spelling of Provot's name (i.e., *Provot* and not *Provost* as in most histories), see Harris (1909, p. 261), although H.H. Bancroft, cited by Auerbach (1943, p. 125, footnote **), "vouches for this spelling [*i.e., Provost* and not *Provot*], confirmed by an assurance from Stella M. Drumm, Librarian, Missouri Historical Society, St. Louis, Missouri." I have not pursued this matter any further, as it is not directly relevant to the topic of this book. Harris (1909) also gives a short summary, with portrait, of this great mountain man's life on p. 258–262.

[332]Kit Carson's contributions to Frémont's success and hence to a scientific mapping of the American West have been immense. For a biography of this remarkable man, see Carter (1968).

[333]Now known to be Upper Cretaceous marine sedimentary rocks (King and Beikman, 1974). See Hall's (1845, p. 296) geological report for the second expedition, where these rocks were identified as Cretaceous only.

[334]This is the latest Miocene-Pliocene Kortes Formation containing sparse Clarendonian (Middle-Upper Miocene) land-mammal age fossils (see Flanagan and Montagne, 1993, especially p. 588–589 and fig. 9).

[335]These are the Archaean granites of ages ca. 2.6 Ga (Frost and Frost, 1993, p. 63).

[336]Early Proterozoic mafic magmatism (Snoke, 1993, fig. 3).

[337]For the exploration history of the Sierra Nevada of California, see Farquhar (1965) and Moore (2000). Moore also gives an outline of its geology that is intelligible to the educated layman.

[338]They went from the Upper Tertiary (mainly continental Pliocene deposits) to the Upper Cretaceous formations along the Arkansas Valley (King and Beikman, 1974). Along their route, they found that the strata dip more steeply eastward than does the topographic surface, so they were traveling down-section.

[339]Hall's error is surprising because he mentioned bones of herbivorous mammalia from the same rocks that Captain Stansbury had sent him (Hall, *in* Stansbury, 1852, p. 402).

[340]For the history of the construction of the first Transcontinental Railroad in the United States, see the monumental study by Bain (1999); however, the surveys occupy slightly more than three pages in his immense tome. The geographer Vance's (1995) superbly illustrated account gives a lot more of the geographical background of the first Transcontinental Railroad in the United States. For a shorter history, see Holbrook (1947). For the history of the Pacific Railroad Surveys, see Viola (1987, chapter entitled "The Great Reconnaisance"), Goetzmann (1991, ch. 7; 1993, ch. 8) and Schubert (1980, ch. VI). For the contribution of the surveys to geology, see Goetzmann (1991, p. 305–326). For the scientific exploration of the western United States in the middle nineteenth century, see also Wallace (1955).

[341]A colorful and quarrelsome personality, a sort of European James Hall in character, Marcou was not Swiss, as Goetzmann writes (Goetzmann, 1991, p. 287; 1993, p. 317. Goetzmann is inconsistent with respect to the nationality he assigns to Marcou; on p. 325 of his 1991 book, he calls him a Frenchman), but a Frenchman from Salins in the French Jura, educated in Besançon and Paris. He carried many of the biases of his Parisian education to North America. For Marcou's life and contributions, see the papers cited in Sarjeant's *Geologists and the History of Geology,* v. III; for his American contributions specifically, also see the scattred references to him in de Margerie (1952, 1954).

[342]For Gibbs, see Goetzmann (1991, p. 317).

[343]Beckwith had to furnish the reports because the commander of the survey, Captain John W. Gunnison, was "barbarously massacred" along with seven others on 26 October 1853 in the valley of the Sevier River, including the party's botanist F. Creuzefeldt, by the savage "Pah Utah Indians" as he ran out of his tent with the intention of declaring their friendly intent (Beckwith, 1854a, p. 9; Goetzmann, 1991, p. 285)! The geologist and physician, James Schiel, was one of the few who were able to get away. Gunnison's noble memory is perpetuated by his countrymen in the naming of Gunnison Island of the Great Salt Lake and the town of Gunnison, Utah. For a portrait of Gunnison, see Schubert (1980, p. 100).

[344] When Schiel delivered his report, the Permian System had not yet become common knowledge, having been proposed by Sir Roderick I. Murchison in 1841 in a letter to Professor Fischer de Waldheim, the ex-President of the Soci-

ety of Naturalists of Moscow. For a reprint of this letter, see Mather and Mason (1939, p. 247–249).

345For the quotation of the passages referred to by Lieutenant Beckwith, see above p. 171.

346By "Putrid sea," Lieutenant Beckwith is here referring to the Sivash Sea in the Ukraine, on the southeastern coast of the Sea of Azov. In Russian, it is known as the *Gniloye More*. It is a salt lagoon, with the deepest point barely –1.5 m, separated from the Sea of Azov by the 110-km-long Arabat Tongue (with which Beckwith compares the paleo-peninsula he is describing). The Genichesk Strait maintains the communication of the Sivash Sea with the Sea of Azov.

347To see to what remarkable degree Lieutenant Beckwith's intuition was confirmed by subsequent geological work in the Great Basin, compare Gilbert (1875, ch. III; 1882; 1890), Gilbert and Howell (*in* Wheeler {undated, map 1, entitled "Restored Outline of Lake Bonneville"}), King (1878, p. 488–529 and the "Analytical Geological Map of the Exploration of the Fortieth Parallel—VI. Lakes of the Glacial Period" between p. 528–529). For the results of the modern phase of research, see especially Morrison (1965; entire Great Basin area) for progress up to the 1960s and the papers in part II (Geology and Geophysics) in Gwynn (1980); for the current state of the art, see Machette (1988; for Lake Bonneville and the eastern Great Basin) and Adams et al. (1999; for Lake Lahontan and the western Great Basin). Adams et al. (1999, fig. 1) give the most up-to-date map that I am aware of for the Pleistocene lakes in the Great Basin. Compare that with King's map cited above.

348By "the creek," Schiel means the Sangre de Cristo Creek, "a small stream of clear, cold water" (Schiel, *in* Beckwith, 1854a, p. 37). The locality Schiel is describing is near the El Sangre de Cristo Pass roughly at 37°36′ N and 105°20′ W (for the description of the area, see Beckwith, 1854a, p. 37–38). See the "*Map N°3 From Santa Fe Crossing to the Coo-che-to Pass*" (*in* Anonymous, 1859, route near the 38th and 39th parallels—Beckwith's Report, v.. II) and the colored lithographs by J.M. Stanley after sketches by R.M. Kern, showing panoramas of the pass (*in* Beckwith, 1854a, facing p. 37, 40).

349Here Schiel inserts the following footnote: "The cambrian system [*sic*], as distinguished from the silurian system [*sic*] by its age and organic remains, is not recognized any longer by geologists. Comp. Murchison, in Quarterly Journal Geology [*sic*], soc. [*sic*] VIII, 1852. Murchison's Siluria, 1854."

350It is 0.04°.

351This observation has been marvelously corroborated and led to fruitful speculations in modern times concerning the origin of the great western swell of the United States territory by Gordon P. Eaton (1987). Eaton appropriately named the ridge forming the crest of the swell in the southwestern United States the *Alvarado Ridge* after the first civilized man who saw it, namely one of Coronado's captains, Hernando de Alvarado (see above).

352Blake translates Marcou's "les gypses" as "gypsiferous formation" following his own preferred terminology. In my rendering, I adhere to Marcou's original.

353Here Marcou was wrong. The evaporitic and clastic rocks he so enthusiastically assigned to the Triassic are now known to be Leonardian and early Guadalupian (middle Permian) rocks (Johnson et al., 1988, especially figs. 8I and 8J).

354Between 102° W longitude and the Rocky Mountain front in New Mexico, Marcou's age assignment has been corroborated (see King and Beikman, 1974). It differs from his "Upper Silurian" assignment (with a small area of "Lower Silurian" just to the east of Spanish Peaks) seen on his first map of the United States that had been drawn on the basis of the literature. Compare Marcou (1853) map with Marcou (1858) folding colored map and Marcou (1888, p. 31–35).

355William Phipps Blake (1826–1910), a native of New York state, was a graduate of Yale (1852) and also worked with James Hall (Goetzmann, 1991, p. 317). His geological background was thus strongly shaped by the Yale-New York Survey traditions (Anonymous, 1910; Keyes, 1939).

356His observations revealed to him the thickest "Triassic" was located just east of the Sandia Mountains though! But farther east he had not seen the base of the "Triassic."

357Blake's footnote: "Resumé. Report of Lieutenant A. W. Whipple, H. Doc. 129, p. 46." For a full reference, see Marcou [1854]. Here, Blake does not give a complete statement of Marcou's views. In the Whipple Resumé just referred to, Marcou had interpreted an unconformity to exist between the Triassic and the Jurassic and thus a pre-Jurassic phase of mountain-building (Marcou, [1854], p. 47) and a pre-Upper Cretaceous unconformity and, thus, what Marcou believed to be an end-Jurassic phase of mountain-building (Marcou [1854], p. 46; 1856, p. 169). Marcou's map shows why he thought there was an unconformity between the Triassic and the Cretaceous (e.g., northeast of the Sierra Madre {the Zuni Mountains}, north-northwest of Camp 68, he actually mapped such an unconformity: see Fig. 96A). Because his map has no Jurassic symbol, it is not possible to see his cartographic justification for the post-Jurassic unconformity. In his cross section, both his Neocomian (i.e., Lower Cretaceous) and his White Chalk (i.e., Upper Cretaceous) in places sit directly on the Triassic, thus indicating a pre-Neocomian and possibly a pre-Upper Cretaceous phase of erosion that swept away the Jurassic; and in other places, also the Neocomian, which are indicated to be present elsewhere where erosion had not destroyed them. It is not easy to disentangle from his graphic depictions what Marcou was thinking at the time in terms of the tectonic history of the southeastern United States because he did not leave a detailed and complete description of his interpretations. But, as he was not in the habit of changing his opinions, one could consult Marcou (1888, p. 43–56) for the tectonic development in the late Jurassic and the early to medial Cretaceous.

358Thomas Antisell (1817–1893) was an Irishman (and not a Scotsman as claimed in Goetzmann, 1991, p. 308; as in Marcou's case, Goetzmann is inconsistent and calls him an "Irish surgeon" on p. 317) born in Dublin, where he was educated as a physician and chemist. In 1848, Antisell came to New York. In 1854, he entered government service as surgeon and geologist for Lieutenant Parke's survey. During the Civil War, he entered the Union Army and functioned as a surgeon both in the field and in the Harewood Hospital. After the war, he settled in Washington, D.C., and worked in the Patent Office and in the Department of Agriculture as a chemist. In 1871, he went to Japan with the Secretary of Agriculture, Horace Capron, to provide technical assistance to the Japanese. He returned to Washington, D.C., and died there. Antisell also taught at Georgetown University, where his archive (which contains material concerning his geologic investigations in California) is now kept. For more information, visit the following Web site: www.library.georgetown.edu/dept/speccoll/cl137.htm.

359See Goetzmann (1991, p. 378, footnote 14) for the uncertainty of the amount of money.

360For sources on von Egloffstein, see Miller (1970, p. 152, footnote 16). Concerning sources on Möllhausen, I confine myself to Miller's two papers (1972a, 1972b) and the dissertation from which they resulted (Miller, 1970), plus the Möllhausen entry in Henze (1993, p. 508 and the references he cites). As to Möllhausen's scientific interests and accomplishments, Miller wrote: "Möllhausen's scientific accomplishments and interests never reached the level of acumen associated with the Duke [*Paul Wilhelm Friedrich von Württemberg (1797–1860); for sources of information about the Duke's life, see Miller, 1970, footnote 17 on p. 21 and footnote 20 on p. 23–24)*] or Humboldt. He is remembered today as an author and artist. Möllhausen was something of a zoologist, a geologist, a botanist and an ethnologist. But primarily he was a diarist and an artist" (Miller, 1970, p. 213). Möllhausen made three trips to America. The first was a private trip, commenced in 1849 when he was only 24 years of age, but he had the good fortune of meeting the Duke of Württemberg in 1851, whom he accompanied from the Mississippi River valley to Fort Laramie, when the expedition had to be given up owing to adverse weather conditions and the hostility of the natives. Möllhausen then spent a few weeks with the Oto and Omaha Indians. Returning south on the Mississippi River, he was reunited with the Duke and returned to Berlin in 1852 (see Miller, 1970, p. 24–78). He returned to the United States to join in 1853 the Whipple Survey and met Ives there (for Möllhausen's record of this journey, see Möllhausen {1858}; this book was immediately translated into English by Mrs. Percy Sinnett and published by Longmans, Brown & Green in London under the title *Diary of A Journey from the Mississippi to the Coasts of the Pacific with a United States Government Expedition* in 2 volumes: see Miller, 1970, p. 15, footnote 4). As Goetzmann (1991, p. 380) rightly points out, that Möllhausen took the trouble of returning to the United States a third time to join the expedition Ives was to lead says much in favor of Ives. The Ives expedition was Möllhausen's last journey in the New World (Möllhausen, 1861; for an unpublished English translation of this important book, see Miller, 1970, p. 150, footnote 13). After Möllhausen returned home, he turned novelist and published more that 100 novels on life in the American West, until his death in 1905, to earn the nickname of "the German Cooper." The last edition of the *Brockhaus Enzyklopädie* (1991, v. 15, p. 23) lists the following as Möllhausen's more important novels: *Der Flüchtling* (1861, 4 vols.), *Der Halbindianer* (1861, 4 vols.), *Das Mormonenmädchen* (1864, 6 vols.), *Die Mandanenweise* (1865, 4 vols.), *Der Hochlandspfeiffer* (1868, 6 vols.), and *Die Kinder des Sträflings* (1876, 4 vols.). See also von Humboldt (1858a) and Goetzmann (1991, p. 310, footnote 10).

[361]Note (1) the distinct echo of Buache and Pallas, despite von Humboldt's best efforts to show that the great central plateau of Asia, the fountain-head of the central continental plateau idea of Buache (1761), was myth, and (2) Dana's descriptions of North America having two grand swellings along the oceanic margins and a lower "great central area of the continent," a vast plain, scarcely affected by important tectonic events (Dana, 1847b, p. 98).

[362]Now known as the Diamond Creek Pluton, forming a part of the Precambrian Ruby Intrusive Complex, which in turn is a part of the Zoroaster Plutonic Complex. The composition of the Diamond Creek Pluton is tonalitic (Babcock, 1990). Möllhausen gives the most convenient graphic summary of the stratigraphy of the mouth region of the Diamond Creek on the basis of Newberry's work:

Limestone with fossils Siliceous shale Limestone Sandstone with no fossils	Lower Carboniferous
Limestone Sandstone Slate with some fossil corals	Devonian
Red and white sandstone with no fossils Green and violet-colored slates	Silurian
Red sandstone with no fossils	Potsdam Sandstone
Granite	500 ft.[151.5 m]

(Möllhausen, 1861, v. II, p. 395, endnote 6). Compare this section with the one given in Figure 106 herein.

[363]For the present-day knowledge of the Carboniferous and Permian stratigraphy of the Grand Canyon district, see Blakey (1990), Middleton et al. (1990), Turner (1990), and Hopkins (1990). For a helpful roadlog to see the exposures of these two systems in the canyon area, see Baars (2002, ch. IV). For those who are not geologists, I recommend Redfern's (1980) and Price's (1999) excellent, popular books.

[364]Here, reference is to the Cordilleran ranges that Newberry studied as the geologist for Lieutenant R.S. Williamson's railroad survey from the Sacramento Valley to the Columbia River (Newberry, 1856).

[365]Şengör (1998, p. 82, footnote 115) wrote that Suess was the first to use the term "embryonic fold" and the associated concept of the embryonic development of mountains because he was unaware of Newberry's usage. It seems that the physician Newberry may have been the one who introduced the term embryo into tectonics.

[366]To document the ancient ancestry of the faults in the Rocky Mountain region, Cloos cites an abstract by Rollin Chamberlin (1939). Chamberlin may well have heard from his father, T.C. Chamberlin, of Newberry's ideas and thus may have built a bridge between him and Cloos. But, by the time the younger Chamberlin was being educated, such ideas had become common knowledge.

[367]For Hayden's life and the history of his survey, see Nelson et al. (1981), Foster (1994), Nelson and Fryxell (1997), and Cassidy (2000). For briefer accounts, see Bartlett (1962, part one) and Goetzmann (1993, ch. XIV). For the struggle for directorship of the U.S. Geological Survey between Hayden and his opponent Powell, Jaffe (2000, especially p. 207–226) gives an interesting account from the perspective of the Marsh-Cope feud.

[368]For King's life and the history of his survey, see Wilkins (1988). For briefer accounts, see Crossette (1946, pp. 98-100), Bartlett (1962, part two), Goetzmann (1993, ch. XII). Nelson and Rabbitts (1997), and Schubert (1980, p. 136–140).

[369]Of all the four Great Survey leaders, "Major" John Wesley Powell is by far the most popular. His popularity among non-scientists rests mainly on his adventurous 1869 Grand Canyon trip and, in some part, on his ethnographical studies of North American Indians and now, increasingly more, because of his advocacy of a rational legal basis for land use in the arid American West. Even if he had not undertaken his Grand Canyon trip and written nothing on Indians or on land-use legislation, he still would have had a formidable reputation as a great geologist on the basis of the superb geological work that he did and also enabled others to do. Powell had a great knack for recognizing and attracting talent. He maintained around himself an atmosphere conducive to creativity for the scientists working under his direction, sharing his means, knowledge, and ideas generously with them. Both Gilbert and Dutton pointed out that it was impossible to tell how much of their ideas came out of Powell's brain. As a result, there is a vast literature on, and many biographies of, Powell.

For the minutes of the meeting held in memory of Powell at the Smithsonian Institution on 26 September 1902, see Langley et al. (1902). The Powell biographies that I have read are the following: Dellenbaugh (1909), Darrah (1951), Stegner (1953; this is not a proper biography as it starts with Powell's 1868 expedition and emphasizes his career and its influence on the opening of the American West), Terrell (1969), and Worster (2001). None of these biographies are satisfactory from the viewpoint of the historian of geological ideas. For geologically more informative writings on Powell, see Gilbert (1902, 1903), Davis (1909, 1915), Chorley et al., (1964, ch. 27), Hunt (1969a, 1969b), McKee (1969), and Rabbitt (1969). Also see Crossette (1946, p. 130–132) and Nelson (1996). Watson (1954) is a collection of the reports of Powell's western explorations. Dolnick (2001) is a meticulous account of the first canyon voyage by Powell. Regarding that journey, also see Ferguson (1961). Anderson's (1979) paper on the same journey contributes no new data but gives the unfortunate impression of being written with the sole purpose of detracting from Powell's glory. Anderson (1983) continues in the same vein but presents new data in the form of reminiscences and comments of Powell's comrades of the two voyages. Tikalsky's (1982) analysis of Powell's role and leadership during the two expeditions seems much more balanced and the result is favorable for the Major. For Powell's own account, see Powell (1875), but in that narrative, he conflated the accounts of the first and the second voyages (and was justly criticized for it; see Rabbitt, 1969, p. 20 {it "is not good history"}, and Anderson, 1983). For another narrative of the same journey, with the same defects, see Powell (1895; the best reprint of this now exceedingly rare book {see the description and account of rarity in Catalogue 18 [2002] of the Five Quail Books, Prescott, Arizona, p. 42, item 308} is the 1964 facsimile edition by the Argosy-Antiquarian Ltd., New York). The *Utah Historical Quarterly* (1947, v. 15) includes biographical data on the participants of the first journey.

For an account of the second voyage, see Dellenbaugh (1908[undated]) and *Utah Historical Quarterly* (1948–1949, v. 16-17). For shorter accounts, see Bartlett (1962, part 3), Rabbitt (1980), and Goetzmann (1993, ch. XV). For the anthropologist Powell, see Fowler et al. (1969). For a collection of sundry writings by Powell, see Crossette (1970) and deBuys (2001). For the history of the Powell Memorial on the south rim of the Grand Canyon, see Burggraaf (1997).

[370]Wheeler's life is the least well-publicized among the leaders of the Great Surveys. I am aware of but a single book on him, that by Dawdy (1993). Devoted to exposing a sinister side of Wheeler's activity as the head of one of the Great Surveys, this little book hardly does justice to the scope and quality of the geographical and geological work undertaken by the Wheeler Survey, neither does it give a just appreciation of the geological and geographical results obtained. Dawdy's book does contain a good bibliography, however. For brief and unpartisan accounts, see Bartlett (1962, part four), Goetzmann (1993, ch. XIII), Schubert (1980, p. 140–149), and Karrow (1986).

[371]For the history of the Great Surveys, Bartlett (1962) is the only comprehensive one-volume account with which I am familiar. For informative accounts, see Manning (1967, ch. 1), Rabbitt (1979, ch. 9–13), Viola 1987, chapter entitled "Arsenic and Directions," p. 121–177), and Goetzmann (1993, ch. XII–XV). Also see Smith (1918[1973]) and Schubert (1980, ch. VIII). Gilman (1872, 1973), Davis (1924), and de Margerie (1952, 1954) give sketchy summaries of the geographical work of the Surveys. Chorley et al. (1964, ch. 30) is a summary and evaluation of the geomorphological work of the Great Surveys. Merrill (1904, ch. VII; 1924, ch. VIII) provides much information about the

Great Surveys and their leaders. Powell (1878) is a valuable document comparing the work of the four Surveys and emphasizing the areas of overlap. All the biographies of the leaders of the Great Surveys cited in the four preceding endnotes naturally also give information relating to the Surveys themselves. Schmeckebier (1904) is a catalogue and index of the publications of the four Great Surveys.

[372]It is in that sense that most Europeans learn about the Rocky Mountains. That was how I learned it at school in İstanbul. Now, North American geologists use the designation made popular by Alexander von Humboldt, namely the "Cordillera," for what Powell referred to as the *Rocky Mountain System*. Gilman (1872, p. 117) notes that it was Josiah Dwight Whitney, who proposed the term *Cordillera of the United States* "for all that vast and intricate system of upheavals lying along the western portion of our territory."

[373]Modern studies are in complete agreement with Powell's estimate of the age of Basin-and-Range faulting (cf. Snow and Wernicke, 2000). This is naturally to be understood within the strictures indicated by Dickinson (2002).

[374]For Mallet, see Davison (1927, ch. V, and the obituaries he quotes on p. 66, footnote *). For the controversy with George Poulett Scrope concerning his contraction theory, see Wilding (1996).

[375]For summaries of Daubrée's important studies on metamorphism that influenced Dutton, see Daubrée (1860, 1867).

[376]Here Dutton inserts the following footnote: "The exact quantity of gelatinous silica is not known, and would be difficult to determine in view of the great quantity of water mechanically held in the clots. Of course, it is not intended that the enormous bulk of precipitates of alumina and silica represents the condition of hydrothermal rocks, but rather those precipitates after water is expressed and the pulp condensed."

[377]Dutton here adds the following footnote reporting observations made by Powell's party in the West:

It is the opinion of many observers in the mountain regions of Colorado Territory, that this is a country of light sediments. Mr. A.R. Marvine and Prof. Powell regard this as apparent and not real. The upturned edges of the strata in the "hog-backs" are no doubt thin, but there is good evidence that they thicken rapidly lower down, for they are unconformable throughout; and the general view adopted seems to be that as far as they were thrown down they were turned up at the edges and attenuated again by erosion, the detritus being carried farther out. The evidence of a stupendous wasting and erosion of this country throughout Tertiary time is complete; and as the whole area of deposit was lacustrine, it may well be asked what became of the detritus, if vast bodies of it do not remain there still?

[378]King earlier referred to Sir William Thomson's (later Lord Kelvin) 1876 presidential address to the Section A of the British Association (Thomson, 1876) in relation to the tidal argument. See Oreskes (1999, especially p. 26–30, 66–69, and the references cited therein) for the development of the argument for the rigidity of the earth from geophysical arguments.

[379]It is tempting to see in the fault-related case a variant of the uplift-rifting scenarios similar to those of Élie de Beaumont (in Dufrénoy and Élie de Beaumont, 1841) and Cloos (1939) and their followers. But such an interpretation would be entirely fallacious here. King was looking at faulting during the Laramide orogeny in the U.S. Rockies, which was a shortening event and the faults he was considering were mostly thrust faults and some crustal-scale tear faults delimiting irregularly shaped uplifts and basins. This case is one example of the usefulness of the copeogenic-falcogenic distinction that can at once embrace extensional, strike-slip, and shortening events and the broad, essentially faultless subsidences and uplifts.

[380]I remind the reader that Dutton here uses the term "Rocky Mountain region" in the sense of the North American Cordillera, following Powell's example.

[381]Von Richthofen assumed that propylite was a primary magmatic product. Dutton followed him in that conclusion but noted that it was unnecessary to create a group distinct from andesites and dacites (Dutton, 1880, p. 108–109).

[382]Work on the highlands of the western United States is continuing along many lines of investigation. An enormous distance has been already covered, and most workers now agree that the high topography is of thermal origin. Of the recent literature, I may perhaps quote some which I think are representative of different models or emphases: Bird (1979), Thompson and Zoback (1979), Eaton (1982, 1986, 1987), Beghoul et al. (1993), Parsons et al. (1994), and Spencer (1996). But see Gregory and Chase (1994), Chase et al. (1998), and McQuarrie and Chase (2000). The difference between the Chase group of models and the others is the timing of the uplift. Brian Wernicke and Martha A. House of the California Institute of Technology are hoping soon to date the timing of the incision of the Grand Canyon using the (U-Th)/He thermochronology in apatite (Brian Wernicke, personal communication, 2002) This should resolve a major issue handed over since the time of Dutton. For the geology of the Grand Canyon area, see (in addition to papers and books already cited): Authors' Collective (1969), Watkins (1969), Smiley et al. (1984), the northern Arizona section in Davis and VandenDolder (1987), Elston et al. (1989), and Harris et al. (1997). For the non-geologists, Chronic (1988) is a useful introduction.

[383]Few geologists may be aware of the circumstances under which Dutton's classical paper on isostasy originated. He relates it himself in a footnote at the end of his great paper with the following words: "The following paper was written hastily to occupy a vacant half hour of a meeting of the Philosophical Society without thought of publication. I have yielded however to the kind solicitation of friends to consent to its publication. It contains a rough outline of some thoughts which have worked in my mind for the last fifteen years and which, from time to time, I have discussed at length in unpublished manuscripts and in familiar conversation with my esteemed colleagues" (Dutton, 1892, p. 64). It seems, from Dutton's essay review of the first edition of the Reverend Osmond Fisher's *Physics of the Earth's Crust*, that he had written such unpublished manuscripts and had invented the term isostasy (in the form of isostacy, which is inconsistent with the rules of transliteration of the Greek into English) by 1882: "I have long been convinced that this doctrine [*i.e. the floating of the crust on a yielding substratum*] must form an important part of any true theory of the earth's evolution. In an unpublished paper I have used the terms isostatic and isostacy to express that condition of the terrestrial surface which would follow from the floatation of the crust upon a liquid or highly plastic substratum;—different portions of the crust being of unequal density. Isobaric would have ben a preferable term, but it is preoccupied in hypsometry" (Dutton, 1882b, p. 289, footnote *). I am grateful to Professor Antony R. Orme for drawing my attention to this early publication of the term isostasy.

[384]Eduard Suess is my candidate for the greatest geologist of all times. I can hardly think of anything in our conceptual appartus in global tectonics today that somehow does not go back to him. That so many of his concepts survived the rise of plate tectonics, indeed in some instances helped to bring the rise about, is the best testimony to his greatness. In addition, he was the founder of the science of urban geology. His range of interests reached from paleozoology through stratigraphy, applied geology and hydrogeology to structural geology, seismology, and tectonics. For much of his life he was an active liberal politician, first in local parliaments and then in the imperial Austrian Parliament. He created the international cartel of academies while presiding over the Imperial Academy of Sciences in Vienna. A great humanist, Suess had such a spell-binding effect on his contemporaries that upon his return home from the 1903 International Geological Congress in Vienna, famous for its heated debates on the nappe theory, the well-known French geologist Charles Barrois' only response to those asking about his impressions of the congress was *"j'ai vu Suess!"* (I saw Suess!).

Suess wrote an autobiography which stops at year 1894 (Suess, 1916). It is almost entirely devoted to his public life and very little to geology, although it contains critical passages concerning his intellectual development in geology, but, as Seidl (2000) began to show, with some misremembrances. A major biography on him does not exist (and is badly needed). The long critical study by Tietze (1917) is so full of misunderstandings that it is misleading even for the professional geologist. A semi-popular book about Suess is that by Obruchev and Zotina (1937) in the series "Lives of Remarkable People" published by the Journal and Newspaper Union in Moscow. There is no doubt that this is the best biography and scientific evaluation of Suess in existence. Obruchev was himself a great geologist and geographer and a friend of Suess. The book contains a set of pictures about Suess' private life (including one of his wife, Hermine) and wonderful descriptions of his daily habits of work and recreation that I have seen nowhere else. It is a great pity that this biography has not yet been translated into any western European tongue. Eugen Wegmann's (1981) article on him in the *Dictionary of Scientific Biography* is excellent, flowing out of the pen of the only close and very able collaborator of Suess' only true heir, Émile Argand! Here I list a few of what I consider relevant books and papers on him published after Wegmann's article: Şengör (1982; 1994; 2000a, 2000b), Greene (1982, especially ch. 6, 7), Hamann (1983), Pinneker (1989), Tollmann (1990), and Cernajsek et al. (1999; with a fine bibliography on Suess).

[385]Suess' ideas developed mostly from his own observations in the eastern and southern Alps and in southern Italy. In 1862, when he published his *Der Boden der Stadt Wien* (The Ground of the City of Vienna), he still portrayed the Alps, *à la* Studer, as a symmetrical chain (see especially Suess, 1862, p. 16–17). It was his view of the structure of the Basilicata region in southern Italy on 12 April 1871, during his trip there with Gerhard vom Rath (for a description of the day and the geology seen, see vom Rath, 1871, p. 130–144) that finally convinced him that both the Apennines and the Alps were asymmetric chains (Suess, 1916, p. 233; for a current assessment of the structure of the Basilicata, see Grasso, 2001, especially fig. 16.15, showing how correct Suess' impression was). Only then did Suess find a preference for asymmetric orogens in the writings of Dana and his American predecessors, such as Amos Eaton, Edward Hitchcock, James M. Safford, and the Rodgers brothers (cf. Merrill, 1924, p. 78, 149–151, 218–220, 331–333; see also Gerstner, 1994, especially ch. 8. Of these authors, Suess cites Dana only in the *Entstehung,* and Dana and the Rogers brothers in the *Antlitz*, choosing to confine his references to the most recent publications). This no doubt greatly increased his confidence in the universal validity of his views. However, Suess' model (see Şengör, 1982, 1998) is significantly different from, and far more sophisticated than, Dana's (for a recent assessment of Dana's global tectonics, see Dott, 1997; also see Struik, 1948; Daniels 1968[1994]; Prendergast, 1978, ch. 2; Newell, 1993, 1997; Dott, 1985; and Mayo, 1985, for a part of the American context). However, Oreskes' (1999, ch. 1) recent portrayal of the Dana-Suess differences is entirely misleading (see Şengör, 2003).

[386]Note Prévost's influence! Recently, Seidl (2000, p. 55) pointed out that the unpublished syllabus Suess submitted in 1857 to the then Minister of Education of the Austro-Hungarian Empire, Count Leo Thun-Hohenstein, contained reference to Charles Darwin's studies on coral reefs. As we have seen above, Darwin had explained the origin and development of the coral reefs by subsidence of large areas in the ocean basins. It is very likely that Suess combined this theory with Prévost's, and also possibly with Dana's ideas, to build his own prejudice in favor of models of terrestrial tectonics in which subsidence phenomena were dominant (see Greene 1982, p. 164). Needless to say, Suess was also familiar with Lyell's work, whom he personally met and held in high esteem (Suess, 1904, p. iv). He would thus have known Lyell's preference for the *preponderance of tectonic subsidence over tectonic uplift even in a non-contracting earth*.

[387]But Suess did repeat the possibility of very large-wavelength and low-amplitude doming by lateral shortening, referring to Diener's interpretation (Diener, 1886, p. 398) between the Libanon and the Anti-Libanon. The translation of this passage in the English edition is incorrect, where Suess is made to talk about "a dome of somewhat greater amplitude" (Suess, 1906, p. 552). In the original (Suess, 1888, p. 699), he says: "... eine Wölbung von etwas grösserer Spannweite" (i.e., an uparching of a somewhat greater span).

[388]But also, see his criticism of Prévost's "subsidence-alone" model (Suess, 1875, p. 65). However, most of Suess' contemporaries saw Suess' model as being one of "subsidence-only" (see, for example, Supan, 1911, p. 370; also Greene, 1982, ch. 7; Oreskes, 1999, ch. 1).

[389]i.e., ridge-making.

[390]This must be a slip of the hand of the great author. He means *centrifugal*. Suess probably had open before him Prévost's 1840 paper while writing this passage, where on p. 201 (see the quotation on p. 108 herein) Prévost does talk about centripetal motion as the direction of the terrestrial contraction. The appearance of de Luc's name on the same page of Prévost lends support to my conjecture.

[391]Neither Prévost (1840, p. 201) nor Suess points to the exact places where de Luc discussed these ideas. Those places are the following: de Luc (1798, Vth letter, especially p. 226–227; 1809a, p. 47–48; 1809b, p. 37–38). De Luc's conclusion was obvious even in the earliest formulation of the non-thermal contraction theory by Descartes (1644[1842], articles 43 and 44 {p. 375–376} and fig. 30 and 31 {Figs. 21C, D herein}; also see Dennis, 1982, p. 8–9 and fig. 1 which reproduces Descartes' two figures). Any collapsing solid will naturally give the same result.

[392]Sedimentary rock successions correlatable world-wide. This idea first came about within the deluge-based geological theories such as that of Steno in the seventeenth century or of Woodward (1695, especially p. 71–74). The idea is historically traceable from the Mesopotamian and Biblical deluge myths to these theories and from these directly to the ideas of Abraham Gottlob Werner (see, for example, Dean, 1985; Huggett, 1989, especially ch. 1–7), who first suggested to separate general formations from local formations (e.g., Ospovat, 1960, p. 159–160; 1971, p. 19 and 100; also see Reichetzer, 1812, p. 42 and 63; d'Aubuisson de Voisins, 1819, p. 326–328; and von Leonhard, 1832, p. 192). General formations were those laid down simultaneously throughout the earth's surface "as Werner also seemed to admit" (Ospovat, 1971, p. 100; d'Aubouisson de Voisins, 1819, p. 326; de Humboldt, 1823, p. 1)!

Alexander von Humboldt's great stratigraphic study of 1823, in which he tried to correlate such general formations between the Old and the New Worlds, lent not only renewed credibility to such ideas, but awakened the enthusiasm of geologists to apply the newly developed biostratigraphic methodology to their fortification, although intercontinental correlations based on fossils had begun before him. (In addition to Cuvier's studies on young Eurasian successions on the basis of terrestrial quadruped fossils of the Tertiary and Quaternary strata, see, for example, the correlation of American and European Paleozoic rocks on the basis of invertebrate fossils by Conybeare and Philipps {1822, p. xiii, note *}, which was surprisingly ignored in Marcou's {1853, p. 14–17} review of transatlantic correlations of Palaeozoic strata.) The erection of the geological time scale in the first half of the ninteenth century (Berry, 1987) and its rapid application world-wide, mainly by the activities of the colonial surveys (for the British colonies, see, for example, Branagan and Townley, 1976; Vodden, 1992; Grout, 1995. For British imperial geological policy, see especially Smith, 1989, p. 27–33 and the references in his footnote 4, and Stafford's excellent book on Murchison: Stafford, 1989. For the Russian Empire, see Burde, Strelnikov, Mezhelovsky, and others, 2000. For the French colonies, see Furon, 1955. For Dutch southeast Asia, see Rutten, 1923. For the history of the meager biostratigraphic information from China before Suess began writing the *Antlitz*, see Yang and Li, 1996; Pumpelly, 1866, provides an overview of the state of geological knowledge in China in the mid-nineteenth century. For colonies in South America, see Figueirôa and Lopes, 1994. For the circum-Pacific, see Branagan and McNally, 1994) and the geologists in the United States (Merrill, 1904, 1920, 1924) set the scene in which Suess could pose the problem of the general, or, to use an expression more popular then, "the universal formations" (see Bertrand, 1897, p. xv). However, a Werner-style petrographic correlation that was still popular in the early twentieth century in some quarters (see the rebuttal in Salomon, 1926, and the references therein) is not to be read into Suess' thinking, and neither the mysteriously rapid, allegedly eustatic fluctuations, which still remain popular in the minds of others in our days (e.g., Vail et al., 1977; Haq et al., 1988).

[393]Here the word *formation* is used by Suess in the modern sense of a *system*. (For the modern definition of a system, see Salvador, 1994, p. 81–82.)

[394]This is the Silurian of Murchison. As Suess was writing these lines in the spring of 1883 (Suess, 1916, p. 323), the Cambrian-Silurian debate had not yet entirely died down, although only a few years later, Cambrian began appearing in textbooks. For the history of this debate to place Suess' usage in perspective, see Secord (1986), who writes that the end of the debate was reached as late as 1901 (Secord, 1986, p. 310).

[395]The largest of these cauldron subsidences, Suess thought, produced ocean basins; the smallest produced hinterland depressions behind mountain belts, such as the Po Plain behind the Alps.

[396]I think it is quite wrong to view Suess' discussion of the main kinds of deformation and magmatism in the first part of his first volume as an attack on uniformitarianism as Greene (1982, ch. 6 and 7 *passim*) does following Tietze (1917), de Launay (1905), and some others (cf. Şengör, 1982, 2000a). To the contrary, Suess' discussions in all his publications in which processes are treated, always start with the *now observable phenomena* and proceed to those *the effects of which become exposed only with time* to emphasize the importance of uniformitarianism (see especially Şengör, 2000a, p. 63). This has been his standard approach to all geological problems. When, for example, he studied the environments in which brachiopods lived, he started first with the living forms and their environments, because brachiopods not only show a great persistence through many geological systems, but also exhibit many identical forms in many areas. "These conditions have convinced me that the class of

Brachiopods can only be a starting point for broader geological deductions, when their present external conditions of existence are studied somewhat more closely, because this alone would yield the clues to the apparent abnormalities in their occurrence. It is the purpose of this article to gather together first the observations relating to the occurrence of living brachiopods, however insufficient they may be, and then to apply them to the fossil occurrences" (Suess, 1859, p. 187). In his 1873 study of the earthquakes of lower Austria, to give a tectonic example, Suess started with the 3 January 1873 earthquake, which he himself had experienced: "Notwithstanding the danger of sounding monotonous, I have been meticulous in its reporting *because it constitutes the most secure foundation to interpret the older earthquakes in Lower Austria*" (Suess, 1873, p. 1, italics mine). Greene's (1982, p. 166) point that Suess enlarged the meaning of uniformitarianism beyond Lyell's quietism (Suess now might have said "steady-state geology" after Dott's, 1969, apposite characterization) is a much more accurate description of Suess' intention and actual accomplishment than to say that Suess made catastrophism respectable again. People who did make catastrophism respectable again in the twentieth century did so in explicit opposition to Suess (cf. Kober, 1928, p. 19; and Şengör, 1982).

[397]For sources on the life of this remarkable individual, see the Pettersen entry in Sarjeant's *Geologists and the History of Geology*, v. III, p. 1881.

[398]*Seter* is also used in the English geological terminology. The fourth edition of the *Glossary of Geology* (Jackson, 1997, p. 585) defines it as a "Norwegian term for a wave-cut rock terrace." For more lexicographic information, see Stamp (1962, p. 413).

[399]For a current account of the temple and the geology around it, see Nazzaro's fine guidebook article with an excellent bibliography (Nazzaro, 1995).

[400]For a brief history of glacio-eustasy, see Dott (1992).

[401]*Novum Organum*, Book II, aphorism XXVII (see Anderson, 1960, p. 176): "Again, there is the Old and New World, both of which are broad and extended towards the north, narrow and pointed towards the south" (see also Carozzi, 1970b, 1970c). Oldroyd (1996, p. 176) implies that a regularity and determinism was implicit in Suess' explanation of the shapes of the continents. "The author invited the reader to consider the general configuration of the earth, especially the shapes of the continents and ocean basins (which for Lyellian theory, for example, were quite fortuitous) ..." Those shapes were equally fortuitous in Suess' theory. Nobody could tell where and how exactly the subsident cones creating the ocean floors were to form according to Suess' theory. It is important not to ascribe to Suess any of the regularist and determinist, also catastrophist, views of his immediate predecessors and contemporaries (see Şengör, 1982, 1998, 2000a). This has been done before and has led to much misunderstanding and misrepresentation (e.g., Tietze, 1917, *passim*; De Launay, 1905, p. 85; and, in part following him, Greene in his otherwise excellent account of Suess' ideas: Greene, 1982, ch. 6 and 7, especially p. 178).

[402]In the mountains named by Powell after the great physicist and Secretary of the Smithsonian, Joseph Henry (Powell, 1875, p. 178), who had supported the great Colorado venture and who had also implored his friend James Hall to consider carefully before publishing so absurd a statement on the geological history of the Appalachians (see Joseph Henry's letter to James Hall quoted on p. 129).

[403]From plates 15 and 16 *in* Hunt (1953), it is clear that the "one bed, the Vermilion Cliff sandstone, broken only by erosion" (Gilbert, 1880, p. 75), covering the Lesser Mount Holmes laccolith, corresponds with the Upper Triassic Chinle Formation, though the laccolith cuts down section eastwards into the Shinarump Conglomerate, which is equivalent to lower Chinle. For the details of the stratigraphy, see also Gregory (1917, p. 53–55), Gregory and Moore (1931, p. 62–64), Jackson (1998, fig. 2); also see, Hintze (1988, p. 44). I am much indebted to Professor William R. Dickinson, Dr. Nathan A. Niemi, and especially Mr. Eric M. Horseman for instructing me about the stratigraphy of the Henry Mountains.

[404]French stratigrapher and botanist who studied the Jurassic rocks of Normandy. For a biographical sketch, see Welsch (1908).

[405]For Le Conte's life, the most comprehensive source I know is Stephens (1982).

[406]Émile Argand (1879–1940), professor of geology at the University of Neuchâtel in Switzerland, is, to my knowledge, the only geologist who came very close to Suess in his knowledge of the regional geology of the entire earth and in his understanding of it. A multi-faceted genius who commanded 17 languages, Argand was the geologist who first unraveled the structure of the Alps in a manner very close to our present understanding. His immortal *La Tectonique de l'Asie* (Tectonics of Asia) was an illustration and extension of Wegener's theory of continental drift particulary as it is applied to continental deformation. That work foreshadowed many aspects of our present thinking regarding continental deformation, and that is why Philipp England and Dan McKenzie named in 1982 the dimensionless number specifying a continent's ability to flow under the influence of body forces: *the Argand Number* (England and McKenzie, 1982, 1983). For Argand's life and work, see Carandell (1928), Lugeon (1940), Thalmann (1943), Dubois (1976), Carozzi (1977), and Schaer (1991). Also see my Tethys Lecture (Şengör, 1998, p. 80–93) for an illustrated account with many quotations.

[407]Émile Haug (1861–1927) was the principle inspiration not only behind the main fixist school of the twentieth century, but also (paradoxically) behind many of Argand's early thoughts (see Haug, 1925). For his life, see Lutaud (1958). I give an illustrated synopsis of his main tectonic ideas in Şengör (1998, p. 53–58).

[408]The *Dictionary of Scientific Biography* has no entry for Fisher, which is surprising. He figures neither in the eleventh nor in the most recent edition of the *Encyclopaedia Britannica*. Sargeant's *Geologists and the History of Geology* lists, together with its second supplement, a number of references for his life and work, to which I am able to add only Muir-Wood (1985, p. 27–29) and Oreskes (1999, p. 25–29) which are regrettably unreliable, mixing up ideas published in the first and the second editions or altogether ignoring the important second edition of Fisher's (1889) great classic, *The Physics of the Earth's Crust*.

[409]Fisher later used Sir George Darwin's arguments of internal differential rotation of various shells of the earth generating additional heat to combat Lord Kelvin's estimate of the age of the earth as 24 m.y. (Fisher, 1895). Fisher showed that if internal heat is being generated, it would be impossible to measure the age of the earth using the methods employed by Lord Kelvin. Soon, the discovery of radioactivity resoundingly vindicated this point, even if not Fisher's preferred model of heat generation in the earth.

[410]Fisher, when still a believer in the contraction theory, was originally hostile to von Humboldt's idea:

> If we now consider the second cause capable of producing a difference of radial contraction, viz. a diversity of materials of the globe at the two places in question, it is palpable that this cannot explain *oscillations* of level. For that would require the materials to become changed in their properties from time to time, in a manner highly improbable. Humboldt's suggestion of secular currents in the interior to explain the oscillations of level is directly opposed to the condition of rigidity of the nucleus. In short, it seems that no modification of the theory of difference of radial contraction, arising from cooling merely can be relied upon. (Fisher, 1879, p. 439)

Sigurdsson (1999, p. 7) recently pointed out that it was Count Rumford (Benjamin Thompson, 1753–1814), the soldier, scientist and prolific inventor, who first suggested in 1797 the possibility of convection currents in the earth. For Count Rumford's life and bibliography, see Brown (1979, 1981).

[411]But see Direktion der Deutschen Seewarte (1882, plate 1; 1891, plates 1 and 2), von Richthofen (1902, especially plate III), and Schott (1912, especially plate V).

[412]Faulting was commonly thought to be an accompaniment of epeirogeny in the twentieth century (despite Stille). See, for example, the strong statements the geographer Machatschek made in his influential paper on epeirogeny: (Machatschek, 1918): "Association of large-scale swellings with fracture events now appears to be the the most dominant type of epeirogenic movement" (p. 8); "So, everywhere in Germany epeirogenic movements go hand in hand with fracturing" (p. 12); "Indeed, also most of the epirogenic uplift regions are associated with areas of fracture" (p. 22); "Blankenhorn's and Kober's descriptions from the Syrian block mosaic and the conditions in central Germany show how fractures may be associated, on the one hand with general uplifts, and on the other with sideways compression"(p. 26). The results of the studies of Powell and Dutton from the plateau country must certainly have supported these opinions.

[413]Otto Ampferer was clearly one of the greatest tectonicians of the twentieth century. For his life, see von Klebelsberg (1948), Cornelius (1946–1948), and Götzinger (1947). Davis et al. (1974), Şengör (1977), and Thenius (1980, 1988) discuss his ideas that anticipated many of our present-day concepts. Also see Oreskes (1999, p. 119).

[414]Thomas Mellard Reade (1832–1909), British architect and engineer (see Anonymous, 1909).

[415]See Suess (1914).

[416]Osmond Fisher had thought of compensating continental margin shortening by sub-oceanic extension (see above). Ampferer tried to do both within the orogens, possibly as a result of Suess' emphasis on simultaneous shortening in the externides of a mountain belt and on subsidence and vulcanicity in the internides ("Suess' rule": cf. Şengör, 1993), an interpretation that became very popular in the last decade of the twentieth century for the very same examples, namely the couples of the Tyrrhenian Sea/Appenines, Pannonian Basin/Carpathians, and Aegean Basin/Dinarides Suess had used in 1875 and in 1883!

[417]But in the last chapter of the second volume of the *Antlitz*, Suess pointed out that only eustatic sea-level changes were capable of explaining the *synchroneity* of certain sequences world-wide (Suess, 1888, p. 684), a statement that to this day has retained its validity. That statement Ampferer does not discuss.

[418]Stille dominated tectonic thinking during the first half of the twentieth century. Starting from his doctoral dissertation focussing on the area in the Teutoburg Forest (between the towns of Detmold and Altenbeken in the present-day federal state of Nordrhein-Westfalen in Germany) and gradually expanding into the area of north-central Germany, he developed a world-picture similar to that of James Dwight Dana (and his American successors), being also under the influence of the French master Emile Haug. His world–regional approach has commonly been compared with that of Suess, but Stille found the very same tectonic styles he had recognized in Germany wherever he directed his attention. Only after he considered the tectonics of the Americas and the Pacific Ocean did he enlarge his conceptual basis, and by the 1950s, his interpretations began to converge with those of Suess (dating from the beginning of the century). It was too late. Plate tectonics overtook him, the initial lights of which he saw as a very old man. Stille did not like what he saw because it involved continental drift, but he had the grace to acknowledge that his time had passed (Şengör, 1999b). The best source for Stille's life is Carlé's (1988, p. 103–328) excellent book. In addition to that, see also Lotze (1956), Bederke (1967), Pilger (1967, 1977), Martini (1967), Brinkmann (1970), and Schwan (1986). Lehmann's (1971) critical paper gives an extreme and somewhat personal account of the opposition that formed against Stille's ideas (that is why five German journals of earth science refused to publish it). Şengör (1996) provides a detailed discussion and critique of Stille's philosophy of science. For those who can read Russian, Bogdanov and Khain (1964) is an invaluable source concerning Stille's ideas.

[419]Really *copeogenic*, for Stille's criteria for orogeny includes taphrogeny and keirogeny.

[420]Really *falcogenic*, for Stille would have considered the entire mid-oceanic ridge system as orogenic, as he did in Iceland (Stille, 1939b).

[421]Leopold Kober (1883–1970) was professor of geology at the University of Vienna and was responsible, together with Hans Stille in Germany, for creating the dominant fixist picture of the tectonics of the earth in the twenteith century. The terms *Orogen* and *Kratogen* were coined by him, the latter of which Stille later converted to craton. For his life, see Medwenitsch (1970) and Tollmann (1983). In Şengör (1998, p. 58–63), I give an illustrated summary of Kober's world picture.

[422]Walter Herman Bucher (1888–1965) was, together with Marshall Kay, the most influential tectonician of the United States in the second and the third quarters of the twentieth century. A convinced fixist of the Kober-Stille school (although his doctoral dissertation was completed under a distinguished mobilist, namely Wilhelm Salomon-Calvi), his influence on Maurice Ewing of Columbia University is said to have prevented for years Ewing's acceptance of plate tectonics. Kendall's (1981) article on him in the *Dictionary of Scientific Biography* is regrettably unsatisfactory, omitting many important phases of Bucher's work, such as his role in the second Taconic controversy or his synthesis of the geology of northern South America and of the Caribbean that may have later influenced J. Tuzo Wilson's important work there. Also see Bucher's biographical memoir in volume 40 of the *Scientific Memoirs* of the U.S. National Academy of Sciences.

[423]Josef Chavanne (1846–1902), Austrian geographer, who worked mostly in Africa: in the Congo Basin and in the Sahara, fixed numerous astronomical locations and established barometric heights, which eventually served as an enlarged basis for a new hypsometric map of Africa (Chavanne, 1881). For Chavanne's life and work, see Henze (1978, p. 560, and the sources cited therein).

[424]Although there is no indication of it in Dureau-de-Lamalle's (1807) description of the interior of Africa, that the Sahara was divided into subbasins defined by elevated regions such as Tibesti was already known to Ritter (1822, especially sections 16, 17, 33, 34, 35, 37; and the *Karte von Afrika* … {1822}). His database consisted of very few and unconnected observations, and the middle part of the Sahara still remained unknown until the observations of such travelers as Major Denham, Lieutenant Clapperton, Dr. Oudney (who discovered Lake Chad on 4 February 1823), and René-Auguste Caillié (1799–1828, who later crossed the Sahara), were published beginning with 1826. Ritter also knew, as did his friend Alexander von Humboldt (as I have shown above), that the sub-Saharan Africa was generally a highland and depicted it as such in the map earlier referred to. Ansted (1863) regarded Africa south of the Equator as a plateau on the basis mainly of Livingstone's reports (his p. 85). He considered this plateau of not very significant elevation, but rising towards the coasts. North of the areas explored by Livingstone, Ansted noted that reports had been received from the natives indicating the presence of a *table land* in that direction (his p. 87). Ansted (1863, p. 93) summarized the existing knowledge on the structure of Africa as follows:

> There is the great east and west mountain chain of the Atlas running across the continent near the north coast, and corresponding high ground near the east and west coast. This latter elevation forms a boundary wall not generally more than six thousand feet above the sea, extending towards the south-east and south-west parallel to the seaboard, and converging in the high table land of the Cape. Directly south of the Atlas range is the Great Sahara, which is by no means a complete desert, although being irregularly and poorly supplied with water, it is, on the whole, unfavourable for vegetable and animal life. There are no lofty central mountains of any kind, but in their place a succession of vast plains, south of the equator, which are well watered by an anastomosing system of rivers connected with great sheets of shallow water, varying greatly in dimensions at different seasons.

Ansted (1863, ch. V) gives a very brief summary of the geologically relevant geographical exploration of Africa to 1861. Hibbert (1982) is a convenient summary of the geographical exploration of the Dark Continent between 1769 and 1889. For the history of the geographical exploration of the Sahara, see the excellent short summary account in Durou (1993). Alan Moorehead's two books on the Nile give good summaries of the exploration of the highlands and the rifts of East Africa in the eighteenth and nineteenth centuries, although with a strongly Anglophonic orientation (Moorehead, 1960, 1962; these books had enlarged second editions in 1971 and 1972 respectively, but I did not have a chance to read them). To offset that bias, see Hassert (1941). This last is to be complemented by the excellent bibliographies by Lobitzer (1981 and 1982) and Kainbacher (2002). Of these, Lobitzer's is orientated toward geological research, whereas Kainbacher's toward travel and geographical exploration. The first volume of Numa Broc's *Dictionnaire Illustré des Explorateurs et Grands Voyageurs Français du XIXe Siecle* (1988) mentioned in the preface above is devoted to Africa and is a wonderful, richly illustrated source, complementing the Anglophonic and Germanophonic histories and bibliographies of geological and geographical exploration in Africa. For a fairly complete summary and a helpful bibliography of the knowledge on the geology of Africa between Ansted's summary and the beginning of the twentieth century, see Knox (1905). Mohr (1999) is a bibliography of the geological exploration from 1830 to 1950 of the east African Rift System and the associated highlands, and Kalb et al. (2000) is another, covering the earth sciences for Ethiopia, Eritrea, Somalia, and Djibouti for the interval 1620-1993. Veatch (1935, p. 3–8) gives a short summary of the evolution of thought on the geology of the Congo Basin, one of the grandest falcogenic structures in Africa. For a superb facsimile atlas of the history of cartography of Africa until the end of the eighteenth century, see Klemp (1968).

[425]Kevin Charles Antony Burke (1929–), British geologist who became a naturalized U.S. citizen in 1979. Kevin Burke worked extensively both in east and west equatorial Africa in the 1950s, 1960s, and early 1970s. He was the earliest among the workers in Africa to embrace the plate tectonic model and, on the

basis of it, to propose novel—and still enduring—interpretations of African geology, not uncommonly in a full and acknowledged awareness of the international work that had preceded him well into the beginning of the twentieth century. Many of the "African" models Burke proposed in diverse geological fields ranging from tectonics to geomorphology have since found ready applications on other continents, almost to vindicate Du Toit's motto in the title page of his classic *Our Wandering Continents*: "Africa forms the Key" (Du Toit, 1937).

[426]Erich Krenkel (1880–1964), German geologist, born in Reichenau, Saxony, on 4 December 1880 and died in Frankfurt am Main on 5 May 1964. Krenkel studied law and geology at the universities of Heidelberg, Munich, and Leipzig. In 1912, he became a lecturer (*Privatdozent*) in Munich and was professor between 1926 and 1945 in Leipzig. Krenkel worked in east Africa (he was a sort of German Gregory, but of a higher caliber as a geologist than Gregory) and contributed substantially to our understanding of the tectonics of this continent. His 1922 book *Die Bruchzonen Ostafrikas* (The Fracture Zones of East Africa) not only contains a masterly synthesis of the then existing knowledge, but does so in the framework of an interpretation that remains mostly compatible with the present knowledge. His four-volume *Geologie Afrikas* (Geology of Africa; 1925–1938) remains valuable in its many parts. He later wrote a single volume, *Geologie und Bodenschätze Afrikas* (1957; Geology and the Mineral Resources of Africa), which he regarded as a shortened second edition of the *Geologie Afrikas*. Also see his *Geologie der Deutschen Kolonien in Afrika* (Krenkel, 1939) and *Der Geologische Bau der Deutschen Kolonien in Afrika und in der Südsee* (Krenkel, 1940).

Krenkel's great contributions to the geology of Africa remain undisputed. His unfortunate political involvement during the Nazi period regretably overshadowed his later life, which may be the reason why I failed to obtain a photograph of him. (For the details of Krenkel's life, I am grateful to Dr. Inge Seibold of the Geologenarchiv of the Geologische Vereinigung e. V. in Freiburg i. Br.).

[427]Erich Otto Haarmann (1882–1945) belonged to a wealthy west German family active in the mining business and was himself educated both as a mining engineer and a geologist. His best-known contributions were in tectonics, in which he repeatedly emphasized the primacy and importance of vertical movements. He put together an immense archive of all geological primary documents (mostly manuscripts) and a card-catalogue of all other existing collections that he had found out about. All that archive vanished entirely on 1 March 1943 during a British bombing raid on Berlin. Haarmann never recovered from that loss. He influenced, among others, Hans Cloos, who dedicated to him his great classic on uplift, fracturing, and volcanism (Cloos, 1939). For Haarmann's life and accomplishments, see Anonymous [Hans Cloos] (1942, p. 85–87) and Carlé (1988, p. 442–464).

[428]See Suess (1888, p. 699). Argand (1924, especially fig. 6) certainly implied a similar model for producing the African basins and uplifts. The most recent reincarnation of Diener's model of uplift-building by all-sided compression, with an added component of vulcanicity, that I am aware of, is Chesworth's (1975) suggestion of creating the Massif Central high by flexing the lithosphere of France (which he calls the Gallic Plate) by compressing it between the Carnic Plate (i.e., Apulia) and the Iberian Plate, as presented by Dewey et al. (1973).

[429]See volume 83 of the *Zeitschrift der Deutschen Geologischen Gesellschaft* (1931) for summaries of the individual critiques.

[430]Mikhail Mikhailovich Tetyayev was a loner in many ways in his geological pursuits. A pioneer in appreciating the importance of Mesozoic tectonic movements in central and north-central Asia, he overdid his case and ended up losing peer support. He developed and emphasized the method of using facies differences on platforms for an analysis of vertical motions of the earth's outer rocky rind, which was further developed by his student V.V. Beloussov. For his life, see Orlov (1999, p. 307 and the references there). For a summary and evaluation of his work in tectonics, see Beloussov (1961).

[431]Vladimir Vladimirovich Beloussov was an early associate of Tetyayev and learned much from him and from such other outstanding teachers as Vladimir Ivanovich Vernadsky (1863–1945), the man who developed Suess' concept of the biosphere further and founded the concept of global ecology. Under the influence of Tetyayev and on the basis of his field experience in the Greater Caucasus, Beloussov developed his ideas emphasizing the primary vertical motions of the earth's outer rocky rind and their influence on folding patterns. In many ways, Beloussov's view of tectonism was similar to that of Dutton. His strong opposition to plate tectonics was partly a result of his intellectual background developed strongly within the Kober-Stillean framework (with Bemmelenian overtones) and partly owing to his inability to incorporate the message of the oceans into his tectonic framework. Towards the end of his life, he began incorporating plate tectonic interpretations into his works under "alternative explanations." Beloussov has rendered an important service to geology by unceasingly emphasizing the importance of understanding the subtle differential vertical movements within large cratons. In this, time has vindicated him; hence, the dedication I placed at the head of this present book. For a summary of his life and work, see Interdepartmental Tectonic Committee and Editorial Committee of Geotektonika (1987), Chernov (1989, p. 18–20), and Sholpo (1999, 2000). Also see the brief biographical sketch in Beloussov (1962. p. xi).

[432]For his life, see Foose (1973) and Tollmann (1984); for the history of the development of his theory of undation tectonics, see Havemann (1969).

[433]Hans Cloos was a geologist who combined a comprehensive artistic and scientific intuition with a penetrating ability of observation at all scales much in the manner of Emile Argand, although Cloos' sweep was not nearly as all-embracing as Argand's. First and foremost, Cloos was a structural geologist, although broad enough (and wise enough, because having a palaeontological habilitation on top of a geological dissertation increased his chances of finding employment in a university) to write his habilitation thesis on the Middle Jurassic Ammonites from the Moluccas (Cloos, 1916; what a loss for palaeontology that he was prevented by the onset of World War I from preparing for the printer his drawings of the fossils!). Cloos worked all his life much closer to the outcrop than Argand, but he also delivered broad syntheses, such as his grand uplift-fracturing-volcanism paper discussed herein. He was a great humanist, a decided anti-Nazi, and protector of all sorts of people pursued by the Nazis. Near the end of the World War II, he was saved from Nazi persecution by the invading Allied armies, who offered him the mayorship of the city of Bonn (in the university of which he was professor), which he declined. Only three years after the war, the Geological Society of America decorated him with its highest distinction, the Penrose Medal.

For the life of Cloos, the best guide is his marvellously written and illustrated autobiography (Cloos, 1947), which was translated into English under the title *Conversation with the Earth*, and published by Adolph Knopf, New York. I consider it a must-read for any young geologist. Also see Ketin (1952), von Bubnoff (1953, with a bibliography of Cloos' writings), and Martin (1968).

[434]Although, Cloos also emphasized that "such swells mark erosion margins and facies boundaries in the layers of the past. The coeval change of fabric is small. Therefore, folding of large wavelength (*Großfaltung*) is commonly recognized not in the structure of the crust, but in the concurrent erosion and deposition, i.e. in the stratigraphic record" (Cloos, 1936, p. 397). It is thus clear that he agreed with Stille and with everbody else who thought that large-wavelength structures leave the rock fabric commonly intact. Stille (1940, p. 247, footnote 1) later noted this with approval.

[435]For example, figures 15 and 16 *in* Cloos (1939; herein Fig. 137A and B). Cloos' experimental domes have flank dips of about 10°, which is at least 10 times steeper than any real dome on earth, culminating in rifts (see Şengör, 2001a).

[436]Du Toit (1937, p. 250–253, and especially fig. 253) repeats Taber's arguments in recognition of his contribution. Cloos does not cite Du Toit. Taber's ideas, as enlarged by Du Toit and Cloos, have been recently revived and expanded. For examples, see Buck (1988) and Wernicke and Axen (1988), although the memory of earlier ideas seems to have faded, despite Holmes' attempt to revive them (Holmes, 1965, p. 1049–1050 and fig. 760).

[437]It is interesting to note that the long-reigning contraction hypothesis of mountain-building collapsed for exactly the same reason, except the signs were reversed (and the allegedly contraction-related domes, the eventual collapse of which supposedly was the origin of mountain-building, were closer in diameter to the diameter of the earth than the more modest cymatogenic domes). For one example of a rejection of the contraction theory owing to its geometric insufficiency, see Penck (1920, p. 344–345).

[438]Bucher's book was twice reprinted by Hafner Publishing company in New York: once in 1957 and again in 1964. This shows that his view found a wide audience until a year before the invention of plate tectonics by Tuzo Wilson!

[439]John Tuzo Wilson (1908–1993) invented the most important theory of the earth sciences in the twentieth century, namely plate tectonics. He coined the word *plate* and showed why plates must have three kinds of boundaries, namely, extensional, convergent, and strike-slip on a spherical earth (Wilson, 1965a). He also provided possibly the second most important theory, that of the mantle plumes. Of Canadian and British parentage, Wilson was born and raised in Canada and educated in Canada, Great Britain, and the United States. He did diverse work in geology and geophysics, ranging from geomorphology of the glaciated terrains to the geophysics of ocean floors. For his life and work, see Wilson (1982) and van Rijn (1993).

[440]Wilson's hypothesis strangely resembles the native myth of the arrival of Pélé, the volcano goddess, at Kilauea on the island of Hawaii, her current residence. Pélé was chased out of her native land, Kalakeenuiakane (although the oldest version of this legend does not mention the land where Pélé came from: cf. Westervelt, 1916, p. 4), by her sister Namakaokahai because of jealousy. Namakaokahai's husband, Aukele, described to Pélé a wonderful new land of games and sports and advised her to seek it. Pélé took her youngest and favorite sister, Hiiaka (*Hiiaka-i-ka-poli-o-Pélé:* "Hiiaka in the heart of Pélé") and first arrived at Kauai, the first of the Hawaiian Islands on the west-northwest. It is told that the native volcano god, Ailaau (i.e., "forest-eater") fled upon Pélé's approach. Pélé had a magic digging rod, called Pa-oa, to excavate the sort of residence she liked, a deep fire pit high atop a mountain of lava. However, she found Kauai's mountain rocks too hard to excavate, so she tried those along the shore. But each time she dug a fire pit, the sea goddess, Hina, would rush in to put out the flames and thus would not allow the new volcano goddess a congenial surrounding. Not successful at Kauai, Pélé tried her luck at Oahu, where she dug a large flaming crater at what is now known as Diamond Head, the southern extremity of the island. Here too, the sea goddess eventually chased out the goddess of the flames. So Pélé went to the next island, Maui, and finally to the big island, Hawaii. Her arrival in any one place was announced by earthquakes, volcanic eruptions, thunder, and lightening. In the big island of Hawaii, away from the competition of the sea goddess, Pélé constructed the grand fire pit of Halemaumau inside the southern part of Kilauea's caldera. She also made other homes on the island, such as the great caldera of Mokuaweoweo at the top of the Mauna Loa. Now and then the goddess is seen rushing from one to the other, sometimes in the form of a swift lava stream, sometimes as a ball of lightening (summarized from Frazer, 1919, p. 217; Westervelt, 1916, ch. II; McBride, 1968; Bullard, 1976, p. 12–13; Mullins, 1977). Although the Pélé myth has a number of variants, variations pertain to details and not to the essence of the west-northwest to east-southeast progression of volcanic activity in time.

Of all religions of nature, few represent such an accurate and detailed description of their natural substance as does the myth of Pélé (and of all the deities I know of, only the Olympians are as thoroughly "human" as Pélé). No geologist could read an account of Pélé without recognizing in her person, in her actions, and in her youthful beauty at the top of the mountain, the different aspects of Hawaiian volcanic activity. For example, the description of her home answers perfectly to a crater of Hawaiian-type shield volcano. The color of her fair hair comes, in these islands of black, blue and green, from the golden glare of the glassy filaments of basalt drawn across the air during an eruption, which are known as *Pélé's hair* (the same thing as "rock wool," which is widely used as insulator; for a color photograph, see Takahashi and Griggs, 1987, p. 878). Tear-shaped volcanic glass fragments that owe their shape to solidification while in flight and that rain as lapilli during eruptions are called *Pélé's tears* (for a color photograph, see Takahashi and Griggs, 1987, p. 879). Olivine crystals in the Hawaiian lava flows are locally known as *Pélé diamonds*.

As the brief account above shows, the age progression of the Hawaiian Islands, at least of their volcanic activity, seems not to have escaped the insightful scrutiny of the early Polynesian inhabitants of these beautiful islands. The myth of Pélé's arrival in Hawaii is in effect nothing more than an explanation of their inference of the age relations of the volcanic activity on the islands.

The present critics of the mantle plume theory say, in essence, that the plumes have as much reality as Pélé does (though perhaps more predictive power).

[441]Not published by the Harvard University Press, as incorrectly stated in Tibbets (1992, p. 153, note 68). When questioned, Harvard University Press informs the enquirer that they have never heard of such a book, which has no ISBN number (because it appeared in what is technically a journal). The series in which it appeared is sold at the exorbitant price of U.S. $220 per issue, so Mahmud's book comes to U.S. $660! It may be purchased from the editors by calling the Massachusetts telephone number 1-617-585-8796.

[442]This book was first published in 1908 and reprinted in 1926. It was reprinted again in 1962 by the Yale University Press with a new foreword by William H. Goetzmann. The University of Arizona Press reprinted it yet again but did not put a date on it. It is the Arizona reprint that I have consulted.

[443]This is an English translation by Francis Blagdon (12mo) of the famous book by Pallas entitled *Bemerkungen auf Einer Reise in die Südlichen Statthalterschaften des Russischen Reichs in den Jahren 1793 und 1794*, published by Martini in Leipzig in two quarto volumes (XXXII +516 p. and XXIV +525 p.) in 1799–1801. It is a report of his second great expedition in Russia. A French translation also was published under the title *Voyages Entrepris dans les Gouvernements Meridionaux de l'empire de Russie. Dans les Années 1793 et 1794* by Deterville in Paris in 1805.

[444]This publication gives an account of de Santarem's great Atlas from his own pen. However, the Viscount wrote it near the end of his life in ill health, and consequently it contains many errors. The well-known London antiquarian bookseller, Bernard Quaritch, prepared in 1864 a title-page and a list of contents for de Santarem's 1849(a) Atlas. Twelve copies of the title page and an "Index of Maps" were printed (reprinted in 1908) with the following note. Owing to the great rarity of this document, I reproduce the contents of that note, which also pertains to the 1855 paper by Viscount de Santarem (from the Harvard University Map Collection, catalogue number MB. 952. 6):

> The Vicomte de Santarem published originally, in 1842, a work entitled "Recherches sur la Priorité de la Découverte de la Côte Occidentale de l'Afrique" with an Atlas consisting of 30 plates. He afterwards made this Atlas, (which was in fact unfinished at the time) the foundation of the present great work, which contains 78 plates, and was published at the expense of the Portuguese Government. It was not, however, completed, in consequence of his death in the year 1855. The Maps in this last Atlas are not numbered, except those belonging to the original work, the numeration of which is no longer appropriate. There are frequently several Maps on one sheet or page of the new series, and these have been selected without any principle of sequence or order. There is no list or index for the arrangement of the sheets; but M. de Santarem communicated to the "Nouvelles Annales des Voyages" (1855, v. 2) shortly before his death, a classified list or catalogue of the several *Maps*, and this may serve as a guide for arranging the sheets. The sheets in this copy follow as nearly as possible the natural order of development suggested by the titles to the four great divisions of the work, and take precedence according to the relative priority of the earliest Map in each as numbered in the "Nouvelles Annales." Upon this plan the ensuing Index has been prepred. M. de Santarem speaks of all the Maps enumerated in his list as "published"; but there are a few which cannot be traced. Many inexactituteds will be observed in his descriptions and notices—two or three of the Maps in the list in the "Nouvelles Annales" are repeated more than once. The dates assigned on the maps and in the list differ by a century. It would appear that the list in the "Nouvelles Annales" was very carelessly drawn up (probably in consequence of illness), but it is nevertheless the only guide afforded to the intention of the author. (Also see Sijmons, 1985, for further attempts at ordering de Santarem's maps.)

[445]The authorship appears on the title page as Mrs. Wiliam T. Sedgwick, who was Mary Katrine (Rice) Sedgwick.

Index

C

G

H

I

J

K

L

M

N

O

T